# Handbook of Nondestructive Evaluation

## About the Author

**Charles J. Hellier** founded and was president of Hellier Technical Training and Consulting. After Hellier was acquired in 1998, he remained as president and then was general manager until he retired in 2003. He is currently the vice president of NDT Classroom and the principal of The Summit Group, LLC.

# Handbook of Nondestructive Evaluation

Charles J. Hellier

Third Edition

McGraw Hill

New York Chicago San Francisco Athens London
Madrid Mexico City Milan New Delhi
Singapore Sydney Toronto

**Library of Congress Control Number: 2019957425**

McGraw-Hill Education books are available at special quantity discounts to use as premiums and sales promotions or for use in corporate training programs. To contact a representative, please visit the Contact Us page at www.mhprofessional.com.

**Handbook of Nondestructive Evaluation, Third Edition**



1 2 3 4 5 6 7 8 9 CD 25 24 23 22 21 20

ISBN 978-1-260-44143-7
MHID 1-260-44143-1

**Sponsoring Editor**
Robert Argentieri

**Editing Supervisor**
Stephen M. Smith

**Production Supervisor**
Pamela A. Pelton

**Acquisitions Coordinator**
Elizabeth M. Houde

**Project Manager**
Revathi Viswanathan,
Cenveo® Publisher Services

**Copy Editor**
Surendra N. Shivam

**Proofreader**
Cenveo Publisher Services

**Indexer**
Arc Indexing, Inc.

**Art Director, Cover**
Jeff Weeks

**Composition**
Cenveo Publisher Services

This Handbook is dedicated to the memory of two pioneers in the field of nondestructive testing:

Robert C. McMaster, 1913–1986
Samuel A. Wenk, 1914–1990

Their work served as the foundation for the technology that has become so important to industry. Without the contributions of these "giants," the world would not be the same.

# Contents

# Contributors

**Michael W. Allgaier** Chap. 3
**Michael Clark** Chaps. 4, 5
**John Drury** Chap. 7
**Richard D. Finlayson** Chap. 10
**Richard A. Harrison** Chap. 5
**Charles J. Hellier** Chaps. 1, 4, 6
**Daniel Kass** Chap. 12
**James Mitchell** Chap. 10
**Frederick A. (Fred) Morro** Chap. 11
**Tom Nelligan** Chap. 12
**John W. Newman** Chap. 14
**Emilie Peloquin** Chap. 12
**Robert B. Pond, Jr.** Chap. 2
**George R. Quinn** Chap. 8
**Joseph L. Rose** Chap. 13
**Michael Shakinovsky** Chap. 7
**John Snell** Chap. 9
**Rob Spring** Chap. 9

# Preface

This third edition of the Handbook includes an extensive update with new techniques and methods such as full matrix capture, pulsed eddy current, and shearography NDT. NDE continues to be a growing and dynamic technology with more developments to come.

One may wonder why the title of this Handbook contains the word "evaluation" instead of the generic term "testing" that is usually used in connection with "nondestructive." The *American Heritage Dictionary* properly defines "nondestructive" as "Of, relating to, or being a process that does not result in damage to the material under investigation or testing." The most appropriate definitions of the word "test(ing)" from the same source are "to determine the presence or properties of a substance" and "to exhibit a given characteristic when subjected to a test." There are also several other definitions that do not really apply. "Evaluate," on the other hand, has a definition that seems to be more fitting for the intent of this Handbook: "To examine and judge carefully; appraise." "Evaluation," as defined in ASTM E-1316, is "A review following interpretation of the indications noted, to determine whether they meet specified acceptance criteria."

In reality, these terms have been used interchangeably with other expressions such as "inspection," "examination," and "investigation." In general, all of these terms refer to the same technology, one that is still widely unknown or misunderstood by the general public. And the use of these different terms may have, in fact, contributed to this misunderstanding. Assuming it is acceptable to take some liberties with these definitions, I would like to suggest that an appropriate definition of NDE, NDT, or NDI would be "A process that does not result in any damage or change to the material or part under examination and through which the presence of conditions or discontinuities can be detected or measured, then evaluated."

It is the intent of this Handbook to introduce the technology of nondestructive testing to those who are interested in a general overview of the most widely used methods. There are many excellent reference books on the various methods that can provide additional in-depth information, if desired.

The key ingredient in the NDT process is the practitioner. Many times, NDT personnel are subjected to unfavorable environments and hazardous working conditions. These same individuals are required to complete extensive training programs and fulfill lengthy experience requirements as a prerequisite to becoming certified. And it doesn't stop there. Many codes and specifications require periodic retraining and recertification. Most inspectors/examiners are under constant scrutiny by client auditors or third-party overseers. At times, travel to remote locations is required, resulting in extended periods away from home and long workdays. There should always be that desire to "do it right." Think of the consequences if a serious discontinuity is missed and some type of failure results. Conscientious examiners are concerned and caring individuals. In NDT, there is no room for those who are "just doing their job." It takes a special kind of dedicated person, but the rewards are great! The thought of helping humankind by being involved in a technology that is devoted to making this world a safer place is motivation for many. NDT is an honorable profession for those who are honorable. When NDT practitioners lose their ethics, they have lost everything!

This Handbook has been created by a group of professionals who all believe this to be true. It is our desire that it will be a source of knowledge and reference for many who are interested in this unique and challenging technology. The quest for excellence should be never-ending. As Robert Browning once wrote, "Ah! But a man's reach should exceed his grasp, or what's a heaven for?"

## Acknowledgments

This Handbook is the result of the combined efforts of many. Each contributor spent untold hours in the preparation of their segment and had to persevere through my many phone calls and e-mails. This was a labor of love, and their responses and continued efforts were unending. It will be obvious to the reader that each one has shared their expertise in a professional and comprehensive way. They are the "cream of the crop," and I will always be in their debt.

Many thanks also go to the dedicated proofreaders, including Robert E. Cameron, Sheryl Shakinovsky, and my best friend and wife, Karen E. Hellier.

### Special Acknowledgment

Michael Shakinovsky provided significant support and constant guidance throughout the development of the Handbook. He assisted with the editing and was the coauthor of Chapter 7, Ultrasonic Testing. I shall always be grateful for his efforts and friendship.

### Credits for Third Edition

1. Michael Shakinovsky, Consultant
2. Lawrence Mullins, NxtNdT
3. Jeffrey Monks, The Equipcon Group
4. Karen Smith, Olympus Scientific Solutions
5. Dan Carnevale, Danatronics
6. Martin Theriault, Eddyfi Technologies
7. Melissa Syvrais, Eddyfi Technologies
8. Andreanne Potvin, Eddyfi Technologies
9. Emilie Peloquin, Olympus Scientific Solutions
10. Sam H. Serhan, TecScan Systems
11. Jim Stuckey, TechCorr

### Radiographic Compendium Contributions

1. Jonathan R. Konerza, Sentinel Integrity
2. Jeri Matza, NDT Direct
3. Terry Plasek, Fujifilm

### In Memoriam

George R. Quinn, major contributor to Chapter 8, Eddy Current Testing, passed away on December 5, 2018. He was certified in eddy current and ultrasonic testing, and was an outstanding instructor and leading authority in both methods. Above all, he was a close friend and colleague for over 40 years.

*Charles J. Hellier*

# Handbook of Nondestructive Evaluation

# CHAPTER 1

# Introduction to Nondestructive Testing

## 1.1 What Is Nondestructive Testing?

A general definition of nondestructive testing (NDT) is an examination, test, or evaluation performed on any type of test object without changing or altering that object in any way, in order to determine the absence or presence of conditions or discontinuities that may have an effect on the usefulness or serviceability of that object. Nondestructive tests may also be conducted to measure other test object characteristics, such as size; dimension; configuration; or structure, including alloy content, hardness, grain size, etc. The simplest of all definitions is basically an examination that is performed on an object of any type, size, shape or material to determine the presence or absence of discontinuities, or to evaluate other material characteristics. Nondestructive examination (NDE), nondestructive inspection (NDI), and nondestructive evaluation (NDE) are also expressions commonly used to describe this technology. Although this technology has been effectively in use for decades, it is still generally unknown by the average person, who takes it for granted that buildings will not collapse, planes will not crash, and products will not fail. Although NDT cannot guarantee that failures will not occur, it plays a significant role in minimizing the possibilities of failure. Other variables, such as inadequate design and improper application of the object, may contribute to failure even when NDT is appropriately applied.

NDT, as a technology, has seen significant growth and unique innovation over the past 25 years. It is, in fact, considered today to be one of the fastest growing technologies from the standpoint of uniqueness and innovation. Recent equipment improvements and modifications, as well as a more thorough understanding of materials and the use of various products and systems, have all contributed to a technology that is very significant and one that has found widespread use and acceptance throughout many industries. This technology touches our lives daily. It has probably done more to enhance safety than any other technology, including that of the medical profession. One can only imagine the significant number of accidents and unplanned outages that would occur if it were not for the effective use of nondestructive testing. It has become an integral part of virtually every process in industry, where product failure can result in accidents or bodily injury. It is depended upon to one extent or another in virtually every major industry that is in existence today.

Nondestructive testing, in fact, is a process that is performed on a daily basis by the average individual, who is not aware that it is taking place. For example, when a coin is deposited in the slot of a vending machine and the selection is made, whether it is candy or a soft drink, that coin is actually subjected to a series of nondestructive tests. It is checked for size, weight, shape, and metallurgical properties very quickly, and if it passes all of these tests satisfactorily, the product that is being purchased will make its way through the dispenser. It is common to use sonic energy to determine the location of a stud behind a wallboard. The sense of sight is employed regularly to evaluate characteristics such as color, shape, movement, and distance, as well as for identification purposes. These examples, in a very broad sense, meet the definition of nondestructive testing—an object is evaluated without changing it or altering it in any fashion.

The human body has been described as one of the most unique nondestructive testing instruments ever created. Heat can be sensed by placing a hand in close proximity to a hot object and, without touching it, determining that there is a relatively higher temperature present in that object. With the sense of smell, a determination can be made that there is an unpleasant substance present based simply on the odor that emanates from it. Without visibly observing an object, it is possible to determine roughness, configuration, size, and shape simply through the sense of touch. The sense of hearing allows the analysis of various sounds and noises and, based on this analysis, judgments and decisions relating to the source of those sounds can be made. For example, before crossing a street, one can hear a truck approaching. The obvious decision is not to step out in front of this large, moving object. But of all the human senses, the sense of sight provides us with the most versatile and unique nondestructive testing approach. When one considers the wide application of the sense of sight and the ultimate information that can be determined by mere visual observation, it becomes quite apparent that visual testing (VT) is a very widely used form of nondestructive testing.

In industry, nondestructive testing can do so much more. It can effectively be used for:

1. Examining raw materials prior to processing
2. Evaluating materials during processing as a means of process control
3. Examining finished products
4. Evaluating products and structures once they have been put into service

Nondestructive testing, in fact, can be considered as an extension of the human senses, often through the use of sophisticated electronic instrumentation and other unique equipment. It is possible to increase the sensitivity and application of the human senses when used in conjunction with these instruments and equipment. On the other hand, the misuse or improper application of a nondestructive test can cause catastrophic results. If the test is not properly conducted or if the interpretation of the results is incorrect, disastrous results can occur. It is essential that the proper nondestructive test method and technique be employed by qualified personnel, in order to minimize these problems. Conditions for effective nondestructive testing will be covered and expanded upon later in this chapter.

To summarize, nondestructive testing is a valuable technology that can provide useful information regarding the condition of the object being examined once all the essential elements of the test are considered, approved procedures are followed, and the examinations are conducted by qualified personnel.

## 1.2 Concerns Regarding NDT

There are certain misconceptions and misunderstandings that should be addressed regarding nondestructive testing. One widespread misconception is that the use of nondestructive testing will ensure, to a degree, that a part will not fail or malfunction. This is not necessarily true. Every nondestructive test method has limitations. A nondestructive test by itself is not a panacea. In most cases, a thorough examination will require a minimum of two methods: one for conditions that would exist internally in the part and another method that would be more sensitive to conditions that may exist at the surface of the part. It is essential that the limitations of each method be known prior to use. For example, certain discontinuities may be unfavorably oriented for detection by a specific nondestructive test method. Also, the threshold of detectability is a major variable that must be understood and addressed for each method. It is true that there are standards and codes that describe the type and size of discontinuities that are considered acceptable or rejectable, but if the examination method is not capable of disclosing these conditions, the codes and standards are basically meaningless. Another misconception involves the nature and characteristics of the part or object being examined. It is essential that as much information as possible be known and understood as a prerequisite to establishing test techniques. Important attributes such as the processes that the part has undergone and the intended use of the part, as well as applicable codes and standards, must be thoroughly understood as a prerequisite to performing a nondestructive test. The nature of the discontinuities that are anticipated for the particular test object should also be well known and understood.

At times, the erroneous assumption is made that if a part has been examined using an NDT method or technique, there is some magical transformation that guarantees that the part is sound. Codes and standards establish minimum requirements and are not a source of assurance that discontinuities will not be present. There are acceptable and rejectable discontinuities that are identified by these standards. There is no guarantee that all acceptable discontinuities will not cause some type of problem after the part is in service. Again, this illustrates the need for some type of monitoring or evaluation of the part or structure once it is operational.

Another widespread misunderstanding is related to the personnel performing these examinations. Since NDT is a "hands-on" technology, the qualifications of the examination personnel become a very significant factor. The most sophisticated equipment and the most thoroughly developed techniques and procedures can result in potentially unsatisfactory results when applied by an unqualified examiner. A major ingredient in the effectiveness of a nondestructive test is the personnel conducting it and their level of qualifications. This will be addressed in greater detail later in this chapter.

## 1.3 History of Nondestructive Testing

Where did NDT begin? There are those who would answer this question by referring to the account of the creation of the heavens and the earth in *Genesis:* "In the beginning, God created the heavens and the earth and He *saw* that it was good" (Fig. 1-1). This is a theme that has been used from time to time when discussing the history of nondestructive testing. Seeing that the "heavens and the earth were good" has been identified as the first nondestructive test—a visual test!

**FIGURE 1-1** Earth from space. (*Courtesy of Library of Congress.*)

It is impossible to identify a specific date that would indicate exactly when nondestructive testing, as we know it today, began. In ancient times, the audible ring of a Damascus sword blade would be an indication of how strong the metal would be in combat. This same "sonic" technique was used for decades by blacksmiths (Fig. 1-2) as they listened to the ring of different metals that were being shaped. This approach was also used by early bell-makers. By listening to the ring of the bell, the soundness of the metal could be established in a very general way. Visual testing, while not "officially" considered a part of the early NDT technology, had been in use for many years for a wide range of applications. Heat sensing was used to monitor thermal changes in materials, and "sonic" tests were performed well before the term "nondestructive testing" was ever used.

Table 1-1 lists some of the key events in the chronology of NDT and the individuals who were mostly responsible for these developments. Certainly there were many other individuals who have made significant contributions to the growth of NDT, but it is impossible to name them all.

From the late 1950s to present, NDT has seen unprecedented development, innovation, and growth through new instrumentation and materials. The ability to interface much of the latest equipment with computers has had a dramatic impact on this technology. The ability to store vast amounts of data with almost instant archival capability has taken NDT to a level once only imagined, yet NDT technology is still in its infancy. This chronology will continue to grow as exciting new challenges present themselves through technology expansion and unique material developments. The quest to detect and identify smaller discontinuities will not end until catastrophic failures can no longer be related to the existence of material flaws.

**FIGURE 1-2** Early blacksmith. (*Courtesy of C. Hellier.*)

The roots of nondestructive testing began to take form prior to the 1920s, but the majority of the methods that are known today didn't appear until late in the 1930s and into the early 1940s. Much of the latter developments came about as a result of the tremendous activity during World War II. In the 1920s, there was an awareness of some of the magnetic particle tests (MT) and, of course, the visual test (VT) methods, as well as X-radiography (RT), which at that time was primarily being used in the medical field. In the early days of railroading, the forerunner of the present-day penetrant test (PT), a technique referred to as the "oil and whiting test," had been widely used. And there were also some basic electrical tests using some of the basic principles of eddy current testing (ET). The sonic or "ringing" method, as well as some archaic gamma radiographic techniques using radium as the source of radiation, were both used with limited success. From these roots, NDT technology has evolved to encompass the many sophisticated and unique methods that are in use today. (See Table 1-2 for a comprehensive overview of the major NDT methods.)

Prior to World War II, design engineers were content to rely on unusually high safety factors, which were usually built or engineered into many products, such as pressure vessels and other complex components, of that time. As a result of the war effort, the relationship of discontinuities and imperfections relative to the useful life and application of a product or system became a concern. In addition, there were a significant number of

| | |
|---|---|
| BC (approx.) | Visual testing becomes the first NDT method when God creates the heavens and earth and "sees" that it is good! |
| 1800 | First thermography observations by Sir William Herschel |
| 1831 | First observation of electromagnetic induction by Michael Faraday |
| 1840 | First infrared image produced by Herschel's son, John |
| 1868 | First reference to magnetic particle testing reported by S. H. Saxby, by observing how magnetized gun barrels affect a compass |
| 1879 | Early use of eddy currents to detect differences in conductivity, magnetic permeability, and temperature initiated by E. Hughes |
| 1880–1920 | "Oil and whiting" technique, forerunner of present-day penetrant test used for railroad axles and boilerplates |
| 1895 | X-rays discovered by Wilhelm Conrad Roentgen |
| 1898 | Radium discovered by Marie and Pierre Curie |
| 1922 | Industrial Radiography for metals developed by Dr. H. H. Lester |
| 1927–1928 | Electric current induction/magnetic field detection system developed by Dr. Elmer Sperry and H. C. Drake for the inspection of railroad track |
| 1929 | Magnetic particle tests/equipment pioneered by A. V. deForest and F. B. Doane |
| 1929 | First experiments using quartz transducers to create ultrasonic vibrations in materials were conducted by S. Y. Sokolov in Russia |
| 1930 | Practical uses for gamma radiography using radium were demonstrated by Dr. Robert F. Mehl |
| 1935–1940 | Penetrant techniques developed by Betz, Doane, and deForest |
| 1935–1940s | Eddy current instrument developments by H. C. Knerr, C. Farrow, Theo Zuschlag, and Dr. F. Foerster |
| 1940–1944 | Ultrasonic test method developed in the United States by Dr. Floyd Firestone |
| 1942 | First ultrasonic flaw detector using pulse-echo introduced by D. O. Sproule (United Kingdom) |
| 1946 | First portable ultrasonic thickness measuring instrument, the Audigage, was introduced by Branson |
| 1950 | Acoustic emission introduced as an NDT method by J. Kaiser |
| Mid-1950s | First ultrasonic testing immersion B and C scan instruments developed by Donald C. Erdman |

**Table 1-1** Chronology of Early Key Events in NDT

catastrophic failures and other accidents relating to product inadequacies that brought the concern for system and component quality to the forefront. Some of the improvements in fabrication and inspection practices can be attributed to boilers (Fig. 1-3) and some of their early catastrophic failures.

One such failure occurred on a sunny and unseasonably warm day in Hartford, Connecticut, in March of 1854. People were just returning to their offices and shops after lunchtime. At about two o'clock in the afternoon, a man stepped into the engine room of the Fales and Gray Car Works and began a conversation with the operating engineer. Just about that time, the boiler exploded with tremendous force (Fig. 1-4). The explosion destroyed the boiler room and an adjoining blacksmith shop, and it severely damaged the main building. As a result of this dramatic boiler explosion, nine people were killed

| Method | Principles | Application | Advantages | Limitations |
|---|---|---|---|---|
| Visual testing (VT) | Uses reflected or transmitted light from test object that is imaged with the human eye or other light-sensing device | Many applications in many industries ranging from raw material to finished products and in-service inspection | Can be inexpensive and simple with minimal training required. Broad scope of uses and benefits | Only surface conditions can be evaluated. Effective source of illumination required. Access necessary |
| Penetrant testing (PT) | A liquid containing visible or fluorescent dye is applied to surface and enters discontinuities by capillary action | Virtually any solid nonabsorbent material having uncoated surfaces that are not contaminated | Relatively easy and materials are inexpensive. Extremely sensitive, very versatile. Minimal training | Discontinuities open to the surface only. Surface condition must be relatively smooth and free of contaminants |
| Magnetic particle testing (MT) | Test part is magnetized and fine ferromagnetic particles applied to surface, aligning at discontinuity | All ferromagnetic materials, for surface and slightly subsurface discontinuities; large and small parts | Relatively easy to use. Equipment/material usually inexpensive. Highly sensitive and fast compared to PT | Only surface and a few subsurface discontinuities can be detected. Ferromagnetic materials only |
| Radiographic testing (RT) | Radiographic film is exposed when radiation passes through the test object. Discontinuities affect exposure | Most materials, shapes, and structures. Examples include welds, castings, composites, etc., as manufactured or in-service | Provides a permanent record and high sensitivity. Most widely used and accepted volumetric examination | Limited thickness based on material density. Orientation of planar discontinuities is critical. Radiation hazard |
| Ultrasonic testing (UT) | High-frequency sound pulses from a transducer propagate through the test material, reflecting at interfaces | Most materials can be examined if sound transmission and surface finish are good and shape is not complex | Provides precise, high-sensitivity results quickly. Thickness information, depth, and type of flaw can be obtained from one side of the component | No permanent record (usually). Material attenuation, surface finish, and contour. Requires couplant |
| Eddy current testing (ET) | Localized electrical fields are induced into a conductive test specimen by electromagnetic induction | Virtually all conductive materials can be examined for flaws, metallurgical conditions, thinning, and conductivity | Quick, versatile, sensitive; can be noncontacting; easily adaptable to automation and in situ examinations | Variables must be understood and controlled. Shallow depth of penetration, lift-off effects, and surface condition |
| Thermal infrared testing (TIR) | Temperature variations at the test surface are measured/detected using thermal sensors/detectors/instruments/cameras | Most materials and components where temperature changes are related to part conditions/thermal conductivity | Extremely sensitive to slight temperature changes in small parts or large areas. Provides permanent record | Not effective for detection of flaws in thick parts. Surface only is evaluated. Evaluation requires high skill level |
| Acoustic emission testing (AE) | As discontinuities propagate, energy is released and travels as stress waves through material. These are detected by means of sensors | Welds, pressure vessels, rotating equipment, some composites and other structures subject to stress or loading | Large areas can be monitored to detect deteriorating conditions. Can possibly predict failure | Sensors must contact test surface. Multiple sensors required for flaw location. Signal interpretation required |

**TABLE 1-2** Major NDT Methods—A Comprehensive Overview

**FIGURE 1-3** Old boiler. (*Courtesy of C. Hellier.*)

immediately and 12 died later. In addition, more than 50 were seriously injured. This boiler was almost new—in service for less than 1 month. It was manufactured by a reputable, well-experienced boiler manufacturer. It should be emphasized again that at this time, boilers were being made with unusually high safety margins. In fact, many of the early boilers were fabricated before the principles of thermodynamics were fully understood. This boiler failure in Hartford, Connecticut, was ultimately determined to

**FIGURE 1-4** Boiler explosion. (*Courtesy of C. Hellier.*)

have been caused by an excessive accumulation of steam. Based on a hearing that was held to determine the cause and to establish blame, the jury offered suggestions as to what could be done to prevent or minimize such accidents in the future. Their suggestions included the following:

- Initiation of regulations to prevent careless or inexperienced people from being in charge of boilers.
- Safety inspections to be made on a regular basis by authorized municipal or state representatives.
- Boilers should be placed outside the factory buildings.
- Boilers should be prohibited from operating at higher temperatures than would be consistent with safety.

This was a significant turning point in the importance and progress of inspection and NDT. Ten years later, in 1864, Connecticut passed a Boiler Inspection Law. This law required an annual inspection of every boiler and resulted in the issuance of a certificate if the boiler was satisfactory or, if it weren't, the boiler would be retired from service. Another benefit that resulted from this early boiler explosion was the founding of the Polytechnic Club in 1857. Basically, 12 men who had an interest in boilers met periodically and studied the problems relating to steam boilers.

During those early days of boilers, there were many other dramatic failures. One of the most memorable in history involved a steamship named *Sultana*, a Mississippi side-wheeler with two tall stacks (Fig. 1-5). On April 27th, 1865, she was steaming along above Memphis when three of her four boilers exploded. The actual cause for this catastrophic explosion was never determined. The *Sultana* usually carried

**FIGURE 1-5** The *Sultana*. (*Courtesy of Library of Congress.*)

about 375 passengers, but that day the boat was jammed from stem to stern with almost 2,200 passengers, mostly union soldiers who had just been released from confederate prisons following Lee's surrender at Appomattox. Eyewitness accounts of this disaster reported that the side-wheeler had burned to the water line within 15 minutes and the death toll, although not precise, was estimated to be between 1,200 and 1,600. Depending upon the exact number, this could have been the worst disaster in marine history. In fact, there is a possibility that there were more lives lost as a result of this explosion than there were when the *Titanic* sank in 1912 with a loss of 1,517 lives.

In the spring of 1866, the Connecticut legislature approved an act of incorporation of the Hartford Steam Boiler Inspection and Insurance Company (Fig. 1-6). This is significant because the premise upon which this company was founded was to provide insurance for boilers. Inspection was required as a prerequisite to issuing an insurance policy. On February 14th, 1867, Policy Number 1 was written on three horizontal tubular boilers for a face value of $5,000. The premium was $60. In 1911, the first Boiler Code Committee was formed and the first code was developed during the years of 1911 to 1914 and first published in 1915. Visual testing (VT) was the initial method of nondestructive testing. Surely, the introduction and application of the Boiler Code has had a major impact on the growth and application of nondestructive testing over the years since its inception and first publication.

In 1920, Dr. H. H. Lester, the dean of industrial radiography, began his work at the Watertown Arsenal in Boston, Massachusetts. Figure 1-7 illustrates the lead-lined exposure room of his original X-ray laboratory as it looked in 1922. Dr. Lester was directed to develop X-ray techniques for the examination of castings, welds, and armor

**FIGURE 1-6** Hartford steam boiler advertisement. (*Courtesy of C. Hellier.*)

**Figure 1-7** Dr. Lester's X-ray laboratory. (*Courtesy of C. Hellier.*)

plate to basically improve the quality of materials used by the Army. Even though William Conrad Roëntgen had discovered X-rays some 27 years earlier, not much had been accomplished in applying X-rays for materials evaluation, due primarily to the low energy of the early X-ray units. Dr. Lester's laboratory had equipment that was archaic by today's standards, but his work and that early equipment served as the foundation for future development of the radiographic test method using X-ray sources.

The next key development in the history of nondestructive testing was also due to a catastrophe—a major train derailment. This resulted in the electric current induction/magnetic field detection system that was developed by Dr. Elmer Sperry and H. C. Drake (Fig. 1-8). The primary use of this method was to detect discontinuities in railroad track. From this development came the basic principles upon which the Sperry Rail Service was founded. This ultimately resulted in more advanced railroad track test cars. Tracks are still being inspected today using similar principles. This makes Sperry Rail Service the oldest continuously operated NDT service group in the United States. In 1929, the Magnaflux Corporation was formed to promote the use of magnetic principles

**FIGURE 1-8** Early Sperry inspection railcar. (*Courtesy of C. Hellier.*)

for industrial NDT applications. Magnetic particle testing (MT) principles came from early electromagnetic conduction and induction experiments performed by Professor A. V. deForest and F. B. Doane. Some of the early equipment was very limited in its application, but at that time it was a unique and novel technique able to detect surface discontinuities in ferromagnetic materials. The unit pictured in Fig. 1-9 was one of the earliest devices used to examine railroad parts like the shaft being examined. This examination was very crude and would only detect the most significant discontinuities. Some of the early wet horizontal units manufactured by Magnaflux, like the one illustrated in Fig. 1-10, are still used today.

Dr. Robert F. Mehl was instrumental in developing the practical industrial uses of radium for gamma radiography in the 1930s. He was instrumental in expanding the use of radium for the detection of discontinuities in materials that were not possible to be examined with the low energy X-ray equipment in use at that time. In fact, as a result of his early work in gamma radiography, Dr. Mehl gave the first honor lecture (later named in his honor) in 1941. The title of his lecture was "Developments in Gamma Ray Radiography—1928 to 1941." An early illustration of the use of gamma radiography using radium is illustrated in Fig. 1-11, which shows a practice that was commonly known at that time as the "fish pole" technique.

Coincidentally, 1941 was also the year that the American Industrial Radium and X-ray Society, the forerunner of the American Society for Nondestructive Testing as it is known today, was founded. Early X-ray units were only capable of producing low energy X-rays, making exposure times on structures as depicted in Fig. 1-12 extremely long.

**Figure 1-9** Early magnetic particle unit. (*Courtesy of C. Hellier.*)

**Figure 1-10** Early Magnaflux unit. (*Courtesy of CJH Collection.*)

**FIGURE 1-11** Radium "fish pole" technique. (*Courtesy of C. Hellier.*)

**FIGURE 1-12** Early industrial X-ray unit. (*Courtesy of C. Hellier.*)

There were other significant new developments in the 1940s in the area of industrial radiography. The first million-volt X-ray machines were introduced by General Electric, which provided higher energies to permit the examination of thicker material cross sections.

Even though early penetrant testing (PT) techniques, known as "oil and whiting," had been in use since before the turn of the 20th century, the method was not widely used until the addition of visible dyes and fluorescent materials resulting from the research of Robert and Joseph Switzer in the late 1930s and early 1940s. Figure 1-13 illustrates an early fluorescent penetrant system.

Even though the principles of eddy current testing (ET) had their roots in 1831, when Michael Faraday discovered the principles of electromagnetic induction, it wasn't until the 1940s that the full potential of this method was realized. The first recorded eddy current test was performed by E. E. Hughes in 1879; he was able to distinguish the difference between various metals by noting a change in excitation frequency, which basically resulted from the effects of test material resistivity and magnetic permeability. But it was not until the year 1926 that the first eddy current instrument, which was used to measure material thickness, was developed. Through World War II and the early 1940s, further developments resulted in better and more practical eddy current instruments. Figure 1-14 illustrates an early instrument developed by the Foerster Institute. Notice that a "standard" was placed in the "primary" coil and compared to the response from the part in the "secondary" coil. In the 1950s, Foerster also developed advanced instruments with impedance plane signal displays, which made it possible to discriminate between a number of parameters.

Since the beginning of time, it has been known that materials can emit certain noises when they are stressed. For example, when a piece of wood is bent, creaking or "crying" sounds can be heard. As the noise intensity increases, it may serve as a warning that the object is ultimately going to fail or break apart. In the 1950s, the first extensive study of

**FIGURE 1-13** Early fluorescent penetrant unit. (*Courtesy of C. Hellier.*)

**FIGURE 1-14** Early eddy current unit. (*Courtesy of C. Hellier.*)

this phenomenon, which we now call acoustic emission testing (AE), was reported by Dr. Joseph Kaiser in Munich, Germany, in his Ph.D. thesis. Basically, his studies demonstrated that acoustic emission events were caused by small failures in a material that was being subjected to stress. Much of the original work was done in the audible frequency sound range, but today, for obvious reasons, most acoustic emission monitoring is conducted at very high or ultrasonic frequencies. AE has grown significantly, and this method has become a valuable NDT method for determining condition, behavior, and the in-service characteristics of many materials and structures.

The use of high-frequency sound for the detection of discontinuities in materials was also introduced in the 1940s. The efforts of Dr. Floyd Firestone led to the development of an instrument called the Supersonic Reflectoscope, which was introduced in the United States by Sperry Products in 1944 (Fig. 1-15). In other countries, similar efforts were being made. Equipment and instrumentation was also being developed in England, Russia, and Germany.

Since humans can sense variations in temperature, it has always been possible to observe thermal energy changes. The physics of thermography was observed as early as the 1800s. Late in the 19th century, heat radiation was observed and explored, and ultimately, instrumentation was actually developed to measure changes in radiant energy. John Herschel created the first thermal picture in 1840, but the real development of thermal imaging did not occur in major areas of industry until the 1950s and early 1960s. This unique nondestructive test method has seen phenomenal growth and expansion as more and more applications have been developed.

As mentioned earlier, a key period in the history and development of nondestructive testing came during and after World War II. Prior to the war, nondestructive testing was typically considered to be part of the inspection activities of various companies

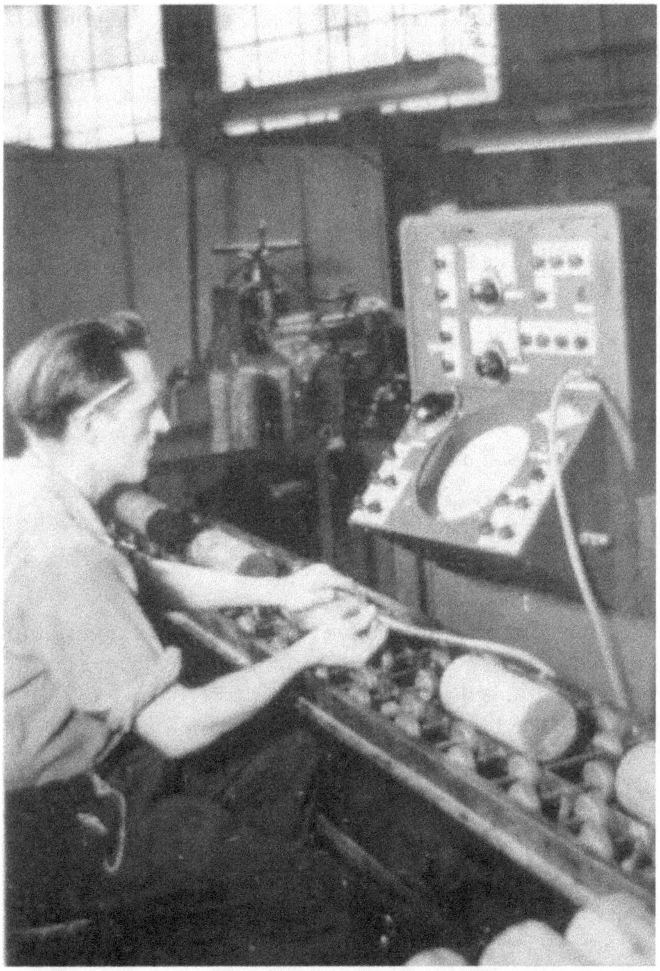

**FIGURE 1-15** Early Sperry ultrasonic unit. (*Courtesy of C. Hellier.*)

where it was being employed. As a recognized technology, it could be said that NDT began with the formation of the American Industrial Radium and X-ray Society (now the American Society for Nondestructive Testing) back in 1941. The evolution of nondestructive testing can be directly related to increased concern for safety, the reduction of safety factors, the development of new materials, and the overall quest for greater product reliability. The changes that have occurred in aerospace, nuclear power, and space exploration have all greatly contributed to the exciting and dynamic changes that have been experienced in this technology.

## 1.4 Nondestructive versus Destructive Tests

Destructive testing has been defined as a form of mechanical test (primarily destructive) of materials whereby certain specific characteristics of the material can be evaluated quantitatively. In some cases, the test specimens being tested are subjected to controlled conditions that simulate service. The information that is obtained through destructive testing is quite precise, but it only applies to the specimen being examined. Since the specimen is destroyed or mechanically changed, it is unlikely that it can be used for other purposes beyond the mechanical test. Such destructive tests can provide very useful information, especially relating to the material's design considerations and useful life. Destructive testing may be dynamic or static and can provide data relative to the following material attributes:

- Ultimate tensile strength
- Yield point
- Ductility
- Elongation characteristics
- Fatigue life
- Corrosion resistance
- Toughness
- Hardness
- Impact resistance

One of the more common destructive testing instruments used for mechanical testing, which is capable of measuring characteristics such as yield point, elongation, and ultimate tensile strength, is illustrated in Fig. 1-16.

Other than the fact that the specimen being examined typically cannot be used after destructive testing for any useful purpose, it must also be stressed that the data achieved

**FIGURE 1-16** Typical tensile testing machine. (*Courtesy of J. Devis Collection.*)

through destructive testing are specific to the test specimen. Another destructive test commonly used to measure a material's resistance to impact is the Charpy test. In this test, a specimen that is usually notched is supported at one end and is broken as a pendulum is released and impacts in the region of the notch. The measure of the material's resistance to impact (or notch toughness) is determined by the subsequent rise of the pendulum (see Fig. 1-17).

Hardness is also an important material characteristic. The hardness test (see Fig. 1-18) measures the material's resistance to plastic deformation. There has always been a minor dispute as to whether this test was nondestructive or destructive, since there usually is an indentation made on the surface of the material. If the hardness test is made without indentation (as is the case when using eddy currents or ultrasonics), it can be considered truly "nondestructive."

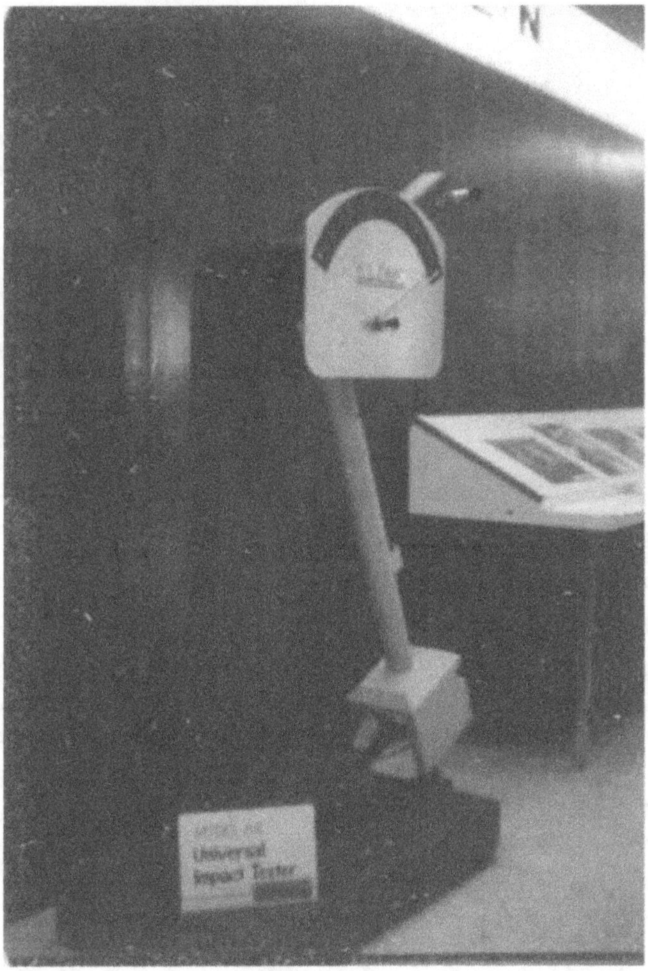


**Figure 1-17** Charpy impact tester. (*Courtesy of J. Devis Collection.*)

**Figure 1-18** Typical hardness tester. (*Courtesy of J. Devis Collection.*)

Although it is assumed in many cases that the test specimen is representative of the material from which it has been taken, it cannot be said with 100 percent reliability that the balance of the material will have exactly the same characteristics as that test specimen. Key benefits of destructive testing include:

- Reliable and accurate data are generated from the test specimen.
- Extremely useful data are generated for design purposes.
- Information can be used to establish standards and specifications.
- Data achieved through destructive testing are usually quantitative.
- Typically, various service conditions are capable of being measured.
- Useful life can generally be predicted.

Limitations of destructive testing include:

- Data apply only to the specimen being examined.
- Most destructive test specimens cannot be used once the test is complete.
- Many destructive tests require large, expensive equipment in a laboratory environment.

Benefits of nondestructive testing include:

- The part is not changed or altered and can be used after examination.
- Every item or a large portion of the material can be examined with no adverse consequences.
- Materials can be examined for conditions internal and at the surface.
- Parts can be examined while in service.
- Many NDT methods are portable and can be taken to the object to be examined.
- Overall, nondestructive testing is cost effective.

Limitations of nondestructive testing include:

- It is usually quite operator dependent.
- Some methods do not provide permanent records of the examination.
- NDT methods do not generally provide quantitative data.
- Orientation of discontinuities must be considered.
- Evaluation of some test results is subjective and subject to dispute.
- While most methods are cost effective, some, such as radiography, can be expensive.
- Defined procedures that have been qualified are essential.

In conclusion, there are obvious benefits for requiring both nondestructive and destructive testing. Each is capable of providing extremely useful information, and when used jointly can be very valuable to the designer when considering useful life and application of the part.

## 1.5 Conditions for Effective Nondestructive Testing

There are many variables associated with nondestructive testing that must be controlled and optimized. The following are major factors that must be considered in order for a nondestructive test to be effective:

1. *The product must be "testable."* There are limitations inherent with each of the nondestructive test methods, and it is essential that these limitations be known so that the appropriate method is applied based on the variables associated with the test object. For example, it would be very difficult to provide a meaningful ultrasonic test on a small casting with very complex shapes and rough surfaces. In this case, it would be much more appropriate to consider radiography. In another case, the object may be extremely thick and high in density, making radiography impractical. Ultrasonic testing, on the other hand,

may be very effective. In addition to the test object being "testable," it must also be accessible.

2. *Approved procedures must be followed.* It is essential that all nondestructive examinations be performed following procedures that have been developed in accordance with the requirements or specifications that apply. In addition, it is necessary to qualify or "prove" the procedure to ensure that it will detect the applicable discontinuities or conditions and that the part can be examined in a manner that will satisfy the requirements. Once the procedure has been qualified, a certified NDT Level III individual or other quality-assurance person who is suitably qualified to properly assess the adequacy of the procedure should approve it.
3. *Equipment is operating properly.* All equipment to be used must be in good operating condition and properly calibrated. In addition, control checks should be performed periodically to ensure that the equipment and accessory items are functioning properly. Annual calibrations are usually required but a "functional" check is necessary as a prerequisite to actual test performance.
4. *Documentation is complete.* It is essential that proper test documentation be completed at the conclusion of the examination. This should address all of the key elements of the examination, including calibration data, equipment and part description, procedure used, identification of discontinuities if detected, etc. These are all key elements. In addition, the test documentation should be legible. There have been cases in which the examination was performed properly and yet the documentation was so difficult to interpret that it cast doubt on the results and led to concerns regarding the validity of the entire process.
5. *Personnel are qualified.* Since nondestructive testing is a "hands-on" technology and depends greatly on the capabilities of the individuals performing the examinations, personnel must not only be qualified, but also properly certified. Qualification includes meaningful training through recognized and approved providers, and defined, documented experience.

## 1.6 Personnel Considerations

### 1.6.1 Introduction

Nondestructive testing effectiveness depends largely on the qualifications of the individuals performing the test. Most tests require thorough control of many variables. The role of personnel qualification has been controversial for several decades. There are many different positions to ensure an NDT practitioner is qualified. The most common approach is to use some form of formal certification by their employer or by a third-party certification organization. The controversies and different positions taken regarding this subject have resulted in the development of several different certification programs.

The term "qualification" generally refers to the knowledge, skills, and abilities of the individual performing the tests, which are achieved through a balanced blend of training and experience. "Certification" is defined as the documentation or testimony of an individual's qualification. Therefore, the processes involved in attaining a certification necessitate that the individual satisfactorily completes certain "levels" of qualification (training combined with experience) as a certification prerequisite. A simple way to relate to this system is to consider the steps involved in becoming a licensed driver. A candidate for a driver's license must go through a series of practical exercises in learning how to maneuver and control a motor vehicle and is required to review and understand the various regulations dealing with driving a vehicle. Once the regulations are studied and understood, and upon completion of actual practice driving a vehicle, the individual is ready to take the "certification examination." The government requires the applicant to pass both written and vision examinations, and also must demonstrate their ability to operate and maneuver the vehicle. Once those steps are completed, the candidate is issued certification in the form of a driver's license. Mere possession of a driver's license does not guarantee there will not be mistakes. It is obvious that there are individuals who have driver's licenses but are not necessarily qualified to safely drive a vehicle. This is often apparent during "rush hour" traffic. It is unfortunate that this same situation occurs in nondestructive testing. Since many individuals are certified by their employers, there are major variations within a given level of certification among NDT practitioners. Countries that have adopted some form of centralized certification do not experience variations to the same degree as those using employer certification approaches.

### 1.6.2 History

One of the earliest references to any form of qualification program for NDT personnel was found in the 1945 spring issue of a journal entitled *Industrial Radiography*, published by the American Industrial Radium and X-ray Society. This organization's name was eventually changed to the Society for Nondestructive Testing (SNT) and, ultimately, the American Society for Nondestructive Testing (ASNT). The original journal, *Industrial Radiography*, is now known as *Materials Evaluation*. An article in that 1945 issue entitled "Qualifications of an Industrial Radiographer" proposed that the society establish standards for "registration" of radiographers using some type of examination, which would result in a certification program. By the late 1950s, the subject of qualification or registration was being discussed more frequently. A 1961 issue of the same contained an article entitled "Certification of Industrial Radiographers in Canada." Then, in 1963, at the Society for Nondestructive Testing's national conference, a newly formed task group presented a report entitled "Recommended Examination Procedure" for personnel certification. Finally, in 1967, ASNT published the first edition of a "Recommended Practice" for qualification and certification of nondestructive testing personnel in five methods (PT, MT, UT, RT, and ET). This first edition was referred to as the 1968 edition of SNT-TC-1A which was a set of recommendations designed to provide guidelines to assist the employer in the development of a process that the document referred to as a "Written Practice." This Written Practice became the key procedure for the qualification and certification of the employer's NDT personnel.

SNT-TC-1A continues to be used extensively in the United States and many other countries. It is probably the most widely used program for NDT personnel certification in the world. Over the years, it has been revised, starting in 1975, and it continues to be revised at approximately 4-year intervals.

### 1.6.3 An Overview of SNT-TC-1A

This program—a "Recommended Practice"—provides for three different levels of personnel certifications. Individuals who are just beginning their NDT careers are usually referred to as "trainees." A trainee is in the process of gaining qualifications to be certified. After completion of formalized training and experience, the trainee is considered to be qualified and with satisfactory passing of examinations, the individual can be certified to Level I. The Level I is qualified to perform specific calibrations, tests, and evaluations for acceptance or rejection in accordance with written instructions or procedures and to record the examination results. A Level I individual should receive instruction and technical supervision from an individual certified to a higher level. After completion of additional training, and experience, the Level I can take additional examinations and be considered qualified to be certified as a Level II.

By definition, a certified Level II individual is qualified to set up and calibrate equipment and to interpret, as well as evaluate, the test results with respect to applicable codes, standards, and specifications. Generally, the Level II will follow procedures prepared and approved by the highest level in this system, a Level III. The Level II may be responsible for providing on-the-job training and guidance to the trainees and Level I personnel. It is expected that the Level II is thoroughly familiar with the various aspects of the method for which certified. A very important qualification requirement for the Level IIs is that they should be able to organize and report the results of the NDT performed.

The highest certification level described in SNT-TC-1A is the Level III. The Level III is a highly qualified individual, capable of establishing techniques, developing procedures, and interpreting codes, standards, and specifications. The Level III should also be capable of designating a particular test method for a given application, as well as specifying the correct technique and procedure to be used. In general, Level IIIs are responsible technically for the NDT operations for which they are qualified and assigned and should also have the capability of interpreting and evaluating examination results in terms of existing codes, standards, and specifications. The Level III should have sufficient practical background in materials, fabrication, and product technology applicable to their job function in order to establish techniques and assist in establishing acceptance criteria where none are available. The Level III should have general familiarity with other major NDT methods and be capable of training and examining Level I and Level II certification candidates.

In summary, SNT-TC-1A is a set of guidelines designed to assist employers in developing their own certification and qualification procedures. Since this program is a set of recommendations, it provides employers wide latitude so that their procedures can be tailored to the products being manufactured and to satisfy the requirements of the client. The application of this program has resulted in some distinct benefits and significant shortcomings.

#### *SNT-TC-1A Benefits*

This program has provided many benefits to the user. It is recognized as the leading recommended practice for the qualification and certification of NDT personnel. Some of the major benefits include:

1. It provides flexibility as interpreted and applied by each individual employer.
2. It requires employers to analyze their position on certification and document it through preparation of their specific written practice.

3. It implies customer responsibility through their acceptance and evaluation of the written practice, which, in turn, can ensure the adequacy necessary to comply with the terms and conditions of the contract or purchase order.
4. Through the written practice or procedure, the employer has an implied responsibility to train and to ensure that their personnel are experienced and competent.
5. It gives the employer excellent guidelines with respect to examinations that can be administered in the certification process.
6. It provides a common foundation that, even with its flexibility, provides an audit path.

#### *SNT-TC-1A Limitations*

There are a number of concerns with the use of recommended practices.

1. A lack of consistency resulting from employers taking advantage of the latitude contained in this recommended practice.
2. The fact that employers certify their personnel assumes a high level of responsibility and competence on the part of the employer. If employers do not understand the recommendations, or do not apply the intent of the recommendations, the effectiveness of their program can be questioned.

There are several paths that an individual can take to achieve employer-based Level III certification, which results in a lack of uniformity within an employer-based certification scheme. An individual can be designated by the employer as a Level III, examined by the employer to become a Level III, or can be examined through the use of an independent outside agency and, based on the results, become certified as a Level III by the employer. This in turn can result in inconsistencies in the Level IIIs.

#### *SNT-TC-1A Summary*

In summary, the limitations and weaknesses of SNT-TC-1A, as an employer-based program, will depend greatly upon the integrity of the employer. The solution to this problem lies with either a standard or code that will establish minimum requirements for certification, or a fully centralized approach to the certification of NDT personnel.

### 1.6.4 ASNT Level III Certificate Program

In 1976, the American Society for Nondestructive Testing initiated a program to issue Level III certificates to individuals through a "grandfathering" process that provided for issuance of the certificates without an examination. A select committee of highly recognized NDT professionals was appointed to review the applications submitted by those who met certain minimum requirements that had been previously established by an ad hoc Level III committee. During a window of 6 months, September 1, 1976, to February 28, 1977, over 1,300 applications were received and reviewed by the select committee. Upon completion of the review, a total of 713 individuals were granted a Level III certificate by grandfathering. Beginning in 1977, ASNT offered Level III examinations in Basic, PT, MT, RT, UT, and ET. Eventually, visual testing (VT), acoustic emission (AE), and infrared (IR) were added. Examinations offered now also include neutron radiography (NR), leak testing (LT), and magnetic flux leakage (ML).

The written examinations administered by ASNT cover the fundamentals and principles for each method as well as questions relating to application and establishment of techniques and procedures. The subjects of interpreting codes and specifications as they apply to each method are also included. In addition to the method examination, a basic examination is also administered that covers topics such as materials, fabrication, product technology, certification programs, and general knowledge of the basic NDT methods. The basic examination is required to be taken and passed only once and is a prerequisite to obtaining a certificate for any method. Once a Level III certificate is issued, the employer may have to administer additional examinations if required by their written practice. Once the conditions of the written practice are met, the employer can certify the individual. If the written practice states that the only condition for certification as Level III is the passing of the ASNT examinations, the employer may then certify with no further examinations other than a vision examination, which is a requirement for all levels. In this context, certification remains the responsibility of the employer.

### 1.6.5 Certification Program Overview

Some commonly used certification programs are briefly described here. There are certain elements, unique to each program, but a review of these programs points to a need for the uniformity and universal endorsement of a central program that will apply consistently to all NDT personnel.

#### *ASNT/ANSI-CP189*

This program is a standard for the qualification and certification of nondestructive testing personnel. Initially issued in 1991, and revised in 1995, 2006, and 2011, it provides minimum requirements for personnel whose job functions require appropriate knowledge of the various principles and application of the nondestructive testing method for which they will be certified. There are similarities between SNT-TC-1A and CP189. The major difference is that CP189 is a standard (not a recommended practice). Since it is still an employer-based program, it requires the employer to develop a procedure for the certification and qualification of their personnel. The possession of an ASNT Level III Certificate is a prerequisite to Level III Certification by the employer.

#### *NAS 410*

This standard was initially published as a military standard, MILSTD-410D, which was created in 1974 and was eventually superseded by MILSTD-410E in 1991. This program was primarily used by military agencies and government agencies, as well as by a number of prime aerospace companies and their contractors. It contained the minimum requirements for the qualification and certification for personnel in nondestructive testing and included provisions for training, experience, and examination. MILSTD-410E has been replaced with a National Aerospace Standard document referred to as NAS 410. The format of NAS 410 is similar in many respects to SNT-TC-1A and CP189. Like CP189, NAS 410 is a standard containing minimum requirements.

#### *NAVSEA 250-1500-1*

This program, which was initiated in the late 1950s, was specifically developed for use in the naval nuclear program. It is a form of central certification requiring candidates to go to an examination site to take examinations that consist of both written and practical parts. They are administered by an independent agency. It is generally believed that this program was developed as a result of the late Admiral Rickover's disdain for recommended practices in the areas of quality assurance and NDT.

### ISO 9712

This international standard was first published in 1992 and revised in 1999 and again in 2005. It has been adopted and is the basis for the certification schemes in over 40 countries. Many believe it is an excellent format for a truly international central certification program. It establishes a system for the qualification and certification of personnel to perform "industrial" nondestructive testing. Certifications under this program are accomplished by a central independent body, which administers procedures for the certification of NDT personnel. There are many countries that have adopted ISO 9712 as a basis for their country's certification program, including Canada, Japan, China, Korea, Russia, some South American countries and most European countries, and others.

### EN 473

EN 473 is European standard for the qualification and certification of NDT personnel and was first issued in January 1993 and revised in 2008. It contained general requirements regarding the examination and certification but did not contain specific technical requirements for the quality level of the examinations. Therefore, additional documents or programs were required to define and describe the examinations. It also included reference to ISO/EN 17024 for the accreditation of personnel certification bodies. Since there were many similarities between ISO 9712 and EN 473, there was widespread consensus to consider combining the two standards. With the introduction of the revised ISO 9712 in January 2013, EN 473 as well as the former version of ISO 9712 were replaced. Even though there are many similarities, the new version of ISO 9712 contains certain changes.

### The ASNT Central Certification Program

The ASNT Central Certification Program (ACCP) was introduced in 1996 and provides for the certification of NDT personnel through the administration of both written and hands-on practical examinations. It is designed to provide a form of centralized certification to those industries and employers who believe there are many benefits to this approach as compared to an employer-based system. Initially, it was not generally endorsed or widely used and was not broadly embraced across industry lines.

At the present time, ASNT is revamping the ACCP ASNT Level II and III programs to align with other global NDT certification schemes, in order to comply with ISO 17024, ISO 9712:2012 and ANSI/ASNT CP106. The goal is to provide industry with personnel who have achieved the highest standard of certification through written and practical examinations and provide those personnel with independent transferable certifications. The program will promote national and international acceptance of NDT certification and support qualification requirements described in SNT-TC-1A and CP-189.

### Other Major Certification Programs

**Certification Scheme for Welding and Inspection Personnel** This has been in use since the early 1970s and applies specifically to personnel involved with the examination of welds. Since its inception, over 20,000 individuals have been certified worldwide Certification Scheme for Welding and Inspection Personnel. This has been in use since the early 1970s and applies specifically to personnel involved with the examination of welds. Since its inception, over 20,000 individuals have been certified worldwide under this program. The administration of this program is the responsibility of The Welding Institute in England.

**Personnel Certification in Nondestructive Testing** This is a worldwide centralized certification scheme that was developed from guidelines established in the 1970s. Since 1985, it has superseded certification programs operated by a number of other organizations and is now one of the most widely recognized schemes in the world. It addresses industry sectors such as aerospace, casting, welding, wrought metals, and railroads. Over 15,000 certificates have been issued globally. It is ISO 9712 compliant.

**The Canadian Certification System** This began in 1960 as a centralized certification program and has remained a basically third-party program administered by the Ministry of the Federal Government of Canada (Natural Resources Canada). This program has implemented the provisions of the Canadian National NDT Standards. Standard CAN/CGSB-48.9712-95 complied with ISO 9712, 1992 edition. In December 2000, it was replaced with CAN/CGSB-48.9712, 2000 edition, which complies with ISO 9712, 1999 edition, and EN 473, 2000 edition. Broad-sector certification is offered in five methods (RT, UT, ET, MT, and PT) to three levels. Three levels of certification in radiography are also available for the aerospace industry.

*American Society of Mechanical Engineers (ASME)* has adopted a somewhat different approach to certification. In the standard ASME Nondestructive Examination and Quality Control and Certification Program—ANDE-1, the focus is first to the needs of the specific industry sector (SIS) for nuclear power related pressure vessels and piping (NSIS). Subsequently, ASME is addressing the non-nuclear pressure vessel and piping sector. As with other programs, ASME will provide a third-party certification based on a central examination; however, the approach to training and experience is markedly different. Unlike other programs, ANDE-1 is based on industry established job-task-analysis (JTA) which layout the knowledge and skill important to NDT performance and these tasks must be successfully completed and documented through use of a qualification checklist or qual-card. Successful documented completion of the qual-card is the necessary precursor to the certification examinations.

Some countries, including France (COFREND), Japan (JSNDI), and others, have centralized certification programs that either comply with, or are patterned after, ISO 9712.

## 1.7 Certification Summary

As evidenced by the number of diverse certification schemes that are in existence worldwide, some would say that NDT personnel certification is in a state of chaos. It has become more evident that a globally recognized, centralized certification program must be adopted and agreed upon by all countries utilizing NDT. A good move in this direction was the harmonization of ISO 9712 and EN 473 into the new ISO 9712 Standard. SNT-TC-1A certification will most likely remain in use for certain industries, but the benefits of central certification far outweigh the benefits and possible cost-effectiveness of employer-based certification programs. The bottom line is that it is essential that NDT practitioners be qualified and that there must be an independent, unbiased system for evaluating their qualifications and establishing their credentials. The technology of NDT and its predicted growth depends on this.

## References

McMaster, R. C. *Nondestructive Testing Handbook.* Ronald Press, New York, 1959.

Murphy, R. V. "Global Impact of International Standard ISO 9712." *5th International Conference on Certification and Standardization in Nondestructive Testing,* September 26–27, Berlin, Germany.

Posakony, G. J. "A Look to the Past—A Look to the Future." Lester Honor Lecture presented at the National Fall Conference of the American Society for Nondestructive Testing, September 27–30, 1976 in Houston, Texas in honor of Dr. Horace H. Lester, NDT pioneer. Published in the December 1976 issue of *Materials Evaluation.*

# CHAPTER 2

# Discontinuities—Origins and Classification

Structural materials are composed of atoms and molecules that ideally possess material continuity extending down into the submicroscopic scale. This connectivity of material is critical for the transfer of loads through a structure and in many cases to provide an impervious barrier to an external environment. Structural materials also tend to be microscopically inhomogeneous, composed of microscopic regions with different chemistries and/or atomic arrangements. The sizes, shapes, dispositions, amounts, and character of these microscopic inhomogeneities strongly influence the properties of engineering materials.

Designers generally assume and require some level of structural continuity, homogeneity, and definition of material properties. However, absolute homogeneity and continuity never exist in any engineering component. Discontinuities must be expected at some level, and only when exceeding a threshold of a standard are discontinuities considered to be defects in a component. Defects are considered to have the potential to challenge the reliable and safe operation of a component.

Spatially sharp departures from material homogeneity and continuity inside a component at any level of magnification are structural discontinuities. Engineering materials always possess some relatively small geometric discontinuities, many of which are acceptable. Examples of discontinuities include voids, inclusions, laps, folds, cracks, chemical segregation, and sudden changes in microstructure.

Sharp transitions in surface continuity and contour are also considered to be discontinuities and may be defects. Geometric surface discontinuities include sharp angles, notches, gouges, scratches, galling, fretting, pitting, and weld undercutting. Surface discontinuities in engineering structures are unacceptable when they degrade the component performance or durability below the expectations of design and when they challenge the operability, reliability, and life of a component.

Identified discontinuities are evaluated as either rejectable or nonrejectable conditions. A primary goal of nondestructive examination of engineering materials is to determine whether or not the continuity and homogeneity of a component are adequate for its use. The evaluation of discontinuities usually is made in reference to a design basis and generally includes a code or rule-based criteria for acceptance and rejection. A complete evaluation of discontinuities includes their location and number in a component, and the shape, size, orientation, and identification of the origin and characteristics.

Codes and standards for defect characterization are necessarily conservative. The determination of material properties of a component in conjunction with the complete characterization of the discontinuities may be used in a fracture mechanics analysis for a reliable prediction of service of a component.

The origin and types of discontinuities depend primarily on the manufacturing processes and the service histories of engineering components. In some cases, the operational environment may either create discontinuities or induce the growth and development of preexisting discontinuities. Discontinuities in structures may originate at any manufacturing step and may be introduced either during the component use, maintenance, or repair.

Discontinuities may be categorized by the stage in processing at which they are introduced. An *inherent discontinuity* is one that is generated in the original production of an alloy material. Discontinuities occurring in the first forming stages from a primary alloy are called *primary processing discontinuities*, and any discontinuities that occur in subsequent forming and finishing steps are called *secondary processing discontinuities*.

The discontinuities that are created during the use of a component are called *service discontinuities*. Discontinuities additionally may be categorized in terms of the forming process that caused them. For instance, discontinuities generated during welding are called *welding discontinuities*. Any discontinuities caused by casting may be called *casting discontinuities*, and discontinuities generated in forging would be *forging discontinuities*.

An understanding of the origin of discontinuities is useful in determining the types and features of discontinuities that may be expected in the examination of a component. Awareness of the characteristics, locations, and orientations of discontinuities is most helpful and sometimes critical in defect detection and evaluation.

Most metallic engineering structures and components are composed of alloys.[1] Alloys are made from metals that are refined from ores.[2] The metals are cast into primary ingots that are remelted and mixed with other elements to make alloys. Alloy ingots are reformed into shapes by melting and casting, mechanical forming, welding, consolidation of alloy powders, and machining. Combinations of forming processes may also be used to shape components. Discontinuities and inhomogeneities may be introduced in the initial winning of metals from ores, in the primary production of alloys, and during any of these subsequent forming steps.

Discontinuities induced prior to use may propagate with the same or different character during operations, and they may also be generated during repair and during use of a component. Discontinuities typically may arise from application and sequences of loads, reactions with chemical and electrochemical environments, or from other environmental conditions. Energetic inputs that create discontinuities in structures include applied and induced loads, ranging from thermal stresses to damage from neutron bombardment in nuclear power reactors. Service discontinuities also can be generated from the microstructural changes that occur over time within some alloys.

[1] *Alloy* is used in the classical sense to mean any mixture of elements that possesses metallic properties.
[2] An *ore* is a mineral aggregate from which elements may be extracted at a profit. The extraction of metals is called *winning*.

## 2.1 Primary Production of Metals

Metals originate from ores that are generally rich in rock and relatively poor in metals. The extraction of metals from ores requires processes that often carry over some of the mineral impurities from the rocks and the chemical additions used in the refinement process. These impurities exist in large part as low-density nonmetallic slags that tend to float to the surface of the molten metal and are purposefully segregated and separated from the molten metal.

Small amounts of the slags are often retained within the metals during primary production and become incorporated in alloys. Slag impurities trapped in alloys are usually specified and controlled below allowable limits. However, occasionally the impurity contents are outside of acceptable limits, and even if within proper amounts on the average, impurities may not be distributed homogeneously within a primary ingot. Slag concentrations are usually found in the pipe regions of ingots. This localization of slag may be carried over into secondary forming operations to generate stringers and laminations. Slag inclusions within a structure are discontinuities called nonmetallic inclusions.

Although there is usually material continuity throughout structures that contain inclusions, the physical properties of the inclusions are deficient compared to the alloy matrix. Inclusions are usually weak and brittle in ambient conditions, and these inclusions may be soft and weak at high temperatures. Inclusions that are planar in shape may act as geometric notches within a material.

Inclusions in steel often are composed of the silicates and sulfides that carry over from the iron ore, the dissolution of refractory systems, and purposeful additions that are made to refine steel from its ores. Many of the nonmetallic inclusions are plastic at high temperatures and capable of being deformed during hot working of steels. These included particles will elongate in the direction of flow of the steel when steel is shaped at relatively high temperatures. Elongated inclusions in a steel structure are called stringers (Fig. 2-1). Stringers in adequately high number will

**FIGURE 2-1** Silicate stringer in cold drawn bar of steel (235 ×). [From *The Making, Shaping and Treating of Steel*, 10th Edition, U.S. Steel (1985).]

result in property differences as a function of direction, a property called *geometric anisotropy*, in the steel. Such steel will have a fibrous structure like wood and will be stronger along the elongated fibrous structure direction and weaker across the elongated structure. Rupture and cracking along the weak planes of this structure during subsequent forming or use generates a discontinuity called *lamellar tearing*. Lamellar tearing has the appearance of a split-wood grain structure.

Manganese is added to steel during primary production, in part to react with excess sulfur that might be in the steel, because manganese has a greater chemical affinity for sulfur than iron. However, if manganese is not present in adequate quantity to balance out the sulfur, brittle iron sulfide will form in the steel at the boundaries of the first grains that form. Manganese sulfide is generally in the benign form of rounded particles. These spherically shaped discontinuities are accepted in controlled amounts because of their random distribution through the structure and the rounded nature of the phase boundaries.

The inhomogeneities of inclusions common in steels and in other alloys are generally acceptable if the size and total inclusion content is relatively small and the distribution is adequately random.

## 2.2 Casting

Metals and alloys are transformed from a liquid to a crystalline solid by the extraction of thermal energy from a melt during casting in a mold. Casting is usually a step in the primary production of metals and in the initial production of alloys. Casting is also the starting point for many discontinuities in engineering components. Casting of alloys results in the change of a metallic liquid solution into a crystalline solid alloy. Virtually all of the engineering alloys are crystalline solids.[3] During casting processes, crystalline solids form inhomogeneously[4] from very small crystals on the relatively cold walls of a casting mold. The crystallization of a solid begins at discrete locations called nucleation sites, and alloy crystals grow from these sites into the alloy melt. These separated and individual crystals extend, consuming the melt, until they connect with each other and form a solid. In a single-phase solid, each of the individual crystals is a grain, and the place of the meeting of two grains is a grain boundary. Adjacent grains in single-phase material differ from each other only in the orientations of their pattern of atoms. The size, shape, and spatial and geometric disposition further characterize grains in a structure.[5]

The growth of grains in castings is sometimes directional and the most rapid growth tends to proceed in the direction of maximum heat dissipation. An extreme condition of this directionality in castings is called *ingotism* (Fig. 2-2). Directionality of grain structure may create difficulties in ultrasonic examinations, as is experienced in cast stainless steel piping.

[3]There is a class of engineering material called metallic glass that currently has special electronic applications. These are alloys that have some characteristic metallic properties and an amorphous atomic structure.

[4]Inhomogeneity of casting nucleation is the local and independent formation of small crystals starting at the liquidus temperature for most alloys and at the solidification temperature for metals.

[5]The spatial disposition of grains includes on occasion the nonrandom distribution of orientations of the grains. This material property of a structure is called "texture," and it is a measure of the anisotropy or directionality of material properties in an alloy.

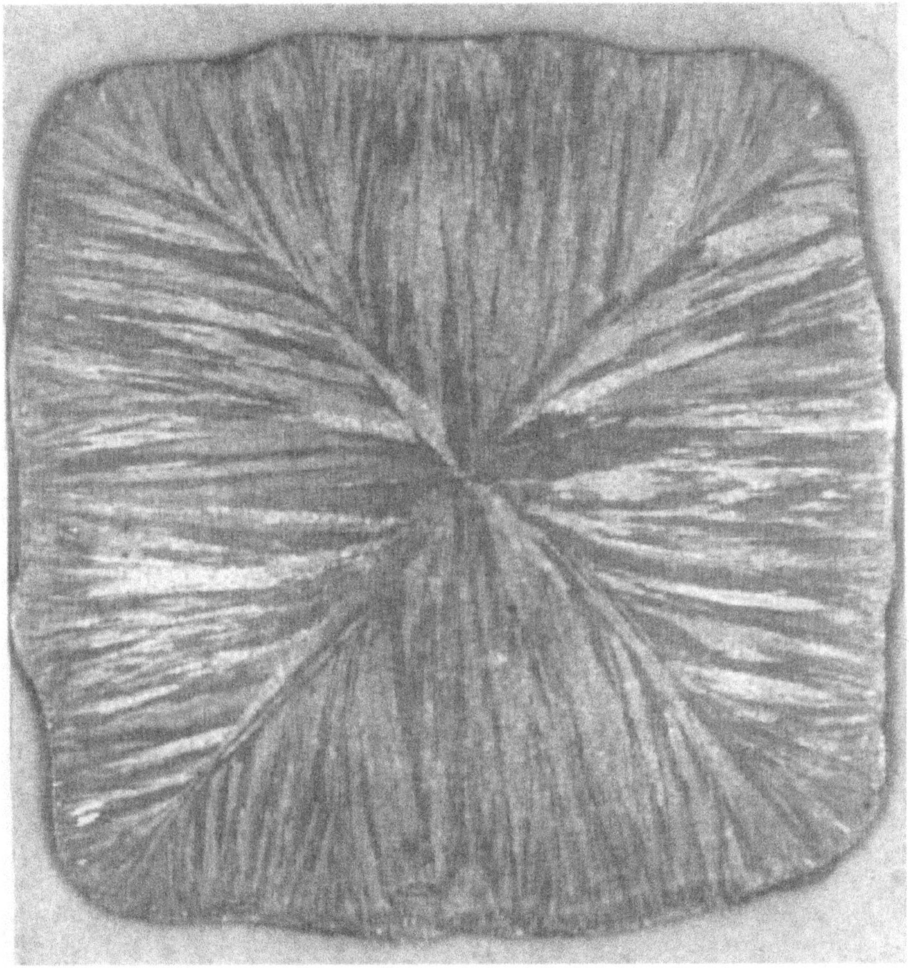

**FIGURE 2-2** Ingotism in cross section of stainless steel billet. [From *Liquid Metals and Solidification*, ASMI (1958).]

When metals and most alloys are solidified, the crystalline structures are denser than the liquids from which they came. Consequently, the casting volume is less than the melt volume after the liquid changes to a crystalline solid. This phenomenon is called solidification shrinkage. A natural consequence of solidification shrinkage occurs when liquid is entrapped within the crystalline shell of a solidifying casting. Unfilled spaces in the cast solid are created by the shrinkage of the included melt. These resulting voids and small holes in a casting are called shrinkage voids or shrinkage porosity. A central depression called "pipe" is formed when the shrinkage is concentrated in the top of an ingot (Fig. 2-3). Pipe often contains concentrations of slag and must be cropped from the ingot prior to secondary fabrication.

Solidification shrinkage also may induce cracking of the casting. This cracking may occur at relatively high temperatures when the solidified material is relatively weak.

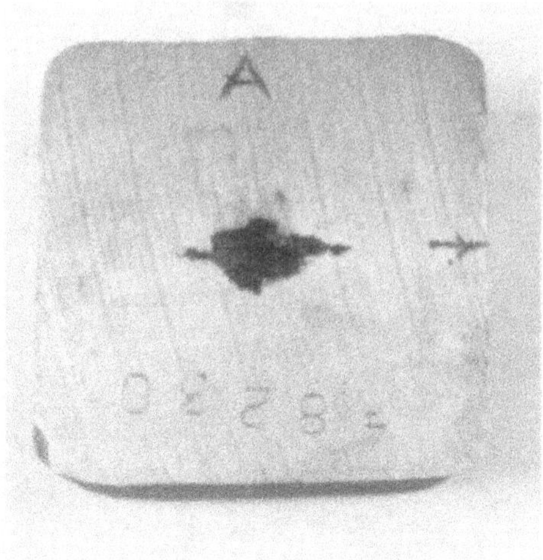

**FIGURE 2-3** A cross section through a billet showing pipe from solidification shrinkage. (*Courtesy of C. Hellier.*)

The cracking caused by thermal contraction of the cast solid also may occur at weak locations within the microstructure of an alloy, typically at grain and phase boundaries. Forms of casting discontinuities include pipe, voids, porosity, microporosity, interdendritic porosity, slag, and cracks.

Shrinkage voids and porosity are simply unfilled macroscopic regions of a casting that have been created by the entrapment of melt within the solidifying casting. Voids usually are found at the last regions of a casting to solidify, and occasionally they appear on the surfaces of castings. Shrinkage voids are potentially deleterious, weakening structures by reducing the continuity of load-bearing material.

Shrinkage porosity is a distribution of a number of small voids. These small voids are usually localized in clusters and exist in the regions of final solidification. Shrinkage porosity extends from the macroscopic to the microscopic scale. Porosity also reduces the load-bearing section of a component. If porosity is adequately limited in amount and uniformly distributed, it may have a negligible effect on the strength of a component. Extensive porosity can degrade the mechanical properties of a component.

Porosity also presents a potential problem when it provides a path for the breach of a component fluid boundary. Interconnected pores may readily provide leakage paths. Surface porosity may also reduce the fatigue resistance of a component. Microporosity may be found in the specific form of interdendritic porosity. This occurs when the last fluid to solidify is contained between the arms of crystalline growth forms called dendrites. The dendrite arms are spaced usually in a regular array, and the arm spacing is dependent on the solidification rate. The dendritic forms extend during solidification and the final form of the structure is a grain or multiphase structure. The fine interdendritic porosity will remain as a residue of the casting process. This form of discontinuity is rarely adequate in size, number, or distribution throughout a structure to challenge component integrity.

Gas evolution from melts may also cause porosity in cast structures. Gases are more soluble in liquid alloys and metals than in their corresponding crystalline solids. During the crystallization process, the gas is released from the growing crystal into the

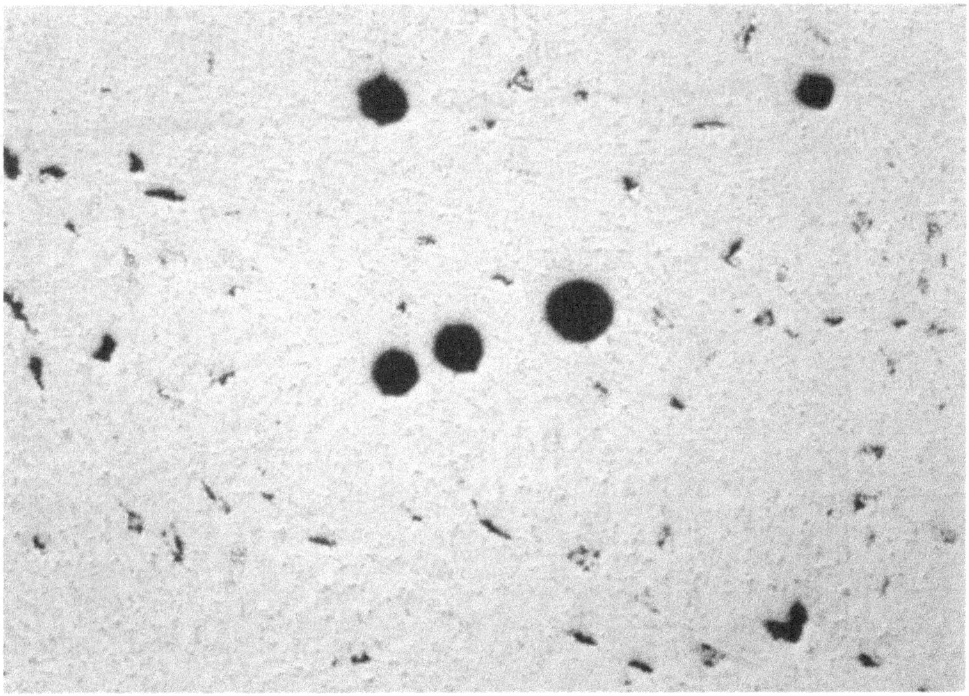

**FIGURE 2-4** Gas porosity in aluminum alloy die casting. (*Courtesy of R. B. Pond, Jr.*)

remaining liquid. If the gas is released in the interior of the casting, it may be entrapped and exist in the form of gas voids or gas porosity (Fig. 2-4). Gas porosity has some characteristics that differ from those of shrinkage porosity. Unlike shrinkage porosity, gas develops in the solidifying melt with pressure. The gases will try to move within the melt to lower-pressure regions. This occasionally leads to elongated void structures called wormholes. These elongated pores form as the gas tries to move away from the solid–liquid interfaces in the casting. This elongated structure of pores may provide a ready leakage path through the boundary of the casting. Small breaches of the surface by gas porosity are called pinholes.

Gas porosity, like shrinkage porosity, will decrease the load-bearing capacity of a component. Dissolved gas emanating near the surface of a component may cause deformation and can even cause splitting of the surface material. The appearance and name of this form of surface and near-surface discontinuity is "blistering."

There are occasions when a casting mold may contain regions that are not completely filled. These discontinuities in structure are called casting cavities. This condition is caused by blockage of the melt from parts of the mold cavity due to early solidification within a feed passage.

Shrinkage and the stresses arising from shrinkage after casting may be adequate to rupture a casting, as shown in Fig. 2-5a. Tensile stresses introduced by shrinkage of a hot part against the constraints of a mold may be adequate to cause local cracking of the component at relatively high temperatures. This is called solidification shrinkage cracking, which may occur through the crystal grains of an alloy or around the boundaries of grains

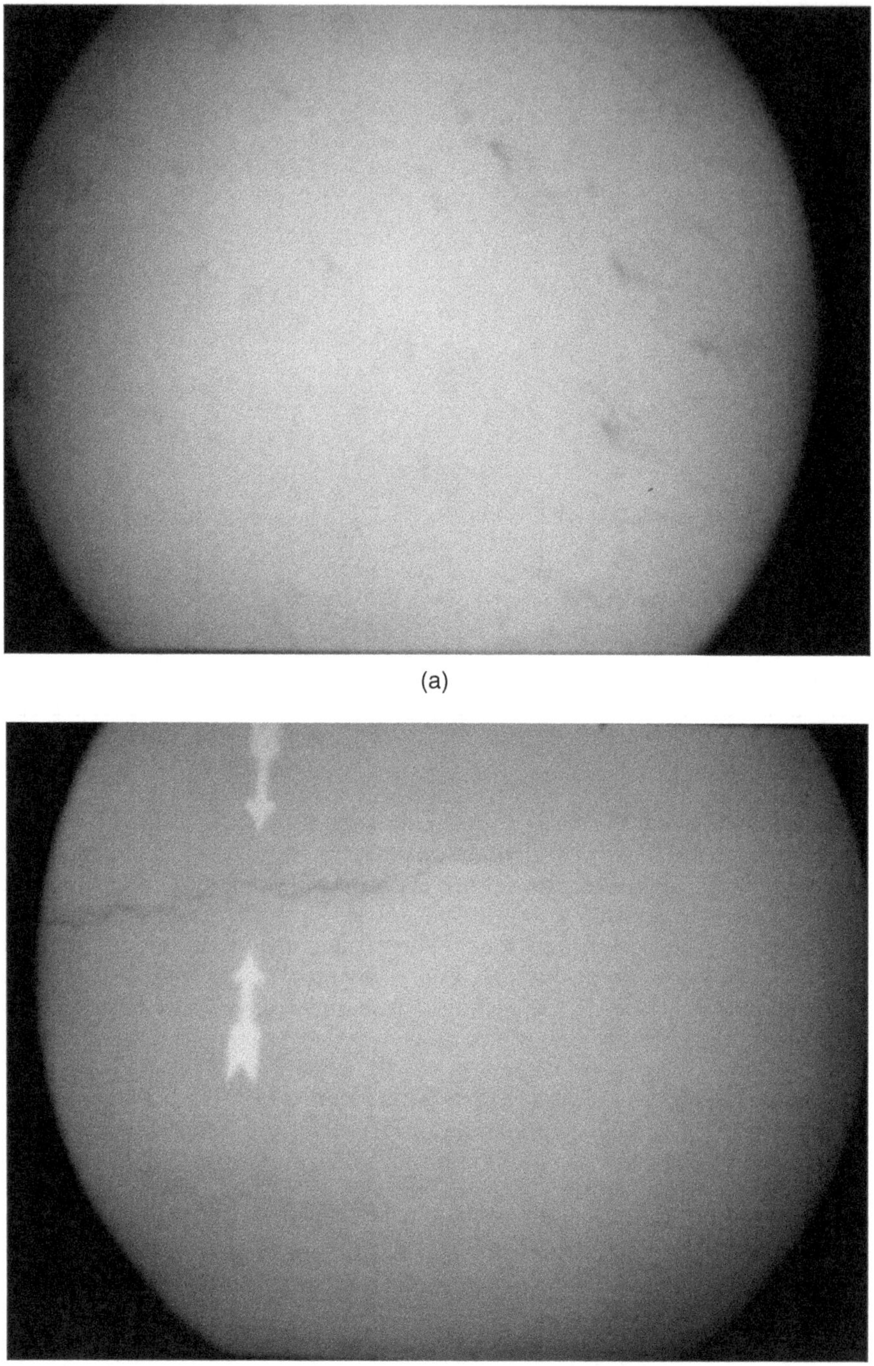

(a)

(b)

**Figure 2-5** (*a*) Radiograph of shrinkage in a casting. (*b*) Radiograph of hot tear in a casting. (*Courtesy of C. Hellier.*)

and solid phases. Shrinkage cracks may later propagate during subsequent heating and mechanically forming in a component and cause failure.

Hot cracking and hot tearing may also occur due to tensile stresses during hot mechanical forming of a component as seen in Fig. 2-5b. Hot cracking usually is considered to be internal cracking; hot tearing involves cracks open to the surface of a casting. These solidification and hot cracking discontinuities are deleterious to service and are unacceptable.

Inclusions of foreign objects in casting typically occur when pieces of refractory are broken off into the melt. The refractory may come from the primary production process or from refractories used in casting molds and pouring crucibles. There is another form of discontinuity characteristic of casting called a cold shut. This is a discontinuity on the surface of a casting caused by a stream of liquid metal solidifying on and not fusing with a previously solidified part of the component. A cold shut may also refer separately to a plugging of a channel in a mold by early solidification, which then prevents the entire mold cavity from filling, resulting in casting cavities.

A scab is a surface discontinuity that has a rough and porous texture, usually with an underlying cavity caused by a refractory inclusion near the surface. Scabs are found more commonly in thin sections of castings.

A mold parting line may give rise to a geometric discontinuity on the casting called a casting seam. A mold parting line seam is a potential indicator of a casting process.[6] Such a linear surface discontinuity, if too deep, may be the location of a fatigue crack during operation of the component.

An inherent cause of microscopic inhomogeneity in materials comes from phase separation in alloy solids. A phase is a region of matter that is homogeneous and physically distinct from its surroundings, independent of the size, form, or disposition of the phase. Most engineering alloys are composed of structures that are mixtures of numerous microscopic phases that are different from each other in chemistry and/or crystal structure. These structures are a consequence of the natural separation of phases that occurs over discrete regions of temperature and chemistry, with tendencies as shown in equilibrium phase diagrams.

In metallurgical systems, an equilibrium-phase diagram is an empirical map of the phases in alloys that exist in equilibrium as a function of temperature and composition. Although most engineering alloys generally are composed of many different phases and characteristic geometries of the combinations of phases called constituents, the principle of chemical-phase separation may be seen by example. A most important equilibrium binary-phase diagram is that of the iron-iron carbide (iron-carbon) system that is the basis of our understanding of steel and cast iron metallurgy (Fig. 2-6). This diagram provides insights to the heat treatment of steels.

An equilibrium-phase diagram is an empirical, graphical representation of the transitions of phase as a function of alloy composition and temperature. Binary diagrams (a map for an alloy of two components) are easily interpreted. The composition in percentage of the two components and a temperature of that alloy is a point on the phase diagram. That point will reside either in a single- or in a two-phase region.[7] If the alloy at temperature resides in a two-phase region, only the boundaries of that region exist in

[6] There are casting processes, such as investment casting and continuous casting, which leave no mold parting lines. Parting lines may also be dressed out of a component after casting.

[7] The point may also rest on a line that contacts three phases that are in equilibrium for a two-component system.

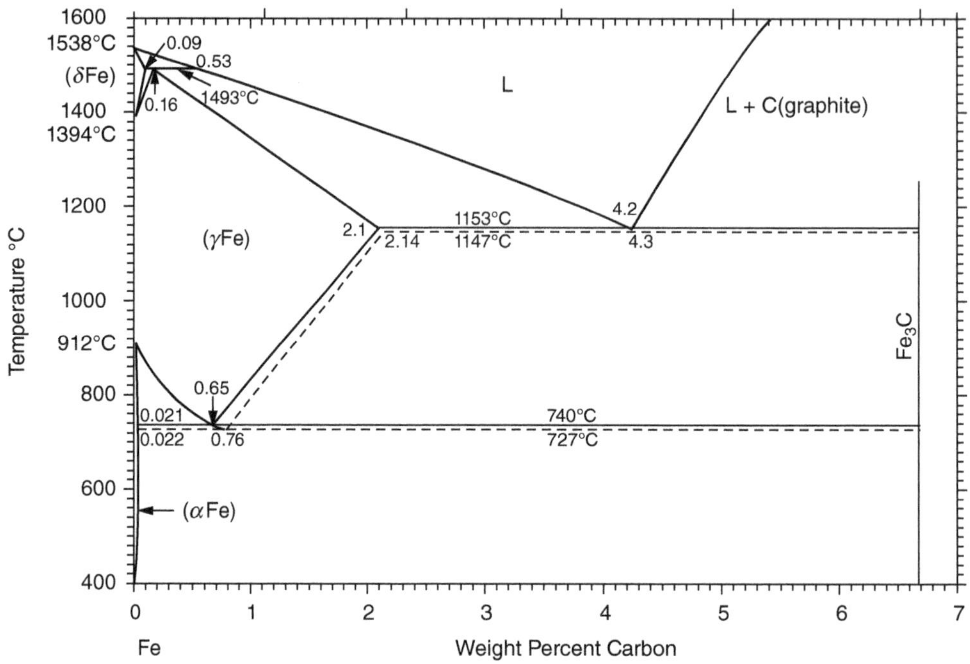


**FIGURE 2-6** The equilibrium binary phase diagram for iron-carbon. [Courtesy of ASM Volume 3, *Alloy Phase Diagrams* (2002).]

the material. A chemical separation of the phases occurs in the two-phase region. Those two phases are different in chemistry from each other with the constraint that the average chemistry of the alloy is the original composition. If an alloy at temperature is in a single-phase region of the diagram, then the chemistry of the phase is the average chemistry of the alloy. In many engineering systems, the pressure either is not a variable or has little influence on phases that are produced. This simplifies the rules of phase equilibrium for metallic systems. We can best explore this phenomenon with examples.

Iron is an element with an allotropic transition, and this element will display a rearrangement of atoms in its condensed crystalline structure at a definite temperature. Above 910°C iron exists in a crystalline form called face-centered-cubic (FCC). Many elements and alloys have this crystal structure. In iron, this form is also called *austenite* and *gamma iron*. Below 910°C, iron is found in a different crystal structure called body-centered-cubic (BCC), which is variously called *ferrite* and *alpha iron*. Only exactly at the transition temperature does alpha and gamma iron coexist for pure iron.

When carbon is introduced to iron, austenite and ferrite will exist over a range of temperature (Fig. 2-6). At any given temperature within the two-phase region of austenite and ferrite, the carbon content of the two phases will be fixed by the boundaries at that temperature. When the temperature of an iron carbon reaches 727°C, nearly pure ferrite is in equilibrium with austenite of 0.76 weight percent carbon and also with a compound $Fe_3C$, iron carbide. As heat is extracted from the iron-carbon alloy at 727°C, the austenite decomposes to ferrite and iron carbide (cementite). If the decomposition occurs at a relatively slow rate, the microstructure is a characteristic lamellar mixture of ferrite and carbide called pearlite. A micrograph of 0.72 percent carbon steel is shown in Fig. 2-7.

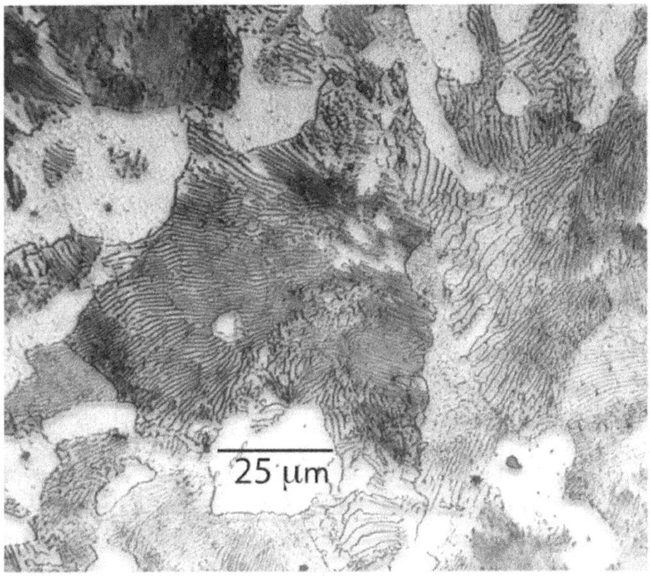


**Figure 2-7** Pearlitic microstructure of the iron-carbon eutectoid. (*Courtesy of Professor Robert Pond.*)

Microscopic mixtures of numerous solid phases are the rule for engineering alloys. The physical properties of the different phases are not the same. The separation of the different solids on a microscopic scale does not necessarily represent a structural discontinuity of concern. However, the sizes, shapes, and distributions of alloy solid phases characterize the microstructure, which is critical in determining the physical properties of the alloy. Macroscopic segregation in the casting operation can give rise to inhomogeneities that are deleterious in an alloy. Macroscopic chemical segregation is a condition that often occurs in very large alloy castings, such as ingots. The segregated glassy material formed during casting is called slag. Relatively large local aggregates of slag challenge the mechanical strength of an alloy.

There are also changes in alloy microstructure that occur at relatively high temperatures where a nonequilibrium structure changes toward an equilibrium structure. Graphitization of carbon-molybdenum steels is a good example of this phenomenon. The carbides in this alloy steel will decompose to iron and graphite at high temperatures and the location of the decomposition will favor higher tensile stressed regions. Therefore, stressed regions of these steels will change their microstructure over time and precipitate weak particles of graphite. These graphitic regions may eventually initiate cracks.

## 2.3 Cracks

A planar breach in continuity in a material is called a crack. Usually, a crack is envisioned as a physical breach of a material that had previously been continuous. Forces from forming operations or usage of components are generally the cause of cracks. Cracks from mechanical forming operations may be internal such as seen in some forging bursts, and cracks also may emanate from surfaces.

Cracking may occur during casting operations. Solidification cracking may occur when a cast component with reentrant angles solidifies and then cools and shrinks while it is restrained by the mold. Metallic materials are weaker at high temperatures, and the forces from differential contraction may be enough to cause cracking.

Cracking may also be induced during operations. Most commonly cracks arise from cyclic tensile stresses. These cracks are fatigue cracks. Cracking may also be induced by corrosive effects in a phenomenon called stress corrosion cracking.

Additionally, the evolution of gas in the interior of a casting at high temperatures may cause pressure that is adequate to crack the interior of the casting. There are also crack-like discontinuities that are not formed by forces. These are disruptions in continuity of a component that are caused by the overlaying of two surfaces that are not joined, except along a boundary. An example of a crack-like discontinuity would be lack of fusion in a welding operation. The surface of the weldment is not fused to the base material substrate, and this discontinuity is geometrically similar to a crack.

Lack of fusion, like a crack, can initiate fatigue cracking during component use. Lack of penetration in a welding operation, generally at the root of a weld, may also provide a discontinuity in a structure that has crack-like geometry. For instance, incomplete penetration in an autogenous weld used to join pipe will yield two mating base material surfaces that are a crack-like discontinuity in the joint (Fig. 2-8).

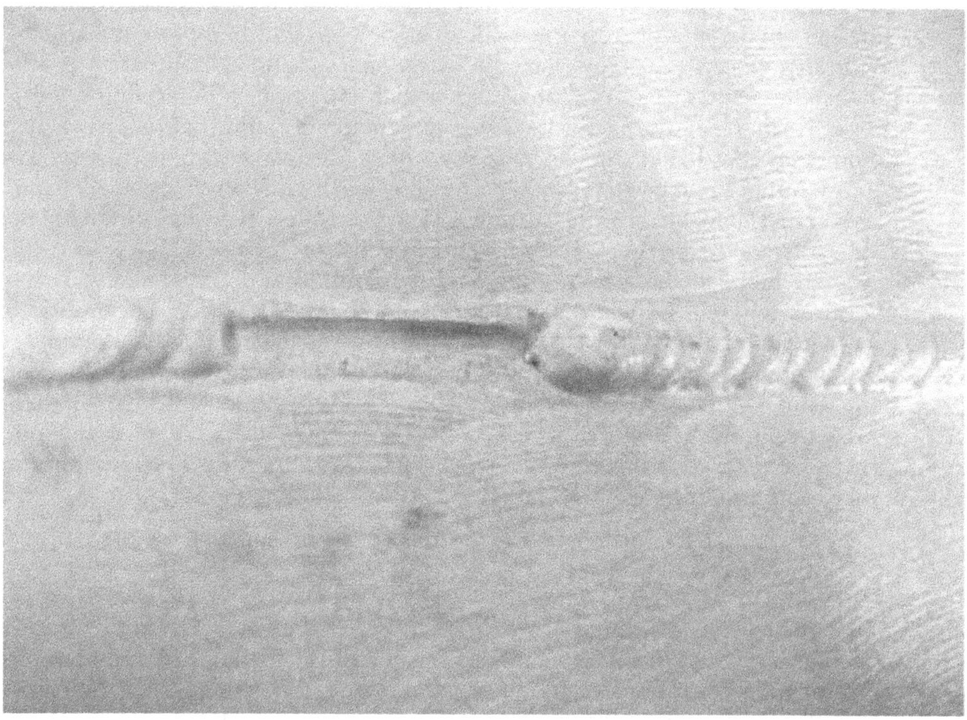

**FIGURE 2-8** Lack of penetration in weld. (*Courtesy of C. Hellier.*)

## 2.4 Welding Discontinuities

There are many welding processes, and each may give rise to discontinuities that are common to casting and occasionally unique to the welding process. Inclusion of nonmetallic material is common to welding processes that use protective glasses called slags. Welding over slag-covered surfaces may trap slag inside the weldment. This type of weld discontinuity occurs in shielded metal arc (SMAW), submerged arc (SAW), and flux core arc welding (FCAW) processes.

Slag inclusions in welding have the same effect as the slag inclusions in cast components. Slag weakens a structure by limiting its load-bearing capability. If rounded in shape, the major effect of slag is a reduction in load-bearing material. Generally, there are allowable amounts of slag above which the slag condition is rejectable. Welding discontinuities include solidification cracks that are caused by the casting operation.

There is a stress condition that arises in welding operations due to the differential expansion and contraction of the base material. This is caused by the unrestrained expansion of the base material into the melt during welding. After solidification of the weldment, restraint against the contraction of the base material by the solid bridge across the joint induces stresses as the weld cools. The differential expansion and subsequent restraint on cooling may give rise to forces that exceed the ultimate strength of either the base material, weldment, or heat-affected zone. This may cause deformation, cracking, fracture, and at the least, residual stresses.

These damage effects are more pronounced in thick wall material and may be minimized by adequate preheating of the base material. The preheating reduces the differential in expansion and contraction during welding. Post–weld heat treatment (PWHT) is sometimes additionally required to relieve the residual stresses in a weld and to make beneficial changes in microstructure.

Other cracks in welds are classified according to their location or the conditions of occurrence. For instance, cracking that is created in the base material adjacent to the fusion zone of a weld is called under bead cracking. Cracking occurring in the process of post–weld heat treatment is called reheat cracking. Discontinuities called lack of fusion are found in local unfused regions between beads in a weld or between the base material and the weld. Incomplete melting of a substrate in a welding process causes lack of fusion in the weldment. Any crack-like discontinuities are potential initiation sites for fatigue cracks and they are usually disallowed by code.

A microstructural change that occurs in welding is called the heat-affected zone (HAZ). The HAZ is the unmelted region of base material adjacent to the weldment that has a microstructure that is altered by the high temperature of the welding operation (Fig. 2-9). This HAZ has effectively been through a locally applied heat treatment during the welding process.

In the case of ferritic steels, the microstructural change is largely due to the allotropic transformation[8] of the steel. This transformation occurs as a function of temperature–time profiles in the HAZ, and the microstructure follows the results of routine heat treatment of a steel. If the transformation occurs at too rapid a rate, because of rapid cooling from the heat sink of the base material, then the heat-affected zone may form a metastable structure called martensite. Martensite is hard and brittle and prone to crack

[8]Allotropic transformation is the change in crystal structure, in this case, in a specific temperature range.

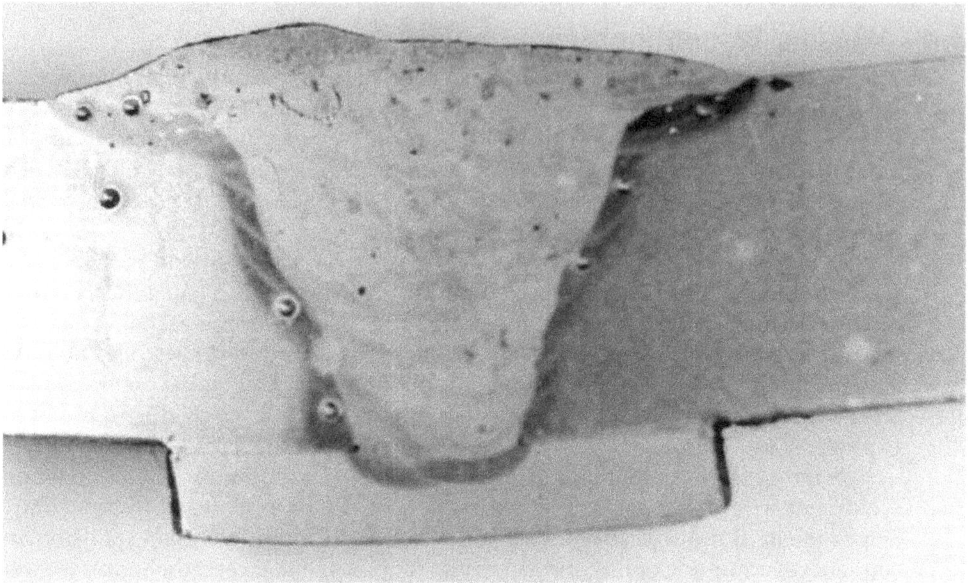

FIGURE 2-9 Cross section of a weld showing heat-affected zone as a dark band adjacent to the weld. (*Courtesy of R. B. Pond, Jr.*)

under quenching stresses and operational loads. Martensite formation tends to occur in medium- and high-carbon steels and in high-strength ferritic alloy steels. Mitigation may require preheating the base material to reduce the rate of cooling so as to reduce or eliminate the martensite transformation. Proper post–weld heat treatment will decompose the martensite into a benign microstructure of ferrite and carbides.

Other microscopic and material property changes may occur in the HAZ of different alloys. For instance, an alloy that is strengthened by work hardening will soften in the HAZ from a weld. Heat-affected zone properties may result in some damage mechanisms being available. For instance, the HAZ of AISI 304 stainless steel with relatively high carbon content has a microstructure that may be susceptible to stress corrosion cracking.

Similar to casting operations, welding may introduce gas into the molten weldment that subsequently is released, resulting in gas wormholes and porosity. There is also the possibility of damage due to incorporation of hydrogen atoms into a steel weldment. This damage typically occurs when water and organic molecules are split apart and disassociated in the high-energy welding process. The melt will incorporate hydrogen that after solidification will exist in nonequilibrium concentrations in the solidified weld.

The hydrogen atoms migrate in the crystal structure until by chance two come close to each other and combine, forming molecular hydrogen. This molecule will not be mobile and will act as a sink for other hydrogen atoms. After a time, there will be multiple small hydrogen bubbles formed in the interior of the crystalline solid. The hydrogen will typically precipitate at inclusion boundaries and grain boundaries.

If the alloy is relatively of low strength and ductile, the material around the gas bubbles will plastically deform to accommodate the damage. If the alloy is very strong, it will resist the formation of the bubbles and the first accommodation to stress may be cleavage and fracture of the crystalline structure. This damage is called hydrogen cracking or delayed cracking and it occurs in the interior of the welded component, usually in the HAZ.

This mode of damage requires time to occur and is the reason for mandatory requirements for a waiting period prior to nondestructive examination for some high-strength steels.

Because of the sensitivity of hydrogen damage in high-strength steels, sources of hydrogen must be controlled. Requirements are found for dryness and cleanliness of the base material and for use of low-hydrogen consumables for the filler materials and fluxes.

There are geometric discontinuities associated with welding. One of these is the formation of a ditching at the weld toe. The weld toe is the edge of the fusion zone on the surface of a base material. This discontinuity is called undercut, and it has the effect of creating a "notch" effect on the surface (Fig. 2-10).

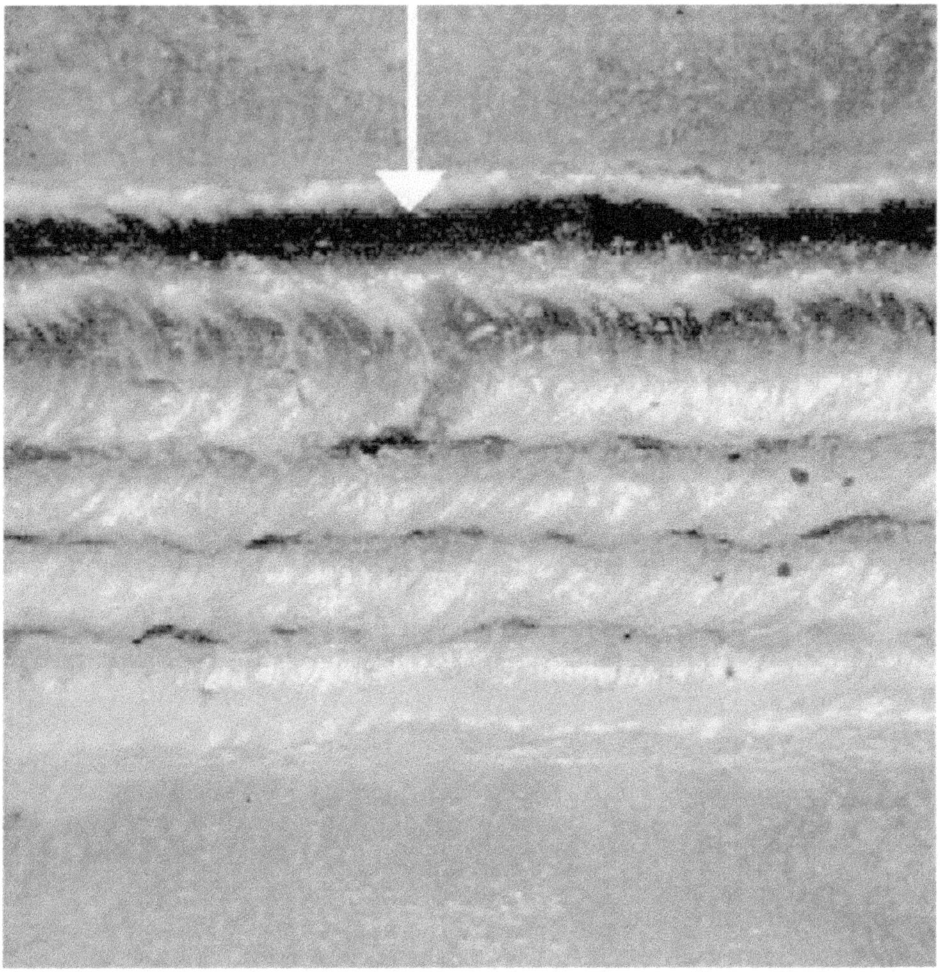

**FIGURE 2-10** Undercut at the edges of weld beads in a multiple pass weld. (*Courtesy of C. Hellier.*)

Excessive crown reinforcement of a weldment usually results in a discontinuous geometric transition with the base material. The geometric discontinuity at the toe of the weld in this case will be a region of stress concentration. When this condition exceeds a design limit, it is called excessive convexity. The same type condition may occur when the weld crown is concave. The concave discontinuity is called "excess concavity" when it exceeds a design or code criterion.

A joint that is incompletely filled represents a geometric departure from design. This condition is called lack of penetration. This may provide a geometric discontinuity in addition to inadequate load-bearing material in the weld joint. The condition of inadequate filling of a joint also exists in brazing, where it is called lack of fill.

Brazing is a technique that joins components by capillary action of and metallurgical bonding with a lower melting point filler material that is drawn into a narrow space between the parent materials. The flow of the brazing alloy is dependent on the gap, surface cleanliness, and temperature of the base materials. Inadequate coverage of the brazing alloy in the joint weakens the joint.

There also are brazing conditions of inadequate bonding called unbond. Inadequate cleanliness and contamination of the base material surfaces, incorrect brazing materials, and inadequate brazing temperature may cause unbond.

## 2.5 Discontinuities Resulting from Plastic Deformation

Forging is a process that forms an alloy by shaping it in a die under compressive loading. The flow of the metal is responsive to distributed loads, and the ability of the metal to flow is dependent on its temperature. The higher the temperature, the easier the plastic deformation and the greater the ductility.

It is possible that the stress resulting from a forming operation may exceed the strength of the material, causing the material to break apart. In a forging operation, this load-induced cracking is called a burst. A burst may be entirely internal or the cracking may extend to the surface of the component (Fig. 2-11).

The volume of a body is approximately constant when a component is mechanically formed. During the flowing of the alloy, material occasionally will lap over itself due to surface flow instability. The folded material will not fuse if the temperature is low and if the surface is contaminated with dirt. This type of discontinuity is called a fold or a lap. A long straight lap may be called a seam. Laps and folds have crack-like characteristics and they are considered to be rejectable discontinuities.

There are geometric discontinuities caused by pressing scale and debris left in forging dies into the surface of a component. These indentations are called scale pits.

**Figure 2-11** Multiple forging bursts in a steel blank. (*Courtesy of C. Hellier.*)

A crack that transects a section is called a split. Splits may occur in forging and in other forming techniques. Rolling, swaging, spin forming, and extrusion may give rise to cracks, laps, burst, splits, scale pits, etc.

## 2.6 Corrosion-Induced Discontinuities

Electrochemical corrosion of alloys may be thought of in terms of a battery. This form of corrosion requires an anode, a cathode, an electrolyte, and an electrical connection between the cathode and the anode. The degradation generally occurs on the anode of the cathode–anode couple. Many operational environments provide the opportunity for electrochemical corrosion and resulting discontinuities.

A common surface discontinuity caused by corrosion is pitting (Fig. 2-12). Pitting is a localized attack on a surface on which specific regions are anodic to the remaining surface and the attack, once started, persists at those areas. The pitting usually contains the associated debris of corrosion products that provides an

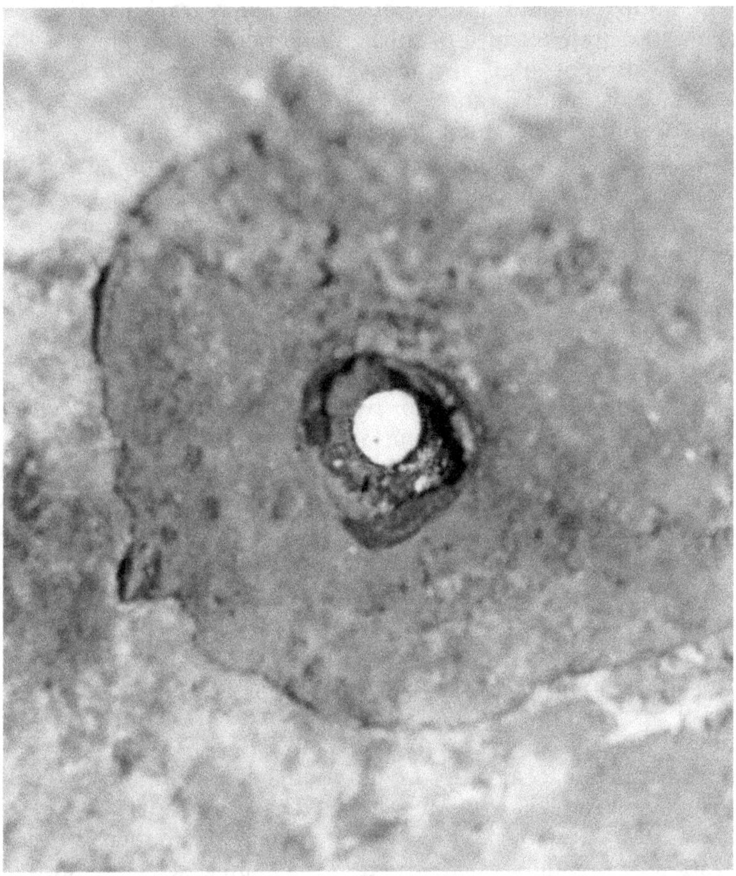

**FIGURE 2-12** Corrosion pit that penetrated a pipe wall (5x). (*Courtesy of R. B. Pond, Jr.*)

environment that assists in the continued local attack. Pitting behavior is a product of the alloy and the chemistry of the electrolyte. There are many combinations of alloy and electrolyte that will not result in pitting.

The forms of pits are variable and range over a wide spectrum of geometry. Pits may be conical in form; they may mushroom beneath the surface or take tortuous paths (a form called wormhole corrosion). Pitting damage is potentially deleterious. In addition to loss and penetration of the material, the rough surface of the pitting is a ready initiation site for fatigue cracking in components that are subjected to cyclic tensile loads. Additionally, the fatigue resistance of an alloy is often lowered in the electrolyte. An electrolyte that causes pitting will also be aggressive with respect to lowering the fatigue resistance of the component, and fatigue cracks may emanate from the pitted regions.

There are crack discontinuities that may be induced by corrosion. A cracking mechanism known as stress corrosion cracking (SCC) requires a metal or alloy within a range of microstructural conditions to be under tensile stress in one of a limited number of specific electrolytes for it to exhibit this behavior. Any alloy or metal may be susceptible to stress corrosion cracking under special conditions.

Stress corrosion cracks that run around the grain and phase boundaries of an alloy are known as intergranular stress corrosion cracking (IGSCC). They may also run across the grains (trans- or intragranular cracking). The cracking may be mixed mode, consisting of inter- and transgranular cracking. In most cases, stress corrosion cracks have many branches (Fig. 2-13). These cracks will begin on the surface and may be deleterious, causing penetration of the wall of the component and serving as a site of fatigue propagation in cyclic tensile loading conditions. Stress corrosion cracking condition is unacceptable.

A corrosion attack at the grain boundaries occurring over broad areas is called intergranular attack (IGA). The character of IGA often includes a checkerboard pattern of cracking on a surface. Although IGA does not usually penetrate the wall of a component, it does provide a ready initiation site for fatigue.

There is a mode of cracking discontinuity that is induced at the cathode of a corrosion cell. This is called hydrogen embrittlement. The electrochemical reaction at the cathode includes the reduction of hydrogen ions from the electrolyte. The atom of hydrogen that is formed on the surface of alloys is small enough to dissolve into the alloy. Over time, the hydrogen will migrate throughout the component. In the same way as embrittlement of high-strength material welds, the hydrogen will form molecular hydrogen at inclusions and grain and phase boundaries. The pressures associated with the included hydrogen gas are computed to be extremely high.

Ductile materials will plastically deform in the region of a bubble to accommodate its growth, but high-strength materials may resist plastic deformation and crack before they flow. The cracking is often by cleavage of grains in the material. The cracking may occur at the boundaries of grains. This internal cracking may occur over substantial regions of a component and cause a weakening of the component as well as producing initiation sites for fatigue crack propagation.

The fracture surfaces of hydrogen-induced cracking are typically shiny and faceted. These small cracks are called flakes. There are situations where corrosive attack will cause strips of material to disengage from a component surface. Elongation and segregation of the alloy microstructure may cause selective subsurface attack. This discontinuity is called delamination.

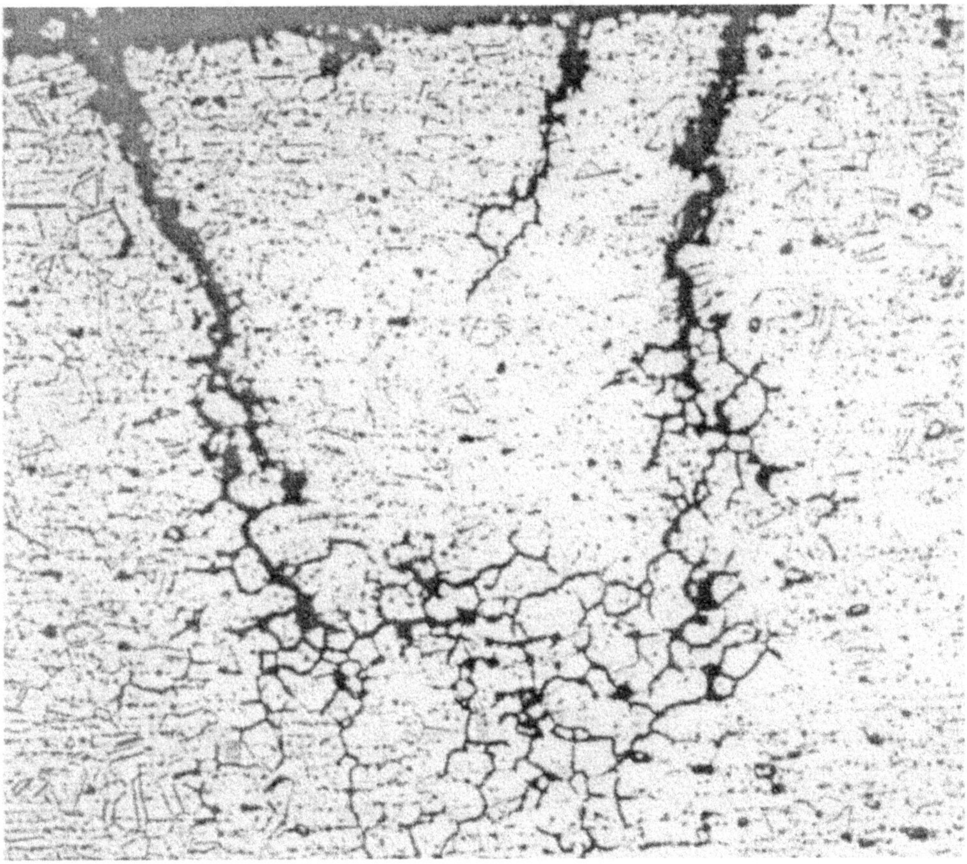

**FIGURE 2-13** Stress corrosion cracking in nickel alloy (200x). (*Courtesy of R. B. Pond, Jr.*)

## 2.7 Operationally Induced Discontinuities—Fatigue Cracking

Alloys that are subjected to cyclic tensile loading may exhibit surface cracks after a critical number of load cycles. This damage may occur even though the component has maximum tensile stress far below the yield stress for the material. Cracking will often initiate at a discontinuity in the form of a discontinuous geometric change or inhomogeneity in the alloy. Once begun, a fatigue crack usually will propagate irreversibly, a small amount with each cycle of tensile load. Eventually, the fatigue crack will grow large enough to cause a catastrophic failure of the component.

Fatigue cracks have unique characteristics. A fatigue crack will propagate on the plane(s) of maximum tensile stress. Often, the distributed tensile load is perpendicular to the planar surface of the material. In these situations, the fatigue cracking will be relatively flat. Usually, each cycle of cracking induces a variation in stress state local to the tip of the crack. Large excursions in load cause the crack to run slightly out of plane during the stress cycle. The consequence of this is a sequence of concentric ridges that

**FIGURE 2-14** Beach marks in a drill rod fractured by fatigue. (*Courtesy of R. B. Pond, Jr.*)

are called beach marks (Fig. 2-14). The beach marks radiate from the initiation site of the cracking and extend to the point where the crack extends catastrophically due to ductile failure or brittle failure.

Another characteristic of fatigue is macroscopic plastic deformation with no change of shape associated with the crack propagation. For this reason, fatigue cracking preserves the shape of the component and fatigue cracks are often hard to see, requiring augmented inspection techniques for detection.

Conditions of thermal cycling may cause stresses due to differential contraction and expansion of the cooled and heated surface compared to the bulk of a component. There is the possibility in this condition of the creation of thermally induced fatigue cracking. These cracks usually occur over an area and they are often in a checkerboard pattern. Thermal cracks are initiators of crack propagation due to other operational stresses.

In grinding operations on steels, it is possible to raise the temperature of a surface layer high enough to cause an allotropic transformation to generate martensite.

The associated thermal stresses can give rise to a cracking pattern that is called crazing. This is often associated with an oxide-induced coloration due to the high-temperature condition. Craze cracks are also initiation sites for fatigue.

## 2.8 Operationally Induced Discontinuities—Creep

High-temperature operation of alloys for long periods of time may give rise to an operationally induced cracking called creep cracking. This condition may occur even if the stresses are relatively low. The cracks are usually preceded by discontinuities in the form of creep voids, which are small and distributed. The creep voids grow in size and number until cracking links these voids. The catastrophic failure due to high-temperature creep is usually a relatively low ductility failure characterized by a thick-lipped rupture (Fig. 2-15).

**FIGURE 2-15** Thick lip fracture from creep in a power plant superheater tube. (*Courtesy of R. B. Pond, Jr.*)

## 2.9 Operationally Induced Discontinuities—Brittle Fracture

Brittle fracture occurs with limited ductility. Brittle fracture is usually catastrophic in nature and nearly always emanates from a discontinuity. Some materials such as ceramics and glasses are inherently brittle, and some alloys may be made brittle due to their process history and operating environment.

The formation of martensite in steel is an example of embrittlement caused by microstructural changes in an alloy. This is usually caused by an allotropic phase transformation to face-centered cubic austenite at relatively high temperature followed by rapid cooling and a reversed phase transformation to body-centered cubic ferrite at relatively low temperatures. Martensite is a hard and brittle material that is very flaw sensitive. Thermally induced decomposition of martensite to ferrite and iron carbide in a characteristic mixture called tempered martensite restores toughness and ductility to the steel.

Ferritic steels will change toughness and ductility as a function of temperature, becoming "glass-like" below a critical transition temperature. The critical temperature is a function of steel alloy chemistry and microstructure and it ranges from far below the freezing point of water to hundreds of degrees centigrade.

In all cases, brittle fracture will begin at a discontinuity in a structure. For a given discontinuity in a material, there is a critical static stress that must be exceeded before a crack may propagate. Conversely for a given material, environment, geometry, and stress, there is a critical flaw size, shape, and orientation that will propagate. This is a basis for categorizing crack-like discontinuities as critical in size.

There are conditions of precipitation of phases at grain boundaries of alloys that cause the grain boundaries to be brittle, and brittle fracture in those cases will tend to be confined to the envelope of the grain boundaries. Fracture about the grain boundaries has a characteristic appearance called a "rock-candy" fracture surface (Fig. 2-16).

**FIGURE 2-16** "Rock candy" fracture surface of a grade-8 bolt cracked by hydrogen embrittlement. *(Courtesy of R. B. Pond, Jr.)*

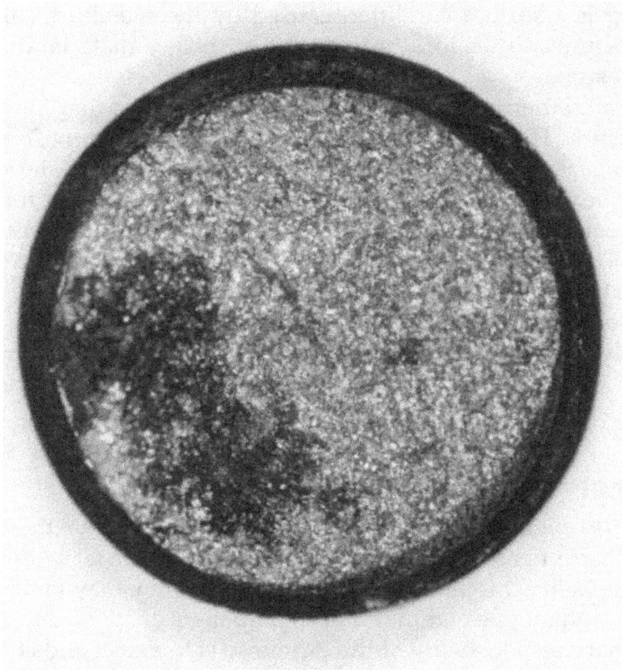

**FIGURE 2-17** Brittle fracture surface of fastener. Bright flecks are planar cleavage facets of ferrite in the steel. (*Courtesy of R. B. Pond, Jr.*)

In other cases, like for stress corrosion cracking and hydrogen embrittlement, the brittle cracking may occur through the grains on crystallographic planes of weakness. This gives a specular, faceted appearance to the brittle cracked surface (Fig. 2-17).

## 2.10 Geometric Discontinuities

Geometric discontinuities are often involved as sites of cracking and failure. These discontinuities may be created by deficiencies in the design, manufacture, operation, and repair of a component. Sharp transitions in surface contour will be locations of stress concentration, and under cyclic loading conditions these locations likely will be the first to crack due to fatigue.

An open and sharp relief in the surface of a component is called a notch. The dimensions and the root radius of a notch are critical characteristics of the extent of stress concentration that will exist under operational loads.

Gouges are usually considered to be relatively shallow grooves or cuts on a surface. Gouges may be caused by plastic deformation or by cutting action. A cut in a surface is a discontinuity similar to a gouge, except that it is usually caused by the removal of material from the surface by interaction with other components, similar to machining. Water flow and high-velocity gas flow (erosion), abrasive particle wear, mechanical wear, local plastic deformation, and local chemical action may cause gouges.

Galling is a surface condition caused by the metallurgical bonding of surfaces under pressure and the local tearing out of surface material during motion between the surfaces.

There are other machining-related discontinuities that have specific and unique characteristics. These include excessive undercut, sharp radius, scoring, scratching, and burring. Fretting is a wear condition that occurs during operations when two surfaces repeatedly rub in a reciprocating motion. The debris from the surface interactions and oxidation products are included between the surfaces and act as a grinding medium. The result of this motion is a roughened surface that invariably will have reduced fatigue resistance.

Surface finish is not usually considered to be a discontinuity. However, surface finish strongly influences the fatigue resistance of a material, and for given environmental conditions, smoother surfaces provide a greater resistance to fatigue cracking.

## 2.11 Summary

Challenges to engineering-material integrity largely involve the discontinuities in components. Disruptions in continuity may be either internal or on the surface. Discontinuities may be macroscopic or microscopic, and they may limit the strength, ductility, toughness, and fatigue endurance of a component.

A primary responsibility of the examiner is to detect and characterize discontinuities in a component. The location and type of discontinuity is dependent on the fabrication and operation history of a component. Examiners should have the advantage of understanding the relationships of product form and manufacturing and operational histories to the consequent discontinuities in a component.

Understanding the origin of discontinuities in engineering components should result in efficiencies and completeness in nondestructive examinations and enhancements in quality of examination results.

## 2.12 Glossary of Metallurgy and Discontinuity Terms

**Anisotropy** The material condition in which material vector properties change with direction.

**Annealing** Any treatment of metals and alloys at relatively elevated temperatures for the purpose of altering the properties of the material. The types of changes include softening, reducing residual stress, and recrystallizing.

**Artifact** An indication originating from a source other than a discontinuity that resembles an indication from a discontinuity.

**Billet** A solid, semi-finished, round or square product that has been formed by hot working (e.g., forging, rolling, and extrusion).

**Blister** A surface or near-surface discontinuity in metal and cast alloy components caused by gas evolution during casting. Small blisters are called "pinheads" or "pepper blisters."

**Blowhole** A discontinuity in the form of a hole in a casting or a weld that is caused by gas entrapment during solidification.

**Brazing** Joining of metals and alloys by bonding; the alloys have liquidus temperatures above 800°F (below the liquidus temperatures of the materials being joined).

**Brittle Cracking** The propagation of a crack that requires relatively little energy and results in little or no plastic deformation.

**Brittleness** The material quality that leads to fracture after little or no plastic deformation.

**Burr** Ragged edge on a part usually caused by a machining or grinding process.

**Burst** Fissure or rupture caused by improper rolling or forging.

**Capillary Action** The movement of liquid within narrow spaces that is caused by surface tension between the liquid and the substrates. The mechanism that is used to fill or penetrate a joint in soldering and brazing.

**Cast Structure** The microscopic and macroscopic distribution of grains and chemistry that is characteristic of a casting.

**Casting Discontinuities** Discontinuities that are generated in casting operations.

**Casting Shrinkage** The reduction in component volume during casting. The reductions are caused by (1) liquid shrinkage, which is the reduction of volume of the liquid as it cools to the liquidus temperature; (2) solidification shrinkage, which is the total reduction of volume of the alloy through solidification; (3) the shrinkage of the casting as it cools to room temperature.

**Charpy Test** An impact fracture test that characterizes the energy required to break a standard specimen. The test machine is a pendulum hammer.

**Chatter** The vibration of a tool in a grinding or cutting operation. Also, the wavy surface on a machined component caused by chattering of the working tool.

**Checks** Multiple small cracks on the surface of components caused during manufacturing.

**Cleavage** The splitting of a crystal on a crystallographic plane of relatively high density.

**Cold Shut** A discontinuity on the surface of a casting caused by the impingement of melt (without fusion) within a part of a casting.

**Cold Working** Plastically deforming a material at relatively low temperatures, resulting in creation of dislocation defects in the crystal structure.

**Columnar Crystal Structure** Elongated grains that are perpendicular to a casting surface.

**Constituent** A characteristic geometric arrangement of microscopic phases.

**Corrosion** The electrochemical degradation of metallic materials.

**Corrosion Fatigue** An acceleration of fatigue damage caused by a corrosive environment.

**Crack** A breach in material continuity in the form of a narrow planar separation.

**Crater** A local depression in the surface of a component caused by excessive chip contact in machining or arc disturbance of a weldment.

**Creep Cracks** Cracking that is caused by linking of creep voids near the end of tertiary creep.

**Creep Voids** Small voids that form in the third stage of creep.

**Crevice Corrosion** The loss of surface material in a crevice subjected to an electrolyte.

**Decarburization** The loss of carbon from the surface of a ferrous alloy because of high-temperature oxidation of carbon.

**Defect** A component discontinuity that has shape, size, orientation, or location, such that it is detrimental to the useful service of the part.

**Dendrite** A tree-like crystal structure that forms in some casting and vapor deposition crystallization.

**Ductility** The ability of a material to deform plastically without fracturing.

**Exfoliation** A corrosion degradation mode that causes layers parallel to the surface of an alloy to be separated and elevated due to the formation of corrosion product.

**Fatigue** Progressive cracking leading to fracture that is caused by cyclic tensile loading in the range of elastic stress, eventually initiating small cracks that sequentially and irreversibly enlarge under the action of fluctuating stress.

**Flakes** Short internal fissures in ferrous materials caused by stresses produced by evolution of hydrogen after hot working. Fractured surfaces containing flakes with bright and shiny surfaces.

**Folds** Discontinuities composed of overlapping surface material.

**Forging Discontinuities** Discontinuities that are created in forging operations.

**Fracture** A break, rupture, or crack large enough to cause a full or partial separation of a component.

**Fretting** Low-amplitude reciprocal motion between two component surfaces under pressure causing surface roughness.

**Gas Holes** Holes created by gas escaping from molten metal during solidification.

**Gas Porosity** Minute voids distributed in a casting that are caused by the release of gas during the solidification process.

**Geometric Discontinuity** A sharp change in surface configuration that may be specified or an unplanned consequence of manufacture.

**Gouge** A groove cut in a surface caused by mechanical, thermal, or other energy sources.

**Grain** An individual crystal in a polycrystalline material.

**Grain Boundary** The narrow zone of material between crystals of differing orientation.

**Grinding Cracks** Shallow cracks formed in a surface of relatively hard materials because of excessive grinding heat.

**Gross Porosity** Pores, gas holes, or globular voids in weldments or castings that are larger and in greater number than is acceptable in good practice.

**Heat-Affected Zone** The portion of base material that was not melted during brazing, cutting, or welding whose microstructure and physical properties have been altered by the heat of the joining operation.

**Hot Cracks** Cracks that are caused by differential contraction between a casting and its mold. These may be branched and scattered through the interior and on the surface of the casting.

**Hot Tear** A relatively large fracture formed in the interior or on the surface of a cast component due to restricted contraction.

**Hydrogen Embrittlement** Low ductility caused by cracking of the interior of a component due to the evolution and precipitation of hydrogen gas.

**Inclusion** Usually a solid foreign material that is encapsulated in an alloy. Inclusions composed of compounds such as oxides, sulfides, or silicates are referred to as nonmetallic inclusions.

**Incomplete Fusion** Failure of weldment to fuse with the base material or the underlying or adjacent weld bead in a weld.

**Incomplete Penetration** Fusion into a joint that is not as full as dictated by design.

**Inherent Discontinuity** A discontinuity that is generated in the primary production of a material.

**Lack of Fusion** Failure of a weldment to fuse with the base material in a weld.

**Lamellar Tearing** The typical rupture of a material that is weakened by elongated slag inclusion. The crack surfaces have the appearance of wood fracture.

**Lamination** Separation or structural weakness, usually in plate that is aligned parallel to the surface of a component. This may be caused by the elongation during plastic forming of segregates that are the result of pipe, blisters, seams, and inclusions.

**Lap** A surface discontinuity appearing as a seam caused by the folding over of hot alloy fins, ears, or corners during forging, rolling, or other plastic forming without fusing the folds to the underlying material.

**Macroshrinkage** Casting discontinuity that is detectable at magnifications less than 10 times and that is the result of voids caused by solidification shrinkage of contained material.

**Metallurgical Notch** A material discontinuity, usually involving hardness that has the geometric characteristics of a crack.

**Microfissure** A microscopic crack.

**Microsegregates** Segregates that are microscopic in size.

**Microshrinkage Cracking** Microscopic cracks that are caused by solidification shrinkage.

**Microshrinkage Porosity** Microscopic pores that are caused by solidification shrinkage.

**Nonmetallic Inclusion** A slag or glass-like inclusion in an alloy.

**Notch** A sharp reentrant on a surface of a component that causes a local concentration of stress.

**Phase Diagram** A descriptive map that shows the existence of equilibrium phases, usually as a function of alloy composition and temperature.

**Phase Diagram—Binary** A phase diagram between two components.

**Phase Diagram—Ternary** A phase diagram between three components; the representation uses a three-dimensional representation of data. The interior of an equilateral triangle base providing unique location representing all the compositions possible between the three components.

**Pitting** Forming of small cavities in a surface by corrosion, arcing, wear, or other mechanical means.

**Primary Processing Discontinuities** Discontinuities that are generated in the first forming steps of an alloy.

**Scale Pits** Surface discontinuities in the form of local indentations caused by the capture and rolling of scale into the surface of an alloy.

**Scratches** Small grooves in a surface created by the cutting or deformation from a particle or foreign protuberance moving on that surface.

**Secondary Processing Discontinuities** Discontinuities that are generated in the secondary and finishing forming steps.

**Segregation** Nonuniform distribution of alloying elements, impurities, or phases in alloys.

**Service Discontinuities** Discontinuities that are generated by service conditions.

**Slag** An aggregate of nonmetallic glasses that is found in primary metals production and in welding processes. Slag may be used as a covering to protect the underlying alloy from oxidation.

**Split** A rupture in a component, usually open to the surface.

**Stringers** Nonmetallic inclusions that are elongated by a hot forming process.

**Voids** A cavity inside a casting or weldment that is usually caused by solidification shrinkage.

**Weld Undercut** A groove at the toe or root of a weld caused by melting of the base material.

**Welding Discontinuities** Any breaches in material continuity that are generated in welding processes.

## 2.13 Discontinuity Guide

The following table provides examples of discontinuities. These examples are designed as a reference and to provide the reader with a representation as to how some of the conditions described herein and in other chapters can appear. All discontinuities are different. When there is doubt regarding the classification and type of discontinuities, this guide may help. Following the table is a collection of figures that illustrate some of the discontinuities (Figs. 2-18 to 2-28).

**Discontinuity Guide**

| Nondestructive Test Methods | Visual | Fluorescent Penetrant | Visible Penetrant | Wet DC Magnetic Particle | Dry DC Magnetic Particle | Dry AC Magnetic Particle | Eddy Current | Thermal Infrared | Radiography | Straight Beam Ultrasonics | Angle Beam Ultrasonics |
|---|---|---|---|---|---|---|---|---|---|---|---|
| **Types of Discontinuities** | **Surface and Near Surface Methods** | | | | | | | | **Subsurface** | | |
| *Process Category* | | | | | | | | | | | |
| *Inherent* | | | | | | | | | | | |
| Laminations | P | U | U | U | U | U | U | U | U | A(I)[c] | U |
| Pipe[d] | U | U | U | U | U | U | U | U | A(2) | A(1) | A(3) |
| Seams | P | A(4) | A(5) | A(1) | A(2) | A(3) | P | P | U | U | P[b] |
| Stringers | U | U | U | A(1) | A(2) | A(3) | P | U | P | P | P[a] |
| *Mechanical Forming Processes* | | | | | | | | | | | |
| Flare or Split | U | U | U | U | U | U | U | U | U | A(1) | A(2) |
| Forging Bursts | P | A(2) | A(3) | A(1) | A(4) | A(5) | P | P | P | P | P |
| Forging Laps | P | A(2) | A(3) | A(1) | P | P | P | U | U | P | P |
| Rolling Lap (Seam) | P | A(2) | A(3) | A(1) | P | P | P | U | U | P | P |
| *Casting Process* | | | | | | | | | | | |
| Casting Cold Shuts | P | P | P | P | P | P | P | U | A(1) | P | P |
| Casting Shrinkage Cracks | U | U | U | P | P | U | U | U | A(1) | A(2) | P |
| Gas Porosity | U | U | U | U | P | U | U | U | A(1) | A(2) | P |
| Hot Tears | U | U | U | P | P | U | U | U | A(1) | A(2) | P |
| Scabs | P | A(3) | A(4) | A(1) | A(2) | A(5) | P | P | U | U | P |
| Shrinkage Porosity | P[e] | P[e] | P[e] | U | U | U | U | U | A(1) | A(2) | P |
| Inclusions | U | P[e] | P[e] | P[e] | P[e] | P[e] | U | U | A(1) | A(2) | P |

(*Continued*)

| Nondestructive Test Methods | Visual | Fluorescent Penetrant | Visible Penetrant | Wet DC Magnetic Particle | Dry DC Magnetic Particle | Dry AC Magnetic Particle | Eddy Current | Thermal Infrared | Radiography | Straight Beam Ultrasonics | Angle Beam Ultrasonics |
|---|---|---|---|---|---|---|---|---|---|---|---|
| **Types of Discontinuities** | **Surface and Near Surface Methods** | | | | | | | | **Subsurface** | | |
| *Secondary Process* | | | | | | | | | | | |
| Grinding Checks | P | A(2) | A(3) | A(1) | P | P | P | U | U | U | P |
| Machining Tears | P | A(2) | A(3) | A(1) | U | U | U | U | U | U | U |
| Plating Cracks | P | A(2)$^{e}$ | A(3)$^{e}$ | A(1) | P | P$^{e}$ | P | U | U | U | U |
| Quench and Heat Cracks | P | A(2) | A(3) | A(1) | A(4) | A(5) | P | P | P | U | P |
| *Welding/Joining* | | | | | | | | | | | |
| Crater Cracks | P | A(2) | A(3) | A(1) | P | P | P | U | U | P | P |
| Dense Inclusion | U | U | U | U | P | U | U | U | A(1) | P | U |
| Excessive Concavity | A(1) | U | U | U | U | U | U | U | U | U | U |
| Excessive Convexity | A(1) | U | U | U | U | U | U | U | U | U | U |
| Incomplete Penetration | P | U | U | U | U | U | U | U | A(1) | U | A(2) |
| Lack of Fusion | U | U | U | P | P | U | U | U | A(2) | P | A(1) |
| Porosity | P | P | P | P | U | U | U | U | A(1) | P | U |
| Slag Inclusion | U | U | U | P | A(2) | U | U | U | A(1) | P | P |
| Subsurface Cracks | U | U | U | U | U | U | P | U | A(1) | A(1) | A(1) |
| Surface Cracks | P$^{e}$ | P$^{e}$ | P$^{e}$ | U | P | P | P | U | A(2) | U | A(1) |
| Underbead Cracks | U | U | U | U | U | U | P | U | A(1) | A(1) | A(1) |
| Undercut | A(1) | U | U | U | U | U | U | U | U | U | P |

| | | | | | | | | | | |
|---|---|---|---|---|---|---|---|---|---|---|
| *Brazing* | | | | | | | | | | | |
| Lack of Fill | U | U | U | U | U | U | U | U | U | A(1) | U |
| Unbond[f] | U | U | U | U | U | U | U | U | U | A(1) | U |
| *Service* | | | | | | | | | | | |
| Brittle Overload Crack | P | A(5) | A(4) | A(2) | A(1) | A(3) | P | U | P | P | P |
| Corrosion Pitting | P | A(1) | A(2) | U | U | U | P | U | P | P | U |
| Fatigue Cracking | P | A(5) | A(4) | A(2) | A(1) | A(3) | P | P | P | P | P |
| Fretting | A(1) | U | U | U | U | U | U | U | U | U | U |
| Graphitization | U | U | U | U | U | U | P | U | U | U | A(1) |
| Metallurgical Notch | U | U | U | U | U | U | P | U | P | U | P |
| Overload Cracking | P | A(5) | A(4) | A(2) | A(1) | A(3) | P | U | P | P | P |
| Segregation | U | U | U | U | U | U | P | U | A(1) | A(2) | A(3) |
| Stress Corrosion Cracking | P | A(1) | A(2) | P | U | U | A(3) | U | P | U | P |
| Wear | A(1) | U | U | U | U | U | U | U | U | U | U |

Key:
U = Unsatisfactory; P = Possible; A(1) = First order of preference; A(2) = Second order of preference; A(3) = Third order of preference; A(4) = Fourth order of preference
[a]Inspection of billet possible.
[b]Possible using surface waves.
[c]Assuming it is completely internal.
[d]May be seen on ends by penetrant or magnetic particle.
[e]If open to surface.
[f]Thermal tests possible.
***Note:*** Acoustic emission testing (AET) has not been included in this guide because this method applies only to those discontinuities that propagate under applied loads.

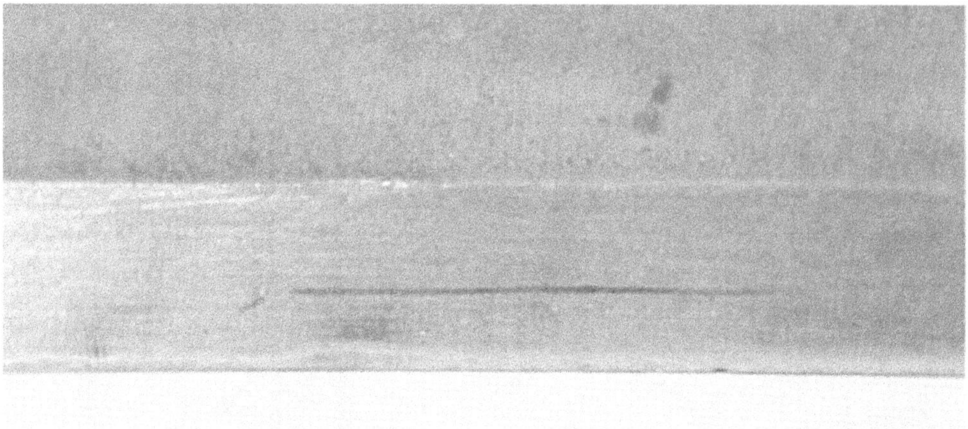

**Figure 2-18** Lamination in plate edge.

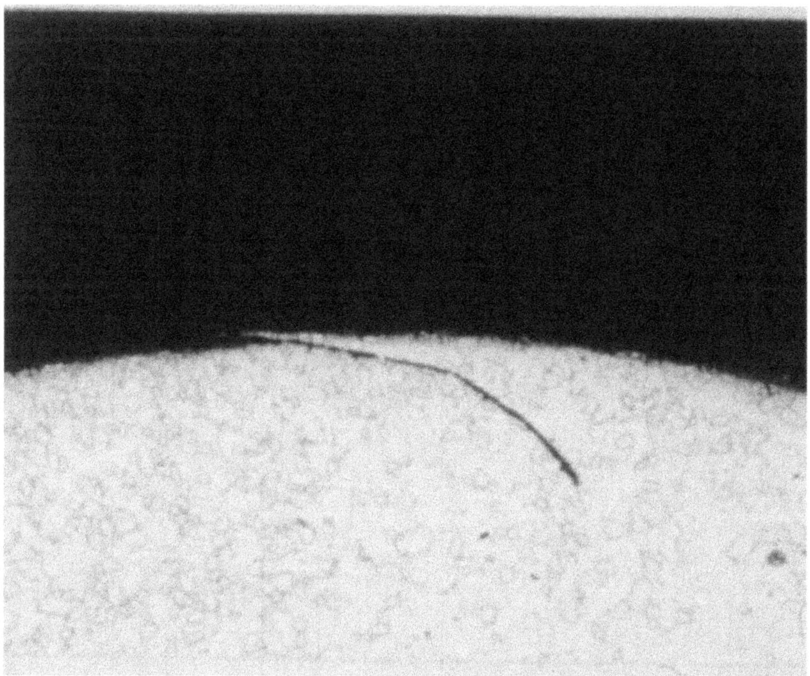

**Figure 2-19** Forging lap.

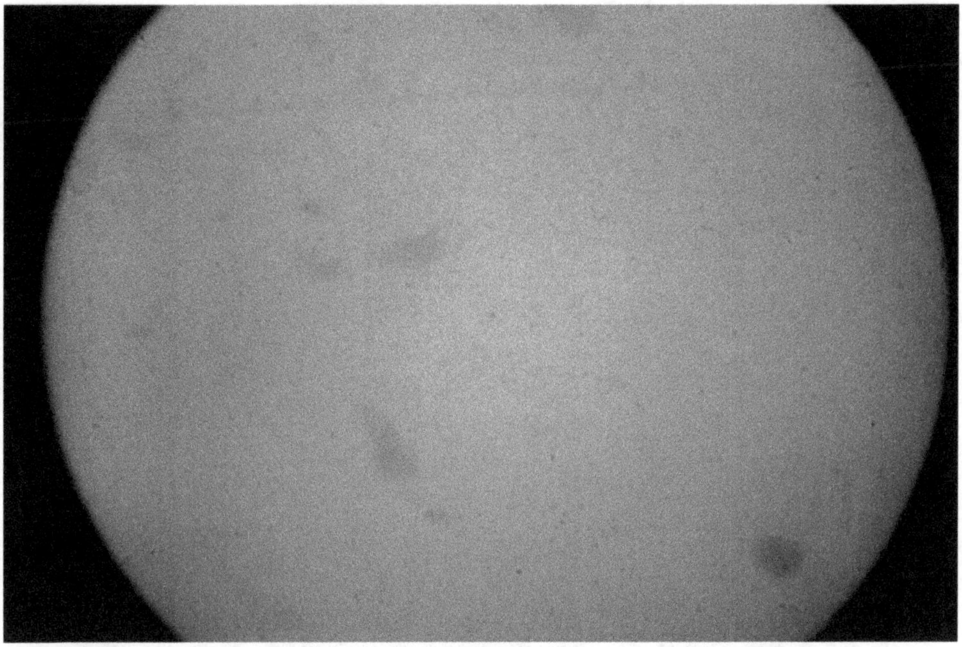

**FIGURE 2-20** Sand and slag inclusions in a casting.

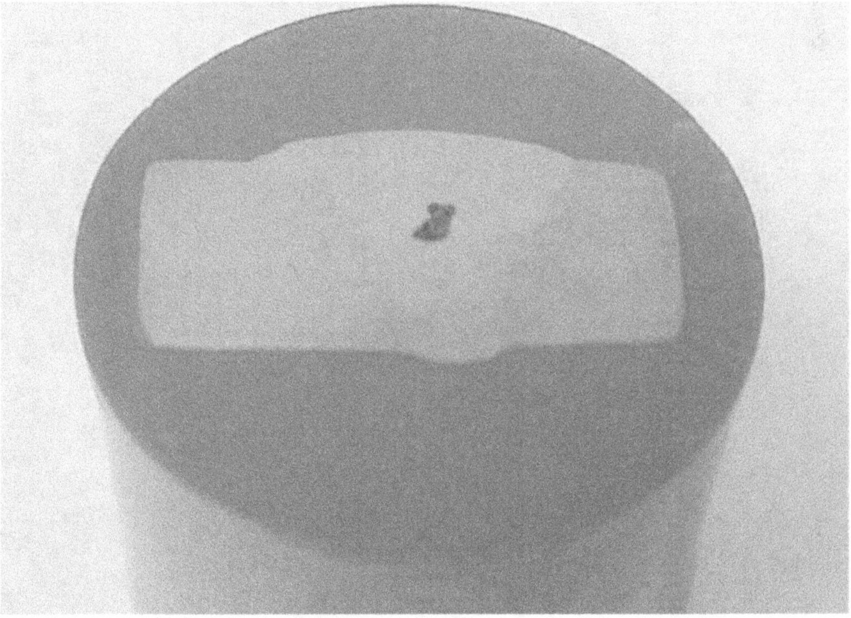

**FIGURE 2-21** Slag inclusion in weld cross section.

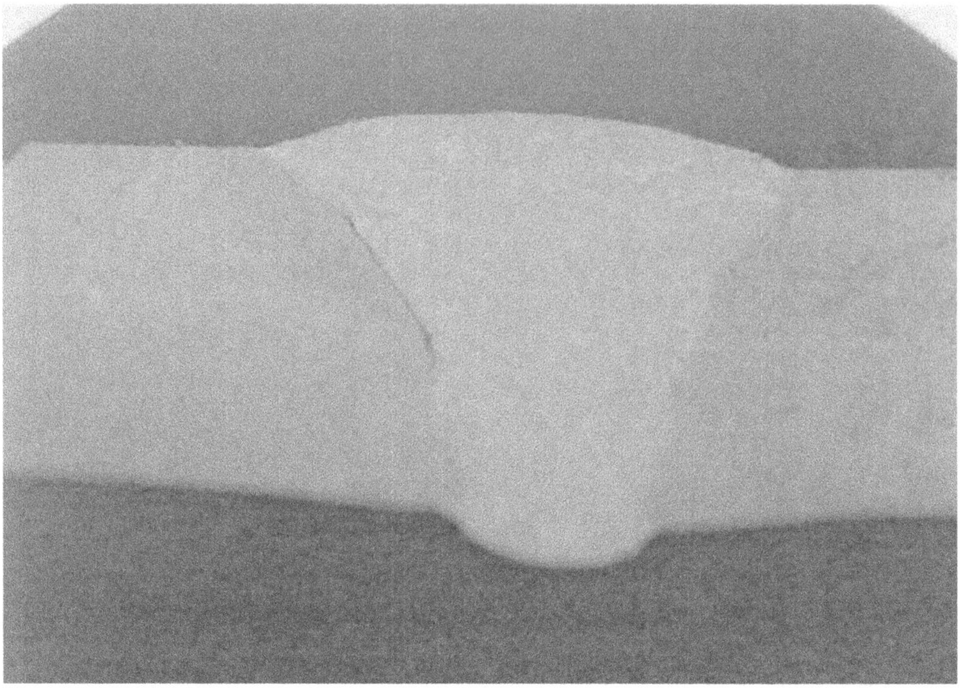

FIGURE 2-22 Lack of fusion.

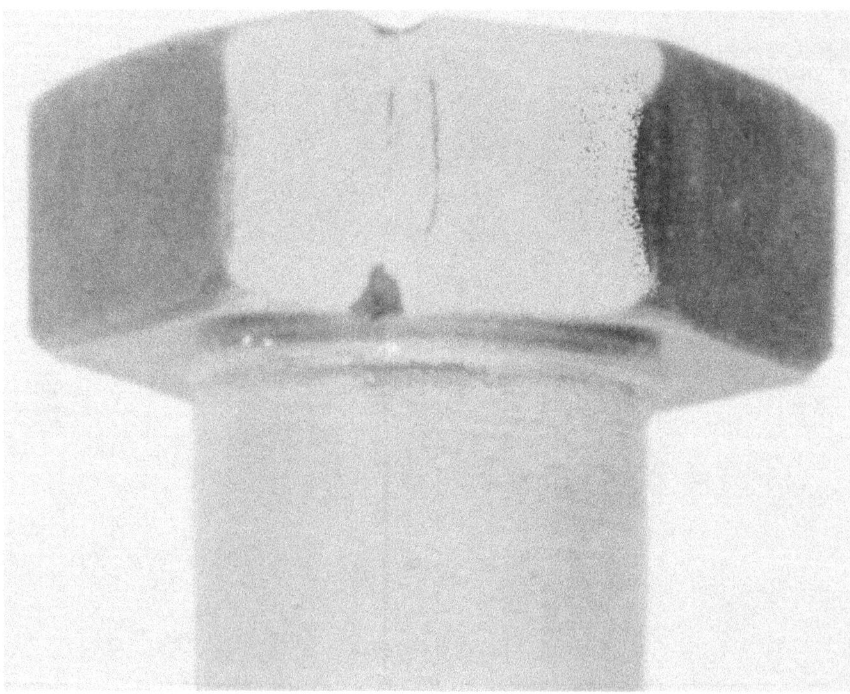

FIGURE 2-23 Seam in bolt.

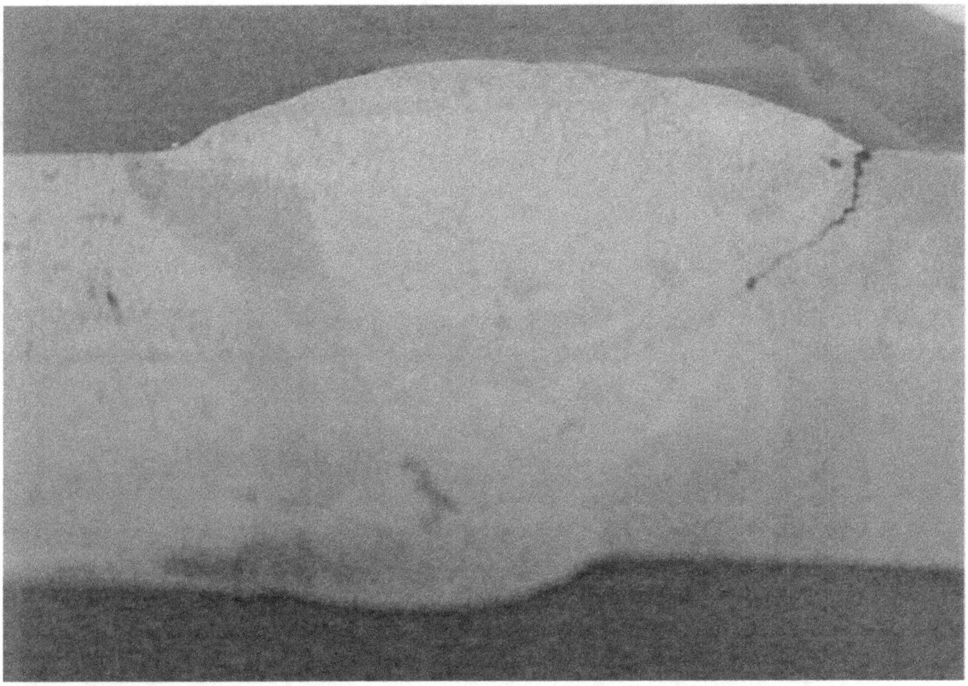

**Figure 2-24** Toe crack in weld.

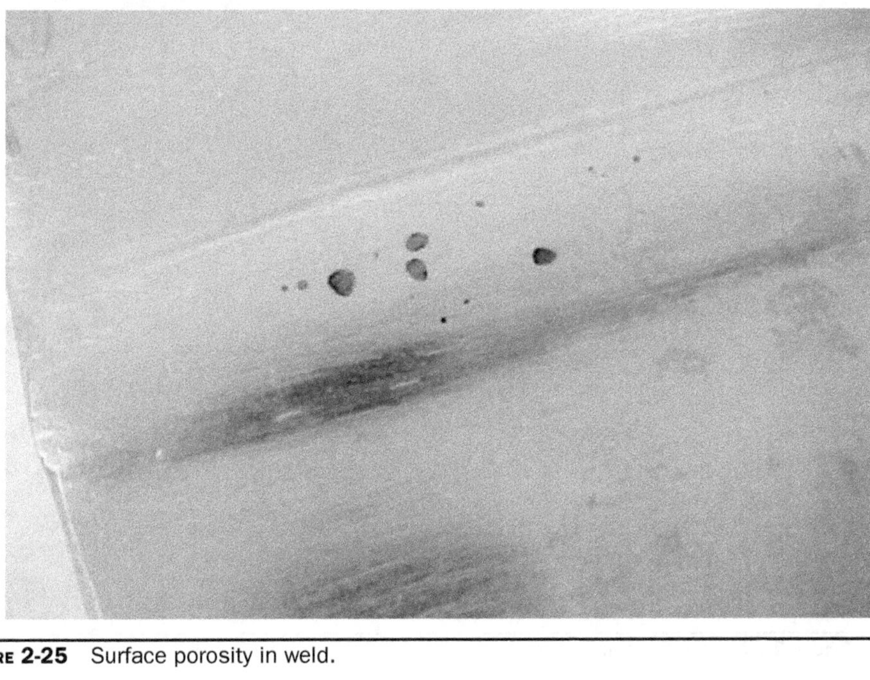

**Figure 2-25** Surface porosity in weld.

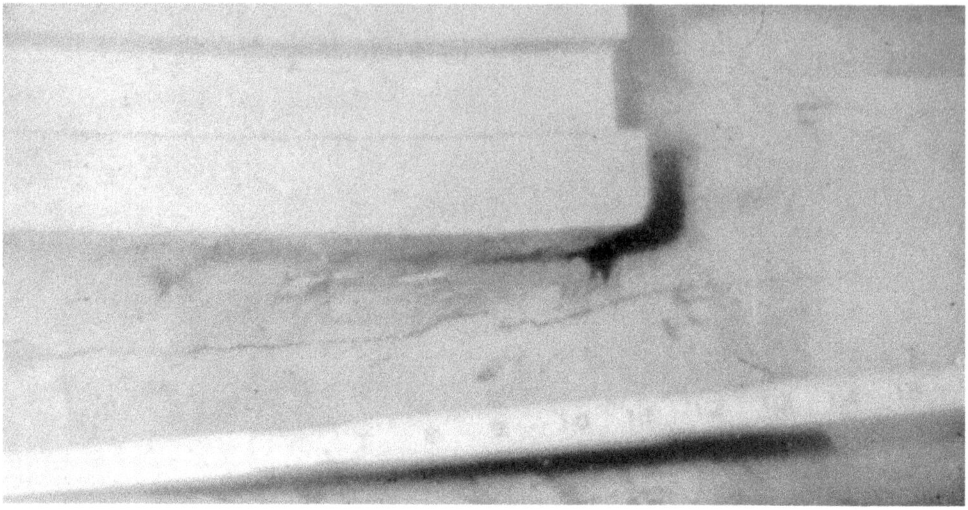

**FIGURE 2-26** Crack between two welds.

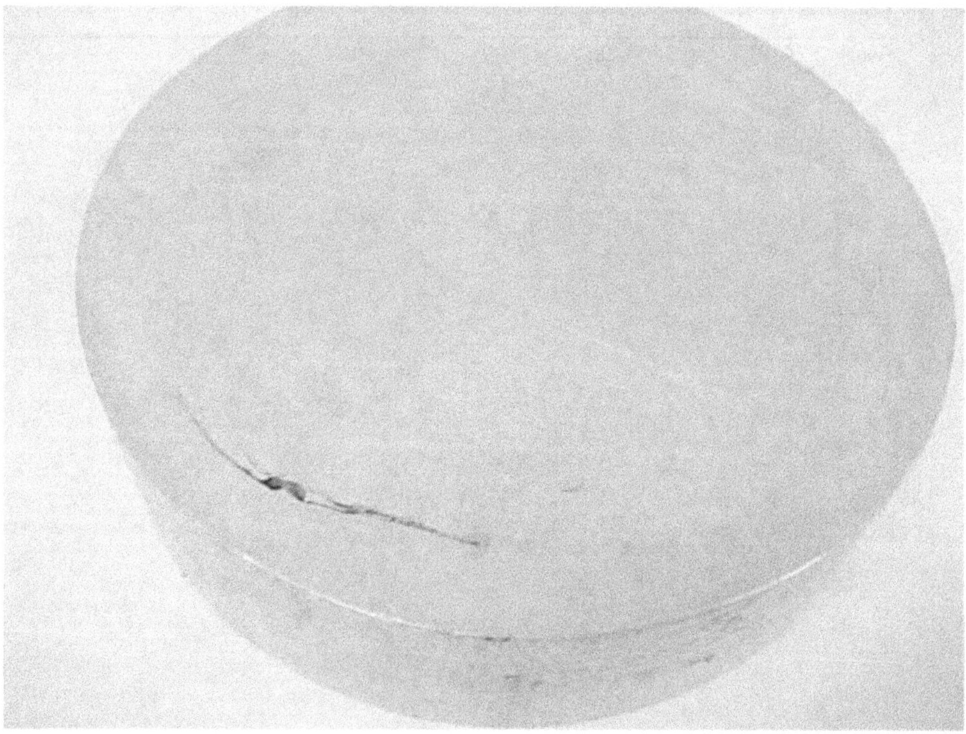

**FIGURE 2-27** Burst in bar.

**FIGURE 2-28** Crack in bolt.

# CHAPTER 3
# Visual Testing

## 3.1 History and Development

Visual examination or testing (VT) is a method of nondestructive testing that has been neglected for much of its industrial application life. VT was the first nondestructive test (NDT) method used in the nondestructive testing industry but was one of the last methods to be formally acknowledged.

In 1854 in Hartford, Connecticut, a boiler at the Fales and Gray Car works explodes, killing 21 people and seriously injuring 50. Within a decade, the state of Connecticut passed a law requiring annual inspection (in this case visual) of boilers. Since then, visual inspection has always been the presumed first inspection before every nondestructive testing method. The American Welding Society (AWS) was founded in 1919 as a multifaceted, nonprofit organization with a goal to advance the science, technology, and application of welding and related joining disciplines. AWS Website AWS has been known for decades to have a 40-hour course to prepare inspectors to qualify as a Certified Weld Inspector (CWI), which is a visual inspection mode of examination of welds.

In the early 1980s, the need for the development of a training program for visual examination was prompted by the American Society of Mechanical Engineers. ASME Section XI—Rules for Inservice Inspection of Nuclear Power Plant Components—addressed this technology with a program that included 120 hours of formal training.

The program was designed to qualify personnel as visual examiners. Examination personnel scrutinizing the general condition of components were to meet the American Society for Nondestructive Testing, Recommended Practice No. SNT-TC-1A.

Visual examination of components for general mechanical and structural conditions was to satisfy the requirements of ANSI N45.2.6. This standard was codified via the U.S. Federal Regulations, "Title 10, Code of Federal Regulations, Part 50," requiring nuclear power plants to meet certain requirements for licensing. ASME sectored the visual examination into four categories based on the scope of inspection. The categories are classified as VT-1, VT-2, VT-3, and VT-4. VT-1 addresses the condition of a component, VT-2 the location or evidence of leakage, VT-3 the general mechanical and structural conditions of components and their supports, and VT-4 (which has been eliminated) focused on the conditions relating to the operability of components or devices.

Performance requirements for NDT are referenced in ASME Boiler and Pressure Vessel Code, Section V—Nondestructive Examination. Direct and remote visual testing is described in Article 9 of Section V—Visual Examination. Direct visual testing is defined as using "visual aids such as mirrors, telescopes, cameras, or other

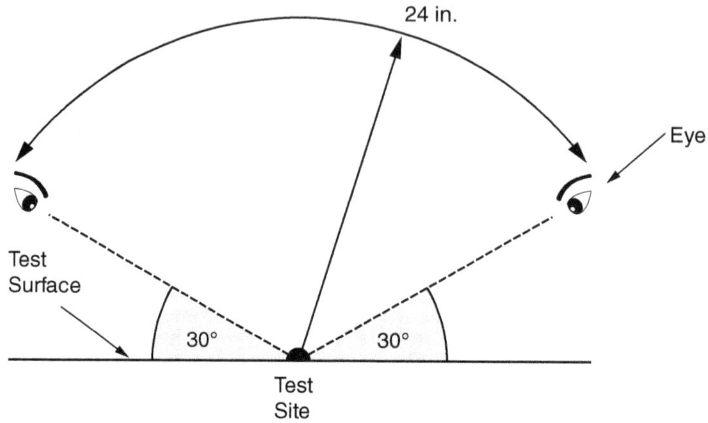


**FIGURE 3-1** Minimum angle for typical visual testing. (*Courtesy of Mistras Group, Inc.*)

suitable instruments." Direct visual examination is conducted when access allows the eye to be within 24 inches (610 mm) of the surface to be examined, and at an angle not less than 30° to the surface to be examined. This is illustrated in Fig. 3-1.

Remote visual testing is divided into three categories: borescopes, fiberscopes, and video technology. These have been developed chronologically. Borescopes, also referred to as endoscopes, were originally used to inspect the bores of rifles or cannons utilizing a hollow tube and mirror. The second generation of the endoscopes included a relay lens system in a rigid tube. This upgraded the image. Due to its rigid structure, endoscopes are limited to mostly straight-line access, as is depicted in Fig. 3-2.

Later innovations corrected this limitation by providing flexibility to the endoscopes. By 1955, the introduction of glass fiber bundles and fiber-optic image transmission enabled the development of the fiberscope. Medical researchers experimented with different techniques in fiber-optic image transmission during this period.

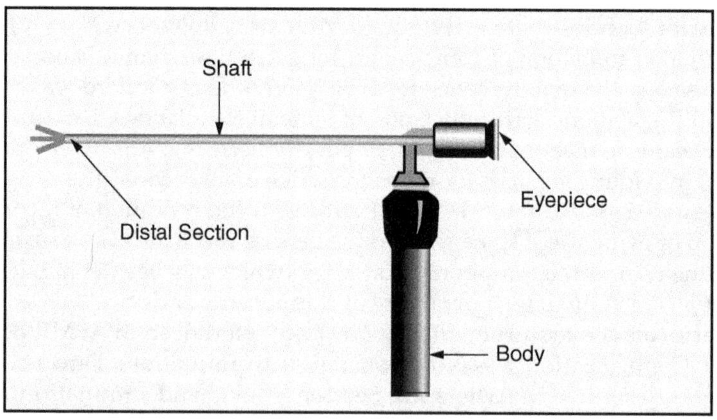


**FIGURE 3-2** Rigid borescope. (*Courtesy of Mistras Group, Inc.*)

Imaging with fiber-optic bundles decreased the clarity of the image transmission compared with the rigid lens systems of borescopes; however, this was a small price to pay for the opportunities it presented. The flexibility of the bundle opened up previously inaccessible areas to remote visual inspection, providing a more versatile tool to be used in industrial situations. This often eliminated the need to dismantle equipment for inspection.

A typical fiber-optic borescope is illustrated in Fig. 3-3.

The evolution of the endoscope continued as the problems of eye fatigue associated with the use of endoscopes and fiberscopes prompted the development of various "add-on" cameras or closed-circuit TV cameras that allowed for the display of images on a monitor. Add-on camera systems have been replaced due to their bulky exterior. Smaller, solid-state imaging sensors, some of which are known as charge-coupled devices (CCDs), are replacing them. Figure 3-4 provides a basic demonstration of how a CCD works.

This new generation of CCDs stimulated a new wave of videoendoscope technology. Small in diameter with high-resolution images, this new technology increased the range of industrial endoscopy applications. The physical size of the CCD, as well as its ability to allow for electronic image processing and its other advantages broaden the application possibilities. The CCD's ability to record images as electronic files allows a wide range of image management, manipulations, and retention or retrieval options. With the advent of digital storage technology, the recording of images on other permanent media enhances the systems versatility.

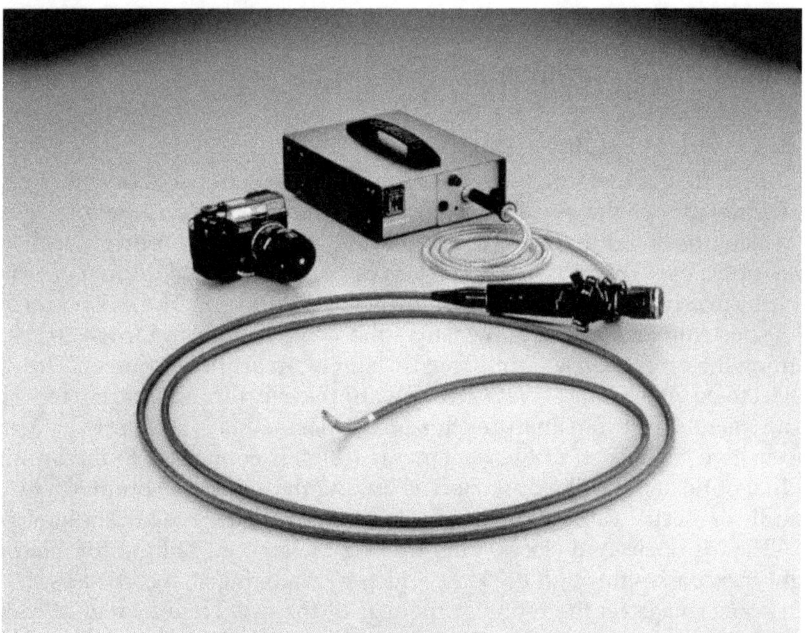

**FIGURE 3-3** Typical fiber-optic borescope. (*Courtesy of Olympus Industrial, with permission.*)

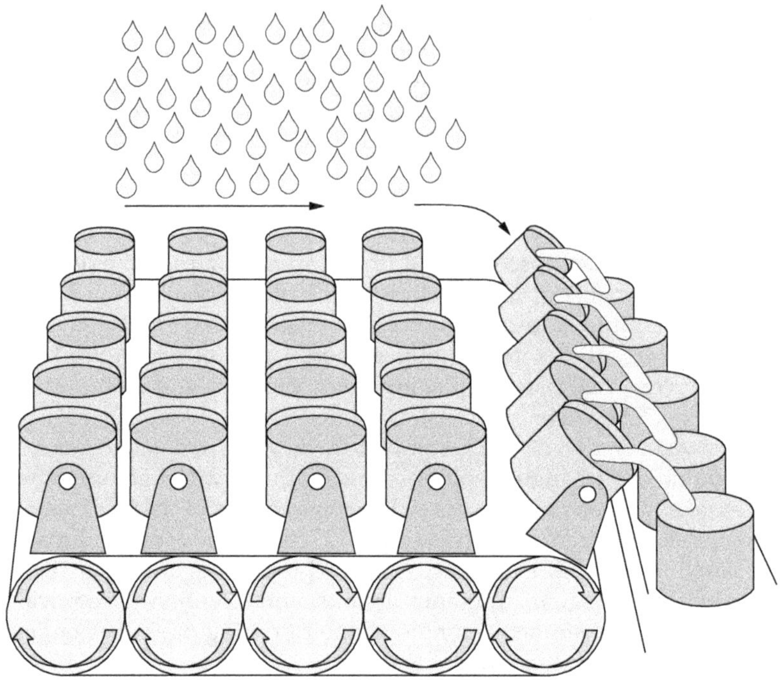

FIGURE 3-4 How charged coupling works. (*Courtesy of Mistras Group, Inc.*)

## 3.2 Theory and Principles

### 3.2.1 Object Factors

In order to understand the physics of vision, it is necessary to first consider the characteristics of the eye. The eye can be compared to a radiation detector. Different wavelengths of light travel through the lens and reach the retina, which is located at the back of the eye. The rods and the cones of the retina in the human eye can sense wavelengths from about 400 nm up to approximately 760 nm. The eye performs the function of a spectrum analyzer that measures the wavelengths and intensity, as well as determining the origin of the light (from the sun or an artificial source). The light strikes the object to be viewed and is reflected toward the eye, through the lens and onto the retina as an image. The brain analyzes this image. The retina is similar to an array of tiny photosensitive cells. Each of these elements (cells) is connected to the brain through individual optic nerves. The optic nerves linking the eye to the brain can be compared to a bundle of electric cables. The major components of the eye are shown in Fig. 3-5.

The iris opens and closes, thus varying the amount of light reaching the retina. The light then passes through the lens, which by changing shape, focuses the light and produces the image on the retina at the rear of the eye. Here a layer of rods and cones is found. The neurological connection from the rods and the cones passes through the rear of the eye via the optic nerve, which transmits the neurological signals to the brain.

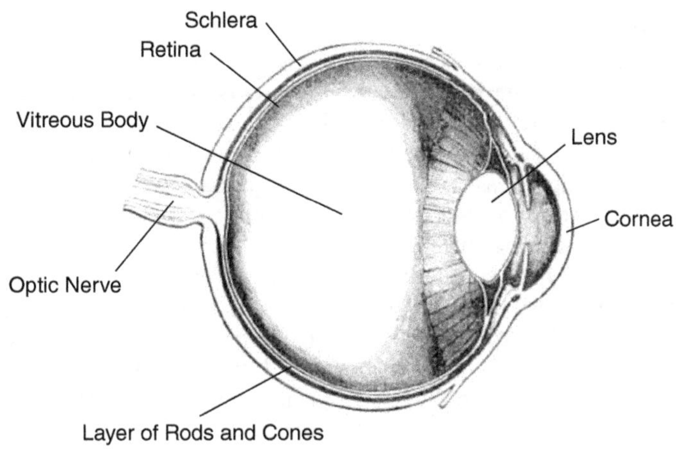


**FIGURE 3-5** Major components of a human eye.

The brain processes the signals as perceptions of colors and details that vary in light intensity and color. It is necessary for a certain minimum level of light to be present before the eye can produce an image. This level is known as the "intensity threshold." Contrast is something that shows differences between images placed side by side. Lighting requirements are frequently expressed in terms of ratios, due to the eye's ability to perceive a percentage of change rather than an absolute change in brightness. The retina will only retain an image for a certain amount of time. This varies according to the size of the object and speed at which it is moving. The main limitations of vision are intensity threshold, contrast, visual angle, and time threshold. This is what makes background to object ratio of brightness important.

Visual acuity is the ability to distinguish very small details. For example, as the distance from the eye to the object increases, two lines that are close together appear as one heavy, dark line. The normal eye can distinguish a sharp image when the object being viewed subtends an arc of one-twelfth of a degree (five minutes), irrespective of distance from the eye to the object. The smallest discernible object could be as small as one minute.

White light contains all colors. Newton proved that color is not a characteristic of an object, but rather that various wavelengths of light are perceived as different colors by the eye. Color can be described as having three measurable properties: brightness, hue, and saturation. The color of an object, ranging from light to dark, emitting more or less light, is known as brightness. Different wavelengths give us different perspectives of colors; this is known as hue. How green something is as opposed to white is how saturated it is with green wavelengths reflecting off the object.

In most NDT personnel qualification and certification programs, visual acuity examinations are a requirement for certification. The visual inspector's natural visual acuity must be examined. The Jaeger (J) test, for example, is used for near-distance visual acuity. It consists of an English language text printed on an off-white card. The parameters for near-distance visual acuity are described in personnel certification and qualification programs. Visual acuity requirements will vary depending upon the needs of specific industries.

### 3.2.2 Light

Light wavelengths are measured in angstrom units ($10^{-10}$ meters) and currently are also measured using the nanometer (nm), which is $10^{-9}$ meters. Brightness of light is an important factor in test environments. The apparent brightness of a test surface depends on the intensity of the light and the reflectivity of the surface reflecting the light to the eye. Excessive brightness will interfere with the ability to see and will "white out" the object. Inadequate light can cause excessive shadows and result in insufficient light being reflected from the surface, preventing observation of the surface attributes. Some codes require a minimum intensity of 15 foot-candles (fc) for general visual testing and a minimum intensity of 50 fc for critical and fine detail viewing. The Illumination Engineering Society requires 100 to 300 fc for critical work. The inverse square law governs the intensity of light noted or measured. It states that luminance ($E$) at a point on a surface varies directly with the luminous intensity of the source ($I$) and inversely as the square of the distance ($d$) between the surface and the source (see Eq. 3-1 and Fig. 3-6).

$$E = \frac{I}{d^2} \tag{3-1}$$

where:
$E$ = luminance
$I$ = intensity of the source
$d$ = distance between the source and the surface

This equation is accurate within 0.5 percent when $d$ is at least five times the maximum dimension of the source, as viewed from the point on the surface.

#### Cleanliness

The amount of light that reaches the eye from an object is dependent on the cleanliness of the reflecting surface. In visual testing, the amount of light may be affected by distance, reflectance, brightness, contrast or the cleanliness, texture, size, and shape of the test object.

Cleanliness is a basic requirement for a successful visual test. Opaque dirt can mask or hide attributes, and excessively bright surfaces cause glare and prevent observation of the visual attributes.

#### Brightness

Excessive brightness within the field of view can cause an unpleasant sensation called glare. Glare interferes with the ability to see clearly and make critical observations and judgments.

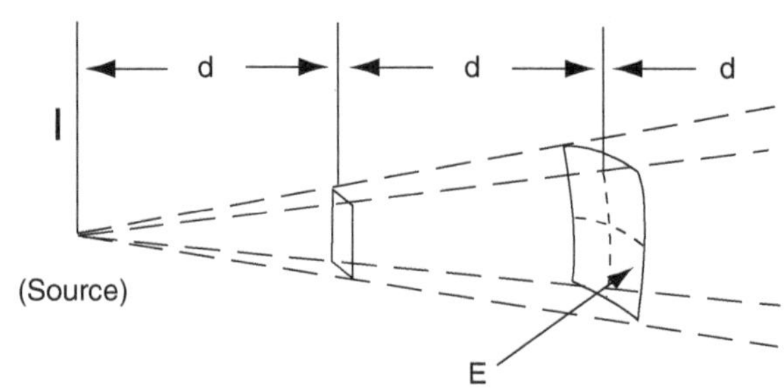


**FIGURE 3-6** Inverse square law.

#### Surface Condition

Scale, rust, contaminants, and processes such as milling, grinding, and etching may affect the ability to examine a surface. This will be further discussed in the section relating to tools, equipment, and accessories.

#### Shape

The shape of an object can influence the amount of light reflected to the eye, due to various angles that can determine the amount of light that will be reflected back to the eye.

#### Size

The size of an object will determine the type of scan pattern that may be used to view 100 percent of the object or it may determine that some magnification is necessary to get a closer view of details otherwise unobservable.

#### Temperature

Excessive temperature may cause distortion in viewing due to the heat wave effect. Most are familiar with the heat waves coming off a desert, resulting in a mirage; this is known as "heat wave distortion." In a nuclear reactor, the image of underwater components is frequently distorted due to heat waves rising from the core that can interfere with the view from an underwater camera as it scans a particular component during visual examination. Excessively smooth or shiny surfaces may cause "white out" as previously mentioned.

#### Texture and Reflectance

One of the greatest variables in viewing an object is the amount of light that is reflected from it, and the angle at which the light strikes the eye. Excessive rust or roughness can cause diffusion of the light and limit the light returning to the eye. This can easily be corrected by increasing the amount of light or improving the surface condition of the object under examination.

### 3.2.3 Human Factors

#### Environmental

The amount of light required for a visual test depends on several factors, such as speed or detailed desired reflection from background, and other inspection variables. These inspection variables include physiological processes, psychological states, and the inspector's experience, health, and fatigue. All of these factors contribute to the accuracy of a visual inspection. One of the key factors in viewing a lighted object is the difference (contrast) between the light on the object and the background. A contrast ratio of 3:1 between the test object and the background is desirable. If the background is dark, a ratio of 1:3 is recommended between the test object and lighter surroundings, 3 being the most intense light in both cases.

Psychological factors can also affect a visual inspector's performance. Surrounding colors and patterns can have an effect on the inspector's attitude. Dark walls can absorb up to 50 percent of the light. A high contrast on the pattern being inspected can cause eye fatigue. It is recommended that blue colors be utilized and brilliant colors avoided.

### Physiological

The act of seeing something is not a passive activity. The observer must be active in keeping track of what is going on. Constant eye shifting back and forth from one location to another or scanning a large area at a rapid speed causes the muscles in the eye to fatigue. If the eye isn't focusing quickly when changing directions, the image can be lost altogether. Any fatigue on the part of the observer can result in reduced efficiency and accuracy in interpreting the visual data.

### Psychological

Individuals can be in various psychological states. They can be suffering from tensions, emotions, and other influences. These may influence the appraisal and ability to visualize an object and may also influence performance of a visual task. The intention of a viewer may affect perception. A great deal of information is potentially available immediately after viewing, but if one does not expect to find certain attributes, one may well overlook the physical evidence that has been viewed and not perceive them. One of the ways to overcome this is to know ahead of time what to expect, the attributes of what is to be seen, and what the greatest possibility is of these attributes existing.

### Perception

The ability of the eye to sense a variety of views is not constant. If one is well rested in starting out an inspection activity, perception can be greater than when one is fatigued.

Grossly changing light levels may cause painful glare. After a long period of relative darkness, normal light may seem painful. (Normal light is that which is comfortable to the eyes.) Sudden exposure to full sunlight can cause discomfort. It may take up to 30 minutes for the iris to adjust and regain normal vision. As the iris becomes tired and the muscles that adjust the lens become fatigued due to age, overuse, drugs, disease, or emotions, vision can be greatly affected. Vision examinations are usually administered annually to provide assurance that the inspector meets the requirements for performing VT. Daily influences external to the inspector, such as emotions, drugs, excessive light, and inadequate sleep, can cause temporary loss of visual acuity.

Another influence on perception is appearance. The two lines represented in Fig. 3-7 appear to have different lengths due to the perceptions created by the "Vee" extensions on each end. In fact, the two lines are identical in length; the brain perceives the lengths differently.

### Visual Angle and Distance

Visualize two pencils representing two parallel lines. If these are perpendicular to the incident angle of view, two pencils can be clearly seen. As the two pencils rotate

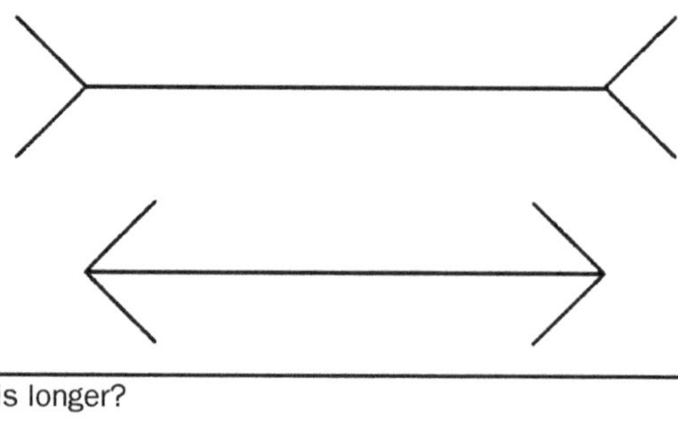

**Figure 3-7** Which line is longer?

and become relatively parallel to the incident view of the eye, the two pencils appear to become one, since one is now behind the other. The quantitative ability of a person to resolve objects is determined in a practical manner from the distance of the object to the eye and the angle of separation of two points that can be resolved by the eye. This is known as resolving power. For the average eye, the minimal resolvable angular separation of two points on an object is about one minute of arc (1/60th of one degree). This means that at about 12 inches (300 mm) from the test surface, the best resolution to be expected is 0.0035 inches (0.09 mm); at 24 inches (600 mm), the best anticipated resolution is about 0.007 inches (0.18 mm). The best result from a visual test is obtained when the object is brought close to the eye and a large visual angle is obtained. There is a limitation to distance; the eye may not sharply focus if it is nearer than 10 inches (254 mm) to an object. Therefore, direct visual tests are usually best performed at a distance of 10 to 24 inches (250 to 610 mm). Also of importance is the viewing angle between the line of sight that the eye makes with the test site and the plane of the test surface. As described previously, two pencils can become one as the viewing angle of the eye changes. For practical consideration, this angle should not be less than 30° of view off the plane of the surface under inspection, as shown in Fig. 3-1.

## 3.3 Equipment and Accessories

### 3.3.1 Direct Visual—Eyes

The most important instrument in visual testing is the human eye (refer to Fig. 3-5). As the camera replicates the action of the human eye, it can serve as a model for discussing the eye's components. Light emanates from a source and strikes an object. Light reflects off the object and travels through a medium, usually air or other transparent or translucent materials such as water, glass, and crystals. The light incident on the human eye enters the lens, after which it is focused on the retina. There the cones and rods transform electromagnetic radiation energy into neurological signals that are transmitted to the brain through individual optic nerves. Analogous to focusing the lens of a camera by changing the depth of field and angle, the lens in the eye is elastic. It changes shape as the eye muscle contracts or relaxes. The iris opens (like the aperture opening in a camera) to allow different quantities of light into the eye. The light first passes through the lens and is focused on the retina. The image focus depends on the shape of the lens as it is changed by the variations in tension of the eye muscle. The light rays then pass through the fluid in the eye and strike the rear of the eyeball at the retina (properly focused).

This image-recording medium (the retina) is equivalent in function to the film in a camera. The image is temporarily retained (up to 1/6 of a second) before being passed to the brain. The neurological connection from the rods and cones in the retina goes through the rear of the eye via the optic nerve. The optic nerve transmits the neurological signals to the brain, which processes them as perceptions of colors and details that vary in light, intensity, and hue. Through this neurological connection, the brain translates the signals into three-dimensional perception.

The unique "equipment" known as the eye can suffer damage and deterioration over time. Or it can be inherently defective. The lens can become inflexible with age and not be able to change shape so as to focus. The lens itself can become defective due to cataracts, stigmas, scars, or scratches that distort or block light transmission.

The recording medium (the retina) can become detached due to sharp blows, genetic predisposition, or disease. The ability of the eye to perceive color may have been limited or abnormal since birth (color blindness). The optical nerve system can become damaged like a wire that is shorted or broken. Chemical influences such as the effects of alcohol or hallucinogens, brain cell deterioration, or neurological impairment can affect the processing system in the brain.

The process of seeing is even more analogous to a CCD, in which light striking small chips, causes electrons to accumulate, resulting in the transmission of signals to a computer where they are reassembled onto chip arrays that yield an image for the human eye to perceive. This is discussed in more detail later in this section.

To summarize, the eyes are the most important and essential components in the entire VT system. It would be impossible to perform examinations without them. They also play a key role in all other methods of NDT, including evaluation prior to, during, and most importantly, after completion of the performance of the examination.

### 3.3.2 Direct Visual Aids

The eye may need assistance when visualizing the detail of the vast variety of surfaces that are normally accessible to the direct view. Enhancement of the view can be achieved through magnification. A change of angle can be achieved through reflective mirrors. Both of these aids can enhance views, and their removal restores direct view again. Hence, their use and the option to return to unaided viewing results in "direct viewing with an aid."

#### *Magnifier*

The magnifier can increase the image size of the viewed object. The power of the magnification is expressed as follows:

$$\text{Magnifying power} = 10 \div \text{focal length (in inches)}$$

The following is an example of determining the magnification power of a lens: A piece of paper is held in one hand and a magnifying lens in the other. The image of a light source (e.g., a candle) is visible on the paper. After focusing the image on the paper by moving the lens toward and away from the paper, measure the distance from the center of the lens to the paper. This distance is the focal length of the lens. Divide 10 by the focal length distance in inches. The resultant quotient is the magnifying power of the lens. For example, assume the focal distance was 2 inches; the power of the lens is 10 divided by 2 inches, which equals a magnification power of five (5×). (See Fig. 3-8.)

Another unit of measurement for the magnifier is the diopter. This is a measurement of the refractive power of lenses equal to the reciprocal of the focal length in meters. A "five-diopter" lens has the magnification power of five times. The principal limitation to the amount of magnification is the depth of field. As magnification increases, the depth of field decreases. This is why a microfocused view of a small object, such as a bee on a flower, yields an image with the bee in focus and an out-of-focus background. It may be so out of focus that it is sometimes blurred out and not distinguishable.

Two common visual magnification devices are handheld lenses and pocket magnifiers or microscopes. Handheld lenses with a frame and handle may contain one lens or multiple fold-out lenses. They are generally plastic (acrylic) or glass. Normal sizes range from 1/2-inch to 6 inches. The other common magnifying device is the pocket microscope.

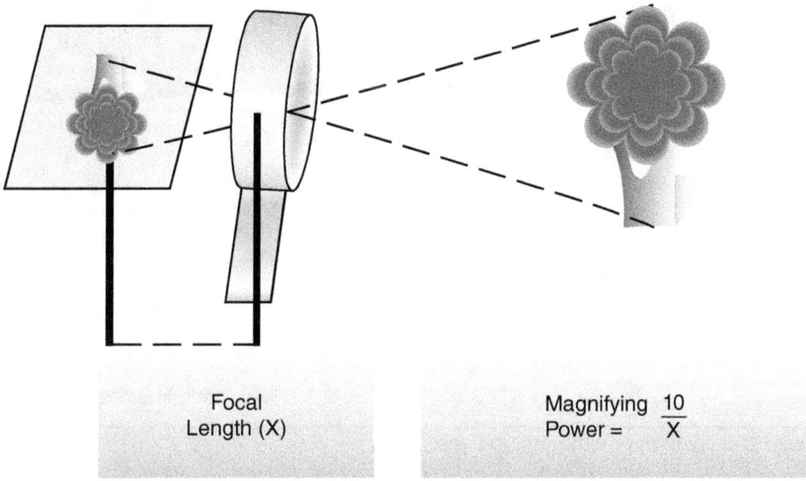


**Figure 3-8** Magnifying power calculations. (*Courtesy of Mistras Group, Inc.*)

Small-diameter tubes (usually 6 inches in length) are fitted with 1/2-inch-diameter lenses. Light is available through cut-outs in the tube or through translucent tube ends. Due to the higher magnification ranges of 25× to 60×, the depth of field and field of view are extremely limited. The larger the diameter of the lens, the lower the magnifying power. Therefore, the practical limit is 20× with a focal distance of 1/2 inch (10/0.5 inch = 20×).

### *Light Sources (Direct)*

When employing magnification devices, additional light sources are usually required. Several lighting devices are available that permit light to be concentrated on a small spot. The most common of these is the handheld flashlight. This light source is usually held at some angle to, and within inches of, the specimen to be examined. The actual light intensity (foot-candles or lumens) at the surface to be examined is dependent on distance, light angle, light bulb wattage, and battery strength. Another common source of auxiliary light is the "drop light," which is usually a 100-watt light bulb encased in a protective cage with a reflective shield connected to a length of electric cable. Again, the actual light intensity is dependent upon angle, distance, and wattage. A light meter can be placed on the specimen surface, and the actual light intensity measurement can be made. As discussed, when the distance from the light source to a surface is doubled, the light intensity is decreased to one-fourth of the original intensity. This is seen in Eq. 3-2, which is expressed differently from Eq. 3-1:

$$\frac{I_1}{I_2} = \frac{D_2^2}{D_1^2} \tag{3-2}$$

where:

$I_1$ = intensity at first location
$I_2$ = intensity at second location
$D_1$ = distance to first location
$D_2$ = distance to second location

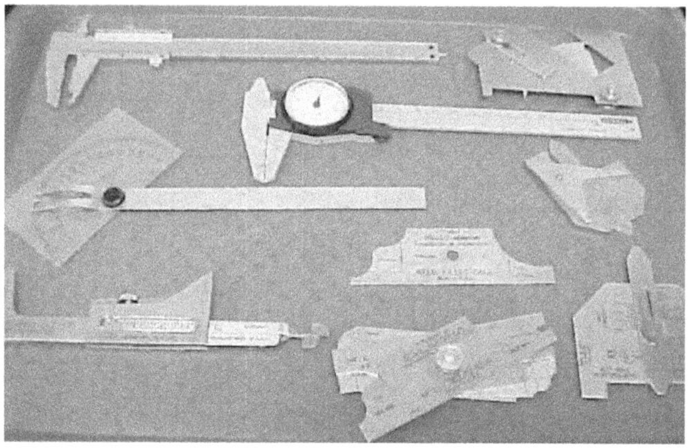

**Figure 3-9** Various measuring gauges.

### *Measuring Devices*

There are multitudes of measuring devices available for different applications (see Fig. 3-9). For the sake of brevity, only a few will be discussed here. The direct visual inspection method is frequently augmented by the use of several common tools used to measure dimensions, discontinuities, or range of inspection. Among these are linear-measuring devices, outside-diameter micrometers, inside-diameter (ID)/outside-diameter (OD) calipers, depth indicators, optical comparators, gauges, templates, and miscellaneous measuring devices.

The most common linear-measuring device is the straightedge scale. Typical scale lengths are 6 inches (15 cm) and 12 inches (30 cm) long. Additionally, 25-foot to 50-foot tape measures are frequently utilized. The 6-inch pocket scale is frequently misused by placing the thumb on one edge of the specimen and abutting the end of the scale against the soft and pliable thumb flesh. This is inherently inaccurate when a movable surface is the beginning point for a measurement. Preferably, the first whole unit of measurement (metric or imperial) should be aligned with one edge of the specimen and the other edge read for total distance. Caution: One must then remember to subtract the first unit of measurement from the total measurement noted.

Micrometers are used to measure outside or inside diameters. They are very accurate measuring devices commonly utilized to measure to the nearest one-thousandth of an inch. Micrometers are available that can be used to measure to an accuracy of one ten-thousandth of an inch. The OD micrometer is made up of several parts, including the anvil, stirrup, spindle, sleeve, thimble, internal screw, ratchet, and knurled knob. It is worth noting that the essential component is the internal screw. It has been turned to have 40 threads per inch. One divided by 40 results in 0.025 inch of spindle movement per 360° turn of the thimble. The micrometer may be outfitted with attachments for different applications. For example, a 0.2-inch ball may be attached to the spindle to accommodate the inside radius of a pipe when measuring pipe wall thickness. Additionally, a pointed attachment may be utilized when measuring pits or very localized variations in wall thickness. Today many micrometers are available with digital readouts. The art of reading the scales is less and less necessary.

The vernier caliper is a variation of the basic caliper. A caliper is usually used to measure the outside diameter of round objects. A common application is the measurement of remaining stock on a part in a machine lathe. If transfer calipers are used, the calipers are placed with the two contact points touching opposite sides of a round object. The calipers are removed and compared to the distance between the contact points as measured on a linear scale. Vernier calipers ID and OD capabilities, depending on which set of points or jaws are utilized. The handle that forms part of the stationary jaw has one or two inscribed scales. There is also a vernier slide assembly that is integral with the movable jaw. The correct OD or ID set of calipers or jaws makes contact with the specimen. The user reads the vernier scale zero mark upon the stationary scale for the whole unit measurement. The vernier scale marks are matched with the stationary scale marks and the best match-up of lines are utilized to read the linear distance to the nearest one-thousandth of an inch or millimeter as the case may be. These two basic forms of calipers have since been replaced with direct read-out dial or digital calipers. Some digital gauges can be interfaced with a laptop or handheld computer so that a large number of readings may be stored.

Today, most calipers are digital read-out making the vernier scale less often used. (See Fig. 3-10.)

Depth indicators (dial indicators) are frequently used to measure surface discontinuity depths. Examples of discontinuities are pits, corrosion, or wastage. Verification of dimensions can also be conducted. In either case, zero depth must first be verified. The dial indicator or the digital read-out indicator is placed on a flat surface and zeroed. It is then moved over the depression. The spindle movement indicates the depth of the depression. When using the older dial indicator, the sweep hand rotations must be counted if the movement exceeds one 360° sweep. If a digital indicator is used, one must note the minimum and maximum point to which the indicator or its extensions are to range, e.g., 0 to 1 or 4 to 5, etc. It is a common mistake not to count the number of rotations or not to note the beginning and end range of the indicator, resulting in only the last increment of measurement being accurate.

Optical comparators are frequently used in machine shops where close tolerance measurements are desired. A comparator produces a two-dimensional enlarged image of the object on a large smoked (frosted) glass screen. The reflected light or background lighting is utilized to cast a magnified image onto the glass screen. This image is compared with a template of the object to check for dimensional accuracy.

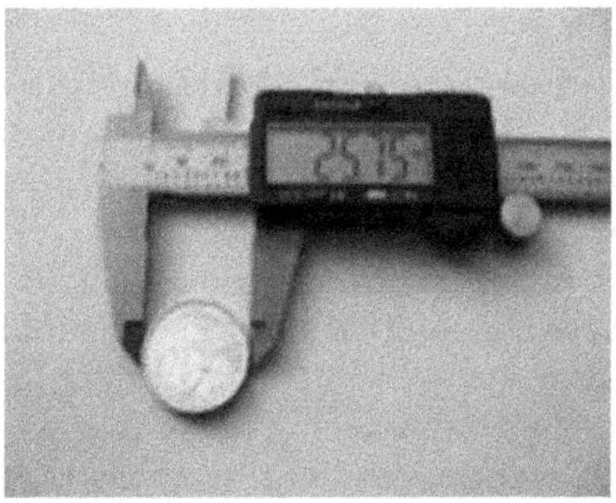

**FIGURE 3-10** Digital caliper.

The welding, fabrication, and construction industries use templates extensively to measure fillets, offset, mismatch, weld reinforcement, undercut, and other dimensional attributes. A common application is a template with minimum and/or maximum dimensions made from sheet metal stock. The actual weld or base metal is compared to the template to determine "go" or "no-go" status. Accurate and actual measurements would still require linear measuring devices.

Miscellaneous measuring devices come in many sizes and configurations. Snap gauges, feeler gauges, radius gauges, temperature gauges, pitch gauges, and diameter gauges are just a few. Some applications could have a gauge, custom made for a specific need or requirement.

### 3.3.3 Remote Visual (Indirect)

Whenever the eye cannot obtain a direct, unobstructed view of the specimen test surface without use of another tool, instrument, or device, a "remote visual" examination is performed.

Recall that a direct visual examination is an examination that can usually be made when access is sufficient to place the eye within 24 inches (610 mm) of the surface to be examined and at an angle not less than 30° to the surface to be examined (see Fig. 3-1). Most codes permit the use of mirrors and magnifying lens to assist with direct visual examinations. A remote visual examination can be defined as an examination that uses visual aids such as mirrors, telescopes, borescopes, fiber optics, cameras, or other suitable instruments.

#### *Borescopes*

One of the oldest applications of remote visual examination is the inspection of gun barrels. The failure of a gun barrel, be it the earliest mortar siege gun of medieval times or the precision machined barrel of a modern rifle, is catastrophic to the gunner at the very least. Two obstacles had to be overcome in the inspection of gun barrels: access to the area to be inspected, and a provision of a light source to provide adequate illumination in order to see the conditions of interest on the inside surface. Small rigid borescopes containing a series of lenses provided the answer (see Fig. 3-11).

Small lamps were used at the far end to provide sufficient light. Since the original applications utilized a lens train to access the bore of a rifle, the term "borescope" was the original and lasting term for the device. From this technology evolved the glass-fiber bundle (referred to as a fiber-optic borescope). This device transmits both the light to illuminate the inspection surface as well as the light reflected off the object back to the viewing end (see Fig. 3-12).

Both the rigid and the fiber-optic borescopes are capable of providing access to small openings. The rigid borescope utilizes a lens train; this is called the lens-optic technique. The fiber-optic borescope utilizes a fiber bundle with a lens at each end; this is called the fiber-optic technique.

A more common means of transferring light to the examination surface is with the use of a light source transmitted through a fiber bundle. The reflective image from the object is returned to the eye via a second fiber "image" bundle. Both bundles are made up of individual fibers that are "clad" with a glass material with a different refractive value sufficient to reflect the light from the outside diameter of each fiber back through the core and down the length of the glass fiber.

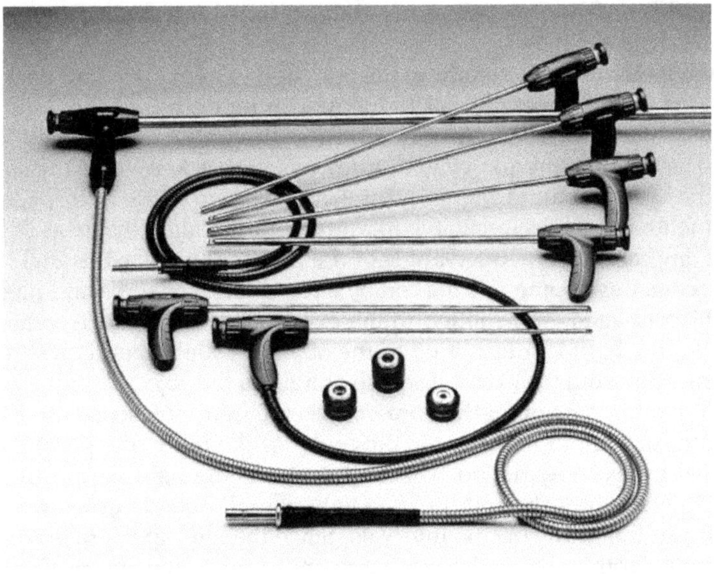

**Figure 3-11** Small rigid borescope. (*Courtesy of Olympus Industrial, with permission.*)

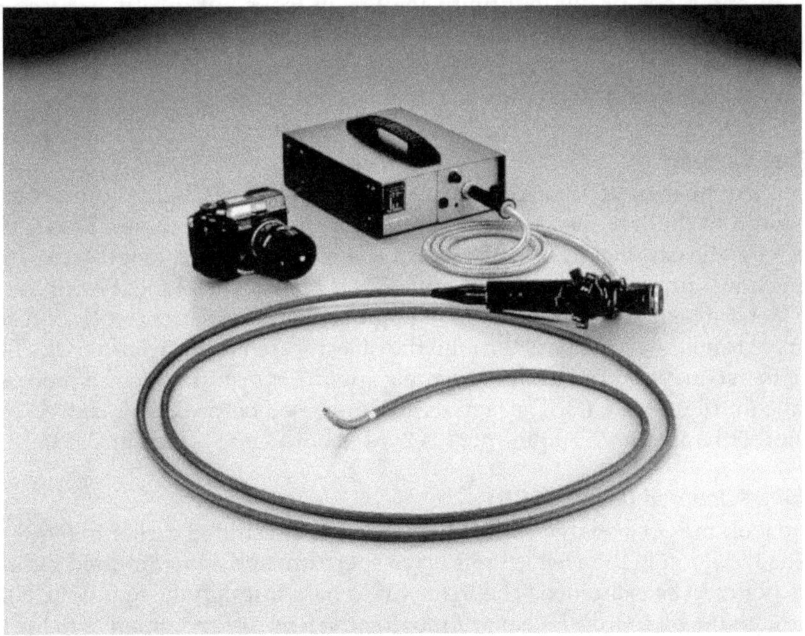

**Figure 3-12** Fiber-optic borescope. (*Courtesy of Olympus Industrial, with permission.*)

Another method of delivering the light to the test site is with the use of light-emitting diodes (LEDs).

Both methods of transmitting the light and image down and back in a bore result in variations in the direction that the view can be delivered. Looking straight ahead is known as the direct view. An angle off the straight-ahead but still forward-looking is known as the fore-oblique view. A sideways look is known as the side view. Anything more than 90° past the straight-ahead view is known as the retrospective view. Different manufacturers will designate the forward, straight-ahead view as 0° or alternatively as 90°. In any case, the included angle of the resultant view is the "field of view."

A recent development is the combination of the lens-optic and fiber-optic techniques. The object image is transmitted to the eye along a rigid tube containing the lens-optic train. But the light is transmitted to the object through fiberglass surrounding the tube, bringing light from the external source of light to the object.

Yet another version of the borescope is the mini-borescope. A single solid fiber diffuses ions in a parabolic arc from the center of the rod to the periphery of the rod, with a graded index of refraction. This still captures the light within the rod and passes it down the length of the rod. But the light actually bends down the length of the rod, forming an image at specific intervals. Since the lens aperture is so small, the lens has an infinite depth of field, eliminating the need for a focusing mechanism.

An interesting physical fact can be observed when using borescopes. A wide field of view reduces the magnification but results in greater depth of field. This can be observed when looking down a great distance during a tube inspection. The image less than a quarter inch away will be magnified. But the image from a quarter inch to one foot away, and theoretically to infinity, may be in focus but greatly reduced in image size. Conversely, a narrow angle field of view produces higher magnification but results in a shallow depth of field. The image at one point may be in focus, but at a short distance in front and behind that point, the image is out of focus.

#### *Fiber Cameras*

Like the fiber-optic borescope, the fiber camera uses light from an external source brought to the test site via the fiber-optic bundle (light guide). This fiber bundle that delivers the external light to the object site is limited in its length. An articulated (four-directional tip) bundle can be manufactured up to 40 feet long. Fiberglass bundles up to 45 feet can be made without articulating tips. Unlike the fiber-optic borescope, the light waves from the object pass through the objective lens and strike a CCD. The light waves are then converted to electronic signals and are transmitted to the processor in the central control unit (CCU). The signals are processed before being viewed on the monitor. Another name for this equipment is "video borescope" (see Fig. 3-13).

#### *Charge-Coupled Device (CCD)*

The workings of a charge-coupled device were illustrated earlier in Fig. 3-4. Electromagnetic radiation in the visible light wave spectrum and some infrared radiation reflect off the object to be examined. The light waves pass through the lens, which can be focused through the mechanical means of moving the lens distance relative to the solid-state silicon chip or CCD. The CCD comprises thousands of light-sensitive elements arrayed in a geometric pattern of rows and columns (a matrix). Each single microchip contains a single point or picture element known as a pixel. Each chip is struck by light waves. The intensity or quantity of light striking the chip determines the amount of electrons that are

(a)

**Figure 3-13** (*a*) Video borescope. (*Courtesy of Olympus Industrial, with permission.*)

(b)

**Figure 3-13** (*b*) Video borescope. (*Courtesy of GE Industrial Technology, with permission.*)

generated in the silicon sandwich (chip). This accumulation of electrons in proportion to the amount of light that strikes that pixel is periodically passed on as packets. These packets of electrons are transferred from one pixel site to its adjacent site, just as water is moved from one bucket to the next in a bucket brigade.

The force required to move these electrons is produced in the pixels by means of a voltage that alternately turns on and off. As this "clock voltage" alternates, the electrons move from one site to the other. After the last pixel, the electrons are converted to a voltage proportionate to their number. This voltage is amplified, filtered, and then applied to the input of a video monitor. The differential in electrical charge is what is processed down the wire at the other end, where the processor is located. By assignment, the pixel is coded as one of the primary colors (red, green, or blue) and is reassembled as electronic signals that are sent to a color monitor. The application of CCD images to flexible endoscopes during the 1970s permitted development of a solid-state scope that relied on electronics rather than fiber optics for image transmission. The CCD image processor produces a brighter, higher-resolution image than does the conventional fiberscope. This advanced microelectronic technology enables the CCD to function as a miniature TV camera able to record and display images with great clarity on a video monitor.

### *CCD Cameras*

**Modern Remote** Visual inspection allows us to utilize a variety of cameras based on CCD image processor technology. Remote inspection equipment ranges from instruments that can access areas through very small openings from 3.9 mm up to cavernous rooms. Very small orifices in jet engines, tubes, pipes, culverts, vessels, tanks, rooms, and open areas can all be inspected by the right video equipment by remote technology.

Digital video borescopes, Pan-Tilt-Zoom cameras, robotic pipe crawlers, and freewheeling robotic platforms can all perform remote visual inspection applications.

A variety of video borescopes from basic cordless and no options to reconfigurable probe options with a variety of recording medium options all exist.

Video borescope probes can be used in openings as small as 3.9 mm (0.152 inch).

The total distance available for inspection is limited by the fiber bundle transmitting the light. If the bundle is flexible enough to make sharp turns, it may not be stiff enough to probe long distances. Stiff probes don't bend around corners well.

Once images are captured electronically, many software programs either "on-board" or remotely processed can provide major advancements image manipulation, measurement, retention, and retrieval. Moving from video analog images to digital images has allowed significant advances in image acquisition, dimensional measurement, and data management for remote visual inspection (RVI). It has become standard to measure attributes through specialized video probe tips. These measurement techniques include comparison, shadow, stereo, and 3D measurements. All of these measurement techniques are available for the video borescope instruments.

Another variation of the CCD camera is the one developed for the examination of boiler tubes (see Fig. 3-14).

The limiting factor in the fiber-optic cameras is the distance the fiber bundle can be extended. There is also a practical limit as to how long the fiber bundle can be manufactured. Also, the longer the fiber bundle, the greater the possibility of breaks and disconnections. The solution to these problems is to place the light source at the object end via electrical wire. Since boiler tubes have inside diameters of several inches,

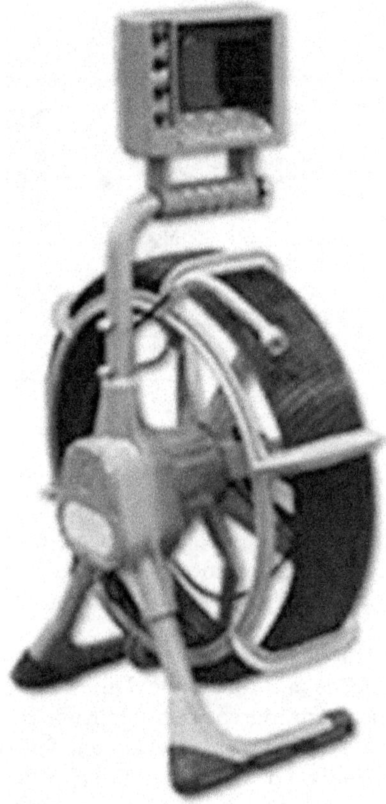

**FIGURE 3-14** CCD tube cameras. (*Courtesy of GE Industrial Technology, with permission.*)

the logistics of assembling a ring of light can be realized and cable lengths of up to 100 feet in length can be attained. The same elements of a lens and CCD transmitting the signal back to the CCU (processor) and on to the monitor remain, but now multiple wires traverse the length of the cable—one with power for the light source and another to transfer signals from the CCD to the processor. No fiber bundles are necessary.

Robotic crawler tractors can carry an assortment of lighting arrangements and cameras with fixed and multi-directional or positioned cameras. Pan and tilt alone can change the elevation of the view from a mobile platform. The pipe sizes capable of being inspected range from 2 inches nominal pipe size (NPS) up to 60 inches NPS. Additionally, tracking in small rooms, tanks, and other large cavities (cisterns, underground storage tanks, underground or separate rooms) can all be inspected remotely (see Fig. 3-15).

Robotic crawlers can inspect enclosed spaces such as air ducts, piping, vessels, and tanks with a range of up to 1,500 feet. Most robotic crawler systems need at least a 6-inch-diameter opening to operate. There are several variations that provide the camera with a platform from which to operate. Mobility can be gained by mounting cameras onto robotic self-propelled platforms or crawlers. A power wheeled system or tractor with sufficient weight can overcome great amounts of friction from the cable, but can only enter an opening larger than the tractor. The larger the tractor, the longer

**FIGURE 3-15** Robotic crawler CCD cameras. (*Courtesy of GE Industrial Technology, with permission.*)

the distance that can be traveled. However, access to smaller openings is limited. Smaller tractors can access smaller openings but cannot travel as far.

Pan, tilt, and zoom cameras are CCD cameras on a pole with without mechanized gears to allow panning, tilting, and motorized lens to allow zooming. They are typically used for large area inspections such as tanks or vessels.

**CCD Tank Camera with Pan, Tilt, and Zoom** Pan, tilt, and zoom cameras may be placed on a pole or dangled from the end of a cable. Direct view on LCD monitors may be done with recording options spanning various electronic recording media exist.

Adding a zoom lens to focus on objects very far away requires great versatility in adjusting the lens assembly. Focusing at greater distances creates more problems to be overcome. First, light, which decreases to one-fourth in intensity for each doubling of the distance, must be greatly increased. Second, motor controls for the remote movement of the more complicated zoom lens assembly must be dealt with. Third, motor controls for panning and tilting must be designed to cover a large inspection area with the possibility of limited camera movement (see Fig. 3-16).

**FIGURE 3-16** CCD tank camera with pan, tilt, and zoom lens. (*Courtesy of GE Inspection Technologies, with permission.*)

The lights now have to target two possible zones of illumination due to the longer viewing distances. Spot lamps or flood lamps could be used. The typical light source is two 35-watt bulbs with reflectors. The spot lamp will produce 8,500 candle power with a 10° beam spread. The flood lamp will provide 15,000 candle power with a 38° beam spread. These can become very hot and dangerous in hazardous environments.

The motor controls for the zoom lens are bulkier because the lens assembly is now more complicated. The lens components in these assemblies are moved to change the relative physical positions, thereby varying the focal length (FL) and angle of view through specified ranges of magnification. Zooming is a lens adjustment that permits seeing detailed close-up images of a subject or a broad overall view. The angular field of view (FOV) can be made narrower or wider, depending on the setting. This ability to zoom in or out gives the impression of camera movement, even though there is no movement at all. This is done with variable focal length and multiple lens assemblies that go from wide angle to telephoto while the image remains in focus. To achieve this effect, three groups of lens elements are needed:

1. Front-focusing objective group
2. Movable zoom group
3. Rear stationary relay group

If the zoom lens is well designed, a scene in focus at the wide-angle (short focal length) setting remains in focus at the narrow-angle (telephoto) setting and everywhere in between.

Pan and tilt mechanisms allow the camera to be pointed in almost any direction. The pan and tilt controller has a two-axis drive capability for panning (side to side) and tilting (up and down). The addition of the pan and tilt gears and drives make for a heavy head that needs additional strength in the support fixture. To ensure spark-free and low or noncombustible situations, the components must all be housed in an inert gas environment. Additionally, the internal pressure of the fixture housing must exceed the external pressure to ensure that explosive gases do not enter the housing (where sparks are possible). The housing must be airtight to prevent sparks from exiting the housing into combustible environments.

The zoom lens with pan and tilt capabilities offers a great advantage over the fixed focal length lens in that many different fields of views can be obtained with one lens. An additional point of interest regarding the zoom lens is that the typical magnification limit via optical adjustment is 12:1. Commercially available units can attain magnification up to 24:1. This doubling of magnification is attributable to computer enhancement. It can be noted during zooming that the magnification is smooth up to the 12:1 point, and then a slight "jolt" is noted as the computer enhancement takes over. At the maximum magnification, slight pixel squares may be barely visible to the eye when viewed on the monitor.

### 3.3.4 Image Recording

Video cassette recorders and video tape recorders have been replaced by the ability to save images on DVDs, flash media, and memory cards.

Regardless of the camera, monitor, or fixturing mechanisms, the image is usually recorded review, evaluation, and record retention. The medium of the future may well be the digital laser disc or digital tape medium.

This is one area of technology in which the consumer market has driven the commercial market. Electronic real-time digital format is the norm for image retention, transfer, and retrievability. This includes flash memory and solid-state memory cards. Technology has advanced, with memory now available in gigabytes, signal compression, and increased speed of processing.

### 3.3.5 Monitors

The number of pixels in a monitor does not relate directly with the resolution of the image of the object. The display resolution of a digital television or display device is the number of distinct pixels in each dimension that can be displayed. It can be an ambiguous term especially as the displayed resolution is controlled by all different factors in cathode ray tube (CRT) and flat panel or projection displays using fixed picture-element (pixel arrays) (see Fig. 3-17 for monitor).

One use of the term "display resolution" applies to fixed-pixel-array displays such as plasma display panels (PDPs), liquid crystal displays (LCDs), digital light processing (DLP) projectors, or similar technologies, and is simply the physical number of columns and rows of pixels creating the display (e.g., 1,920 × 1,080). A consequence of having a fixed grid display is that, for multiformat video inputs, all displays need a "scaling engine" (a digital video processor that includes a memory array) to match the incoming picture format to the display.

Note that the use of the word resolution here is a misnomer, though common. The term "display resolution" is usually used to mean pixel dimensions, the number of pixels in each dimension (e.g., 1,920 × 1,080), which does not tell anything about the resolution of the display on which the image is actually formed. Instead, resolution properly refers to the pixel density, the number of pixels per unit distance or area, not total number of pixels. In digital measurement, the display resolution would be given in pixels per inch.

The 1951 USAF Resolution Test Target is typically used for visible camera or sensor contrast tests and calibration. This is one calibration criterion for display resolution determination (see Fig. 3-18).

### 3.3.6 Digital Cameras

Digital cameras use the CCD as the means of capturing images. These images are stored temporarily on floppy discs or memory chips, or are archived permanently. The bitmap

**FIGURE 3-17** Monitor. (*Courtesy of Dell.*)

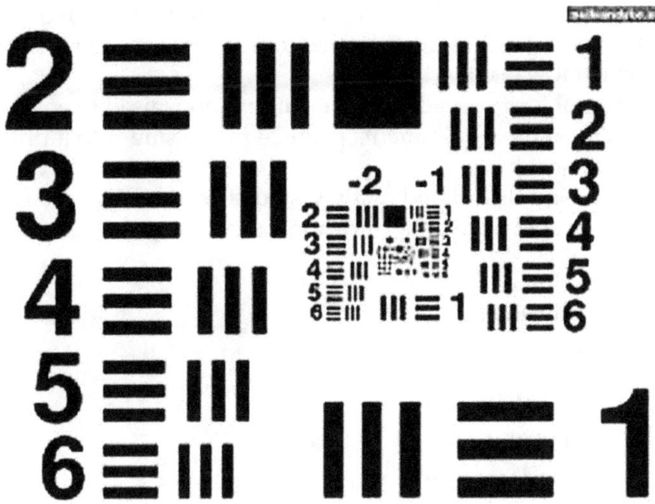

**Figure 3-18** 1951 USAF Resolution Test Target. (*Courtesy of Electro Optical Industries, Inc. Santa Barbara, CA.*)

or other file format can then be downloaded directly to a personal computer, attached to e-mail, stored on web pages, or otherwise transferred for viewing. Some cameras utilize removable media cards for short-term or permanent archiving.

Typical resolution may vary. CCD pixel arrays with 640 × 480 to 1,280 × 960, or 1,280 × 1,024, 1,536 × 1,024 up to 1,920 × 1,080 are available. Many cameras feature LCD screens built into the camera for on-the-spot viewing of the image, thus allowing retakes as required. Reusable storage cards can provide image-capturing capability limited only by the number of storage cards available.

Once an image is downloaded to a personal computer, various image programs allow image manipulations. Enlargements, segmenting, framing, changing brightness or color, as well as other options allow a wide range of changes to be made. Image printouts on color printers are producing near-photographic quality prints.

## 3.4 Applications and Techniques

### 3.4.1 General Applications

Visual examinations and other nondestructive test methods cover the spectrum of examining materials from raw product form to the end of their useful lives. Initially, when raw material is produced, a visual examination is conducted to locate inherent discontinuities. As the material is further transformed through the manufacturing process, a product results. At this stage, the visual examination method is used to find discontinuities that are produced during the primary processing steps. When the product is further developed into its final shape and appearance, the secondary processes that give the product its final form can also introduce new discontinuities. Finally, the product is placed into service and is subject to stresses, corrosion, and erosion while performing its intended function. The process concludes when the material has reached the end of its useful life and is removed from the source. At every stage, the visual examination method is applied using various techniques to ascertain the physical condition of the material that became the component, system, or structure serving the needs for which it was intended.

After material is produced, visual examination is used to ensure that a product will meet the specification requirements prior to processing into a product form for use in its intended service.

The technology associated with visual testing (VT) and remote visual testing (RVT) includes a spectrum of applications, including various products and industries such as:

- Tanks and vessels
- Buildings
- Fossil-fuel power plants
- Nuclear power plants
- Turbines and generators
- Refinery plants
- Aerospace
- Transportation
- Infrastructure

Tanks and vessels usually contain fluids, gases, or steam. Fluids may be as corrosive as acid or as passive as water, either of which can cause corrosion. Tank contents are not always stored at high pressure. Conversely, vessels usually contain substances under substantial pressure. This pressure, coupled with the corrosive effects of fluids and thermal or mechanical stresses, may result in cracking, distortion, or stress corrosion of the vessel material.

Buildings also serve as a source for a myriad of RVT applications. These applications include location of clogged piping; examination of heating and cooling (HVAC) heat exchangers; and looking for cracking, pitting, blockages, and mechanical damage to the components. Structural damage that may be present in the support systems, beams, flooring, or shells, such as cracking, corrosion, erosion, or warpage, can also be detected.

Fossil-fuel power plants have piping, tubing, tanks, vessels, and structures that are exposed to corrosive and erosive environments, as well as to other stresses. These components may require RVT.

Turbines and generators, existing at both fossil-fuel and nuclear power plants, are vulnerable to damage due to high temperatures, pressures, wear, vibration, and impingement of steam, water, or particles. Accessing the small openings and crevices to reach damaged turbine blades becomes a very tedious job and a serious challenge, but the effort of performing remote inspections through limited-access ports reduces the need and cost of downtime and disassembly of major components.

VT and RVT technologies and techniques are used in nuclear power plants as well. Water used for shielding and cooling is exposed to both ionizing radiation and radioactive surface contamination. The use of water as a coolant and radiation shield in a nuclear environment places additional requirements on RVT evaluation. The equipment must not only be waterproof but also tolerant of radioactive environments.

Due to process requirements in refineries, the containment of pressure and temperature is a necessity of paramount importance, as is the containment of hazardous materials. These same materials can be a source of corrosion to piping, tanks, vessels, and structures, all of which are in constant need of monitoring.

### 3.4.2 Standards, Codes, and Specifications

The discussed applications all require VT and RVT to detect surface anomalies. A common source for material specifications is the American Society for Testing and Material (ASTM) standards. ASTM was founded in 1898 and is a scientific and technical organization formed for "the development of standards on characteristics and performance of materials, products, systems and services; and the promotion of related knowledge." At this time, the annual ASTM standards fill 75 volumes and are divided into 16 sections.

An ASTM standard represents a common viewpoint of producers, users, consumers, and general interest groups intended to aid industry, government agencies, and the general public. Two metal material sections are Section 1—Iron and Steel Products and Section 2—Nonferrous Metal Products. These standards provide guidance on the material conditions that must exist in order to be considered satisfactory for use.

Additionally, when material is fabricated and formed into a product, other standards, specifications, and codes delineate visual testing requirements. The American Society of Mechanical Engineers (ASME) publishes the ASME Boiler and Pressure Vessel Code, Sections I through XI. The material section is Section II. The design sections are I, III, IV, VIII, and X. Section V contains the methodology for nondestructive examination, and Section IX addresses welding and brazing. Section VI deals with heating boilers, Section VIII is pressure vessels, and Section XI, in-service inspection of nuclear power plant components. Scattered throughout these sections are visual examination requirements.

The American National Standards Institute (ANSI) is a consensus-approval-based organization. ANSI's B31.1—Power Piping and B31.7—Nuclear Power Piping also provide visual examination requirements for materials, fabrication, and erection.

The American Petroleum Institute (API) has developed approximately 500 equipment and operating standards relating to the petroleum industry that are used worldwide. An example is the API standard for Welding of Pipelines and Related Facilities (API 1104).

The aerospace and military standards are being replaced with the more commonly accepted industry codes and standards.

### 3.4.3 Visual Detection of Discontinuities

Chapter 2 addresses and illustrates many of the following discontinuities in greater detail.

#### *Inherent Discontinuities*

The VT technique most often used to detect inherent discontinuities is direct visual. The human eye assisted by measuring devices, auxiliary light sources, visual aids (e.g., magnifiers and mirrors), and the recording media of photographs and sketches are most commonly used for this application.

**Applications** Raw material may be checked for inherent discontinuities and compared with the requirements of material specifications for acceptance. Frequently, the first examinations performed on new materials and products are dimensional checks. Attributes such as thickness, diameter, out of round, roughness or smoothness are examples of the first round of visual checks made on new materials. Checking for discontinuities that interrupt the normal physical structure of the part is also conducted at the material fabrication stage. The inspector might look for blemishes or imperfections that are generally superficial and could be located by observing a stain, discoloration, or other variations on the surface. Discontinuities on the surface could indicate other anomalies in the material. Discontinuities or lack of some element that is essential to the usefulness of the part should be of importance to the inspector. Any of these attributes would be evaluated against the specified code, acceptance criteria, standard, or customer requirements. The following is a typical sequence of observation steps an inspector or examiner might follow for a general condition visual examination:

1. The observation of an anomaly—something different, not normal, or irregular.
2. The visual indication is evaluated and determined to be relevant or not.
3. The relevant indication is compared to the appropriate standard, code, specification, or customer requirements.
4. The relevant indication is judged acceptable, not recordable, recordable, or unacceptable and rejected.

Discontinuities may start out as something small. They may result from corrosion, a slight scratch, or other discontinuities inherent in the material such as porosity, cracks, or inclusions. These seemingly harmless anomalies may develop into cracks due to stress concentration under varying loads and propagate with time. Eventually, there may no longer be sufficient solid material left to carry the load.

A defect is a discontinuity of such size, shape, type, number, or location that results in a condition that will cause the material to fail in service or not be used for its intended purpose.

**Ingots** Common discontinuities detected visually in the ingot are cracks, scabs, pipe, and voids.

Cracks may occur longitudinally or in the transverse plane in the ingot. Although cracks may be observable in the ingot stage, it is more likely that they will appear in subsequent processing operations.

Scabs are surface conditions on ingots caused by the hot molten metal splashing on the mold surface, solidifying, and oxidizing such that it will not re-fuse to the ingot. Again this condition may be more detrimental to the product at later stages of processing.

Pipe is a cavity formed by shrinkage during the last stage of solidification of the molten metal. It is an elongated void with a "pipe like" shape and occurs in the center of the ingot. This discontinuity can result in a lamination after the ingot is rolled into plate or sheet. If the ingot is processed into bars, blooms, or billets, the pipe will become elongated as a result of the processing and generally remain in the center.

Voids are produced during the solidification of the ingot, caused by insufficient degassing. The gas, which becomes entrapped, often forms spherical cavities.

Nonmetallic inclusions in ingots are chiefly due to deoxidizing materials added to the molten steel in the furnace, ladle, or ingot mold. Oxides and sulfides constitute the bulk of nonmetallic inclusions.

All of these discontinuities can occur during the pouring of molten metal into the ingot mold. The resulting discontinuities are the same as those found in the processing stage when pouring metal into a mold and creating a casting.

**Techniques** The types of anomalies found in ingots are typically gross in nature. These can generally be found with direct visual examination techniques. A magnifying lens, auxiliary light, and possibly a mirror should suffice for this application.

#### *Primary-Processing Discontinuities*

The visual technique most often used to detect primary-processing discontinuities is the direct visual technique. Measuring devices, auxiliary light sources, visual aids (e.g., magnifiers and mirrors), along with the recording media of photographs and sketches, are the most common techniques used for this application.

**Applications** Materials are formed into a final or near-final shape as a result of the various primary processes. The forming processes of wrought products include forging, rolling, drawing, extruding, and piercing. Discontinuities caused by these methods of forming the product shape are known as primary-processing discontinuities. When steel ingots are worked down into usable sizes and shapes such as billets and forging blanks, some of the inherent discontinuities described earlier may appear. These primary-processing discontinuities may ultimately become defects. Working the metal down into useful forms such as bars, rods, wire, and forged and cast shapes will either change the shape of existing discontinuities inherent to the product or possibly introduce new ones.

**Forging** Typical forging discontinuities include forging bursts, laps, and cracks. Bursts are internal or external forging discontinuities. Forging bursts appear as scaly, ragged cavities inside the forging. (*Note:* Internal bursts are more likely to be detected with the use of volumetric inspection methods such as radiography or ultrasonics. For additional information, refer to Chapter 2.)

Forging laps are folded flaps of metal that have been forced into the surface of the part during the forging process. The lap can vary from shallow to very deep on the forging surface. A lap can appear to be tight and irregular, straight, crack-like and linear, or even wide and U-shaped. Forging laps are difficult to detect without magnification.

Cracks in a forging are usually referred to as "forging bursts." Many forging bursts open to the surface are detectable visually; however, the smaller ones may require other methods such as PT or MT.

**Technique** The direct visual technique is best suited for this application. This can be best accomplished by examination using a 5× to 10× magnifier with an auxiliary light source held at an angle so as to cast a shadow on any irregularities.

**Rolling** Rolling of product forms can create a variety of discontinuities. The more common ones are seams, stringers, cracks, and laminations. These can appear in product forms such as bars, rods, channels, beams, angle iron, and strip metal.

Seams and stringers generally appear on the outside surface of the product parallel to the direction of rolling and are very difficult to detect visually. Seams contain nothing other than entrapped oxides. A seam is usually produced when two unbonded surfaces are pressed together during the rolling process. Stringers are inclusions that get "strung out" during rolling and may be observable if exposed to the surface. Complex structures may be formed from plate or rolled shapes, making some surface indications difficult to detect. Additionally, internal discontinuities such as laminations are best detected using UT.

**Techniques** The direct visual technique is best suited to detect these conditions in this application. This can be best accomplished by examination using a 5× to 10× magnifier with an auxiliary light source held at an angle so as to cast a shadow on any irregularities.

Indirect visual examination with a borescope or fiberscope may be appropriate for viewing internal areas of complex product forms. The limiting factors will be access and distance.

**Drawing, Extruding, and Piercing** Drawn, extruded, or pierced products may exhibit similar types of surface discontinuities. Any irregularities on the surface that are gross in nature could result in problems during the forming process. Common observable anomalies may be scabs, slugs, or scoring.

Scabs are loosely adhering material that can be removed from the final product. These can result from dirty dies that press in the loose foreign material during the drawing process or from oxide layers that are not bonded to the material.

Slugs are usually larger pieces of material that are firmly impressed into the product form and may not come off easily.

Scoring appears as deep scratching of the surface of the product by foreign material that may be adhering to the inside of the die or piercing mandrel during processing.

**Technique** The direct visual technique is best suited to detect these conditions. This can be best accomplished by examination using a 5× to 10× handheld magnifying lens with an auxiliary light source held at an angle so as to cast a shadow on any irregularities.

**Castings** Casting is a primary process. Castings may exhibit various types of discontinuities that are inherent in the metal forming process, including nonmetallic inclusions (slag), hot tears, gas, unfused chills and chaplets, cold shuts, and misrun. Some of these discontinuities may also occur internally. The casting process can produce some very complex shapes. Access for visual examination can be a real challenge.

Inclusions exposed to the surface may be akin to the "tip of the iceberg" in that the visible area may be merely a small percentage of the inclusion contained within the casting. The visual examiner can only detect and measure the exposed area of the discontinuity.

Hot tears are more likely to be visible on the surface, since they occur at locations where changes in thickness occur. Hot tears are the result of different coefficients of expansion and contraction at a change of thickness. The abrupt change from thick to thin material sections provides a stress riser. The different rates of cooling (contraction) could cause a rupture of the metal under stress at these junctions.

Gas is a spherical void condition. It may occur both internally and at the surface of the casting. Surface gas typically has a hemispherical shape, since if exposed to the surface an incomplete sphere is trapped in the casting. There may be one or many of these anomalies distributed randomly or in clusters at any location.

Unfused chaplets and chills will be visible only if they protrude through the surface. Remember that the chaplet is usually metal with a low melting point that is used to support the internal mold cores until the molten metal enters the mold and supposedly melts the chaplet. The examiner may note the "footprint" of the metal form that is exposed to the surface. The chill can be likened to a nail with a head. The "head" may be all that the examiner observes, with the remainder of the "nail" encased in the casting.

Cold shuts could result from balls of molten splattered metal that adhere to the mold wall prior to the remainder of the molten metal rising over it and entrapping it on the inside of the mold and thus the outside of the casting. It can appear as a circle around a ball of trapped metal. Cold shuts also result when two regions of the casting do not metallurgically fuse; one surface begins to solidify before the remainder of the molten metal forms around or over the already solidified metal. This condition is generally not detectable by standard VT techniques.

Misrun is a surface condition in which the casting mold does not completely fill with molten metal, thereby leaving a noticeable depression, which is readily detectable with VT.

**Techniques** Either the direct visual or remote visual techniques are necessary for examining castings for exposed inclusions—hot tears, surface voids, unfused chaplets or chills, cold shuts, and misrun. Accessible surfaces can best be examined by using a 5× to 10× handheld magnifying lens with an auxiliary light source held at an angle so as to cast a shadow on any irregularities. The surfaces that are internal to the casting may require examination using a rigid borescope or flexible articulating end fiber-optic borescope, the latter being more versatile. Either one would bring the remote surface under close examination and provide light to the examination site. Extra care is required to ensure full coverage and proper orientation when performing internal exams. Knowing where the discontinuity is located and accurately measuring its dimensions are requirements for proper evaluation of the indication. If a visual record is desired, a fiber-optic video camera may be appropriate. The distance to the area of interest as well as the size of the access port are important factors to be considered when choosing the proper instrument. Additionally, it is difficult to obtain accurate measurements at the remote site. Some instruments have been outfitted with detachable scales that are visible in the field of view. Other instruments have remote measuring devices that rely on shadow casting and geometric calculation techniques. These devices require calibration on surfaces similar to the orientation plane of the examination surface in order to be accurate.

### *Metal Joining Processes*

Metal joining processes include a number of welding and allied processes. Each major process must be considered for unique as well as common discontinuities. Metalworking industries generally use soldering, brazing, and a broad variety of welding processes. A generic definition of the basic welding process is "a materials joining process that produces coalescence of materials by heating them to suitable temperatures, with or without the use of filler metal."

Soldering utilizes filler material that becomes liquid at temperatures usually not exceeding 450°C (±840°F). The solder is heated and distributed between two surfaces of a joint by capillary action.

Brazing produces coalescence of materials by heating them to a suitable temperature and using a filler metal that becomes liquid above 450°C (±840°F). The filler metal is distributed between the closely matching surfaces of the joint by capillary action. In either case, when completed, the only visible signs of completed wetting, and hence

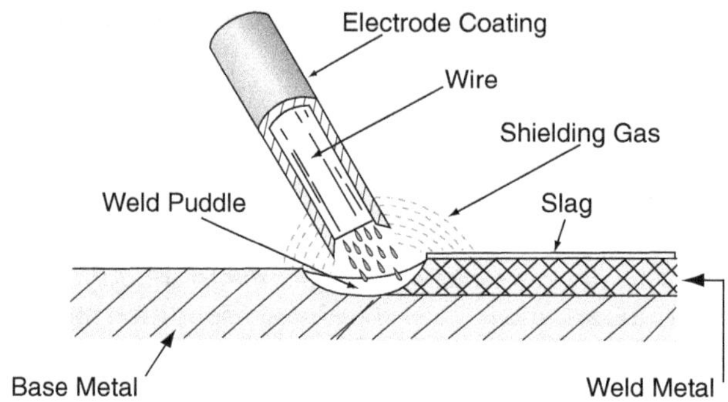


**Figure 3-19** Shielded metal arc welding process (SMAW).

bonding, is the micro fillet at the intersection of the two materials. The normal configuration for joining is a lap joint of two materials overlapping each other. When heated and wetted with filler material, the resultant fillet should be uniform the entire length of the joint.

Arc welding produces coalescence of metals by heating them with an arc, with or without the application of pressure, and with or without the use of filler metal. Shielded metal arc welding (SMAW; Fig. 3-19), submerged arc welding (SAW; Fig. 3-20), gas tungsten arc welding (GTAW; Fig. 3-21), and gas metal arc welding (GMAW; Fig. 3-22) all have two things in common. Each utilizes an electric arc as the heat source for melting the base and filler metal, and each provides a means of shielding the arc to block out harmful elements found in the air (oxygen).

SMAW and SAW both use the combustion of flux as the means of consuming oxygen at the weld site, thus preventing the oxidation (burning) of the metals being joined. The common by-product of burning flux is slag.

GTAW and GMAW both utilize inert gas cupped around the weld site to eliminate the oxygen portion of the fire triangle (oxygen, heat, and fuel) and thus prevent the metals from burning at high temperatures. The electrode bearing the heat for GTAW is the tungsten electrode. In this process, the electrode is not consumed. However, deposits of

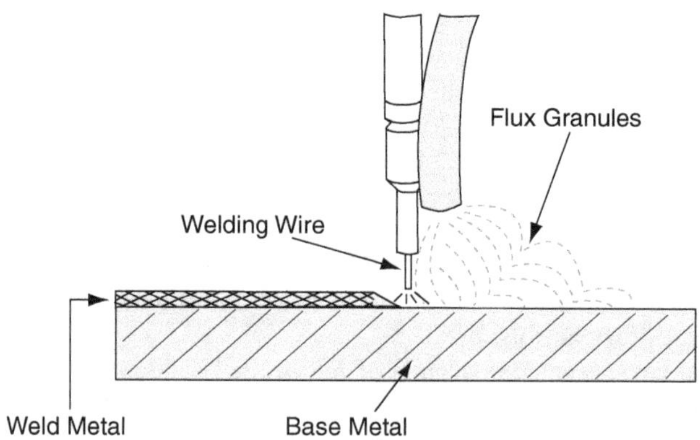


**Figure 3-20** Submerged arc welding process (SAW).

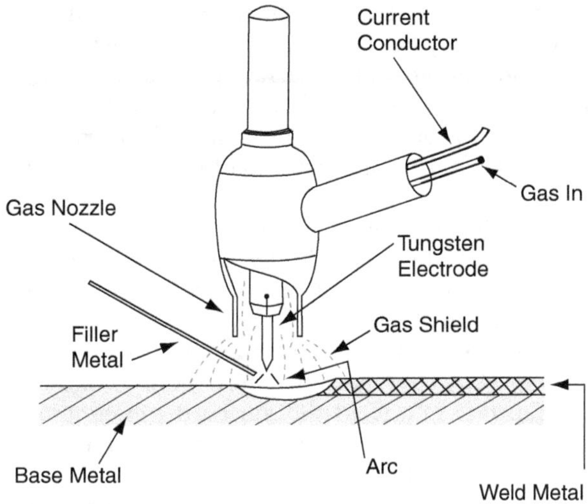


**FIGURE 3-21** Gas tungsten arc welding process (GTAW).

tungsten may flake off and remain in the weld. This is visible only if they remain on the surface of the weld during welding. GMAW uses a consumable electrode of filler metal. Therefore, tungsten cannot be deposited in the weld.

Welding processes are akin to a microcasting process. Therefore, many discontinuities found in the primary casting process can also be found in the welding processes. Porosity, nonmetallic inclusions, and cracks are common to both casting and welding processes.

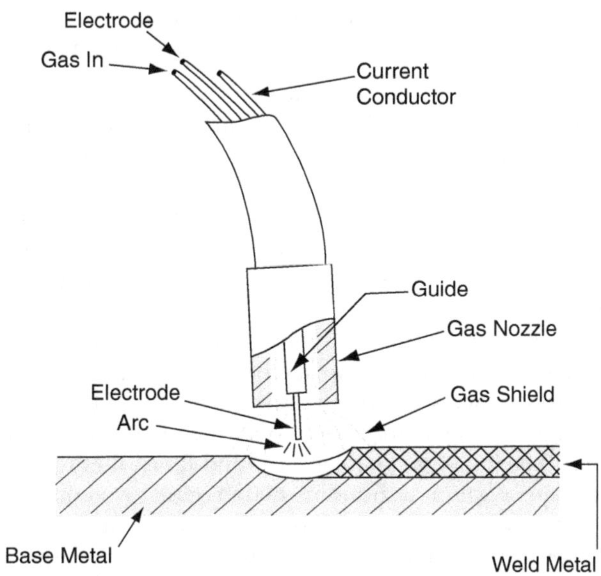


**FIGURE 3-22** Gas metal arc welding process (GMAW).

Porosity may occur internally or at the surface of the weld. As with the casting process, gas porosity may have a hemispherical shape if it is open to the surface. There may be from one to a number of these, randomly distributed or in clusters, at any location. Variations in the shape of the porosity may be called by descriptive names such as wormhole porosity, cluster, and random or linearly aligned porosity. They all originate from trapped gases as the liquid metal solidifies upon cooling, thus forming spherical voids in the metal.

Nonmetallic inclusions commonly referred to as slag (burned flux in the case of welding processes) may be trapped internally or at the surface of the weld face. This condition may be sporadic, isolated, or clustered in groups. Slag is something included in the metal that is not metallic.

*Note:* Only the processes that utilize flux can have significant slag residue from the burning of the flux. Any base metal may have nonmetallic inclusions or impurities that may rise to the surface of a weld site during the melting of the metal. However, this quantity of "slag" is slight and not to be confused with the major source of nonmetallic inclusions, which is burned flux.

Cracking, with all its possible shapes, forms, and degrees of severity will occur whenever metal relieves itself of stresses by rupturing. The various orientations, locations, and appearances cause the cracks to have many names that uniquely identify them. Crater cracks are found in the depressed cup of the terminal puddle made at the last arc location of the weld bead. Transverse cracks are oriented perpendicular to the longitudinal axis of the weld. Longitudinal cracks are parallel to the long axis of the weld. Microcracks could occur anywhere but are very small by definition. Heat-affected zone (HAZ) cracks occur at the toe of the weld and may exist in the weld and/or the base metal, but are limited to the zone adjacent to the toe of the weld. All cracks are characterized as being crack-like in appearance, that is to say, irregular, jagged, and linear in overall disposition.

Each weld process may have weld discontinuities unique to that process. Since only GTAW has a tungsten nonconsumable electrode, it is the only process that can have tungsten inclusions. The other weld processes use consumable material as filler.

**Techniques** The direct visual technique is best suited for detecting most external weld discontinuities. This can be best accomplished by examination using a 5× to 10× magnifier with an auxiliary light source held at an angle so as to cast a shadow on any irregularities.

Additionally, linear measuring devices are commonly employed to measure each discontinuity. Dimensional and profile checks are the most common direct visual examinations for welds. These visual examinations would be conducted at fit-up, in process, and after welding.

During prewelding and fit-up, many dimensions should be checked. Root opening, which is the separation between the members to be joined, and root face lands should also be checked. Groove face surfaces on each member of a butt joint must be checked. The single bevel angle on one member of the two members and the groove angle between the two members to be joined have tolerances and limits that require verification. The weld joint may have penetration minimums that necessitate confirmation. The members to be joined typically have minimum thickness measurement requirements. The inside diameter of butt pipe joints has diameter tolerances, land dimensions, and counterbores that must also be verified.

During welding, controls are frequently evaluated by the visual inspector, e.g., temperature control with the use of a pyrometer, temperature crayons, or other temperature indicators. Distortion or movement tolerances are checked during the welding process.

Welding process controls such as amperage, temperature, weld wire, material certifications, and welder qualification may also be performed by the visual examiner, the quality control inspector, quality verifier or, in this era of Total Quality Management (TQM) and other quality control philosophies, by the welders themselves. Other weld attributes that can be evaluated visually in the process include bead and weave pattern. Excessive weave patterns may induce uneven heat sinks, creating a skin effect that could lead to brittle layers of weld metal that may crack in service. Interpass cleaning is necessary to prevent entrapment of nonmetallic inclusions. Porosity in the intermediate weld beads may propagate throughout the weld cross section with subsequent bead deposits.

Another imperfection that may occur during the welding process prior to completion is lack of fusion. Since lack of fusion is defined as nonfusion between weld beads or between the weld and the parent metal, it is not visible to the eye in the final weld. During deposit of each weld layer, sharp crevices that may lead to lack of fusion on the next weld deposit can be detected and either removed or ground into a radius, or more heat can be applied to melt to the bottom of the crevice. Detection of lack of sidewall fusion usually requires a volumetric test (i.e., UT or RT).

When dealing with completed welds, many dimensions and weld attributes are inspected for compliance with the weld specification. Final cleaning, weld size, and profile dimensions are also checked. Weld face, length, width, and thickness must meet dimensional requirements found on the drawing. The weld should also be examined to detect possible surface discontinuities. Convexity and concavity affect the weld thickness. The legs and actual throat dimensions of fillet welds must be measured. Welds that are accessible from both sides should be checked for adequate penetration. Partial penetration welds cannot be adequately detected by visual means unless cross-sectioned, which defeats the nondestructive aspect of visual examination. Stringer weld beads can be counted, lengths measured, distances between welds measured, and occurrence of surface discontinuities observed.

Final weld examinations include visual inspection for the following dimensional and profile attributes:

*Misalignment*. Excessive offset observed on the inside or outside diameters of a pipe or the misalignment of the near or opposite sides of butt welds.

*Underfill*. Insufficient weld filler material to fill the weld groove face. The weld surface is below the base metal surface (see Fig. 3-23).

*Undercut*. Base metal melting without sufficient metal deposit to bring the weld metal level with the base metal. Emphasis is on the melting of base metal and insufficient filler metal being added (see Fig. 3-24).

*Overlap*. Weld metal extends over the base metal at the toe of the weld without fusion to the base metal (see Fig. 3-25).

*Excessive reinforcement*. Weld reinforcement that exceeds the specification limits (see Fig. 3-26).

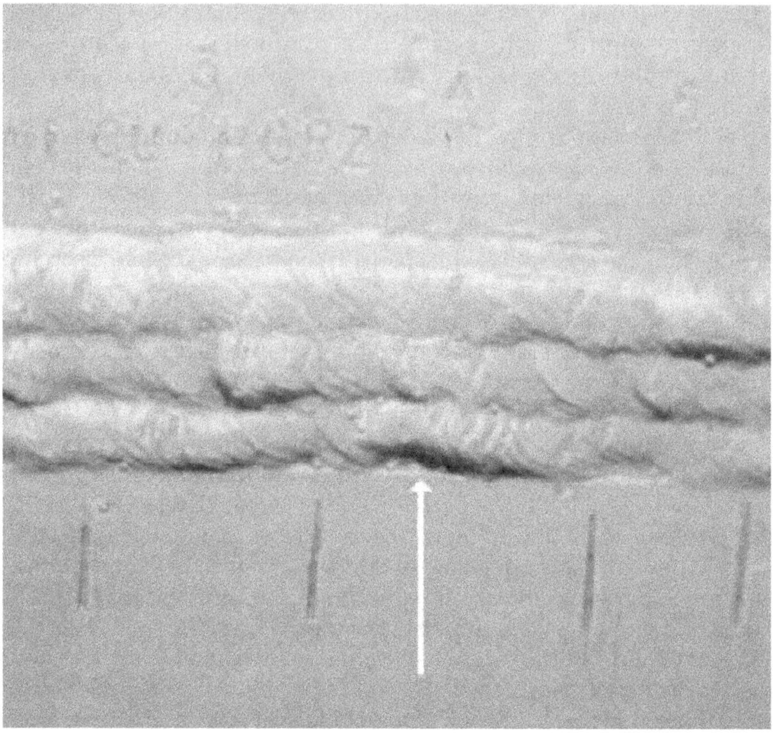

**Figure 3-23** Underfill.

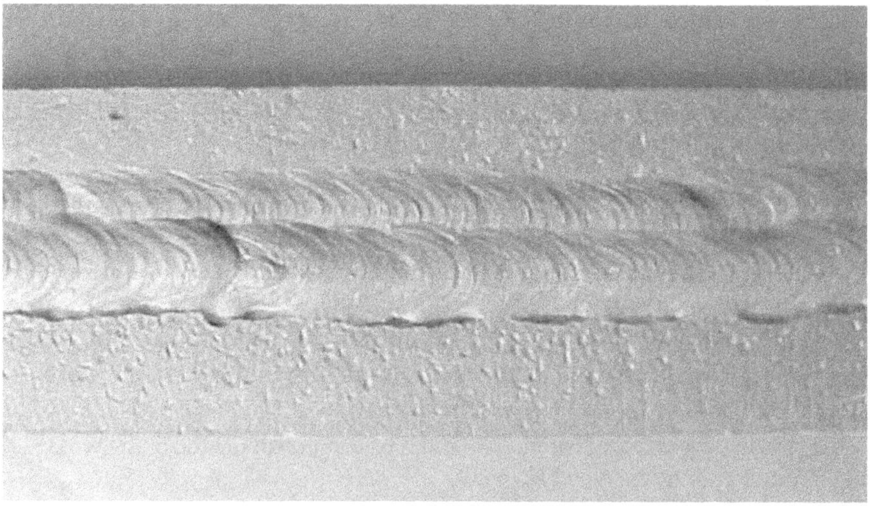

**Figure 3-24** Undercut.

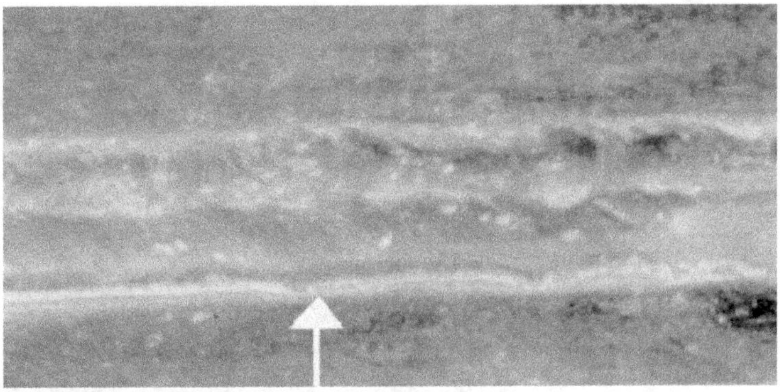

**FIGURE 3-25** Overlap (or rollover).

*Concavity.* A depression on the inside diameter (root concavity) of a pipe, or near the far side of plate that results in less than specified weld thickness and is less than the base material thickness of the thinnest member (see Fig. 3-27).

*Insufficient throat in a fillet weld.* A depression on the face of a fillet weld that results in the actual throat being less than that specified.

*Insufficient leg in a fillet weld.* Less leg dimension in the fillet weld than that specified.

*Incomplete penetration.* This is the absence of penetration of the root pass. It is only observable by VT when the opposite side from the weld is accessible. It is, in fact,

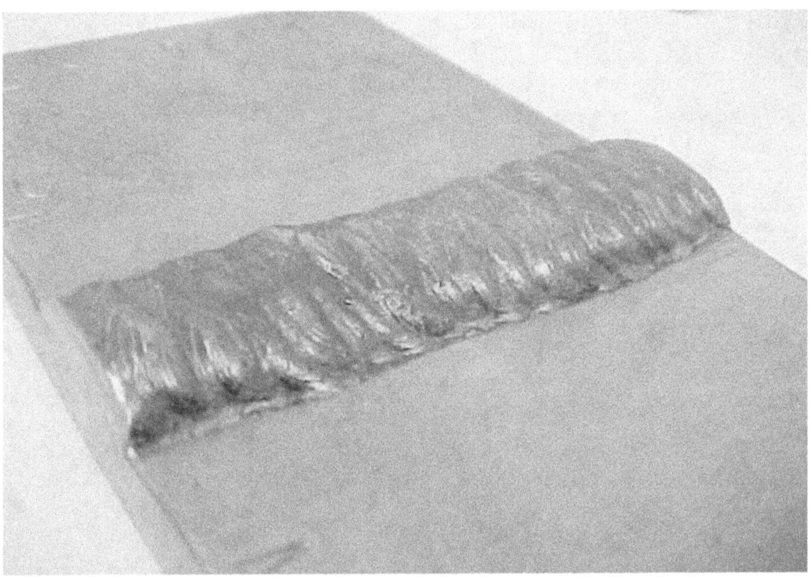

**FIGURE 3-26** Excessive reinforcement.

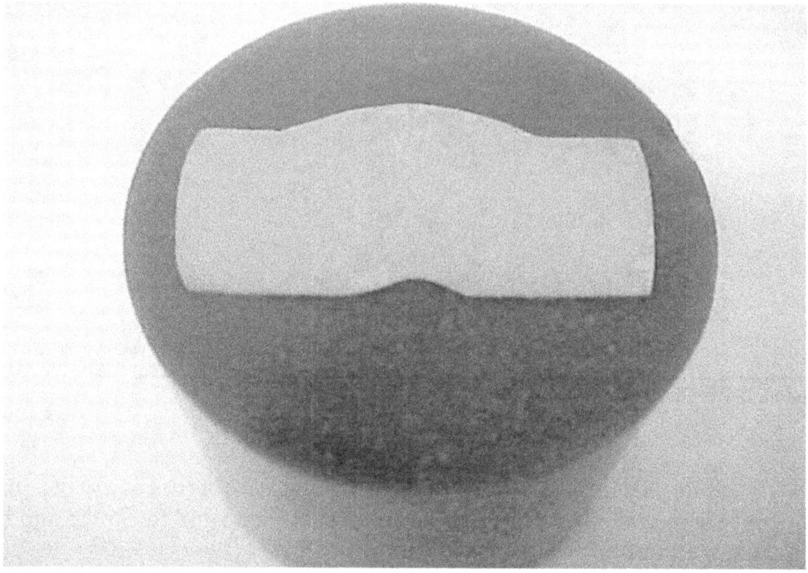

**Figure 3-27** Concavity.

the original edge of the plate or pipe to be joined (at least one side) still visible and not totally fused or melted (see Fig. 3-28).

*Burn-through.* Excessive heat resulting in the loss of some metal, leaving behind a cavity that can vary in size and can extend from one side of the weld to the other.

**Figure 3-28** Incomplete penetration.

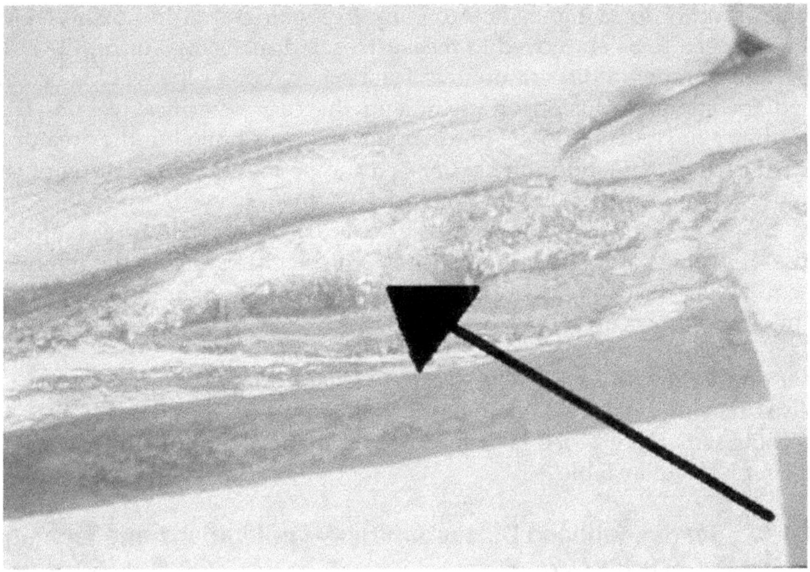

**Figure 3-29** Weld crater.

*Weld craters*. Depressions at the end of the weld, representing the last puddle in the weld bead (see Fig. 3-29).

*Crater cracks*. As mentioned above, cracks that occur in the weld crater. These cracks are due to thinner portions of the weld in the crater cooling first and becoming rigid. The main weld cools later and shrinks away from the crater portion of the weld. This results in residual stresses in the weld relieving themselves via rupture of metal or cracking. The crater crack often has a star-like appearance and is not easily detected with conventional VT techniques.

*Arc strikes*. Arc welding that strikes the base metal, usually in an effort to initiate an arc to start welding another bead. These arc strikes melt localized portions of the base metal to shallow depths. The differentials in material hardness frequently result in stress risers, yielding microcracks at the time of welding or thereafter in service.

*Spatter*. Globules of material expelled from the weld puddle or from the consumable electrode during welding. The localized melting of the skin of the base material contacted by the molten balls of excess metal can also cause local stress risers. Again, these differentials in material hardness frequently result in stress risers, yielding microcracks at the time of welding or later in service.

**Techniques** Visual examination of welds and other metallurgical joining processes can best be performed with the direct visual technique. This application requires a variety of mechanical aids, such as templates, gauges, and measuring tools, to maximize efficiency and ensure consistency of the results. The first overview should be accomplished by examination using a 5× to 10× magnifier with an auxiliary light source held at an angle in order to cast a shadow on any irregularities. Additionally, linear measuring devices need to be employed to measure each dimensional or profile attribute and the size and number of discontinuities. The root area of a pipe weld is occasionally examined for complete fusion or melting in the case of critical welds. This application requires a flexible fiber-optic borescope that is positioned in the area of the root weld. This visual examination can ensure that the penetration or melting of the root is adequate, in lieu of performing radiography.

On some occasions, the root insert can be observed during real-time welding to observe welder progress and adequacy. This is hazardous to the remote visual instrument, as the heat can destroy the fibers or lens if they are too close to the weld root while welding occurs.

#### *Measuring Devices for Welding Process Visual Examination*

Measurements and examinations of welds are conducted before, during, and after welding with the use of a large array of mechanical aids, including, but not limited to, the items listed in Table 3-1.

### 3.4.4 Service-Induced Discontinuities—Applications and Techniques

Service-induced discontinuities are the results of material deterioration during use. Wear, erosion, corrosion, and loss of integrity through fatigue all may lead to the failure of a component while performing its intended function. Wastage or general material loss can occur when the size of a component has been reduced such that the cross-sectional area is no longer able to support the designed load. Mechanical fatigue due to cyclic loading and thermal fatigue due to temperature fluctuations frequently lead to excessive stresses, cracking, and loss of integrity. Stress combined with a corrosive environment frequently results in stress corrosion cracking.

| Linear Measuring | General Gauges | Weld Gauges | Optical Aids | Image Recorders | Misc. | Light Sources |
|---|---|---|---|---|---|---|
| Steel rule & scale | Depth gauge | Cambridge weld gauge | Reticule magnifier | Video camera | Combination square | LED flashlight |
| Tape | Thread gauge | Fillet weld gauge | Microscope (pocket) | Digital camera | Temperature gauge | Incandescent flashlight |
| Vernier caliper | Pitch gauge | Template gauge | Telescope | | Level and plumb line | High-intensity spotlight |
| Dial caliper | Thickness gauge | Hi-Lo gauge | Optical comparator | | Surface comparator | Halogen lamps |
| Digital caliper | Diameter gauge | Weld height gauge | | | | Xenon lamps |
| Dial indicators | Snap gauge | | | | | |
| Micrometers | Radius gauge | | | | | |

**Table 3-1** Mechanical Aids Listed by Functional Category

### Wear

Wear can be defined as the undesired removal of material caused by contacting surfaces through mechanical action. In many respects, wear is similar to corrosion. The difference is that wear is mechanical in origin and corrosion is chemical in origin. The different types of wear are abrasive, erosive, grinding, gouging, adhesive, and fretting.

Abrasive wear has the appearance of a multitude of small scratches or grit marks that generally result in a local uniform reduction of material thickness. It occurs when one surface rolls or slides under pressure against another surface.

Erosive wear occurs when particles in a fluid or other carrier slide and roll at a relatively high velocity against a surface. Each moving particle contacting the surface removes a minute portion of the surface. The cumulative effect over a long period of time is significant and can produce a high degree of material loss. A "scalloped" appearance with wave patterns over a general area can indicate erosive wear.

Grinding wear occurs primarily when particles under high pressure cut very small grooves at relatively low speed across a metal surface. High-pressure, low-speed operation is characteristic of metal turning against a dull cutting edge. A series of parallel scratches, either curved or straight, resulting in reduced material thickness, would indicate grinding wear. Gouging wear is caused by high pressure that lifts fragments from a metal surface. When hard abrasive products are crushed, battered, or pounded under high pressure, rapid deterioration of the contact surfaces can be expected. Grooves with a buildup of material at the deeper end of the trough are good indications of gouging.

Adhesive wear can be characterized as microwelding. The microwelding or adhesive property causes a stationary component to lose particles of material that adhere to a moving component. Other terms for adhesive wear are scuffing, galling, and seizing. Adhesive wear frequently starts out on a small scale but rapidly escalates as the two sides alternately weld and tear metal from each other's surfaces.

Fretting wear is similar to adhesive wear in that microwelding occurs on mating surfaces. The difference is that adhesive wear is related to moving surfaces and fretting wear is related to stationary surfaces that unintentionally move due to applied stresses. Minute motions do occur, but the cyclic motion is extremely small in amplitude. It is, however, enough to cause microwelding and enough to cause fretting on both surfaces. These "welded" areas do not remain joined, and with subsequent movement of the surfaces, they break off. The particle debris resulting from this process causes wear. Other terms for fretting wear is fretting corrosion, false brinelling, friction oxidation, chafing fatigue, and wear oxidation.

Visual techniques for examining materials for wear include magnification, auxiliary light, and surface comparators. The surface patterns, depth of wear, knowledge of the part's function, and hardness all play a role in effective evaluation. The biggest obstacle to overcome is to gain access to the surfaces where the wear is occurring. By definition, two surfaces in close contact could cause wear. Therefore, disassembly is required for direct visual examination.

### Corrosion and Erosion

Corrosion can be defined as the deterioration of a metal resulting from electrochemical reactions with its environment. There are many types of corrosion (at least two may be acting simultaneously).

Galvanic corrosion is caused by physical differences between contacting metals or a metal and its environment. The corroding material is known as the "anode" and the noncorroding complimentary material is known as the "cathode." When corrosion takes place, three conditions are necessary for the electrochemical reaction to occur. (1) Two different materials are needed. (2) They must meet in electrical contact with an electrolyte or electrically conductive liquid or paste. (3) When the anode corrodes, such as in the operation of a battery, hydrogen is released and energy in the form of an electric current is present. The continued electrochemical reaction causes more galvanic corrosion and additional material loss. To break the cycle, one of these three elements must be removed.

Uniform corrosion is the most common form. In the case of iron or steel, it is better known as rusting. Uniform corrosion is the result of microscopic galvanic cells in the surface of the metal. Because of local chemical differences, impurities, and alloying intermetallics within the metal, there are microscopic anodes and cathodes ready to cause corrosion if an electrolyte such as moisture is introduced. The corrosion is uniform only on the macroscopic level.

Crevice corrosion is a kind of galvanic corrosion that is caused by differential oxygen concentration cell corrosion, or what is also known as "poultice corrosion." Normally, metal may be expected to corrode when exposed to high levels of oxygen. However, a crevice between two surfaces or metal under a poultice of moist debris is more likely to corrode than the exposed metal. This occurs because there is little oxygen within the crevice or under the poultice. The metal there is anodic and it corrodes. Areas exposed to higher oxygen content are cathodic and do not corrode. Concealed metal at the edge of a joint or under debris tends to pit and eventually becomes perforated through the metal thickness. The gas bubbles under a coated (painted) surface of steel can be attributed to crevice corrosion.

An erosion corrosion form of material loss is particularly common in pumps or impellers. This form is known as cavitational erosion. A liquid stream can destroy the protective surface film on a metal. This is done through a change in pressure on the trailing edge of pump or impeller blades with an associated creation of gas bubbles. The subsequent collapse or implosion of the bubble can cause a great deal of accelerated metal loss over time. The collapsing cavities implode on the metal surface with compressive stresses estimated at thousands of pounds per square inch. With constant repetition of the pounding, the fatigue mechanism progresses until pits form in the material surface at these locations.

Corrosion may enter the picture if surface films are formed on virgin metal in the pit when the system is at rest. The surface film is rapidly destroyed by the high compressive forces when the motion resumes and the cycle starts again.

### *Fatigue Cracking*

Loss of integrity through fatigue cracking can be caused by many mechanisms. Fatigue can cause the failure of a material or component under repeated, fluctuating stresses having a maximum value less than the tensile strength of the material. In general, fatigue strength is a function of the ultimate tensile strength. There are three stages of fatigue: (1) initial damage, (2) crack propagation in the component, and (3) failure resulting from sectional reduction. The cracking mechanisms of mechanical fatigue and thermal fatigue are as follows.

Mechanical fatigue depends on the nature of the stress, how the stress is applied, the rapidity of the stress fluctuation, the stress magnitude, the fabrication procedures utilized, and the environment in which the item is used. The higher the tensile strength, the more susceptible the part is to mechanical failure. Surface finish is important when considering mechanical fatigue. The more polished the surface, the more fatigue-resistant the item is. When fatigue cracks occur, they characteristically are relatively straight and nonbranching. The cracks are transgranular (across the grains) when viewed metallographically.

Thermal fatigue is differentiated from mechanical fatigue by the nature of the load causing the fatigue. In thermal fatigue, there are temperature differentials within a component, which induce significant strains and stresses. Differential expansion may be caused by parts with different coefficients of expansion, i.e., parts within one component having different temperatures or connected to components having different temperatures. The cyclic rate of thermal fatigue is typically low. Thus, it is often but not exclusively a high-temperature phenomenon. High thermal stresses can lead to failure in a few cycles.

Thermal fatigue is often initiated where there is a change in section thickness. This is due to different amounts of heat that is absorbed and the resultant temperature gradient. Examples of components that may suffer thermal fatigue are superheater tubes, pump casings, turbine blades, and piping. Failure is often caused by a single crack propagating through the cross section of the part. Occasionally, the failed piece may display a crack that is partially through the sectional area. This will exhibit a color resulting from oxidation. This occurs prior to the break, which exhibits a newer cracked surface.

**Techniques** In general, fatigue cracks are open to the surface. Examining components such as pumps, valves, piping, vessels, and heat exchangers for wear, erosion corrosion, and loss of integrity through fatigue can best be performed with the direct visual technique. Access may frequently be limited and remote visual techniques may be required.

The first overview should be attempted by direct visual examination using a 5× to 10× handheld magnifying lens with an auxiliary light source held at an angle so as to cast a shadow on any irregularities. Additionally, linear measuring devices may be needed to measure each observable discontinuity. Internal examinations of valves, pumps, or components may require flexible fiber-optic borescopes, zoom cameras, crawler devices with miniature cameras, and handheld or guided fixture devices.

### 3.4.5 Corrosion, Erosion, and Fatigue Applications

The locations of corrosion and erosion are usually inside pressure containers. Common examples are pumps, valves, piping, vessels, and heat exchangers.

Pumps function to move liquid from one place to another. This is achieved by imparting energy to the medium, generally by means of converting electrical to mechanical energy. In the process, resistance is encountered in the form of friction. Friction occurs from both the medium moving in the pressure boundary and from the moving parts of the pump. Gravity is also a force that has to be partially overcome.

In many cases, the medium being moved may have corrosive properties. Erosive wear is common in all cases. All of these factors lead to deterioration of the internal pressure boundary and the internal parts of a pump. Access to the areas to be examined usually requires some disassembly, if not the total breakdown of the pump for inspection. Frequently, the primary access to pumps "in situ" is either through the suction or through discharge ends of the pump. Occasionally, disassembly provides access through the pump shaft opening.

**Techniques** A common access technique is with the use of a crawler, which can be a tractor motor device with mounted camera and trailing electrical wire. This approach would be acceptable if the suction or discharge pipe were nearly horizontal. If the access pipe is vertical or very steep and the gravity force exceeds the friction of the tractor wheels, a different mechanism is needed. It may be a centering device with expanding and contracting spider arms. Sequential expansion and contraction of alternating sets of centering arms may be mechanically or hydraulically operated to allow the device to climb up or down a pipe.

Once access to the impeller blade or pump vanes is achieved, a zoom camera or a camera with an articulating head can be manipulated to view the area of interest. The leading and trailing edges of the blades and vanes are very important areas to inspect. Additionally, the pump casings should be examined for signs of washout (erosion) or pitting. Pits or ridges can reduce efficiency and accelerate casing wear. Impellers should be checked for cracking. Diffuser vanes should be examined for wear or erosion. Shaft sleeves should be checked for scoring. Bearings should be checked for wear or damage. Shafts should be checked for corrosion, wear, and signs of washout.

Valves function to stop or start flow, regulate flow, and control the direction of flow. Short of catastrophic failure, the common problems with valves are erosion and corrosion (as with pumps). Different types of valves have different locations that should be focused on for examination.

Gate valves use linear movement to start or stop flow. The upstream seat of the wedge (gate) and downstream body seat are the most likely places for erosion and wear. The sealing surfaces behind removable seat rings are good locations to expect erosion. The walls of the packing and stuffing box are a probable location of corrosion. The stem should have a smooth surface to allow sealing contact with the stuffing material. Scores, scratches, pitting, and erosion could all allow leakage.

Globe valves regulate flow and employ rotational movement. The sealing surfaces of the disc and body of the globe valve should be checked for cracks, scratches, galling, pits, indentations, or erosion that may prevent disc-to-body sealing. Examination of these areas would require either upstream or downstream pipe openings and an articulating fiber-scope or video camera head, assuming that the valve and its openings are large enough to accommodate access. The valve stem should have a smooth surface to allow sealing contact with the stuffing material. Checking the stuffing box wall would require stem and stuffing removal.

Swing check valves function to control the direction of flow, and as the name implies, the disc "swings." As fluid pushes against the disc, it opens to allow flow. When back-pressure is present, the disc reverses direction and seals against the body seal, preventing reverse flow. The areas to be inspected are the seating surfaces of the disc and seat ring. Additionally, the swinging gate rotates on a hinge. If access is possible, the hinge pin should be checked for wear or distortion, as should the seating surfaces. Swing check valves frequently have access port or bonnet caps that provide access from the top of the valve.

An articulating fiberscope is the most appropriate instrument for the examination of valves. Limited success could be achieved with a rigid borescope. Marginal success may be achieved using a flashlight and mirror. Only very large swing check valves could accommodate a miniature camera in conjunction with a video monitor or LCD.

Ball and plug valves function to start or stop flow. They have rotational movement. Openings from the up or downstream side could allow access for a rigid or flexible borescope to check the ball and seating surfaces for physical damage, wear, or corrosion that may cause leakage. As with other valves with stems, the stuffing box and stem should be checked for scratches, pits, wear, or erosion that might cause leakage.

An articulating fiberscope would be an effective device for the examination of ball and plug valves. Limited examination may be achieved with the use of a rigid borescope. Some success may be achieved with the use of a flashlight and mirror.

Butterfly valves function to start and stop flow in addition to regulating flow. The movement of the valve is rotational. The primary area for signs of physical damage, wear, or corrosion is the horizontal plane at the centerline. It is here that the plastic seal may be subjected to high flow rates and the most wear and erosion. As with the other valves with stems, the stuffing box and stem should be checked for scratches, pits, wear, or erosion that might cause leakage. Limited view may be achieved with the use of a flashlight and a mirror. An articulating fiberscope would be an appropriate instrument to use. A limited examination may be achieved with the use of a rigid borescope.

Diaphragm valves function to control or regulate flow. Their movement is linear. The sealing surface is a flexible diaphragm with a wear-resistant coating or layer sealing against the body of the valve. Most importantly, the diaphragm must be checked for evidence of aging and cracking. An articulating fiberscope would be the most appropriate instrument. Limited success may be achieved with a rigid borescope. A flashlight and a mirror may also be useful.

Most of the discussion regarding piping systems to this point has been related to the pumps and valves in the piping system. This is where the energy is imparted to the component or changed in the liquid medium. It is in these areas where energy is transformed or transferred that damage to the internals of the component or part within the system usually occurs. Evidence of the failure to contain a liquid can be observed externally by signs of leakage. Piping systems are designed to contain specified pressures during normal operation.

Testing these systems usually requires achieving certain pressures at certain temperatures and for specified amounts of time. The more critical the function of the system, the higher the safety factors that are used. When holding times and pressures have been determined, a pressure test can be conducted. First, the inspector needs to be familiar with the complete scope of the impending test. Pumps, pressure gauges, lighting, drawings, temperature gauges, and layout familiarity are necessary. A checklist for the inspector would comprise a list of possible locations where evidence of leakage may be found. Exposed surfaces, surrounding areas and components, floor conditions, conditions of equipment or components, and drainage systems should be part of the checklist. Evidence of leakage could range from drips of water to a spraying torrent. Access to "out of the way" piping systems is needed to prevent any zone of interest from being overlooked.

### Remote Visual Techniques

**Leakage** Remote visual examination with the use of binoculars and a spotlight may allow access to areas not easily observed by direct visual examination. Evidence of leakage may be noted by observing stains. These stains may be red or rust discoloration due to oxidation of ferrous materials, white crystals from liquid containing boric acid, yellow stains from sulfur-containing liquid, and green stains from copper-containing liquid.

Gravity will take water to the lowest point on vertical piping. Horizontal runs of pipe could drip at any low point. If the piping is covered with insulation, time must be allowed to pass to let the leakage penetrate through the insulation. A drop of water is very uniform in size. Since it is the smallest unit of evidence of leakage, being able to see a drop of water at the examination location would be the indication upon which to qualify the remote (using binoculars) visual method. The drop of water should be visible at the maximum distance to be inspected with a given light source. Light intensity is most often measured with a calibrated light meter.

In summary, leakage examination should be systematic and scheduled. The normal system pressure should be achieved or exceeded for the examination. A minimum amount of time should pass to allow evidence of leakage to be observed. Leak testing provides a good technique to discover a mechanical problem and has the advantage of testing the entire system at one time.

### Component Supports

Component supports have different configurations. Hangers, snubbers, restraints, and other component supports are all designed to support piping systems. Detailed knowledge of each component is required prior to performing a thorough examination. A brief description of each type of component support will be followed by some general inspection techniques, as well as a discussion of the conditions that should be observed in order to ensure proper functioning of the supports.

Hangers usually consist of a rod attached to a component by a welded "ear" or clamp. They usually are simple in design, with few parts and connections.

Spring supports can be of the variable spring-loaded type or the constant-weight type. They usually carry the weight from above and support the pipe in tension.

Pipe supports typically support the pipe from below. They can be in contact with the pipe in the form of a saddle, clamp, or tee. As with overhead supports, they can be variable spring load or constant load. They are installed vertically, carry the weight from below, and support the pipe in compression.

Restraints typically are rigid and anchor the piping system to a building structure. The purpose of a restraint is to prevent movement in one axis. Multiple restraints can prevent movement in several directions. They are typically installed horizontally to vertical runs of pipe but may be oriented in any direction. They are usually attached to the pipe by means of clamps. The other end of the restraint is anchored to the concrete or welded to I-beams. They prevent severed pipes from whipping and rupturing other pipes.

Snubbers are vibration dampers that are installed on components to minimize the effects of dynamic loading. They are attached to the piping by clamps and to a building or structure by anchor bolts or welded connections. The snubber provides structural stability through mechanical or hydraulic action. The snubber allows normal movement during operations but restrains, restricts, or prevents movement during dynamic loading events.

**Techniques** The examination of hangers, snubbers, restraints, and component supports generally is done by direct visual examination. On occasion, indirect or remote visual examination is required due to access limitations. In general, the inspector is looking for signs of mechanism failure or future failure. This could include inadequate construction practices, physical damage in service or since installation, signs of overload, signs of corrosion, or signs of fatigue.

Generally, hangers, snubbers, restraints, and component supports are checked to determine their general mechanical and structural conditions. The inspector checks for loose parts, debris or abnormal corrosion products, wear, erosion, corrosion, and the loss of integrity at bolted or welded connections.

Specific conditions that should be observed include:

Inadequate construction practices that may be evidenced by discontinuities in the base material may be in the form of roughly cut or deformed threads in rods, bolts, or nuts; bolts, nuts, and rods not tightened or torqued properly; wrong-sized parts; missing washers, cotter pins, or locking nuts; and rough, excessive, or inadequate welds.

Physical damage may be evidenced by bent, twisted, or deformed clamps, rods, or beams; warped, discolored, or burned parts (due to welding); missing nuts, pins, load scales, or identification numbers; cracked, sheared, or broken pins, rods, bolts, or welds; chipped concrete; and elongated bolt holes, eyes, swivels, or turnbuckles.

Overload may be evidenced by cracked welds, broken pins, sheared rods, or stripped threads; deformed mounting or saddle plates; cracked concrete at embedded bolts and plates; twisted or bent parts; and elongated bolt or pin holes.

Corrosion may be evidenced by light or heavy layers of rust on exposed surfaces; shallow or deep pits; buildup of corrosion products or loss of material at openings and crevices; loss of rivets on load scales; frozen rollers or bearings; loss of threads on bolts, rods, and nuts; and areas of eroded material.

Fatigue can be evidenced by cracked or broken parts. These could appear as thin lines or gaps running through or adjacent to welds, gaps or openings in threaded roots, and gaps or openings in bars, rods, or pipes adjacent to an encircling clamp.

#### *Pressure Vessels*

Vessels or tanks are designed to contain liquids or gases while maintaining their integrity at specified design pressures. Visual examinations of vessels are performed to determine the condition of the critical inner surfaces, such as the high-stress points at the junction of the vertical and horizontal surfaces, at nozzle penetrations, at construction weld joint intersections, and transitions between thin and thick sections. Less apparent is the importance of the intersection of air and product, where crevice corrosion (discussed earlier) could be found. Direct visual inspection may be performed where access is possible. The typical large vessel contains product. Internal equipment or components may block access, as may high-pressure gases or liquids, or hazardous materials that offer biological or radioactive dangers to humans. Hence, it is seldom possible to examine the surfaces of pressure vessels by direct visual means within 24 inches and at an angle greater than 30° from the plane of inspection. More commonly, remote visual is performed with binoculars, periscopes, closed circuit television, fiberscopes, or zoom cameras mounted on fixtures that aid access.

Nuclear reactor vessels are inspected in accordance with the Internal Vessel Visual Inspection (IVVI) program. Simply stated, the internal components of the reactor vessel are inventoried. The type of inspection required for each component and surface area is identified as to its importance and the type of visual examination to be performed. For the more important structures, components and stress points receive close inspection with sensitive calibration requirements. For example, component integrity is checked for fine cracks. A 0.001-inch-diameter wire is used to prove adequate visual system resolution.

For a less stringent examination, the observance of a 1/32-inch black line on the 18 percent gray card or a sheet of metal can be used as verification of resolution. An example of a gray card is illustrated in Fig. 3-30.

The intent of this inspection is to look for general integrity of the part or component. Location of distortion, missing parts, debris, general wear, major cracking, and discoloration are the main objects of this examination.

**Techniques** Nuclear reactor vessels are inspected using a vidicon CCTV camera that can tolerate high levels of radiation without burning out. The CCD camera has become more radiation-tolerant for use in high-radiation environments. At this time, the vidicon

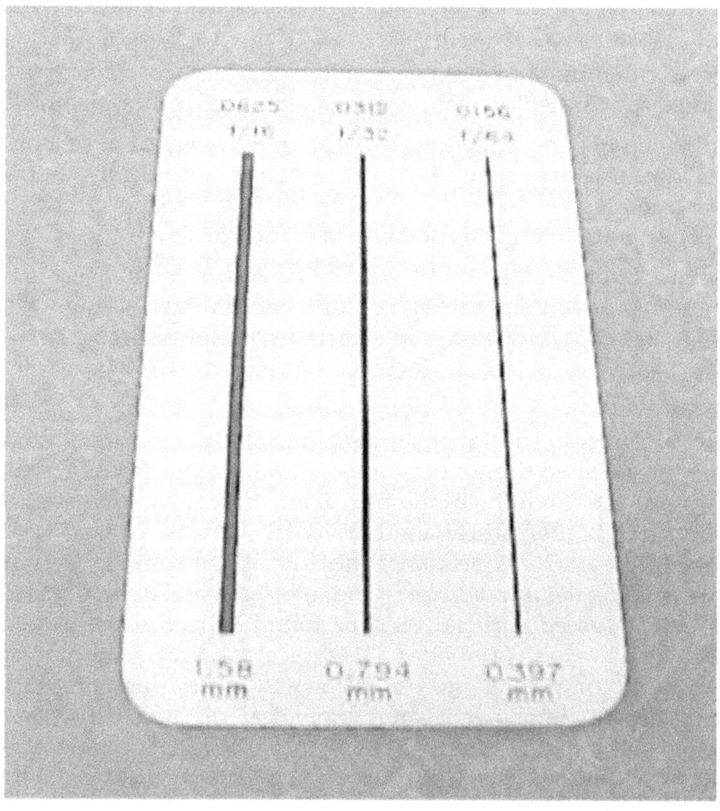


**FIGURE 3-30** Example of gray card.

camera is still attached to long cables that are dangled from a bridge into 40 feet of water. The water provides radiation protection for examination personnel. With an extra control line attached, the skilled operator can access very restricted areas. Like the strings used by a puppeteer, the extra line articulates the small-diameter camera into very tight spaces, allowing awkward but necessary angles of view to be achieved. Alternatively, zoom cameras are attached to long aluminum poles that are placed in central areas to provide a stable base from which the camera can zoom in and out with minimal movement and maximum flexibility.

Storage tanks for petrochemical products are frequently inspected to assess their condition for continued service. Tanks that are several stories tall and hundreds of feet in diameter are inspected through access ports at the top. Smaller tanks on railroad cars and highway tractor-trailers are also inspected for the same reasons. In these cases, a camera on a pole may be attached to a gimble apparatus from the tank access port and manipulated to view as much area as possible. Alternatively, if the tank is free of product, the camera may be mounted on a tripod atop the tank and lowered into the tank. If access to the bottom of the tank is possible, it may be placed on the floor. From there it can be panned, tilted, and zoomed to achieve maximum coverage for inspecting the physical condition of the tank or vessel.

Another remote technique is to send in a crawler robot with the pan, tilt, and zoom camera attached. Tank or vessel configuration is very important for determining the specific technique most suitable for remote visual examination.

Heat exchangers, boilers, and steam generators provide more unique inspection challenges. The open space usual in a tank or empty vessel is now replaced with tubes, baffles, tube supports, deflector plates, and other structural supports. Lens-optic borescopes, fiber-optic borescopes, and crawlers with pan, tilt, and zoom cameras may enable access to the heat exchanger internals. Boilers are often the victims of overheat failures.

Short-term, severe overheating often leads to distortion and failures. Less extreme overheating leads to accelerated scaling or corrosion until failure results. Toward the end of their useful life, tubes may show signs of creep cracks by observable changes in dimensions.

Distorted boiler tubes and enlarged diameters may be early indicators of immanent creep failure.

## 3.5 Evaluation and Reporting of Test Results

Evaluating and reporting of visual test results are dependent on the applicable codes, standards, specifications, contract requirements, and customer needs. Some general criteria apply to all examinations. The first requirement to be established is to determine the scope of the inspection. The second requirement is to decide "how" to inspect an object. The third requirement is to evaluate the results in accordance with a known standard. The fourth requirement is to accurately and legibly report the results.

Visual examination has long been an integral part of construction, fabrication, and manufacturing. As mentioned in Chapter 1, the Book of Genesis in the Bible mentions the first nondestructive test, and in particular, the first visual examination. This belief can be documented in the quote, ". . . and God saw that it was good." Apparently the acceptance criterion was in the eye of the beholder, in this case, the Creator. Industrial applications can be categorized into various groups, such as energy, petrochemical, aerospace, transportation, and construction.

The American Welding Society's Structural Welding Code for steel requires that welds meet certain workmanship standards. Welded structures such as buildings, tubular structures, and bridges must be welded in accordance with specific requirements regarding the material, welding processes, and joint designs. Once the application is established, welding parameters are specified in the workmanship requirements. The weld quality requirements are described in the workmanship section. The section on Inspection spells out the qualifications of Certified Weld Inspectors (CWIs). The tasks of the CWI include ensuring that the materials, weld procedures, welder qualification, and records of inspection all meet code requirements. The inspector keeps a record of qualifications of all welders, welding operators, and tackers; all procedure qualifications or other tests that are made; and other information that may be required. Generally, the CWI performs weld inspections in accordance with a procedure that contains the AWS Code requirements.

The American Society for Mechanical Engineers (ASME), Boiler and Pressure Vessel Code, Section V, Nondestructive Examination, Article 9, Visual Examination lists the following elements to be included in a visual examination procedure:

1. How visual examination is to be performed
2. Type of surface conditions available
3. Method or tools for surface preparation, if any
4. Whether direct or remote viewing is used
5. Special illumination, instruments, or equipment to be used, if any
6. Sequence of performing examination, as applicable
7. Data to be tabulated, if any
8. Report forms to be completed

The procedure should also contain or reference a report of the approach used to demonstrate that the examination procedure was adequate (procedure qualification).

In general, a fine line, 1/32 inch or less in width, or some other artificial flaw located on the surface to be examined or a similar surface is considered satisfactory evidence that the procedure is adequate. To prove the procedure, the line or artificial flaw should be located in the most unfavorable location on the area being examined. It is common to list acceptance criteria for various product forms in the procedure. The product forms should be specified in order to reduce the number of different visual procedures needed. In any case, substituting equipment with that of a different manufacturer or changing the details of the test parameters may require procedure requalification.

### 3.5.1 Reports

Depending on specific code requirements, a written report may be necessary. The report may contain the following for general applications:

1. Date of test
2. Examination procedure used
3. Illuminators and illumination requirements

4. Instruments, equipment, and tools
5. Test results
6. Inspector's signature and certification level

When specific product forms are expected to be examined repeatedly, it may be appropriate to design a report with attributes specific to the examination.

A report of the results of a weld examination may include the following items:

1. Surface finish
2. Discontinuities
3. Undercut
4. Overlap
5. Concavity
6. Convexity
7. Weld dimensions
8. Alignment
9. Evidence of mishandling
10. Arc strikes
11. Other appropriate observations

A report of the results of bolting examinations may include the following items:

1. Discontinuities
2. Necking down
3. Erosion
4. Crevice corrosion

A report of the results of valve examinations may include the following items:

1. Surface markings.
2. Stem conditions.
3. Seating surface cracks, scratches, galling, etc.
4. Guide surface clearances, wear, galling, corrosion, misalignment.
5. Internal body surface erosion, wear, steam cutting.
6. Stem-to-wedge connection for wear and corrosion.
7. Diaphragms (if applicable) checked for aging, damage, cracking.
8. Ball joints (if applicable) should be checked at bearing surfaces and position stops for wear.
9. Hinge pins (if applicable) should be checked for wear, clearances, and misalignment.
10. Stuffing boxes (if applicable) should be checked for steam cutting, corrosion, and mishandling.

A report of the results of pump exams may include the following items:

1. Erosion
2. Corrosion
3. Pitting
4. Leaching
5. Broken or loose parts
6. Debris
7. Shaft alignment

A report of the results of hanger, support, snubber, and restraint examinations may include the following items:

1. Galled or scraped areas of pipe surface (wear)
2. Loose nuts, pins, bolts, and locking pins
3. Oversized and undersized gaps and clearances
4. Distortion of the restraint
5. No freedom of movement or locked-up conditions
6. Extensive or deep corrosion
7. Cracked welds or broken parts
8. Stripped threads
9. Elongated bolt holes
10. Necked-down bolts or rods
11. Sheared rods
12. Debris
13. Loose parts
14. Erosion

*Note:* Snubbers have special performance requirements unique for each application. These functional requirements exceed the normal visual examiner's scope of inspection.

General surface conditions are not the limit of VT-3 examinations in the ASME Section XI environment. Rather stroke, speed, distance, and operability are inspected too. Not applicable to VT-3. Used to be covered under VT-4, which is no longer used for visual examiners in ASME Section XI.

A report of the results of vessel exams may include the following items:

1. Loose parts
2. Debris
3. Abnormal corrosion products
4. Wear
5. Erosion

6. Corrosion
7. Loss of integrity (broken or cracked)

Reports of visual examination results may be supplemented by special recording media or other permanent electronic record forms. Whenever questionable conditions or indications are encountered, some form of a permanent record should be considered. Standard or digital cameras with sufficient resolution to discern and record the attribute or condition of interest are excellent means of accomplishing this. Digital cameras usually include special computer programs that allow the image to be imported directly into the electronic report. Videotape recordings are useful for large areas that are being examined.

This approach is especially helpful in providing evidence of coverage and the examination technique applied. Video camera recordings may also permit a picture frame to be "captured" and recorded directly on the electronic media report. Historically, sketches are simple and can be very helpful in recording examinations and providing location data regarding various discontinuities. Sketches have the disadvantage of incorporating the inspector's perception and subsequent misinformation that would not be in a photograph.

Sketches made manually also may not be legible. Photographs and electronic images should include, where possible, a scale in order to permit the comparison of size and distance.

Any disorientation of angle between the scale and the line of sight (sometimes referred to as "parallax") could distort the true dimensions of an attribute, discontinuity, or condition. The use of a scale will usually provide a better perspective and result in better accuracy.

### 3.5.2 Acceptance Criteria

Acceptance criteria for welds will be as designated by the applicable code or specification and will usually include the following discontinuities:

1. Cracks
2. Incomplete penetration
3. Crater pits and cracks
4. Arc strikes
5. Undercut (dimension will be specified)
6. Surface porosity (usually defined by a maximum single size or some formula of aggregate amount in a total length of weld)
7. Slag (surface)
8. Spatter
9. Burn-through or melt-through
10. Overlap and rollover
11. Lack of fill
12. Excessive reinforcement

Specific acceptance criteria can be found in various codes, specifications, and standards. In many cases, they will be specified as part of the contract requirements.

Records of inspections or examinations should also include statements indicating part acceptance (if applicable), and include appropriate documentation to support findings. Inspection report forms should be complete, legible, concise, and signed and dated. If corrections are necessary, they should be made with a single line through the entry error, initialed, and dated. All reports, results, and/or drawings, sketches, photos, electronic files, digital video discs (DVD), or videotapes should be listed and retained in accordance with contractual requirements or codes and included with the data sheet.

## 3.6 Advantages and Limitations

### 3.6.1 Direct Visual Testing

As has been discussed, direct visual testing is a test in which the eye must be within 24 inches of the test surface and the angle of the line of site must be no less than 30° from the plane of the test surface. Direct visual examination has an advantage over other techniques in that it generally affords the clearest view of the inspection surface. The direct view will result in the processing of the full spectrum of light wavelengths that are available to the eye and the brain to form an image. All the color, shadows, textures, and visual attributes that are generally in existence are observed directly by the inspector. The limitation is that artificial enhancement of the image is not always possible and access restrictions may limit the view. Environmental extremes that are hazardous to inspection personnel may limit the viewing time or visualization of the image altogether.

#### *Photographic Cameras*

An advantage of photography is that it is an excellent record-keeping technique. It can be used very effectively in conjunction with written records or a voice recorder. The most commonly used devices may still be the single lens reflex and macro cameras. Digital cameras provide exceptional resolution and detail. The digital camera allows multiple images to be stored and downloaded to another camera, TV monitor, or computer storage media. Once filed, images can be retrieved and imported into a report. The limitation may be the quality of the image, i.e., the degree of resolution and clarity, depending upon the equipment used. There may also be limitations to the available storage space or memory. This may limit the quality or quantity of the stored images.

#### *Film-Based Records*

Resolution, contrast, and light sensitivity of film-based photographic records all depend on the grain size of the emulsion. In order to achieve greater detail, the use of film that has a finer-grain size will be necessary. The decrease in grain size results in a film having a slower speed. This means that the film will require an increased exposure to light to record the image. Exposure is a function of light and time. A higher-speed film will have larger grains. This results in the reduction of the amount of light that is necessary to turn the film base into a latent image. This equates to a reduction in exposure. The latent image will appear as a negative image after it is chemically processed. The negative image is transferred to light-sensitive paper as a positive image. The result is a positive paper-based photograph. The American and international standard for film speed is expressed in ASA numbers. The European or German standard is expressed in the DIN number. Color film comparisons are listed in Table 3-2.

| Slow | Medium | Fast |
|---|---|---|
| < 32 ASA (16 DIN) | 64–125 ASA (19–22 DIN) | 160 > (23 DIN) |

**Table 3-2** Comparative Color Film Speeds

The three most common film speeds used for nonprofessional photographs are ASA 100, 200, and 400. The finest grain and slowest speed combination for snapshot photography with ASA 100 is considered as medium speed in the range of available film. Microphotography requires more light than is available from a consumer-type flash attachment and a much finer-grain film than that which is used by the amateur photographer.

### *CCD-Based Records*

Section III discussed the concept of the charge-coupled device (CCD). This is reviewed briefly in the following paragraph.

Photons (packets of energy that behave like particles), also known as light, are reflected from an object. They then pass through the lens of the camera and fall into the collection region of the picture element or pixel. Here electrons are freed from the semiconductor's crystal lattice. Thus, the more light that falls on the pixel, the more electrons that are liberated. Electrical fields trap and isolate the mobile electrons in one pixel from those of other pixels until the electrons can produce an indication of the amount of light that fell on the pixel. Electrons are passed from one CCD to another. This is analogous to a bucket brigade moving water. A tiny color filter over each pixel allows photons of only one color to pass through into the pixel's collection region. This pixel is assigned that color during the reconstruction of the image on the LCD monitor. The array of light-sensitive picture elements, or pixels, each measuring five to 25 microns across, make up the array matrix. The low-end digital camera for nonprofessional use typically has an array of 640 × 480 pixels. At the time of writing, arrays up to 16 megapixel cameras are commercially available. Top-of-the-line cameras are available for professional use. These comprise arrays that have millions of pixels.

### *Photographic Techniques*

Whether the recording medium is emulsion-coated acetate-based film or CCD matrix arrays, the lens creates an image by focusing the light rays from an object on a plane behind the lens. An important concept to understand is depth of field. Depth of field can be defined as the overall range of focus apparent in a photograph. The principal plane of focus is the single plane through the subject that is actually in focus. The distance in front of and behind the principal plane of focus that is in focus is known as the depth of field.

When working at higher magnifications, this effect becomes even more significant. For this application, the lens diaphragm is used to control the principal plane of focus or depth of field.

If the lens diaphragm opening is reduced, portions of the image that were blurred appear sharper because areas of the object both in front of and behind the principal plane become focused. If the opening is further reduced, this effect is greater. Increasing or reducing the diaphragm opening controls the depth of field. The offsetting factor when increasing the depth of field in this manner is that there is less light reaching the film or CCD array.

A balance between adequate light and optimum depth of field must be achieved. When using a typical 55-mm lens, the depth of field or area of focus extends farther behind the principal plane of focus than it does in front of it. Therefore, the best depth of field can be obtained by focusing at approximately one-third into the region or area of the object or area of interest. This way, the two-thirds region behind the optimum focus will also be in focus. As magnification is increased (by using 90-mm or 120-mm lenses), the reverse is true. This is why macrophotography, e.g., of a bee on a flower, may yield a sharply focused image of the flower or the bee while the background may appear out of focus.

#### *Fiber-Optic Cameras and Video Borescopes*

Video borescopes with CCD cameras are available with diameters as small as a quarter inch and pixel arrays with 410,000 pixels coupled with micro lens technology.

The four-way articulating tip is available in up to 20 feet lengths. Without the articulating tip, lengths of 45 feet can be achieved. The camera control unit (CCU) can fit under the 7-inch diagonal color monitor and can control light intensity, gain, white balance, and shutter speed. The light source can be provided with 150-watt halogen lamps or optionally with 300-watt Xenon lamps.

This instrument is excellent for examining the inside surfaces of tubing found in heat exchangers, feedwater heaters, condensers, and steam generators. It would be especially useful for short distances in small-orifice access scenarios, such as with pumps, valves, engines, small-diameter pipes, and short-distance views of turbines. It would be a poor choice for large-space inspections, such as tanks, vessels, large-diameter pipes, or anywhere where detail at some distance was desired and an additional light supply could not be provided. The small diameter of the lens can yield near-infinite focus, but inadequate light would limit the view.

#### *CCD Cameras*

Camera equipment specifically developed for boiler tube inspection application (see Fig. 3-12) utilizes a half-inch color CCD with 420,000 pixels, a micro lens, a 460 horizontal video line (HVL) monitor for SVHS video, and a light source made from various light head options. These light attachments can produce from 9.6 to 80 watts of illumination.

The elimination of the fiber bundle to transmit light to the site makes the electrical wire and frequency of the electric current the limiting factor for length rather than the limits of how long a fiber bundle can be manufactured. The half-inch-diameter cable can be manufactured in 100-foot-long lengths. Depending on light head diameters, tube diameters down to a half inch can be accessed. This type camera is best suited for boiler tubes but can do an adequate job on larger-diameter pipes or small vessels. The larger pipe diameters reduce the ability of the camera head to rest anywhere but on the bottom of the pipe, restricting the viewing options. Additionally, pushing a cable does not work well for any distance if the space is sufficient to allow snaking and curling of the cable. Therefore, centering devices and push rods are necessary to alleviate these problems. There remains the problem of articulation. The wide angle of view, as would be achieved with a "fisheye" lens, allows for a large downfield view and limited side field examination. This makes it suitable for many applications where examination for general conditions is the goal. Drain lines, oil lines, steam and process lines, headers, and vessels can all be internally inspected with the miniature camera with light head to lengths of 100 feet.

### CCD Camera with Pan, Tilt, and Zoom Lens

The pan, tilt, and zoom (PTZ) camera, as illustrated in Fig. 3-13, is designed for multipurpose applications that require objects at major distances to be brought under closer examination.

A one-third inch CCD chip is joined with a 12:1 or 24:1 lens train for zoom capabilities. The 12:1 lens can yield greater than 460 horizontal video lines of image. The 24:1 lens can yield more than 380 horizontal video lines of image. The 12:1 lens and the 24:1 lens both utilize variable lens diameters from 0.216 inches (5.4 mm) to 2.6 inches (65 mm).

A spot or flood lamp is needed for adequate light at the greater distances for which these cameras are employed. The standard cable length is 100 feet, but lengths up to 500 feet can be manufactured. A pan range of ± 175° is available. A tilt range of 253° is achievable. This type of camera and light arrangement can be mounted on a tripod, tractor crawler, push pole, or any other suitable platform to examine a wide variety of surface configurations.

This visual examination system is particularly useful in hazardous environments and confined spaces. Tank and vessel applications are probably the most common. Inspections of steam headers, sumps, manifolds, pipe system supports, and large-diameter piping (including sewer and water utility lines) are also common, as are petrochemical, refinery, power generation, and engineering services industry applications. The transportation industry requires infrastructure inspection of structures such as bridges, shafts, and tunnels.

The nuclear power and medical industries may need contaminated or hostile environment areas surveyed. Other industrial surveillance and law enforcement applications may include security and terrorist bomb investigations using remote means.

Other applications in the nuclear industry include the inspection of vast surface areas of reactor containment structures made of concrete or steel. A single location of a camera with a 24:1 zoom lens can inspect concrete walls and roofs for spalling, cracking, defoliation, and other concrete deterioration anomalies much more efficiently than placing an inspector hundreds of feet in the air and covering thousands of square feet of surface manually.

It is interesting to note the difference between 12:1 and 24:1 zoom capability. Moving the lens assembly optically performs the zoom effect up to 12:1 magnification. The zoom effect from 12:1 up to 24:1 is achieved by computer enhancement. Portions of the image are expanded and digitally magnified to create enlargements, providing greater detail for viewing and interpretation. When observing the zoom effect of this camera, the transition is smooth and seamless up to 12:1. Beyond 12:1 magnification, a slight "jiggle" in the image may be observed. From 12:1 to 24:1, the appearance of squares for each pixel is barely visible, but detectable. This is due to the digital magnification process.

### Delivery Systems

Previous discussions of remote visual systems have made some reference to delivery systems. The following paragraphs list these by category. Some typical means of moving a camera from one point to another include fiberglass push rods, small-, medium-, and large-wheeled tractors, mobile platforms with tracks, and various crawlers with retrieval attachments. Applications can vary considerably with regard to the access size or opening, length of travel, environment, magnification, and lighting requirements. The basic elements of a remote visual delivery system are the camera, light, tractor, cable, mobility control unit, pan, tilt, and zoom controls, and monitor.

A small camera can be affixed to push rods. Limitations may be encountered when multiple changes in direction are required; i.e., the push rod soon binds up against multiple turns in the pipe, shaft, or opening.

A steerable-wheeled tractor can negotiate many turns with a camera attached for showing the way. The smaller the opening in which the tractor must operate, the greater the limitation to the payload that can be delivered. For example, a two-inch diameter pipe-crawling tractor may only weigh 3.5 pounds, thereby gaining access to a three-inch nominal pipe size. But it may only be able to drag along 100 meters of quarter-inch-diameter cable. Multiple friction drag or contact points may reduce this to a much shorter distance.

Additionally, inclined angles of the shaft, pipe, or structure would increase the load and decrease the access distance. Remote-controlled delivery systems may eliminate the need for cable but would require radio contact at all times.

A larger tractor that would fit into a 4-inch-diameter pipe may be able to carry a 16-pound load. This may include 100 to 200 meters of cable. Larger tractors may carry a 24-pound load and 200 or more meters of cable. The advantage of larger tractors and accompanying larger wheels would be the ability to traverse adverse terrain and steeper inclines.

The unmanned rovers that landed on Mars in the mid-1990s demonstrated that multijointed and complex designs can be designed to fit into small packages. They may appear frail but they were able to cross over rocks approaching a 2:1 ratio. Generally, the heavier the structure, the more the friction that can be generated; thus, the heavier the load, the rougher the terrain that can be negotiated.

Any of these remote systems can incorporate a retrieval system. The three-pronged fork and tine is one approach. Two-pronged pincers resembling a crab claw is another.

Each additional function requires more cables, lines, and articulating capabilities.

#### *Monitors*

Video monitors are the last link in the video system chain. They are the display devices. In simple terms, the monitor reverses the electronic coding of the video camera and returns the signals to visible displays. The standard monitor since the invention of TV has been the cathode ray tube (CRT). Solid-state and liquid crystal displays (LCD) have mostly replaced the CRT, allowing for thinner and smaller monitors. In particular, the miniature camera with lens and CCD designed for boiler tubes has been outfitted with LCDs. The detail and nature of the indications being sought in tubes, pipes, and small vessels do not necessarily require higher resolution. The LCD display is less expensive and more rugged in the field.

### Conclusion

Generally, the main limitation to visual testing is access. The image of the object must be delivered to the eye. That image is always of the surface of an object. Visual testing is capable of examining the surface of an object unless the material is translucent. Remote visual testing advances are being driven today, as in recent years, by consumer demand and improvements in video technology. The challenge remains to understand fully "what" the inspector is examining and "how" the image is delivered to the eye. As designers make the image-gathering package smaller and smaller, the limitations of access will be further reduced. Applications in the field of medicine have been influencing the industrial field for years. Military applications including drones and robotic devices should continue to bring innovations to the technology of remote visual testing.

## 3.7 Glossary of Key Terms

This section provides definitions of the terms associated with visual examination or testing and the nomenclature associated with visual equipment. Most terms relating to discontinuities are defined in Chapter 2 and within the text of this chapter.

**Angle of Field** The greatest angle in between two rays coming from the object through the objective lens and into the optical system.

**Angstrom** Unit of length, equal to 0.1 nanometer.

**Aspect Ratio** In television, the ratio of the frame width to the frame height.

**Borescope** An industrial scope used to transmit images from inaccessible interiors for visual testing. Borescopes are so called because they were originally used in machined apertures and holes, such as gun barrel bores. There are both flexible and rigid, fiber-optic, and geometric light (lens-optic) borescopes.

**Brightness** The attribute of visual perception in accordance with which an area appears to emit more or less light.

**Burned-In Image** An image that persists in a fixed position in the output signal of a camera tube after the camera has been turned to a different scene.

**Charge-Coupled Device (CCD)** A solid state image sensor. CCDs are widely used inspection systems because of their accuracy, high-speed scanning, and long service life. Semiconductors, when struck by photons, accumulate electrons that are passed on from one charge-coupled device to another. Each pixel represents one color or another. Each pixel and its color is reconstructed on a monitor to form an image.

**Code** A standard enacted or enforced as a law.

**Color** Sensation by which humans distinguish light of different intensities (brightness) and wavelengths (hue).

**Cone** In biology, a retinal receptor that dominates the retinal response when the luminance level is high and provides the basis for the perception of color.

**Contrast** The range of light and dark values in a picture or the ratio between the maximum and minimum brightness values.

**Corrosion** Loss or degradation of metal as a result of chemical reaction.

**Crevice Corrosion** A kind of galvanic corrosion caused by differences in metal ion concentrations in neighboring portions of the corrodent.

**Depth of Field** In photography, the range of distance over which an imaging system gives satisfactory definition when its lens is in the best focus for a specific distance. (See also Depth of Focus.)

**Depth of Focus** The region in front of and behind the focused distance within which objects still produce an image of acceptable sharpness or resolution when viewed by an observer having normal vision, or an observer whose vision is corrected to normal. In the case of a fixed-focus system, this parameter is often called the Depth of Field (DOF).

**Direct Viewing** Viewing of a test object in the viewer's immediate presence. The term is used in the fields of robotics and surveillance to distinguish conventional from remote viewing.

**Distal** In a manipulative or interrogation system, of or pertaining to the end opposite from the eyepiece and farthest from the person using the system. Also objective or tip.

**Dust-tight** So constructed that dust will not enter the enclosing case.

**Dustproof** So constructed or protected that dust will not interfere with successful operation.

**Endoscope** Device for viewing the interior of objects. From the Greek words for inside view, the term endoscope is used mainly for medical instruments.

**Erosion** Loss of material or degradation of surface quality caused by friction or abrasion from moving fluids or particles.

**Explosion-proof, Intrinsically Safe, Purged and Pressurized** Constructed in a manner to prevent the surrounding atmosphere from being exploded by the operation of, or the results from, operating the item so classified.

**Fiber Optics** An array of flexible glass or plastic fibers that has the capability of transmitting light (random array) or an image (coherent array) axially through the fiber bundles.

**Flux (luminous)** The intensity of light per unit area of its source.

**Foot-candle** A unit of illuminance when the foot is taken as the unit of length. It is the illuminance on a surface one square foot in area on which there is a uniformly distributed flux of one lumen, or the illuminance at a surface, all points of which are at a distance of one foot from a uniform source of one candle.

**Fovea Centralis** A small depression near the center of the retina, constituting the area of most acute vision.

**Frame (in Television)** The total area, occupied by the picture, that is scanned while the picture signal is not blanked.

**Geometric Distortion** Any aberration that causes a reproduced picture to be geometrically dissimilar to the perspective-plane projection of the original scene.

**Horizontal (Hum) Bars** Relatively broad horizontal bars, alternately black and white, that extend over the entire picture. They may be stationary or may move up or down. Sometimes referred to as a "Venetian-blind" effect, they are caused by the approximate 60-cycle interfering frequency or one of its harmonic frequencies.

**Image** A reproduction of an object produced by light rays. An image-forming optical system gathers a beam of light diverging from an object point and transforms it into a beam that converges toward another point. If the beam converges to a point, a real image is produced, which can be projected upon a screen, a film plane, video camera tube, or semiconductor array.

**Interlaced Scanning** A scanning process in which the distance from center to center of successively scanned lines is two or more times the nominal line width, and in which the adjacent lines belong to different fields.

**Lumen** The unit of luminous flux. It is equal to the flux through a unit solid angle (steradian) from a uniform point source of one candle, or to the flux on a unit surface, all points of which are at unit distance from a uniform point source of one candle.

**Object** The figure seen through or imaged by an optical system that may contain natural or artificial structures, or may be the real or virtual image of an object formed by another optical system.

**Peripheral Vision** The seeing of objects displaced from the primary line of sight and outside the central visual field.

**Pixel** A lighted point on the screen of a digital image. The image from a conventional computer monitor is an array of over 256,000 pixels, each of which has a numerical value. The higher the number for a pixel, the brighter it is. Formerly called picture element.

**Raster** A predetermined pattern of scanning lines that provides substantially uniform coverage of an area.

**Reference Standard** Workpiece or energy source prepared according to precise instructions by an approved agency for tests and calibrations requiring precise and consistent measurements. (In the case of photometry, the standard is a light source.)

**Reflection** A general term for the process by which the incident flux leaves a surface or medium from the incident side, without change in frequency. Reflection is usually a combination of regular and diffuse reflection.

**Remote Viewing** Viewing of a test object not in the viewer's immediate presence. The word remote previously implied either closed-circuit television or fiber-optic systems remote enough so that, for example, the eyepiece and the objective lens could be in different rooms. High-resolution video and digital signals can now be transmitted around the world with little loss of image quality. Compare with Direct Viewing.

**Resolution** An aspect of image quality pertaining to a system's ability to reproduce objects, often measured by resolving a pair of adjacent objects or parallel lines. See also Resolving Power.

**Resolution (Horizontal)** The amount of resolvable detail in the horizontal direction in an image. It is usually expressed as the number of distinct vertical lines, alternately black and white, that can be seen in a distance equal to image height. This information usually is derived by observation of the vertical wedge of a test pattern. An image that is sharp and clear and shows small details has good or high, resolution. If the picture is soft and blurred and small details are indistinct, it has poor or low resolution. Horizontal resolution depends upon the high-frequency amplitude and phase response of the system, the transmission medium, and the image monitor, as well as the size of the scanning spots.

**Resolution (Vertical)** The amount of resolvable detail in the vertical direction in a picture. It is usually expressed as the number of distinct horizontal lines, alternately black and white, that can be seen in a test pattern. Vertical resolution is fundamentally limited by the number of horizontal scanning lines per frame. Beyond this, vertical resolution depends on the size and shape of the scanning spots of the pickup equipment and picture monitor and does not depend upon the high-frequency response or bandwidth of the transmission medium or picture monitor.

**Resolution Threshold** Minimum distance between a pair of points or parallel lines when they can be distinguished as two, rather than one; expressed in minutes of arc.

**Resolving Power** The ability of vision or other detection system to separate two points. Resolving power depends on the angle of vision and the distance of the sensor from the test surface. Resolving power is often measured using parallel lines. Compare with Resolution.

**Retained Image (Image Burn)** A change produced in or on the target that remains for a large number of frames after the removal of a previously stationary light image and yields a spurious electrical signal corresponding to that light image.

**Retina** The tissue at the rear of the eye that senses light.

**Rod** Retinal receptor that responds at low levels of luminance, even down below the threshold for cones. At these levels, there is no basis for perceiving differences in hue and saturation. No rods are found in the fovea centralis (see definition).

**Simple Magnifier** A device having a single converging lens.

**Specification** A set of instructions or standards invoked by any organization to govern the results or performance of a specific set of tasks or products.

**Standard** Document to control and govern practices in an industry or application, applied on a national or international basis and usually produced by consensus. See also Reference Standard.

**Tip** In casual usage, the distal or objective end of a borescope.

**Trace** Line formed by electron beams scanning from left to right on a video screen to generate a picture.

**Video** Pertaining to the transmission and display of images in an electronic format that can be displayed on a cathode ray screen.

**Videoscope** Jargon for video borescope. See Borescope, Video.

**Visibility** The quality or state of being perceivable by the eye. In many outdoor applications, visibility is defined in terms of the distance at which an object can be just perceived by the eye. In indoor applications, it usually is defined in terms of the contrast or size of a standard test object, observed under standardized viewing conditions.

**Vision** Perception by eyesight.

**Visual Acuity** The ability to distinguish fine details visually. Quantitatively, it is the reciprocal of the minimum angular separation in minutes of two lines of width subtending (extending under or opposite to) one minute of arc when the lines are just resolvable as separate.

**Visual Angle** The angle subtended by an object or detail at the point of observation. It usually is measured in minutes of arc.

**Visual Field** The locus of objects or points in space that can be perceived when the head and eyes are kept fixed. The field may be monocular or binocular.

**Visual Perception** The interpretation of impressions transmitted from the retina to the brain in terms of information about the physical world displayed before the eye.

Visual perception involves any one or more of the following: recognition of the presence of something (object, aperture, or medium); identifying it; locating it in space; noting its relation to other things; identifying its movement, color, brightness, or form.

**Visual Testing** Method of nondestructive testing using electromagnetic radiation at visible frequencies.

**Waterproof** So constructed or protected that water will not interfere with successful operation. This does not imply submersion to any depth or pressure.

**Watertight** Provided with an enclosing case that will exclude water applied in the form of a hose stream for a specified time as stated in the following note.

*Note:* A common form of specification for watertight is: "So constructed that there shall be no leakage of water into the enclosure when subjected to a stream from a hose with a one-inch nozzle and delivering at least 65 gallons per minute, with the water directed at the enclosure from a distance of not less than 10 feet for a period of 5 minutes, during which period the water may be directed in one or more directions as desired."

## Bibliography

American Society for Nondestructive Testing. *Nondestructive Testing Handbook*, 3rd ed, Vol. 9, *Visual Testing*. American Society for Nondestructive Testing, 2010.

American Society for Materials Handbook, Section 3.03 Nondestructive Testing, ASTM, 100 Barr Harbor Drive, West Conshohocken, PA, Publication Code 01-030396-22 (1996).

American Society for Testing and Materials. *1999 Annual Book of ASTM Standards, Section 3, Metals Test Methods and Analytical Procedures, Volume 03.03, Nondestructive Testing*. Philadelphia, The American Society for Testing and Materials, 1999.

American Welding Society. *Welding Inspection Handbook*, 3rd ed. American Welding Society, Miami, Florida, 2000.

Anderson, Robert Clark. *Inspection of Metals, Vol. 1 Visual Examination.* American Society for Materials, 1983.

GE Industrial Technology Website http://www.ge-mcs.com/en/inspection-technologies.html, Remote Visual Inspection Product Overview, 7-31-2011. The Illuminating Engineering Society, of North America. *IES Lighting Handbook: Reference Volume.* New York, The Illuminating Engineering Society, of North America, 1984.

Inoue, Shinya. *Video Microscopy*. New York, Plenum Press, 1989.

Lorenz, Peter G. *The Science of Remote Visual Inspection, Technologies, Applications, Equipment*. New York, RVI Olympus Corporation, Lake Success, 1990.

Lorenz, Peter G. *Expanding Technology Adds Value to Visual Testing and Remote Visual Inspection Procedures, Materials Evaluation*, September, ASNT, Columbus, OH, 1998.

Olympus Industrial Website http://www.olympus-ims.com/en/, video borescopes brochures, 5-8-2012.

*Visual Examination Technology Containment Inspection*, Level II IWE/IWL, 2nd ed. American Society for Metals, 1997.

Wikipedia http://en.wikipedia.org/wiki/Pixel, definition of pixel, 5-8-2012.

# CHAPTER 4

# Penetrant Testing

## 4.1 Introduction

Penetrant testing (PT) is one of the most widely used nondestructive testing (NDT) methods for the detection of surface discontinuities in nonporous solid materials. It is almost certainly the most commonly used surface NDT method today because it can be applied to virtually any magnetic or nonmagnetic material. PT provides industry with a wide range of sensitivities and techniques that make it especially adaptable to a broad range of sizes and shapes. It is extremely useful for examinations that are conducted in remote field locations, since it is extremely portable. The method is also very appropriate in a production-type environment where many smaller parts can be processed in a relatively short period of time. This method has numerous advantages and limitations that can be found in Section 4.10, Advantages and Limitations, of this chapter.

## 4.2 History and Development

Although the exact date of the first "penetrant" test is unknown, it is generally believed that the earliest tests were performed in the late-19th century and were primarily limited to the examination of various railroad parts, such as axles and shafts. Even though it was a very rudimentary method, it was capable of revealing fairly large cracks in metallic parts by using what is referred to as the "oil and whiting" method. It is interesting to note that the oil and whiting method employed the same processing steps that are used today with current penetrant testing procedures. In this early penetrant method, the part to be examined was cleaned and then submerged in dirty engine oil. The oil that was used in those days came from large locomotive engines and was very heavy. It was generally diluted with kerosene or alcohol so that it would be thin enough to penetrate surface discontinuities. It seemed that the dirty oil worked the best in the presence of a discontinuity, since it provided a dark, oily stain on the test surface. After saturation with oil, the part was allowed to drain. During the draining time, known today as the "dwell" time, the thinned oil would penetrate surface cracks. After the excess oil was removed from the surface with a solvent, the part was coated with "whiting," which consisted of a chalk-like powder suspended in alcohol. The oil that was entrapped in the void would then "bleed" out into the whiting and a dark, oily stain indicated the presence and location of a discontinuity.

Sometimes the whiting used was a whitewash material similar to that used to wash and paint fences, hence the expression "oil and whiting" method. It is also believed that talcum powder was used in those early days as a developer to help bring the oil back out to the surface so it could be observed. As can be expected, there were many problems with this early technique. There was a general lack of consistency, since there were no established procedures or standards and the dwell and development times were pretty much left up to the judgment of the user. The materials varied in content and use, and only the grossest type of discontinuities could be detected.

The oil and whiting method had a place in the early examinations of railroad parts, but its use began to diminish with the introduction of magnetic particle testing (MT) in the 1930s. Its decline was most notable from 1925 into the mid-1930s. At this time, many parts that were considered to be critical were made of ferromagnetic materials and the advent of MT techniques provided a much more reliable and repeatable method for the detection of surface discontinuities. By the mid-1930s, the use of aluminum and other materials that could not be magnetized was increasing and it was quite apparent that there had to be another nondestructive test method for detecting discontinuities in these nonferromagnetic materials. Certainly, at that time, internal discontinuities could be detected using X-ray techniques, but many of the X-ray techniques in use were not capable of revealing smaller, tight discontinuities at the surface. It follows that there was an obvious need for a method that would be sensitive enough to detect these small surface discontinuities.

The early pioneers Carl Betz, F. B. Doane, and Taber deForest, who worked for the Magnaflux Corporation at the time, were experimenting with many different types of liquids and solvents that might fulfill this need for surface discontinuity detection. These early techniques that were tried used brittle lacquer, electrolysis, anodizing, etching, and various color-contrast penetrants. Some of these early approaches were discarded, with the exception of the anodizing process. During this period, the anodizing process was used for detecting cracks in critical aluminum parts, generally associated with aircraft. This resulted in the publication of a military specification, MIL-I-8474, on September 18th, 1946. The title of this specification was "Anodizing Process for Inspection of Aluminum Alloys and Parts."

Years before the publication of this military standard, Robert Switzer had been working with fluorescent materials, primarily for producing fluorescent advertisements for movie theaters. He and his younger brother, Joseph, were pioneers in the early use and development of various fluorescent materials. As early as 1937, Robert Switzer became aware of a problem that a local casting company was having with parts for the Ford Motor Company. There was a large batch of aluminum castings that were found to contain a number of discontinuities that were not observed until after the surfaces of the castings had been machined. Switzer recalled how the different fluorescent materials he was familiar with were capable of clinging to surfaces and fluorescing when observed using ultraviolet (black) light. He thought that this material would be appropriate for detecting the surface cracks that had been uncovered by machining. He was able to obtain a number of samples and began experimenting with the various fluorescent pigments that he was developing. He found that although these pigments were unique in clinging to a person's hands and other porous-type materials, they were not very effective for the detection of very fine surface cracks. Switzer continued to try different combinations and eventually introduced them into various liquids that would be used to carry the fluorescent pigment into the discontinuities. His work was ultimately successful, and he applied for a patent in August of 1938. It is interesting to note that he asked his brother, Joseph, to share the patent-filing fee, which amounted to $15.00, and Joe declined. Robert then proceeded to file the patent on his own and, ultimately, found information about the Magnaflux Corporation, which at that time was still developing and pioneering the magnetic particle inspection method. He decided to investigate the company and contacted one of their sales representatives in New York City to discuss the possibility of expanding their surface discontinuity efforts to include a fluorescent penetrant.

Some of the early attempts to demonstrate the fluorescent penetrant method to Carl Betz and others from the Magnaflux Corporation were not successful. Lack of success in these early demonstrations was probably due to the fact that the castings being examined were supposed to have discontinuities, but those discontinuities probably did not exist. One of the observers, A. V. deForest, who was a pioneer in MT, happened to have a specimen containing known discontinuities that he had used to demonstrate MT, and a fluorescent penetrant test was performed on this specimen. The indications were quite apparent. Naturally, all of the observers were very impressed that the known discontinuities appeared, In addition, other discontinuities that A. V. deForest was not aware of were also observed.

Subsequent to this demonstration, it was found that the castings that were first examined had been peened, thus closing the discontinuities that were at the surface. This demonstration proved unique. The year 1941 became memorable, not only for the patent that was awarded to Switzer that summer, but because it also marked the beginning of World War II. This new test would be widely used in supporting the different products that would be used by the military.

### 4.2.1 Early Penetrant Techniques

Some of the early penetrant processes were quite similar to those in use today. The part would be cleaned with a strong solvent and then, after drying, it was immersed in the penetrant for about 10 minutes. After this penetration, or dwell time, the penetrant was removed, usually with a strong solvent, and the part was wiped until dry and clean. The removal step was usually performed under a black light. According to early accounts, the parts would then be struck with a hammer, which would cause the entrapped penetrant materials to "bleed out" to the surface, at which time the part would be examined under a black light. These early techniques were still quite archaic, and the test results were not very consistent. During the following years, in order to achieve the level of consistency that is essential for quality assurance, many different types of materials were tried and a variety of techniques were attempted. These led to a most unique development: the water-washable or water-removable (perhaps a more appropriate term) penetrant.

Water-rinsable penetrant materials and related equipment were first offered around June of 1942. There was much interest in them, and a number of companies started to use the water-rinsable technique. As a matter of record, the first purchaser of the water-removable penetrant equipment was Aluminum Industries of Cincinnati, Ohio, which would use the equipment for the examination of aluminum castings. Many other applications and uses followed, including the testing of propeller blades, pistons, valves, and other critical aircraft parts. It should be noted that a major step forward in the application of PT occurred when it was included in the maintenance and overhaul programs for aircraft engines. As a result of this application, the PT method of nondestructive testing was on its way to gaining wide acceptance. A patent on the "water-washable" technique was applied for in June 1942 and was issued in July 1945. The early developer compounds were yellowish materials consisting mainly of talc. The wet developer technique began to be used in late 1944 and early 1945.

One of the problems associated with the water-removable technique was the potential for removal of entrapped penetrant from discontinuities as a result of a vigorous water rinse step; this led to concerns about "overwashing." This overwashing was the result of an emulsifier that had been mixed in with the penetrant. In fact, some early penetrants were marketed as promoting "super washability," which resulted in extremely clean surfaces but also produced a greater danger of the penetrant being removed from surface discontinuity openings. The solution to this overwashing concern was to remove the emulsifier from the penetrant and to apply it later. This finally occurred in 1952 when this process was referred to as the post-emulsification, or PE, technique.

The first post-emulsification penetrants were introduced in 1953. With this technique, the emulsifier was applied after the penetrant dwell time and carefully controlled, so that the penetrant in the discontinuities was not emulsified. It would remain in the discontinuity even after the emulsified surface penetrant was removed with a water rinse. Although this added an additional step and an additional liquid to the process, it did provide a higher level of control over the detection of small, shallow discontinuities that may not have detected with the use of a typical water-removable technique containing emulsifiers.

### 4.2.2 Visible Penetrants

The penetrant techniques described to this point were of the fluorescent type. The fluorescent penetrant technique required tanks, a water supply, electricity for the black lights, and a darkened area for the evaluation of indications. In order to permit penetrant tests to be performed in the field and to provide portability, a simpler, visible dye had to be developed. In the 1940s, a Northrop metallurgist named Rebecca Smith developed a visible dye-penetrant approach. Rebecca Smith, who would be later known as Becky Starling, collaborated with Northrop chemists Lloyd Stockman and Elliot Brady, who also assisted in the development of a visible dye penetrant. This was considered necessary for examining critical jet engine parts outdoors, where creating a darkened area necessary for the use of fluorescent penetrants was inconvenient. The development of the visible dye-penetrant technique would take several years; Stockman applied for a patent in March 1949. By this time, there were several choices of penetrants: fluorescent water removable, fluorescent emulsifiable, and a visible dye.

### 4.2.3 Other Developments

In the late 1950s and early 1960s, much work was done to quantify and analyze the various penetrants that were available. By the early 1960s, a variety of techniques were being used with a range of sensitivities that would satisfy the demanding requirements of many industries. Many of the variables associated with the penetrant tests were evaluated and the entire process was improved, so that consistency and sensitivity would be an intrinsic part of the process. Originally, one of the most widely used military standards, MIL-I-6866, was issued in 1950. The standard was canceled in 1964 and replaced by MIL-STD-6866, which was cancelled and replaced by the American Society for Testing and Materials (ASTM) standard E1417 in 1996. Currently, there has been a shift away from military standards and, eventually, all military standards will be replaced with standards developed by the American Society for Testing and Materials. The remainder of this chapter will focus on the materials and techniques that are in use at the present time and will describe the many applications for which this unique NDT method can be effectively applied.

## 4.3 Theory and Principles

The basic principle upon which penetrant testing is based is that of capillary "attraction" or "action." Capillary action is a surface tension phenomenon that permits liquids to be drawn into tight openings as a result of the energies that are present at the surfaces of the openings. In most high school physics classes, the principle of capillary action is demonstrated by placing a glass straw into a beaker filled with colored water. The surface tension associated with the opening of the glass straw, or capillary, causes the liquid level to move to a higher level inside that capillary than the level of the liquid in the beaker. A simple demonstration of capillary action using two glass panels clamped together is illustrated in Fig. 4-1a and b.

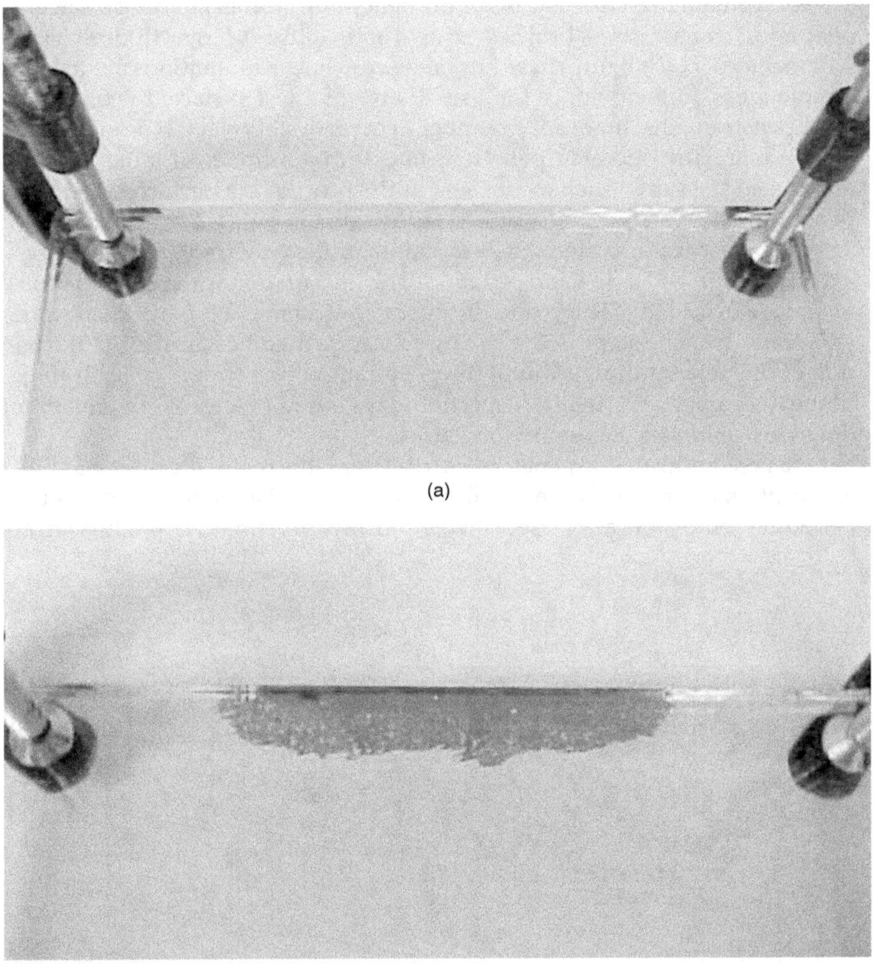

(a)

(b)

**Figure 4-1** Demonstration of capillary action. (*a*) Glass panels clamped together. (*b*) Visible color-contrast penetrant applied to the edge of the panels. (All illustrations in this chapter courtesy of C. J. Hellier.)

One can consider that discontinuities open to the surface behave in much the same fashion as shown by the glass panels in Fig. 4-1b. The liquid used in this example is a typical visible contrast penetrant. The capillary action forces are very strong; in fact, if a penetrant test were being performed on a specimen in an overhead position, the penetrant would be drawn into the opening, against the force of gravity. The capillary force is much stronger than gravity, and the discontinuities will be detected even though they may be in an overhead specimen.

## 4.4 Penetrant Equipment and Materials

### 4.4.1 Penetrant Equipment

Penetrant systems range from simple portable kits to large, complex in-line test systems. The kits contain pressurized cans of the penetrant, cleaner/remover, solvent, and developer and, in some cases, brushes, swabs, and cloths. A larger fluorescent penetrant kit will include a black light. These kits are used when examinations are to be conducted in remote areas, in the field, or for a small area of a test surface. In contrast to these portable penetrant kits, there are a number of diverse stationary-type systems. These range from a manually operated penetrant line with a number of tanks, to very expensive automated lines, in which most steps in the process are performed automatically. The penetrant lines can be very simple, as illustrated in Fig. 4-2.

In this particular system, there is a tank for the penetrant, a tank for the water rinse, a drying oven, and a developer station. The final station is the examination area, which includes a black light. This manually operated system is a typical small water-removable penetrant line. The steps in the testing process would be: cleaning of the parts, application of the penetrant, removal of the penetrant with a water spray, drying, application of the developer, and finally, inspection. This entire process is covered in much greater detail in Section 4.5, Penetrant Procedures.

If a post-emulsifiable penetrant is to be used, the manually operated line will require an additional tank. This tank will contain an emulsifier that will render the surface penetrant removable after a specified emulsification time. Again, this technique will be

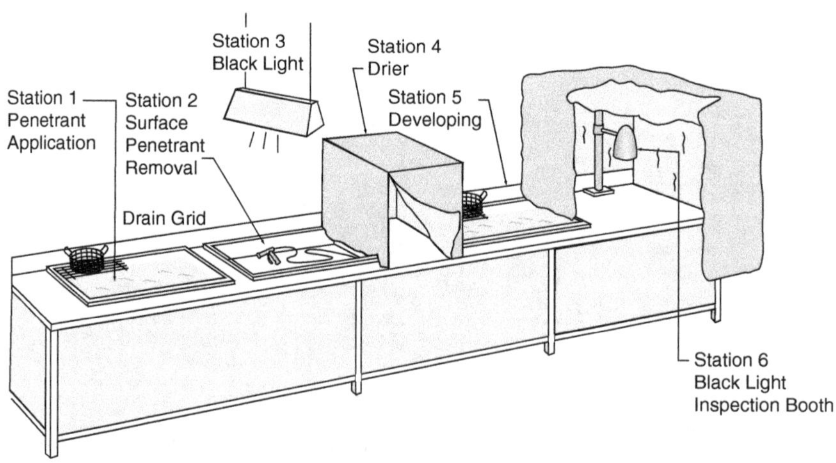


**Figure 4-2** Typical fluorescent penetrant line arrangement.

**FIGURE 4-3** Automated fluorescent penetrant line.

covered in much greater detail later in this chapter. The automatic penetrant lines in use today vary from small, rather simple systems to very large complex lines that are computer controlled. Figure 4-3 illustrates a large automatic penetrant line.

Although the steps in an automated penetrant system have been somewhat mechanized, it is interesting to note that the examinations still must be conducted by inspectors who have been trained and are qualified in the process. The arrangement of these large automated penetrant lines varies with different layouts to permit the most flexibility from the standpoint of processing the parts. Normally, the systems will be arranged in a straight line; however, a U-shape or other configuration may be used to provide more effective use of floor space.

### 4.4.2 Black Lights, Light Intensity Meters, and Test Panels

The *black light* is an essential accessory for fluorescent penetrant inspection. Since the early 1940s, mercury vapor high-intensity lamps were used to generate ultraviolet-A (UVA) wavelengths typically in the range of 315 to 400 nm (3,150 to 4,000 Å) for the newly developed fluorescent penetrant technique. Mercury, being a hazardous material, became more deeply regulated, and there was a general reluctance to use it in the 1980s and beyond. It eventually became obsolete for fluorescent penetrant lighting applications.

FIGURE 4-4 LED black light.

In the mid-1990s, a new technology, micro-power gas discharge (MPXL) lighting, became available for use with fluorescent penetrants as well as fluorescent magnetic particles. This type of lighting provided benefits in the form of the UVA high intensities that were being generated, a warm-up time of less than 15 seconds, and its insensitivity to magnetic fields. Older mercury lights that were shut off required a cooldown period and additional warm-up time prior to using them for inspection again.

In the 2000s, light-emitting diode (LED) technology developed into near full-spectrum wavelength potential (see Fig. 4-4). In addition to generating UV wavelengths with good repeatability, the benefits included:

- Instant-on at full intensity
- Lighter weight equipment
- Low power use
- Battery operation
- Ability to generate purer UVA wavelengths (360 to 370 nm)
- Long life
- Insensitivity to magnetic fields
- Nonfilament lights are more robust to mechanical abuse, dropping, etc.

*Light intensity meters* are used to measure both white light intensities when visible PT is used and black light intensities for fluorescent penetrant techniques. This measurement is necessary to verify code compliance and to ensure that there is no serious degradation of the lights. Some meters are designed to measure both white light and black light intensities.

*Test panels*, including comparator blocks, controlled cracked panels, tapered plated panels, and others for specific industries, such as the Testing and Monitoring (TAM) panel (typically used in aerospace), are employed to control the various attributes of the PT system. They also provide a means for monitoring the materials and the process. This is discussed in Section 4.9, Quality Control Considerations.

In summary, the equipment used will be greatly influenced by the size, shape, and quantity of products that are to be examined. If there are large quantities involved on a continuing basis, the use of an automated system may be appropriate, whereas with small quantities of parts, the use of penetrant kits may be more suitable. The size and configuration of the part will also influence the type of penetrants that will be most appropriate.

### 4.4.3 Penetrant Materials

The various materials that will be used in the different penetrant processes must exhibit certain characteristics. Above all, these materials must be compatible with each other and collectively provide the highest sensitivity for the application. The term "penetrant family" is sometimes used to indicate a group of materials all from the same manufacturer. The intent is to provide a degree of assurance that the different materials will be compatible with each other. There are usually some provisions for using materials outside the "family" if the combination of the different materials can be proven compatible through qualification tests.

The materials used in the penetrant process are classified into four groups. The characteristics for each will be presented in detail. The first group of materials that are essential for a penetrant test are precleaners. The second group of materials, which has the greatest influence on sensitivity, are the penetrants. The third group comprises the emulsifiers and solvent removers, and the fourth group the developers.

#### *Precleaners*

Choosing the correct precleaning material is an essential first step in the penetrant process. No single cleaning method is appropriate for all metals or contaminants. The precleaning method must be capable of removing dirt and oils without causing harm to the article under inspection. The surface of the material under test must be thoroughly cleaned to ensure that all contaminants that may prohibit or restrict the penetrant from entering surface openings are removed. Thorough cleaning is essential if the examination results are to be reliable. Not only does the surface have to be thoroughly cleaned, but openings must be free from contaminants such as oil and water, oxides of any kind, paint, or other foreign materials, which can greatly reduce the penetrant sensitivity.

Typical cleaners are described here.

**Solvents** These are probably the most widely used liquids for precleaning inspection surfaces remote locations or when inspecting a small area of a part. There are a variety of solvents that can be effective in dissolving oil, films, grease, and other contaminants. These solvents should be free of any residues that would remain on the surface. Solvents cannot be used for the removal of spatter, rust, or similar materials on the surface. These must be removed by some type of a mechanical cleaning process (see Section 4.5, Prerequisites, Surface Condition Considerations).

**Vapor Degreasing** Vapor degreasing is an extremely effective method for removing oils, grease, wax, and other non-water soluble contaminants. In a vapor degreaser, solvent is boiled at the bottom of a deep tank. Near the top of the vapor degreaser are cooling coils, which encircle the inside of the degreaser. The coils catch the vapors and condense them back to liquid. Parts are lowered into the vapors but not into the boiling liquid. The oils and non-water soluble contaminants become diluted and drip into the boiling solvent.

**Ultrasonic Cleaning** Ultrasonic cleaning is a very effective method of precleaning. Not only will the contaminants be removed from the surface but also if there are entrapped contaminants in discontinuities and other surface openings, the power that is generated in the ultrasonic cleaning process will usually be effective in breaking up and removing them. The ultrasonic cleaner is used with solvent or water and detergent. A typical small ultrasonic cleaner is illustrated in Fig. 4-5.

**Alkaline Cleaning** Alkaline cleaners used for precleaning are nonflammable water solutions that, typically, contain specially selected detergents that are capable of removing various types of contamination.

**Steam Cleaning** In some rare instances, steam may be used to remove contaminants from the surface. Although very effective in removing oil-based contaminants, this is not a widely used technique.

**Water and Detergent Cleaning** There are various devices that utilize hot water and detergents to clean part surfaces. This technique depends largely upon the type of contamination that is present on the test surfaces. Usually, if parts are covered with oil or grease,

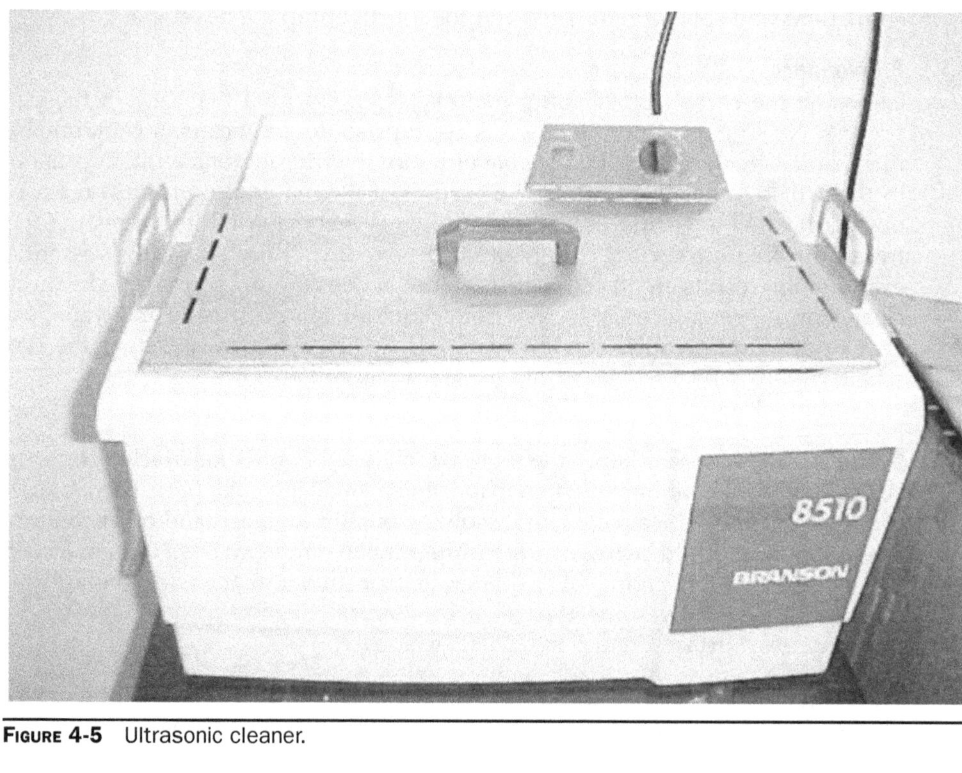


**Figure 4-5** Ultrasonic cleaner.

the contaminants will not be satisfactorily removed from the surface with this cleaning technique.

**Chemical Cleaning** Chemical cleaning techniques usually involve etchants, acids, or alkaline baths. This precleaning approach is primarily confined to softer materials, such as aluminum and titanium, where prior mechanical surface treatments, such as machining or grinding, could possibly have smeared metal over discontinuity openings. Both acid or alkaline liquids are usually effective in the removal of rust and surface scale; however, a slight amount of the surface material is also removed, so this process must be very carefully controlled. Steps must be taken to ensure the complete removal of these liquids from all surface openings.

### *Penetrants*

The most important characteristic that affects the ability of a penetrant to penetrate an opening is that of "wetability." Wetability is a characteristic of a liquid and its response to a surface. If a drop of water is placed on a very smooth, flat surface, a droplet with a very pronounced contour will result, as shown in Fig. 4-6c. Although water is a liquid and is "wet," its wetting characteristics are not good enough to make it an effective penetrant.

The "contact angle," θ, is measured from a line drawn in an arc from its junction point with the surface to the opposite surface (see Fig. 4-6). If that same droplet of liquid is emulsified, such as would be the case with the addition of a small amount of liquid soap, the droplet will tend to flatten out and the contact angle will be somewhat decreased (see Fig. 4-6b). In the case of a liquid penetrant, its wetting properties are so great that it will, essentially, lie almost flat on a smooth surface and the contact angle will be very low, as in Fig. 4-6a. Therefore, the penetrants with the lowest contact angles will have the best wetability and provide good penetrability for a given material. Two other important characteristics of the penetrant are the dye concentrate and the viscosity. Dye concentrate has a major and direct influence on the "seeability" or sensitivity of the penetrant material, assuming that the wetting properties and the penetrability are of a very high level. The dye concentrate makes the penetrant more visible to the human eye. In some early penetrants, different-colored dyes were tried, including blue, yellow, and green, but it seemed that the red dye resulted in the best response to visible observation. In fact, the term "contrast ratio" is generally used to express the seeability of a penetrant. If a white penetrant were used on a white developed background, the contrast ratio would be 1:1. In other words, there would be no contrast between the penetrant and the background. In the case of a red penetrant, the contrast ratio is said to be 6:1, making that contrast very noticeable on a white background surface. The dye concentrate is also important in a fluorescent penetrant. The contrast ratio of a fluorescent penetrant is said by many references to be 40:1, compared to the 6:1 of a visible red dye. In reality, the contrast ratio of the fluorescent penetrant being viewed under a black light in a virtually dark room would essentially be the same as would exist if there were a single candle in a perfectly pitch-black room. The contrast ratio in this case would be exceptional and, for this reason, the fluorescent penetrant produces a much higher degree of seeability or sensitivity as compared to the visible dye penetrants.

Viscosity is defined as the state or quality of being viscous. Liquids with higher viscosity values are thicker than those with lower ones. Although viscosity is an important characteristic of a penetrant, it is not as influential as the wetting characteristics

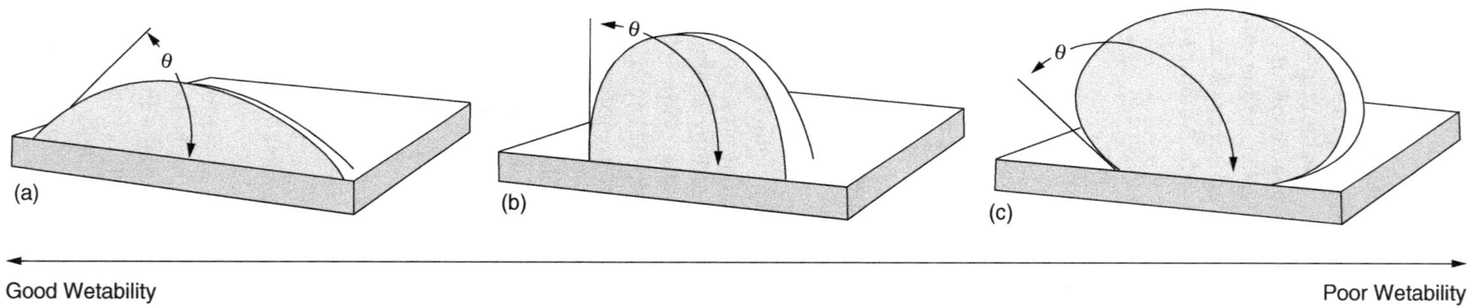


**Figure 4-6** Wetability characteristics.

and dye concentrate. If the penetrant has good wetting characteristics and exceptional dye concentration, it will still provide a meaningful examination, even if the viscosity is high. The difference in viscosity will influence the actual dwell time, or the amount of time that it will take for the liquid to effectively penetrate a given surface opening.

There are other characteristics that effective penetrants should possess. In addition to being able to penetrate small surface openings, they must also:

- Be relatively easy to remove from the surface during the removal step
- Be able to remain in the discontinuities until they are withdrawn during the development step
- Be able to bleed from the discontinuities when the developer interacts with it and have the ability to spread out in the developer layer
- Have excellent color and the ability to be displayed as a contrasting indication in order to provide the sensitivity that is necessary
- Exhibit no chemical reaction between the penetrant materials and the test specimen
- Not evaporate or dry rapidly

In addition, they should be nonflammable, odorless, and nontoxic; possess stability under conditions of storage; and be cohesive, adhesive, and relatively low in cost.

In summary, the most important characteristics of a penetrant when performing a penetrant test are (1) capillary action, (2) wetting characteristics, (3) dye concentrate, and (4) viscosity.

### Emulsifiers/Removers

The purpose of the emulsifiers used in penetrant testing is to emulsify or break down the excess surface penetrant material. In order for these emulsifiers to be effective, they should also possess certain characteristics, including:

- The reaction of the emulsifier with any entrapped penetrant in a discontinuity should be minimal in order to ensure that maximum sensitivity is achieved.
- The emulsifier must be compatible with the penetrant.
- The emulsifier must readily mix with and emulsify this excess surface penetrant.
- The emulsifier mixed with the surface penetrant should be readily removable from the surface with a water spray.

### Solvent Removers

Solvent removers are used with the solvent removable technique and must be capable of effectively removing the excess surface penetrant. There are a number of commercially available solvents that make excellent removers. These solvents should readily mix with the penetrant residues and be capable of removing the final remnants from the surface. They should also evaporate quickly and not leave any residue themselves. It is essential that the removers not be applied directly to the surface, since they are also good penetrants. Spraying or flushing the part with the solvent during the removal step is prohibited by many specifications. Even so, there are still users who insist on performing this unacceptable practice in order to "thoroughly" remove the surface penetrant. When using the visible color-contrast penetrants, a slight trace of pink on the cloth or paper towel will indicate that the removal is adequate. For fluorescent penetrants, slight traces of the fluorescent penetrant as observed under the black light will also indicate the proper level of removal.

### Developers

There are four basic types of developers:

1. Dry developer
2. Solvent-based developers, also referred to as "spirit" or nonaqueous
3. Wet developers suspended in water
4. Wet developers that are soluble in water

Developers have been described as "chalk" particles, primarily because of their white, chalk-like appearance.

In order for the developers to be effective in pulling or extracting the penetrant from entrapped discontinuities, thus presenting the penetrant bleed-out as an indication that can be evaluated, they should possess certain key characteristics. They should:

- Be able to uniformly cover the surface with a thin, smooth coating
- Have good absorption characteristics to promote the maximum blotting of the penetrant that is entrapped in discontinuities
- Be nonfluorescent if used with fluorescent penetrants
- Provide a good contrast background that will result in an acceptable contrast ratio
- Be easily applied to the test specimen
- Be inert with respect to the test materials
- Be nontoxic and compatible with the penetrant materials
- Be easy to remove from the test specimen after the examination is complete

There are other types of developers that are used on rare occasions. These are referred to as strippable, plastic film, or lacquer developers. They are typically nonaqueous suspensions containing a resin dissolved in the solvent carrier. This developer sets up after application and is then stripped off the surface, with the indications in place. It can then be stored and maintained as part of the inspection report.

## 4.5 Penetrant Procedures

Selecting the correct technique for penetrant testing is very important. Prior to performing the examination, a procedure should be developed and qualified. When preparing the procedure, the following should be considered:

- The requirements of the code, specification, or contract
- The type and size of discontinuity that is anticipated

- The surface condition of the test specimen
- The configuration of the part
- The quantity of parts to be examined
- Systems and equipment that are available

### 4.5.1 Prerequisites

Prior to any penetrant test, certain prerequisites have to be addressed.

#### *Temperature*

Penetrant materials are influenced by temperature variations. Most codes and specifications require that the test part and the penetrant materials be within a specified temperature range, typically from 40°F (4.4°C) to as high as 125°F (51.6°C). The part and the penetrant materials must fall within the specified temperature range. If the test part or the penetrant is extremely cold, the penetrant becomes very thick and viscous, which will affect the time it will take to penetrate the discontinuities. If the test surface or penetrants are high in temperature, some of the more volatile constituents may evaporate from the penetrant, leaving a thick residue that will not effectively penetrate the discontinuities.

#### *Environmental Considerations*

Since some of the solvent cleaners and removers used with penetrant testing can be somewhat flammable, it is essential that the penetrant test be performed in an area where there are no open flames or sparks that may tend to cause the penetrant materials to ignite. Typically, penetrant materials have relatively high flash points, but some of the cleaner/remover solvents could ignite when exposed to sparks or open flames. Also, some of the solvents may give off fumes. Therefore, penetrant testing should be performed in an area where there is adequate ventilation.

#### *Lighting*

There must be adequate lighting in the examination area, especially during the time when the evaluation is performed.

#### *Surface Condition Considerations*

Surfaces to be examined having coatings such as paint or plating, or extremely rough conditions, must be addressed. If the surface contains scale and rust, some type of strong mechanical cleaning process is required. Many codes and specifications do not permit the use of some mechanical cleaning techniques, such as grit-blasting, shot-peening, or sandblasting, since these processes tend to peen over the test surface, potentially closing a crack or other surface discontinuity. If wire brushing is used to remove scale or rust, it should be done with extreme care for the same reason. If extreme pressure is applied to a grinding wheel or power wire brush, it is possible to cause a smearing of the metal on the surface. Many codes and specifications require that mechanical cleaning methods, if used, such as grit- or sandblasting be followed with a caustic chemical cleaning step in order to remove smeared material that may prevent penetrant from entering a surface connected discontinuity.

FIGURE 4-7 Precleaning of a weld surface.

### 4.5.2 The Penetrant Procedure

#### *Precleaning*

After addressing the prerequisites, it is necessary to remove all contaminants from the surface and, after the surface has been cleaned, all evidence of any residues that may remain (see Fig. 4-7). After precleaning, it is essential that the precleaners evaporate and that the test surface be totally dry prior to application of the penetrant. This will prevent contamination or dilution of the penetrant in the event that it interacts and becomes mixed with the precleaner. Water is most common contaminant to a penetrant examination. If a discontinuity is filled with water, the penetrant will not be able to enter.

#### *Penetrant Application*

The penetrant can be applied to the surface of the test part in virtually any effective manner, including brushing (Fig. 4-8), dipping the part into the penetrant, immersion, spraying, or just pouring it on the surface. Figure 4-9 shows a water-removable fluorescent penetrant being applied with an electrostatic sprayer. The key is to ensure that the area of interest is effectively wetted and that the penetrant liquid does not dry during the penetration or *dwell time,* which is the period of time from when the penetrant is applied to the surface until it is removed. The codes and specifications give detailed dwell times that must be followed. It is quite common to have a dwell time of 10 to 15 minutes for many applications.

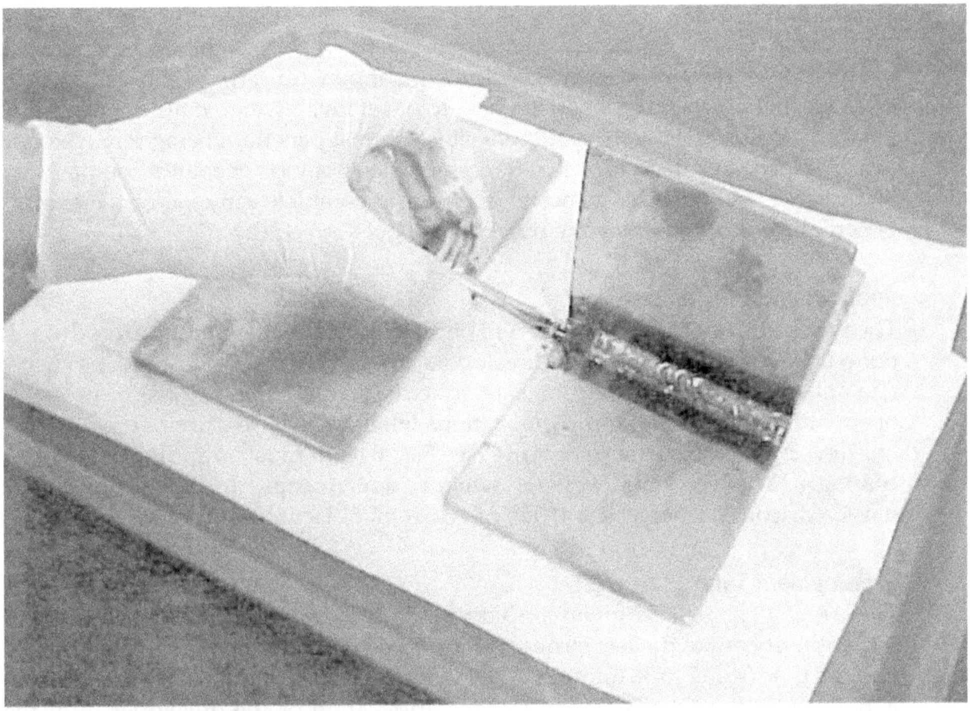

**Figure 4-8** Application of a visible penetrant.

**Figure 4-9** Water-removable fluorescent penetrant being applied with electrostatic spray.

### Penetrant Removal

In this step, the excess surface penetrant is removed from the test specimen surface; the method of removal depends on the type of penetrant that is being used. There are three techniques for excess surface penetrant removal: water, emulsifiers, and solvents. Figure 4-10 illustrates excess surface visible contrast penetrant being removed with a solvent-dampened cloth. Removal of fluorescent penetrants is usually accomplished under a black light. This provides a means of ensuring complete removal of the excess surface penetrant while minimizing the possibility of over-removal.

### Application of Developer

The type of developer to be used will be specified in the penetrant procedure. As mentioned above, the four types of developers are dry, nonaqueous, aqueous suspendable, and aqueous soluble. The entire test surface or area of interest must be properly developed, although there are rare applications where developers are not used. A nonaqueous developer is applied by spraying (see Fig. 4-11). It must be applied in a thin, uniform coating. Thick layers of developer, whether nonaqueous, dry, or aqueous, can tend to mask a discontinuity bleed-out, especially if that discontinuity is small and tight.

### Development Time

The developer must be given ample time to draw the entrapped penetrant from the discontinuity out to the test surface. Many codes and specifications will require a development time from 7 to 30 minutes and, in some cases, as long as 60 minutes. Development is defined as the time it takes from the application of the developer until the actual evaluation commences. It is recommended that the surface be observed immediately

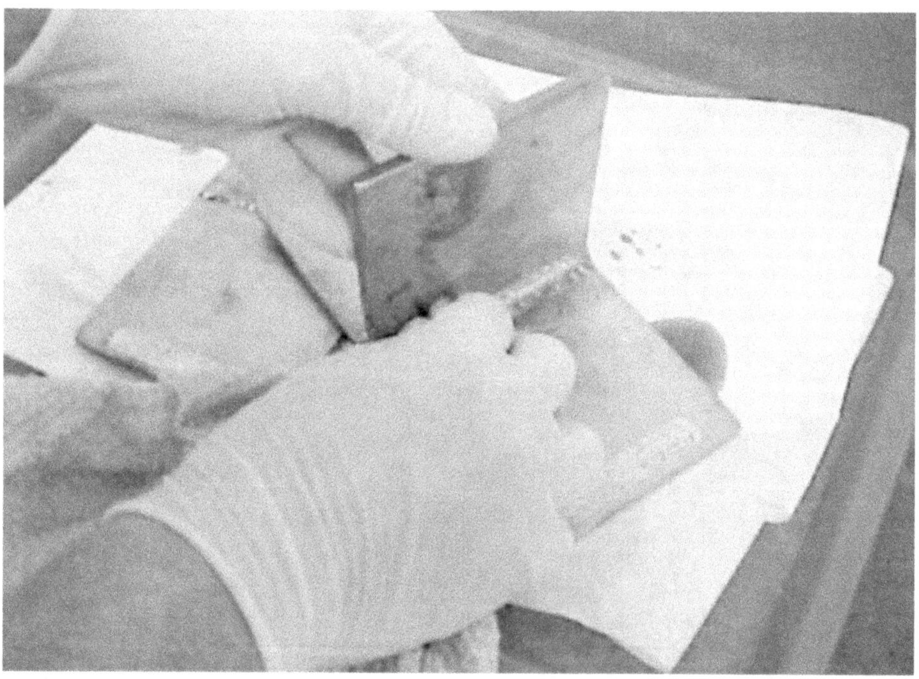

**Figure 4-10** Removal of excess surface penetrant.

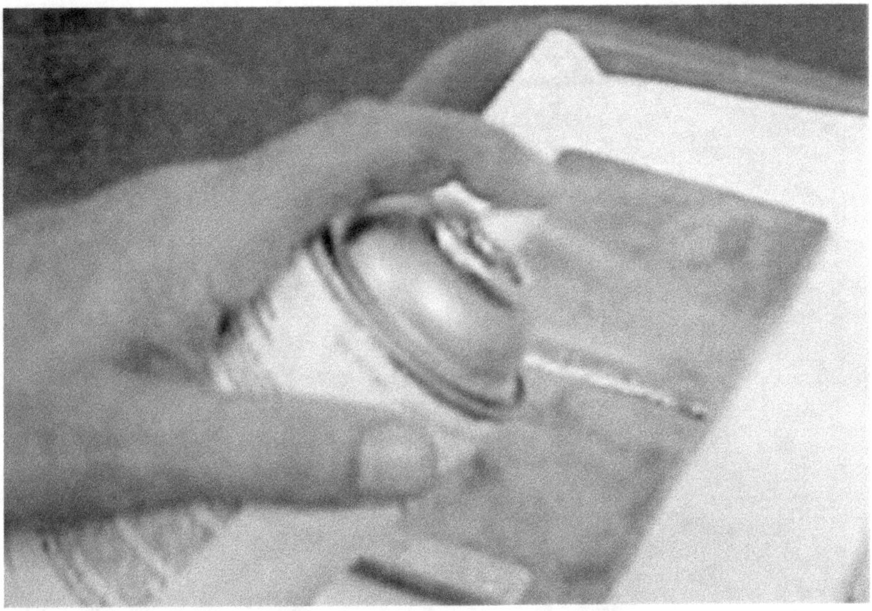

FIGURE 4-11 Application of a nonaqueous developer.

after the application of the developer to assist in the characterizing and to determine the extent of the indication(s).

### *Interpretation*

Upon completion of the development time, the indications from discontinuities or other sources that have formed must be interpreted. A visible contrast penetrant bleed-out is illustrated in Fig. 4-12, and fluorescent penetrant indications are shown in Fig. 4-13. Bleed-outs are interpreted based primarily on their size, shape, and intensity (see Section 4.7, Evaluation and Disposition).

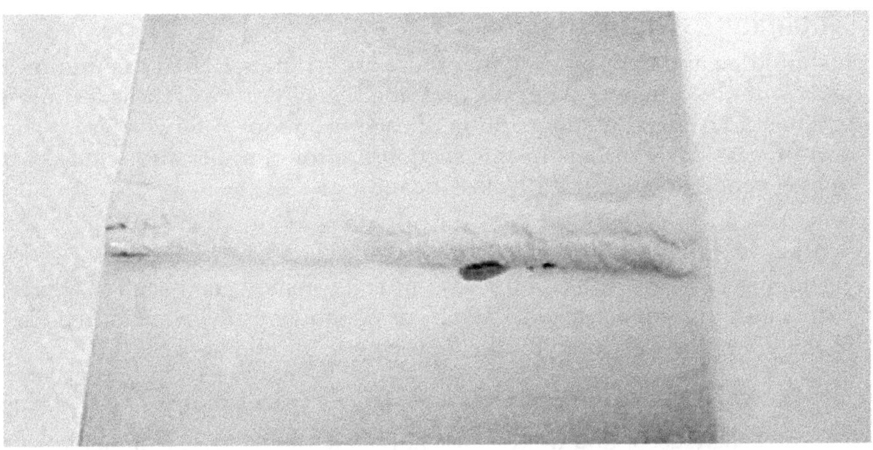

FIGURE 4-12 Indication of a bleed-out.

**Figure 4-13** Fluorescent penetrant indication of a crack.

***Post-Cleaning***

After the part has been evaluated and the report completed, all traces of any remaining penetrant and developer must be thoroughly removed from the test surface prior to it being placed into service or returned for further processing.

## 4.6 Techniques and Variables

It should be apparent by now that there are a number of PT techniques that can be used with the different materials described. A summary of these techniques is listed in Table 4-1. A detailed description of each technique follows. (***Note:*** The technique and process designations in this section are for simplification and do not directly relate to code or specification classifications.)

### 4.6.1 Technique I, Process A (I-A)

Technique I Process A uses a fluorescent water-removable penetrant that can be used with either dry, aqueous suspendable, or nonaqueous developers (see Fig. 4-14). This technique is generally used for the following applications:

1. When a large number of parts or large surface areas are to be examined.
2. When discontinuities that are not broad or shallow are anticipated.

| Technique | Process | Materials |
|---|---|---|
| I<br>(Fluorescent) | A<br>(Fig. 4-14) | Water-removable penetrant; dry, aqueous suspendable, or nonaqueous developer |
| | B<br>(Fig. 4-15) | Post-emulsifiable penetrant; lipophilic emulsifier; dry, aqueous, or nonaqueous developer |
| | C<br>(Fig. 4-16) | Solvent-removable penetrant; solvent cleaner/remover; dry or nonaqueous developer |
| | D<br>(Fig. 4-17) | Same as I B except the emulsifier is hydrophilic |
| II (Visible,<br>color contrast) | A<br>(Fig. 4-14) | Water-removable penetrant; aqueous or nonaqueous developer |
| | B<br>(Fig. 4-15) | Post-emulsifiable penetrant, emulsifier, and aqueous or nonaqueous developer |
| | C<br>(Fig. 4-16) | Solvent-removable penetrant; solvent cleaner/remover, aqueous or nonaqueous developer |

**TABLE 4-1** Penetrant Technique Classification Summary

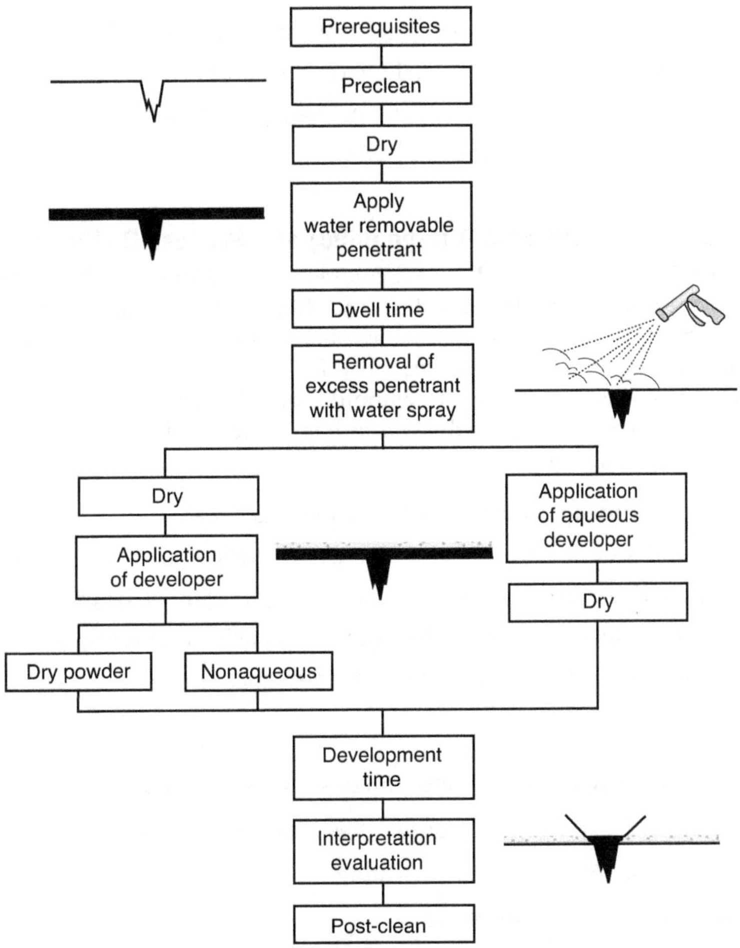


**FIGURE 4-14** Water-removable technique (Process I-A or II-A).

3. When parts to be examined have complex configurations such as threads, keyways, or other geometric variation.
4. When the parts to be examined have surfaces that are rough, such as with sand castings or as-welded conditions.

Advantages:

1. Higher sensitivity.
2. Excess penetrant is easily removed with a coarse spray.
3. Easily adaptable for large surfaces and large quantities of small parts.
4. The cost is relatively low.

Limitations:

1. A darkened area is required for evaluation.
2. Under- or over-removal of penetrant material is possible.
3. Water contamination can degrade the effectiveness of the penetrant.
4. Not effective for broad or shallow discontinuities.
5. Dryers are required (usually) when using developers.
6. This technique is usually not portable.

### 4.6.2 Technique I Process B (Lipophilic) and Process D (Hydrophilic)

Technique I, Processes B and D use a fluorescent post-emulsifiable penetrant, a lipophilic (L) or hydrophilic (H) emulsifier, and dry, aqueous, or nonaqueous developers. (Figures 4-15 and 4-17 illustrate overviews of Processes B and D, respectively.) The materials used are very similar to those described for Technique I Process A, except that these penetrants are not water-removable without emulsification. A lipophilic or hydrophilic emulsifier must be used after the dwell time has expired. This technique is generally used in the following situations:

1. When a large quantity of parts must be examined
2. When discontinuities that are broad and shallow are anticipated
3. For the detection of stress cracks or intergranular corrosion
4. For the detection of small discontinuities such as grinding cracks
5. Applications requiring higher-sensitivity techniques

Advantages:

1. High sensitivity for the detection of smaller discontinuities
2. For broad or shallow discontinuities (when they are expected)
3. Adaptable for high-quantity testing
4. Not easily affected by acids
5. Less susceptible to over-removal than Technique I-A

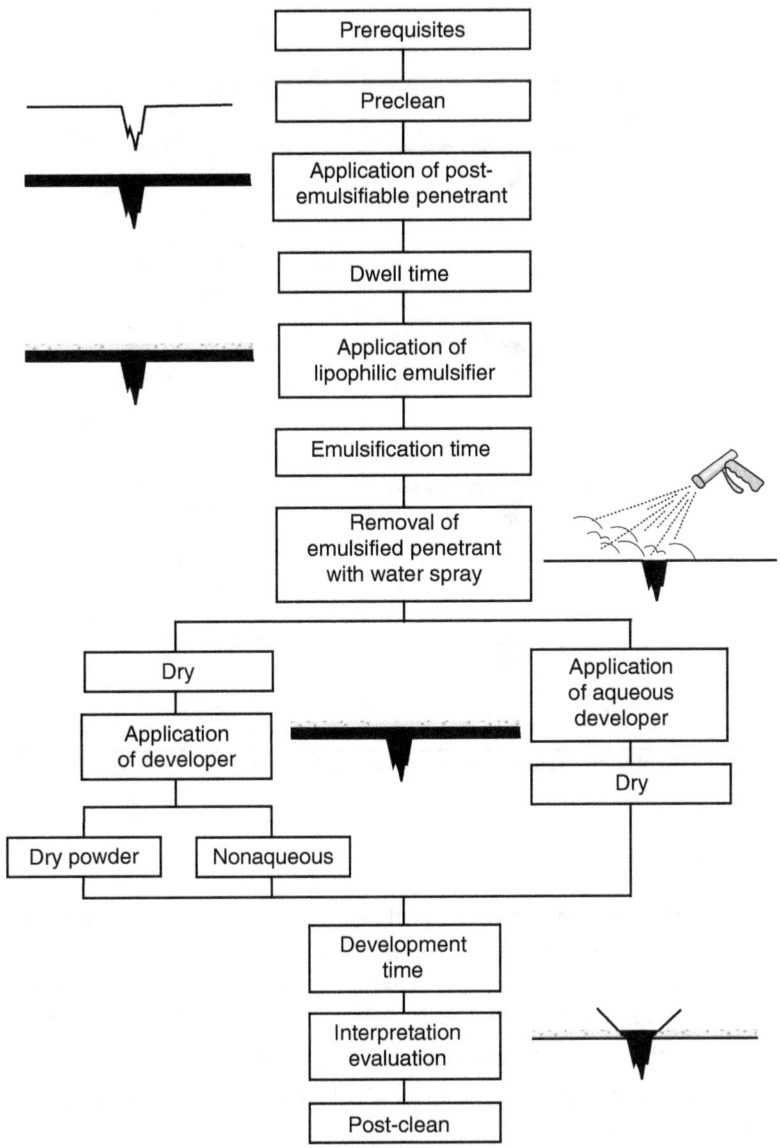


**FIGURE 4-15** Post-emulsifiable technique (lipophilic) (Process I-B or II-B).

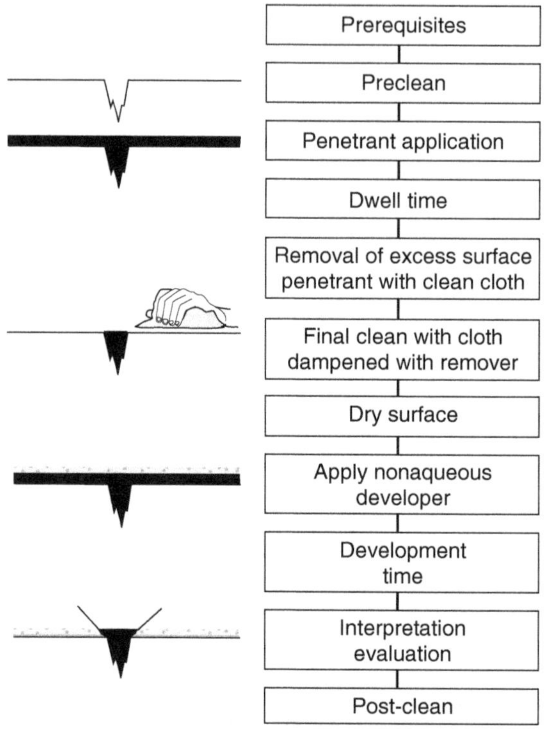


**FIGURE 4-16** Solvent-removable technique (Process I-C or II-C).

Limitations:

1. This technique has an additional step, which requires an emulsifier. Therefore, more time and material are necessary.
2. It is not as effective for parts with complex shapes (e.g., threads) or rough surfaces, as is Technique I-A.
3. The emulsification time must be closely controlled.
4. As with Technique I-A, it requires drying prior to the application of dry or nonaqueous developers.
5. It is usually not portable.

### 4.6.3 Technique I, Process C (I-C)

Technique I Process C uses a fluorescent penetrant, which is solvent-removable, a solvent cleaner/remover, and a nonaqueous developer. The excess surface penetrant is first removed with a dry cloth, followed by cleaning with a cloth dampened with a solvent remover. (Figure 4-16 illustrates an overview of Technique I-C.) This process is generally used when removal with water is not desirable due to part size, weight, surface condition, water availability, or when a heat source is not readily available for drying.

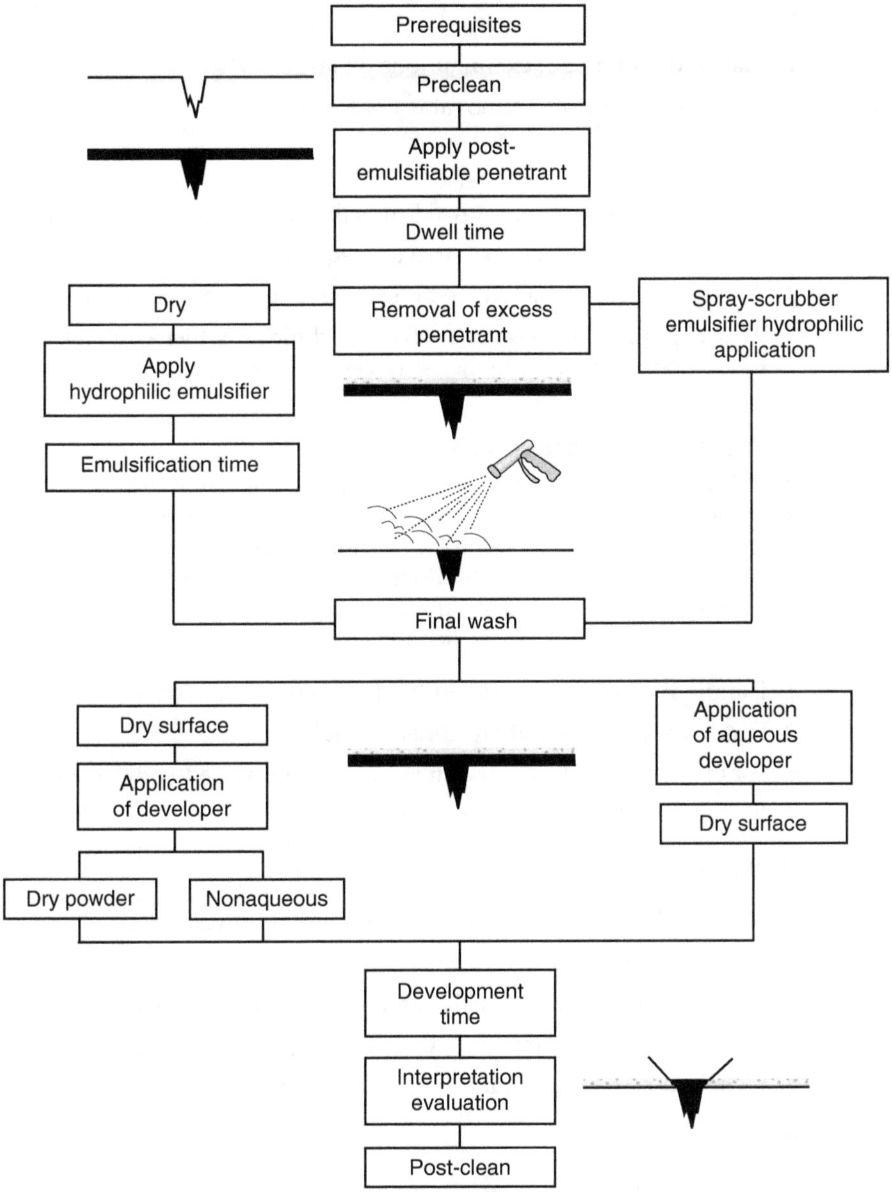


**FIGURE 4-17** Post-emulsifiable technique (hydrophilic) (Process I-D).

Advantages:

1. Can be used for spot examinations on large parts
2. Effective when water removal is not feasible

Limitations:

1. The use of solvent for removal limits this technique to smaller areas.
2. A black light and darkened area are required.
3. The sensitivity can be reduced if excessive remover is applied.
4. A "background" may occur with this technique, which could affect the contrast ratio, especially with rougher surfaces.

### 4.6.4 Technique II Process A (II-A)

Technique II Process A uses a visible color-contrast, water-removable penetrant, and an aqueous or nonaqueous developer. Dry developer is not usually used with Technique II penetrants. Some specifications, in fact, do not permit the use of dry developers with Technique II penetrants. Figure 4-14 illustrates Technique II-A.

The penetrant contains an emulsifier, making it water removable. This technique is generally used for the following applications:

1. Examinations of a large quantity of parts or large surface areas
2. For discontinuities that are generally tight
3. For the examination of parts with threads, keyways, and other complex geometries
4. For parts with generally rough surfaces

Advantages:

1. No black light or darkened area is required for evaluation.
2. It is relatively quick and inexpensive.
3. The excess penetrant is easily removed with a coarse water spray.
4. It is effective for the examinations of a large quantity of parts.
5. It can be used for rough surfaces, keyways, threads, and other complex geometries.

Limitations:

1. Its sensitivity is inferior to Technique I-A.
2. Penetrant can be over-removed.
3. Water contamination can degrade the effectiveness of the penetrant.
4. It is not usually effective for the detection of broad or shallow discontinuities.

### 4.6.5 Technique II Process B (II-B)

Technique II Process B uses a visible color-contrast, post-emulsifiable penetrant, an emulsifier, and an aqueous or nonaqueous developer. The materials used (except for the penetrant) are very similar to those described for Technique I Process B (as illustrated in Fig. 4-15). An emulsifier (usually lipophilic) is applied to the surface penetrant after the dwell time to make it water removable. Technique II Process B is generally used for the following applications:

1. When a large quantity of parts must be examined.
2. Whenever lower sensitivity than that achieved with Technique I is acceptable.
3. When broad and shallow discontinuities are anticipated.

Advantages:

1. No black light or darkened area for evaluation is required.
2. Broad or shallow discontinuities may be detected.
3. Useful when there are large quantities of parts to be examined.
4. This technique is not as susceptible to over-removal, as are the Process A penetrants.

Limitations:

1. The additional step of an emulsifier requires more time and additional material.
2. It is not as effective for parts with a complex geometry (e.g., threads), as is Process A.
3. The emulsification time is very critical and must be closely controlled.
4. Drying is required if nonaqueous developers are used.

### 4.6.6 Technique II, Process C (II-C)

Technique II Process C uses a visible, color-contrast, solvent-removable penetrant, a solvent cleaner/remover, and an aqueous or nonaqueous developer. Figure 4-16 illustrates this technique.

The excess penetrant is not water-removable and must be removed with a solvent remover. This technique is widely used for field applications when water removal is not feasible, or when examinations are to be conducted in a remote location.

Advantages:

1. This technique is very portable and can be used virtually anywhere.
2. It can be used when water removal is not possible.
3. Black lights or darkened evaluation areas are not required. Evaluation is done in visible light.
4. It is very adaptable for a wide range of applications.

Limitations:

1. The use of solvent to remove excess surface penetrant limits the examinations to smaller areas and parts without a complex geometry.
2. Sensitivity is reduced when an excessive amount of remover is used during the removal step.
3. Excess penetrant removal is difficult on rough surfaces, such as sand casting and as-welded surfaces, and usually results in a "background."

4. This technique has a lower level of sensitivity compared to Technique I penetrants.
5. It is more "operator-dependent" due to the variables involved in the removal step.

### 4.6.7 Summary

Of all the techniques described in Section 4.6, the most widely used fluorescent penetrant technique is I-A (water-removable). Technique II-C is the most widely used visible color-contrast penetrant.

## 4.7 Evaluation and Disposition

After the penetrant process has been completed and the indications are observed and recorded, the final step will be to establish whether or not these conditions are acceptable or rejectable. The size of the indication can usually be related to the amount of penetrant entrapped in the discontinuity. The larger the discontinuity volume, the greater the amount of penetrant that will be entrapped and, therefore, the larger the bleed-out after development. The shape of the indication is important because it relates to the type or nature of the discontinuity; e.g., a crack or lack of fusion will show up as a linear bleed-out rather than a rounded one. A linear indication, by most codes and specifications, is defined as a bleed-out whose length is three times or greater than its width. The intensity of the bleed-out gives some evidence as to how tight the discontinuity is. A broad shallow-type discontinuity will tend to be diluted, to an extent, by the remover liquid, and not be as brilliant as a bleed-out from a very tight discontinuity. It is essential that corrective action be taken to remove or repair the discontinuity if it is deemed to be rejectable. In most cases, a crack or other serious discontinuity will be cause for the rejection or scrapping of the part. Repairs to discontinuities will often be accomplished by grinding. A recommended technique to ensure complete removal of an indication after grinding is to merely reapply the developer. This usually verifies whether the discontinuity has been removed, since the bleed-out will reappear if it has not (see Fig. 4-18).

This process should be repeated, i.e., grinding, reapplication of developer, and then grinding again until no further bleed-out occurs. At that point, the area that has been ground out must be reexamined, following the penetrant procedure from the beginning, to ensure that, in fact, the discontinuity has been totally removed. It is further recommended that the grinding be performed in the same direction as the longest dimension of the discontinuity. This is to minimize the possibility of smearing the material over the discontinuity. After the repair is completed, the repaired surface must be reexamined. During the evaluation process, it is necessary that a suitable light source be used. For visible penetrants, a light source of 100 foot-candles is common, although codes and specifications may require different light intensities. A black light is used for evaluation of fluorescent penetrants, as seen in Fig. 4-19. An intensity of between 800 to 1,200 pW per square centimeter is typically required.

Penetrant indications must be recorded. For recording purposes, a number of satisfactory techniques can be used, including photographs, hand sketches, and the transparent tape lift-off technique. Photographic techniques employed for recording visible penetrant indications are quite standard. When photographing indications under black light conditions, specialized exposures, and filters may be necessary when using photographic film. Digital cameras are usually quite exceptional for recording both visible and fluorescent penetrant indications. In addition, the digital

**Figure 4-18** Initial grind-out of an indication.

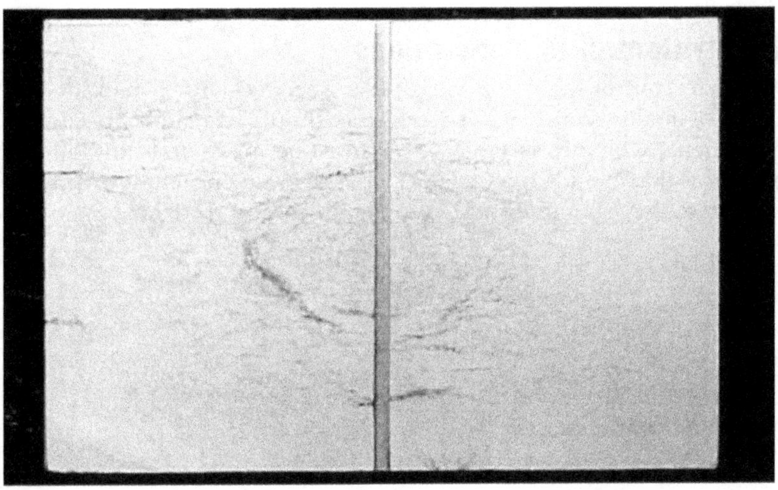

**Figure 4-19** Comparator block.

image can be observed immediately after exposure. Hand drawings, when used with test reports, should be prepared with as much accuracy and detail as possible. The transparent tape lift-off technique is usually very effective for visible dye penetrants, since the tape can be placed directly on the developed indication, lifted off, then placed on the test report. This will provide a means of displaying the actual bleed-out indication with its true size and intensity. This technique is not usually effective with fluorescent penetrants, unless the test report is to be evaluated under a black light.

## 4.8 Penetrant Testing Applications

Penetrant testing is extremely versatile and has many applications. It is used in virtually every major industry and for a wide variety of product forms. Industries that widely use penetrant testing techniques include:

Power generation, both fossil- and nuclear fueled

Petrochemical

Marine/shipbuilding

Metalworking, including foundries and forging shops

Aerospace

Virtually all of the various welding processes and metals-joining industries

Another unique application of penetrant testing is for the detection of through-wall leaks. With this application, penetrant is applied to one surface, for example, of a tank, and developer applied to the opposite surface. If there is an unimpeded through leak, the penetrant will follow that path and be exposed on the opposite developed surface as an indication. Some leak tests, such as those for complex components and piping systems that must be evaluated to determine the source and location of a leak path, are conducted using fluorescent tracers.

## 4.9 Quality Control Considerations

In order to control the many variables associated with penetrant testing, there are a number of quality control (QC) checks that should be made periodically. The applicable requirements will specify which ones must be made and how often. There are three major areas that include the various QC issues, and once these are identified as essential or required, they should be added to the procedure. The three major categories are:

1. Material checks
    - New
    - In-use
2. System checks
3. Equipment checks

### 4.9.1 Material Checks—New

All incoming or new penetrant materials should be verified to ensure compliance with specifications. Some codes and specifications contain unique requirements regarding the presence of contaminants in the PT materials such as chlorine, sulfur, and halogens. Some codes and standards include analytical tests to determine the amount of these contaminants, but most of the penetrant manufacturers will provide material certifications that specify the amount in that particular batch of materials. Each can or container of the penetrant will have the batch number clearly marked to make it easy to correlate with the certifications. These certifications should be maintained in the QA files for future reference and for ready access in the event of an audit.

### 4.9.2 Material Checks—In-Use

It is good practice, and in some cases a requirement, that penetrant materials be checked periodically to ensure that they have not degraded and are still performing satisfactorily. These checks may include, but are not limited to, the following:

1. Contamination of the penetrant
2. Water content in the penetrant (for water-removable penetrants)
3. Water content in lipophilic emulsifier
4. The condition of the dry developer (fluorescent penetrant carry-over, etc.)
5. Wet developer contamination
6. Aqueous developer concentration (using a hydrometer)
7. Hydrophilic emulsifier concentration

The results of these checks should be entered in a logbook to permit periodic reviews of the condition and the trends of the materials as they are being used.

### 4.9.3 System Checks

Probably the most effective overall checks of the performance of penetrant systems involve the periodic use of panels or samples containing known discontinuities (see Fig. 4-20). As mentioned earlier, a number of cracked panels are available for this purpose, including plated, single thickness; plated, tapered to provide for a range of crack depths; and industry-specific panels. Prior to use, these panels should be thoroughly

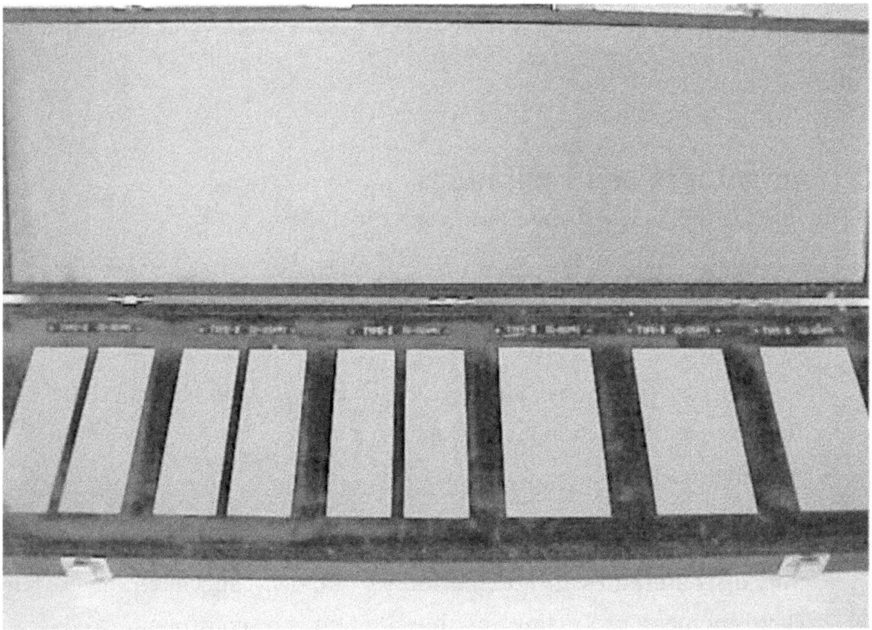

**Figure 4-20** Cracked plated panels.

cleaned, preferably in an ultrasonic cleaner to ensure that there are no residual penetrants remaining that could affect their response.

There are other systems checks that may be waived if the system performance check with the panels is satisfactory. They include tests for:

1. Penetrant brightness
2. Penetrant removability
3. Penetrant sensitivity
4. Emulsifier removability

### 4.9.4 Equipment Checks

There are routine checks that should be made on the equipment that is used in the PT process. They include, but are not limited to, the following:

1. Black lights—for intensity, filter condition, bulb, and reflector.
2. Light meters, black and white—for functionality and calibration status. Some codes require that these meters be periodically calibrated; this is usually done by the meter manufacturer.
3. The temperature of the drying oven—to ensure that it is within the specified range.
4. The water used for removal—temperature and pressure.
5. Pressure gauges, when compressed air is used for application of the penetrant, emulsifiers, and developers.

Table 4-2 lists key items considered by some codes and standards as necessary for maintaining control over some performance variables encountered in penetrant testing.

## 4.10 Advantages and Limitations

The major advantages of penetrant testing include:

Portability.

Cost (inexpensive).

Sensitivity.

Versatile—virtually any solid nonporous material can be inspected.

Effective for production inspection.

Nondestructive.

The limitations include:

Only discontinuities open to the surface of the test specimen can be detected.

There are many processing variables that must be controlled.

Temperature variation effects.

| Test | Frequency |
|---|---|
| System performance | Daily |
| Penetrant contamination | Daily |
| Developer contamination (aqueous: soluble and suspension) | Daily |
| Developer condition (dry) | Daily |
| Water spray pressure | Each Shift |
| Water spray temperature | Each Shift |
| Black light intensity | Daily |
| Black light reflectors, filters, and bulbs | Daily |
| Inspection area cleanliness | Daily |
| Emulsifier concentration (hydrophilic) | Weekly |
| Penetrant sensitivity | Weekly |
| Fluorescent brightness | Quarterly |
| Penetrant removability | Monthly |
| Emulsifier removability | Monthly |
| Emulsifier water content (lipophilic) | Monthly |
| Drying oven temperature | Quarterly |
| Light meter calibration | Semiannually |

**TABLE 4-2** Recommended Tests and Frequency of Performance

Surface condition and configuration.

Surface preparation is necessary.

The process is usually messy.

## 4.11 Glossary of Penetrant Testing Terms

**Adhesion** The tendency of the penetrant to adhere to the surface.

**Angstrom Unit (Å)** A unit of length that may be used to express the wavelength of electromagnetic radiation (i.e., light). One angstrom unit is equal to 0.1 nanometers (1 nm = $10^{-9}$ m).

**Background** The surface of the test part against which the indication is viewed. It may be the natural surface of the test part or the developer coating on the surface.

**Black Light** Electromagnetic radiation in the near-ultraviolet range of wavelength (315–400 nm, 3,150–4,000 Å).

**Black Light Filter** A filter that transmits near-ultraviolet radiation while absorbing other wavelengths.

**Bleed-out** The surfacing of penetrant entrapped in discontinuities to form indications.

**Blotting** Developer soaking up penetrant from a discontinuity to accelerate bleed-out.

**Carrier** A liquid, either aqueous or nonaqueous, in which penetrant examination materials are dissolved or suspended.

**Clean** Free of contaminants.

**Cohesion** The intermolecular action by which the elements of a penetrant are held together.

**Contaminant** Any foreign substance present on the test surface or in the examination materials that adversely affects the performance of penetrant materials.

**Contrast** The difference in visibility (brightness or coloration) between an indication and the background.

**Contrast Ratio** The ratio of the amount of light reflected or emitted between a penetrant and a background (usually the test surface). The contrast ratio for visible color-contrast penetrants is 6:1; it is 40:1 for fluorescent penetrants.

**Developer** A material that is applied to the test surface to accelerate bleed-out and to enhance the contrast of indications.

**Developer, Aqueous** A suspension of developer particles in water.

**Developer, Dry Powder** A fine, free-flowing powder used as supplied.

**Developer, Liquid Film** A suspension of developer particles in a vehicle that leaves a resin/polymer film on the test surface after drying.

**Developer, Nonaqueous** Developer particles suspended in a nonaqueous vehicle prior to application.

**Developer, Soluble** A developer completely soluble in its carrier (not a suspension of powder in a liquid) that dries to an absorptive coating.

**Development Time** The elapsed time between the application of the developer and the examination of the part.

**Drain Time** That portion of the dwell time during which the excess penetrant or emulsifier drains from the part.

**Drying Oven** An oven used for increasing the evaporation rate of rinse water or an aqueous developer vehicle from test parts.

**Drying Time** The time required for a cleaned, rinsed, or wet developed part to dry.

**Dwell Time** The total time that the penetrant is in contact with the test surface, including the time required for application and the drain time.

**Emulsification Time** The time that an emulsifier is permitted to remain on the part to combine with the surface penetrant prior to removal.

**Emulsifier** A liquid that interacts with an oily substance to make it water-removable.

**Emulsifier, Hydrophilic** A water-based liquid used in penetrant examination that interacts with the penetrant oil, rendering it water-removable.

**Emulsifier, Lipophilic** An oil-based liquid used in penetrant testing that interacts with the penetrant oil, rendering it water-removable.

**Etching** The removal of surface material by chemical or electrochemical methods.

**Evaluation** A review to determine whether indications meet specified acceptance criteria.

**False Indication** A response not attributed to a discontinuity or test object condition (usually caused by faulty or improper NDT processing).

**Family** A complete series of penetrant materials required for the performance of a penetrant examination; usually from the same manufacturer.

**Foot-candle (fc)** The illumination on a surface 1 $ft^2$ in area, on which is uniformly distributed a flux of 1 lm (lumen). It equals 10.8 $lm/m^2$.

**Immersion Rinse** A means of removing surface penetrant, in which the test part is immersed in a tank of either water or remover.

**Nonrelevant Indication** An indication caused by a condition related to the test object shape or design that entraps penetrant but is not rejectable.

**Over-emulsification** Excessive emulsification time that may result in the removal of penetrant from discontinuities.

**Over-removal** Too long and/or too vigorous use of the remover liquid that may result in the removal of penetrant from discontinuities.

**Penetrant** A solution or suspension of dye (visible or fluorescent) that is used for the detection and evaluation of surface-breaking discontinuities.

**Penetrant Comparator Block** An intentionally flawed specimen having separate but adjacent areas for the application of different penetrant materials so that a direct comparison of their relative effectiveness can be obtained.

**Penetrant, Post-emulsifiable** A penetrant that requires the application of a separate emulsifier to render the excess surface penetrant water-washable.

**Penetrant, Solvent-removable** A penetrant so formulated that most of the excess surface penetrant can be removed by wiping, with the remaining surface penetrant traces removable by further wiping with a cloth or similar material lightly moistened with a solvent remover.

**Penetrant, Visible** A penetrant that is characterized by an intense color, usually red.

**Penetrant, Water-removable** A penetrant with a built-in emulsifier.

**Post-cleaning** The removal of residual penetrant test materials from the test part after the penetrant test has been completed.

**Post-emulsification** The application of a separate emulsifier after the penetrant dwell time.

**Precleaning** The removal of surface contaminants from the test part so that they will not interfere with the examination process.

**Solvent Remover** A volatile liquid used to remove excess penetrant from the surface being examined.

**TAM Panel** A controlled test panel for checking system performance.

**Viscosity** The property of a fluid that presents a resistance to shearing flow (measured in centistokes).

**Wetability** The ability of a liquid to spread over and adhere to solid surfaces.

## References

Lovejoy, D. *Penetrant Testing—A Practical Guide.* Chapman and Hall, 1991, 2–6 Boundary Row, London, UK.

Metals Test Methods and Analytical Procedures, Vol. 03, Nondestructive Testing. American Society for Testing and Materials International 2011 Annual Book of Standards. 100 Barr Harbor Drive, PO Box C700, West Conshohocken, PA, 19428-2959.

Nondestructive Evaluation and Quality Control, Vol. 17. *ASM International Handbook,* 1992. Metals Park, Ohio.

# CHAPTER 5
# Magnetic Particle Testing

## 5.1 History and Development

### 5.1.1 Magnetism Discovered

The ancient Greeks, originally those near the city of Magnesia, and also the early Chinese knew about strange and rare stones (possibly chunks of iron ore struck by lightning) with the power to attract iron. A steel needle stroked with such a "lodestone" became "magnetic" and around the year 1000 A.D., the Chinese found that such a needle, when freely suspended, pointed north–south.

### 5.1.2 Early Physicists Develop the "Basics" of Magnetism

In the late 1700s and the early 1800s, several exiting new discoveries were made in the field of Physics that paved the way for today's magnetic particle testing (MT) technology.

In the 1700s, Charles Coulomb, a French physicist, discovered that "the magnetic forces of attraction and repulsion are directly proportional to the strength of the poles and inversely proportional to the square of the distance from them" (the inverse square law). He also invented the magnetoscope and magnetometer, which are devices for measuring the Earth's magnetic field strength.

Until 1821, only one kind of magnetism was known; the one produced by iron magnets. Then, Danish scientist Hans Christian Oersted, while demonstrating to friends the flow of an electric current in a wire, noticed that the current caused a nearby compass needle to move. Physicist Andre-Marie Ampere, who concluded that the nature of magnetism was quite different from what everyone had believed, studied this new phenomenon in France. It was basically a *force between electric currents:* Two parallel currents in the same direction *attract*, in the opposite direction they *repel*. This proved that magnetic fields exert an influence on current flow.

### 5.1.3 Phenomenon Discovered Leading to Inspection Principles

In the late 1800s, it was observed that a compass needle deflected when it passed over a crack in a magnetized cannon barrel. In the 1920s, it was discovered that iron filings on parts held in a magnetic machining chuck produced patterns showing the magnetic lines of flux. Closer investigation revealed patterns that corresponded to discontinuities within the parts.

This is probably the earliest documented application of the magnetic particle method for detecting discontinuities in engineering components. This event was recorded in the journal, *Engineering*.

### 5.1.4 Development of Current MT Techniques

Development of the means to establish the magnetic flux in test objects ran in two distinctly separate paths on either side of the Atlantic Ocean. Most American developments used DC storage batteries to produce magnetism in components, whereas the European developments were based on alternating current. These initial preferences are still present in many of the standards and codes, although the DC storage batteries have been replaced with full-wave rectified current.

Advances were also made in the detection media. Instead of using iron filings, iron oxides and iron powders became much more popular, and colored coatings were added to increase the visibility of the indications formed. Later still, fluorescent coatings were applied to the particles, which greatly increased the sensitivity of the inspection process when viewed under "black light" or ultraviolet illumination.

### 5.1.5 The Present

Today's range of magnetic particle testing techniques is vast and encompasses portable, transportable, fixed, and semiautomatic inspection devices. These use permanent magnets, electromagnets using either AC or DC, or a combination of both. The detection media are available as dry powder or as a suspension in a liquid carrier. They are supplied in many colors so as to provide a contrast with the test surface background color and are also available as fluorescent particles for maximum sensitivity.

## 5.2 Theory and Principles

### 5.2.1 Introduction

Magnetic particle testing (MT) is a nondestructive testing (NDT) method for detecting discontinuities that are primarily linear and located at or near the surface of ferromagnetic components and structures. MT is governed by the laws of magnetism and is therefore restricted to the inspection of materials that can support magnetic flux lines. Metals can be classified as ferromagnetic, paramagnetic, or diamagnetic.

*Ferromagnetic* metals are those that are strongly attracted to a magnet and can become easily magnetized. Examples include iron, nickel, and cobalt.

*Paramagnetic* metals such as austenitic stainless steel are very weakly attracted by magnetic forces of attraction and cannot be magnetized.

*Diamagnetic* metals are very slightly repelled by a magnet and cannot be magnetized. Examples include bismuth, gold, and antimony.

*Only those metals classified as ferromagnetic can be effectively inspected by MT.* In order to understand MT, one should have a basic understanding of magnetism and electromagnetism.

### 5.2.2 Principles of Magnetism

#### *Polarity*

Many of the basic principles of magnetism can be deduced by simple observation of the behavior of a magnetized rod and its interaction with ferromagnetic materials, including other magnetized rods. If the rod is suspended at its center, it will eventually align itself with the Earth's magnetic field so that one end points to geographic north and the other end to the south. If the north-pointing end is identified, it will be found that it is always this end that points north. By convention, this end of the rod is called the "north-seeking pole," usually abbreviated as "north pole," and the other end is called the "south pole."

#### *Magnetic Forces*

When the north pole of one magnetized rod is placed close to the south pole of another, it will be observed that they attract one another. *The closer they come together, the stronger the force of attraction.* Conversely, if two north poles or two south poles are placed close together, they will repel each other. This can be summarized as *like poles repel, unlike poles attract.* One way of defining the phenomenon of magnetism could be: "a mechanical force of attraction or repulsion that one body has upon another," especially those that are ferromagnetic.

#### *Magnetic Field*

The simple observations of attracting and repelling indicate that some force field surrounds the magnetized rod. Although invisible, this force field is clearly three-dimensional because the attraction or repulsion can be experienced all around the rod. A two-dimensional slice through this field can be made visible by placing a sheet of plain white paper over a bar magnet and sprinkling ferromagnetic particles onto it. The particles will collect around the lines of force in the magnetic field, producing an image such as the one shown in Fig. 5-1.

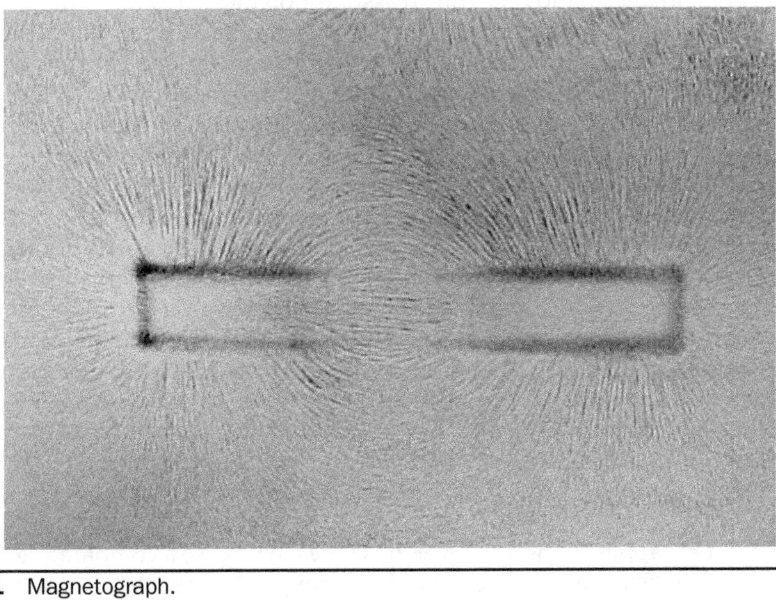

**FIGURE 5-1** Magnetograph.

This image is called a "magnetograph," and the lines of force are referred to as lines of "magnetic flux." Lines of magnetic flux will flow in unbroken paths that do not cross each other. Each line of force forms a closed loop that flows through and around the magnet. The word "flow" suggests some sort of direction of movement and by convention this direction is said to be the direction that would be taken by a "unit north pole" placed at the north pole of a bar magnet. A unit north pole is an imaginary concept of a particle with no corresponding south pole. Such a particle would be repelled by the magnet's north pole and attracted to the south pole. In other words, magnetic flow is from its *north pole to its south pole through the air* around the magnet and in order to complete the magnetic circuit, flow will be from the south pole to the north pole within the magnet.

### Flux Density

The flowing force of magnetism is called *magnetic flux*. The magnetograph image does not show the direction of flux flow, but it can be seen from the magnetograph that the *area of maximum flux concentration (flux density) is at the poles*. Flux density is defined as "the number of lines of force per unit area." The unit area referred to is a slice taken perpendicular to the lines of force. Flux density is measured in Gauss or Tesla, the Tesla being the current unit, and flux density is given the symbol $\beta$ (beta).

### Magnetizing Force

The total number of lines of force making up a magnetic field determines the strength of the force of attraction or repulsion that can be exerted by the magnet and is known as the *magnetizing force* and given the symbol "*H*."

### Magnetic Permeability

A German physicist, Wilhelm Weber, postulated a theory about a material's ability to generate or concentrate a magnetic flux field. This theory became known as the magnetic domain theory and relies on an assumption that a magnetic domain is the smallest independent particle within a material that still exhibits a north and south pole (i.e., it has polarity). In an unmagnetized material, these magnetic domains are arranged in a random (haphazard) direction such that their magnetic fields cancel each other out when considering the total magnetic field exhibited by the material.

When a magnetic force is applied to the material, the domains will tend to align themselves with the magnetizing force so that the domains' north poles point in one direction while the south poles point in the opposite direction. The material will now exhibit an overall polarity, which equates to the sum of all of the magnetic domains combined. A flux field will exist around and through the material, as depicted in the sketch in Fig. 5-2.

The ease with which the domains align themselves is called "permeability," which is expressed with the symbol $\mu$ (mu). Materials in which the domains align easily under low magnetizing forces are said to have high permeability. To determine absolute values for permeability, the flux density produced is divided by the magnetizing force applied. Stated mathematically this becomes

$$\mu = \frac{\beta}{H}$$

For practical applications, however, it is far easier to use a comparative measure of permeability, which is determined simply by comparing a material's permeability to that

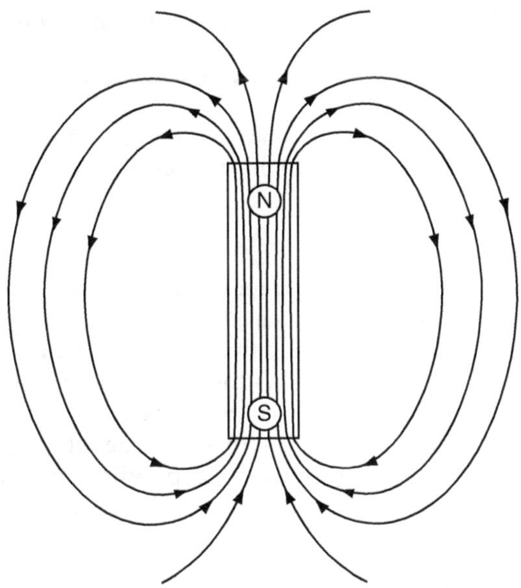


**FIGURE 5-2** Magnetograph sketch showing polarity and flux direction.

of a vacuum. This produces a response that is referred to as *relative permeability* and is given the symbol $\mu_r$.

Permeability in magnetic theory can be compared with *conductivity* in electrical theory. Relative permeability can then be compared with the IACS (International Annealed Copper Standard) conductivity scale, in which all conductive materials have their conductivity expressed as a percentage of that of copper. Ferromagnetic materials will have relative permeability values far in excess of one, which is the base value for a vacuum (sometimes referred to as *unity*). They can have values of several hundred times a vacuum's permeability.

Paramagnetic materials in which the domains resist alignment even under high magnetizing forces are said to have low permeability. Paramagnetic materials will have a relative permeability value of slightly greater than 1; for example, 1.003 for some stainless steels.

The third class of metals consists of the diamagnetic materials. The magnetic domains of diamagnetic materials will rotate to positions that are 90° to the direction of the applied magnetizing force and will create a very slight repulsion to the magnetizing force.

Diamagnetic materials will have a relative permeability value slightly less than 1; for example, 0.9996. Paramagnetic and diamagnetic materials are usually referred to as *nonmagnetic* because of their slight or absence of reaction to a magnet.

### Magnetic Reluctance

"Reluctance" in a magnetic circuit is the equivalent of resistance in an electrical circuit. The magnetic flux will always follow the path of least magnetic reluctance, which is usually through a ferromagnetic material. The factors affecting reluctance are:

1. The length of the magnetic circuit ($\lambda$)
2. The cross-sectional area of the magnetic circuit ($A$)
3. The permeability of the magnetic circuit ($\mu$)

The reluctance, $R$, of a given magnetic circuit can be described mathematically as

$$R = \frac{\lambda}{\mu A}$$

Another way of referring to the phenomenon of a material having low permeability is by stating that the material has high reluctance (the domains are reluctant to align themselves).

### Magnetic Saturation

The lines of force in a magnetic field repel adjacent lines flowing in the same direction. As the flux density increases, the force of repulsion increases. For a given material there is a maximum value for flux density that can be sustained. Upon reaching this value, the material is said to be *saturated*. As the flux density is increased toward saturation, the reluctance of the material increases and the permeability decreases toward that of a vacuum. At saturation, any further increase in magnetizing force finds that the path of least reluctance is now through the air surrounding the material and the excess flux extends out into the air.

### Hysteresis

Hysteresis describes how the flux density ($\beta$) in a magnetic material varies as the magnetizing force ($H$) is varied. A graphical representation of how flux density increases with an increase in the magnetizing force is shown in Fig. 5-3.

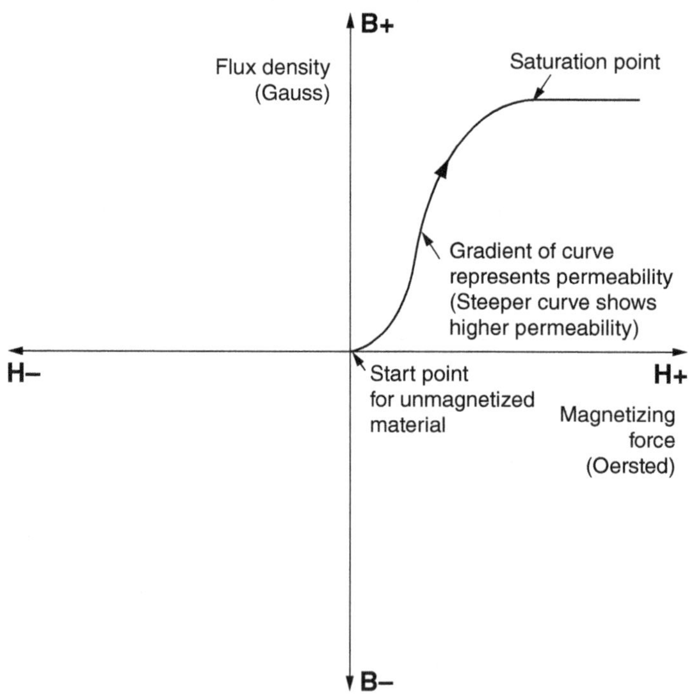


**FIGURE 5-3** Hysteresis curve—flux density due to initial magnetization of ferromagnetic material.

From the graph it can be seen that a ferromagnetic material produces a steep initial "virgin curve" due to the relatively small amount of magnetizing force required to produce a high flux density. The flux is being concentrated by the material's permeability. The steepness of the initial virgin curve is therefore a measure of the material's permeability.

Also noticeable from the graph is the fact that the ferromagnetic material curve changes in gradient toward the top and will actually flatten out completely at one point. From this point onward, any increase in magnetizing force will not produce an increase in flux density. As mentioned above, this condition is called "saturation." The reason for the change in gradient is due to the reduced amount of magnetic domains available for alignment. The total number of domains present is fixed; therefore, as more of these become locked in alignment due to the magnetizing force, there will be fewer "fluid" domains to create the permeability effect. The relative permeability of a material becomes less as it becomes more magnetized.

The graph shown in Fig. 5-3 displays a positive magnetizing force that produces an increasing positive flux density. If the effects of reducing the magnetizing force and plotting the resultant reduction in flux density are considered, even more information about the properties of ferromagnetic materials is gained, as shown in Fig. 5-4.

As seen in Fig. 5-3, increasing the magnetizing force will increase the flux density until the saturation point is reached. If the magnetizing force is decreased, it will produce a corresponding decrease in flux density. The curve produced by this interaction will deviate from the "virgin curve" produced during the initial magnetization.

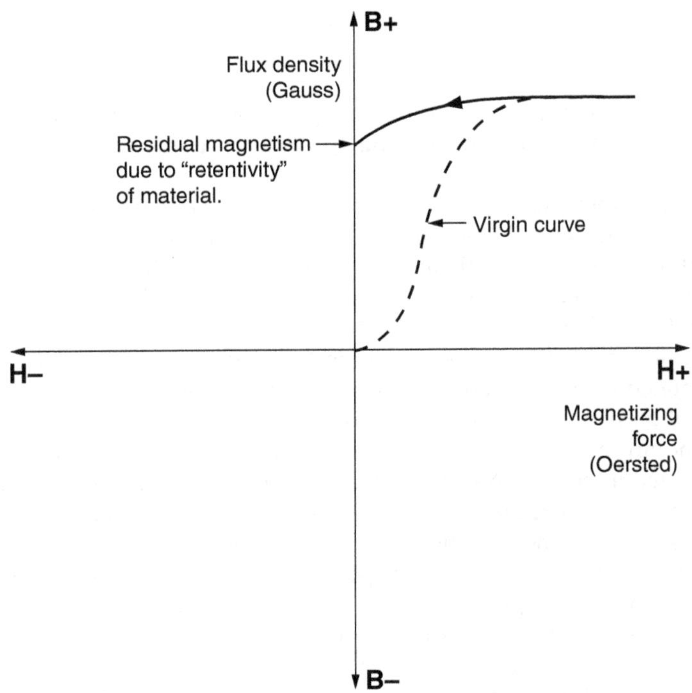


**FIGURE 5-4** Hysteresis curve—residual magnetism.

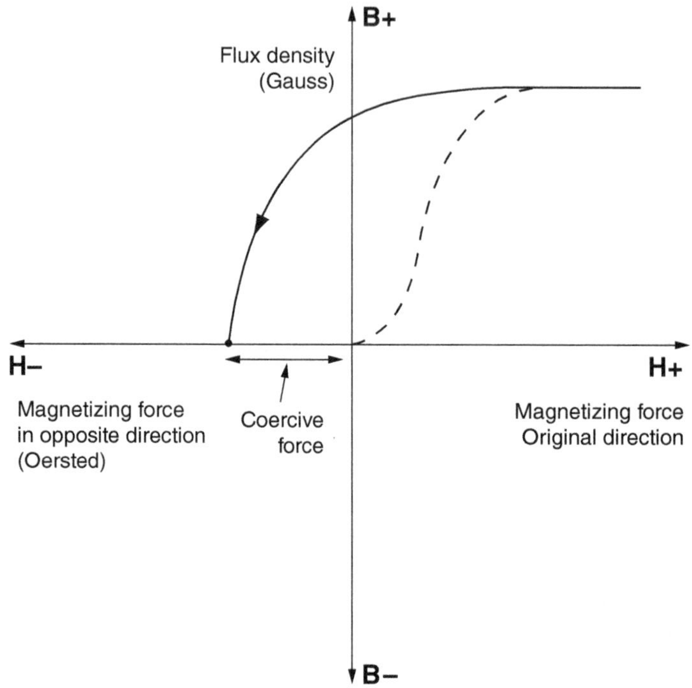


**FIGURE 5-5** Hysteresis curve—coercive force.

As depicted by the dotted line in Fig. 5-5, a point is eventually reached where the magnetizing force is no longer being applied, but there still is a positive flux density in the material.

This phenomenon is due to the *retentivity* of the material, which "retains" some of the flux density and is also referred to as a *remnant field* or *residual magnetism*. The amount of residual magnetism will be determined by the retentivity of the material and the amount of magnetizing force initially applied. The retentivity value is highest among the hardened ferromagnetic materials.

Consider what has happened at the magnetic domain level. The domains were initially rotated from their random rest positions by the initial magnetizing force to a position of alignment. When the magnetizing force stops, some of the domains rotate back to a random orientation. Some of the domains, however, remain aligned, resulting in residual magnetism. These domains will remain aligned until a force applied in the opposite direction causes them to rotate back and will actually rotate some of the domains into alignment in the opposite direction.

In Fig. 5-5, the effects of applying this opposing magnetizing force can be observed. As this negative force is increased (a force with opposite polarity), the flux density is reduced. The force required to reduce the net flux density to zero is called the *coercive force*.

In Fig. 5-6, the effects of increasing the opposite magnetizing force beyond the coercive force can be seen. The material exhibits a flux density flowing in the

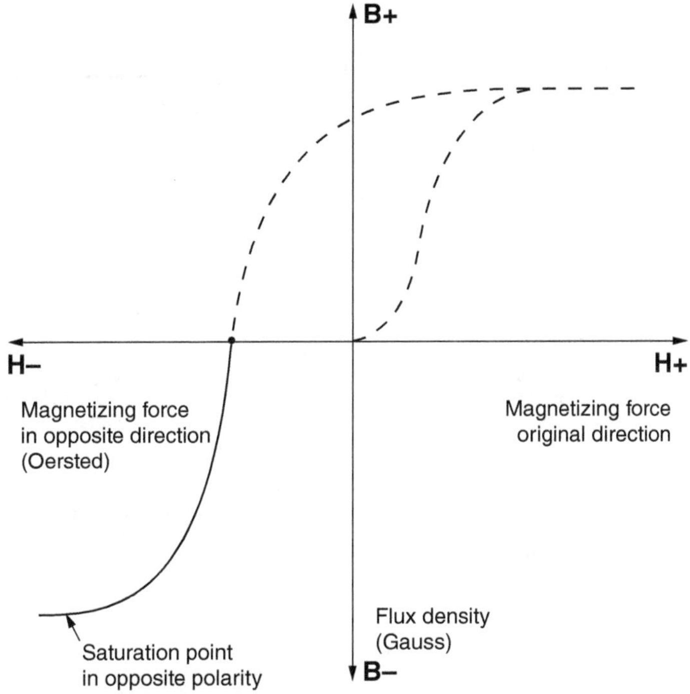


**FIGURE 5-6** Hysteresis curve—saturation point with opposite polarity.

opposite direction, which is expressed in the graph as flux density having a polarity opposite to the original. It is also measured in Gauss. The magnetic domains are now rotated such that their north and south poles are aligned 180° out of phase from the initial alignment. A condition is once more achieved whereby all of the domains become aligned in this manner, and saturation occurs in the opposite polarity when all of the domains are aligned.

If this "opposite" magnetizing force is now reduced back to zero, a reduction of the flux density will be observed, but again, some flux density will be present at the point when there is no magnetizing force applied (Fig. 5-7).

To eliminate this field and bring the flux density to zero once again, magnetizing force in the original direction must be applied. Again, the amount of force required to achieve zero flux density is called "coercive force" (Fig. 5-8).

A further increase in the magnetizing force will produce an increase in flux density until saturation is once again achieved in the original direction, thereby completing the loop known as the *hysteresis loop* (Fig. 5-9). The hysteresis loop can be used to display several of the material's properties. For example, permeability is indicated by the gradient of the virgin curve; reluctance can also be determined from this gradient; a steep curve indicates high permeability, whereas a shallow curve indicates high reluctance.

The residual magnetism is indicated by the flux density value at a magnetizing force of zero. It can be seen in Fig. 5-9 that once the virgin curve has been used to

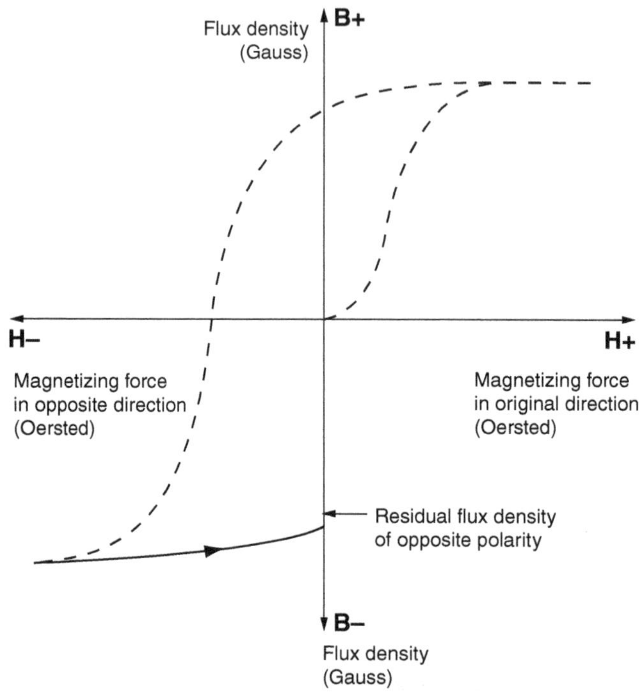


**FIGURE 5-7** Hysteresis curve—residual magnetism with opposite polarity.

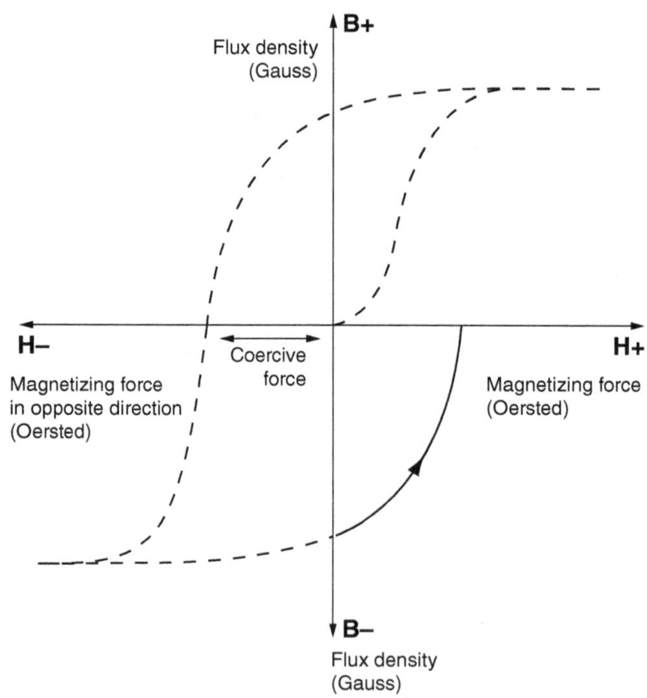


**FIGURE 5-8** Hysteresis curve—coercive force with opposite polarity.

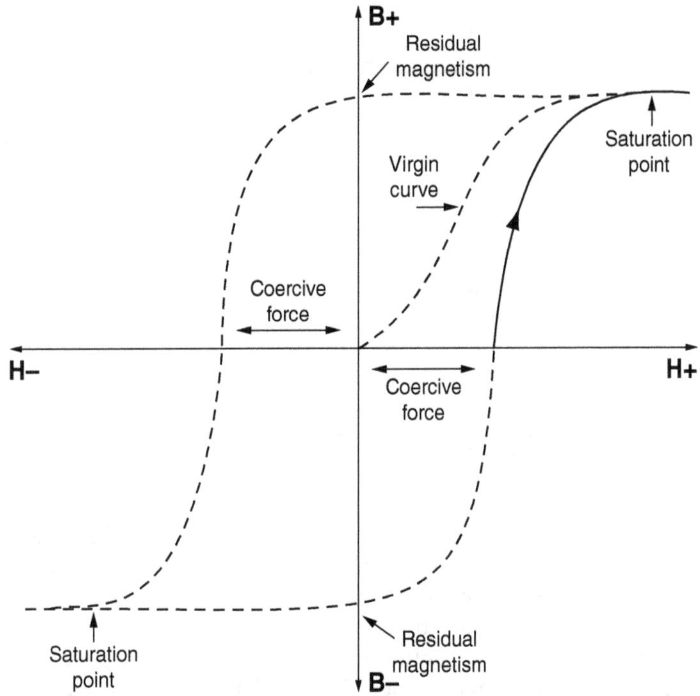


**Figure 5-9** Hysteresis curve—completed hysteresis loop.

produce saturation, the graph continues to follow the S-shaped hysteresis loop and will never return to zero, due to the residual magnetism. This condition will remain until the material is demagnetized.

### Magnetization

In the creation of a hysteresis loop, the increase and decrease of magnetizing force were made possible by positioning a powerful permanent magnet closer to or farther away from the object being magnetized. The "negative magnetization" or magnetization in the opposite direction was accomplished by reversing the orientation of the powerful magnet, such that the opposite pole was closest to the object. Magnetism can also be produced by electrical means. If an electrical current flows through a conductor, a magnetic flux will be produced that flows around the conductor and also in the air surrounding that conductor. This flux flows in a circular direction at 90° to the direction of the electric current flow, as shown in Fig. 5-10.

### Electromagnetic Field Direction

By convention, the actual direction of the flux lines relative to the current flow can be illustrated if the conductor is held in one's left hand. Consider that the electron flow due to the electric current is in the same direction in which the thumb is pointing. The magnetic flux will flow around the conductor in the same direction in which the fingers are pointing (Fig. 5-11).

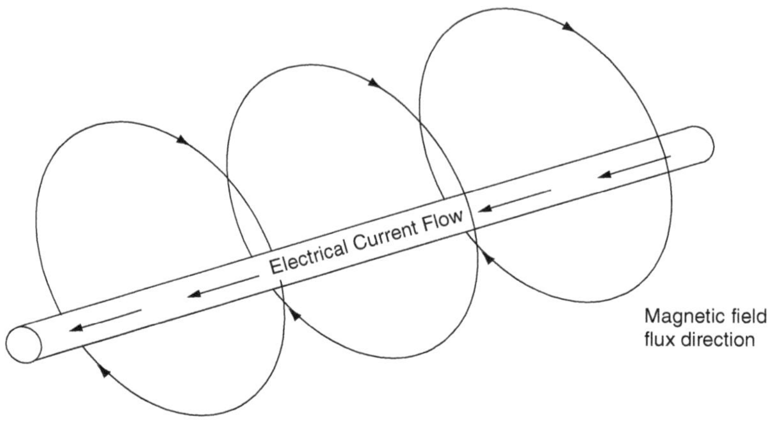


**Figure 5-10** Flux lines around a current-carrying conductor.

This "left-hand rule" applies only if it is considered that the *electron flow* is from negative to positive. This is a relatively recent belief. The original *electric current flow,* concept given by French physicist Ampere, was that the current flowed from positive to negative. If this were the case, then the "right-hand rule" would apply.

If the conductor were the actual ferromagnetic test object, concentrated circular flux lines would be produced in the object. The magnetization of a part in this way is referred to as "direct" magnetization. Current is passed directly through the part, creating the circular flux field.

If a nonmagnetic conductor were formed into a "loop," the magnetic flux circulating in the air around and through the loop would be as shown in Fig. 5-12. It can be seen that at the center of the loop, the flux is basically linear. This effect becomes more pronounced if the conductor is wound into several loops close together so as to form a coil.

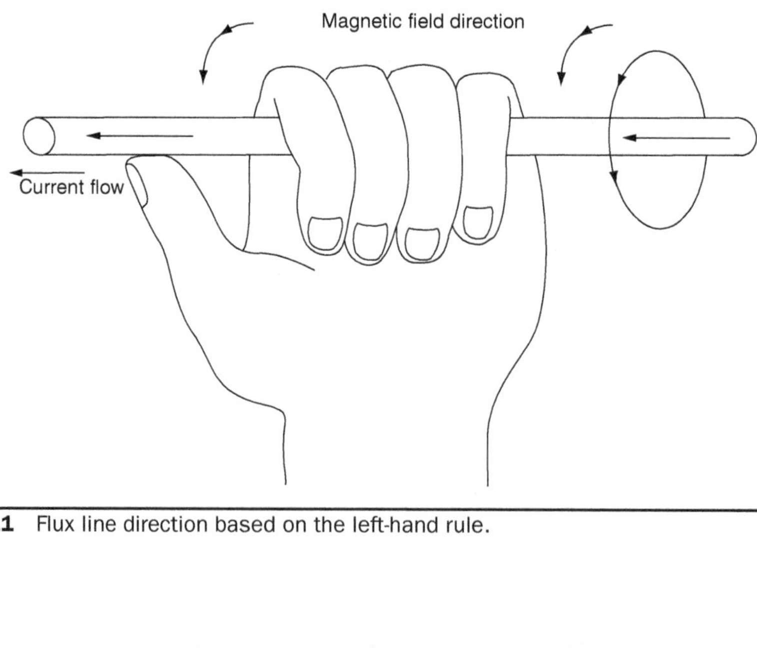


**Figure 5-11** Flux line direction based on the left-hand rule.

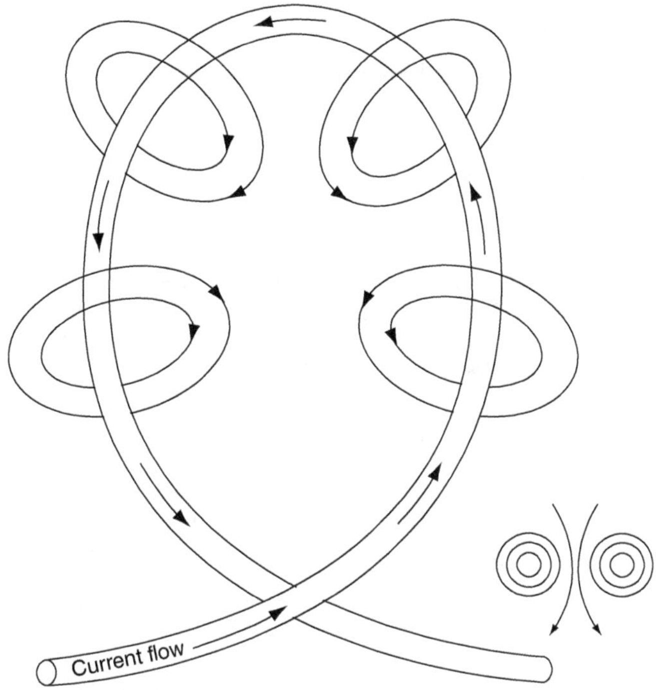


**Figure 5-12** Current flowing in a coil showing flux lines produced.

If the turns of the coil are close enough to each other, the individual flux fields from each loop will join together to create a field for the whole coil (see Fig. 5-13).

When a ferromagnetic object is placed inside this coil, flux lines are induced into the object, resulting in a longitudinal flux field in the object in approximately the same direction as the coil axis. Magnetization of a part within a coil is accomplished by magnetic induction and is referred to as "indirect" magnetization. It is therefore possible to produce both circular and longitudinal magnetic flux fields in the test object using electromagnetism.

### 5.2.3 Detection of Discontinuities

#### *Distorted Fields*

The lines of force in the internal field of a magnetized material will tend to distribute themselves evenly through the material, provided that the material is homogenous. The presence of a discontinuity presents an interruption to the field and an increase in reluctance. The lines of force prefer the path of least reluctance and will therefore redistribute themselves in the material by bending around the discontinuity. The field becomes "distorted" by the discontinuity.

#### *Leakage Fields*

As a discontinuity gets larger, the remaining metal path in the part becomes more restricted and the magnetic flux approaches saturation for that part of the material.

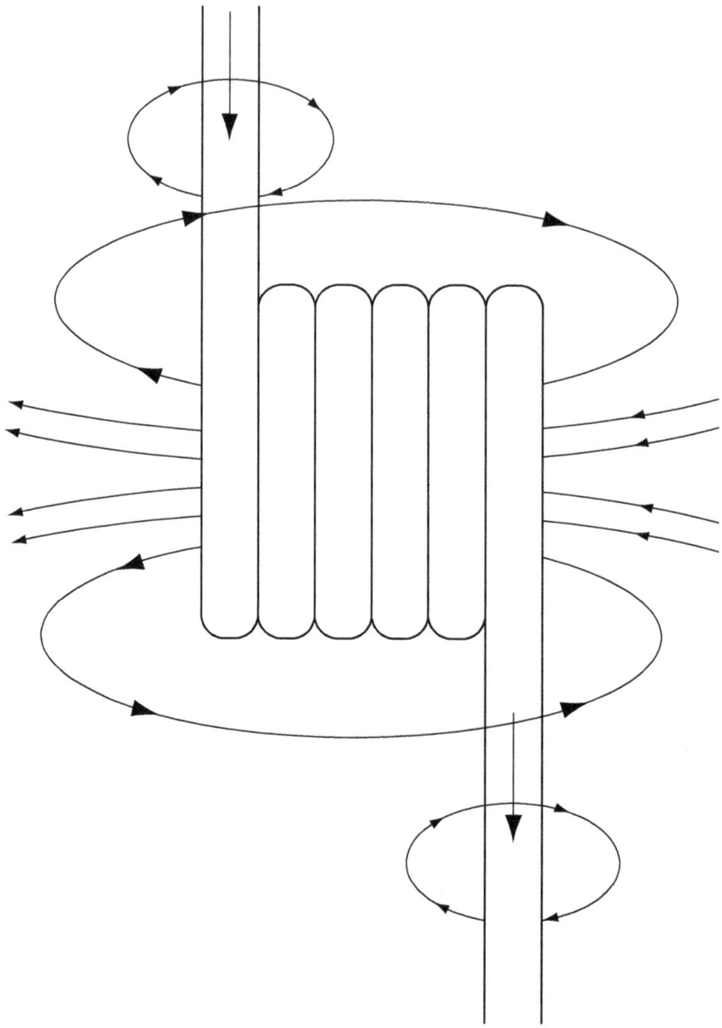

**Figure 5-13** Current in a multiturn coil and combined flux lines produced.

Some of the magnetic lines of force then find that a path through air or across the discontinuity presents a lower reluctance than the remaining metal. As a result, some flux lines "break out" of the surface of the metal into the air. This is called a *leakage field*. It is interesting to note that a leakage field may exist both at the near surface and also at a remote or hidden surface.

In order to create a leakage field, the discontinuity must interrupt the lines of force in the material. A narrow discontinuity oriented parallel to the flux lines will not create a leakage field. In order to produce a leakage field, a discontinuity must interrupt the field usually considered to be within 45° to the perpendicular. It follows that in order to detect a discontinuity with any orientation, the part must be magnetized in at least two directions, 90° to one another.

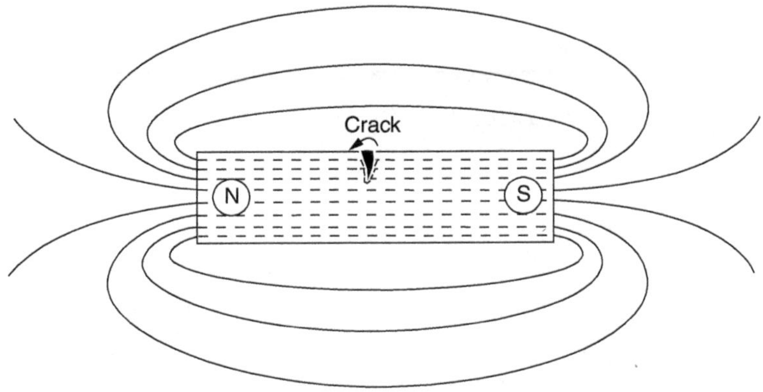


**FIGURE 5-14** Flux distortion and leakage field at a crack.

### *Making the Leakage Field Visible*

When a flux leakage field occurs, a north and south pole will be created at that location (see Fig. 5-14). It has already been established that the maximum flux density will occur at the poles. Therefore, whenever a discontinuity disrupts the flux lines and flux leakage occurs, an area of high flux density will be produced. When ferromagnetic particles are applied to the surface of the test object, they are strongly attracted to this flux leakage area and will form an accumulation of particles, producing a visible indication of that discontinuity (see Fig. 5-15). The creation of an indication relies on the flux lines being distorted sufficiently to produce flux leakage.

### *Part Geometry*

Some aspects of the part's geometry may also produce flux leakage. For example, threads, keyways, or other abrupt changes in section thickness will cause flux leakage and produce indications referred to as "nonrelevant" (see Fig. 5-16).

### *Control of Magnetization*

The dimensions of the discontinuity in relation to the component thickness, together with the strength of the applied magnetizing force, will determine the strength of the

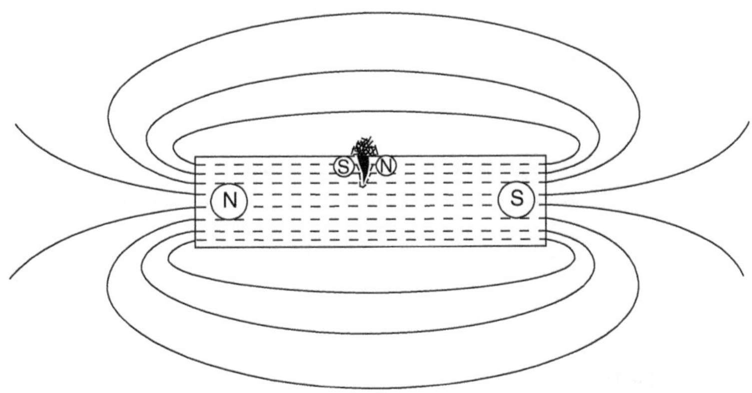


**FIGURE 5-15** Particles applied.

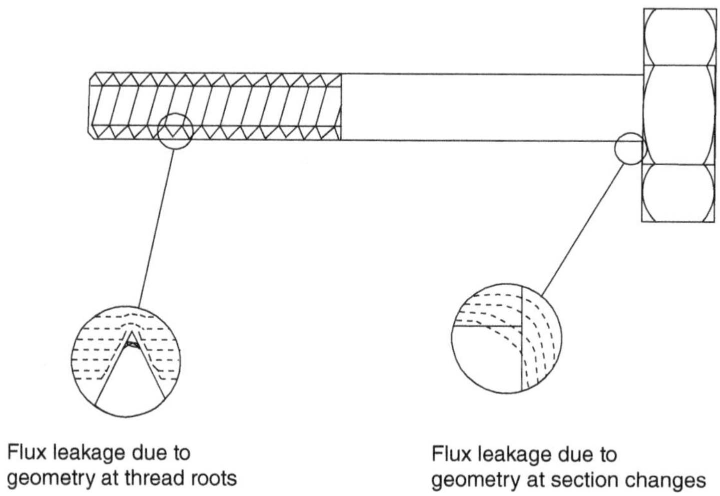


**Figure 5-16** Flux leakage at threads and geometric changes.

leakage field. When the component has a complex geometry, no one value of magnetizing force will be suitable if nonrelevant indications from section changes are to be avoided. It is therefore necessary to control the magnetizing force to suit the geometry of the part. With electromagnetism, there is far greater control than with permanent magnets. By varying the applied current, the magnetic flux density produced can be controlled to minimize nonrelevant indications while still complying with inspection requirements.

## 5.2.4 Types of Electrical Current

### *Direct Current (DC)*

Current produced from a battery is constant in amplitude and direction and is called "direct current" or DC. A graphical representation of DC field is illustrated in Fig. 5-17.

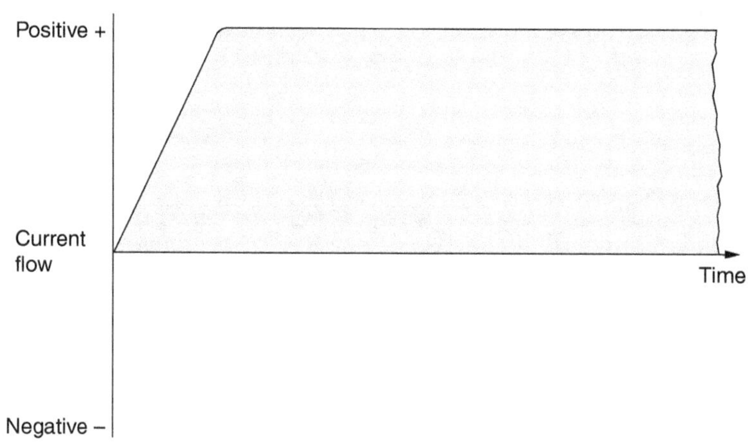


**Figure 5-17** Graphical display of direct current waveform.

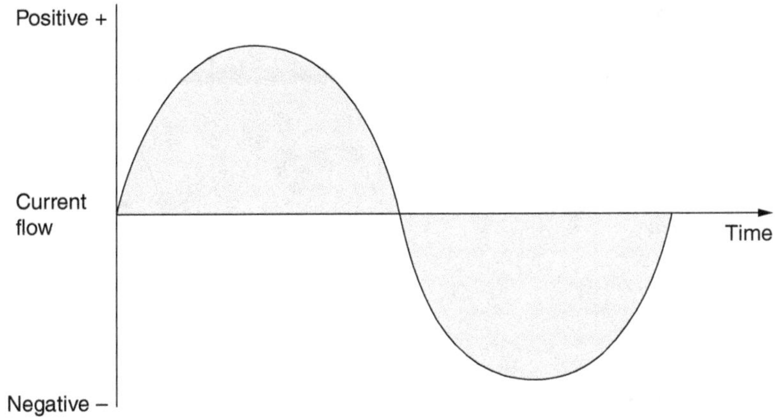


**Figure 5-18** Graphical display of alternating current waveform.

### Alternating Current (AC)

Current provided from a conventional electrical outlet constantly varies in amplitude and also reverses in flow direction (polarity) many times a second. It begins at zero and flows in one direction, rising from zero to maximum flow, then decreases back to zero before reversing in flow direction, rising from zero to a maximum flow in the opposite direction before falling back to zero once again to complete a full cycle. The actual number of cycles per second is called *frequency*. The frequency of AC power in the United States is 60 cycles per second or 60 hertz (1 hertz = 1 cycle per second). A graphical representation of this cycle is shown in Fig. 5-18. This type of current is referred to as alternating current because it is continuously alternating in direction.

### Half-Wave Rectification

This alternating current can be modified in several ways. Its frequency can be altered by using an oscillating circuit (although this is seldom done in MT) or the "negative" portion of the current flow can be removed to allow the current to flow only in one direction. This is called *half-wave rectified alternating current* (HWAC), and its effects on the waveform can be seen in Fig. 5-19. Obviously, if half of the electrical current is removed, this will reduce the magnetic flux produced.

### Full-Wave Rectification

One other alternative is to reverse the polarity of the negative portion of the wave to make it also positive. This is called *full-wave rectified alternating current* (FWAC), and it causes little or no loss of the original AC wave's power (see Fig. 5-20).

### Electrical Power

The power developed in an electrical circuit is measured in watts (W) and is the product of the applied voltage ($V$) and the resulting current ($I$):

$$\text{Watts} = V \times I$$

One of the by-products of the current developed is heat, which may damage metals if the current is applied for an extended period of time. At the same time, the flux

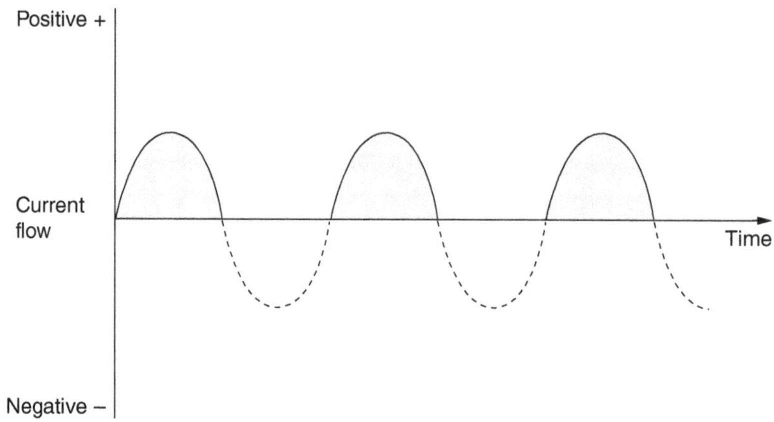


**FIGURE 5-19** Graphical display of half-wave rectified (HWAC) current.

density achieved with electromagnetism is proportional to the current flowing. These two factors need to be balanced when selecting the magnetizing parameters for MT. For safety reasons, most MT equipment operates at a low voltage (typically 3 V), and so the current is the greater contributor to the power in the circuit.

The electrical current amplitude is measured in amperes (amps). With DC, the amplitude is constant; however, with any of the sinusoidal wave forms produced by AC, half-wave or full-wave rectified, the "peak" amperage produced at the maximum flow rate at the peaks of the waves will only apply for an instant. The average current includes the zeros, the peaks, and all points in between. The average value for an AC wave or full-wave rectified AC will be approximately 70 percent of the peak value. This value is referred to as the *root mean square* (RMS). It can be seen that the power developed at 1,000 amp DC will be higher than at 1,000 amp AC. With half-wave rectified AC, the difference becomes much more pronounced because half of the AC wave is not used; therefore, the actual output will only be approximately 35 percent of that produced by DC.

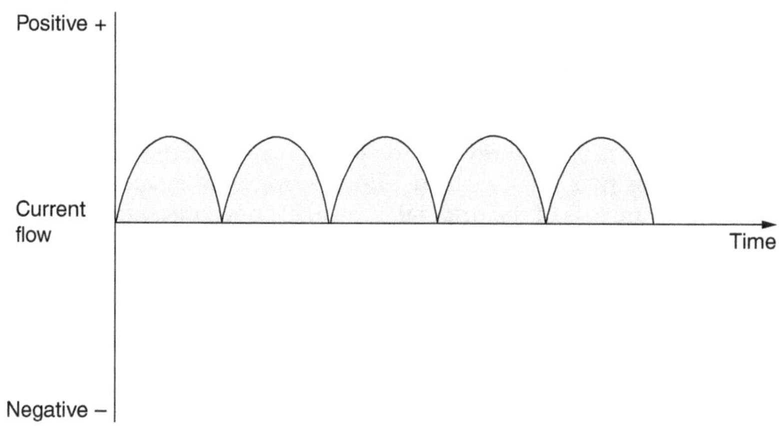


**FIGURE 5-20** Graphical display of full-wave rectified (FWAC) current.

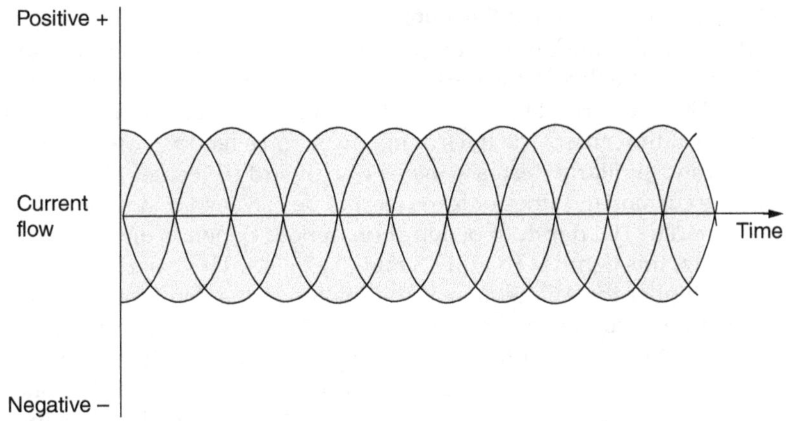


**FIGURE 5-21** Graphical display of three-phase wave AC.

Care should be exercised when setting amperage values based on meter readings. Some meters indicate peak values, and some will indicate RMS values.

With three-phase current, when the AC wave is one-third into its cycle, a second wave is generated; and when the second wave is one-third into its cycle, a third wave is generated. The result is three AC waves being generated simultaneously, with each one overlapping the others, thus providing more current flow and a smoother distribution. See Fig. 5-21.

Early versions of MT equipment required banks of batteries to produce pure direct current. These were cumbersome and long charge cycles to produce relatively few "shots" before recharging again were necessary. Equipment with DC capability relies on modifying AC current into DC by full-wave rectification. This is further enhanced by using smoothing capacitors to slow down the rate of decay of the peaks and help to "fill in" the troughs. If three-phase current is used with full-wave rectification, this smoothing of the waveform and current flow is improved to a point where almost pure DC results (see Fig. 5-22).

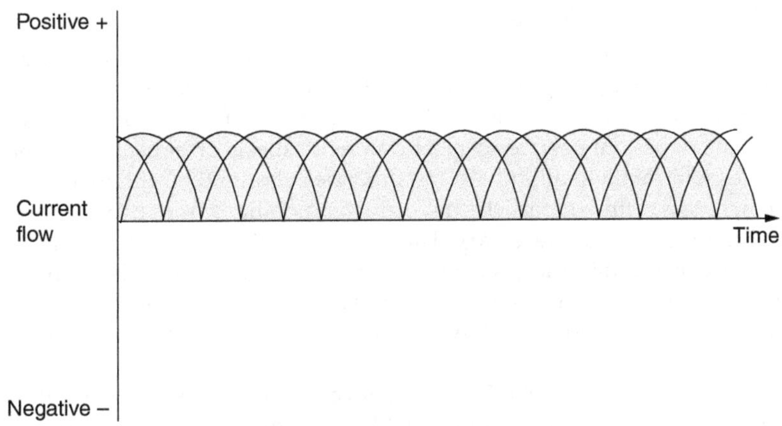


**FIGURE 5-22** Graphical display of three-phase full-wave rectified AC.

#### *The Influence of AC and DC Currents*

In addition to the amount of flux produced, the different currents used will affect the distribution of the flux in the part.

With DC, the current tends to use the entire cross section of the conductor, decreasing to zero at the center. The lines of magnetic force tend to spread through the material thickness and similarly decrease in density toward the center. With any of the AC wave forms, the current and lines of force tend to be confined to a region close to the surface of the material. The depth of penetration of both depends on the frequency of the AC. For 60 hertz, this depth is roughly 3 mm (0.125 in.). This phenomenon is called the *skin effect*. The higher the AC frequency, the shallower the depth of penetration.

This difference in flux distribution can be used to an advantage. If it is desired to detect discontinuities slightly deeper in a part, then DC will result in a greater penetration of the flux. If the goal is to detect surface discontinuities, the concentration of the flux at the surface due to the skin effect will enhance detection. It is for this reason that DC is preferred for the detection of manufacturing induced discontinuities, which may occur anywhere in the part. The deeper the discontinuity is below the test surface, the broader and less distinct the indication becomes. To enhance the contrast it is necessary to use dry particles, since the particles are larger than those used in a suspension. The larger particles are also less mobile and more reluctant to gather at the region of the flux leakage, so that some form of external agitation may be required (blowing on the surface or tapping the component). Also, the amperage required to achieve sufficient flux density using DC will be higher than for AC and this will increase the heating effect. Before deciding to use DC current for deeper penetration, it may be worth considering whether another NDT method would be more suitable.

AC current is preferred for the detection of surface and near-surface discontinuities, particularly during in-service inspection, where discontinuities, by nature, tend to be surface-breaking (e.g., fatigue cracks). For a given current, the flux density near the surface is higher when using AC and this means that lower currents than those required for DC can be used. This, in turn, produces less heat. Another advantage is that an AC field, by its cyclic nature, tends to vibrate the particles, thereby increasing their movement toward the area of flux leakage. Most users regard magnetic particle testing as a method for the detection of surface and near-surface discontinuities only.

### 5.2.5 Detection Media

The detection media used for MT are basically similar to the ferromagnetic particles used in producing a magnetograph; however, they are of a higher level of refinement. Their size and shape are carefully controlled to produce a variety of different sizes and shapes. Different shapes are desired to meet different aspects of indication formation. Elongated particles will rotate to align with the flux lines, due to their increased length-to-diameter ratio, whereas rounded particles have greater mobility to move to the areas of flux leakage.

Different sizes are necessary. The smaller particles have the ability to accumulate at small discontinuities and produce a sharper indication. This is often useful in identifying the nature of the discontinuity. For example, a crack indication will have a jagged appearance. Larger particles have the ability to bridge across larger discontinuities and increase contrast.

Particles may be applied in a dry form or may be suspended in a liquid such as kerosene or a similar petroleum distillate. They may also be suspended in a water carrier containing suitable additives such as wetting agents and antifoam liquids.

The particles used in the wet suspensions are usually iron oxides as opposed to iron filings, due to their lighter weight and greater ability to remain in suspension. The disadvantage of using iron oxides is the reduced permeability of the oxides compared to iron filings. This results in a lesser response to the weaker leakage fields produced by subsurface discontinuities. A distinct advantage of using suspended particles is their greater mobility in the liquid, which improves the detection of small surface discontinuities. Whenever the wet suspension is used, it is essential that periodic checks be made to determine the particle concentration in the liquid carrier. This is referred to as the "settling test" and will be discussed in detail at the end of Section 5.4.

The visual impact of indications is also assisted by the addition of a coating to the particles to provide a greater contrast with the test surface. This results in a significant increase in "seeability" or sensitivity. The coatings fall into two distinct groups, color contrast and fluorescent.

Color contrast coatings are available in several colors, including blue, red, gray, and black. See Fig. 5-23 for examples of dry particles. The choice of color depends upon the color of the test surface and the degree of contrast to be provided. If suitable contrast cannot be achieved, another possibility is to coat the part with a specially formulated white contrast paint, which is quick drying and easily removed after inspection by using a solvent. This will create a good contrast between the black particles in a liquid carrier and the white background (see Fig. 5-24).

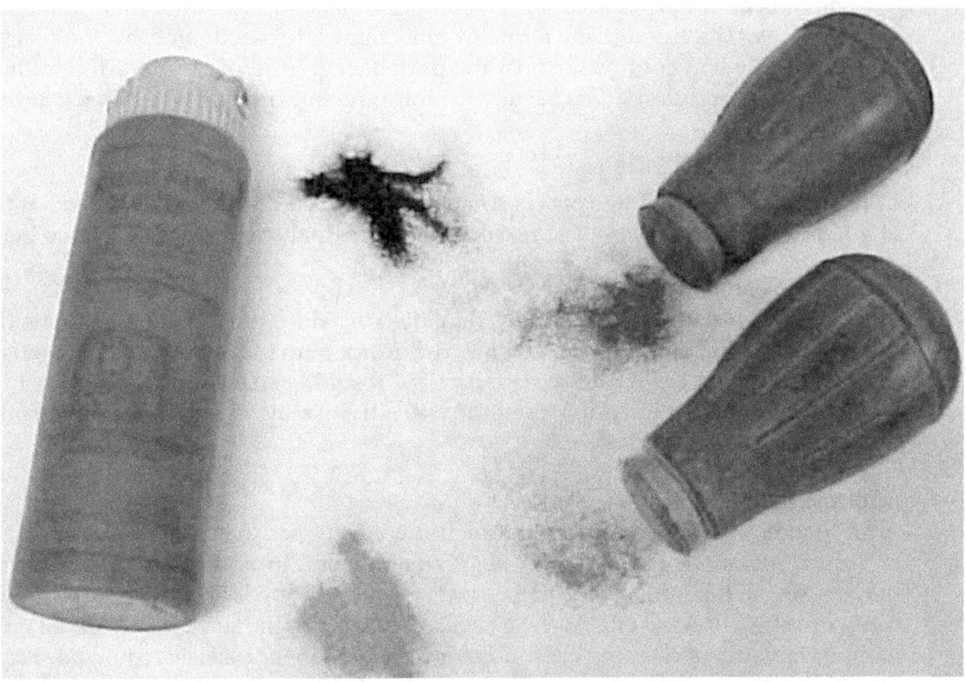

**FIGURE 5-23** Examples of dry particles.

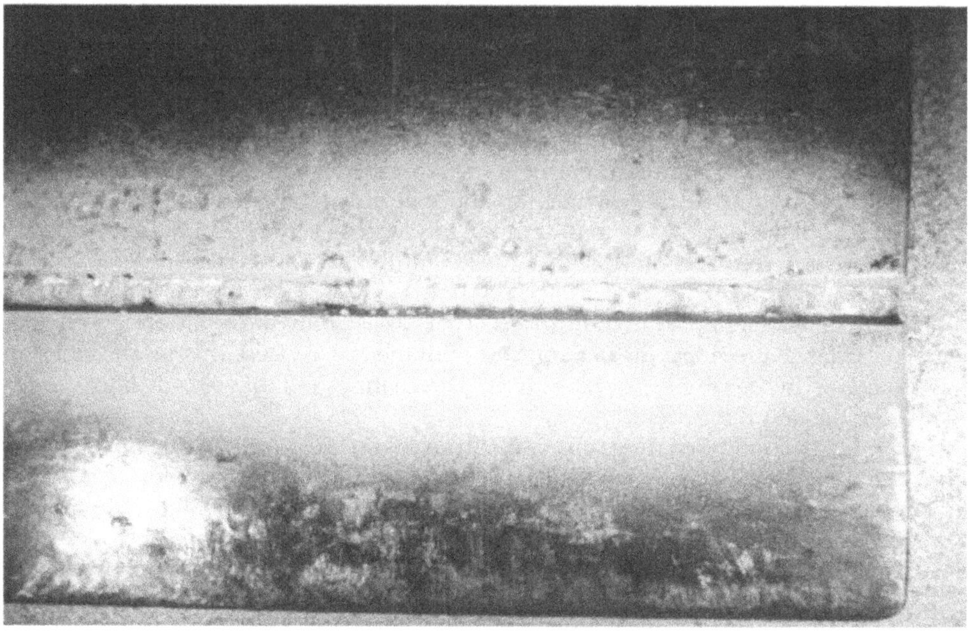

FIGURE 5-24 Weld with white contrast paint.

The fluorescent particles are viewed under black light illumination in a darkened viewing area. The fluorescent particles emit light when activated by the black light, providing excellent contrast against the dark background of the part due to the darkened viewing conditions. (See Section 5.3 for more information about black lights.)

## 5.2.6 Demagnetization

Demagnetization of the test part is sometimes necessary before, during, and at the end of an inspection, for a variety of reasons. The principal reasons are described below.

### *Before and During the Test*

If the test part has been left with a residual field in a different direction to the field about to be applied, the resulting field will be the vector sum of the residual magnetism and the newly applied field. The direction of the resulting field will therefore not be precisely in the direction intended; therefore, sensitivity may be reduced. Demagnetization is recommended in this case.

### *After the Test*

A residual field in a component being returned to service may have a detrimental effect on future operations of the part or nearby equipment. In an aircraft or ship, a compass may be affected. High-speed rotating parts may induce eddy currents in nearby structures, causing a braking effect on the part and heating of the surface. A residual field may hamper subsequent cleaning operations and ferromagnetic particles could be attracted to the object, causing premature wear. If the part is to be welded after inspection, the residual field may deflect the weld arc and make it difficult to control the weld deposit.

#### Demagnetization Process

In order to effectively demagnetize a part, two requirements must be met:

1. The polarity must be successively reversed.
2. The field strength (flux density) must be decreased.

This means that each successive current application will be in the opposite direction to the previously applied force and will be at a lower level. It should be below the saturation point, but must be above that required to produce a higher flux field than retained. This will produce successively lower residual fields at each application until an acceptably low field exists. It should be noted that although total demagnetization will not result, the field can be brought to an acceptable level.

These two requirements can be met using a reversing DC field; however, AC is preferred, due to its inherent reversing nature. The most satisfactory method of demagnetization using AC is to gradually reduce the applied current to zero or, more commonly, to pass the test part through an AC demagnetizer. See Fig. 5-45. After demagnetization, a field indicator similar to the one in Fig. 5-42 is used to establish the effectiveness of the demagnetization. Many codes and specifications require that no more than 2 Gauss remain in the part after demagnetization.

One way to completely demagnetize is by heating the part to a temperature above its curie point and allowing it to cool without any magnetizing force acting upon it. If this can be accomplished, the residual field will be totally removed and the part will be totally demagnetized. The curie points for steels range from 720°C (1,296°F) to 800°C (1,440°F). For most NDT applications, however, this approach is not practical.

## 5.3 Equipment and Accessories

Different MT equipment is available for production line and field-type applications. To facilitate the varied applications, several basic groups of test equipment are available. They are classified as "stationary," "mobile," or "portable" units.

### 5.3.1 Stationary Units

A stationary unit is referred to as a wet horizontal unit and usually has capabilities for producing longitudinal and circular fields. Some units will also be capable of demagnetizing the parts, although this is usually accomplished with a separate demagnetizer. Figure 5-25 illustrates a typical wet horizontal stationary unit. This unit has a fixed headstock and a sliding tailstock. The part is placed between these and the tailstock is adjusted to just slightly longer than the length of the part. When the tailstock is locked in this position, a foot-switch controlled pneumatic headstock closes and grips the part tightly to permit the current to flow, thereby producing a circular magnetic field. Figure 5-26 illustrates the inspection of a gear shaft using the direct magnetization circular wet fluorescent technique.

If hollow parts such as tubular products or ring-shaped parts are to be inspected, a central conductor technique can be used. This involves the use of a bar of high electrical conductivity, typically copper or aluminum, that is gripped between the headstock and tail-stock. When current is passed through the conductor bar, a circular magnetic field will be induced in the component. It should be noted that the flux density in the part

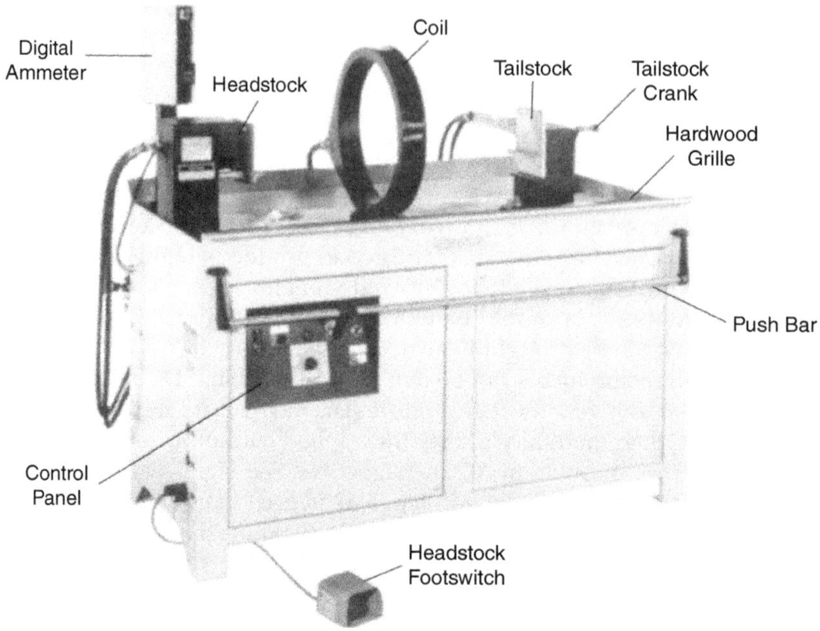


**Figure 5-25** Wet horizontal unit.

will be highest at the inner surface. Advantages of this technique are that several components may be inspected simultaneously and there is no risk of overheating the parts or arcing between the unit and the part(s) (see Fig. 5-39, central conductor/threader bar), since this is an indirect magnetization technique.

For longitudinal magnetization techniques using coils (as in Fig. 5-25), the part can either remain in this position if a centralized coil position is used or it can be placed on

**Figure 5-26** Wet horizontal system inspection of a gear shaft using wet fluorescent particles.

the lower inside surface of the coil. The ratio of coil inner diameter to component outer diameter, or cross-sectional area ratios, will determine which position should be used. This will vary based on specification requirements.

The amperage can be adjusted using a variable self-regulating current control. An adjustable timer controls the current duration. Demagnetization is accomplished using either the coil in the wet horizontal unit or a separate AC coil demagnetizer. The stationary units are very versatile in their range of capabilities; however, the main limitation is their inability to be taken to a field location. For this, "mobile" or "portable" equipment must be used.

### 5.3.2 Mobile and Portable Units

The main difference between mobile and portable equipment is in their ability to be moved. A mobile unit must be transported to a location, and then moved around on its own wheels or castors (see Fig. 5-27). It can weigh up to 1,000 lb. (500 kg). A portable unit can usually be carried by one person and will probably weigh no more than 50 lb (25 kg) (see Fig. 5-28).

Despite their size difference, both of these types of units can accomplish similar inspections with the use of various accessories. The power outputs of these units vary considerably. As expected, the mobile unit will have a much higher output than the portable unit. Some of the mobile units can produce as much as 6,000 amps, whereas the portable units are usually limited to a maximum of 1,500 amps.

Common accessories include clamps and prods for producing circular magnetization. Great care must be exercised when using prods or clamps to ensure that good contact is maintained with the test specimen, otherwise electrical arcing can result. This can produce localized areas of surface damage and potential stress risers in the part.

Cable-wrapped coils and rigid coils are used with the mobile and portable units for producing longitudinal magnetization. Both mobile and portable units usually have selectable AC or DC positions, which also enables demagnetization to be accomplished.

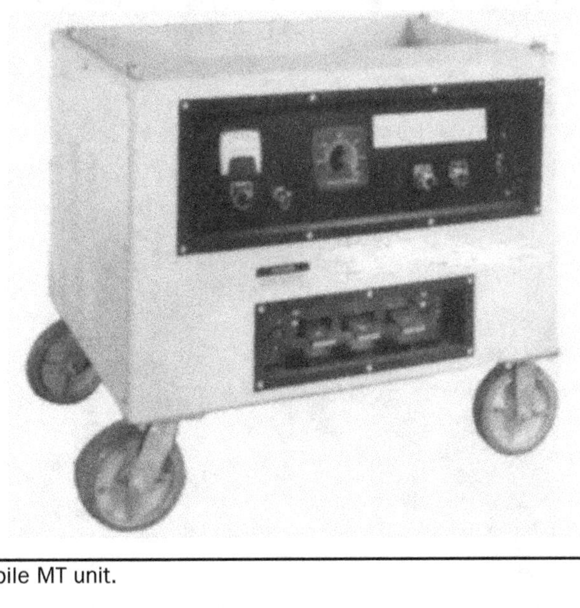

**FIGURE 5-27** Mobile MT unit.

FIGURE 5-28 Portable DC prod unit.

### 5.3.3 Electromagnetic Yoke

Even greater portability can be achieved by using an electromagnetic yoke, which is referred to as an AC yoke or contour probe (see Fig. 5-29). This unit can be used with AC and is also available in a battery pack version, which further increases its portability by eliminating the need for an AC power source. Many yokes have articulating legs to

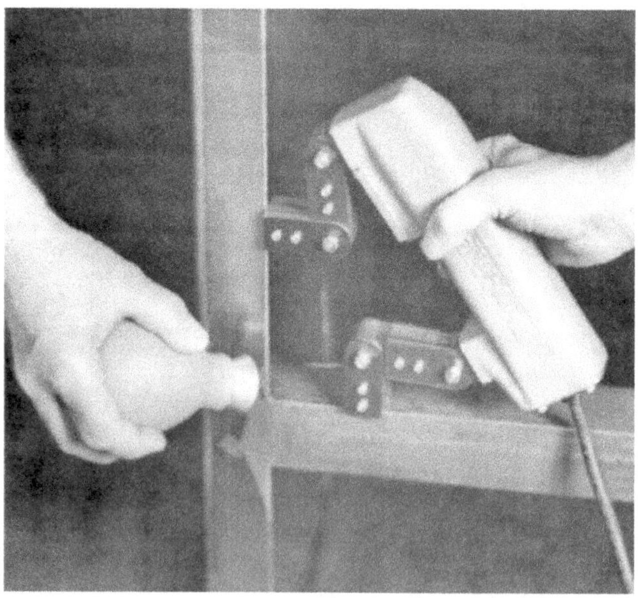

FIGURE 5-29 AC yoke.

facilitate various inspection area profiles. These yokes produce only longitudinal magnetization; repositioning is required to achieve flux line orientation in at least two 90° opposing directions.

### 5.3.4 Permanent Magnets

Small permanent magnets can also be used to produce localized longitudinal magnetization, but there are some limitations to their use (see Fig. 5-30). Due to their strong fields, particles tend to become attracted to the legs more readily than the test surface. At times, the use of these magnets can become unwieldy.

### 5.3.5 Black Lights

The use of fluorescent particles requires a black light (see Fig. 5-31). One type of black light contains a regulated current ballast transformer to limit the current drawn by the arc in the mercury vapor bulb. The light produced may also contain some white light and harmful UV radiation. It is therefore essential that a correct black light filter be used. This filter will allow the relatively harmless portion of the ultraviolet spectrum to pass through with a center wavelength of 365 nanometers (nm). The condition of the filter must be regularly checked to ensure that white light or harmful UV is not present.

### 5.3.6 Flux Direction Indicators

Flux direction can be determined by using one of several different indicators, as detailed below.

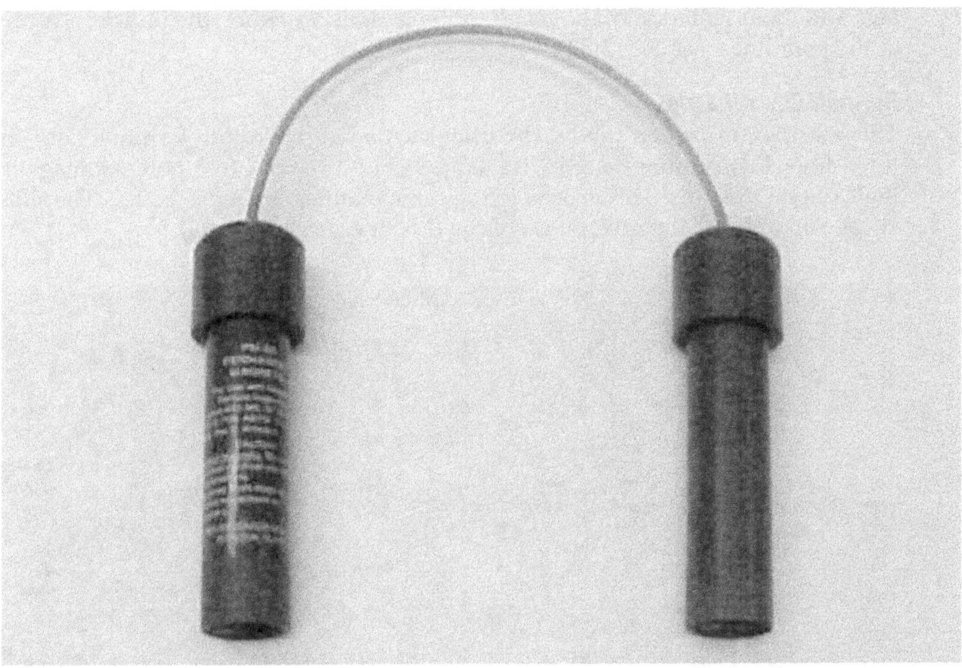

**FIGURE 5-30** Permanent magnets.

FIGURE 5-31 Black light.

### Pie Gauge

This is one of the most commonly used devices in MT. It consists of eight low-carbon steel segments, furnace brazed together, to form an octagonal flat plate that is then copper plated on one side to hide the joint lines. The gauge is placed on the test specimen during magnetization, with its copper face up. The particles are applied to this face and the orientation of the resultant field is displayed by the indications produced at the joint lines (see Fig. 5-32).

### Burmah Castrol Strips

These strips consist of a sandwich structure made up of a thin ferromagnetic foil with three fine slots running lengthwise sandwiched between two nonferromagnetic foils that conceal the slots. This is used in the same manner as the pie gauge. The slots in the strips will indicate the orientation of the flux lines (see Fig. 5-33).

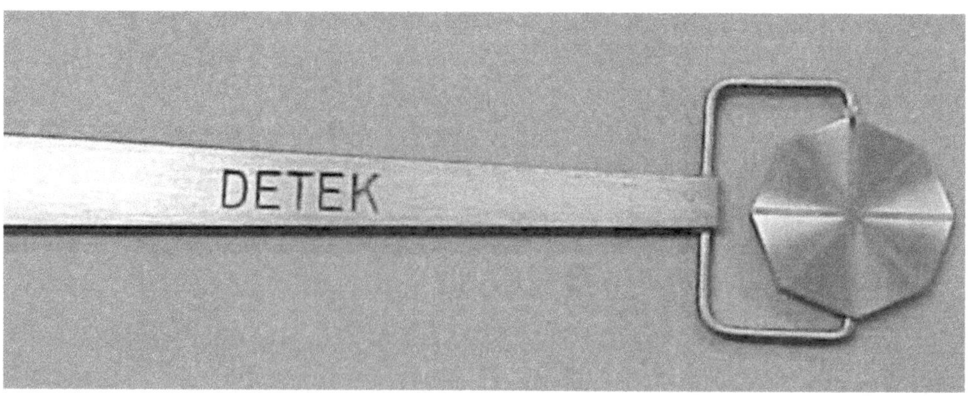


FIGURE 5-32 Pie gauge.

**FIGURE 5-33** Castrol magnetic flux indicators.

#### *Quantitative Quality Indicator (QQI)*

This is a recent development of the foil concept and consists of a thin metallic foil, but instead of having slots in only one direction; the slots form circles or crosses, thereby making them able to indicate flux line direction without repositioning the indicators. The portion of the circle or the leg(s) of a cross determine which flaw orientation should be detected (see Fig. 5-34).

### 5.3.7 AC Demagnetizing Coils

Separate coils are used when the demagnetization cannot be accomplished on the MT equipment. Many portable and mobile units have provisions for stepped reduction in current. The demagnetizing coil is often operated at a fixed current value and the reduction in field strength is achieved by passing the part(s) through the coil and beyond the coil along the central axis for a distance typically two to three times the length of the part(s).

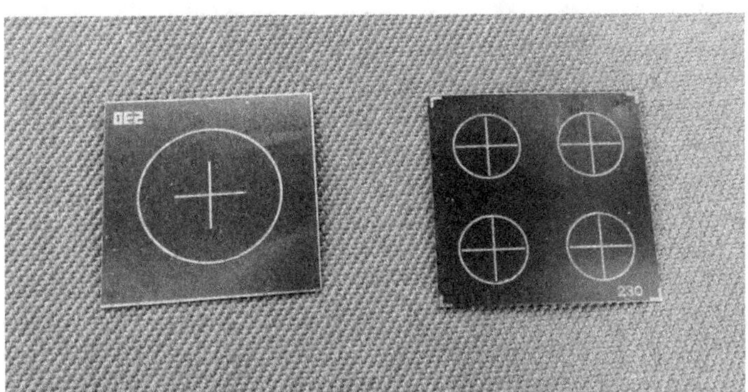


**FIGURE 5-34** Quantitative quality indicators.

This is effective for removing most of the residual fields because the residual vector of each cycle rapidly swings around to longitudinal. Thereafter, as the distance between the coil and the part increases, the field reduces until it is close to zero.

## 5.4 Techniques

Magnetic particle testing techniques can be categorized into several key groups. They are explored here.

### 5.4.1 Magnetization Techniques

There are several factors that will dictate which magnetization technique is chosen for a particular test. The primary factor is the probable direction of anticipated discontinuities. The direction can usually be determined for fatigue cracks by studying the component loading and stress points, but in most other cases, the direction can be entirely unpredictable. Whether the direction is known or not, the component must be inspected using two field directions perpendicular to one another. This can be achieved using one or more of the magnetizing techniques described here.

The next factor in the order of importance is the strength of the field to be applied, which will be discussed in Section 5.5, Variables. The remaining factors to be considered include the end use of the part, the location of the part (in a factory, at a remote site, or in situ in a structure), and the susceptibility of the part to cracking due to heating or arc burns.

### 5.4.2 Magnetic Flow Techniques (Indirect)

The "indirect" or magnetic flow technique is defined as one in which the test part becomes part of the magnetic circuit by bridging the path between the poles of a permanent magnet or electromagnet. Figure 5-35a illustrates permanent magnets used in this way. The magnets can be turned 90°, as shown in Fig. 5-35b.

The AC yoke is an example of an electromagnet that is used in much the same way. They can also be used as magnetic flow devices, as shown in Fig. 5-36a and b. The main advantage of the indirect technique is that the risk of arc burning of critical components does not exist. Also, the use of permanent magnets or yokes can be very convenient for in situ inspections in confined spaces or remote locations.

### 5.4.3 Current Flow Techniques (Direct)

The "direct" technique is defined as one in which the magnetizing current flows through the part, thereby completing the electrical circuit. This is accomplished by placing the part between the "heads" of a stationary unit such as the one shown in Fig. 5-25 or by using "prods," as shown in Fig. 5-28.

The magnetic field formed with the direct technique is at right angles to the current direction of the current. In the case of a round bar held between the heads in a wet horizontal unit, the magnetic lines of force circulate around the bar as the current flows from one end to the other. This technique produces "circular magnetism." The flux line pattern in a flat plate when prods are used is illustrated in Fig. 5-37.

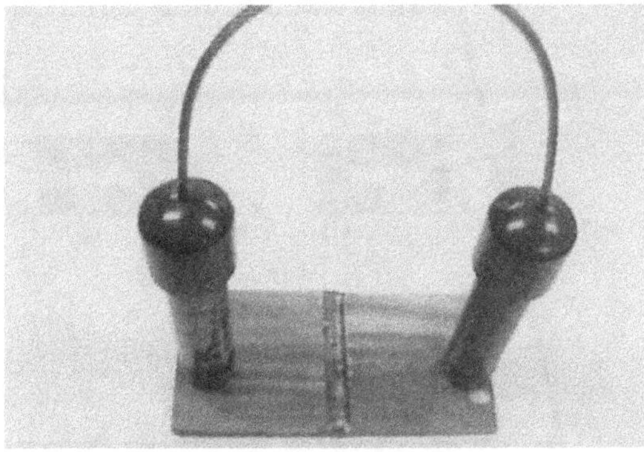

(a)

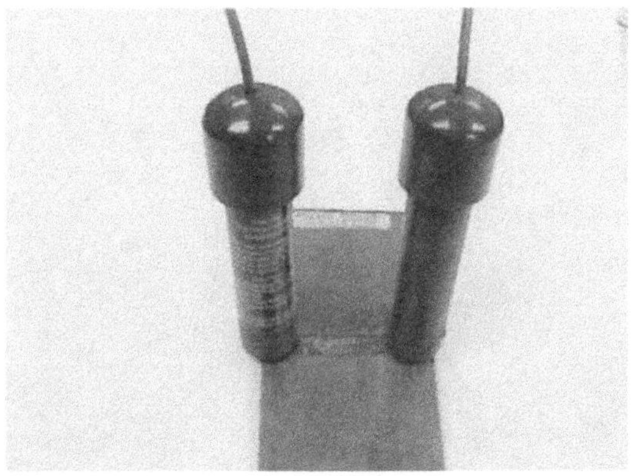

(b)

**Figure 5-35** (*a*) Magnetic flow technique. (*b*) Magnetic flow technique at 90°.

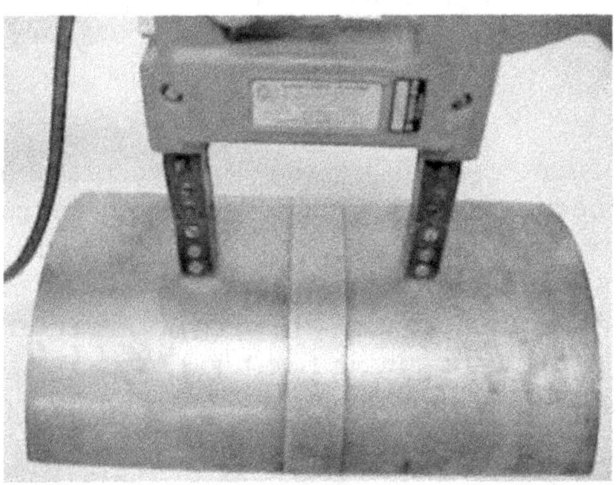

(a)

(b)

**Figure 5-36** (*a*) Electromagnet. (*b*) Electromagnet at 90°.

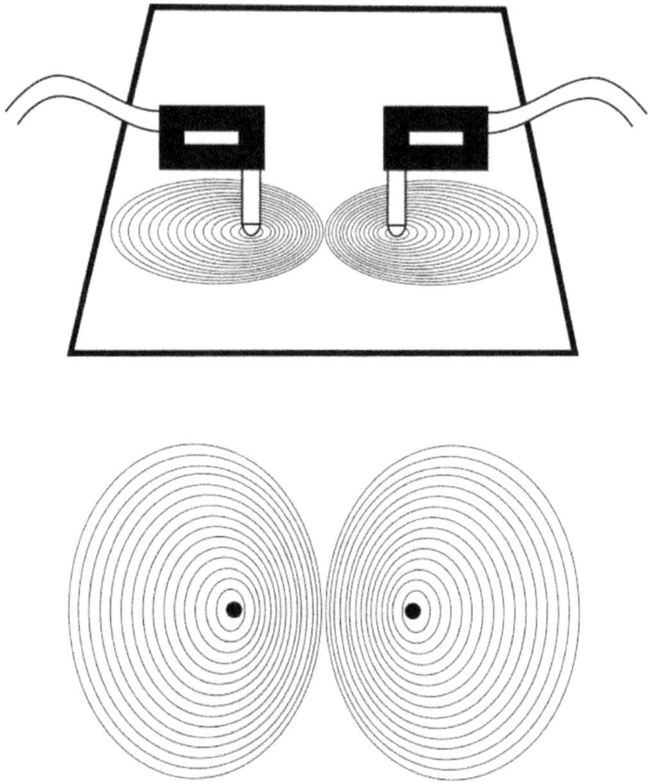

FIGURE 5-37 Sketch of flux lines—DC prod.

One of the dangers in using the direct technique is that contact head or prod can cause a burn in the part if the high current is passing through a small contact area. For this reason, the contact faces on the heads should be flexible, and the tips of the prods should have a low melting point in order to spread the thermal load.

### 5.4.4 Coil Technique

Figure 5-38 shows a cylindrical component inserted into the windings of a coil. This produces longitudinal magnetism in the part as the current flows through the coil and creates a longitudinal field through and around the coil. This is referred to as the "longitudinal magnetization" technique. It has the advantage of not causing thermal damage, since a current does not pass through the part (indirect). In wet horizontal units, it is usual to have a fixed coil, usually containing five turns; with mobile and portable units, it is more usual to wrap the current-carrying cable around the part to form a coil.

### 5.4.5 Central Conductor Technique

"Central" conductor is sometimes a misnomer because the opening through which the nonmagnetic conductor is positioned is not always central. The technique is sometimes

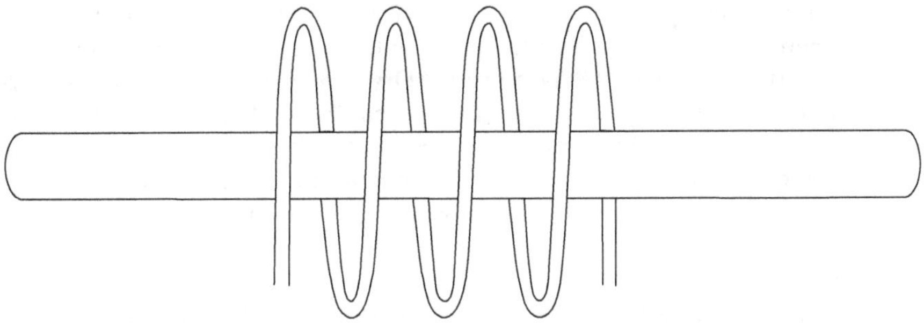

**FIGURE 5-38** Coil technique.

referred to as the "threader bar" technique to show the distinction. Figure 5-39 is a sketch of a tubular component being tested using this technique. Like the coil technique, it is also indirect because the current passes through the conductor and not the part. This technique produces circular magnetization and can be used to inspect the insider and outside surfaces of the part.

### 5.4.6 Continuous and Residual Techniques

There are four techniques that can be used to inspect parts using these methods of magnetization. They are:

1. Dry continuous
2. Dry residual
3. Wet continuous
4. Wet residual

There are advantages and limitations associated with each technique and careful selection must be made depending upon the application. The term "continuous" is used

**FIGURE 5-39** Central conductor/threader bar.

when the magnetic particles are applied while the current is still flowing. "Residual" is used when the material has sufficient retentivity to allow application of the magnetic particles after the current has ceased. "Dry" means that the magnetic particles are applied in fine particle form. "Wet" means that the particles are applied suspended in a liquid carrier. The particles may be suspended in kerosene or other similar petroleum distillates or may be suspended in water containing specially formulated additives. A brief description of the key parameters of each technique follows.

### Dry Continuous

This technique uses dry particles that are applied while the magnetizing force is on. The particle application must cease before the current flow ceases. The use of dry particles is useful for detecting slightly subsurface discontinuities, since the particles have higher permeability compared to the particles in a wet suspension. This higher permeability is needed to detect the weaker leakage fields produced by slightly subsurface discontinuities. Applying the particles during the magnetization provides maximum sensitivity, since the flux density will always be at its maximum during the current flow.

One disadvantage of dry particles is their relatively poor mobility when used with DC current. Mobility can be improved by gently tapping the test part or by using half-wave rectified or AC current. The pulsed nature of the half-wave rectified and AC current creates a vibratory effect on the particles, causing them to "dance" toward the leak age fields. This technique is used extensively for field inspections using prods or yokes. Care must be exercised when removing the excess particles from the test surface to minimize any effect on the particles held by discontinuities. In this respect, the current should also be flowing while the excess particles are being removed with low air pressure.

### Dry Residual

This technique also uses dry particles but differs from the previous technique in that the magnetizing current is applied for a brief period, which produces a residual magnetic field in the part (DC is preferred). The dry particles are applied after the magnetizing force has ceased. The retentivity of the material produces a residual magnetic field, which results in leakage fields at discontinuities open to the surface. This technique is suitable only with materials having high retentivity. This must be determined prior to selecting this technique.

There are further disadvantages when compared to the continuous techniques. The leakage fields produced will be weaker, thereby reducing the sensitivity. Small and subsurface discontinuities are difficult to detect. The lack of the magnetizing current's vibratory effect during particle application will also reduce sensitivity due to the decrease in particle mobility.

The main advantage of this technique is that multiple parts or batches can be magnetized simultaneously, and then examined individually after particle applications. This technique would only be used when a lower sensitivity of high-retentivity parts is acceptable.

### Wet Continuous

This technique uses particles suspended in a liquid carrier and is applied simultaneously with the magnetizing current. Compared to the dry particles, the suspended particles are generally of a lower permeability, which makes this technique less favorable for the detection of slightly subsurface discontinuities.

There are advantages and limitations of the different liquid carriers. Kerosene and petroleum distillates are more expensive and may produce some health and flammability problems, but they also help to lubricate the sliding mechanisms on the wet horizontal unit and do not constitute a corrosion source for the equipment or the parts being inspected. Water, on the other hand, is inexpensive, readily available, poses no health or flammability hazards, but is a source of corrosion. Using corrosion-inhibiting additives can reduce the corrosion risk, although it can never be completely eliminated.

The greatly improved mobility of the particles makes this technique very suitable for the detection of small surface discontinuities. Additionally, the suspension will adhere to complex shapes better than dry particles, due to the surface tension combating the effects of gravity. These advantages combined with the high flux densities of the continuous technique provide the maximum sensitivity for surface discontinuities. When using this technique, the suspension flow should cease before the current stops, otherwise indications may be washed away.

#### *Wet Residual*

This technique also uses suspended particles, but the suspension is applied after the magnetizing force has been stopped. As previously detailed, the residual method will be of a lower sensitivity than the continuous method, but has the advantage of increased inspection speed, due to multiple part or batch inspection possibilities.

#### *Color and Fluorescence*

Each of the previous techniques may be further subdivided into color contrast or fluorescent particles techniques depending on the availability of equipment and materials, type of discontinuity, and sensitivity requirements. Some inspection requirements will mandate the use of fluorescent particles, due to the increased sensitivity provided by the higher contrast ratio.

### 5.4.7 Summary of Technique Choices

Obviously there are many options and choices to consider when developing the most suitable technique for the part to be inspected. In order to provide the optimum inspection approach, some of the key areas to be evaluated include:

- Material type
- Part configuration
- Dimensions
- Surface condition
- Type and location of discontinuities anticipated
- Code and specification requirements
- Customer-specific requirements

Once these are known, the choice of technique can be made. These key items should be considered:

- Type of current (AC, HWDC, FWDC)
- Wet suspension or dry particles

- Visible or fluorescent particles
- Direct or indirect magnetization
- Continuous or residual
- Stationary or portable equipment

The best combination can be determined through qualification by validation through the use of controlled test specimens.

### 5.4.8 Quality Control Considerations

In order to maintain consistency and high level of control over the MT process, it is highly recommended (and in many cases required) that certain checks be completed on the system. Probably the best ongoing quality check is to periodically inspect a controlled test specimen with known discontinuities. The test specimens should be carefully maintained and the periodic checks recorded.

A suspension concentration test, sometimes referred to as a "settling test," should be performed daily. This test is usually performed with a 100 mL ASTM pear-shaped centrifuge tube (see Fig. 5-40). A sample of the suspension (after thorough agitation) is deposited directly into the tube from the hose dispenser and allowed to settle, typically for a period of 30 minutes. To assist in the settlement, the tube may be "run" through the demagnetizer. The purpose of this check is to ensure that the proper concentration of particles is being maintained in the liquid carrier. The applicable specifications will detail the concentration range. Typical ranges are:

For visible particles, 1.0 to 2.4 mL/100 mL

For fluorescent particles, 0.1 to 0.4 mL/100 mL

A settling test tube in a stand is illustrated in Fig. 5-40.

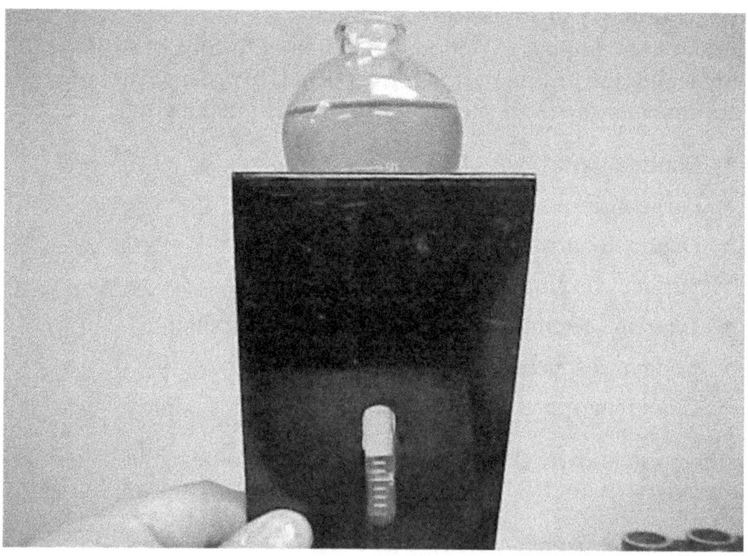

**FIGURE 5-40** Settling test.

**FIGURE 5-41** Ketos ring.

The Ketos ring seen in Fig. 5-41 is a device made of tool steel and is designed to show the effectiveness of the MT field and the relative penetration based on the number of holes that display indications. Both wet suspension and dry particles are used, and general practice calls for the following:

| | Number of Hole Indications | |
|---|---|---|
| **Amperage (FWDC)** | **Wet Suspension** | **Dry Particles** |
| 1,400 | 3 | 4 |
| 2,500 | 5 | 6 |
| 3,400 | 6 | 7 |

As with all the quality checks, it is essential that results of the ring test be recorded.

A final check for the presence of residual magnetism should always be made using a simple field indicator such as the one in Fig. 5-42.

Other QC checks may include:

- Suspension contamination
- Ammeter accuracy
- Timer performance
- Black light intensity
- White light intensity
- Ambient visible light
- Water in oil-based suspension liquid
- Fluorescent particle brilliance

The key to successful MT and control of the system is to periodically perform the appropriate checks and maintain a detailed record of them in a logbook.

**FIGURE 5-42** Field indicator.

## 5.5 Variables

As mentioned in previous sections, there are many variables to consider when deciding upon the best inspection technique for a particular application. A listing of some of the considerations and options follows.

### 5.5.1 Technique Considerations

Customer requirements—specifications, standards, or drawing requirements

Equipment availability—AC/DC, stationary, portable, black light, etc.

AC power availability—needed for all except permanent yokes or magnets

Part size requirements—e.g., will it fit into a wet horizontal unit?

Part location—will it require portable equipment?

Types of discontinuities anticipated—orientation and location

Retentivity of test material—must be high if residual technique is to be used

### 5.5.2 Options

#### *AC or DC (HWDC or FWDC) Magnetizing Current*

Use AC for surface discontinuities or those that would be expected during in-service inspection. Use DC for deeper penetration and possible detection of slightly subsurface discontinuities.

#### *Continuous or Residual*

Use the residual technique for inspection of highly retentive materials, especially where large batches are to be inspected. Use the continuous technique for all other applications, especially where higher sensitivity is required.

### *Wet or Dry Particle Application*

Wet particles have better mobility, but generally lower permeability. Use wet for small surface discontinuities or service-induced discontinuities. Use dry for slightly subsurface discontinuity detection.

### *Color Contrast or Fluorescent Particles*

Color contrast does not require a darkened viewing area or black light illumination; however, the sensitivity will be lower than that of fluorescent particles. Use color contrast for field inspections where highest sensitivity is not required. Use fluorescent where maximum sensitivity is essential.

### *Stationary, Mobile, or Portable Units*

In a production or maintenance facility, the versatility of the wet horizontal unit makes it the ideal choice; however, many applications involving immovable or large test objects eliminate this possibility, resulting in the need for mobile or portable units.

### *Amperage to Be Used*

The amperage to use will be determined based on specification and standard requirements and the test part dimensions.

## 5.5.3 Amperage Selection

Many international standards give tables of current values based on the type of current available in the magnetizing equipment, the sensitivity required, and the component geometry and material. In the absence of a specified standard, the following general "rules of thumb" have been widely used.

**For direct circular magnetization** of basically round sections, use a range of 300 to 800 amps per inch diameter (12 to 32 amps/mm). This is usually 500 amps/inch (20 amps/mm) based on peak value for full-wave rectified current. Values above 500 amps/inch diameter (20 amps/mm) are only recommended for low-permeability materials or for slightly subsurface discontinuity detection.

**Example 5.1** For a 1.5-in.-diameter bar, 9 in. long, 1.5 in. × 500 amps/in. = 750 amps.

**For central conductor circular magnetization** of hollow round components with **conductor bar placed concentrically** with the component, the same current values stated above would apply, based on the outer diameter or inner diameter, depending on which surface is to be examined.

**Example 5.2** Tubular component, 2 in. outside diameter, 1.5 in. inside diameter, using a 1-in.-diameter central conductor (looking for OD discontinuities), 2 in. outside diameter × 500 amps/in. = 1,000 amps.

**For central conductor circular magnetization of hollow round parts, with the conductor bar placed against the inner surface of the component,** the same formula should be used for calculating the direct current amperage. However, instead of basing the calculation on the outer diameter, it should be based on the diameter of the central conductor plus two times the wall thickness of the part. The area that is considered to be effectively magnetized using this technique is a distance around the inner diameter equating to four times the central conductor diameter. The part should be rotated and reinspected as often as necessary to achieve a minimum of a 10 percent overlap between each examination area.

**Example 5.3** Outside diameter 2 in., inside diameter 1.5 in., 1-in.-diameter central conductor, wall thickness 0.25 in.:

Central conductor = 1.0 in.

2 × wall thickness; 2 × 0.25 in. = 0.5 in.

Part diameter = 1.5 in.

Part diameter × 500 amps/in. = 1.5 in. × 500 amps/in. = 750 amps

**For longitudinal magnetization using a rigid coil,** when the part cross-sectional area (cross-sectional area is found by squaring the radius and multiplying by $\pi$) is **less than 10 percent of the coil inner cross-sectional area (low fill factor coil),** the number of "ampere turns" to be used can be found using one of the following formulas:

If part is placed against inside of coil, ampere turns
= 45,000 ÷ (part length) × (part diameter)

If part is placed centrally in the coil, ampere turns
= 43,000 × coil inner radius (6 × length/diameter ratio) – 5

*Note 1.* The length-to-diameter ratio must be in the range of 2 (minimum) to 15 (maximum). Parts shorter than twice the diameter should not be inspected using the coil technique. Parts with a length greater than 15 times the diameter should be tested using a number of "shots" until the entire length is covered.

*Note 2.* The "ampere turns" value must then be divided by the number of turns in the coil to determine the amperage to be used.

**Example 5.4** Inspecting a 1.5-in.-diameter round bar with a total length of 9 in. using a 5-turn coil with an internal diameter of 12 in. Determine if the coil is a low fill factor by comparing area ratios:

Area of part = 0.75 in. (radius) squared × $\pi$ = 1.767 in.$^2$

Area of coil = 6 in. (radius) squared × $\pi$ = 113 in.$^2$

Ratio of areas = 1.767 ÷ 113 = 1.56% (low fill factor)

Amperes to be used if part positioned against inside of coil:

$$= 45{,}000 \div \frac{9}{1.5}$$

$$= 45{,}000 \div 6$$

$$= 7{,}500 \text{ ampere turns}$$

$$\text{Divide by 5 turns} = \frac{7{,}500}{5} = 1{,}500$$

Amperes to be used if part positioned centrally in coil:

$$= \frac{43{,}000 \times \text{coil radius}}{6 \times (\text{length/diameter ratio})} - 5$$

$$= \frac{43{,}000 \times 6}{(6 \times 9 \text{ in.}/1.5 \text{ in.}) - 5}$$

$$= \frac{258{,}000}{(6 \times 6) - 5}$$

$$= \frac{258{,}000}{(36) - 5}$$

$$= \frac{258{,}000}{31}$$

$$= 8{,}322 \text{ ampere turns}$$

$$\text{Divide by 5 turns} = \frac{8{,}322}{5} = 1{,}664 \text{ amps}$$

**For longitudinal magnetization using a coil** (either rigid or wrapped cable), where the cross-sectional area of the part is at least one-half (50 percent) of the coil inner cross-sectional area (high fill factor coils), the number of amperes to be used can be calculated by using the following formula:

$$\text{Ampere turns} = 35{,}000 \div \frac{\text{part length}}{\text{part diameter}} + 2$$

Again, the actual amperage setting is determined by dividing the "ampere turn" value by the number of turns.

**Example 5.5** Inspecting a 1.5-in.-diameter round bar with a total length of 9 in. using a 5-turn coil with an internal diameter of 2 in. The fill factor is determined by dividing the coil area by the part area:

Area of coil = 1 in. $r^2 \times \pi = 3.142$ in.

Area of part = 0.75 in. $r^2 \times \pi = 1.767$ in.

Ratio of areas = 1.767 divided by 3.142 × 100 = 56.23% (high fill factor coil)

$$\text{Ampere turns} = 35{,}000 \div \frac{9}{1.5} + 2$$

$$= 35{,}000 \div 6 + 2$$

$$= 35{,}000 \div 8$$

$$= 4{,}375 \text{ ampere turns}$$

$$\text{Divide by five turns} = \frac{4{,}375}{5} = 875 \text{ amps}$$

**For longitudinal magnetization using a coil where the part cross-sectional area is between 10 and 50 percent of the coil inner cross-sectional area** (intermediate fill factor coils), the number of amperes can be calculated by using the formula:

Ampere turns = [(Ampere turn value for high fill factor coil) × 10 – (coil cross-sectional area/part cross-sectional area) ÷ (8)] + [(ampere turn value for low fill factor coil) × (coil cross-sectional area/part cross-sectional area – 2)] ÷ 8

**Example 5.6** Inspecting a 1.5-in.-diameter round bar with a total length of 9 in. using a 5-turn coil with an inside diameter of 4 in., the fill factor ratio is determined by dividing the coil inner diameter by the part outer diameter:

Area of the coil = 2 in. $r^2 \times \pi = 12.566$ in.$^2$

Area of the part = 0.75 in. $r^2 \times \pi = 1.767$ in.$^2$

Ratio = 1.767 ÷ 12.566 × 100 = 14.06% (intermediate fill factor coil)

From Example 5.5 (high fill factor coil), amperage for this part would be 875 amps. From Example 5.4 (low fill factor coil), ampere turns for this part would be 1,500 amps. Insert the above values into the formula to give

$$875 \times \frac{10 - 7.11}{8} + \frac{1{,}500 \times (7.11 - 2)}{8} = \frac{875 \times 2.89}{8} + \frac{1{,}500 \times 5.11}{8}$$

$$\frac{2{,}528.75}{8} + \frac{7{,}665}{8} = 316 + 958 = 1{,}274 \text{ amps}$$

*Note:* Amperages to be used are the calculated amperages ± 10%.

**For circular magnetization using prods,** the amperage to be used is usually determined by the prod spacing and part thickness. For part thicknesses up to ¾ in. (19 mm), 90 to 115 amps per inch (3.5 to 4.5 amps per mm) of prod spacing should be used. For part thicknesses over ¾ in. (19 mm), 100 to 125 amps per inch (4 to 5 amps per mm) of prod spacing should be used. The prod spacing will usually be between 2 in. (50 mm) and 8 in. (200 mm).

**For longitudinal magnetization using an AC yoke,** the yoke should be proven to having a field strength capable of lifting a dead weight of at least 10 lb. (45 N) with a 2 to 4 in. (50 to 100 mm) spacing between the legs.

When using a DC yoke or a permanent magnet, it should be proven to have sufficient lifting capability to lift a dead weight of 40 lb. (135 N) with a 2 to 4 in. (50 to 100 mm) spacing between the legs. Alternatively, a weight of 50 lb. (25 N) with a leg spacing of 4 to 6 in. (100 to 150 mm) spacing can be used. It should be noted that there may be different weight requirements depending upon the applicable code or specification.

*Note:* Some codes require that the leg spacing for the lift test be the same as to be used during inspection.

### 5.5.4 Caution!

Although the above formulas will produce exact values for required amperages, it should be mentioned that the correct amperage requirements vary based on the applicable specification. Some specifications take into account the variations in flux density due to the differences in permeability between the different steels and also consider the type of current used and the type of meter used to measure the current.

There are specifications that recommend the use of a handheld Gauss meter (such as in Fig. 5-43) held against the part surface to determine if sufficient flux density is being generated in the part. This practice does not take into consideration the permeability variations within the part. A part with a high permeability will concentrate more of the flux within itself, producing the high-density flux necessary for acceptable sensitivity. The flux density in the air surrounding the part will be low as a result of this. A low reading will therefore be produced on the Gauss meter. If the part were to be replaced with a part with lower permeability, the flux density within the part would not be as high and might not be sufficient to produce reliable indications. However, the reading on the Gauss meter would be higher due to the higher flux density in the air surrounding the part.

In summary, although code and specification requirements are very detailed, the practice of determining if sufficient flux is being generated is not an exact science. The ideal way to determine if sufficient flux is being generated in the part is to have an identical part with an actual discontinuity of the type, size, and location as those being sought. If this

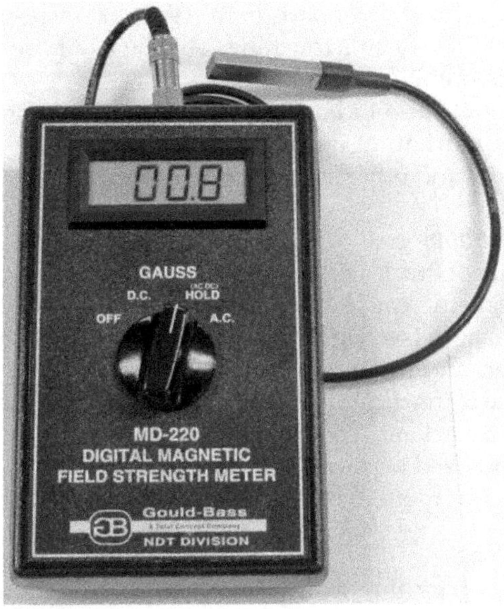


**Figure 5-43** Handheld Gauss meter.

can be reliably found, then the probability of finding an identical or larger discontinuity in the test part is very high.

## 5.6 Evaluation of Test Results and Reporting

The evaluation of test results is the most critical stage in the magnetic particle inspection process and greatly relies on the qualifications of the inspector. The ability of an inspector to observe an indication depends upon the correct application of the proper techniques, and using equipment and accessories that are functioning correctly. Another consideration is the inspector's ability to perform the test correctly. Competence should have previously been demonstrated by satisfying qualification requirements. The qualification process should also include a vision examination to prove visual acuity and ability to distinguish the colors associated with the MT technique to be used.

### 5.6.1 Classification of Indications

Once detected, the indications should be classified as either "false," "nonrelevant," or "relevant" before final evaluation.

#### *False Indications*

False indications can be produced by improper handling and do not relate to the part's condition or use. An example is "magnetic writing." This is typically produced by the formation of indications at local poles that are created when the part comes in contact with another magnetized part prior to or during inspection. This can be eliminated by demagnetization and repeating the inspection.

Magnetic writing is most likely to occur when using the residual method, through poor handling that allows the individual parts to touch. The continuous technique may require the demagnetization of parts before the next inspection to preclude the possibility of magnetized components touching. This type of false indication can be eliminated through careful handling.

Other sources of false indications may be caused through the use of excessively high magnetizing currents or inadequate precleaning of the parts to remove oil, grease, corrosion products, and other surface contaminants.

#### *Nonrelevant Indications*

These are the result of flux leakage due to geometrical or permeability changes in the part. Examples of geometric causes include splines, thread roots, gear teeth, keyways, or abrupt section changes (see Fig. 5-16). A concern with these conditions is that they may also be stress risers and could be the origin for fatigue-induced cracks. These conditions are therefore some of the most critical; the possibility that one of these nonrelevant indications can conceal a crack must be considered. Other potential sources of nonrelevant indications include localized permeability changes in the part, which may be due to localized heat treatment or variations in hardness, and may also occur at the fusion zone of a weld.

#### *Relevant Indications*

These are produced by flux leakages due to discontinuities in the part. When these discontinuities are not in compliance with a code, they are classified as rejectable. If they meet the acceptance criteria, they are considered to be acceptable discontinuities. Discontinuities that do not permit the part to be used for its original purpose or can potentially cause the part or fail are classified as defects.

### 5.6.2 Visual Appearance

*Surface* discontinuities will provide sharp distinct indications, which very closely resemble the shape and size of the discontinuity producing the leakage field (see Fig. 5-44).

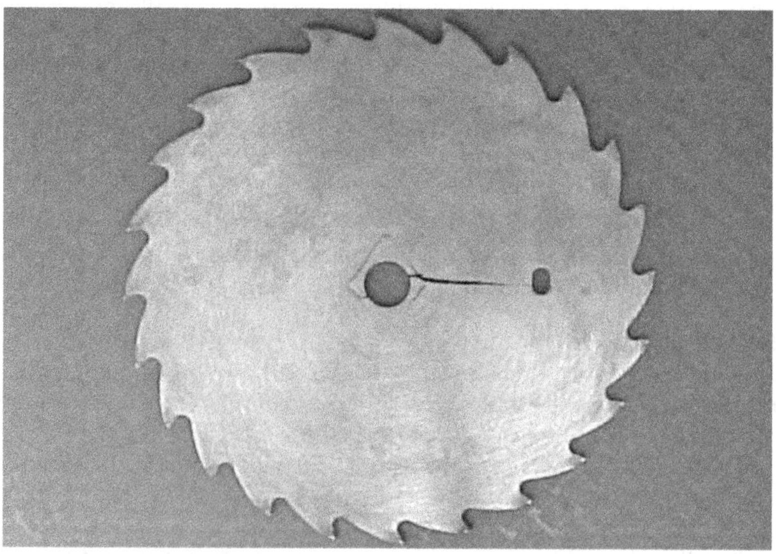

**Figure 5-44** Indication of a quench crack in a saw blade.

It is not feasible to describe every possible type of discontinuity and its appearance when using the MT method. However, a detailed description of discontinuities and their appearances and causes is contained in Chapter 2.

*Subsurface* discontinuities will produce weaker, diffuse, and broad indications, which become less well defined as their depth below the surface increases.

### 5.6.3 Depth Limitations

It is not possible to define at exactly what depth a discontinuity can be detected because of the many variables present. It has been claimed that subsurface discontinuities have been detected as deep as 0.240 in. (6 mm) from the surface.

Sensitivity varies greatly with depth below the surface. A surface crack will usually be easily detected providing the correct current and flux line direction are employed. It is reasonable to expect reliable detection of cracks on the order of at least 0.040 in. (1 mm) in length.

### 5.6.4 Reporting

When the parts have been inspected and all indications evaluated, it will be necessary to prepare a report detailing the results of the test and, if applicable, the size, location, and orientation of discontinuities found. This report may vary considerably from company to company, but as a minimum, it should meet customer requirements and should typically include the following data:

1. Contract and customer
2. Inspection company
3. Date of inspection
4. Inspector's name and qualification and certification level
5. Report number or identification
6. Applicable codes, specifications, or procedures, including type and technique of inspection
7. Acceptance criteria
8. Component description, part number, and serial number
9. Flux line direction with respect to discontinuity orientation
10. Other identification details as requested in the contract; for example, batch or order number
11. Material batch number (particles, liquid carrier, etc.)
12. Results of inspection, including recording of indications
13. Signature or inspection stamp of inspector

A detailed, concise report will enable future evaluations by other inspectors. There are several ways to record the indications as noted in the list below:

1. A descriptive written report including significant dimensions and indication location can be complied.
2. A photograph may be taken under the correct viewing conditions. Black light illumination in a darkened environment works fine if the correct exposure

FIGURE 5-45 Box demagnetizer. (*Courtesy of Magnaflux Corp.*)

(usually several seconds) is used and a camera is mounted on a tripod. It may be necessary to vary the exposure time several times to ensure that the best lighting and exposure parameters are met.

3. A free-hand sketch may be used to supplement the data. Again, key dimensions must be included. Unfortunately, not everyone is a good artist, and the quality and usefulness of sketches can be questionable.
4. A piece of transparent tape may be used to lift the indication from the test surface if using dry particles. When peeled off, the tape will retain the shape and size of the indication through the adherence of the particles to its adhesive layer. The tape can then be applied to the report or other suitable background material to render the indication more visible.
5. Aerosol-based, strippable lacquer may be applied in several thin layers, allowing each layer to dry before applying the next, until sufficient thickness exists, allowing the solidified, flexible film to be peeled off the part.
6. Magnetic rubber inspection may be used to create a permanent record of indications. Magnetic rubber inspection involves using a two-part liquid rubber kit consisting of:
   (a) Room vulcanizing temperature (RVT) rubber supplied in a liquid form. This liquid rubber also contains ferromagnetic powder.
   (b) A catalyst, which when mixed will cause the mixture to solidify at a controlled rate.

When mixed together, the rubber solution is poured onto the area of inspection and a magnetizing force applied during the "cure" time. The cure rate should be slow enough to allow the ferromagnetic powder to migrate to the flux leakage fields, but not so slow as to delay the inspection longer than necessary. Magnetic rubber materials with curing ranges from 5 minutes to 4 hours are available. When the rubber is solidified, it can be peeled off and the indications created by the flux leakage can be observed and retained as a record.

The magnetic rubber inspection technique has uses beyond recording indications. It can also be used to inspect areas and surfaces that are inaccessible for standard MT inspection, such as inside blind fastener holes, particularly threaded holes. The rubber can be poured into the hole and a permanent magnet applied across the hole during the cure. After solidification of the rubber, the solid (but, flexible) rubber plug is removed from the hole and the indications from within the hole can be viewed on the outer surface of the plug. A small stick or other device can be cast into the plug to facilitate easy removal.

## 5.7 Applications

The magnetic particle test method is effective for the detection of surface and slightly subsurface discontinuities in ferromagnetic parts. It can be used as an inspection tool at all stages in the manufacture and end use of a product.

### 5.7.1 Stages of a Product's Life Cycle

Discontinuities can occur in any of the following stages of the product's life cycle and can be classified accordingly.

#### *Inherent*

As the metal cools and solidifies to produce the original ingot, numerous discontinuities can be introduced. Inherent discontinuities that can be detected by magnetic particle testing include nonmetallic inclusions (sometimes referred to as "stringers") and seams located at the surface.

#### *Primary Processing*

When the original material has become solid, it must be worked and formed to produce a rough-shaped product. During these processes, the part is said to be in its "primary processing stage." Typical primary processes include forging, casting, rolling (hot and cold), extrusion, and drawing. During these processing operations, discontinuities may be produced or existing discontinuities may be modified. Examples of primary processing discontinuities detectable by magnetic particle testing include:

- Forging bursts
- Forging laps
- Rolling laps, seams, and stringers
- Rolling seams
- Laminations (at the edges of plates and sheets)
- Casting shrinkage (at the surface)
- Casting inclusions (at the surface)
- Casting cold shuts (at the surface)
- Casting hot tears (at the surface)

#### *Secondary Processing*

After rough shaping the metal in the primary processing stage, it is further refined and shaped to produce finished products. This stage is referred to as "secondary processing" and can include such processes as machining, grinding, plating, and heat treatment. These secondary processes may also produce discontinuities or change the appearance of existing ones. Examples of secondary processing discontinuities detectable by magnetic particle testing include:

- Quench and heat cracks
- Grinding cracks (or checks)
- Machining tears
- Plating cracks

#### *Welding*

When parts are joined together by a welding process, numerous discontinuities may be created. Examples of the welding processes are found in Chapter 2. The following is a list of welding-related discontinuities that may be detected by magnetic particle testing when at or close to the surface:

- Cracks (longitudinal, transverse, and crater)
- Lack of fusion
- Incomplete penetration (if accessible)
- Entrapped slag
- Inclusions
- Overlap or "cold" lap

#### *Service*

When the part is put into service and subjected to stresses and environments that may be detrimental to its structure, service discontinuities may result. Those that are detectable by magnetic particle testing include:

- Fatigue cracks
- Stress corrosion cracks
- Static failure cracks (overstressed structures)

### 5.7.2 Types of Components

Examples of materials, structures, and parts that may be inspected using the magnetic particle test method include:

- Ingots
- Billets
- Slabs
- Blooms
- Bar stock
- Sheet

- Rod
- Wire
- Castings
- Shafts (plain, threaded, or splined)
- Welds
- Bearings and bearing races
- Nuts and bolts
- Gears
- Cylinders
- Discs
- Forgings
- Tubular products
- Plate

### 5.7.3 Industrial Sectors

Magnetic particle testing has many applications throughout industry including but not limited to the following sectors:

- Petrochemical
- Construction
- Aircraft and aerospace
- Automotive
- Defense
- Nuclear
- Transportation
- Shipping (marine)

## 5.8 Advantages and Limitations

### 5.8.1 Advantages

The following are advantages of magnetic particle testing as compared to alternative NDT methods:

1. Test results are virtually instantaneous, in that indications will form within one or two seconds of particle application. No developing or processing times are involved.
2. Permanent records of indications can be produced using photography, magnetic rubber, or transparent tape techniques.
3. MT can be applied in situ, without the need for an AC power supply, by using permanent magnets or battery-powered yokes.

4. Indications are easy to interpret.
5. The indications formed by the particles closely represent the shape and type of the discontinuity.
6. Training and experience requirements prior to becoming certified are significantly less stringent than for UT, RT, or ET, since MT is a relatively simple process.
7. MT equipment can be much less expensive than other NDT equipment. Depending on the degree of automation or scale of operation, it may also be more economical than many other NDT methods.
8. Virtually any size or shape of component can be inspected.
9. Inspections can be performed during all stages of manufacturing.
10. Test part surface preparation is less critical than with penetrant testing.
11. MT can be used to inspect through metallic and nonmetallic coatings or plating with some techniques. It should be noted, however, that a reduction in sensitivity will occur as the thickness of the coating increases. Maximum coating thickness should be established through qualification tests or stipulated in customer specifications or code requirements.
12. There are no known personnel hazards associated with the process because the magnetic fields generated are of short duration; however, the usual electric shock, manual lifting, and chemical (petroleum distillate) precautions apply. Additionally, the parts may become heated during the process if high-amperage current is applied for an extended period.
13. Many parts can be inspected simultaneously if using the residual magnetism technique.
14. MT can be automated for certain production line applications.

### 5.8.2 Limitations

The following are limitations of magnetic particle testing as compared to other NDT methods:

1. It is only effective for the examination of ferromagnetic materials.
2. Discontinuity detection is limited to those at or near the surface.
3. Demagnetization may be required before, between, and after inspections.
4. Discontinuities will only be detected when their major axis interrupts the primary flux lines. This necessitates inspection in more than one direction to assure discontinuity detection regardless of orientation.
5. Some magnetic particle testing techniques may cause damage to the part as a result of arcing or localized overheating of the parts (for example, when using DC prods).
6. Paint and/or coating removal is necessary from localized areas on the part to facilitate good electrical contact when using direct magnetization techniques.
7. Uniform, predictable flux flow through the parts being tested may not be possible due to complex shapes.
8. Nonrelevant indications due to abrupt changes in component profile or local changes in material properties may make interpretation difficult.

## 5.9 Glossary of Key Terms

**Alternating Current (AC)** Electric current that flows through a conductor in a back and forth manner at specific intervals. It provides the best sensitivity for the detection of discontinuities at the surface.

**Background** The general appearance of the surface on which discontinuities are being sought.

**Black Light** Electromagnetic radiation in the 320- to 400-nanometer wavelength range, which is invisible to humans. Also see Ultraviolet Light.

**Burning** Local overheating of the component at the electrical contact area arising from high resistance and the production of an arc.

**Carrier Fluid or Liquid** The fluid in which magnetic particles are suspended to facilitate their application.

**Circumferential Magnetization (Circular Magnetization)** Magnetization that establishes a flux around the periphery of a part.

**Coil Method** A method of magnetization in which part or all of the component is encircled by a current-carrying coil. (The use of this term is usually restricted to instances in which the component does not form part of a continuous magnetic circuit for the flux generated.)

**Current Induction Method** A method of magnetizing in which a circulating current is "induced" into a ring component by the influence of a fluctuating magnetic field that links the component. Also known as indirect magnetization.

**Demagnetization** The process by which a part is returned substantially to the near unmagnetized state.

**Demagnetizing Coil** A coil of wires carrying alternating current that is used for demagnetization.

**Diffuse Indications** Indications that are not clearly defined; e.g., indication of subsurface discontinuities.

**Direct Current (DC)** Electric current that flows through a conductor in only one direction at all times. DC from a battery source has been phased out in favor of "rectified" forms of AC for surface and subsurface discontinuity detection.

**Dry Particle Technique** The application of magnetic particles in dry form (without the use of a liquid carrier).

**Ferromagnetic** Having magnetic permeability, which can be considerably greater than that of air and can vary with flux density.

**Fluorescence** The ability to absorb electromagnetic radiation with a wavelength outside the human visible spectrum (black light) and reemit electromagnetic radiation within the white light spectrum, usually in the yellow to green range.

**Flux** Invisible lines of magnetic force.

**Flux Density $\beta$** Magnetic field strength per unit volume within a ferromagnetic test part; expressed in "Gauss." (See also Tesla.)

**Flux Field Penetration** The ability to establish and drive high-density magnetic lines of force into the test part.

**Full-wave Rectified Current (FWDC)** Electric current that flows through a conductor in one direction only with an increased rate of pulsation surges and drops at specific intervals. FWDC is recommended for effective surface and subsurface discontinuity detection when using wet-suspension techniques.

**Half-wave Rectified Current (HWDC)** Electric current that flows through a conductor in one direction only with pulsating surges and drops at specific intervals—hence the name half-wave. It is most effective for surface and subsurface discontinuity detection when using the dry particle technique due to the vibratory effect produced on the particles.

**Hysteresis Loop (and related terms)** See Section 5.2.

**Linear Indication** Any indication having a length dimension at least three times greater than its width (as defined by some codes).

**Longitudinal Magnetization** Magnetization in which the flux lines are oriented in the part in a direction essentially parallel to its longitudinal axis.

**Magnetic Domains** Ferrous material atoms or molecules, normally represented as small bar magnets with north and south poles.

**Magnetic Particle Inspection** A nondestructive test method that provides for the detection of linear, surface, and near-surface discontinuities in ferromagnetic test materials.

**Magnetic Poles** Those parts of a magnet that are the source of the external magnetic field.

**Magnetic Writing** Spurious indications arising from random local magnetization.

**Magnetism** A form of energy directly associated with electrical current and characterized by fields of lines of force.

**Magnetizing Force (*H*)** Magnetic field strength per unit volume in air, measured in Oersteds.

**Mercury Vapor Lamp** A bulb used for producing ultraviolet radiation. (See also Ultraviolet Light and Black Light.)

**Particle Mobility** The ability to impart activity or motion to the magnetic particles applied to the test surface.

**Permeability** The ease with which a material can be magnetized. The ability of a material to conduct magnetic lines of force.

**Prods** Handheld electrodes attached to cables that transmit the magnetizing current from the power supply to the part under examination.

**Reluctance** The opposition of a material to conduct magnetic lines of force.

**Residual Magnetism** The magnetism remaining in a part after the magnetizing force has ceased.

**Retentivity** The ability of a material to retain magnetism following magnetization.

**Tesla** An expression of flux density. 1 tesla = $10^4$ Gauss or 1 weber/meter$^2$.

**Ultraviolet Light** Electromagnetic radiation in the 200- to 400-nanometer wavelength range. The 200- to 320-nanometer wavelength range contains harmful UV radiation, which can damage the skin tissue or the eyes. It is essential that this be filtered out using the correct black light filter (see also Black Light).

**Weber** 1 weber = $10^8$ lines of force. *Note:* 1 meter$^2$ = 10,000 cm$^2$; therefore, there are 10,000 flux lines/cm$^2$ (1 Gauss).

## References

Lovejoy, D. *Magnetic Particle Inspection—A Practical Guide.* Chapman and Hall, 1993, 2–6 Boundary Row, London, UK.

Metals Test Methods and Analytical Procedures, Vol. 03, Nondestructive Testing. American Society for Testing and Materials International 2011 Annual Book of Standards. 100 Barr Harbor Drive, PO Box C700, West Conshohocken, PA, 19428-2959 USA.

Nondestructive Evaluation and Quality Control, Vol. 17. *ASM International Handbook,* 1992. Metals Park, Ohio.

# CHAPTER 6

# Radiographic Testing

## 6.1 History and Development

The history of radiographic testing (RT) actually involves two beginnings. The first commenced with the discovery of X-rays by Wilhelm Conrad Roentgen in 1895 and the second with the announcement by Marie and Pierre Curie, in December of 1898, that they had demonstrated the existence of a new radioactive material called "radium."

### 6.1.1 The Discovery of X-rays

One of the true giants in radiography is the man who discovered X-rays—Wilhelm Conrad Roentgen (Fig. 6-1). Roentgen was born on March 27, 1845, in Lennep, Germany. His birth house is still in existence and is presently a magnificent museum containing many of the artifacts that he used in his early days of experimenting. It also includes many radiographic devices and interesting radiographs that had been taken over the years as a result of his discovery.

Roentgen was educated in Utrecht and Zurich and ultimately became a Professor of physics at Strasbourg in 1876, Giessen in 1879, and Wurzburg in 1888, where his famous discovery took place. He later became a Professor of physics in Munich in 1899.

There are various accounts as to how X-rays were actually discovered during those days in his experimental laboratory at the University of Wurzburg (Fig. 6-2). Most accounts agree that the discovery took place on November 8, 1895, while he was working in his laboratory. He apparently had been working in a semidarkened laboratory room and was experimenting with a vacuum tube, referred to as a Crooke's tube, which he had covered with black photographic paper in order to better visualize the effects that the cathode ray tube produced. As he was experimenting that day, he observed that as the tube was energized, a cardboard coated with barium platinocyanide happened to be lying on a bench not too far from the tube. Even though no visible light escaped from the Crooke's tube because of the dark paper that encompassed it, Roentgen noted that the barium platinocyanide screen fluoresced. He also observed that as he moved the cardboard with the barium platinocyanide coating closer to the tube, it fluoresced with an even higher intensity, which proved that some unknown ray was emanating from the tube. Another account indicates that the cardboard contained a letter "S," which a student had painted on it with barium platinocyanide salts, and the glowing image was actually in the form of that letter "S." Regardless of which of these accounts is factual, it remains that this discovery that occurred on that November day in 1895 had a major impact, not only on industrial nondestructive testing but also on the world of medicine. As a result of this discovery, Roentgen began conducting further experimentations

**FIGURE 6-1** Wilhelm Conrad Roentgen. (*Courtesy of Charles J. Hellier.*)

that involved taking radiographs of different objects in his laboratory, including a wooden box containing metal weights, a lock on the laboratory door, and a double-barreled shotgun with different size pellets inside (Fig. 6-3). He also X-rayed other materials, such as playing cards, a book, wood, and even some thin sheets of metal. He also observed while he was experimenting that as he placed his hand between the tube and the cardboard containing the fluorescent salts, a faint shadow outline of the parts of his fingers appeared. The actual structure of his bones within his living flesh was observed. Following this came the most historic demonstration of his new discovery, when his wife, who was apparently concerned about all the time he was spending away from home while at the laboratory, came for a visit. It was during this visit that Wilhelm Conrad Roentgen took an X-ray radiograph of his wife's hand (Fig. 6-4).

Notice in that early radiograph, that the bones produce a lighter image, since they were higher in density than the surrounding flesh. History does not give any further details as to why Roentgen X-rayed his wife's hand rather than his own. Some suggest that even in those early days, Roentgen was aware of the biological effect that radiation may have had, but that is pure conjecture. In the first days of his discovery, he referred to this unknown ray as "X light," but many of his fellow scientists referred to them as "Roentgen rays." In fact, the expression "Roentgen rays" was quite popular into the early 1900s. In 1902, the term that Roentgen designated for this discovery, "X-rays," began to take hold.

Obviously, Roentgen was very involved with this remarkable finding and on December 28, 1895 (less than two months after first observing X-rays and making the X-ray of his wife's hand), he presented his first official paper entitled, "On a New Kind of Rays" to the Wurzburg Physical Medical Society. He obviously did not celebrate

(a)

(b)

**Figure 6-2** University of Wurzburg. (*a*) Then and (*b*) now. Author is standing (and pointing) outside Roentgen's laboratory with Dr. Dietmar Henning. (*Courtesy of Charles J. Hellier.*)

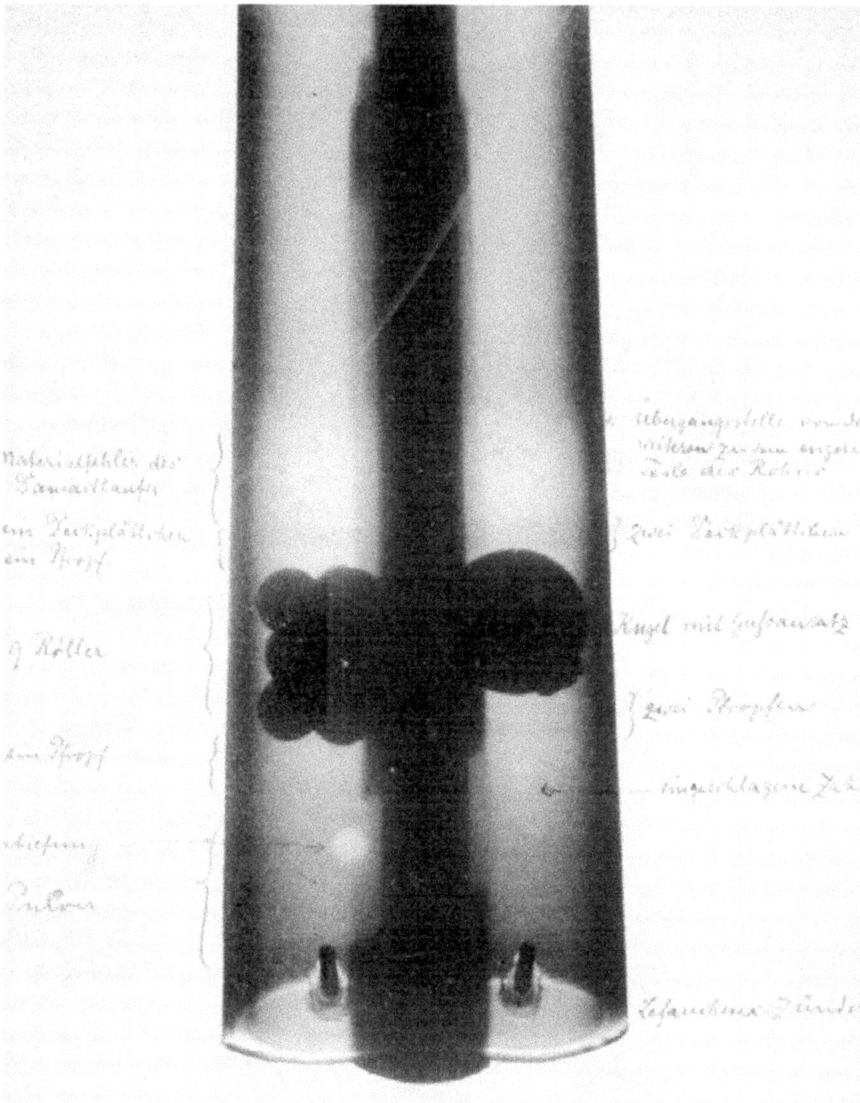

**FIGURE 6-3** Early Roentgen radiograph of a shotgun.

New Year's Day because, on January 1, 1896, he mailed reprints of this paper together with copies of the X-ray of his wife's hand and other radiographs he had taken to many of his colleagues. As a result of this, the very first news of this discovery appeared in the Sunday edition of the *Vienna Presse* on January 5, 1896. From there, it was wired to London and then on to newspapers around the world. The earliest recorded newspaper article in the United States was in the *Milwaukee Journal* on January 7, 1896. Even in those days, before e-mail and satellite transmission of information, the good news traveled fast. In fact, the *Milwaukee Journal* account was published before it was announced in Roentgen's own hometown newspaper, *The Wurzburg General Anzeiger*,

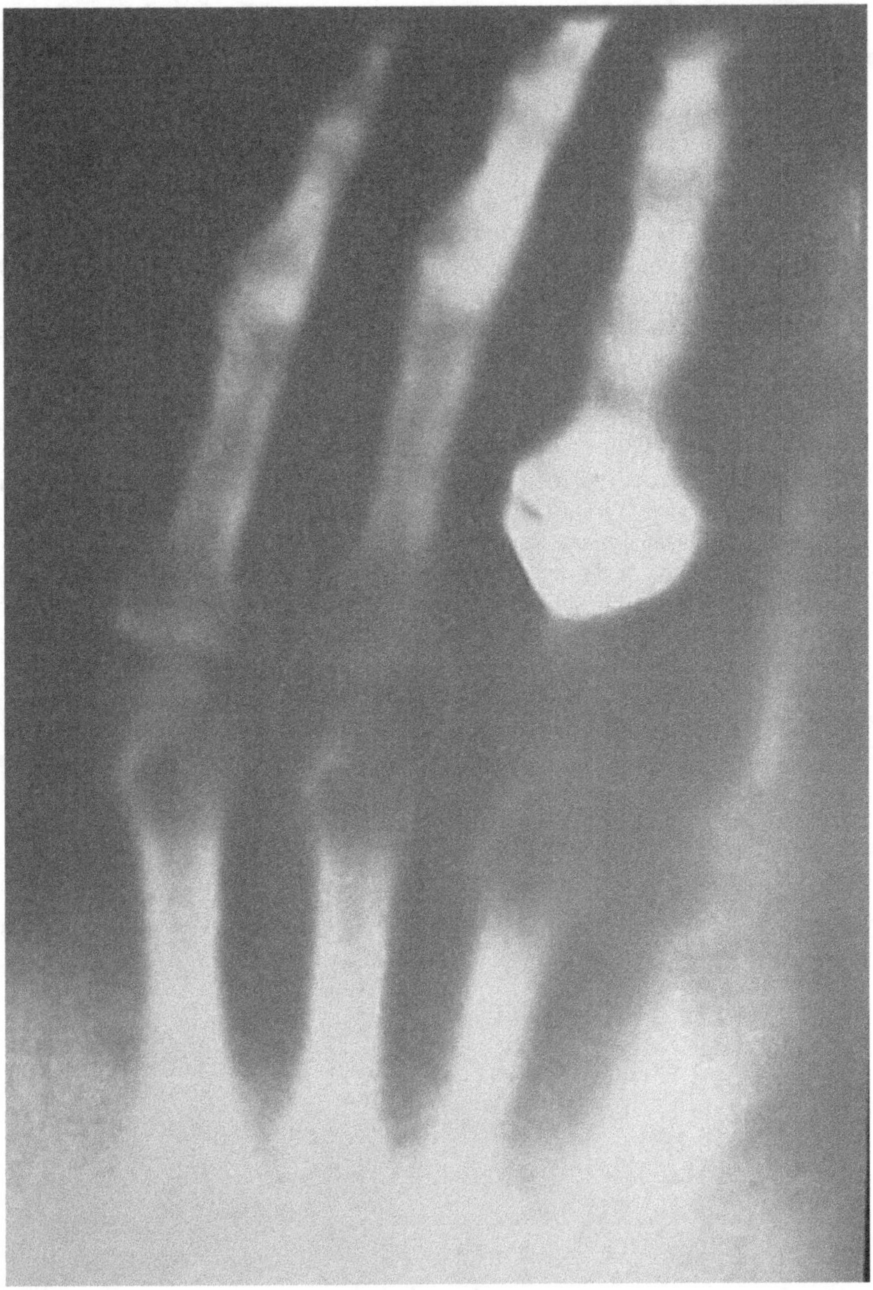

**Figure 6-4** X-ray of Mrs. Roentgen's hand.

which finally reported the event on January 9, 1896. Roentgen made his first public presentation before the Wurzburg Physical Medical Society on January 23, 1896, and after the lecture, Roentgen took an X-ray of the hand of a famous anatomist named Kolliker who proposed that the new discovery be called Roentgen's rays.

The first recorded use of X-rays in the United States was on February 8, 1896, when a young man by the name of Eddie McCoughey from Hanover, New Hampshire, was X-rayed for a broken wrist. That first American X-ray was taken by Dr. Frost and his wife, who was the first head nurse at the Dartmouth College Medical Center.

One might wonder what this all has to do with nondestructive testing, since medical X-ray examinations cannot, theoretically, be considered nondestructive. But if it weren't for this early discovery, many of the early industrial X-ray experiments would have never occurred. In 1901, Roentgen was awarded the first Nobel Prize in science for his work with X-rays (Fig. 6-5).

There is so much more to Roentgen's life and how science recognized him, but his desire was not to be in the limelight. He rejected all of the different commercial offers relating to his discovery. There were false reports that it was really his laboratory assistant who was responsible for first observing X-rays, and he had taken all the credit for the discovery. In his later years, Roentgen withdrew from public life. He died on February 10, 1923, of carcinoma of the rectum and was buried beside his wife in the family grave in Giessen. Another remarkable fact regarding Roentgen is that he refused to patent this discovery, so that the world could freely benefit from his work. At the time of his death, he was nearly bankrupt as a result of the inflation that followed World War I.

In the first 20 years that followed Roentgen's famous discovery, X-rays were primarily used in the medical arena by doctors who were developing medical X-ray

**FIGURE 6-5** Nobel Prize for Physics awarded to Roentgen in 1901. (*Courtesy of Charles J. Hellier.*)

techniques, looking inside the human body while the patient was still alive. In 1912, Laue, Knipping, and Frederick made the first successful experiment using the diffraction of X-rays. Almost immediately after this experiment, W. D. Coolidge of General Electric Co. invented the Coolidge tube, which allowed a much greater intensity of X-rays to be produced and permitted uniform control of the radiation. On the nondestructive testing scene, Dr. H. H. Lester began experimenting with different materials in the early 1920s at the Watertown Arsenal in Watertown, Massachusetts. This laboratory is illustrated in Chapter 1, Fig. 1-7.

While some work continued, the importance of X-rays in NDT did not really become fully known, nor the technology widely utilized, until World War II. The importance of industrial radiography was reflected in the first name of the technical society now known as the American Society for Nondestructive Testing. It was founded in 1941 under the name of The American Industrial Radium and X-ray Society. The developments, triggered by World War II, resulted in significant innovation and invention with radiographic equipment that produced more intense beams and higher energies. Figure 6-6 depicts a "mobile" X-ray unit that was capable of producing energies of up to 250 kV. Certainly, the works of Horace Lester and William Coolidge were significant; however, other contributions were made by early scientists and engineers, such as Gerald Tenney, James Bly, Donald O'Connor, Edward Criscuolo, Daniel Polansky, Samuel Wenk, and many others too numerous to name.

**FIGURE 6-6** Early "mobile" 250 kV X-ray unit. (*Courtesy of Charles J. Hellier.*)

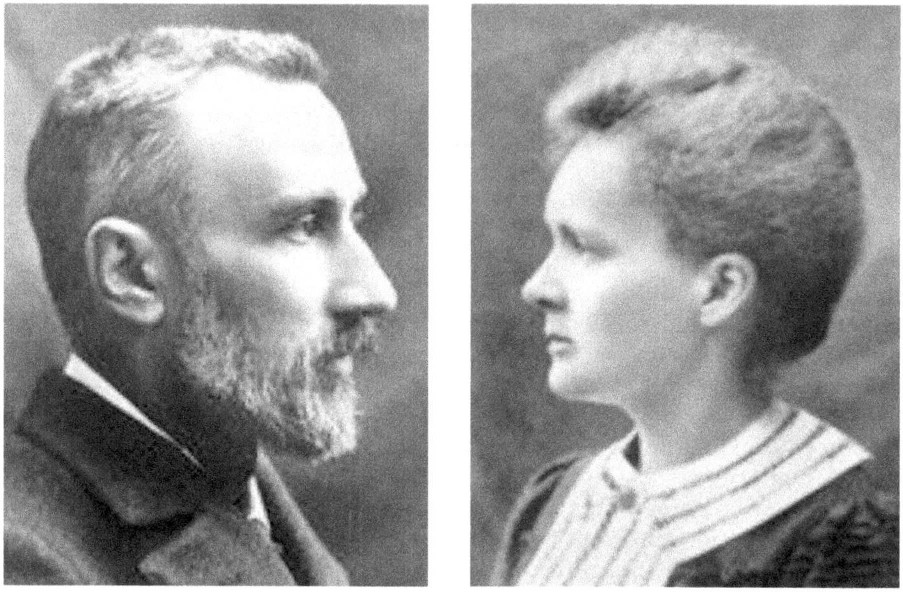

(a) (b)

**Figure 6-7** (*a*) Pierre and (*b*) Marie Curie.

### 6.1.2 The Discovery of Gamma Rays

There are many accounts about Marie and Pierre Curie (Fig. 6-7) and there is no doubt that these two physicists led very interesting, unique lives. Marie Sklodowska, as she was known before she was married, was born in Warsaw on November 7, 1867. When she was 24, she traveled to Paris to study mathematics and physics. While there, she met Pierre Curie, who was 35 years old, 8 years older than Marie. He was an internationally known physicist and a very serious scientist who was dedicated to spending his life doing scientific work. They married in July of 1895 and began to work together on various scientific experiments.

One of their early observations was that thorium gave off the same rays as uranium. They also discovered that there were some reactions coming from materials containing bismuth and barium. When Marie took away a small amount of bismuth, a residue with a much greater activity remained. At the end of June 1898, the Curies had obtained a sufficient amount of this substance and proved it was about three hundred times more active than uranium.

They also extracted a substance from pitchblende, similar to bismuth, which they claimed contained a metal never known before. It was suggested that this new, highly active material be called polonium after the country of Marie's origin. In those early days of experimenting, the term "radioactivity" was introduced. Finally, on December 26, 1898, they produced evidence that there was a very active new material that reacted and behaved chemically, almost like pure barium. At that point, they suggested the name radium be used for this new element.

In order to provide proof that this element did exist, they would have to produce it in a sufficient amount and to determine such characteristics as its atomic weight. In order to do this, they would need a great amount of the pitchblende from which the radium could

**FIGURE 6-8** Laboratory where the Curies extracted radium from pitchblende. (*Courtesy of Charles J. Hellier.*)

be extracted. They were able to obtain several tons of pitchblende and the intensive, laborious work of separating the radium from the pitchblende began. Obviously, it was very hard work and Marie Curie was quoted as saying, "Sometimes I had to spend a whole day stirring a boiling mass with a heavy iron rod nearly as big as myself. I would be broken with fatigue at day's end." She worked in a shed with an earth floor (see Fig. 6-8), which had a glass roof and did not provide adequate protection against rain. In the summer, it was like a hothouse. And yet, history records that this is where they spent the best and happiest years of their lives. While they were conducting all these experiments, both Pierre and Marie were still teaching, so their days must have been very tiresome and long. Ultimately, from the several tons of pitchblende, they were able to isolate one decigram of almost pure radium fluoride and, at that time, determined the atomic weight of the radium to be 225. Marie presented this work in her doctoral thesis on June 25, 1903. As a result of the discovery of this new element, Marie Curie was awarded the Nobel Prize. She was the first woman to receive this award.

Working closely with radium, which is highly radioactive, and the gas radon, a by-product, took its toll. Pierre Curie had scarred and inflamed fingers from holding small glass tubes containing radium salts and solutions. He actually conducted medical tests on himself by wrapping a sample of the radium salts on his arm for up to 10 hours and then evaluated the results of the exposure, which was in the form of a burn, day by day. After many days, a scar still remained and Pierre suggested that the radium might one day be used for the treatment of cancer. Fifty years after those days of experimentation, the presence of radioactivity was still observed in the laboratory where the Curies had worked so hard, requiring extensive clean-up.

Marie also had started to notice the effects of radiation exposure. Her fingers were cracked and scarred. Both Pierre and Marie were beginning to experience signs of fatigue.

They obviously had no idea as to the effect of radiation on the human body at that time. Contrary to the expected, Pierre Curie did not die as a result of his exposure to radiation. He was killed when he was run over by a horse-drawn wagon in Paris on April 19, 1906. This left Marie, who was only 38 years old, alone with two daughters, aged 9 and 2 at the time. Ultimately, Marie Curie would die of leukemia on July 4, 1934.

The work of the Curies marked the beginning of many other developments that would ultimately lead to the radioactive sources that are primarily used in industrial radiography today, namely, iridium 192 and cobalt 60. The scientific community owes much to the Curies for their early sacrificial work, especially considering the damage that they personally suffered as a result of early exposure to radium.

Before the Curies began their work in isolating and identifying radium, the scientist Becquerel discovered that certain radiations were given off by uranium ore. He was responsible, in some respects, for directing and motivating the Curies to begin their work with radium. Another key scientist from England, Rutherford, was one of the scientists who identified the radiation emitted by certain radioactive elements. He also developed the theory that elements had isotopes and was responsible for identifying the existence of the neutron.

With the advent of nuclear reactors capable of generating high neutron intensities, the possibility of producing artificial isotopes became a reality. Iridium 191 and cobalt 59, both elements existing in nature, are quite stable. When exposed to thermal neutrons, the stable isotopes capture a thermal or slow neutron, becoming one unit heavier in mass. With the addition of this neutron, the iridium 191 becomes iridium 192 and cobalt 59 becomes cobalt 60. Both isotopes are unstable and therefore radioactive. More details about these two artificially produced radioactive isotopes will be presented in Section 6.2. Other key developments in the area of radioactive isotopes involve the equipment used to encapsulate the radioactive isotopes and the projectors or "cameras" that store the isotopes, and from which they are operated.

## 6.2 Theory and Principles

There are many excellent references that contain in-depth information regarding the principles of radiation and radiography. It is not the intent of this chapter to cover principles and theory in great depth but to provide the reader with an understanding of the basics, so that the entire process of producing radiographs with a radiation source will be comprehended. Whether the radiation is emitted from an X-ray tube or a gamma ray source, there are some essential components that apply to the process of radiographic testing.

The first component is the source of radiation. X-rays emanate from an X-ray tube and gamma rays are emitted by a radioactive isotope. The second component is the test specimen that is to be examined. The third includes the development of the technique. The fourth involves the taking of the radiograph and the processing of the film. The final component, which is extremely critical, is the interpretation of the radiographic image.

### 6.2.1 Characteristics of Radiation

There are certain unique characteristics relative to radiation that must be understood in order to realize the physics and variables involved with producing a radiograph with the use of a radiation source. These characteristics apply to both x-radiation and gamma radiation. Recall that the only difference between x- and gamma radiation is their origin. X-rays are produced by an X-ray tube and gamma rays come from a radioactive source that is disintegrating. The following characteristics apply to the radiation that will be used in the nondestructive examination of materials:

### 1. Radiation Is Absorbed and Scattered by Material

There are four common absorption processes that influence the amount of radiation that passes through a part. The four absorption processes are:

1. Photoelectric effect
2. Rayleigh scattering
3. Compton scattering
4. Pair production

The *photoelectric effect* is that process in which a photon of low radiation energy (less than 500 kV) transfers all of its energy to an electron in some shell of the material atom. The energy may simply move an electron from one shell to another, or if there is energy above that required to interact with the orbital electron in the material, it will impart kinetic energy and the electron will be ejected from the atom.

Another name for *Rayleigh scattering* is "coherent scattering" and it is a direct interaction between the photon and the orbital electrons in the atom of the material. However, in this case the photon is deflected without any change or reduction of the kinetic energy of the photon or of the energy of the material atoms. In this process, there are no electrons released from the atom. It is estimated that Rayleigh scattering accounts for no more than about 20 percent of the total attenuation.

*Compton scattering* is a direct interaction between photons in the 0.1 to 3.0 MeV energy range and an orbital electron. In this case, when the electron is ejected from the material atom, only a portion of the kinetic energy of the photon is used. The photon then scatters in a different direction than the direction from which it came and actually emerges with a reduced energy and, therefore, a longer wavelength. Compton scattering varies with the atomic number of the material and varies, roughly, inversely with the energy of the photon.

*Pair production* is an attenuation process that results in the creation of two 0.51 MeV photons (as a result of annihilation of the electron-positive pair) of scattered radiation for each high-energy incident photon that is at or above an energy of 1.02 MeV. The two 0.51 MeV photons travel in different directions, causing the production of electromagnetic radiation through interaction with other material particles. Energies exceeding 1.02 MeV result in additional kinetic energy being applied to the pair of particles.

These four forms of attenuation or absorption are depicted in Fig. 6-9.

Total absorption, therefore, is the combined sums of the four different types of absorption.

### 2. Radiation Penetrates

The variables relating to the penetration of the radiation can be expressed with the term "half-value layer." The half-value layer is defined as the thickness of a specific material that will reduce the radiation intensity to one-half of that entering the part. If the initial radiation intensity is 100 roentgens (100 R), a material that is exactly one half-value layer will reduce that 100 R to 50 R. Another half-value layer thickness will reduce that 50 R to 25 R, and so on. If this is carried forward, the radiation never reaches zero. Another term

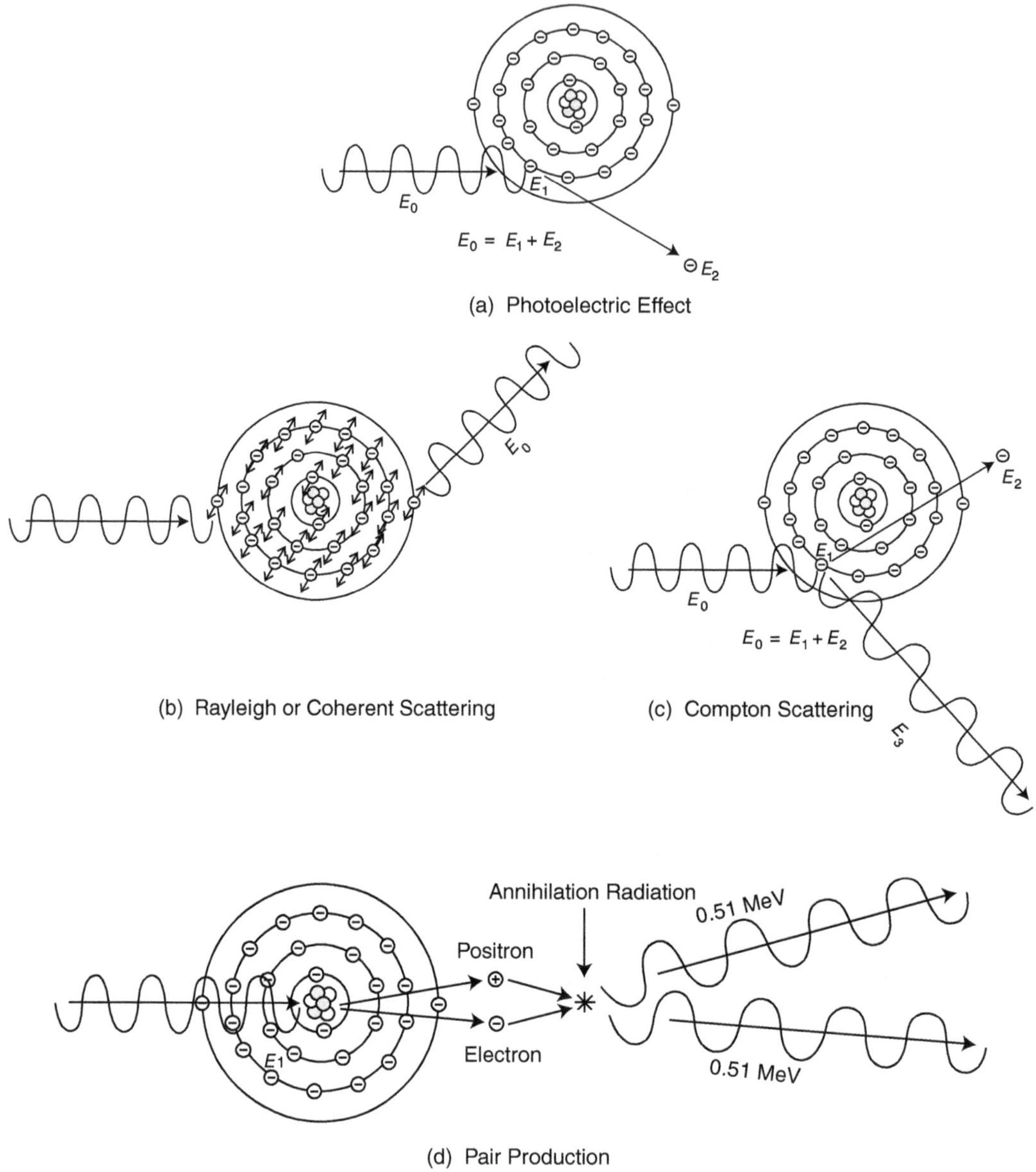


**FIGURE 6-9** Forms of scatter and absorption.

used is the "tenth-value layer." The same principle as the half-value layer applies, except that the thickness is somewhat greater and reduces the initial radiation to one-tenth on the opposite side. The factors involved with these half- and tenth-value layers are:

- *Energy.* The higher the energy, the thicker the half- or tenth-value layers. This is supported by the fact that the higher energy radiation produces shorter wavelength radiation, resulting in better penetration. The half- and tenth-value layers are especially useful for calculating the thickness of shielding when designing an enclosure or room for radiography. (See Table 6-1 for examples of half- and tenth-value layers.)

- *Material type.* The greater the material density, the thinner the half-value layer. Material that is low in density, such as aluminum or titanium, will allow more radiation to pass through, and there will be less absorption and scattering. Materials with higher density, such as steel and lead, provide a much greater chance of interaction because of their higher atomic number. With higher-density materials, there will be more absorption and the half-value layer thickness will be considerably less than with the lower-density materials.
- *Thickness.* As mentioned, the half-value layer thickness is specific for a given material and energy. As the thickness of the material increases, the amount of absorption and scatter increases and the amount of radiation that passes through that thickness decreases.

Figure 6-10 illustrates the half-value layer principle, using two different energies with the same test specimen. When there are discontinuities present in the material, the density through the cross section in that area of the test material will also vary, thereby affecting the amount of radiation that passes through the part. It is for this reason that radiography

| | Half-value | | Tenth-value | | Half-value | |
|---|---|---|---|---|---|---|
| X-ray (kV) | Lead (inches) | Concrete (inches) | Lead (inches) | Concrete (inches) | X-ray | Steel (inches) |
| 50 | 0.002 | 0.2 | 0.007 | 0.66 | 120 kV | 0.10 |
| 70 | 0.007 | 0.5 | 0.023 | 1.65 | 150 kV | 0.14 |
| 100 | 0.009 | 0.7 | 0.028 | 2.31 | 200 kV | 0.20 |
| 125 | 0.011 | 0.8 | 0.035 | 2.64 | 250 kV | 0.25 |
| 150 | 0.012 | 0.9 | 0.039 | 2.79 | 400 kV | 0.35 |
| 200 | 0.020 | 1.0 | 0.065 | 3.30 | 1 MeV | 0.60 |
| 250 | 0.032 | 1.1 | 0.104 | 3.63 | 2 MeV | 0.80 |
| 300 | 0.059 | 1.2 | 0.195 | 3.96 | 4 MeV | 1.00 |
| 400 | 0.087 | 1.3 | 0.286 | 4.29 | 6 MeV | 1.15 |
| 1,000 | 0.315 | 1.8 | 1.04 | 5.94 | 10 MeV | 1.25 |
| 2,000 | 0.393 | 2.45 | 1.299 | 8.60 | 16 MeV+ | 1.30 |
| Isotopes | | | | | | |
| Iridium 192 | 0.190 | 1.9 | 0.640 | 6.20 | | 0.60 |
| Cesium 137 | 0.250 | 2.1 | 0.840 | 7.10 | | 0.68 |
| Cobalt 60 | 0.490 | 2.6 | 1.620 | 8.60 | | 0.87 |
| Selenium 75 | 0.0394 | 1.18 | 0.187 | 3.54 | | |

**TABLE 6-1** Half- and Tenth-Value Layers

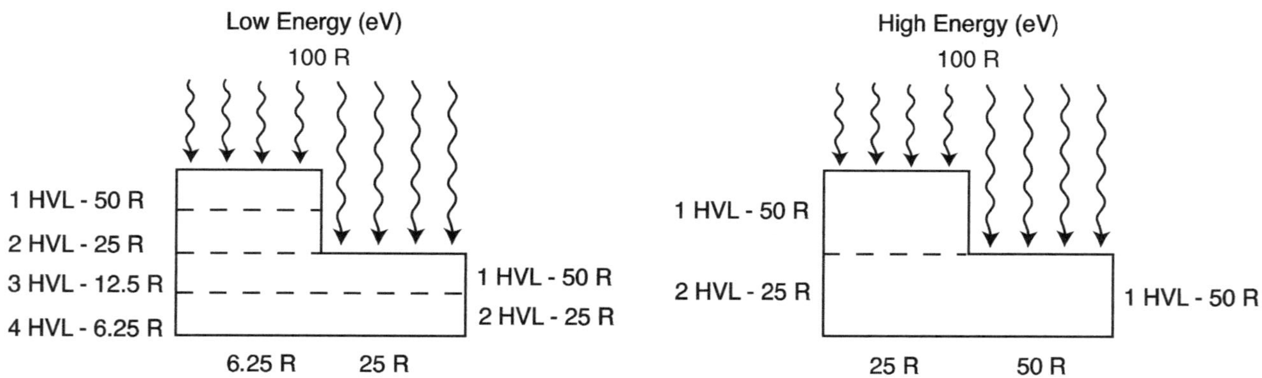


**FIGURE 6-10** Half-value layers.

is so effective for the detection of discontinuities, especially those that result in the reduction of material cross section. The densities of the different discontinuities vary greatly from the base material. If, for example, a weld or casting contains gas pores or other void-type discontinuities, the radiation will pass through these areas with relative ease, creating a higher exposure effect on the radiographic film, resulting in a darker region on the finished radiograph.

### 3. Radiation Travels in Straight Lines and at the Speed of Light

The speed of light is 186,300 miles per second or 299,800 km per second.

### 4. Radiation Exhibits Energy

On the electromagnetic spectrum, the wavelengths of x- and gamma rays are much shorter than that of visible light. Recall that as the radiation energy increases, shorter wavelengths are produced that provide greater penetration through the material being examined.

### 5. Radiation Ionizes Matter

Ionization is a change in the electrical nature or characteristics of matter. Electrons are displaced or knocked out of orbit, thereby changing the electrical balance. This, coincidentally, is what causes the greatest concern to humans. When radiation passes through living tissue, the cells are affected—electrically changed—and damage will result.

### 6. X- and Gamma Radiation Is Not Particulate

Radiation has no mass, and even though x- and gamma rays behave like particles, they are actually weightless. To describe the effect of radiation as it interacts with matter, the radiation is sometimes referred to as photons, which is another way of saying high-speed energy traveling at the speed of light that behaves like particles; but in fact, there are no particles in x- or gamma radiation.

### 7. Radiation Has No Electrical Charge

X- and gamma radiations are not affected by either strong electrical or magnetic fields.

### 8. X- and Gamma Radiation Cannot Be Focused

If x- or gamma radiation is directed toward a glass lens, the radiation would not focus like light does. In fact, the radiation passing through that lens would be absorbed to a greater extent in the thicker portion of the lens and more radiation would pass through the thinner regions of the lens.

### 9. Humans Cannot Sense X- or Gamma Radiation

Radiation cannot be seen, tasted, or felt. It has no odor and if humans are subjected to radiation exposure, the effects are not realized for a period of time. In other words, the body goes through a latent period before the harmful effects of the radiation exposure are evident.

### 10. Radiation Causes Fluorescence in Some Materials

Some minerals and salts will fluoresce when subjected to radiation. Fluoroscopy utilizes a fluorescent imager that converts some of the radiation to light, which provides a visible real-time image for viewing. When used to take a radiograph with film, the light that is emitted from the fluorescent screen creates a much higher exposure effect on the film that is adjacent to it, thereby significantly reducing exposure.

#### *11. Intensity of Radiation*

Radiation is expressed in roentgens per hour (R/hr), roentgens per minute (R/min), or milliroentgens per hour (mR/hr). The intensity decreases with distance. This is a function of the inverse square law, which states that the intensity of the radiation varies inversely with the square of the distance. As the distance from the radiation source is doubled, the intensity is decreased to one-fourth. This will be discussed in more detail in Section 6.4, Variables.

#### *12. Radiation Produces Chemical Changes in Exposed Film*

The film's emulsion is chemically changed through ionization as radiation interacts with it. If it weren't for this characteristic, it would not be possible to produce a radiographic image on film.

### 6.2.2 Principles of X-Radiography

X-rays are produced in a vacuum tube when high-speed electrons, which are negatively charged, are attracted by a positive potential in the anode and collide with a target material. The electrons are produced when a filament, usually tungsten, is heated to incandescence. The resulting electrons "boil" off and collide with the target material in the anode.

A higher current applied to the filament will result in a greater number of electrons and, therefore, higher radiation intensity. The energy of the radiation is a function of the applied voltage, or kilovoltage to the anode. This causes an increase in velocity of the electrons and, as higher voltage is applied, higher-energy radiation results. The energy of the radiation is its single most important characteristic. It is directly related to the wavelength and to the ability of that radiation to penetrate. The higher the energy, the shorter the wavelength, and the greater the penetration. The target in the anode in most X-ray tubes is typically made of tungsten. Tungsten has a high atomic number, which makes it an ideal material to interact with the high-speed electrons. It also has a high melting point, 5,880°F (3,249°C), which is essential since the target is subject to high temperatures as a result of electron impingement. Since almost all of the kinetic energy from the electrons is transformed or converted into heat, the anode must have the capability of dissipating that heat. In fact, 97 to 99 percent of the energy conversion results in the generation of heat and approximately 1 to 3 percent is converted to X-rays.

In summary, as the energy is increased, radiation with shorter wavelengths results in greater penetration. On the cathode side of the tube, which contains the filament and a focusing cup, the number of electrons increases with an increase of applied milliamperes, which results in an increase in intensity of the radiation. Figure 6-11a illustrates several typical X-ray tubes and a cross-sectional sketch of a tube can be found in Fig. 6-11b.

The target size in the anode is very important, since this affects the image sharpness of the object being radiographed. This is directly related to the definition of the final image. The smaller the target or radiation source, the sharper the resultant radiographic image.

As mentioned above, the energy of the radiation is expressed in voltage or kilovoltage, with one electron volt being the amount of voltage that is necessary to energize and displace one electron. The common expression for X-ray energy is in units of kiloelectron volts (keV), but this is shorted to the kV. A kilovolt is equal to one thousand volts. The majority of industrial radiographic techniques use an energy range from about 100 kV to 400 kV. The intensity of the radiation is expressed in milliamperes (mA) (one milliampere = one one-thousandth of an ampere). In a general way, the energy of the radiation can also be described as the "quality" of the radiation and the intensity (in milliamperes) as the

(a)

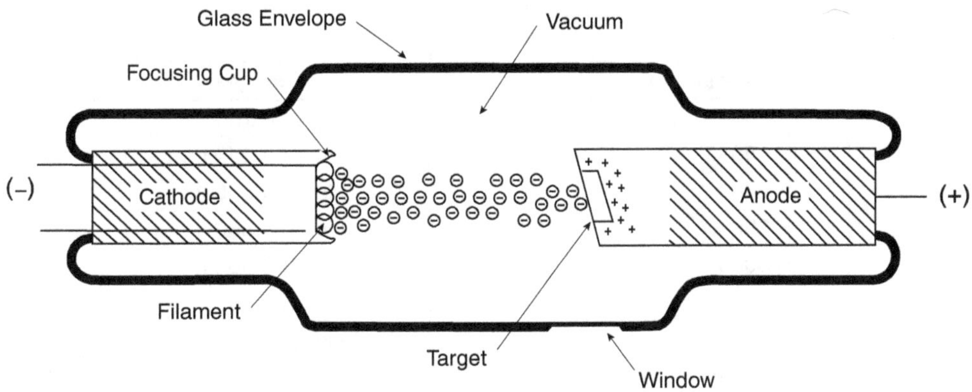


(b)

**FIGURE 6-11** (*a*) X-ray tubes and (*b*) sketch of cross section sketch. (*Courtesy of Charles J. Hellier.*)

"quantity." Energies above the 400 kV range are considered to be high-energy x-radiation. Equipment producing high-energy X-rays include:

- X-ray tubes with a resonance-type transformer
- Betatrons

- Linear accelerators
- Van de Graaff generators

The older of these high-energy X-ray units are the *resonant transformer* X-ray units, which were developed in the 1940s. Their energy range is from 250 kV to 4 MeV. The X-ray tube is symmetrically located within the transformer. The electrons are accelerated to very high velocities by electrodes and focused by a coil as they enter an extension chamber. It is possible to penetrate up to 8 in. of steel with this equipment.

*Betatrons* accelerate the electrons in a circular path by using an alternating magnetic field and typically produce energies in the range of 10 to 30 MeV. The tube is doughnut-shaped and the electrons increase their velocity with each orbit until they reach a very high energy. They are eventually guided from their circular path to a target and X-rays are generated.

*Linear accelerators* produce high-velocity electrons by the use of radio frequency energy coupled to a waveguide and have energies from 5 up to about 25 MeV. These systems contain an electron gun, an accelerator, and a waveguide. The electrons are accelerated along the waveguide by an electrical field.

The *Van de Graaff generators,* also known as electrostatic generators, produce X-ray energies from about 500 kV to as high as 6 MeV. The systems consist of high-speed belts that travel about 4,000 feet/minute and build up electrical charges that are carried to a high-voltage terminal. This results in a high electrostatic charge that is used to accelerate the electrons through the tube to the target.

Table 6-2 contains a list of the different energy levels produced by X-ray sources and approximate thickness penetration ranges that would be typical for steel.

| X-ray Energy | Thickness Range in Steel |
|---|---|
| 100 kV | up to 0.100 in. |
| 150 kV | up to 0.250 in. |
| 200 kV | 0.250 in. to 0.500 in. |
| 250 kV | 0.330 in. to 0.850 in. |
| 300 kV | 0.350 in. to 2.00 in. |
| 400 kV | 0.500 in. to 3.00 in. |
| 1.0 meV | 0.800 in. to 4.0 in. |
| 2.0 meV | 1.00 in. to 5.0 in. |
| 6.0–10.0 meV | 3.00 in. to 16.0 in. |
| 10.0–20.0 meV | 10.0 in. to 24.0 in. |
| 15.0–30.0 meV | 12.0 in. to 30.0 in. |
| Sources | |
| Selenium 75 | 0.125 in. to 1.50 in. |
| Iridium 192 | 0.500 in. to 1.50 in. |
| Cesium 137 | 0.650 in. to 2.0 in. |
| Cobalt 60 | 1.0 in. to 5.0 in. |

**Table 6-2** Radiation Energies and Practical Thickness Ranges

| Source | Symbol | Atomic Number | Atomic Weight | Isotope | Half-life | Energy | Emissivity |
|---|---|---|---|---|---|---|---|
| Cesium | Cs | 55 | 132.91 | 137 | 30 years | 0.66 meV | 4.2 R/C hr @ 1 ft |
| Cobalt | Co | 27 | 58.9 | 60 | 5.3 years | 1.17 meV<br>1.33 meV | 14.5 R/C hr @ 1 ft |
| Iridium | Ir | 77 | 192.2 | 192 | 75 days | 0.61 meV<br>0.21 meV | 5.9 R/C hr @ 1 ft |
| Radium | Ra | 88 | 226 | 226 | 1,602 years | 2.2 meV<br>0.24 meV | 9.0 R/C hr @ 1 ft |
| Selenium | Se | 34 | 78.96 | 75 | 120 days | 97 keV<br>401 keV | 0.595 R/C hr @ 1 meter |

**TABLE 6-3** Gamma Ray Source Data

### 6.2.3 Principles of Gamma Radiography

As mentioned earlier, the two most commonly used gamma ray sources in industrial radiography are iridium 192 and cobalt 60. Other sources that have been used to a limited extent in the industrial area are thulium 170 and cesium 137, and of course, radium 226. Table 6-3 contains a summary of these sources with characteristics and data regarding their half-life, energy, and penetration capabilities in steel.

There are certain characteristics regarding radioactive isotopes that should be understood. The first is energy. Energy of a radioactive isotope is unique to that particular, specific radioactive material. Some of these radioactive isotopes actually give off a spectrum, or a number of unique energy bundles. For sources, energy never changes for a given radioactive isotope. Radioactive isotopes decay with time. With the decay comes reduction in activity or quantity. The radioactive half-life is the amount of time that it takes for a specific radioactive isotope to decay or be reduced to one half of its original activity. Refer to Table 6-3 for examples of isotope half-lives. Recall that with an X-ray source, the intensity or quantity of radiation is a function of the applied milliamperes. With a radioactive isotope, the activity, or the amount of radiation that is given off by a source, is expressed in curies or becquerels. One curie is equal to $3.7 \times 10^{10}$ disintegrations per second and one becquerel is equal to one disintegration per second.

One way to understand disintegration is to visualize the atom of a radioactive isotope as having one neutron too many in the nucleus. In time, the nucleus literally breaks apart or disintegrates. In that act of disintegration, a bundle of radiation with a unique energy is given off. In addition, other by product materials are created. When there are 37 billion of these disintegrations occurring in a unit time of one second, there is one curie of that radioactive isotope. Again, the half-life is the time it takes for the source activity to be reduced to one-half of what it was originally. So, if an iridium 192 source is used when new and the activity is 100 curies, at the end of a half-life, which for iridium is approximately 75 days, the activity would be reduced to one-half, or 50 curies. If the source goes through another half-life, it will reduce to 25 curies; another half-life to 12½ curies; and so on. In other words, it does not matter how many half-lives the source goes through; it never reaches absolute zero.

For cobalt 60, the half-life is 5.3 years. Its energy is higher than iridium. For iridium 192, there are seven principal energy packets produced during the disintegration process ranging from just over 0.2 to about 0.6 MeV (approximately 200 to 600 kV). With cobalt 60, there are two discrete gamma energies (sometimes expressed as photon energies), 1.17 and 1.33 MeV. With this higher energy, it is possible to examine materials that are somewhat thicker or denser. Another factor to consider when using radioactive isotopes is their specific activity. The specific activity of a source is expressed in curies per gram. The greater the number of curies for a given physical size, the higher the specific activity. The major benefit of a high specific activity source is the improved definition that will result from the high activity (curies) and small physical size.

## 6.3 Radiographic Equipment and Accessories

### 6.3.1 X-ray Equipment

In general, industrial X-ray equipment can be categorized by energy groups. Those X-ray units that produce energies up to about 125 kV are considered low energy; from 125 up to 400 kV, medium energy; and those systems that produce radiation energies above 400 kV are considered high energy. As mentioned above, the majority of examinations performed today are done in the medium energy range, from 125 kV up to about 400 kV. It is the most widely used for a broad range of applications, and this chapter will focus primarily on that energy range.

When energy such as 300 kV is mentioned, it usually denotes the maximum energy output for that particular unit. For any given X-ray unit, the energy can usually be varied from a low energy up to the maximum energy for which the tube is rated. As discussed, X-rays are produced in a vacuum tube, by heating a filament that boils off negatively charged electrons. They are attracted to a positively charged anode, where there are very complex collisions and interactions that produce mainly heat, with only a small amount of the energy converted into X-rays.

Industrial X-ray systems come in different sizes, shapes, and duty cycles. There are X-ray units in cabinets that are ideal for radiographing smaller parts made of low-density material or with limited thickness. These cabinet units typically produce lower energies, usually up to about 150 kV, such as the one in Fig. 6-12.

There are X-ray units that are portable, with gas-filled tube heads. These are lightweight and are relatively easy to transport for field radiography. Many of the stationary units are self-rectified and continuous duty, such as the *constant potential units*, which can operate continuously for long periods of time. The gas-filled tubes operate with partial duty cycles, such as 50 percent. This means that the X-ray tube can operate for a specific time with an equal amount of time permitted for cool-down before the next exposure. Figure 6-13 illustrates a 300-kV X-ray tube head and controller.

#### *Gamma Ray Equipment*

As mentioned above, the first gamma ray source used in industrial radiography was radium. Figure 6-14 illustrates the "fish pole" technique in which the radium was transported inside a capsule on the end of a line that was attached to a fishing pole. The radiographer would place the "hook" on the end of the line through an eyelet

FIGURE 6-12 X-ray cabinet. (*Courtesy of Charles J. Hellier.*)

on the capsule, then remove it from the lead storage pig and carry it to the area where the object to be radiographed was located, as illustrated.

The major component in a gamma ray system is the exposure device, such as the one depicted in Fig. 6-15. These exposure devices have shields that contain an S-tube in which the iridium source is maintained inside a stainless steel capsule and is securely locked in place when not in use. The shield material is usually made from depleted uranium, unlike earlier gamma cameras that used shielding made of cast lead. Depleted uranium is slightly heavier than lead and has unusually good radiation

**Figure 6-13** X-ray tube head with controller. (*Courtesy of GE.*)

**Figure 6-14** The "fish pole" technique. (*Courtesy of Charles J. Hellier.*)

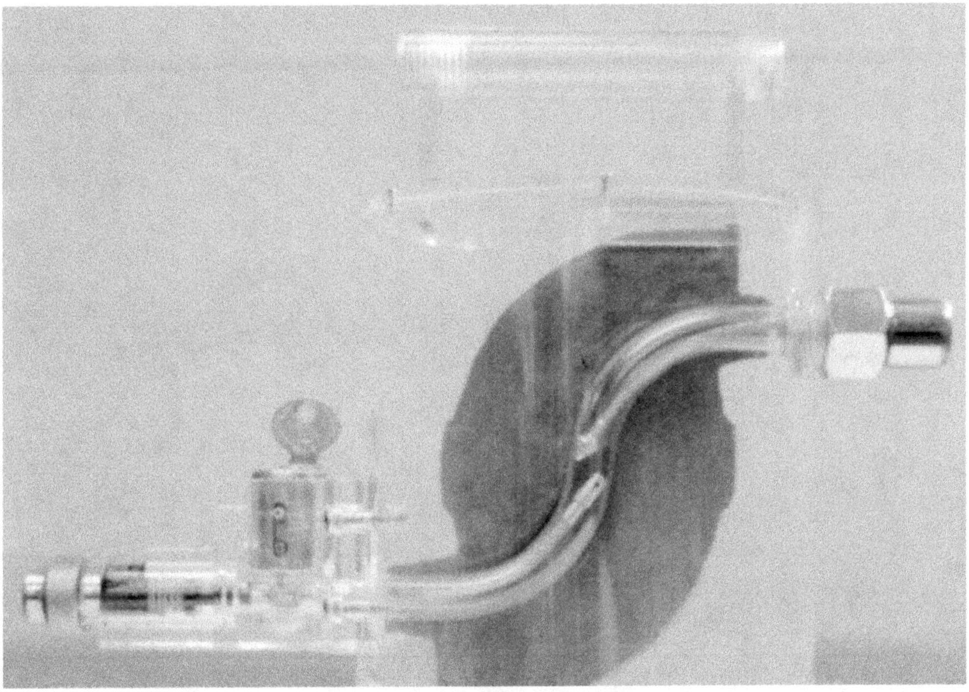

(a)

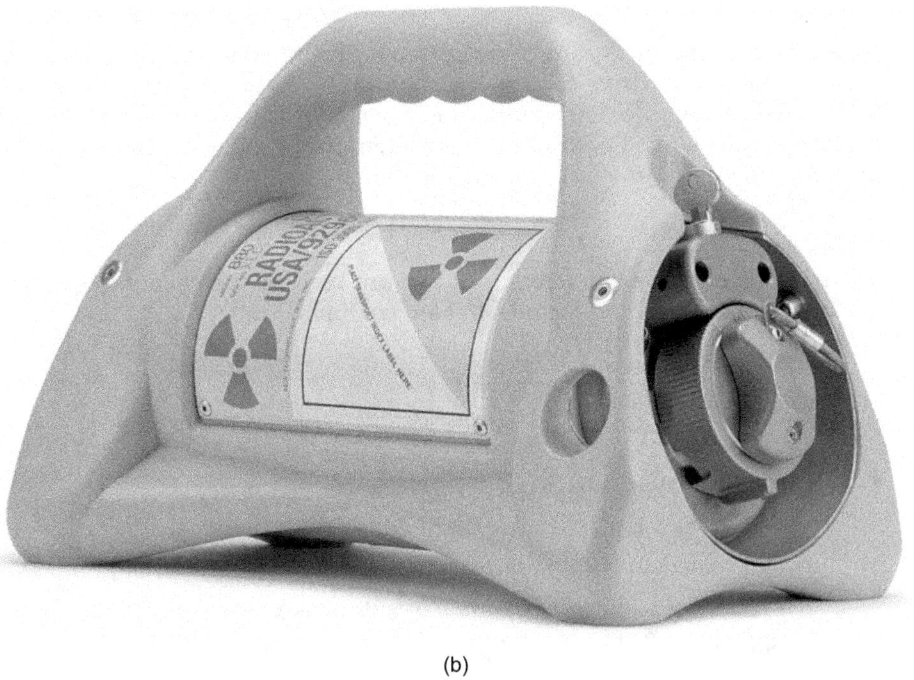

(b)

**Figure 6-15** (*a*) Gamma ray camera cutaway. (*b*) 150 curie Ir192 capacity camera. [(*a*) *Courtesy of Charles J. Hellier*; (*b*) *used with permission of Sentinel NDT*.]

**FIGURE 6-16** Gamma ray exposure setup. *Note:* The darkroom truck should not be in this position when exposures are taken otherwise unexposed film in the darkroom will be exposed. (*Courtesy of Charles J. Hellier.*)

shielding capabilities. A gamma ray set-up is illustrated in Fig. 6-16, showing the camera with the source drive control, the source tube that is positioned in a source stand just above the pipe weld that will be radiographed. The radiographer "cranks" out the source to the extended position. The predetermined exposure time is allowed to elapse and then the source is cranked back into the camera.

### Accessories

There are many different accessories that are used in conjunction with taking radiographs. These accessories include, but are not limited to:

- Film and cassettes
- Lead screens
- Lead numbers and letters
- Penetrameters (image quality indicators)
- Densitometers
- High-intensity film illuminators
- Processing equipment (manual or automatic)
- Film hangers (for manual processing)
- Collimators (especially for use with radioactive sources)

## 6.4 Variables

Of all the nondestructive testing methods described in this book, radiography certainly has the most variables. These variables include:

1. Energy[1]
2. Exposure time
3. mA (X-ray) or curies (gamma ray)
4. Material type and density
5. Material thickness
6. Type of film
7. Screens used
8. Film processing (procedure development time and temperature, etc.)
9. Film density
10. Distance from the radiation source to the object
11. Distance from the object to the film
12. Physical size of the target (X-ray) or source (gamma ray)

In order to control these variables so that the benefits can be maximized for each one, a technique chart should be used. Unfortunately, there are still many radiographs taken by the "trial and error" technique. Some radiographers will take three, four, or more exposures using different techniques, then, after processing the film, will decide which image looks best. The best way to produce a high-quality radiograph every time is through the use of exposure charts. An example of an exposure for a 270-kV X-ray machine is illustrated in Fig. 6-17. Notice that all of the variables are addressed in this chart. The use of this chart is quite simple. It can be easily understood by following a few logical steps. (*Note:* This chart is for demonstration purposes only and will not necessarily work for other X-ray machines of similar energy ratings.)

Step #1. Verify the material type (steel in this case) and find its thickness on the horizontal axis.

Step #2. Project a straight line vertically from that thickness up to the top of the technique chart and notice that there are a number of energies that can be used. If the material were one inch thick, for example, the energies that could be used would range from 160 kV to a maximum of 270 kV. It would not be advisable to use the lowest energy, since the exposure time would be unusually long. If, for example, 160 kV were to be used, the exposure time would be 50 minutes if a tube current of 10 mA were used (30 M or 30 thousand milliampere seconds divided by 10 mA = 3,000 seconds divided by

[1]Recall that energy is a function of the applied kilovoltage for X-rays and can be varied up to the maximum for a specific X-ray tube. The energy is fixed with a specific gamma ray source. Also recall that energy is the most important variable, since it affects the contrast (to be discussed later), and determines the thickness of material that can be effectively examined. It also influences the amount of scatter, which affects the definition or sharpness of the image.

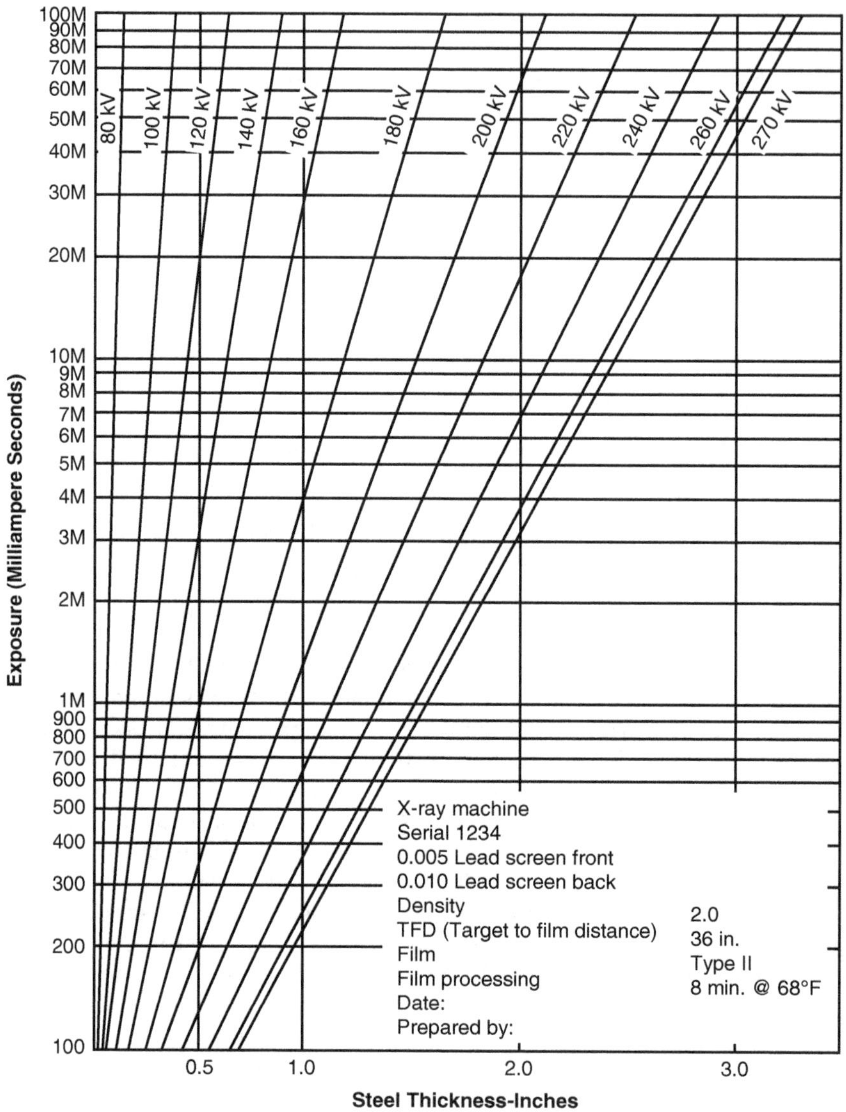


**FIGURE 6-17** Example of an X-ray technique chart.

60 = 50 minutes). This is done by projecting a line horizontally from the thickness/kV intersect point to the vertical. The other problem with this technique is that a greater amount of scatter will result due to the longer wavelengths that occur at the lower energies coupled with the long exposure time. Scatter reduces the clarity or sharpness of the image. If the highest energy were used (270 kV), the exposure time would be 22 seconds (220 milliampere seconds divided by 10 mA = 22 seconds). This would not be advisable, since high-energy techniques result in lower-contrast images. Contrast is the difference in density between two adjacent regions of the part, which are different in thickness or material density. So, it is not advisable to use the lowest or the highest kV

based on the technique chart. The obvious choice would be to select energy somewhere in the middle. For this example, either 200 or 220 kV would be good choices.

Step #3. As explained in Step #2, once the energy is selected, a line is projected to the left vertical axis, where the exposure time and intensity are combined in units of milliampere seconds (mAs). Again, mA is the current applied to the filament that controls the quantity of radiation. Time, in this case in units of seconds, and mA are inversely proportional values. If, for example, a radiograph was taken with an exposure of 5 mA at 240 seconds (4 minutes), the exact same image would be produced if 10 mA were used for 120 seconds (2 minutes), a reduction in exposure time of 50 percent. In each case, there is an exposure time/intensity value of 1,200 milliampere seconds (20 milliampere minutes). So, at the intersection of 1 in. and the 200 kV curve, project over to the vertical axis, and the exposure time in milliampere seconds will be 1,400 mAs (2 minutes 20 seconds), and at 220 kV, the exposure time would be about 650 mAs (1 minute 5 seconds). The exposure was calculated using a tube current of 10 mA. Notice that with just an increase of 20 kV, there is virtually a 50 percent reduction in exposure time. The resultant image quality would be virtually the same using either of these kV settings.

A similar exposure chart for iridium 192 is illustrated in Fig. 6-18. As with the X-ray technique chart, this is an example and may not provide exact results. Notice that there are no energy curves since energy is a fixed value for a given gamma ray source. In this technique chart, there are three different film types having different speeds. The way to establish the correct exposure time is similar to that of the X-ray technique chart. The first step (assuming the material is steel) is to project the material thickness vertically until it intersects with the film type being used. Then draw a line horizontally to the vertical axis. From that axis, the exposure factor (EF) is determined. Using the equation at the top of the chart and knowing the number of curies of iridium 192, an exposure time in minutes can be easily calculated.

**Example**

Thickness = 1.5 in.

Type II film will be used

$D$ (source to film distance) = 36 in. or 3′

$S$ = 60 curies

$T$ = exposure time in minutes (for a film density of 2.0)

Projecting vertically up the chart, the point of intersection with the Type II film is then projected horizontally. The EF (exposure factor) is 50.

$$T = \frac{50 \times (3')^2}{60}$$

$$T = \frac{405}{60}$$

$$T = 7.5 \text{ minutes}$$

If the material to be examined is other than steel, exposure times can be easily calculated by using the radiographic equivalence factors (REFs). Some REFs are illustrated in Table 6-4. (*Note:* It should be mentioned that these factors are approximate but will be sufficient to enable a useable technique to be established.) In order to obtain an equivalent

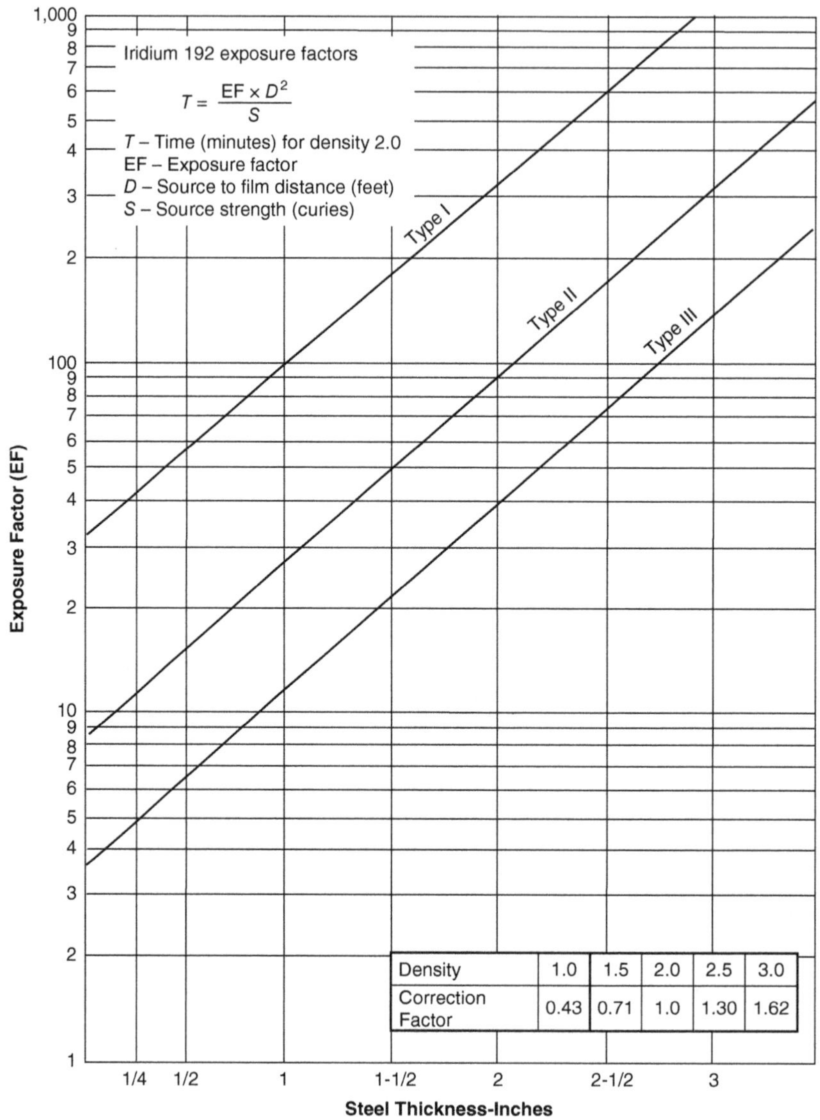


| Density | 1.0 | 1.5 | 2.0 | 2.5 | 3.0 |
|---|---|---|---|---|---|
| Correction Factor | 0.43 | 0.71 | 1.0 | 1.30 | 1.62 |

**FIGURE 6-18** Gamma ray technique chart.

thickness of a material other than steel, the REF should be multiplied by the material thickness. The relevant answer will be the equivalent thickness of steel. Once this thickness is known, the technique chart can be used as if the material were steel.

**Example**

Material is aluminum

Thickness = 5.0 in.

Energy to be used is 220 kV

| | X-rays (kV) | | | | | | | Gamma Rays | | |
|---|---|---|---|---|---|---|---|---|---|---|
| | 50 | 100 | 150 | 220 | 400 | 1,000 | 2,000 | Ir-192 | Cs-137 | Co-60 |
| Magnesium | 0.6 | 0.6 | 0.05 | 0.08 | | | | 0.22 | 0.22 | 0.22 |
| Aluminum | 1.0 | 1.0 | 0.12 | 0.18 | | | | 0.34 | 0.34 | 0.34 |
| Titanium | | 8.0 | 0.63 | 0.71 | 0.71 | 0.9 | 0.9 | 0.9 | 0.9 | 0.9 |
| Steel | | 12.0 | 1.0 | 1.0 | 1.0 | 1.0 | 1.0 | 1.0 | 1.0 | 1.0 |
| Copper | | 18.0 | 1.6 | 1.4 | 1.4 | 1.1 | 1.1 | 1.1 | 1.1 | 1.1 |
| Zinc | | | 1.4 | 1.3 | 1.3 | 1.1 | 1.0 | 1.1 | 1.0 | 1.0 |
| Brass | | | 1.4 | 1.3 | 1.3 | 1.2 | 1.2 | 1.1 | 1.1 | 1.0 |
| Lead | | | 14.0 | 12.0 | | 5.0 | 2.5 | 4.0 | 3.2 | 2.3 |

*Note:* Aluminum is taken as the standard metal at 50 kV and 100 kV and steel at the higher voltages and with gamma rays.

**Table 6-4** Radiographic Equivalence Factors

The REF obtained from the chart for aluminum at 220 kV is 0.18 in. So:

$$\text{REF}_{\text{AL}} \times \text{material thickness} = \text{equivalent thickness in steel}$$

$$0.18 \times 5.0 \text{ in.} = 0.9 \text{ in.}$$

Therefore, a technique for 0.9 in. of steel from the technique chart can be used for the 5.0 in. of aluminum.

In the event there is reason to use a film other than Type II, film characteristic curves (Fig. 6-19) can help to determine the new exposure time. (*Note:* These characteristic curves are examples only. Film manufacturers can provide accurate curves that can be used to calculate precise correction values.) Notice that for gamma rays, the technique chart (Fig. 6-18) has provisions for three different films.

**Example** If it is desired to obtain a radiograph with better definition, a change from the Type II film to a Type I would be appropriate. The 2.0 density (vertical axis) intersects with the Type II film at 1.9 (horizontal axis), which is the log relative exposure. Continuing on the 2.0 film density line over to the Type I film will result in a log relative exposure of approximately 2.55 or a difference of 0.65 in the log relative exposure. The inverse log of 0.65 is 4.5 (rounded off). The new exposure time would be calculated by multiplying the correction factor of 4.5 by the original exposure time used to achieve the 2.0 film density. If the original exposure was 3 minutes, the new exposure time required to achieve a film density of 3.0 would be 13 minutes, 30 seconds.

This film characteristic curve approach can also allow a given density to be changed to any desired density. For example, on the technique chart for X-rays, the resultant film density will be 2.0. For many codes and specifications, a film density of 2.0 is the minimum permitted. It would be much more desirable to produce a radiograph with a higher density. Higher densities produce better radiographic contrast. So, in order to achieve a film density that is different than the one stipulated on the technique chart, the film characteristic curves can be used.

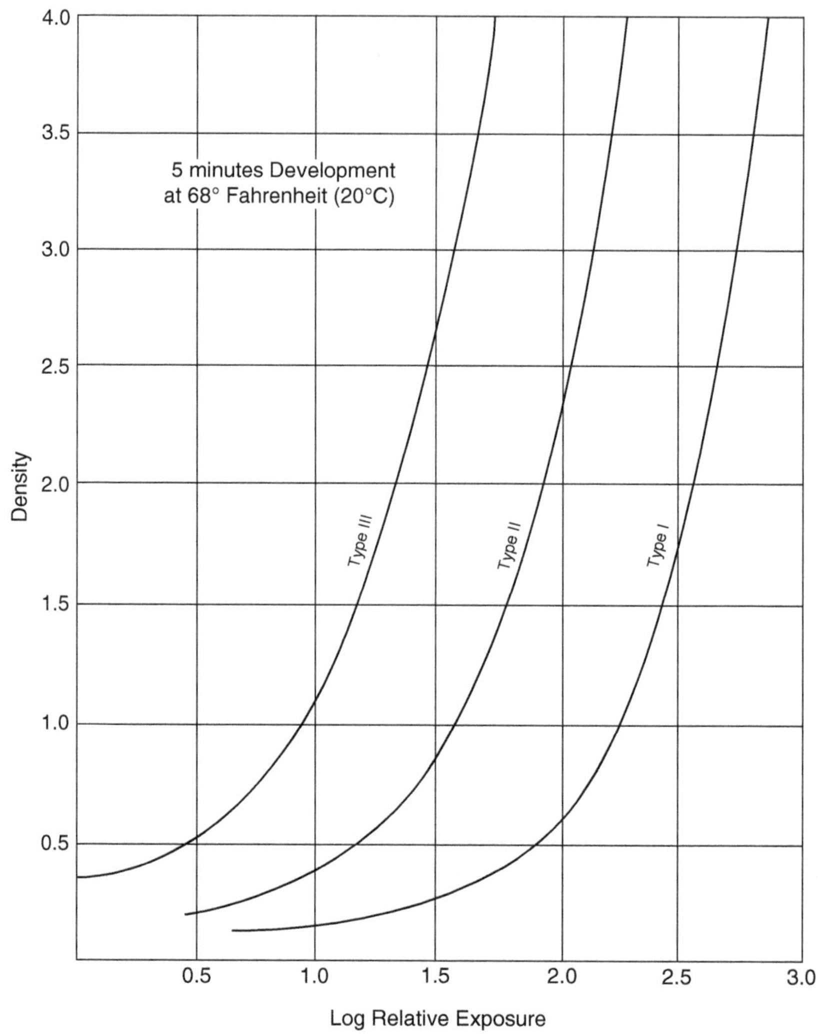


**FIGURE 6-19** Film characteristic curves.

**Example** Using a Type II film, a density of 3.0 is desired instead of the 2.0 as indicated on the technique chart (Fig. 6-19). As in the previous example, the log relative exposure for a Type II film with a density of 2.0 is 1.9. The 3.0 density intersects the Type II film at approximately 2.1. The difference between the 2.1 and 1.9 log relative exposures is 0.2. The inverse log of 0.2 equals 1.6 (rounded off). If the original exposure time were calculated from the technique chart to be 3 minutes (for the 2.0 density), that time would be multiplied by the correction factor of 1.6, for a new exposure time of 4.8 minutes. If all other variables are the same as with the original exposure, a film density of 3.0 will be achieved if the new exposure time of 4 minutes, 48 seconds is used. It should be noted that correction factors are included on the technique chart for iridium 192.

In addition, a specific type of film with one density can be changed directly to another film type with a different density using the same approach.

**Example** If the Type II film and the film density as indicated on the X-ray technique chart is to be changed to a different film (Type I for example, with a density of 2.5), it can be accomplished as in the previous examples. As seen, the Type II film with the 2.0 density has a log relative exposure of 1.9. The Type I film with a 2.5 density has a log relative exposure value of approximately 2.65. The difference in log relative exposures is 0.75 and the inverse log is about 5.6. Multiplying the original exposure time by the correction factor of 5.6 will give a suitable exposure time to achieve a film density of 2.5 with Type I film.

Another variable to be considered is the source to film distance (SFD). Occasionally, the source to film distance is also referred to as the target to film distance (TFD). The TFD generally applies when using an X-ray source and the SFD applies when radioactive isotopes are used. This distance is very critical and plays a major role in establishing the correct technique. Notice on the X-ray technique chart that the TFD is 36 inches. There may be occasions when the 36 inch TFD is not possible because of geometric limitations and a shorter distance may be necessary. On the other hand, the TFD may be increased, which would result in an image with better definition. There is a mathematical relationship between the exposure time and the distance. The technique chart, if used properly, will provide the correct exposure time at a distance of 36 inches. If a different distance is used, a simple equation that gives the relationship between time and distance should be used. The results are very precise and enable the radiographer to essentially use any TFD (X-ray) or SFD (gamma ray) that is appropriate for the object that is being radiographed. The equation for this calculation is:

$$\frac{T_1}{T_2} = \frac{D_1^2}{D_2^2}$$

where:

$T_1$ = original exposure time derived from the technique chart
$T_2$ = new exposure time
$D_1$ = original distance
$D_2$ = new distance

**Example** A new distance of 48 in. is desired to achieve better definition. The original exposure time at 36 in. was 3 minutes. The exposure time required at the new distance can be calculated as follows:

$T_1 = 3$ minutes

$D_1 = 36$ in. (3′)

$D_2 = 48$ in. (4′)

$T_2 = \times$

$$T_2 = 3 \text{ minutes} \times \frac{(4)^2}{(3)^2}$$

$$T_2 = 3 \text{ minutes} \times \frac{16}{9}$$

$T_2 = 5.33$ minutes or 5 minutes and 20 seconds

The other variable that all radiographic personnel must be aware of is the inverse square law. This is important as it applies to technique development but also, more importantly, as it relates to safety. As mentioned, the intensity of radiation varies inversely proportional to the square of the distance and is expressed mathematically in the following equation:

$$\frac{I_1}{I_2} = \frac{D_2^2}{D_1^2}$$

where:
$I_1$ = intensity at distance $D_1$
$I_2$ = intensity at distance $D_2$
$D_1$ = initial distance
$D_2$ = new distance

### 6.4.1 Image Quality

By optimizing the variables that have been discussed, it is feasible to achieve a radiographic image with a high-quality level. The goal of the radiographer should be to achieve the highest possible quality image of the object that is being radiographed. The quality of the radiographic technique is established by comparison with the image of an "image quality indicator" (IQI), also referred to as penetrameter, on the completed radiograph. There are two major types of penetrameters in use today: the shim and wire types. The "shim" or hole type contains a lead identification number at one end and three holes of different diameters. The thickness of the shim-type penetrameter is based on a percentage of the material thickness that is being radiographed. The most common penetrameter thickness is 2 percent of that thickness. The three holes as seen in Fig. 6-20a have different diameters, which are based on the penetrameter thickness.

The smallest hole located in the center is referred to as the 1T. The "T" in the penetrameter designation represents the penetrameter thickness. The 1T hole, therefore, will have a diameter equivalent to the thickness of the penetrameter, which again is usually 2 percent of the thickness of the material being radiographed. The hole diameter at the end opposite to the lead identification numbers is referred to as the 2T hole. Its diameter will be two times the penetrameter thickness. The largest hole is the 4T hole and that diameter will be four times the penetrameter thickness. The outline of the penetrameter and the image of the holes on the finished radiograph establish the quality level for that radiographic technique. It is unfortunate that the hole sizes are, at times, used as a comparison to estimate the size of discontinuities that are detected. This is not the intent of the penetrameter. For example, if the diameter of a 2T hole is a 0.020 in., a discontinuity of that size will not necessarily be detected. Penetrameter holes are very precisely manufactured into the penetrameter shim. They have sharp edges. Discontinuities are quite different from these holes. The penetrameter is used to establish the quality level of the radiographic technique and should not be compared to discontinuities. The typical designations for radiographic quality levels (RQLs) using the shim-type penetrameters will be a combination of two numbers and a letter. For example, the most common quality level is 2-2T. This designation means that a penetrameter whose shim is 2 percent of the material being radiographed must display the 2T hole (which is two times the thickness of the penetrameter) clearly and distinctly on the radiograph. When the 2T hole is clearly discernible, the 4T hole, which is larger, should also be discernible. The 1T may, or may not, be clearly visible. If a 2-1T quality level is specified, the 1T hole must be clearly discernible in a 2 percent thick penetrameter shim. A 2-4T quality level will require the 4T-hole image to be discernable in the radiograph. Although the standard penetrameter shim thickness is typically about 2 percent of the material being radiographed, a 1 percent shim thickness may also be specified. This means that the penetrameter thickness is 1 percent of the material being radiographed. This designation is used when a higher level of quality is required in the radiographic technique. A listing of common shim-type penetrameter designations is found in Table 6-5.

(a)

(b)

**Figure 6-20** (*a*) Hole-type penetrameter on a shim and (*b*) wire penetrameters. (*Courtesy of Charles J. Hellier.*)

The second most commonly used image quality indicator is known as the "wire" penetrameter. The wire penetrameter has been in use for decades in Germany and other European countries. In fact, the first wire penetrameter was developed in Germany and was known as the DIN (Deutsche Industrie-Norm) type penetrameter. Typical ASTM wire penetrameters are illustrated in Fig. 6-20b. Notice that they contain a series of six wires encased in a clear plastic holder with lead identification numbers. Each of the six wires is of a slightly different diameter. There are four general wire sets described in Table 6-5.

Both shim and wire type IQIs must be placed on the source side of the object being radiographed whenever possible. There may be occasions when it is impossible to put the penetrameter on the source side due to geometric considerations or accessibility and, in those cases, some codes permit them to be placed on the film side, between the object being radiographed and the film. When a film side penetrameter is used, the required penetrameter designation is less than stipulated for the source side, since the quality level is somewhat easier to achieve, due to the closeness of the penetrameter to the film.

There are a number of other different designs and types of penetrameters; however, the shim and wire types are the most widely used. There is also a step-hole type that is used, primarily in France. There are codes today that permit the use of either the shim-type penetrameter or the wire-type penetrameter and their equivalence has been proven. Table 6-6 makes a comparison between the shim-type penetrameter and the wire-type penetrameter for the purpose of comparable quality levels.

The importance of achieving the highest possible quality level cannot be stressed enough. Those variables affecting quality level are listed in Table 6-7.

Note that the two major factors that influence sensitivity are contrast and definition. Contrast is the difference in density on two adjacent regions of the radiograph. Definition is the sharpness, or clarity, of the image as has been mentioned. Contrast is affected primarily by energy. Other factors such as film type, differences in thickness of the specimen, density of the test object, and the degree of scatter radiation will also influence contrast. Placing a filter at the source of radiation also affects contrast. On the other hand, the sharpness of the object's image is mostly controlled by geometric considerations such as the size of the source of radiation, the source to object distance, and the object to film distance. The type of screens that are used and their contact with the film will also affect sharpness. In fact, many codes stipulate that the unsharpness, or lack of definition in a radiographic technique, be calculated by using the unsharpness equation:

$$Ug = \frac{F \times t}{d}$$

where:

$Ug$ = geometric unsharpness

$F$ = the physical size of the target or source of radiation (*Note:* For some codes, the physical size is defined as the maximum projected dimension, i.e., the diagonal.)

$t$ = the distance from the source side of the test object (area of interest) to the film

$d$ = the distance from the source to the top of the test object (area of interest)

At times, this equation may be used improperly. The unsharpness equation should be used to verify or to ensure that the image sharpness meets the various code and specification requirements. Often, radiographers will use this equation to calculate the absolute minimum source to film distance that can be used and still meet the code required unsharpness values. In fact, in far too many cases the goal is to just "barely meet" the

| Wire Penetrameter | | | |
|---|---|---|---|
| Set A | | Set B | |
| Wire Diameter, in. (mm) | Wire ID No. | Wire Diameter, in. (mm) | Wire ID No. |
| 0.0032 (0.08)* | 1 | 0.010 (0.25) | 6 |
| 0.004 (0.1) | 2 | 0.013 (0.33) | 7 |
| 0.005 (0.13) | 3 | 0.016 (0.4) | 8 |
| 0.0063 (0.16) | 4 | 0.020 (0.51) | 9 |
| 0.008 (0.2) | 5 | 0.025 (0.64) | 10 |
| 0.010 (0.25) | 6 | 0.032 (0.81) | 11 |
| Set C | | Set D | |
| Wire Diameter, in. (mm) | Wire ID No. | Wire Diameter, in. (mm) | Wire ID No. |
| 0.032 (0.81) | 11 | 0.10 (2.5) | 16 |
| 0.040 (1.02) | 12 | 0.126 (3.2) | 17 |
| 0.050 (1.27) | 13 | 0.160 (4.06) | 18 |
| 0.063 (1.6) | 14 | 0.20 (5.1) | 19 |
| 0.080 (2.03) | 15 | 0.25 (6.4) | 20 |
| 0.100 (2.5) | 16 | 0.32 (8) | 21 |

| Shim (Hole) Penetrameters | | | | |
|---|---|---|---|---|
| Penetrameter Designation | Penetrameter Thickness | 1T Hole Diameter | 2T Hole Diameter | 4T Hole Diameter |
| 5 | 0.005 | 0.010 | 0.020 | 0.040 |
| 7 | 0.0075 | 0.010 | 0.020 | 0.040 |
| 10 | 0.010 | 0.010 | 0.020 | 0.040 |
| 12 | 0.0125 | 0.0125 | 0.025 | 0.050 |
| 15 | 0.015 | 0.015 | 0.030 | 0.060 |
| 17 | 0.0175 | 0.0175 | 0.035 | 0.070 |
| 20 | 0.020 | 0.020 | 0.040 | 0.080 |
| 25 | 0.025 | 0.025 | 0.050 | 0.100 |
| 30 | 0.030 | 0.030 | 0.060 | 0.120 |
| 35 | 0.035 | 0.035 | 0.070 | 0.140 |
| 40 | 0.040 | 0.040 | 0.080 | 0.160 |
| 45 | 0.045 | 0.045 | 0.090 | 0.180 |
| 50 | 0.050 | 0.050 | 0.100 | 0.200 |
| 60 | 0.060 | 0.060 | 0.120 | 0.240 |
| 80 | 0.080 | 0.080 | 0.160 | 0.320 |
| 100 | 0.100 | 0.100 | 0.200 | 0.400 |

*The 0.0032 wire may be used to establish a special quality level if agreed upon between the purchaser and the supplier.

**TABLE 6-5** Shim- and Wire-Type Penetrameter Data

| Shim (Hole)-Type Penetrameter Designation | Wire Diameter Equivalence to Hole Diameter | | |
|---|---|---|---|
| | 1T | 2T | 4T |
| 5 | — | — | — |
| 6 | — | 0.004 | 0.006 |
| 8 | 0.0032 | 0.005 | 0.008 |
| 10 | 0.004 | 0.006 | 0.010 |
| 12 | 0.005 | 0.008 | 0.013 |
| 15 | 0.006 | 0.010 | 0.016 |
| 17 | 0.008 | 0.013 | 0.020 |
| 20 | 0.010 | 0.016 | 0.025 |
| 25 | 0.013 | 0.020 | 0.032 |
| 30 | 0.016 | 0.025 | 0.040 |
| 35 | 0.020 | 0.032 | 0.050 |
| 40 | 0.025 | 0.040 | 0.063 |
| 50 | 0.032 | 0.050 | 0.080 |
| 60 | 0.040 | 0.063 | 0.100 |
| 70 | 0.050 | 0.080 | 0.126 |
| 80 | 0.063 | 0.100 | 0.160 |
| 100 | 0.080 | 0.126 | 0.200 |

**Table 6-6** Penetrameter Comparison

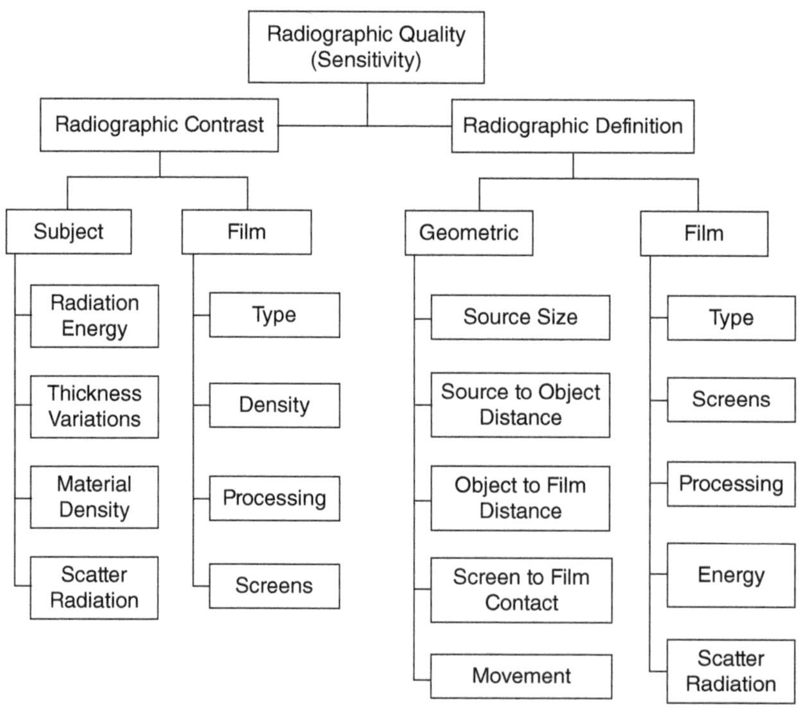


**Table 6-7** Radiographic Quality Variables

quality level requirements. It seems that the variables that directly reduce the quality of the image are those that some radiographers will employ to take the radiographs in the shortest practical time. For example, it should be obvious that when the source to film distance is greater, the image will be sharper, but this requires a longer exposure time. Many radiographers will use the absolute minimum source to film distance to reduce the exposure time. The highest energy available with an X-ray unit should not be used because contrast is reduced. In too many cases, the higher energy is chosen and the resultant radiograph does not have that high-contrast image that makes interpretation much more meaningful. Radiographic film is another area where high quality is sacrificed for speed. Fast film, which results in images with noticeable graininess, is used in preference to the slow film in order to achieve the shortest exposure time. The slow, fine-grain film, which requires longer exposure times, produces a much sharper image with better definition. When all these factors are combined (a fast film with a short source to film distance, and the highest energy used), the resultant image will fall far short of the sensitivity goal that can be achieved. The goal should be to always use a technique that will provide the highest possible quality image.

Lead screens have a major effect on the quality level, primarily affecting definition. Lead screens serve two very important functions in radiography: they intensify the primary radiation and they absorb "soft" or secondary scatter radiation. Lead screens are typically 0.005 inches or 0.010 inches in thickness. Based on the energy of the radiation, either will be used. One lead screen will be placed on top and the other beneath the film, inside a light-tight cassette. The purpose of the top lead screen is to intensify the primary radiation that passes through the object. This intensification occurs as a result of the emission of secondary electrons from the interaction of the photons with the atoms in the lead. This intensification actually creates a sharper, denser image on the film because of this secondary emission. The lead screens also serve to absorb scatter radiation. These lead screens easily absorb scatter, which consists of long-wavelength, lower-energy radiation, whereas the primary radiation passes through with relative ease. The primary source of scatter radiation is from the object itself, and results when the radiation passes and interacts with the material atoms. Again, radiation will penetrate the object, be scattered, and be absorbed. With lead screens, the scattered radiation is absorbed to a great extent and the primary beam is intensified.

The lead screens under the film also improve the image quality by absorbing backscattered radiation, which comes from objects such as floors and tables, underneath the film cassette. Lead screens are effective for both X-rays and gamma rays with energies above approximately 125 kV. When lower energies are used, the lead screens tend to absorb a portion of this lower-energy radiation rather than intensifying it. For gamma radiography using iridium 192, it is quite common to use 0.010-inch lead screens on the top and underneath the film.

Any type of contaminant or blemish on the lead screen surfaces, such as the lower screen in Fig. 6-21, will cause corresponding images of that condition on the radiographic film, so it is extremely important that these lead screens be kept in excellent condition. They should be virtually mirror smooth and should be inspected periodically to ensure that there are no conditions that would cause artifacts. It is recommended that a small, unique identification number be scratched into the lead at one of the corners. In the event that there is an artifact, it will be easy to see that number as an image on the film, facilitating the identification and removal of the faulty lead screen.

**FIGURE 6-21** Lead screens (screen on bottom is unacceptable). (*Courtesy of Charles J. Hellier.*)

Other types of screens used in radiography are the fluorescent types. Fluorescent and fluorometallic screens result in a significant reduction in exposure time since the radiation causes visible light to be emitted that greatly increases the exposure effect on the film. The major problem with fluorescent screens is that light diffusion occurs, which reduces the definition of the image. It is, however, an excellent way to radiograph test specimens that are either extremely thick or dense, where conventional radiography with lead screens would be impossible. Quality levels that would meet most code requirements are difficult to achieve with the use of fluorescent screens. This again is due to the light diffusion of the fluorescent material in the screens.

## 6.5 Techniques and Procedures

The most effective technique used in radiography is one in which the radiation passes through a single thickness and the film is in contact with the surface opposite the source side. This technique is referred to as the "single wall exposure, single view technique" (Fig. 6-22). The radiation passes through one wall of the object (a single thickness), and that thickness is evaluated or "viewed."

The second most common technique is referred as the "double wall exposure, single view technique." In this case, the radiation passes through two walls but only that area closest to the film is evaluated. This technique is depicted in Fig. 6-23.

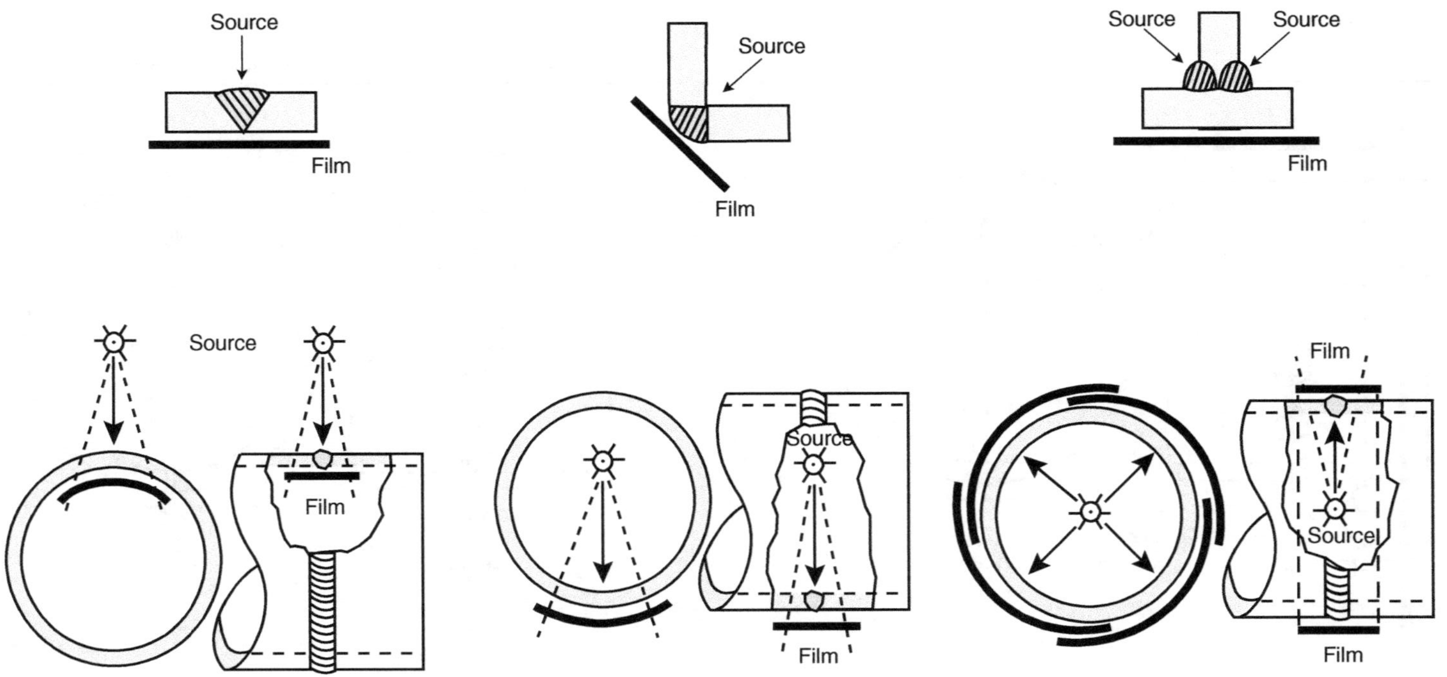


**Figure 6-22** Single wall exposure, single view technique.

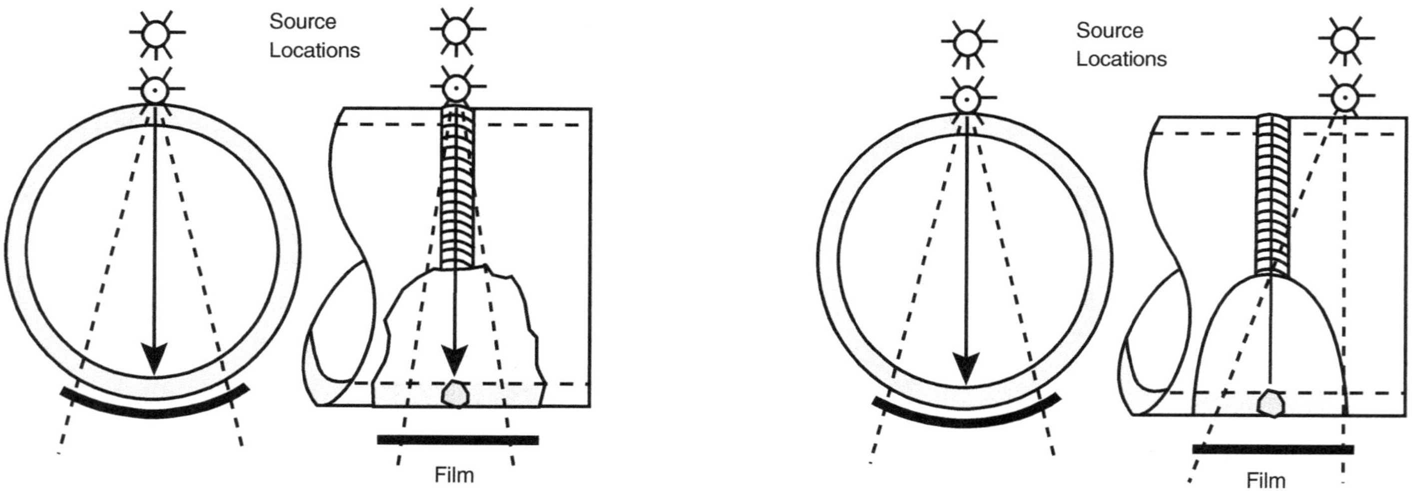


**FIGURE 6-23** Double wall exposure, single view technique.

The third radiographic technique requires the radiation to pass through both walls of the object and both walls are evaluated. This technique is referred to as the "double wall exposure, double view technique" and is usually restricted to parts with small diameters, typically equal to or less than 3.5 in. This technique is illustrated in Fig. 6-24. In this technique, the source of radiation may be positioned directly over the area of interest, thus superimposing the top portion with the region directly under it. As an alternative, the source may be offset by an angle of approximately 15° in order to observe both the top and bottom walls. When the source is offset, the technique is often referred to as the "ellipse" or "elliptical" technique.

In the single wall exposure, single view technique, the penetrameter is placed on the source side of the object (the surface opposite the film). In the double wall exposure, single view technique, the penetrameter may be on the source side of the object or the film side, depending upon accessibility. The double wall exposure, double view technique usually permits the penetrameter to be placed on the source side.

In all of these techniques, the goal is to achieve the highest possible quality level. The penetrameter placement should be as close to the area of interest as possible. If the object being radiographed is a weld, the penetrameter should be as close as possible to the weld, without interfering with the weld image. In the case of castings, it may be necessary to place the penetrameter in the area of interest, depending upon the size and shape of the casting. In other cases, the technique may permit an object of similar radiographic density and thickness to be placed adjacent to the test object. This will serve as a location for the penetrameter to be placed. If the weld has reinforcement, it may be necessary to use a shim of radiographically similar material upon which the penetrameter is placed, as illustrated in Fig. 6-20a. The purpose of the shim is to provide the penetrameter with a through thickness that will closely approximate the weld thickness, including the reinforcement. This is especially important when the codes and specifications require that the density range in the area of interest have a density within a certain percentage of the density through the penetrameter image. More details on this will be found in Section 6.5.3.

With all of these techniques, it is necessary that there be some form of "shooting sketch" that indicates the key aspects of the technique, such as the source location, film location, penetrameter placement, and test part position. This serves as a great aid to the radiographic film interpreter who may not be totally familiar with the radiographic technique used.

One other important aspect of radiography has to do with the source placement. In far too many cases, the radiographic technique is designed to show the best possible image of the penetrameter. It seems that the goal in many cases is to produce a good "picture" of the penetrameter and the essential hole, whereas the goal should be to display the best images of discontinuities, especially those that may not be oriented in a direction favorable to the radiation source. Radiography is extremely sensitive to the orientation of tight planar discontinuities. If a tight planar discontinuity is expected to be at an angle to the source of radiation, it will be difficult if not impossible to detect. These aspects of discontinuity orientation should be taken into consideration when developing the technique. Since the purpose of radiography is to detect discontinuities, the nature, location, and orientation of the expected discontinuities should always be a major factor in establishing the technique.

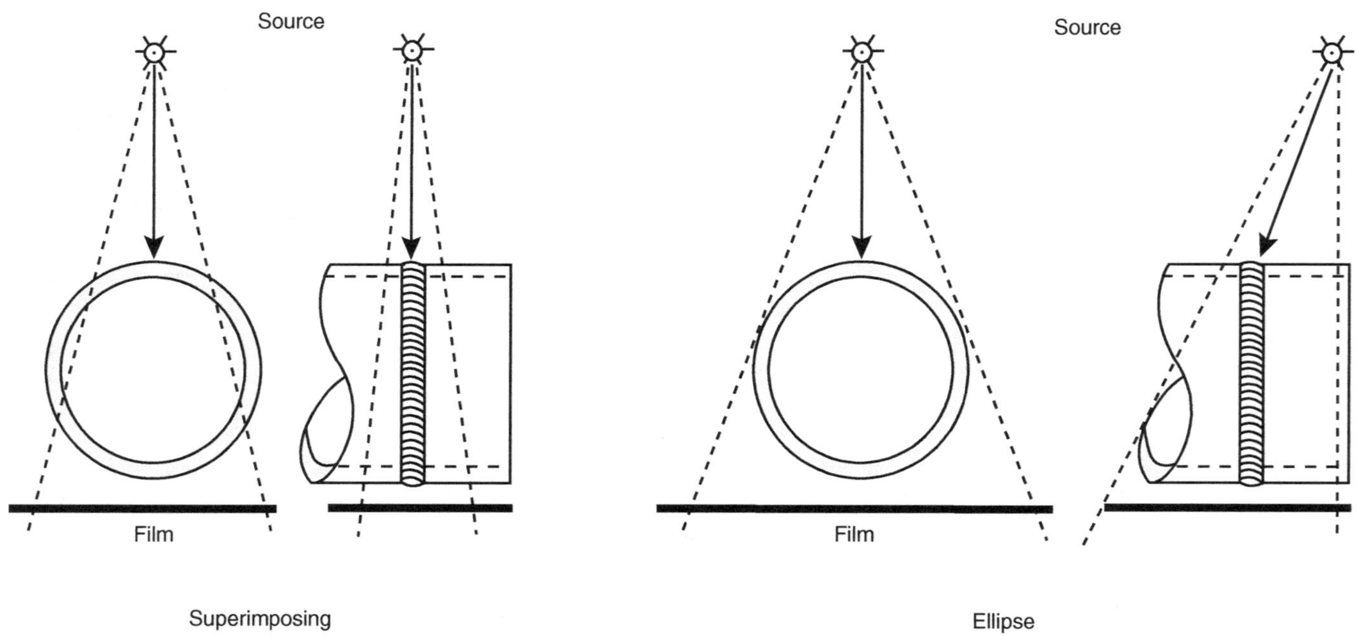


**Figure 6-24** Double wall exposure, double view technique.

### 6.5.1 Procedure

A typical procedure is outlined below to provide guidance, especially when radiography of a specific object is being attempted for the first time. The recommended steps in a radiographic procedure are:

1. Understand the codes, specifications, and customer requirements thoroughly
2. Develop a technique based on the thickness and type of material
3. Prepare a shooting sketch
4. In the darkroom, carefully place the radiographic film in the cassette with the proper lead screens
5. Place the film under the area of interest
6. Ensure that the correct source to film distance is being employed
7. Place the appropriate station markers and identification numbers in the area of interest to ensure easy correlation with a discontinuity if one is detected
8. Set up the exposure parameters
9. Make the exposure
10. In the darkroom, unload and process the film
11. Evaluate the film for artifacts
12. Evaluate the film for compliance
13. Complete a report and store the film

#### *Radiographic Film*

Radiographic film is, by far, the most important permanent record in nondestructive testing. Standard industrial radiographic film contains two emulsions, one on each side of the film base. There are seven layers that make up the cross section of the film. The base, which occupies the vast majority of the thickness, is made from cellulose, triacetate, or polyester. Both sides of this base have a very thin layer that contains an adhesive, and adhering to the adhesive are the layers of emulsion. The emulsion contains many small silver halide crystals, which are suspended in gelatin. This is where the chemical changes take place when the emulsion is exposed to radiation or light. On the two outer surfaces, there is a very thin layer of hardened gelatin, which is designed to protect the emulsion. The emulsion is very sensitive to handling, chemicals, abrasion, and various levels of light exposure, so it is essential that the film be treated with extreme care.

There are four general classifications of industrial radiographic film. *Class I* is described as extra-fine grain, low-speed, with very high-contrast capabilities. This film is generally used for lower-density materials and can be used with or without lead screens. *Class II* is a fine-grain, medium-speed, high-contrast film that is also used for the lower-density materials with low- and medium-energy radiation. This film classification tends to be more widely used than the Class I since it provides very good definition, has fine grain, and is slightly higher in film speed than Class I. It can also be used with or without lead screens. *Class III* is a high-speed film, and therefore requires shorter exposure times. It is typically used for X-rays or gamma rays with higher energies, and can be used with or without lead screens. It is considered a medium-contrast film with high graininess. The *Special Class* includes films with very high definition and very fine grain. There are other classes of wide latitude films that are not used as frequently as the four groups above.

### 6.5.2 Film Processing

After the radiograph has been taken, the film is ready for processing. A typical manual film processing system is illustrated in Fig. 6-25. The first step in manual processing is to carefully place the film onto a stainless steel hanger, which is designed for manual processing, or carefully place it into the feed-in end of the automatic processor (Fig. 6-26). The first step that the film will undergo is development. Developers are alkaline solutions that change the latent or chemically stored image in the radiographic emulsion into a visible image, resulting in various shades of gray or black, depending upon the amount of exposure. Agents in the developer reduce the exposed silver halide to metallic silver. There is also an accelerator, which speeds up the development; a preservative, which prevents the oxidation of the developer; and a restrainer, which helps to control and reduce the chemical fog that can occur during the processing of the film. It is fairly common to use a development time of 5 to 8 minutes at a temperature of 68°F (20°C). Some film manufacturers recommend 4 to 7 minutes at 68°F. If the temperature is higher than 68°F, a shorter development time will be

**FIGURE 6-25** Manual film processing tanks. (*Courtesy of Charles J. Hellier.*)

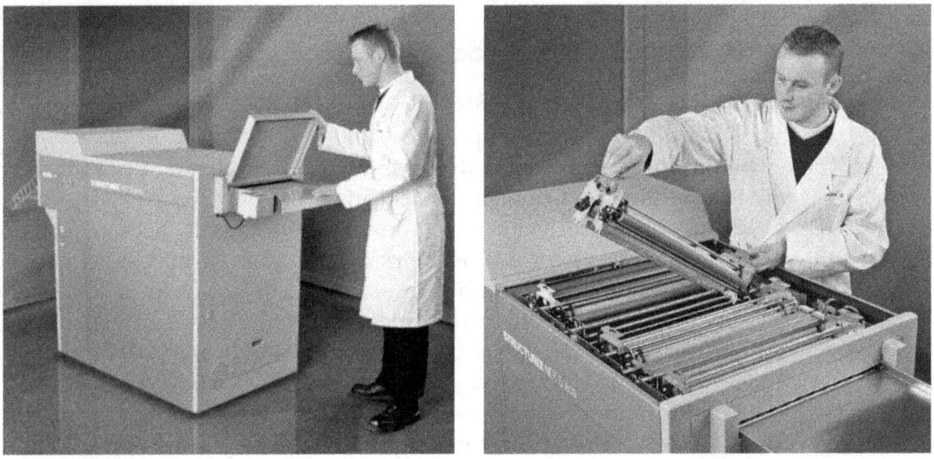

(a) (b)

**FIGURE 6-26** (*a*) Automatic processor and (*b*) automatic processor roller transport system. (*Courtesy of GE.*)

appropriate. On the other hand, if the temperature of the developer is lower than 68°F, a longer development time will be necessary. Just as it is important to follow a procedure for taking the radiograph, it is also important to follow a consistent procedure for the processing of the film. A standard time and temperature for the development should be adopted and remain consistent throughout the radiographic operation.

When the radiographic film is being processed manually, it is necessary to agitate the film while it is in the developer solution. This provides a more uniform development and prevents the formation of air bubbles on the film surface. If the film is not properly agitated, the chemical activity in the development will be inconsistent and a condition called "streaking" may result, characterized by variations in film density.

The next step is to arrest or stop the development. This can be done in one of two ways. The film can be taken out of the developer and placed into a water bath for several minutes, or it can be placed in an acidic solution called stop bath. The stop bath is an excellent way to immediately and consistently arrest the development activity. The acid neutralizes the alkaline developer and protects the acidity of the fixer, which is the next step.

The fixer serves two major functions. First, it clears out the unexposed silver halide crystals remaining in the film and, second, it fixes or hardens the image. The fixer also influences the archival properties of the film. Many times, shortcuts are taken in the fixing process and the film is taken out before it is totally fixed or hardened. In this case, the film may still contain some unexposed silver halide crystals and they will continue to develop on their own. This could ultimately result in the image of the film having a yellowish or brownish appearance. It is good practice to keep the film in the fixer for about 15 minutes, although some film manufacturers recommend at least double the clearing time. Clearing time is the time that it takes for the remaining silver halide crystals to be removed from the emulsion, clearing up the image.

After fixing, the film goes into a water rinse for a period of time, typically 30 minutes. The water should be continually changed in order to remove any remaining traces of the developer or the fixer. If the film is removed from the water rinse prematurely, shortened archival properties can result and the film can eventually turn yellow or brown. At the end of the water rinse time, it is recommended that the film be placed in water containing a wetting agent. This uniformly wets or coats the film so that droplets of water will not form, causing artifacts when the film is dried. It also facilitates consistent drying of the film.

The final step in processing is to dry the film, which is normally done in a warm air recirculating drier designed for this purpose.

The entire film processing procedure (apart from the drying operation) is accomplished under safelight conditions in a darkroom. The darkroom safelights should be tested periodically for film safety. An unexposed film with a coin or some other opaque object on the emulsion should be placed on the worktable. After a period of time that exceeds the longest time that films would be exposed to the safelight, the object should be removed and the film processed. There should be no change in the density of the film in the area of the opaque object. If an image of the object appears on the film, it indicates that the safelight is not "safe" and should be changed.

With automatic processors, the dry-to-dry time is about 10 minutes, compared to over an hour for manual processing. This is because the developer solution is more concentrated and the development temperature is higher, which speeds up the entire process. The automatic processor is the optimal way to process film, since it results in consistent images with fewer film artifacts compared to manual processing. Automatic processors are very expensive and there should be a sufficient quantity of radiographs to be processed in order to make it cost effective. The key to successful film processing with automatic processors is maintenance. The automatic processor must be cleaned often. The roller system must be removed and carefully maintained; otherwise, there will be many problems. Even the most exceptional technique capable of producing the highest quality radiographic images can be rendered meaningless because of problems in processing.

### *Summary*

The first and most important consideration in radiography is the handling of the film. In far too many cases, film artifacts are the cause for the rejection of the final radiograph. Handling the film properly in the darkroom will help to minimize this. The darkroom should have a "wet" side and a "dry" side. The dry side is the area where the film is stored and handled, i.e., loaded into and removed from cassettes. The wet side is where the processor is located, whether it is manual or automatic. The darkroom should be hospital clean at all times. A darkroom that is dirty and untidy is usually a darkroom where film will be scrapped as a result of artifacts.

The five steps in the processing of the film are:

1. Developing
2. Stop bath
3. Fixing
4. Washing
5. Drying

The precautions that should be taken include:

1. Proper control of the development temperature and time
2. Periodic maintenance of the developer and fixer solutions
3. Agitation in the manual system during the development step

4. Maintaining a safelight condition in the darkroom
5. Most important, cleanliness

### 6.5.3 Density

In radiography, film density is defined as the quantitative measure of film blackening as a result of exposure and processing. It can be expressed mathematically as follows:

$$D = \log \frac{I_0}{I_t}$$

where:
$D$ = density
$I_0$ = light incident on the film
$I_t$ = light intensity transmitted through the film

If a film is processed and it has not been exposed to light or radiation, it will be virtually clear. A film that has zero density is totally clear and has a film density reading of zero. If a film is exposed and the resultant film density is one, the amount of light that passes through the film is 10 percent of the incident light. For a film density of 2.0, only 1 percent of the incident light passes through. A film density of 3.0 permits 0.1 percent of the incident light to pass through, a film density of 4.0, 0.01 percent, and so on. Most specifications require that the film density in the area of interest be between 2.0 and 4.0. There are other specifications that differ slightly, but generally, the density range of 2.0 to 4.0 is quite common. Film density is measured with a densitometer (see Fig. 6-27).

It is essential that densitometers be checked prior to each use to assure that the readings are accurate. In the past, film density strips were used to compare known densities to densities in the area of interest. This required a great deal of conjecture and, even

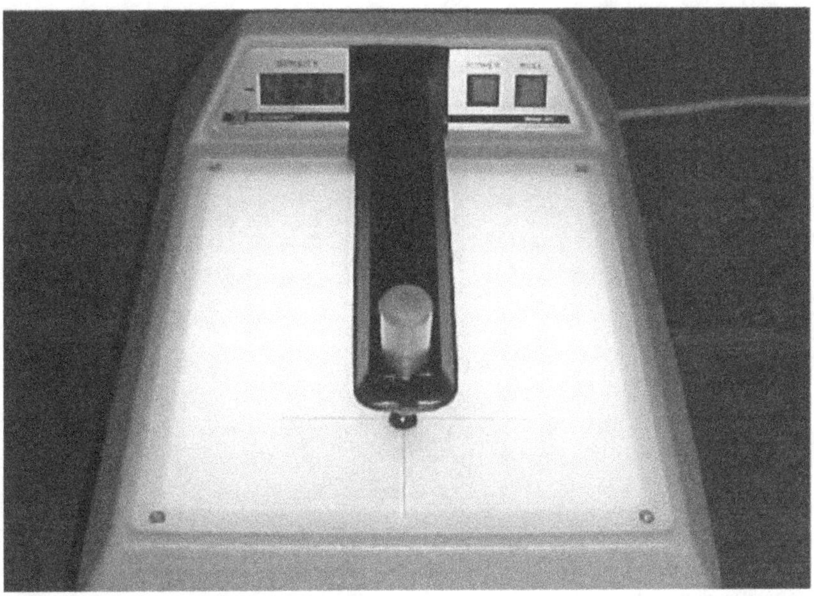

**FIGURE 6-27** Densitometer. (*Courtesy of Charles J. Hellier.*)

though still employed today, the readings are not nearly as accurate as the readings obtained with the densitometer.

As mentioned earlier, the base density of a radiograph is established by taking a density reading through the penetrameter. The density range in the area of interest must not vary greater or less than a percentage of that base density. It is not unusual to have a density range requirement for the area of interest of +30 percent to –15 percent of the base density established through the penetrameter. This, in fact, is good radiographic practice. The quality level of the radiographic technique is established by the penetrameter image and it has been determined that the quality level will also apply to a defined density range. Table 6-8 shows density readings and the +30/–15 percent range. Notice that for a

| Density Reading through Penetrameter | +30% | –15% |
|---|---|---|
| 1.5 | 1.95 | 1.28 |
| 1.6 | 2.08 | 1.36 |
| 1.7 | 2.21 | 1.45 |
| 1.8 | 2.34 | 1.53 |
| 1.9 | 2.47 | 1.62 |
| 2.0 | 2.60 | 1.70 |
| 2.1 | 2.73 | 1.79 |
| 2.2 | 2.86 | 1.87 |
| 2.3 | 2.99 | 1.96 |
| 2.4 | 3.12 | 2.04 |
| 2.5 | 3.25 | 2.13 |
| 2.6 | 3.38 | 2.21 |
| 2.7 | 3.51 | 2.30 |
| 2.8 | 3.64 | 2.38 |
| 2.9 | 3.77 | 2.47 |
| 3.0 | 3.90 | 2.55 |
| 3.1 | 4.03 | 2.64 |
| 3.2 | 4.16 | 2.72 |
| 3.3 | 4.29 | 2.81 |
| 3.4 | 4.42 | 2.89 |
| 3.5 | 4.55 | 2.98 |
| 3.6 | 4.68 | 3.06 |
| 3.7 | 4.81 | 3.15 |
| 3.8 | 4.94 | 3.23 |
| 3.9 | 5.07 | 3.32 |
| 4.0 | 5.20 | 3.40 |

*Note:* Typical acceptable density range for the penetrameter and area of interest is 2.0–4.0. Some specifications may require different ranges.

**Table 6-8** Film Densities

base density reading of 3.0 (i.e., the density through the penetrameter), the maximum density that is considered acceptable in the area of interest would be 3.9 and the minimum density would be 2.55. Most codes and specifications require that even with the application of this range, densities greater than a specified maximum (usually 4.0) and a minimum (usually 2.0) not be exceeded. The density readings in the area of interest should be taken at random in that area. Density readings should not be taken through an area containing a discontinuity. It should be remembered that the density is a measure of establishing that the correct exposure has been employed. The density readings for both the penetrameter and the area of interest should be entered on the radiographic report.

## 6.6 Radiographic Evaluation

The final step in the radiographic process, and by far the most important, is the evaluation of the radiograph. At times, radiographic interpretation has been referred to as a mix of art and science. Science involves all of the physics and principles involved with producing the radiographic image. Interpretation of the image involves art, to an extent. Interpretation of radiographs cannot be taught in a classroom, although that can provide a good foundation. It requires hours of reviewing and understanding the different types of images and the various conditions that are prevalent in industrial radiography. There are individuals who have been evaluating radiographs for decades who will admit that they are still learning how to interpret with each additional radiograph.

After the film density readings are completed, the next step in evaluating radiographs is to take an overall look at the appearance of the radiographic image, as well as the condition of the film itself. This initial look will generally indicate the overall quality that has been achieved. After that, the quality level should be confirmed by observing the image of the IQI and assuring that the essential hole in the shim-type penetrameter, or wire in the wire-type penetrameter, is clearly and discernibly displayed.

The radiographic interpreter should always wear cloth gloves (preferably cotton) when evaluating radiographs. This prevents fingerprints, smears, and smudges from being placed on the surface of the radiograph. After the quality level has been established, it is then appropriate to review the radiographic image for artifacts. There are a large number of artifacts that can appear in a radiograph. Artifacts, or false indications, are classified based on when they were formed, i.e., prior to processing, during processing, or after processing.

The high-intensity illuminator (Fig. 6-28) designed for viewing the radiographs should be in good working condition. Even before the radiograph is initially reviewed, the illuminator should be checked to make sure that the high-intensity light bulbs are working properly and that the viewing area is clean and free from any scratches or artifacts that could be superimposed on the image of the radiograph. The illuminator should be properly designed to dissipate the heat that is created when the high-intensity lights are in operation. There should also be an intensity control to allow the interpreter to increase or decrease the brightness of the light source. It is essential that any light not coming through the radiograph be masked. The light intensity given off from these high-intensity illuminators is very intense and can cause severe discomfort to the eye, so masks are very strongly recommended to assure that the light coming through the film is the only light reaching the eye. Most high-intensity illuminators are operated

**FIGURE 6-28** Interpretation of a radiograph with a high-intensity illuminator. (*Courtesy of Charles J. Hellier.*)

with a foot switch, which permits the interpreter to turn the light on and off while keeping both hands free to handle the film. This is another area where caution is warranted. The light switch should not be activated until the film is in place on the viewing area of the viewer and the light should be switched off prior to its removal.

There seems to be a lack of agreement as to whether magnification should be employed during the evaluation of radiographs. Magnifiers are, in fact, encouraged when they can assist in the proper detection and identification of the different discontinuities. When magnification exceeds a certain power, it can become quite useless, since the grains in the film are also magnified and the images of the discontinuity tend to be diffused.

Radiographs should be viewed in a darkened area. It is advisable to allow the eyes to adapt to the dimly lit conditions prior to radiographic viewing. Recommendations vary regarding the length of dark adaptation time. Experience has shown that periods from one to as high as 15 minutes may be necessary to properly distinguish the density changes within the image. The issues to consider are those such as:

(a) The minimum density changes sought. Small or tight discontinuities may appear as very slight changes in the density of the radiographic image.

(b) If the interpreter has recently been in bright sunlight, longer times may be required for dark adaptation.

For critical interpretation, it may be necessary to mask the lighter parts of the image where the low-density (lighter) areas allow the brighter transmitted light to reduce the interpreter's perception of changes in density.

### 6.6.1 Evaluation for Artifacts

#### *Film Artifacts or False Indications That Are Caused Prior to Processing*

These include:

- Lead screen marks
- Static marks
- Film scratches
- Exposure to light
- Fog due to exposure to low levels of light or aging
- Finger marks
- Pressure marks
- Crimp marks

#### *Artifacts That Are Caused during the Processing of the Film*

These include:

- Pressure marks (from rollers in an automatic processor)
- Chemical and delay streaks
- Pi lines (in automatic processors)
- Chemical spots
- Dirt

#### *Artifacts Caused after Processing*

These are primarily in the form of scratches and fingerprints. Fingerprints will be minimized with the use of cotton gloves.

### 6.6.2 Evaluation for Discontinuities

After the radiograph is evaluated for artifacts, it is finally time to evaluate for discontinuities. Discontinuity conditions that are normally found in welds include those in the following subsections, listed in order of severity.

#### *Cracks*

There are many different types of cracks that are classified by their orientation and location. They will always appear as dark, irregular, linear indications in a radiograph and are the most serious of all discontinuities. They are generally tight and not always detectable by radiography unless their orientation is somewhat in the same plane as the direction of the radiation. The most common types of cracks are those that are oriented longitudinally (along the length of a weld). Others may occur as star-shaped patterns, usually at the end of a weld pass. Some will occasionally be transverse to the length of a weld. Cracks can also appear at the toe of the weld or at the edge of the root pass. Underbead cracks will typically be angular, extending from the fusion zone.

### Lack of Fusion (Inadequate Fusion)

This serious discontinuity results from an absence of metallurgical fusion, either between a weld pass and the base material (weld edge prep) or between two successive weld passes. Lack of fusion is usually very narrow, linear, and tends to be straighter than the crack. At its extremities, the condition "feathers" down to an unusually sharp edge. In many cases, the image of lack of sidewall or lack of fusion appears somewhat straight along one edge and slightly wavy on the other.

### Incomplete Penetration

This discontinuity is an absence of weld metal or an area of "nonfusion" in the root pass of the weld. Its appearance is very straight, dark, linear, and usually "crisp" in sharpness. It can be short but usually has significant length.

### Inclusions (Dense and Less Dense)

Inclusions are basically materials that have been entrapped in the weld that do not belong there. They will have a variety of shapes and dimensions ranging from short and isolated to linear and numerous. The lighter-density inclusions will result in a darker image on the radiograph and the more dense inclusions, such as tungsten, as a lighter image.

### Porosity

When gas is trapped in a weld metal, the void-type condition created is referred to as gas or porosity. Porosity comes in different shapes (globular, tailed, elongated) and distributions (linearly aligned, clustered, isolated, scattered). Porosity will always appear darker, since they are gas filled, and are the easiest of all weld discontinuities to detect.

### Geometric Conditions

There are also geometric conditions that can occur in welds that are observable in a radiograph and should be further addressed by visual examination and dimensional checks. When possible, these conditions should be measured with mechanical gauges. These geometric conditions include the following:

*Concavity.* A concave condition at the root of a weld resulting in a thinner region through the weld cross section. The image of concavity in a radiograph is usually easy to identify since it is a broad indication with a gradual change of density as compared to the abrupt change associated with incomplete penetration.

*Convexity.* Usually considered the opposite of concavity since this condition is a thicker protrusion at the root pass. It is caused by excessive deposit of the root pass and is sometimes referred to as excessive penetration. In extreme cases, the weld metal will "burn through" and may form droplets that are referred to as "icicles." The concern is the abrupt change in contour and the possibility of flow constriction in a pipe due to the protrusion of additional weld metal.

*Undercut.* A depression that occurs at the edge of the weld where it has fused into the base metal at the outer or inner surface. It is like a slight "valley" that continues along the length of the weld and varies in depth. The depth is the measurable dimension that determines its seriousness.

*Underfill.* Sometimes interpreted as undercut since it can appear in the same region of the weld. In fact, it is a different condition that occurs when the weld groove is not completely filled with weld metal. Also, unlike the undercut, the contour of the bottom of this condition is more V-shaped or notch-like as compared to the generally broad gradual configuration of the undercut.

*Overreinforcement.* This condition results from an excessive depositing of weld metal on the outer surface of the weld. The concern is the geometric change that may create a stress riser condition at the outside surface where stresses normally tend to be higher by design.

### 6.6.3 Casting Discontinuities

Discontinuities that occur in castings that can be detected radiographically include the following, listed in order of their severity:

*Hot tears and cracks.* Both serious ruptures or fissures that typically occur in an isolated zone due to the high stresses that build up during the cooling of the casting. Hot tears usually form during initial cooling and cracks later, usually at or near room temperature. On a radiograph, both conditions appear linear and branch-like and are most likely to be in or near an area of thickness change, where the different rates of cooling cause stresses to build up.

*Shrinkage.* Usually in the form of a zone of minute fissures as a result of stresses during cooling. Shrinkage comes in various shapes. Sponge shrinkage has a "spongy" appearance and can be isolated, scattered, or significant in size and density variations. Microshrinkage is feathery in appearance, and the change in density is often quite minor.

*Slag and sand inclusions.* The entrapment of inclusion materials and sand cause these conditions, which will have irregular shapes and variations in density due to the nature of the included matter.

*Gas voids and porosity.* Unlike the inclusions, gas voids and porosity are more uniform, typically globular and dark in appearance. In fact, these discontinuities just look like voids, are normally easy to detect (they are not subject to alignment limitations like cracks), and readily recognizable.

*Cold shuts.* Very tight discontinuities that occur when a surface that has begun to solidify comes in contact with other molten metal as the casting is in the process of being poured. There is usually a thin film of oxide present that prevents total metallurgical fusion. It is a very difficult discontinuity to detect with radiography due to its tight condition and angular orientation.

#### Geometric Conditions

As with welds, there are also geometric conditions in castings that can be observed radiographically. These geometric conditions include the following:

*Misrun.* This condition is actually an absence of metal due to the inadequate filling of the casting mold. It is easily detected by a simple visual test.

*Unfused chaplets*. Metal supports that are strategically placed in the casting mold to support the core. They are designed to be consumed when the molten metal comes into contact with them, and when this does not happen, the circular shape of the chaplet will be apparent in the radiograph.

*Note:* A compendium at the end of this chapter contains radiographs showing examples of false indications, and weld and casting discontinuities.

### 6.6.4 Responsibility of the Interpreter

In order to effectively evaluate radiographs, it is essential that the interpreter be thoroughly familiar with the parts, dimensions, and material, to properly judge whether it is acceptable or not. The interpreter should be thoroughly familiar with the technique that was used to produce the radiograph, how the film was processed, the codes and standards that apply, and acceptance criteria.

If there is a rejectable discontinuity, it is extremely important that the condition be suitably identified as to its exact location. For this purpose, it is good practice for the radiographer to prepare a transparent skin that is overlaid on the radiographic image; then, with a wax pencil, the identification numbers are marked on the skin, as well as the outline and location of the discontinuity. This permits an accurate marking of the discontinuity on the actual part and facilitates precise repair. After repair, the region that originally contained the discontinuity must be reexamined to assure that the discontinuity has been completely removed.

It is essential that the entire evaluation process be entered on a radiographic report form. Report forms differ in format with each individual company. It would be extremely beneficial to the radiographic industry if a consistent, generic report form could be adopted. To this end, sample radiographic report forms for welds and castings are offered (see Fig. 6-29).

## 6.7 Applications

Although the majority of applications in radiographic testing appear to involve welds and castings, it has been effectively applied to many other product forms spanning a wide variety of industries. The major industries include but are not limited to the following:

1. *Power Generation*. A wide range of weld configurations, thicknesses, and materials are used at power plants. Radiographic testing requires radiation sources that are capable of providing energies from the low end of the spectrum to the highest commercially available. Pressure vessels are complex structures that require high-quality initial examinations. The associated piping systems, with their many weld joint geometries, provide continual challenges for the radiographer. Cast valve bodies are usually radiographed at the foundry. In-service examinations offer even greater challenges. There are other conditions the radiographer must deal with, such as accessibility, the environment (temperature; in the case of nuclear power plants, radiation, air quality, circulation, etc.), the transporting and positioning of equipment, the concern

for the security of radiation areas, and the weather when examinations are to be conducted outdoors. There are other nonconventional applications for RT in power plants. There are ongoing concerns regarding the degradation of structures and components as a result of corrosion, vibration, and wear. Many times, the areas are covered with insulation or are very difficult to reach. There are portable RT units that provide for the examination of these conditions and provide contour data through the insulation. When there are similar concerns

Customer ______________________ RT Report # __________ Date ____________

Part ID/ Location ______________________

Weld ID/Location ______________________

**PART DATA**

Base Mat'l Thickness ______________________

Weld thickness ______________________

Material Type ______________________

Welding Process ______________________

Weld Status ______________________

Type of Weld Joint ______________________

Diameter or Weld Length ______________________

RT Procedure __________ Rev. ______

Acceptance Criteria

**TECHNIQUE DATA**

IQI# _____ ☐ s/s ☐ f/s Screen Type __________

Shim Thickness ________ Front Thickness __________

Film Mfr. & Type ________ Inner Thickness __________

Film Size ________ Back Thickness __________

# of Film /Cassette ________

Radiation Type: ☐ Isotope ____ ____ ci. ☐ X-Ray ____kV

Exposure: __________ ci-min /__________ mam

f = ______ t = ______ d = ______ Ug = ______

Film Processing: ☐ Manual ☐ Automatic

**SHOOTING SKETCH**

(Show location and orientation of radiation source, part, film, IQI & location markers)

**INTERPRETATION DATA**

EXPOSURE: ☐ Single wall ☐ Double wall VIEW: ☐ Single wall ☐ Double wall

| Part # | Weld # | Location # | Acc | Rej | Disc. Code | Quality | Density | Artifiacts | Remarks |
|---|---|---|---|---|---|---|---|---|---|
| | | | | | | | | | |
| | | | | | | | | | |
| | | | | | | | | | |
| | | | | | | | | | |
| | | | | | | | | | |
| | | | | | | | | | |
| | | | | | | | | | |
| | | | | | | | | | |
| | | | | | | | | | |
| | | | | | | | | | |

**DISCONTINUITY CODES**

| | | | |
|---|---|---|---|
| POR - Porosity | | | EUC - External Undercut |
| SI - Slag Inclusion | SP - Spatter | DT - Drop Through | IUC - Internal Undercut |
| ESI - Elong. Slag Inclusion | AS - Arc Strike | W - Tungsten Inclusion | UI - Unconsumed Insert |
| IP - Incomplete Penetration | CX. - Convexity | CR(L) - Longitudinal Crack | SO - Surface Oxidation |
| IF - Incomplete Fusion | CV. - Concavity | CR(T) - Transverse Crack | SURF - Surface |
| BT - Burn Through | HiLo - High Low | CR(C) - Crater Crack | UF - Underfill |

Interpreted by: ______________________ NDT Level _____ Date ________

Customer: ______________________ Date ________

(a)

**FIGURE 6-29** (*a*) Radiographic report form for welds.

Customer ______________________
Part ID/ Location ______________________
Casting ID/Location ______________________

PART DATA

Mat'l Thickness ______________________
Material Type ______________________
RT Procedure ____________ Rev. ______
Acceptance Criteria ______________________

RT Report # __________ Page ___ of ___ Date _______

TECHNIQUE DATA

IQI# _____ ☐ s/s ☐ f/s | Screen Type ___________
Shim Thickness _________ | Front Thickness _________
Film Mfr. & Type ________ | Inner Thickness _________
Film Size _____________ | Back Thickness _________
# of Film /Cassette ________
Radiation Type: ☐ Isotope _____ ____ci. ☐ X-Ray ____kV
Exposure: __________ ci-min /__________ mam
f = ______ t = ______ d = ______ Ug = ________
Film Processing: ☐ Manual ☐ Automatic

SHOOTING SKETCH

(Show location and orientation of radiation source, part, film, IQI & location markers)

INTERPRETATION DATA

EXPOSURE: ☐ Single wall ☐ Double wall VIEW: ☐ Single wall ☐ Double wall

| Part # | Casting # | Location # | Acc | Rej | Disc. Code | Quality | Density | Artifiacts | Remarks |
|---|---|---|---|---|---|---|---|---|---|
| | | | | | | | | | |
| | | | | | | | | | |
| | | | | | | | | | |
| | | | | | | | | | |
| | | | | | | | | | |
| | | | | | | | | | |
| | | | | | | | | | |
| | | | | | | | | | |
| | | | | | | | | | |
| | | | | | | | | | |

DISCONTINUITY CODES

| | | | |
|---|---|---|---|
| CR - Crack | SI - Sand Inclusion | SH - Shrinkage | UC - Unfused Chaplet |
| MP - Microporosity | INC - Inclusion | DR - Dross | UCL - Unfused Chill |
| POR - Porosity | DI - Dense Inclusion | HT - Hot Tear | CS - Core Shift |
| WP - Worm Hole Porosity | SS - Sponge Shrinkage | MR - Misrun | SURF - Surface |
| BH - Blow Hole | MS - Micro Shrinkage | COL - Cold Shut | SEG - Segregation |

Interpreted by: ______________________ NDT Level _____ Date ________

Customer: ______________________ Date ________

(b)

**FIGURE 6-29** (*b*) Radiographic report form for castings. (*Continued*)

for the internal conditions of valves, or for the determination of the position of the mechanical components, a higher-energy source may be necessary.

2. *Aerospace.* There are countless uses for RT in the aerospace industry. The critical components in the aircraft engine itself call for thousands of radiographs. The airframe and other related structures depend greatly on radiography for initial and in-service inspection. Water ingress in honeycomb cells can cause problems

due to freezing when the aircraft is airborne. Radiographic techniques can effectively detect this condition in most cases.

3. *Petrochemical.* Radiography is of great importance in the initial construction of petrochemical plants. Considering the many service conditions under which this type of complex operates, establishing a baseline through radiography is of extreme importance. Just as with power plants, in-service inspection is necessary and RT plays a key role. The detection of areas of degradation, corrosion, erosion, wear, and other conditions that develop through the extended operation should be detected and corrected before an unplanned shutdown becomes necessary. Other than the cost of initial construction, unplanned outages are the most costly events in this industry. It should be noted that this also applies to most other major industries.
4. *Nonmetals.* This category includes a wide variety of materials that are inspected by radiography, including plastics, rubber, propellants, ceramics, graphite, concrete, explosives, and many more. Perhaps the electronic industry, because there are some metals involved, also belongs in this industry segment.
5. *Medicine.* There are numerous applications in medicine that do not involve the examination of living creatures. On occasion, foreign objects or contaminants have inadvertently found their way into various drugs and medicines during processing. These elements most usually have densities that are quite different from the products into which they have become included. This makes their examination an ideal application for RT. A prerequisite, of course, must dictate that the product will not be affected by exposure to radiation. When a large quantity of product is involved, real-time radiography may be the most cost-effective approach. This provides for a two-dimensional image to be displayed on a monitor almost immediately, as a result of the conversion of the radiation into an electronic or optical signal.
6. *Law Enforcement and Security.* Radiography is especially adaptable and useful for the detection of hidden contraband and weapons and for the examination of explosive devices such as shells and projectiles. Sealed boxes, envelopes, and other packages have been found to contain devices that were potentially hazardous if they had not been detected. When passing through the security area in an airport, carry-on articles are subjected to real-time radiographic inspection.
7. *Food.* Various food products are examined with some form of RT for the purpose of detecting foreign objects that may have been introduced during processing. In some cases, food items may be inspected for content, distribution of additives (such as nuts, candy, in ice cream), or for quantity. Again, the fastest approach would include fluoroscopy or some other real-time technique.
8. *Objects of Art or Historic Value.* This industry is perhaps the most unusual and most interesting for radiographers. Famous paintings have been radiographed (using low-energy, high-contrast techniques) to disclose hidden paintings or images beneath their outer layer. Mummies, usually of some very prominent ruler or an individual with some historic significance, have been inspected to evaluate the condition of the remains. Statues, vases, old pottery, figureheads

from historic seagoing vessels, and the famous Liberty Bell have all been subjected to radiographic inspections. Entire jet engines, cars, and complicated assemblies have been examined, many times with a single exposure.

### 6.7.1 Summary

The uses of radiography are limited only by the principles inherent in the process and the human mind. There are those that would say that the world of radiography has reached its peak. It is easy to dispute such a narrow opinion when considering the many recent developments and innovations, including industrial computed tomography, real-time radiography, flash (high-speed) imaging, in-motion radiography, microfocus radiography, digital imaging, and the photostimulable luminescence (PSL) process, also known as storage phosphor plates, for capturing radiographic images.

The future of radiography will continue to offer new and exciting techniques that will expand its world of applications and opportunities. Certainly, Wilhelm Conrad Roentgen and Marie and Pierre Curie would have been proud.

## 6.8 Advantages and Limitations of Radiography

### 6.8.1 Advantages

Radiographic testing has many advantages, some of the most significant of which are listed below. It:

1. Provides an extremely accurate and permanent record
2. Is very versatile and can be used to examine many shapes and sizes
3. Is quite sensitive, assuming the discontinuity causes a reasonable reduction of cross- section thickness
4. Permits discontinuity characterization
5. Is widely used and time proven
6. Is a volumetric NDT method

### 6.8.2 Limitations

1. There are safety hazards with the use of radiation devices.
2. RT has thickness limitations, based on material density and energy used.
3. RT can be time consuming.
4. RT is very costly in initial equipment and expendable materials.
5. It is also very dependent on discontinuity orientation.
6. RT requires extensive experience and training of the personnel taking the radiographs and during the interpretation.

### 6.8.3 Safety Considerations

Radiation ionizes matter through the ejection of electrons from their orbit, thereby changing the electrical characteristics of the atoms. It is this process that results in biological effects to humans. Living tissue will be damaged to an extent determined by the dose received. Absorption of radiation by a human is expressed in REM (roentgen equivalent man). It is the product of roentgens or milliroentgens (mR) and the Quality Factor (QF) of the radiation type. Since x- and gamma rays have a QF value of 1, the exposure as measured in Roentgens will be the dose received in units of REM. The age-old rules of protection against radiation exposure involve time, distance, and shielding.

*Time.* The shorter the exposure, the better. Every effort should be made to keep personnel exposure time as close to zero as possible. When performing radiography in the field, zero exposure is virtually impossible. So every attempt should be made to minimize the time of exposure to the body to as short a time as possible.

*Distance.* The further an individual is from a radiation source, the less the exposure. Recall the inverse square law and how radiation intensity varies inversely proportional to the square of the distance. If there is an intensity of 200 mR/hour at a certain distance from a radiation source, simply doubling the distance will decrease the intensity by a factor of one-fourth, or 50 mR. Doubling the distance again will reduce the intensity to 12.5 mR, and so on. Using the inverse square law will permit the radiographer to calculate the approximate exposure at any distance.

*Shielding.* The greater the thickness and density of the shielding material, the less the radiation exposure. This is a function of the half- and tenth-value layers, as discussed in Section 6.2. Lead, of course, makes an excellent shielding material because of its high density.

To summarize, the exposure to radiation should be kept as short as possible, the distance from the radiation as far as reasonable, and the shielding as thick as practical.

There are many excellent references regarding the biological effects of radiation exposure. Unfortunately, there is not a great deal of data to support the effects of radiation on humans. The limited data from the end of World War II and some of the more notable accidents involving radiation have been the primary source for the exposure effects still referred to today. The ALARA (as low as reasonably achievable) concept is the most appropriate way to treat the issue of radiation exposure. If every person working with radiation took every precaution and made every effort to minimize exposures, needless damage would not occur to nearly the same extent as it does.

Since humans cannot sense radiation, it is essential, and in many industries a requirement, that personnel-monitoring devices be used. Such devices include:

1. *Film badges* are small plastic holders that contain a special film, which are worn during the time of working in or near a radiation area. The film is quite sensitive and can detect as low as several mR of low-energy X-rays and up to as high as 2,000 R. They usually contain small metallic filters that enable the type and energy range of the radiation to be identified. After a period of time, they are sent to a laboratory for processing and reading. This is the main source of data that are entered on a lifetime radiographic exposure history report that should be maintained for all radiation workers.
2. *Ionization chambers* are sometimes referred to as pocket dosimeters. They are small, lightweight, and are capable of being read directly from a scale inside the chamber.

3. *Thermoluminescent dosimeters,* referred to as TLDs, are widely used due to their rapid readout response time and broad range, which is quite linear. They can be reused and are excellent under field conditions.
4. *Photoluminescent glasses* are useful down to about 1 R. They build up fluorescent centers when exposed to radiation and emit visible light when viewed with ultraviolet light. The intensity of the light is proportional to the exposure of radiation.

In addition, area monitors should always be employed whenever radiographic activities are in process. Some of the monitors, or survey meters commonly used include:

- Ionization chambers
- Geiger–Mueller (G–M) counters
- Scintillation instruments

(*Note:* Two survey meters are better than one in the event of a failure.)

Radiation should never be taken for granted. One cannot be too cautious when dealing with this potentially hazardous form of electromagnetic energy. And yet, it can be safely and effectively used in radiography by following the rules and being familiar with the regulations. Personnel who do not respect radiation and try to do things their way cause most accidents and overexposures. Radiation does deserve respect!

## 6.9 Compendium of Radiographs

Examples (Figs. 6-30 to 6-47) of a number of conditions (artifacts) and discontinuities are included in the following pages as a guide. It must be stressed that discontinuities are unique and no two are alike. There are many other types that are not covered in this chapter. A thorough knowledge of the materials, how they were processed, the radiographic techniques employed, and an understanding of the variables involved in radiographic interpretation are all essential to effective evaluation.

Figures 6-30 to 6-33 illustrate artifacts; Figs. 6-34 to 6-42 illustrate weld discontinuities; and Figs. 6-43 and 6-45 illustrate casting discontinuities. (*Note:* For examples of radiographs showing other casting discontinuities, refer to Chapter 2: hot tear, Fig. 2-5b; sand and slag inclusions, Fig. 2-20; an additional example of shrinkage, Fig. 2-5a.)

**FIGURE 6-30** Chemical stains. (*Courtesy of TechCorr.*)

**FIGURE 6-31** Light leak. (*Courtesy of TechCorr.*)

**FIGURE 6-32** Pi lines. (*Courtesy of TechCorr.*)

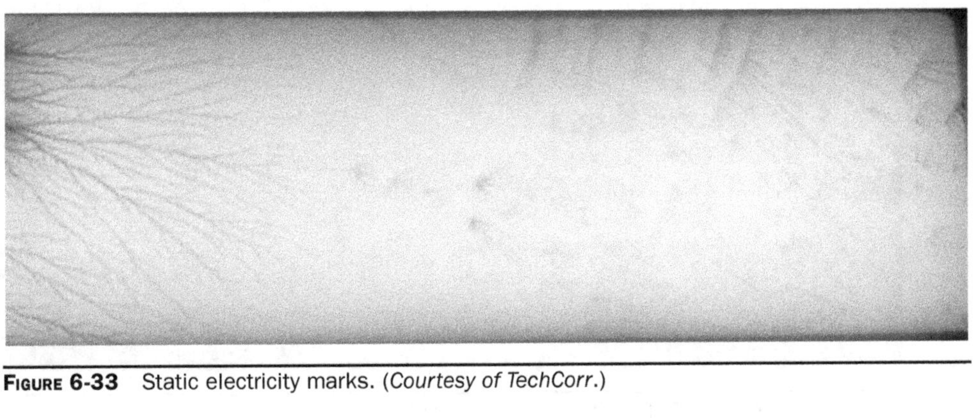

**FIGURE 6-33** Static electricity marks. (*Courtesy of TechCorr.*)

**FIGURE 6-34** Porosity. (*Courtesy of TechCorr.*)

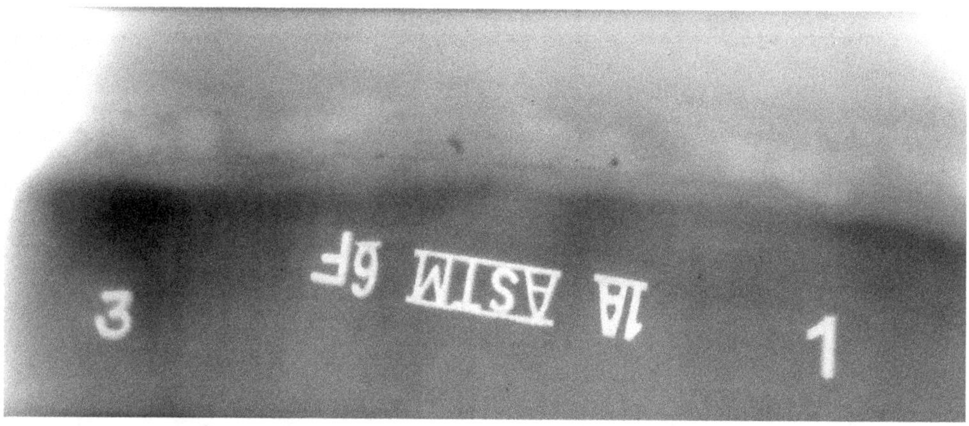


**FIGURE 6-35** Inclusions. (*Courtesy of TechCorr.*)

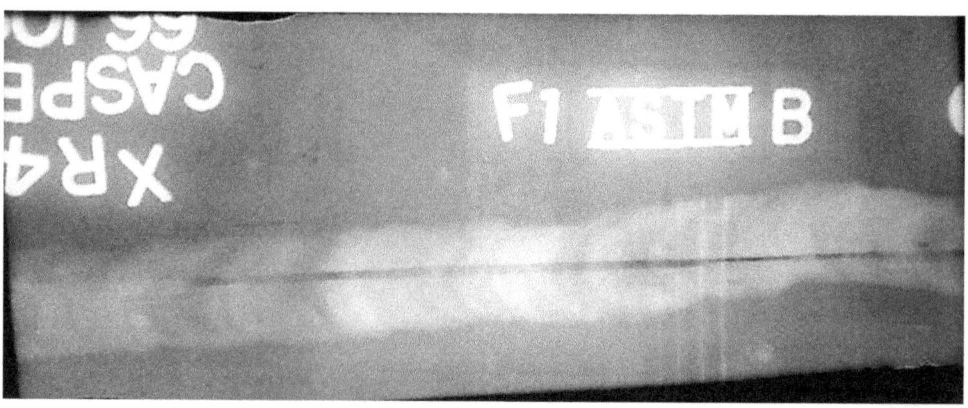


**FIGURE 6-36** Incomplete penetration. (*Courtesy of TechCorr.*)

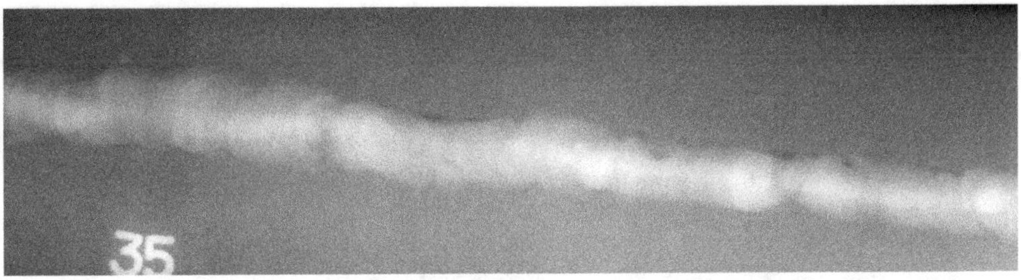


**Figure 6-37** Undercut at edge of weld. (*Courtesy of TechCorr.*)

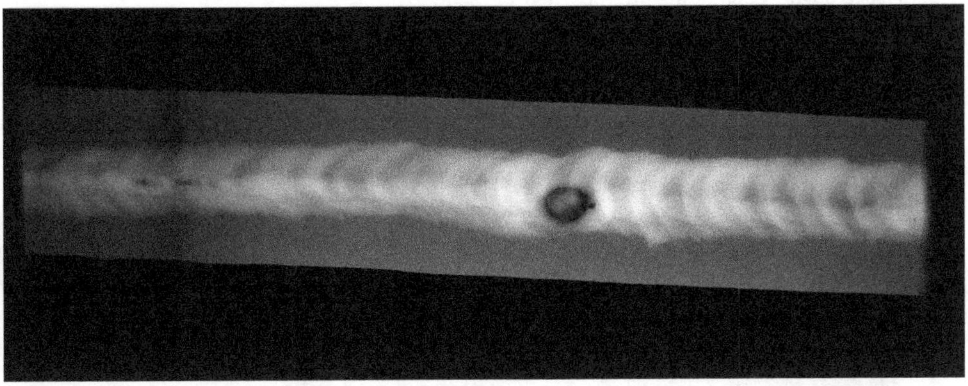

**Figure 6-38** Burn through. (*Courtesy of TechCorr.*)

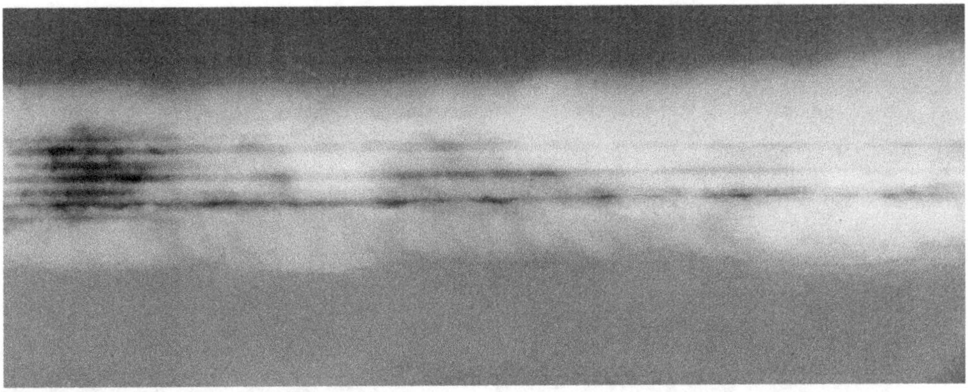

**Figure 6-39** Slugging (weld rods placed in weld groove then welded over). (*Courtesy of Charles J. Hellier.*)

**Figure 6-40** Lack of fusion. (*Courtesy of TechCorr.*)

**Figure 6-41** Tungsten inclusions. (*Courtesy of TechCorr.*)

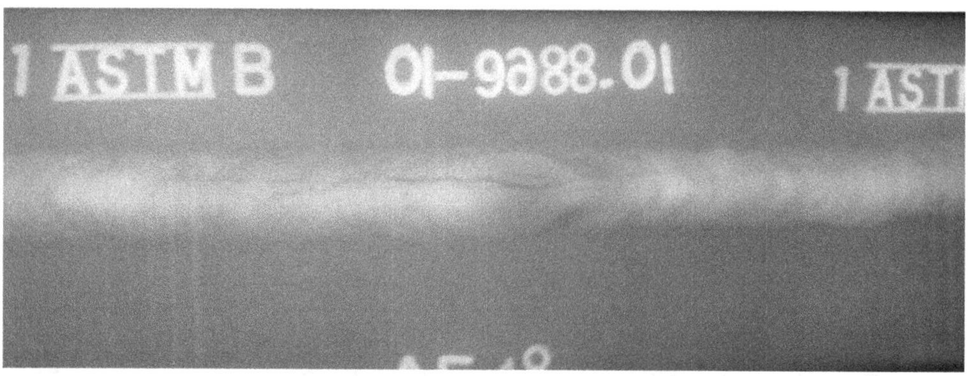


**Figure 6-42** Longitudinal crack. (*Courtesy of TechCorr.*)

**Figure 6-43** Gas porosity in a casting. (*Courtesy of NDT.org.*)

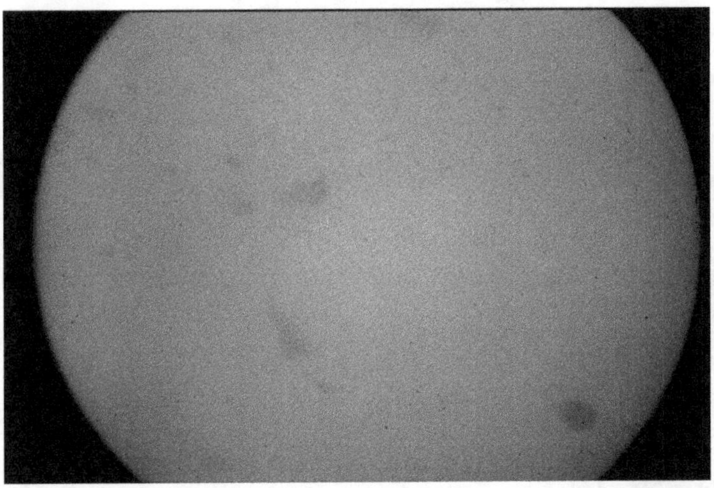

**Figure 6-44** Sand and slag inclusions in a casting. (*Courtesy of Charles J. Hellier.*)

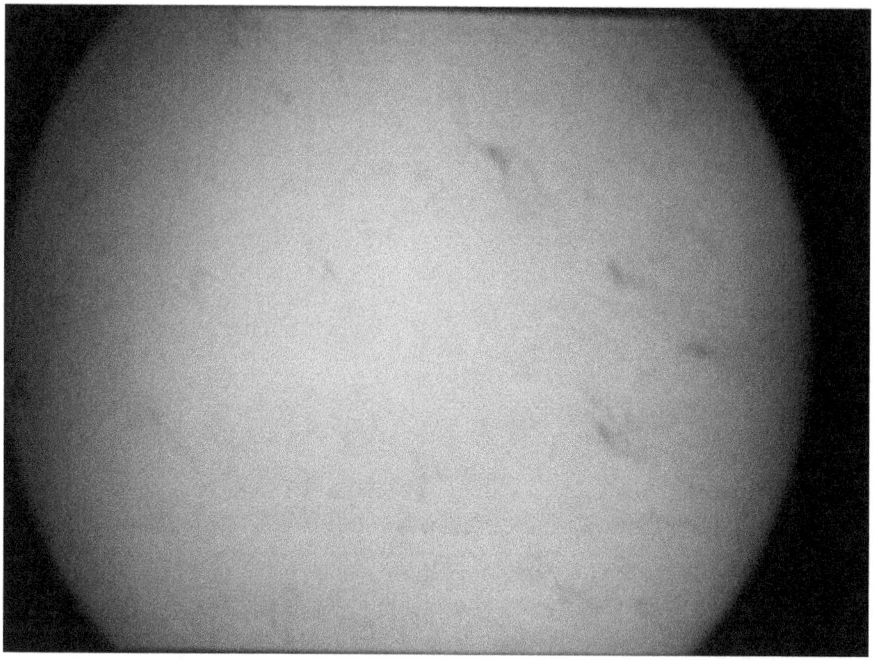

**FIGURE 6-45** Shrinkage in a casting. (*Courtesy of Charles J. Hellier.*)

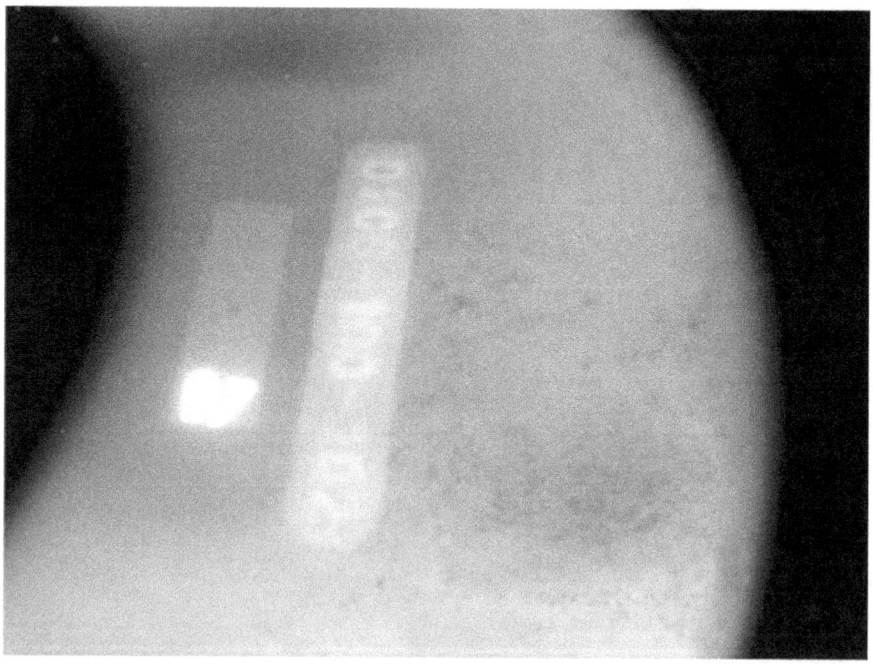

**FIGURE 6-46** Sponge shrinkage in a casting. (*Courtesy of NDT.org.*)

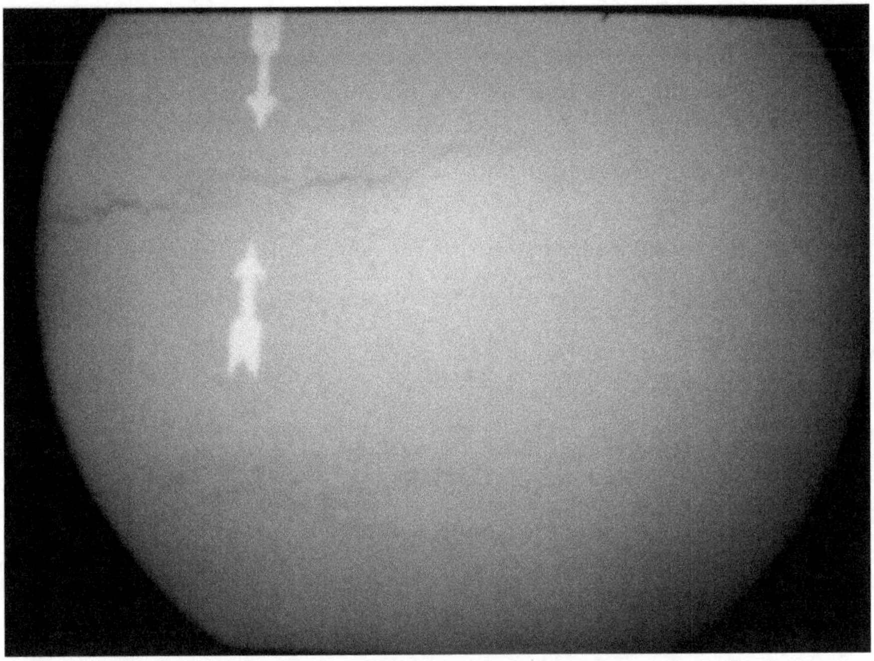

FIGURE 6-47 Hot tear in a casting. (*Courtesy of Charles J. Hellier.*)

## 6.10 Glossary

**Absorption** The process whereby photons of radiation are reduced in number or energy as they pass through matter.

**Angstrom Unit (Å)** A unit of length that is used to express the wavelength of electromagnetic radiation (i.e., light, X-rays, gamma rays). One angstrom unit is equal to 0.1 nanometers (1 nm = $10^{-9}$ m).

**Anode** The positive electrode of a discharge tube. In an X-ray tube, the anode contains the target.

**Artifact** A false indication on a radiograph arising from, but not limited to, faulty manufacture of the film, storage, handling, exposure, or processing.

**Cassette** A light-tight packet or container for holding radiographic film during exposure; may include intensifying or conversion screens.

**Collimator** A device made of radiation absorbent material intended for defining the direction and angular divergence of the radiation beam.

**Composite Viewing** The viewing of two or more superimposed radiographs from a multiple film exposure. The film may be the same or different speeds.

**Constant Potential X-ray Unit** An X-ray system that operates on full-wave rectified current.

**Contrast Sensitivity** A measure of the minimum percentage change in an object that produces a perceptible density/brightness change in the radiographic image.

**Definition, Image** The sharpness of delineation of an image in a radiograph. Generally used qualitatively.

**Densitometer** A device for measuring the density of radiographic film.

**Density, Radiographic** The quantitative measure of film blackening when light is transmitted.

**Duty Cycle** The usable time of a device versus the time that it has to rest when operating "continuously."

**Equivalent Penetrameter Sensitivity** That thickness of penetrameter, expressed as a percentage of the part thickness radiographed, in which a 2T hole would be visible under the same radiographic conditions.

**Exposure Table** A summary of values of radiographic exposures suitable for the different thicknesses of a specific material.

**Film Contrast** A qualitative expression of the slope or steepness of the characteristic curve of a film; that property of a photographic material that is related to the magnitude of the density difference resulting from a given exposure difference.

**Film Speed** A numerical value expressing the response of a radiographic image film to the energy of penetrating radiation under specified conditions.

**Filter** Uniform layer of material, usually of higher atomic number than the specimen, placed between the radiation source and the film for the purpose of preferentially absorbing the lower-energy radiation.

**Focal Spot** In X-ray tubes, the area of the anode that emits X-rays when bombarded with electrons.

**Fog** A general term used to denote any increase in optical density of a processed photographic emulsion caused by anything other than direct action of the image-forming radiation.

**Gamma Radiography** A technique of producing radiographs using a radioactive source emitting gamma rays.

**Geometric Unsharpness** The penumbral shadow in a radiographic image that is dependent upon (1) the radiation source dimensions, (2) the source to object distance, and (3) the object to film distance.

**Graininess** The visual impression of irregularity of silver deposit in a processed film.

**Half-life** The time required for a radioactive isotope to decay to one-half of its original activity.

**Half-value Layer (HVL)** The thickness of a material required to reduce the intensity of a beam of incident radiation to one-half its original intensity.

**Image Quality Indicator (IQI)** A device or combination of devices whose demonstrated image or images provide visual or quantitative data, or both, to determine the radiographic quality and sensitivity. Also known as penetrameter. (*Note:* It is not intended for use in judging size or establishing acceptance limits of discontinuities.)

**Intensifying Screen** A material that converts a part of the radiation into light or electrons and that, when in contact with a recording medium during exposure, improves the quality of the radiograph or reduces the exposure time required to produce an image, or both.

**IQI Sensitivity** The minimum discernible image and the designated hole in the shim-type or the designated wire image in the wire-type image quality indicator.

**Isotope** One of a group of atoms that have the same atomic number (same chemical characteristics) but have a different mass number. The nuclei have the same number of protons but different number of neutrons, resulting in differing values of atomic mass.

**Kinetic Energy** The energy of a body with respect to its motion.

**Latent Image** A chemical change in the film emulsion as a result of a condition produced and persisting in the image receptor by exposure to radiation. It can be converted into a visible image by processing.

**Location Marker** A number or letter made of lead or other high-density material that is placed on an object to provide traceability between a specific area on the radiograph and the part.

**Material Density** A material's mass per unit volume.

**Milliamperes (mA)** The current applied to the filament in the cathode portion of an X-ray tube that controls the intensity of the X-rays.

**Object to Film Distance** The distance between the surface of the source side of the object and the film.

**Photon** A "packet" of electromagnetic radiation.

**Primary Radiation** Radiation coming directly from the source.

**Radiographic Contrast** The difference in density between an image and an adjacent area on a radiograph.

**Radiographic Quality** A qualitative term used to describe the capability of a radiograph to show changes in the area under examination.

**Radiographic Sensitivity** A general or qualitative term referring to the size of the smallest change in a test part that can be displayed on a radiograph.

**Secondary Radiation** Radiation emitted by any substance as the result of radiation exposure by the primary source.

**Source** A machine or radioactive material that emits penetrating radiation.

**Source to Film Distance** The distance between the radiation-producing area of the source and the film.

**Step Wedge Comparison Film** A strip of processed film carrying a stepwise array of increasing photographic density for comparison to radiographs for the purpose of estimating density.

**Subject Contrast** The ratio (or the logarithm of the ratio) of the radiation intensities transmitted by selected portions of the test specimen.

**Target** That part of the anode of an X-ray tube that emits X-rays as a result of impingement of electrons from the filament.

**Transmitted Film Density** The density of radiographic film determined by measuring the transmitted light, usually through the use of a densitometer.

**Tube Current** The current, measured in milliamperes, passing from the cathode to the anode during the operation of an X-ray tube.

## References

Charles J. Hellier and George C. Wheeler. "Radiographic Interpretation." *NDT Handbook,* Section 8, American Society for Nondestructive Testing, 2004.

Charles J. Hellier and Samuel A. Wenk. "Radiographic Interpretation." *NDT Handbook,* Section Eight. American Society for Nondestructive Testing 1984.

"Radiographic Inspection." From *ASM Handbook,* Vol. 17, *Nondestructive Testing and Quality Control,* 2nd ed, May 1992. ASM International.

Radiography and Radiation Testing. *Nondestructive Testing Handbook,* Vol. 4, *Radiographic Testing,* 3rd ed. American Society for Nondestructive Testing, 2002.

# CHAPTER 7

# Ultrasonic Testing

## 7.1 History

The use of sound energy to determine the integrity of solid objects is probably as old as mankind's ability to manufacture objects in pottery and metal. The English language has many words and phrases that illustrate the acceptance of this fact and hint at the way in which sound was used in the past to test for integrity. Expressions such as "the ring of truth" or "sound as a bell" are commonplace in everyday speech to indicate quality, honesty, or good health.

Both phrases allude to the fact that a sharp tap on a solid object will set up a vibration at the natural frequency of the object; that is how a bell or any percussion instrument makes music. Any major disruption of the homogeneity of the object will distort that natural frequency and indicate that there is a problem. The instinct to tap an object is so ingrained into human nature that it probably accounts for all those people who, when viewing a prospective second-hand car purchase, unconsciously kick a tire!

The problem with this simple approach to testing an object is that it takes a relatively large imperfection to cause a significant change in sound for the human ear to detect. This is because the determining factor is the wavelength of the sound wave encountering an imperfection in relation to the size of the imperfection. Wavelength, in turn, depends on the speed of sound in the object and the frequency, or pitch, of the sound wave. Within the human range of audible sound frequencies, the wavelength is rather large in most metals. It wasn't until the ability to generate and detect sound waves at much higher frequencies existed that smaller discontinuities could be detected in metals.

The first steps toward this ability were taken in the 1870s with the publication of Lord Rayleigh's work on sound, "The Theory of Sound." This work explained the nature and properties of sound waves in solids, liquids, and gasses, which led to the development of the techniques that are currently in use in nondestructive testing.

The means for generating and detecting sound waves at frequencies above the audible range followed shortly after with the discovery of the piezoelectric effect by the Curie brothers and Lippmann. In 1880, the Curies found that an electrical potential could be generated by applying mechanical pressure to plates cut in a particular fashion from certain crystals. The following year, Lippmann discovered that the reverse was true and that the application of an electrical signal to these plates caused a mechanical distortion. Naturally occurring crystals of quartz, tourmaline, and Rochelle salt were among those materials displaying the piezoelectric effect.

Over the years, there have been many uses made of this effect, from crystal microphones and gramophone pickups to spark generators for cigarette lighters and, of course, ultrasonic transducers for NDT. However, growth in use of piezoelectricity was slow.

As early as 1912, following the Titanic disaster, it was suggested that sound waves could be used to detect icebergs at sea, an idea that received further stimulation during World War I for the detection of submarines. The pulse echo system developed for this application gave rise to peacetime uses between the two world wars in the fields of hydrographic surveys charting the ocean depths and fishing, where echo sonar was used to detect shoals of fish.

S. Y. Sokolov, in Russia, was the first to suggest using ultrasonic waves to detect discontinuities in metals. In 1929, he described some experiments in which he generated ultrasonic waves in metals including cast iron and steel samples, which were subsequently sectioned. In 1935, he described his design for piezoelectric transducers for generating and detecting ultrasound, including a method of coupling sound to the metal. His method worked through a transmission technique using continuous waves with quartz transducers and mercury as a couplant. An alternating current generator was used to drive the quartz crystal transmitter and vibrations reaching the receiver caused an alternating signal that could be measured.

Much work was carried out using this technique between the wars, particularly in Russia and Germany. However, there was a problem with continuous wave testing and much of the experimental work concentrated on trying to find a solution. The principle of the continuous wave technique was that energy transmitted through the sample would generate a signal of particular amplitude when no obstruction was present. Some of the sound energy would be obstructed by a discontinuity so that the resulting received signal would be weaker.

This principle was fine as long as there was only enough energy to complete one full transition of the sample, as was often the case with large castings, where grain size rapidly attenuated the sound. If, however, there was sufficient energy to set up multiple reflections of the sound within the sample, the reflected waves joined in with later sound waves to produce confusion. If the reflections were in phase with the continuous wave, there was an apparent increase in transmitted energy. On the other hand, if the reflections were out of phase, the signal became weaker, giving a false indication. The same effect could be observed in narrow samples, where the beam could reflect from the side walls. This reflected beam path, being longer than the straight-through path, could also create constructive or destructive interference with the main signal.

This problem limited the development and use of ultrasonic flaw detection until World War II, when several workers on either side of the combating nations adopted the technique used in echo sonar known as pulse echo. In this system, short pulses of sound are transmitted at regular intervals, the transmitted pulses and the resulting echoes being displayed on a cathode ray tube (CRT). The interval between transmitted pulses is arranged to be sufficient to allow all internal reflections to die away before the next pulse is started, thus avoiding the interference effects previously encountered. Both the transmitter and receiver crystals can be positioned on the same surface.

The sweep time of the CRT can be arranged to display one or more transit periods, and the position of the initial pulse and subsequent echoes can be used to determine the depth of the various reflecting surfaces, including discontinuities.

The pulse echo system was devised around 1942. In Britain, the development was attributed to D. O. Sproule, and in the United States to F. Firestone. After the war, the two approaches were compared, together with developments that had taken place in Germany. The main difference was that Sproule had used separate transmitter and receiver elements, whereas Firestone used the transmitter element as a receiver during its quiescent periods. Since there were merits in both approaches under various circumstances, developments since World War II have tended to use either single or dual element transducers.

At this stage of the development of ultrasonic flaw detection, only "straight beam" compression wave techniques, suitable for detecting reflectors parallel to the scanning surface, were in use. Attempts to angle the beam to reflect from surfaces at other orientations gave confusing results because of the existence of mode conversion in solids. With only a small angulation of the beam at the test surface, both compression waves and shear waves were generated in the object being tested. These traveled at different angles of refraction and at different velocities in the sample, making interpretation of the displayed echoes difficult.

In 1947, Sproule described a transducer design that would generate only shear waves in the sample, and this advance allowed the development of ultrasonic techniques for many other types of discontinuities in the welding, aerospace, and foundry industries. Rapid increase in the use and application of ultrasonics followed this development.

During the next 20 years, much of the development of ultrasonic techniques, as opposed to instrumentation, centered on the accurate sizing of the reflectors detected by the beam. Various approaches using beam parameters were tried, with varying degrees of success. Some were intended to estimate actual size, some were intended to assess the minimum theoretical reflector surface area, and some were intended to provide a common "go, no-go" reporting standard. In Britain, the intensity drop technique developed by C. Abrahams used the plotted –6 dB or –20 dB edges of the sound beam to plot the longitudinal and vertical extents of the reflector. In Germany, Krautkramer developed the DGS (distance gain amplitude) system to compare reflector amplitude against known-diameter circular reflectors (flat-bottomed holes) at defined depths for defined transducers. In the United States, signals were compared and reported against a DAC (distance amplitude correction) curve to produce a common reproducible reporting standard.

Each of the above techniques proved to be unsatisfactory for the application of fracture mechanics to determine fitness for purpose. Only the intensity drop technique claimed to assess the critical "through-thickness" dimension and this technique was heavily dependent on operator skill and reflector profile; repeatability of results was poor.

Sproule, in the 1950s, had described the tip diffraction signal originating at the tip of a discontinuity and defined its amplitude as being 30 dB smaller than a corner reflector at the same depth. Whitford, in the 1960s, developed an alternative sizing system to the intensity drop system that he called the "maximum amplitude" system, in which the last maximum, defining the edges of the reflector, was the tip diffracted shear wave echo. In theory, this diffraction signal gave a much more precise location of the limits of the reflector. However, the signal from the diffracted shear wave is weak and difficult for the ultrasonic practitioner to positively identify.

It was Silk, in 1977, who first used the time of flight diffraction (TOFD) technique to display the top and bottom edges of discontinuities in a way that would allow greater accuracy of through-thickness measurement. The method employs angled compression wave transducers, located on the same surface, to both transmit and receive sound. Lateral wave and tip diffraction signals allow accurate triangulation of the top and bottom edges of the reflector. Recent advances in instrumentation allowing real-time and post-inspection analysis of results using computer technology have increased the number of users exploiting this technique.

If physicists have advanced the understanding of the theory and developed more and more ways of using ultrasonics, engineers have been no less productive in improving the instrumentation over the last 50 years. Early ultrasonic flaw detectors used vacuum tubes (valves), needed generated electricity, and were heavy (see Fig. 7.1(a) and (b)). Using quartz crystals, signal amplitude was poor and resolution very poor. After a shaky start, semiconductor technology has produced flaw detectors that are light, very portable, and together with synthetic crystal materials offers performance that is greatly enhanced. Much of this had been achieved by the mid-1970s.

During the 1980s and 1990s, microchips have been incorporated into the flaw detector, allowing the operator to store calibration parameters and signal traces. This, in turn, allows off-line analysis and reevaluation at a later date. Digital technology and the use of LCD display panels instead of CRTs during the 1990s have further reduced the size and weight of the flaw detectors.

(a)

**Figure 7-1(a)** Instruments of the 1950s and 1960s.

(b)

FIGURE 7-1(b) Reflectoscope-steam engine axle inspection.

## 7.2 Theory and Principles

### 7.2.1 Nature of Sound Waves

Sound waves are simply vibrations of the particles making up a solid, liquid, or gas. As an energy form they are therefore an example of mechanical energy, and it follows that, since there must be *something* to vibrate, sound waves cannot exist in a vacuum.

The only human sense that can detect sound waves is hearing, and that sense is restricted to a relatively narrow range of vibration frequencies called "the audible range." It follows that there will be vibration frequencies that are so low or so high that they cannot be detected by the human ear.

The unit of frequency is the hertz, abbreviated as Hz, defined as "one cycle of vibration per second." Sounds below approximately 16 Hz are below the limit of human hearing and are called "subsonic vibrations," and sounds above approximately 20,000 Hz are too high to be heard and are called "ultrasonic vibrations." Between those two values, in the audible range, it is more common to use the term "pitch" to refer to frequency; a high-pitched sound means high audible frequency, and low-pitched means low audible frequency. A piano key pitched at "middle C" is at a frequency of 260 Hz.

Abbreviations are used for high frequencies; 1,000 Hz is shortened to 1 kHz (one kilo-hertz), 1,000,000 Hz becomes 1 MHz (one megahertz), and a billion cycles per second becomes 1 GHz (one gigahertz). In ultrasonic flaw detection, most testing is carried out in the MHz range (0.5 to 25 MHz).

It is fortunate that there are devices called "transducers" that will change sound waves into electrical energy that can be displayed as visual signals on a cathode ray tube (CRT)or liquid crystal display (LCD) screen. This allows all sounds, including those outside the audible range, to be detected and studied. A transducer is defined as a device that will change one form of energy into another, and vice versa. Materials exhibiting the piezo-electric effect are commonly used to both generate and detect sound waves.

### 7.2.2 Vibration and Periodic Motion

A vibration is an example of periodic motion, a term that suggests that the body or particle concerned is undergoing some repetitive change of position with time. Another example is a pendulum swinging back and forth at a steady rate or frequency. To study the essential requirements for a vibration, consider Fig. 7-2. The weight, W, is suspended from a beam by a spring. At rest, two equal and opposite forces are acting on the weight—gravity (G), acting downward, is opposed by the tension (T) in the spring. The weight is

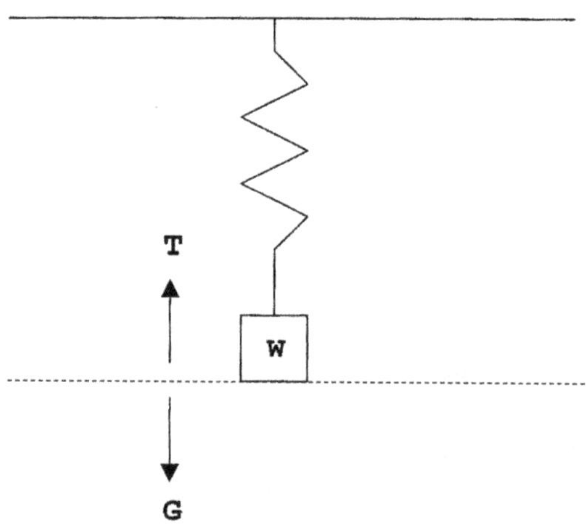


**FIGURE 7-2** Weight on spring extended.

said to be in a state of equilibrium. If the weight is lifted, slackening the spring, then released, gravity will try to restore the weight to its original position. If the weight is pulled down, the tension in the spring will increase; and when the weight is released, this extra tension will try to restore the weight to its original position.

This arrangement provides all the essentials to sustain a vibration. First, there must be something (the weight) to move, and second, there must be a restoring force that will try to counteract that movement or displacement (in this case gravity and the spring).

Imagine that the weight is pulled down from its normal rest position A in Fig. 7-3 to position B, and then released. The extra stretch in the spring will exert a force on W so that it will begin to accelerate back to position A. As it moves, the stretch on the spring decreases until, at position A, both force G and force T are equal again. Since the weight has been accelerating all this time, it has now reached its maximum speed.

Any mass in motion possesses inertia, and this inertia will carry the weight on past the equilibrium position A. But, of course, as soon as it passes A, the spring slackens so that T becomes less than G. In other words, gravity starts to slow the weight down. Eventually, the weight comes to rest at a new position C. Now gravity outweighs the tension in the spring, trying to accelerate W back to position A. On reaching A, inertia ensures that the weight overshoots again and the whole train of events starts again.

If a pen is fixed to the weight and allowed to write against a piece of chart paper pulled at a steady speed past the pen, a trace of the movement of the weight with time, as it bobs up and down, will be drawn, as shown in Fig. 7-4. This trace is typical of all periodic motion and faithfully shows what has happened to the weight.

The steeper the line traced out, the faster is the movement of the weight. At the maximum displacement up or down, the trace is flat, showing that the weight has stopped briefly. As the weight passes through the normal rest position each time, the line is steepest, showing maximum speed. On the trace, $t_1$ represents the weight, traveling upward and passing position A. The next trace position, $t_2$, shows the weight traveling down but again passing A, and $t_3$ again shows the weight traveling upward at position A.

Between $t_1$ and $t_3$ the weight has traveled up to its top limit, down to its bottom limit, and back to where it started. The trace between $t_1$ and $t_3$ has drawn "one cycle" of

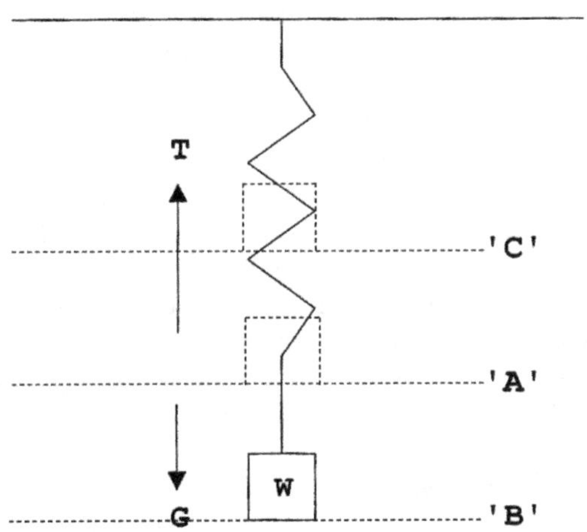


**FIGURE 7-3** Weight on spring extended 2.

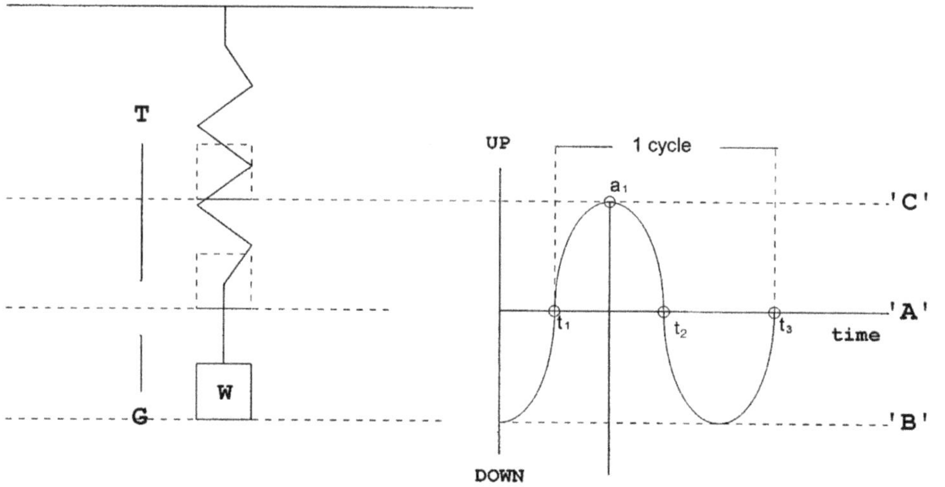


**FIGURE 7-4** Weight and cycle.

the motion of the weight in a set period. If the weight is allowed to carry on bouncing up and down until it eventually comes back to rest, the trace will only go on drawing repeats of that motion, each cycle occupying the same time span. The number of cycles completed in one second has already been defined as the frequency of the vibration.

The maximum displacement of the weight from its normal rest position is shown on the trace as $a_1$ and is known as the amplitude of the vibration. With sound vibrations, frequency is perceived as the pitch of the sound, whereas amplitude is the loudness of the sound.

### 7.2.3 Sound Vibrations

For sound waves in solids, liquids, and gases, the vibrating bodies are the particles making up the substance, and the restoring forces are the elastic bonds holding the substance together. The particles can be imagined to be joined together by springs. If one particle moves toward its neighbor, the spring gets squashed and tends to push the invader back "home." Similarly, if it moves away from its neighbor, the spring gets stretched and the particle is pulled back into place.

Audible sound is an example of a vibration mode called a "compression wave." It travels from the source by a succession of shunting actions from one particle to the next. Each particle vibrates at the frequency of the sound, oscillating to and fro by a distance that is the amplitude or loudness of the sound. As each particle oscillates, it squashes the "spring" to the next neighbor and starts the neighbor oscillating. As the oscillation passes from one particle to the next, and the next, and so on, the sound wave is said to travel or "propagate" through the material. Note that individual particles do not migrate to another place; they only oscillate about a mean position.

### 7.2.4 Modes of Propagation

The type, or mode, of sound wave propagation described above (compression wave) can exist in solids, liquids, or gases. Other modes of vibration can exist, but only in solids.

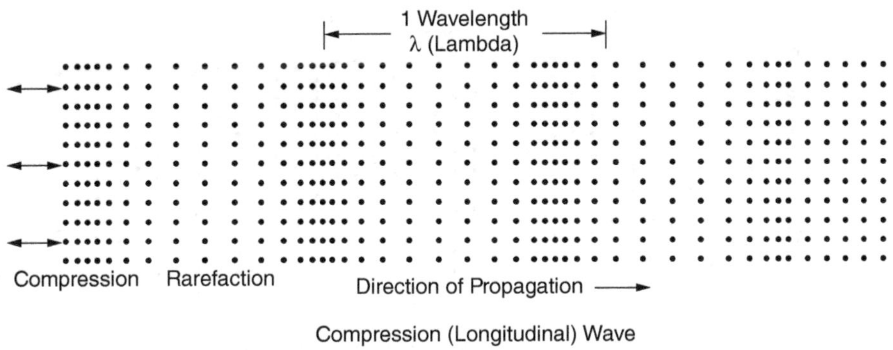


**FIGURE 7-5** Compression wave drawing.

The various ways in which sound can propagate are usually described in terms of the direction of particle motion in relation to the direction in which the sound wave travels. Compression waves can be defined on this basis as waves in which the particle motion is in the same plane as the direction of propagation (see Fig. 7-5).

All three media have forces that bind the particles together to resist squashing or pulling apart (compression or tension). In solids, this is provided by the modulus of elasticity, known as "Young's modulus of elasticity." The pressure of an entrapped gas as it is squashed rises to oppose the squashing force, and the pressure drops if the volume is increased, the partial vacuum applying the restoring force (see Fig. 7-6).

Solids, unlike liquids and gasses, also have rigidity that is a resistance to shear loads. It is the rigidity that has to be overcome when snapping a stick, for instance. The name for this resistance to shear loads in solids is called "the modulus of rigidity,"

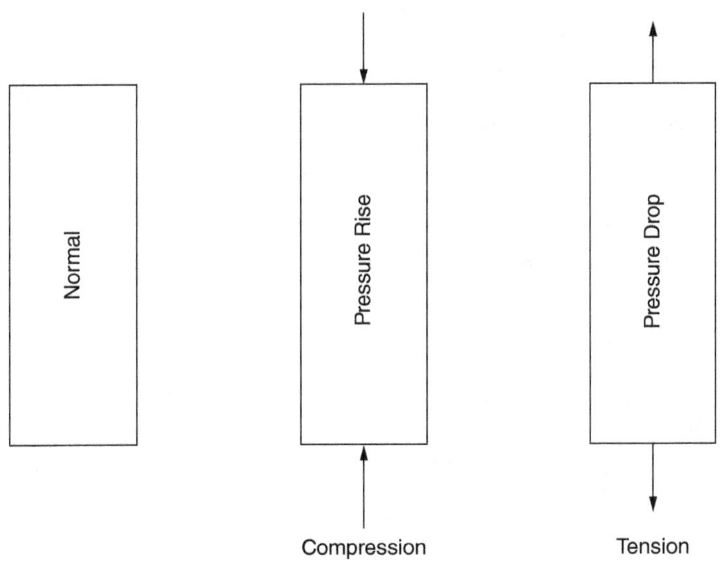


**FIGURE 7-6** Cylinders.

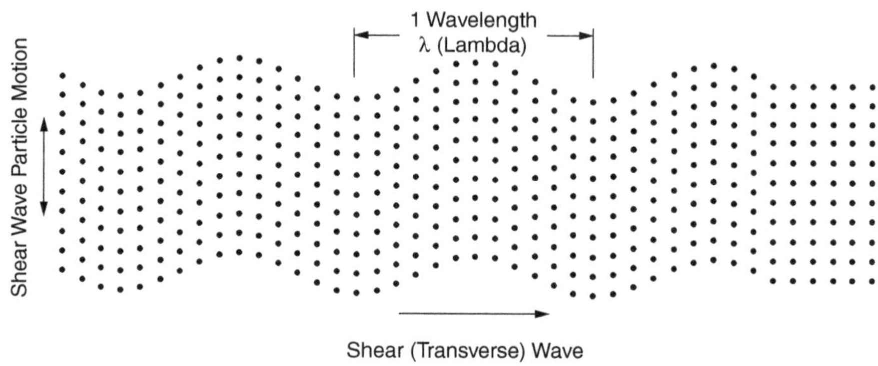


**Figure 7-7** Shear wave drawing.

and it allows sound to propagate in a different way under certain circumstances. This new mode of propagation is known as a shear wave and is defined as a wave in which the particle motion is at right angles to the direction of propagation (see Fig. 7-7).

If a shear wave is set up so that it just skims along the surface of a solid, it again changes mode to one, which is contour following with a peculiar particle motion. This contour following wave is called a surface wave and is defined as a wave in which the particle motion is elliptical, with the major axis of the ellipse perpendicular to the direction of propagation (see Fig. 7-8).

Lamb (plate) waves, like surface waves, propagate parallel to the test surface and have a particle motion that is elliptical. They occur when the thickness of the test material is only a few wavelengths at the test frequency and where the test piece is of uniform thickness. In other words, they can exist best in plate, tube, and wire.

Finally, there is a special type of compression wave that skims along the surface rather like a surface wave and is called a creeping or lateral wave. Its use is described under TOFD techniques.

The four main modes of propagation are the compression wave, the shear wave, the surface wave, and the lamb wave. Each of these has an alternative name that is sometimes used. These alternative names are:

Compression waves are sometimes called longitudinal waves

Shear waves are sometimes called transverse waves

Surface waves are sometimes called Rayleigh waves

Lamb waves are sometimes called plate waves

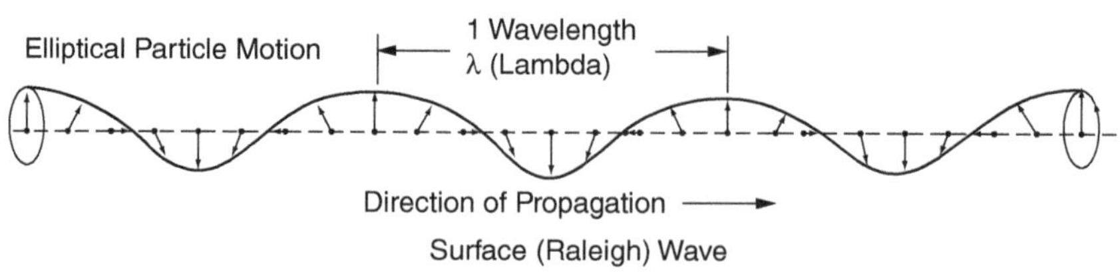


**Figure 7-8** Surface wave drawing.

### 7.2.5 Properties of Sound Waves

#### *Velocity*

Sound travels at different speeds through different materials. This is noticeable when, for example, a railroad worker is observed from a distance striking a rail with a hammer. Since the speed of light is much faster than that of sound, the observer first sees the hammer strike the rail. If the observer is standing close to the rail, he or she next hears the sound of the blow coming from the rail. Finally, the airborne sound resulting from the blow is heard.

This shows us that the speed of sound in the rail is faster than the speed of sound in air. It is true that sound travels faster in liquids than in gasses and faster in metals than in liquids. However, it is also true that sound travels at different speeds in different metals. There is a distinct speed of sound for each material. In ultrasonics, this is called the velocity of sound for that material. This being so, it would be useful to have an understanding of the reasons for the difference.

#### *Factors Affecting Velocity*

The two main factors affecting velocity are the density and the elasticity of the material. To grasp a logical explanation for this, imagine that the molecules of any material are balls whose weight is analogous to the material density. So for lead, the balls would be heavy; they would be lighter for aluminum, and featherweight for air. Also imagine that these balls are joined together by springs representing the elasticity, or strength, of the material. For steel, the springs would be strong; they would be weaker for lead, and very weak for air. With these two concepts in mind, the scene is set.

The speed with which sound propagates through a material depends on how quickly one ball can get its neighbor to take up the vibration; in other words, to pass on the message. To get its neighbor moving, it has to overcome the inertia of that neighbor. Suppose two balls of a given weight are joined by strong springs and two more of equal weight are joined by weak springs. If the first balls in each pair are moved simultaneously, the ball facing the strong spring will quickly build up enough force to overcome the inertia of its neighbor. On the other hand, the ball facing the weak spring will have to move further and thus take longer to build up the equivalent force. From this, the logic tells us that it is reasonable to expect materials with a high value for Young's modulus of elasticity to have a high velocity of sound, as is the case, for example, with steel.

Consider two more pairs of balls. This time the springs are all the same strength but one pair of balls is light and the other heavy. If, again, the first balls in each pair are moved simultaneously, the lightweight pair quickly exchange messages but this takes longer in the heavy pair. Again, the general rule is indicated: the higher the density of a material, the lower the velocity of sound. Lead, for example, has a lower velocity than steel.

Density and elasticity are the dominant factors affecting velocity, but there is another one that plays a relatively minor, but nonetheless significant, role, and it is called Poisson's ratio. It is easy to see that when an elastic band is stretched it also gets thinner. The more it is stretched, the thinner it gets. Poisson's ratio relates the thinning to the stretching and can be calculated by dividing the change in diameter of the elastic band by the change in length.

The velocity of the compression wave for a given material can be calculated from the equation

$$V_c = \sqrt{\frac{E}{\rho} \cdot \frac{1-\sigma}{(1+\sigma)(1-2\sigma)}}$$

where:
$V_c$ = compression wave velocity
$E$ = Young's modulus of elasticity
$\rho$ = material density
$\sigma$ = Poisson's ratio

Shear waves are able to exist in solids and they do not travel at the same velocity as the compression wave in a given material. This is because it is the modulus of rigidity, rather than Young's modulus, that dictates the velocity, and the modulus of rigidity is lower than the modulus of elasticity. This means that the shear wave velocity is always slower than the compression wave velocity in a material. As a rule of thumb, the shear wave velocity is roughly half the compression wave velocity. The velocity can be calculated from

$$V_s = \sqrt{\frac{E}{\rho} \cdot \frac{1}{2(1+\sigma)}} \qquad \text{or alternatively} \qquad V_s = \sqrt{\frac{G}{\rho}}$$

where:
$V_s$ = shear wave velocity
$G$ = modulus of rigidity
$\rho$ = material density
$\sigma$ = Poisson's ratio

Surface (Rayleigh) waves also have their own particular velocity, which is generally taken to be approximately 90 percent of the shear wave velocity.

Although the velocity for each of these modes of propagation can be calculated, it requires a precise knowledge of all the parameters, and these are not usually available to the ultrasonic practitioner. Parameters such as density and strength vary with alloying, heat treatment, casting, rolling, and forging processes, all of which make it difficult to know that the correct values are being used. Instead, it is more normal to carry out a routine called "calibration" during the setting up procedure, for an ultrasonic inspection. In the calibration procedure, the flaw detector time base is adjusted to give a convenient scale against a calibration sample of known thickness and made of the same material as the work to be tested.

### *Wavelength*

The distinction between the oscillating motion of the particles making up a solid, liquid, or gas and the velocity of the sound moving through the substance has already been made. As the particles are completing each cycle of their vibration, the sound wave is moving on in the direction of propagation at the characteristic velocity for that material. It follows that during the time taken to complete one cycle of vibration, the sound wave will move a certain distance depending on the velocity in that material. For gasses with low velocities, that distance is small compared to the distance in metals, which have high velocities. This distance for a given material and sound vibration frequency is called the wavelength.

Wavelength is given the Greek symbol $\lambda$ (lambda) and for any material and sound frequency, can be calculated from the equation

$$\lambda = \frac{V}{f}$$

where:
$\lambda$ = wavelength
$V$ = velocity
$f$ = frequency

**Example 1** Calculate the wavelength of a 5-MHz compression wave in steel, given that the velocity of sound in mild steel is 5,960 meters per second (m/sec).

$$\lambda = \frac{V}{f}$$

$$\therefore \lambda = \frac{5{,}960}{5{,}000{,}000}\text{ meters (m)}$$

$$\therefore \lambda = 0.00192 \text{ m}$$

It would be better to express such a small distance in millimeters (mm) by multiplying the answer by 1,000:

$$\lambda = 0.00192 \times 1000$$

$$\lambda = 1.192 \text{ mm}$$

At ultrasonic frequencies, the wavelength of sound in metals is relatively short and so it is usual to express the wavelength in millimeters. This is done at the start of the calculation by changing the velocity from meters to millimeters per second by multiplying by 1,000.

**Example 2** Calculate the wavelength of a 5-MHz compression wave in aluminum, given that the velocity is 0.252″/$\mu$ sec (6,400 m/sec).

$$\lambda = \frac{V}{f}$$

$$\lambda = \frac{6{,}400 \times 1{,}000}{5{,}000{,}000}$$

$$\lambda = 1.28 \text{ mm}$$

Wavelength is useful in many ways in ultrasonic flaw detection. In the first place, the smallest reflector that can be detected must have a major dimension of at least half a wavelength at the test frequency. If the critical size of the discontinuity that must be detected is known, the knowledge helps with selection of an appropriate test frequency. Wavelength is also used in the calculation of the sound beam shape and the near-field distance. The significance of these will be discussed later.

### *Reflection*

The boundary between one medium and another—for instance, steel to air at the far side of a steel plate—is called an "interface." At an interface, a proportion of the sound may be transmitted to the next medium and the remainder reflected back to the first medium. In the case of a steel to air interface, almost all the energy reflects and virtually none goes into the air. If the steel is under water, so that there is a steel to water interface, 88 percent of the energy is reflected and 12 percent is transmitted into the water. The proportions that will be reflected or transmitted depend upon the properties of the materials on either side of the interface.

In order to understand this phenomenon, consider again the balls and springs. This time, consider a train of tightly packed heavy balls joined by strong springs, representing steel. The steel train leads to a train of widely spaced light balls attached to weak springs, representing the air. If a compression wave is initiated at the start of the steel train, the message is passed on along the train, with each ball limited in its oscillation by its neighbor until the last ball in the steel train. It finds more space and lighter neighbors and is able to move a greater distance from its mean position in the direction of the air. In doing this, it stretches the spring joining it to the last but one steel ball until the spring tension arrests its motion. The last steel ball now starts to accelerate back into position and is so fast when it gets there that it overshoots, crashes into the last but one ball and starts a compression wave going back the other way. This is a reflection.

The two big differences between air and steel are density and elasticity, and these are the factors that decide how much energy is reflected and how much is transmitted at the interface. Each material is given a factor that is used to calculate reflectivity at an interface. This factor is called the "acoustic impedance" and given the symbol $Z$. Acoustic impedance is the product of density and velocity for that material. Stated mathematically

$$Z = \rho \times V$$

where:
$Z$ = acoustic impedance
$\rho$ = material density
$V$ = material velocity

To calculate the percentage of energy reflected at an interface between any two materials, the following formula is used:

$$\text{Reflected energy} = \left(\frac{Z_1 - Z_2}{Z_1 + Z_2}\right)^2 \times 100\%$$

where $Z_1$ and $Z_2$ are the acoustic impedance of the materials on either side of the interface.

**Example 3** Calculate the percentage energy reflected at a steel to water interface, given that the acoustic impedance for steel is 46.7 and that for water is 1.48.

$$\text{Reflected energy} = \left(\frac{Z_1 - Z_2}{Z_1 + Z_2}\right)^2 \times 100\%$$

$$\text{Reflected energy} = \left(\frac{46.7 - 1.48}{46.7 + 1.48}\right)^2 \times 100\%$$

$$\text{Reflected energy} = \left(\frac{45.22}{48.18}\right)^2 \times 100\%$$

$$\text{Reflected energy} = (0.93856)^2 \times 100\%$$

$$\text{Reflected energy} = 0.8809 \times 100\%$$

$$\text{Reflected energy} = 88.09\%$$

**Example 4** Calculate the percentage energy reflected at a steel to air interface, given that the acoustic impedance for steel is 46.7 and that for air is 0.0004.

$$\text{Reflected energy} = \left(\frac{Z_1 - Z_2}{Z_1 + Z_2}\right)^2 \times 100\%$$

$$\text{Reflected energy} = \left(\frac{46.7 - 0.0004}{46.7 + 0.0004}\right)^2 \times 100\%$$

$$\text{Reflected energy} = \left(\frac{46.6996}{46.7004}\right)^2 \times 100\%$$

$$\text{Reflected energy} = (0.99998)^2 \times 100\%$$

$$\text{Reflected energy} = 1.0000 \times 100\% \text{ approx}$$

$$\text{Reflected energy} = 100\%$$

When a beam of sound traveling through a metal sample encounters a discontinuity such as a crack, lamination, void, or nonmetallic inclusion, there is an interface. On one side is the sound metal and on the other the discontinuity. At this interface, some energy will be reflected and some transmitted. If the discontinuity side of the interface is air then the reflection is total; but even for a nonmetallic inclusion, most of the energy will be reflected. This property of sound waves allows for the detection of discontinuities in materials (see Table 7-1).

### Couplant

The property of reflection can also be a problem because if a transducer is simply placed on a part, there must be an air gap, however small. But a solid to air interface creates 100 percent reflection, so the sound goes straight back into the transducer without transmitting into the metal! To overcome this problem there has to be some way to exclude the air using a medium that will match the acoustic impedance of the transducer to the metal. Since this medium must also allow the transducer to be scanned over the surface of the metal, it needs to be a liquid, grease, or paste. Such a substance is called a "couplant."

There are many suitable substances that can be used as couplants, the main criteria being the best possible match and no adverse chemical reaction between the couplant and the metal. Most couplants only allow limited matching because liquids in general have a low acoustic impedance. In immersion testing, the couplant is usually water, which only allows about 12 percent of the energy into the steel, and, of course, only 12 percent of any echoes to pass back across the interface and back to the receiving transducer. Most couplants permit between 10 percent and 15 percent sound transmission. The best of these is glycerin at around 15 percent. Commonly used couplants are:

- Water
- Kerosene
- Oil
- Grease
- Wallpaper paste
- Glycerin
- Special gels designed for the purpose

| Material | Ultrasonic Velocity | | | | |
|---|---|---|---|---|---|
| | Longitudinal | | Transverse (Shear) | | Impedance |
| | in/µs | mm/µs | in/µs | mm/µs | Z |
| **Metals** | | | | | |
| Aluminum 1100-0 | 0.248 | 6.229 | 0.121 | 3.073 | 17.1 |
| Aluminum 2024-T4 | 0.251 | 6.375 | 0.124 | 3.150 | 17.6 |
| Aluminum 6061-T6 | 0.248 | 6.299 | 0.124 | 3.150 | 17.0 |
| Beryllium | 0.507 | 12.878 | 0.350 | 8.890 | 23.5 |
| Brass (70% Cu–30% Zn) | 0.172 | 4.369 | 0.083 | 2.108 | 37.1 |
| Bronze (phosphor 5%) | 0.139 | 3.531 | 0.088 | 2.235 | 31.3 |
| Copper (CP) | 0.187 | 4.750 | 0.092 | 2.337 | 42.5 |
| Gold | 0.128 | 3.251 | 0.047 | 1.194 | 62.6 |
| Hastelloy C | 0.230 | 5.842 | 0.114 | 2.896 | 52.2 |
| Hastelloy X | 0.228 | 5.791 | 0.108 | 2.743 | 47.7 |
| Inconel (wrought) | 0.308 | 7.823 | 0.119 | 3.023 | 64.5 |
| Iron (cast), various alloys | 0.138–0.220 | 3.505–5.588 | 0.087–0.126 | 2.210–3.200 | 24.3–41.2 |
| Lead (94Pb-6Sb) | 0.085 | 2.159 | 0.032 | 0.813 | 23.5 |
| Magnesium, various alloys | 0.215–0.228 | 5.461–5.791 | 0.119–0.122 | 3.023–3.099 | 9.24–10.6 |
| Monel | 0.211 | 5.359 | 0.107 | 2.718 | 47.2 |
| Nickel (CP) | 0.222 | 5.639 | 0.117 | 2.972 | 50.0 |
| Silver (0.99 fine) | 0.142 | 3.607 | 0.063 | 1.600 | 37.8 |
| Steel 1020 | 0.232 | 5.893 | 0.128 | 3.251 | 45.4 |
| Steel 4340 | 0.230 | 5.842 | 0.128 | 3.251 | 45.6 |
| Steel, CRES 300 Series | 0.221–0.226 | 5.613–5.740 | 0.120–0.123 | 3.048–3.124 | 44.6–45.4 |
| Steel, CRES 400 Series | 0.212–0.237 | 5.385–6.020 | 0.118–0.132 | 2.997–3.353 | 41.3–46.3 |
| Titanium, 6Al-4V | 0.243 | 6.172 | 0.130 | 3.302 | 27.3 |
| Zircaloy | 0.186 | 4.724 | 0.093 | 2.362 | 44.2 |
| Zirconium | 0.183 | 4.648 | 0.089 | 2.261 | 30.1 |
| **Polymers** | | | | | |
| Acrylics | 0.105–0.109 | 2.667–2.769 | 0.044–0.057 | 1.118–1.448 | 3.15–3.51 |
| Cellulose acetate | 0.096 | 2.438 | No Shear Component | | 3.19 |
| Nylon | 0.016 | 2.692 | No Shear Component | | — |
| Phenolic | 0.056 | 1.422 | No Shear Component | | 1.90 |
| Polycarbonate | 0.090 | 2.286 | No Shear Component | | 2.71 |
| Polyethylene | 0.105 | 2.667 | No Shear Component | | 2.94 |

**TABLE 7-1** Acoustical Properties of Common Materials

| Material | Ultrasonic Velocity | | | | |
|---|---|---|---|---|---|
| | Longitudinal | | Transverse (Shear) | | Impedance |
| | in/μs | mm/μs | in/μs | mm/μs | Z |
| **Polymers (*Continued*)** | | | | | |
| Polystyrene | 0.094 | 2.388 | 0.045 | 1.143 | 2.52 |
| Rubber (natural) | 0.061 | 1.549 | No Shear Component | | 1.74 |
| Rubber (carbon filter) | 0.066 | 1.676 | No Shear Component | | — |
| Rubber (silicone) | 0.037 | 0.94 | No Shear Component | | 1.40 |
| Teflon | 0.054 | 1.372 | 0.250 | 6.35 | 3.00 |
| **Miscellaneous Solids** | | | | | |
| Alumina ($Al_2O_3$) | 0.427 | 10.846 | No Shear Component | | 43.1 |
| Concrete | 0.167–0.207 | 4.242–5.258 | 0.135 | 3.429 | 12.4 |
| Glass (plate) | 0.227 | 5.766 | No Shear Component | | 14.5 |
| Granite | 0.156 | 3.962 | 0.076 | 1.93 | 10.9 |
| Ice (–16°C) | 0.150 | 3.81 | No Shear Component | | 3.60 |
| Quartz, natural | 0.226 | 5.74 | 0.139 | 3.531 | 15.2 |
| Quartz, fused | 0.219 | 5.563 | 0.302 | 7.671 | 14.5 |
| Sapphire | 0.469 | 11.913 | 0.157 | 3.988 | 47.2 |
| Tungsten carbide | 0.262 | 6.655 | No Shear Component | | 67.6 |
| **Composite Materials** | | | | | |
| Fiberglass (50 v/o) | 0.124 | 3.15 | 0.068 | 1.727 | 6.04 |
| Graphite/Epoxy (60 v/o) | 0.117 | 2.972 | 0.077 | 1.956 | 4.65 |
| Boron/Epoxy (50 v/o) | 0.131 | 3.327 | 0.072 | 1.829 | 6.38 |
| **Liquids** | | | | | |
| Ethylene glycol | 0.064 | 1.626 | No Shear Component | | 1.80 |
| Glycerin | 0.076 | 1.93 | No Shear Component | | 2.42 |
| Oil (SAE 20) | 0.069 | 1.753 | No Shear Component | | 1.51 |
| Water (20°C) | 0.058 | 1.473 | No Shear Component | | 1.48 |
| **Gases** | | | | | |
| Air (20°C) | 0.014 | 0.356 | No Shear Component | | 0.00041 |
| Nitrogen (20°C) | 0.014 | 0.356 | No Shear Component | | 0.00041 |
| Oxygen (20°C) | 0.013 | 0.33 | No Shear Component | | 0.00043 |

Courtesy of NDT Systems, Inc.

**Table 7-1** Acoustical Properties of Common Materials (*Continued*)

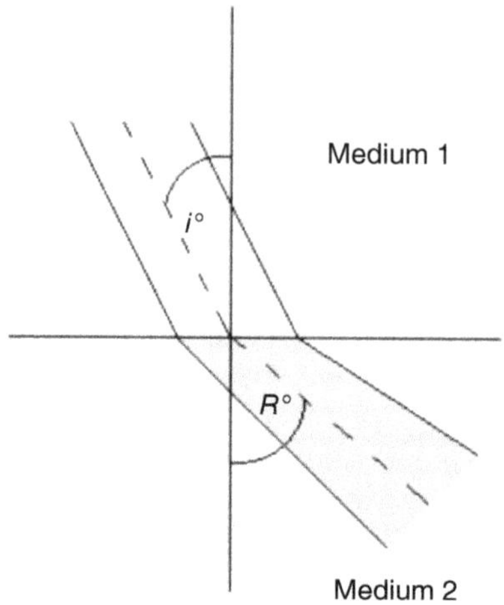


**FIGURE 7-9** Refraction.

### *Refraction and Mode Conversion*

So far only sound entering the metal perpendicular to the surface has been discussed. When the sound is introduced at an angle to the surface called the "angle of incidence," several things may happen, depending on the actual angle of incidence.

Figure 7-9 shows a beam of sound traveling toward an interface at an angle of incidence $i°$ to the perpendicular that is usually called the "normal." The velocity of sound in Medium 1 is $V_1$ and in Medium 2, on the other side of the interface, the velocity is $V_2$.

Assume for this example that $V_1$ is slower than $V_2$, as would be the case if Medium 1 had been water and Medium 2 steel. As the beam travels toward the interface, the whole beam is moving at the same speed until the left-hand edge first strikes the interface. The moment the edge of the beam reaches Medium 2, it speeds up. But the sound still in Medium 1 stays at the old speed. Gradually, as the entire wavefront sweeps the interface, it speeds up until at last the right-hand edge passes across the interface and the entire beam travels on in Medium 2 at the new speed.

During this transition, the beam slues around to a new angle in Medium 2 called the "angle of refraction." It is a bit like somebody passing through a doorway, catching their pocket on the door handle, and being diverted by the braking action on that side. The reason for this refraction is the velocity difference on either side of the interface. Snell's law allows the new angle to be calculated if the two velocities and the angle of incidence are known. Snell's law states that the sine of the angle of incidence divided by the velocity in Medium 1 equals the sine of the angle of refraction divided by the velocity in Medium 2. Stated mathematically:

$$\sin\frac{i°}{V_1} = \sin\frac{R°}{V_2}$$

where:

$\sin i°$ = the sine of the angle of incidence
$\sin R°$ = the sine of the angle of refraction
$V_1$ = the velocity in Medium 1
$V_2$ = the velocity in Medium 2

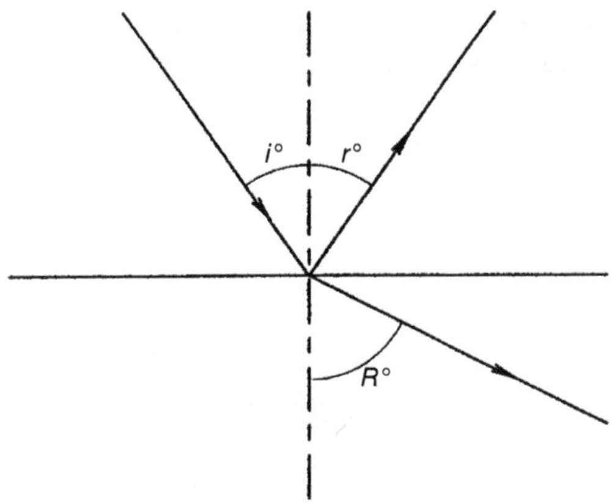


**FIGURE 7-10** Snell's law.

Refraction refers to the transmitted portion of the sound energy at the interface; the proportion of energy reflected is the same as before but the reflected energy leaves the interface at an angle of reflection equal to the angle of incidence, as shown in Fig. 7-10. This diagram shows the angle of incidence ($i°$), angle of reflection ($r°$), and angle of refraction ($R°$).

**Example 5** Calculate the angle of refraction in steel for an incident angle in water of 10° given that the compression wave velocity of sound in water is 1,480 m/sec and in steel 5,960 m/sec.

$$\frac{\sin 10°}{1{,}480} = \frac{\sin R°}{5{,}960}$$

$$\frac{5{,}960 \times \sin 10°}{1{,}480} = \sin R°$$

$$\frac{5{,}960 \times \sin 10°}{1{,}480} = \sin R°$$

$$0.6993 = \sin R°$$

$$R° = 44.37°$$

The relationship between velocity and refraction can be seen in the above example, the compression wave velocity of sound in steel is roughly four times that in water, and the refracted compression wave angle is roughly four times the incident angle.

### Mode Conversion

As the beam of sound is introduced at an angle of incidence to a solid, another phenomenon begins to arise, and that is mode conversion. Although the incident beam is a compression wave, a refracted shear wave begins to develop in the solid as the sound crosses the interface, in addition to the refracted compression wave. For small angles of incidence, the amplitude of the shear wave is small and can be ignored, but as the angle of incidence increases, so does the amplitude of the shear wave. Eventually, both the shear wave and the compression wave are about equal in amplitude. Snell's law shows us that the two modes will not refract through the same angle because the velocity of the shear wave is less than the compression wave.

**Example 6** Calculate the angle of refraction of the shear wave in steel for an incident compression wave of 10° in water, given that the shear wave velocity in steel is 3,240 m/sec.

$$\frac{\sin i^\circ}{V_1} = \frac{\sin R^\circ}{V_2}$$

$$\frac{\sin 10^\circ}{1,480} = \frac{\sin R^\circ}{3,240}$$

$$\frac{3,240 \times \sin 10^\circ}{1,480} = \sin R^\circ$$

$$\frac{3,240 \times 0.1736}{1,480} = \sin R^\circ$$

$$0.3801 = \sin R^\circ$$

$$R^\circ = 22.34^\circ$$

The problem for ultrasonic flaw detection is immediately obvious—two beams traveling at different speeds and in different directions spells chaos! This was a problem that beset the early practitioners until Sproule came up with a solution in 1947. What he did was to get rid of one of the beams. He did this by increasing the angle of incidence until the refracted compression wave refracted to 90°. Any further increase in incident angle leaves only a refracted shear wave. The compression wave is said to have undergone total internal reflection in Medium 1.

The angle of incidence giving a 90° refracted angle for the compression wave is called the "first critical angle." The first critical angle for a water to steel interface is about 15°, and for Plexiglas (Lucite or Perspex) to steel the first critical angle is about 28°. Above the first critical angle of incidence, only a shear wave remains. By suitable choice of an incident angle above the first critical angle, a shear wave beam of any desired angle can be achieved. For immersion testing, the transducer is simply tilted through the calculated angle of incidence in water. For manual scanning, the transducer is mounted on a Plexiglas wedge angled to the desired incident angle.

If the incident angle is increased more and more beyond the first critical angle, eventually the shear wave will also be refracted to 90°. The angle of incidence to achieve this is called the "second critical angle." At the second critical angle, the shear wave undergoes another mode conversion; this time it becomes a surface (Rayleigh) wave, which is the contour-following wave. Any increase in angle of incidence beyond the second critical angle leaves no sound in Medium 2 at all; there is total internal reflection in Medium 1. For water to steel, the second critical angle is about 27°, and for Plexiglas to steel about 58°. Figure 7-11 shows the angle of refraction in steel obtained for increasing angles of incidence in Plexiglas at a Plexiglas to steel interface, and Fig. 7-12 shows the same for a water to steel interface.

### *Reflective Mode Conversion*

Mode conversion also takes place inside a solid when an ultrasonic beam strikes a reflector at an angle of incidence other than perpendicular. Figure 7-13 shows a

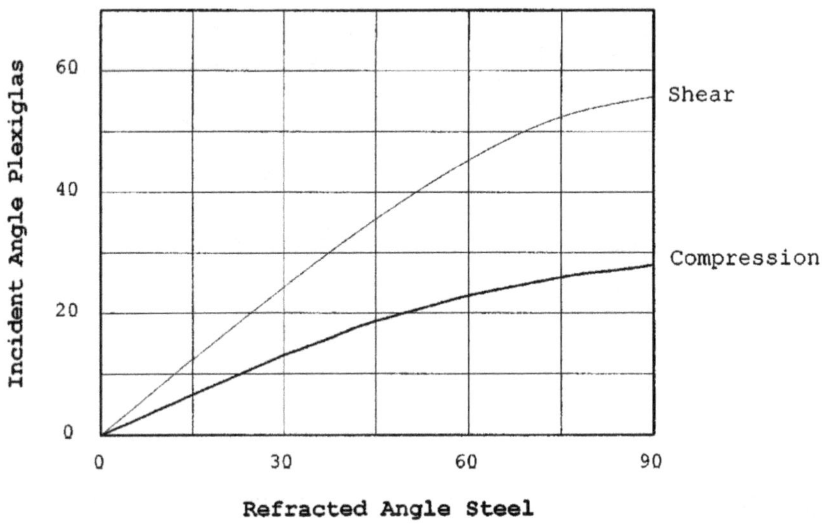


**FIGURE 7-11** Incident versus refracted angles in steel.

compression wave, *C*, striking a steel to air interface at an angle of incidence, *i*°, to the normal. The reflected compression wave, $C_r$, is at a reflected angle *r*° equal to the angle of incidence. However, there is also a mode-converted shear wave, *S*, at an angle derived from Snell's law, *s*°. In ultrasonic flaw detection, this mode conversion can cause confusion, depending on the relative amplitudes of the reflected compression wave, and the mode-converted shear wave.

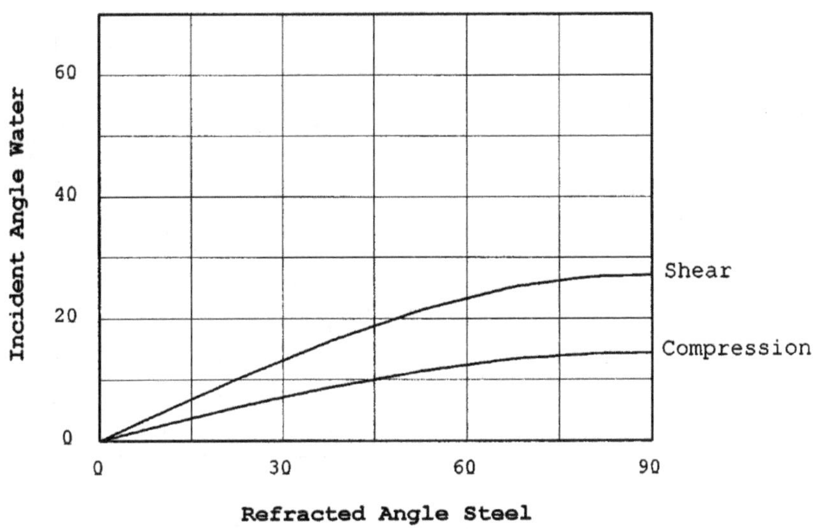


**FIGURE 7-12** Incident versus refracted angles in steel.

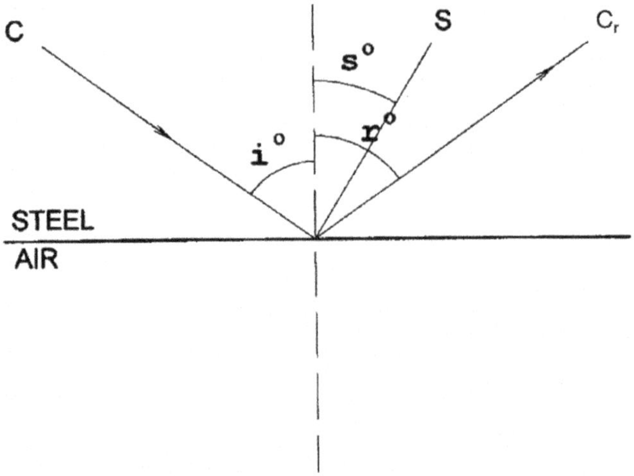


**FIGURE 7-13** Incident and reflected angles.

Figure 7-14 shows the relative amplitudes of the reflected compression wave and the mode conversion shear wave for increasing angles of incidence, of the compression wave at a steel to air interface. It can be seen that at low angles of incidence the shear wave is weak and can be ignored. At an angle of incidence of about 25°, the reflected compression wave and the shear wave are at the same amplitude, and at about 70°, the reflected compression wave is very weak, whereas the shear wave is still very strong.

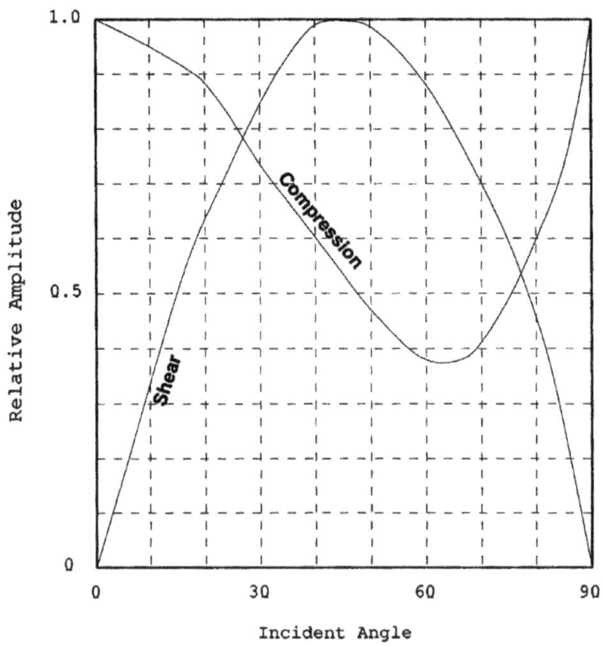


**FIGURE 7-14** Incident angle graph.

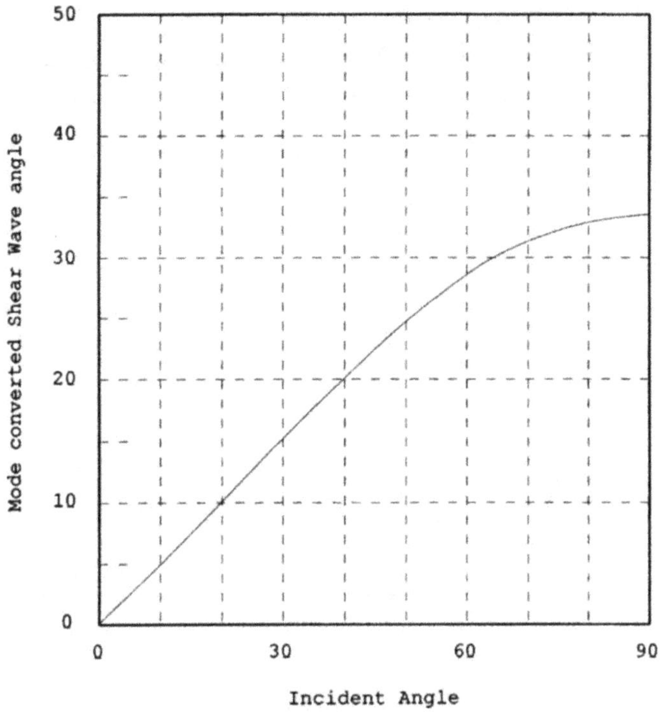


**FIGURE 7-15** Incident angle graph 2.

Figure 7-15 shows the angle of the mode-converted shear wave with respect to the normal for increasing angles of incidence for the compression wave at the steel to air interface. These two graphs show the strength and direction of the mode conversion. The practitioner may encounter these circumstances when carrying out a compression wave (straight beam) test if a reflecting surface is not parallel to the scanning surface.

Mode conversion can also take place when a shear wave meets a reflecting surface, as shown in Fig. 7-16. In this case, the reflected shear wave is again at the same angle as the angle of incidence, but the mode-converted compression wave is at an angle $\alpha°$, which can be calculated from Snell's law.

Figure 7-17 shows the amplitudes of the reflected shear wave and the mode-converted compression wave relative to the amplitude of the incident shear wave.

Figure 7-18 shows the angle of the mode-converted compression wave. It can be seen from the graph that when a shear wave is incident to a reflecting surface at about 30°, less than 10 percent of the shear wave is reflected, but the mode-converted compression wave amplitude rises steeply, and is far greater than the reflected shear wave. Figure 7-18a shows that for this incident angle of shear wave, the mode-converted compression wave will be at an angle of about 65° to the normal. This situation will occur if a 60° angle beam transducer is chosen to examine a weld with a vertical fusion face, such as an electron-beam weld, or the root face of a double "V" weld preparation. As shown in Fig. 7-18b, the shear wave will meet any vertical

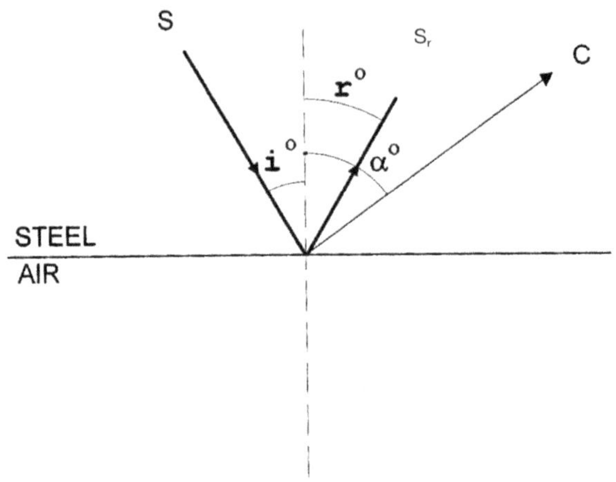


**Figure 7-16** Incident and reflected angles 2.

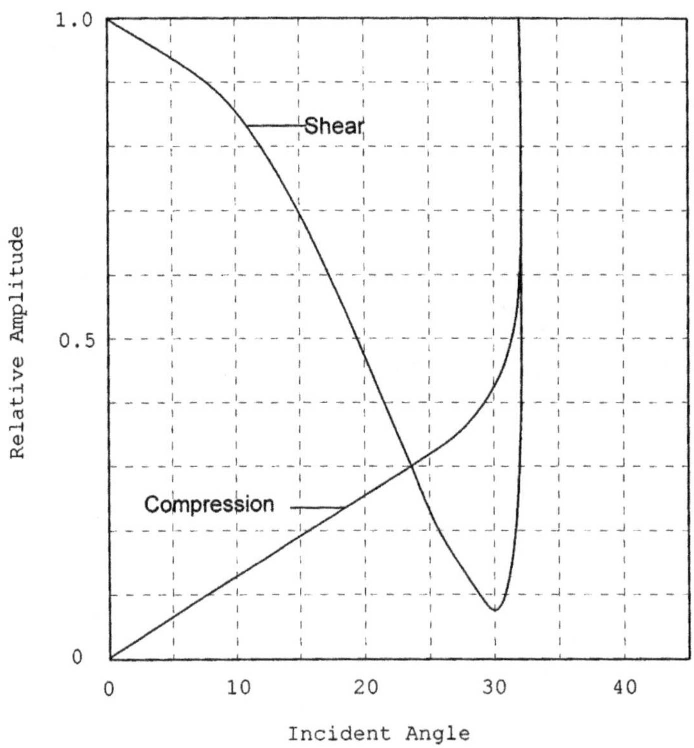


**Figure 7-17** Incident angle graph 3.

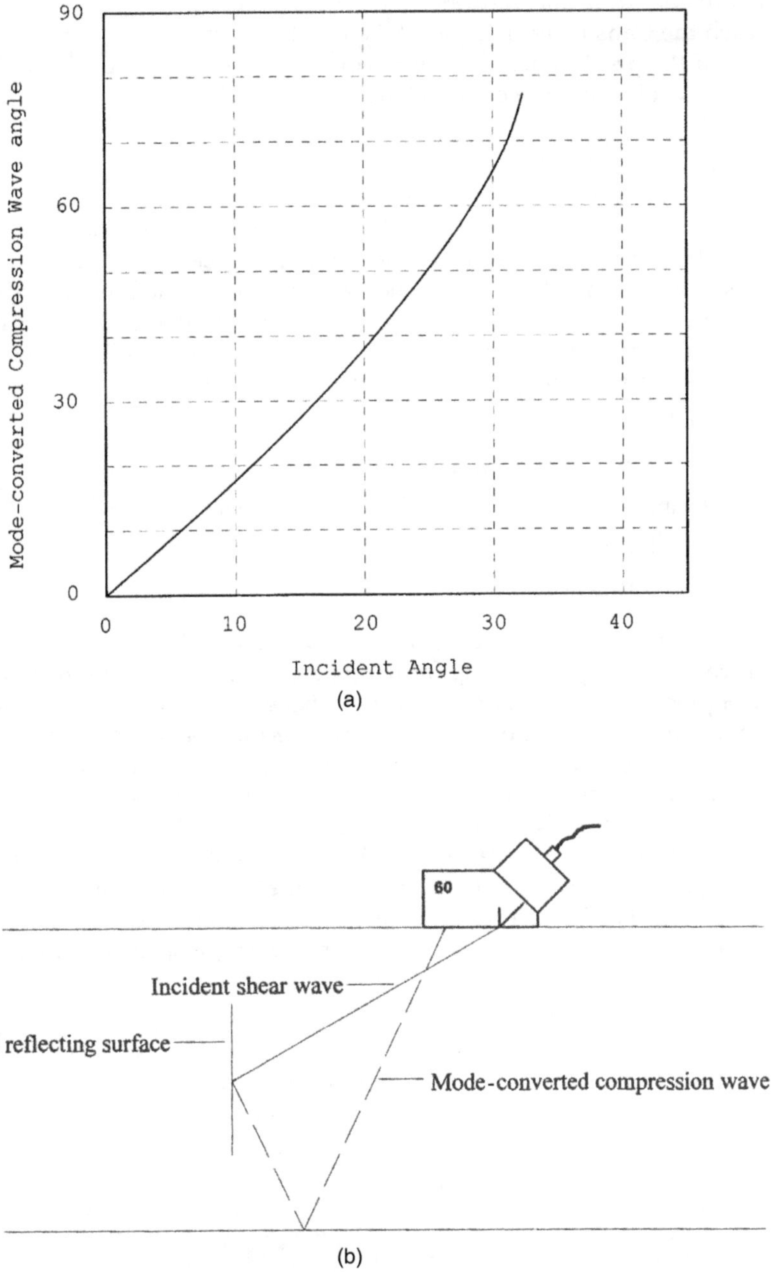


**FIGURE 7-18** *(a)* Incident angle graph 4. *(b)* Mode-converted beam.

nonfusion at an incident angle of 30°, and the strong mode-converted wave will reach the transducer. The path taken by this wave and its velocity will cause confusion if the practitioner is unaware of the problem. Obviously, a 60° beam angle is a wrong choice in those circumstances.

### *Beam Characteristics*

Many of the illustrations so far used have treated the sound as if it were a single ray, but in fact, the sound propagates as a beam. Within the beam, intensity or amplitude of the sound energy varies. The following paragraphs deal with those variations and with the shape of the beam. For convenience, the beam is divided into two distinct zones called the "near field" and the "far field." In these two fields, different mechanisms are at work to vary the sound intensity. The word used to describe what effectively is a gradual loss of sound energy is "attenuation." Attenuation is the combined effect of a number of parameters:

- Interference and diffraction effects
- Interference absorption (friction and heat)
- Interference scatter
- Interference beam spread

Interference and Diffraction Effects. Huygens developed a convenient way of looking at wave energy propagating from a source. He said that a point source was rather like dropping a stone into a pond; the disturbance moves out as an expanding circle on the pond, but from a sound source the circle becomes an expanding sphere—it moves out in all directions. A sphere is a three-dimensional object that is difficult to portray on a sheet of paper, so for this exercise a circle will have to do.

Figure 7-19 shows a point source surrounded by concentric circles representing successive pressure waves of sound frozen in time a short time after the sound starts. The spaces between the circles represent the rarefaction part of each cycle of sound. If the pattern had been frozen a little later in time, the outer circle would have been bigger in diameter. The space between each circle represents one wavelength of the sound

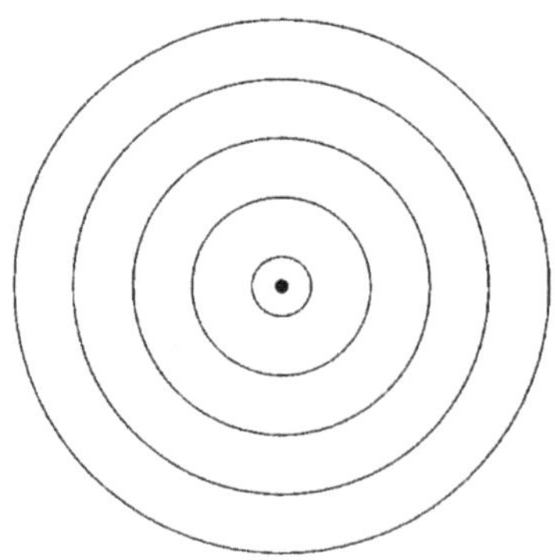

**FIGURE 7-19** Point source.

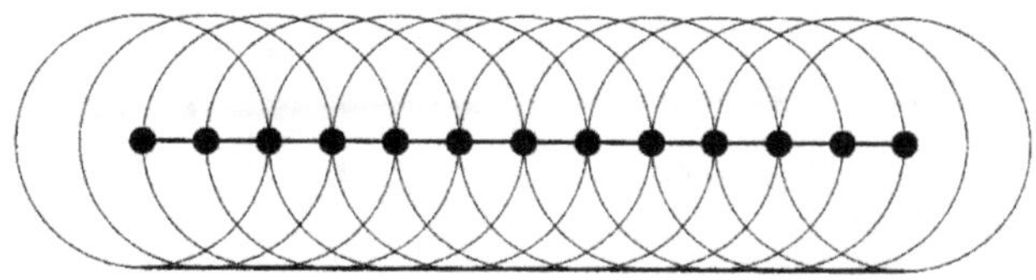

**FIGURE 7-20** Wavefront.

in the material. But an ultrasonic transducer is not a point source, it has a diameter and a surface area, all of which is active. This is called a "finite source" and Huygens said that this could be considered as being made up of an infinite number of point sources. Figure 7-20 shows a finite source with a few of these infinite point sources frozen a short time after the vibration has been started. It can be seen that the wavefronts from the point sources combine to make a united wavefront as the "beam" propagates from the source. But notice how a little bit of the sound is lost around the edges of the source; it is said to "diffract" around the edges; this is one of the energy losses in the near field.

The next source of loss needs a little more explanation. Figure 7-21 shows a finite source again, but this time only the point sources in the center and at the edges are shown, for simplicity. In front of the source is a point "P," which is waiting for sound to arrive. In the diagram, the first pressure wave from the middle of the source has already arrived, but sound from the edges has some way to go yet. "P" is given a gentle nudge in the direction of propagation by the sound from the middle of the source. Figure 7-22 shows the situation a short time later, when the first pressure wave from the edges of the source arrives at "P." The result of these two nudges from either side is again to move "P" in the direction of propagation. But that is without considering what is arriving from the center of the source. In the illustration, the third pressure wave is arriving from the center, resulting in an extra large nudge for "P," representing the combined nudges from the edges and middle. These three simultaneous nudges are called "constructive interference" because the end effect is a local increase in sound intensity. It has happened because "P" is an exact number of wavelengths from both the center and the edges of the source for the frequency of the sound wave. A change in frequency or a shift in the position of "P" might result

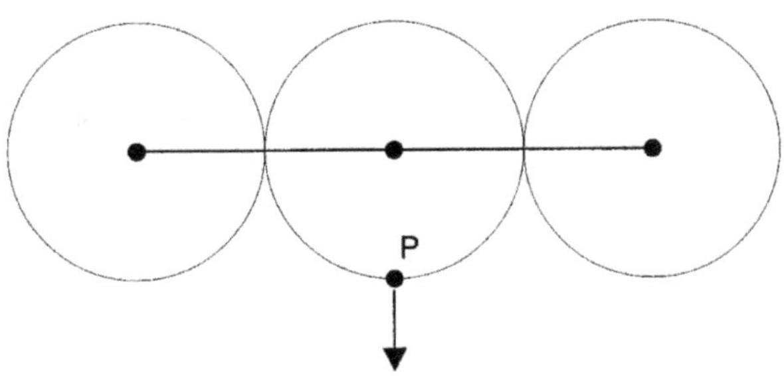


**FIGURE 7-21** Point source 2.

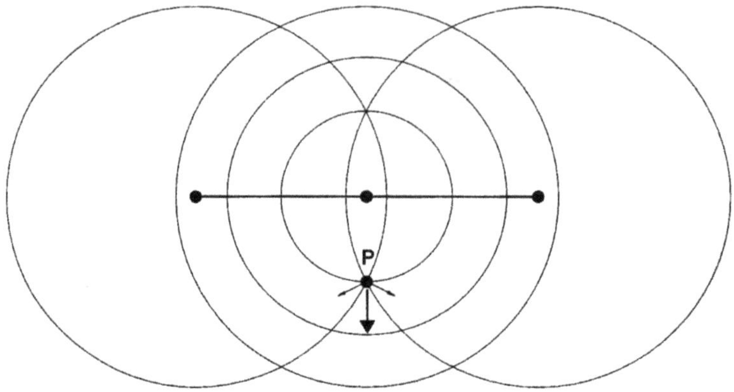


FIGURE 7-22 Point source 3.

in the sound from the center and from the edges arriving at "P" out of phase, as shown in Fig. 7-23.

In this figure, the first pressure waves from the edges have arrived at "P" as a rarefaction arrives from the center. Two forces are pushing and pulling at "P." This is called "destructive interference" and leads to a local reduction in sound energy. For an absolutely pure frequency continuous wave sound, the destruction could be total; in other words, there would be no sound at all at "P." The pulsed, broadband transducers used in ultrasonic flaw detection never quite cancel out.

The reason for this destructive interference is the difference in path length from "P" to the center of the source and to the edges compared to the wavelength. This being so, eventually there will be a distance for "P" in front of the source where the path difference becomes significantly less than a wavelength (see Fig. 7-24) and the interference effects cease. This distance is the end of the "near field." The near-field distance, *NF*, can be calculated from

$$NF = \frac{D^2}{4 \times \lambda}$$

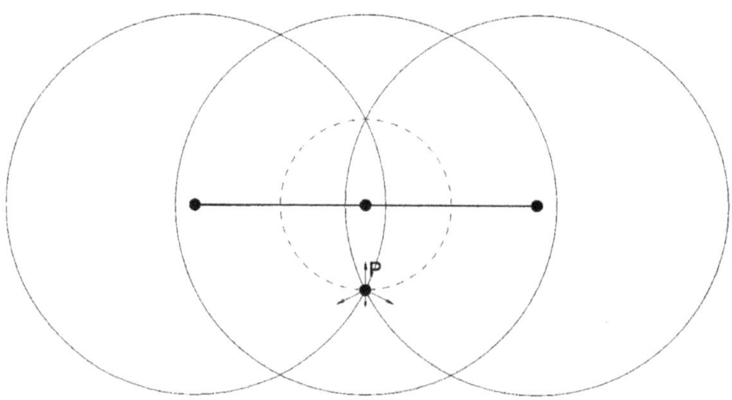


FIGURE 7-23 Point source 4.

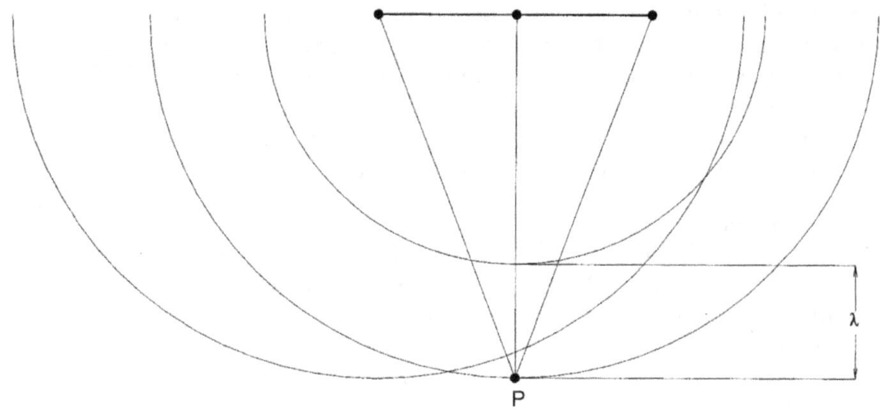


**FIGURE 7-24** Point source 5.

Alternatively, where the wavelength is unknown

$$NF = \frac{D^2 \times f}{4 \times V}$$

where:
$D$ = transducer diameter
$F$ = frequency
$V$ = velocity

**Example 7** Calculate the near-field distance in steel for a 5-MHz compression wave when using a transducer that is 10 mm in diameter.

$$NF = \frac{D^2 \times f}{4 \times V}$$

$$NF = \frac{100 \times 5{,}000{,}000}{4 \times 5{,}960{,}000}$$

$$NF = \frac{5{,}000{,}000}{23{,}840}$$

$$NF = 20.97 \text{ mm}$$

### Absorption

Sound propagates through the vibration of particles of a solid, liquid, or gas and the movement of those particles causes friction and absorbs some of the energy. The rate at which energy is absorbed depends on the material through which the sound is passing and the frequency of the sound. In general, the higher the frequency, the greater the absorption; or put another way, the lower the sound frequency, the further it penetrates into the material.

### Scatter

Sound waves will reflect from interfaces within the material being tested, and grain boundaries in solids are interfaces that may be randomly orientated to the beam. This causes some of the sound to reflect in random directions or "scatter." Very fine-grained material causes very little scatter but coarse-grained material causes considerable scatter.

Scattered energy that does not reach the receiver transducer is "lost" energy. Scattered energy that does reach the receiver is worse! It creates small signals across the time base. This condition is called "noise," "grass," or "hash" and it tends to mask signals from discontinuities.

Both absorption and scatter exist as sources of lost energy in both the near field and the far field. Beam spread is the remaining cause of energy loss affecting the far field.

### Beam Spread

In the near field, the beam is taken to be roughly cylindrical and the same diameter as the transducer crystal. Beyond the near field, in what is called the "far field," the beam spreads out like a cone. The angle of the cone, as shown in Fig. 7-25, can be calculated from

$$\sin = \frac{\theta}{2} = \frac{1.22\,\lambda}{D}$$

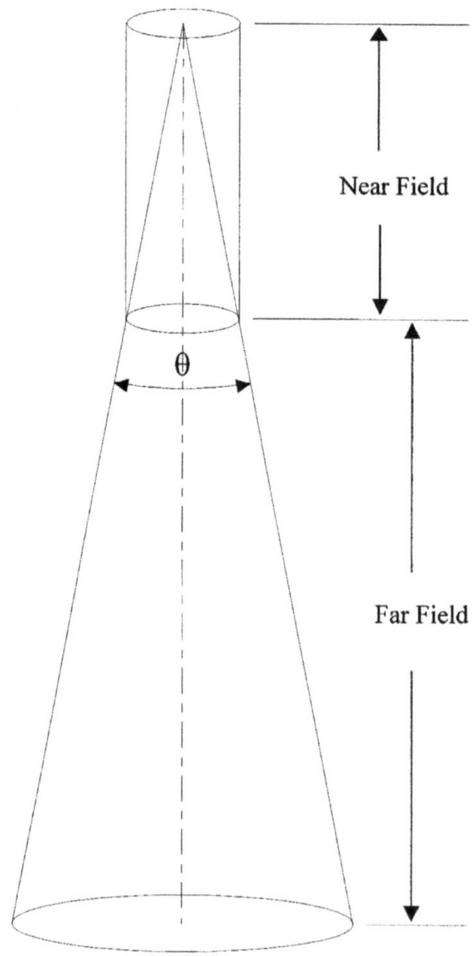


**FIGURE 7-25** Near and far fields.

where:
$\theta/2$ = half angle of beam spread
$\lambda$ = wavelength
$D$ = transducer crystal diameter

The above equation includes the constant 1.22. This calculates the beam spread to the absolute limit of the beam where sound ceases to exist. This is not a practical limit for the ultrasonic practitioner because if sound doesn't exist, it can't be detected or measured. In practice, it is more usual to replace the constant 1.22 with either 0.56 or 1.08. The 0.56 value predicts the limits of the beam where the sound has dropped to one half of the intensity at the beam center. The 1.08 value defines the limits where the sound is one-tenth of that at the beam center.

*Note:* The constants used above (e.g., 0.56, 1.08, and 1.22) are commonly used for calculation of theoretical beam shapes. If the shape of the beam is required for discontinuity sizing purposes, it is more practical to plot the beam shape using a special calibration block, rather than calculate the beam spread (see Intensity Drop Technique on page 390).

**Example 8** Calculate the beam spread angle for a 5-MHz compression wave in steel when using a 10-mm-diameter transducer ($\lambda$ = 1.192 mm).

$$\sin = \frac{\theta}{2} = \frac{1.22\,\lambda}{D}$$

$$\sin = \frac{\theta}{2} = \frac{1.22 \times 1.192}{10}$$

$$\sin = \frac{\theta}{2} = \frac{1.45424}{10}$$

$$\sin = \frac{\theta}{2} = 0.145424$$

$$\sin = \frac{\theta}{2} = 8.36°$$

$$\therefore \theta = 16.72°$$

Figure 7-25 shows the overall beam shape, including the near-field portion.

Figure 7-26 shows the way in which amplitude changes along the beam center. In the near field, there are fluctuations in amplitude because of the interference effects.

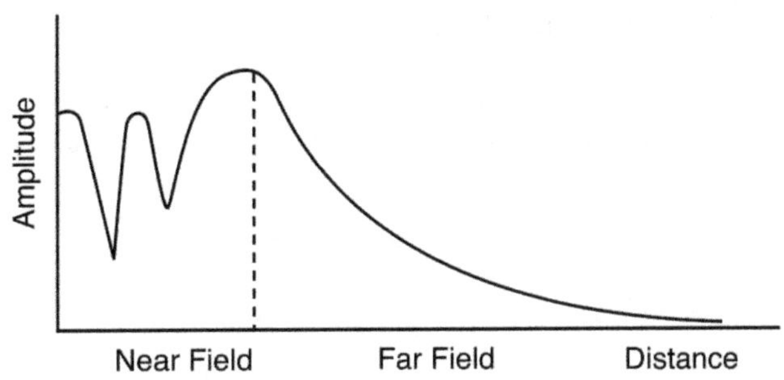


**FIGURE 7-26** Near- and far-field plots.

The last maximum amplitude marks the end of the near field and the beginning of the far field. This is called the $Y_0$ point. In the far field, the intensity can be seen to decay exponentially. From a practical point of view, Fig. 7-26 implies that it is unreliable to use amplitude as an acceptance criterion for flaws detected in the near field. In some applications, this problem is overcome by using a delay line (a Plexiglas column) between the transducer and the work-piece so that the near field is contained within the delay line.

### The Decibel System

In order to compare data, some form of measurement is necessary; e.g., for length, the standard used is either inches or millimeters. Because electric power is proportional to the square of the voltage produced, it can be said that the voltages produced at the transducer are relative to the sound intensity. When measuring sound intensities, the unit of measurement is the bel. The bel is named for Alexander Graham Bell (1847–1922), inventor of the telephone.

The bel being a large unit of measurement, it needs to be broken down into smaller units. These units are called decibels or dB. "Deci" is a prefix that is borrowed from the Latin. It means "one tenth," so a decibel is one-tenth of a Bel. The decibel is a unit used to express the intensity of sound energy. It is equal to 20 times the common logarithm of the ratio of the pressure produced by the sound energy, to a reference pressure. In other words, it is used to express the ratio of the magnitudes of two reflections (signals), each having different magnitudes, equal to 20 times the common logarithm of the voltage or current ratio. In practical terms, if there are two signals on the screen and the difference between these signals needs to be known, it can be calculated. Alternatively, if a signal of known amplitude were to be reduced by a certain percentage, this reduction in gain can be calculated. For example, if a signal height of 100 percent is to be reduced to 10 percent, the reduction in gain can be calculated.

Of course, this can be measured by using the gain control or attenuator on an ultrasonic instrument, assuming that the instrument is linear in the vertical axis (see subsections on linearity in Section 7.3).

To calculate the difference between two signal amplitudes, the following formula is used:

$$\text{dB} = 20 \times \log\left(\frac{A_1}{A_2}\right)$$

where:

$A_1$ = the first percent signal height
$A_2$ = the second percent signal height

**Example 9** Two signals are noted on the screen. The first has an amplitude of 80 percent full screen height (FSH) and the second is noted to be 40 percent FSH. Calculate the difference in dB between the two.

$$\text{dB} = 20 \times \log\left(\frac{80}{40}\right)$$

$$\log 2 = 0.3010$$

$$20 \times 0.3010 = 6.02 \text{ dB}$$

| dB | Ratio |
|---|---|
| 3 | 1.41:1 |
| 6 | 2.00:1 |
| 9 | 2.82:1 |
| 12 | 3.98:1 |
| 14 | 5.01:1 |
| 20 | 10.00:1 |

**TABLE 7-2** Amplitude Ratios

The 80 percent signal is 6 dB greater than the 40 percent signal. It can be seen from this calculation that +6 dB is twice the amplitude.

Conversely, if the 80 and the 40 were inverted, that is, if the 40 were placed over the 80 and divided, the result would be 0.5. The log of 0.5 is the same as the log of 2; however, it calculates to negative or –3,010. If this is multiplied by 20, it can be noted that –6 dB is one half the amplitude.

This means that if the signal amplitude were to increase by 6 dB, the signal height would double. If the signal amplitude were decreased by 6 dB, the signal height would be half of its original amplitude.

The above formula applied to a first signal amplitude of 100 percent and second signal amplitude of 10 percent will result in a difference of 20 dB (see Table 7-2).

Having noted the significance of the pressure (amplitude) differences of reflections, it follows that the same rationale applies to the cross section of the ultrasonic energy "beam." If the beam intensity were shown as a slice across the beam section, it would appear as in Fig. 7-27. Information regarding the actual pressure differences across the beam section is very important, particularly when using transducer movement to investigate or evaluate the characteristics of a discontinuity (see Section 7.3).

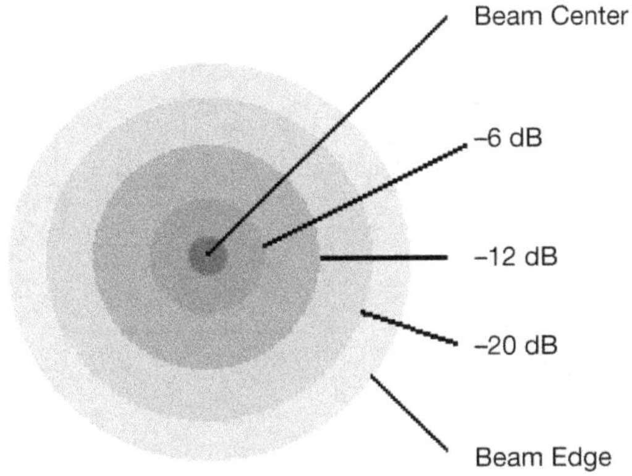


**FIGURE 7-27** Beam slice.

## 7.3 Equipment for Ultrasonic Applications

As with computers, the technology concerning ultrasonic equipment and systems is becoming somewhat transitory. Ultrasonic systems are either battery-operated portable units, multicomponent laboratory ultrasonic systems, or something in between. Whether they are based on modern digital technology or the fast disappearing analog original, "systems" (often defined as instrument plus transducer and cable) basically comprise the following components. The appropriate controls are shown in the ellipses in Fig. 7-28.

1. Transducer
2. Pulser (clock)
3. Receiver/amplifier
4. Display (screen)

To understand how a typical ultrasonic system operates, it is necessary to view one cycle of events, or one pulse. The sequence is as follows.

1. The *clock* signals the *pulser* to provide a short, high-voltage pulse to the *transducer* while simultaneously supplying a voltage to the *time-base trigger* module.
2. The *time-base trigger* starts the "spot" in the CRT on its journey across the screen.
3. The voltage pulse reaches the transducer and is converted into mechanical vibrations (see "piezoelectricity"), which enter the test piece. These vibrations

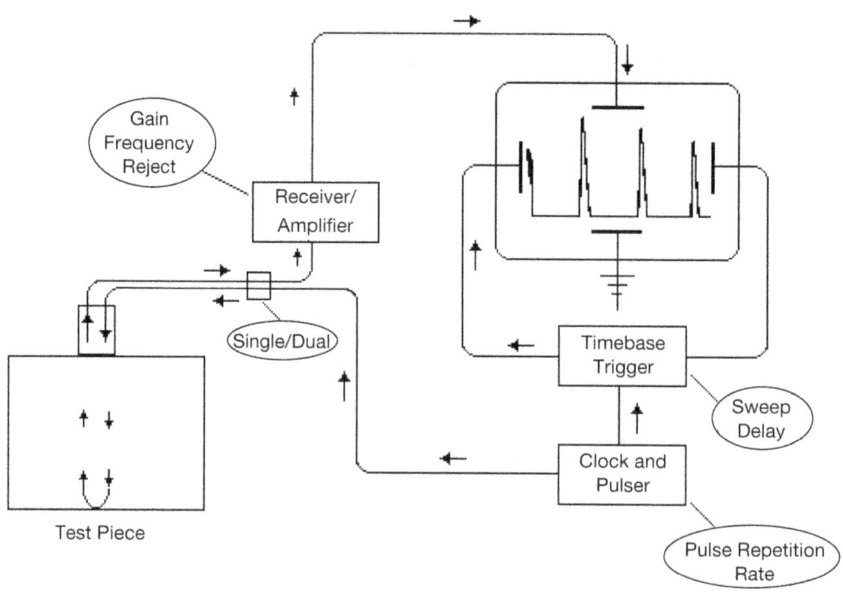


**FIGURE 7-28** Block diagram, analog.

(energy) now travel along their "sound path" through the test piece. All this time, the spot is moving horizontally across the CRT.

4. The energy in the test piece now reflects off the interface (back wall) back toward the transducer, where it is reconverted into a voltage. (The reconverted voltage is a fraction of its original value.)
5. This voltage is now received and amplified by the *receiver/amplifier*.
6. The amplified voltage is sent to the "vertical (*Y* axis) plates" (top and bottom) in the CRT. At this time, the upper *Y* axis plate attracts the spot upward. This motion produces the "signal" on the screen that signifies the time that the energy has taken to make the round trip through the test piece, from the moment the energy leaves the transducer until it is received by the transducer. The spot is set to start its trip at the time the energy enters the test piece. This is manually adjusted by using the *delay* or *zero* control. This step is particularly necessary when using a Plexiglas delay line (see Glossary of Terms).
7. The same "packet" of returning energy has by this time reflected down off the test piece's top interface and now makes a second trip down through the test piece. (The spot continues its horizontal journey across the screen.) The energy reflects once more off the back wall interface and returns again to be received and amplified. The amplifier once again sends the voltage to the *Y* axis plates. The spot is again drawn up toward the upper *Y* axis plate, this time at a "later" position on the time base. This is a "repeat signal" (multiple) that is lower in amplitude because of factors such as attenuation and other losses. The spot is then released to continue its journey across the screen, and the above sequence repeats again and again until the energy in the test piece has been attenuated. The display will show multiple repeat signals, as many as are available in the calibrated time base and according to the amount of amplification (gain) selected. For example, if the screen is calibrated for 5 inches of steel and the test piece is 1 inch thick, there will be five signals on the screen, representing five "round trips" of 1 inch thickness. (The energy has, in fact, traveled 2 inches each trip, one forward and one back, but it is displayed as a series of 1 inch trips on the screen.)
8. The clock now sets off the pulser a second time and the next pulse is produced. The complete scenario is repeated over again, *n* number of times per second. The *n* number of pulses per second is referred to as the *pulse repetition frequency* (PRF) or the *pulse repetition rate* (PRR).

## 7.3.1 Digital Instrumentation

In principle, the above also applies to digital flaw detection instrument (see Fig. 7-29).

### *The Controls and Their Functions*

Instrumentation varies by manufacturer; however, there are three controls that are common to most ultrasonic flaw detection equipment. These controls are sweep (range), delay, and gain.

**Sweep (Range—Coarse and Fine)** These controls usually exist on analog instruments. On digital units, the controls are usually combined into one, designated as "range."

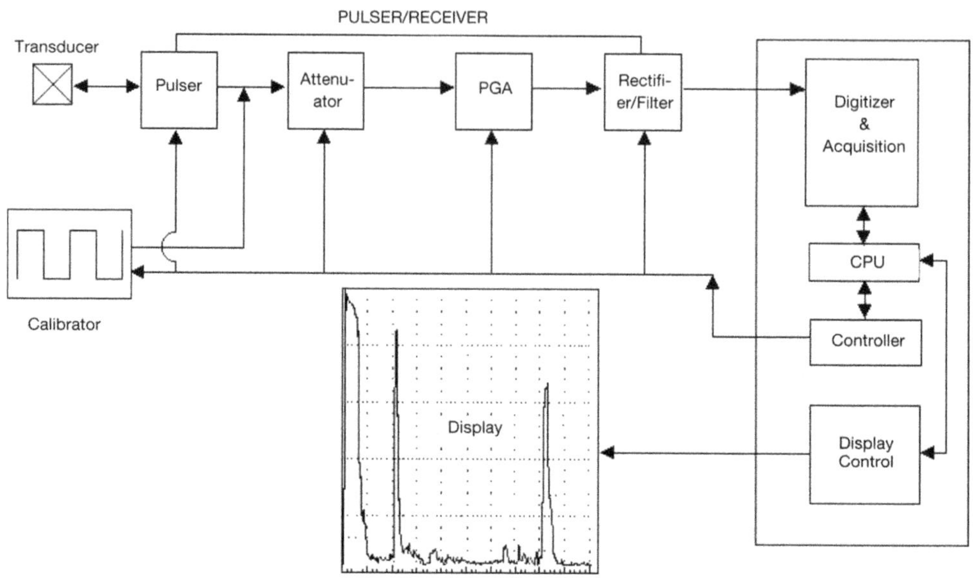


**FIGURE 7-29** Block diagram, digital.

The function of this control is to adjust the speed of the screen spot in order to accommodate displays of varying sound paths. The spot will move slower when the screen is to represent a long sound path, e.g., displaying a time base for a long shaft of steel (slow time base). Conversely, the spot will move very fast across the screen when displaying a few millimeters of steel across the full screen (fast time base).

**Delay (Zero)** This control delays the spot from beginning its journey across the screen. For example, when using a dual transducer or a transducer with a Plexiglas wedge (stand off or delay line), the start time of the spot has to be delayed for an instant to allow the sound to travel through the Plexiglas wedge before it enters the test piece. The display need only show the area from the top of the test piece and onward. Displaying the wedge on the screen is unnecessary and also confusing. When calibrating the system, test piece material zero should appear at screen zero. The delay (zero) control will be used to accomplish this.

**Gain (Attenuator)** Because everything is relative to something else, and to facilitate the ability to make decisions about the nature of the signals that are observed on the CRT when conducting an ultrasonic examination, certain comparisons need to be made. The meaning of a signal on the screen is rather limited unless it is related to something tangible such as another signal from a known reflector.

If the signal from a reflector in a test piece is compared with a signal from a known reference reflector, theoretically (all things being equal), the size of the reflectors can be compared. [*Note:* Signal amplitude in itself does not necessarily indicate reflector (discontinuity) size.]

If accept or reject decisions are made based on signal amplitude, consideration of a signal that saturates the screen may be impossible. That is because it exceeds 100 percent full screen height (FSH). Since the signal is above the viewable screen, it is impossible to make any comparisons unless the signal height is adjusted so that the top can be seen. If the instrument gain were to be adjusted so as to lower this signal until it becomes the same height as our reference signal, the actual reference signal height may be reduced so much that it cannot be quantified; or, worst case, it may be so low in amplitude as not to be visible at all, so signal amplitudes cannot be compared.

The solution to this problem is to employ a "volume" control, much the same as those found in stereo systems. This device is known as either a "calibrated gain control" or, in some cases, an "attenuator." The difference between the two devices is their functionality. This is discussed below. If a numeric value could be attributed to the amount that the signal amplitude (gain) is adjusted, numeric signal height comparisons could be conducted.

Gain controls and attenuator controls operate using similar circuitry. The difference to the user is that when using an attenuator, increasing the attenuation results in diminished signal height. When using the calibrated gain control, increasing the gain level increases the signal height. The differences in signal height and variations in gain need to be known in order to obtain accurate data.

Others may vary between instruments by respective manufacturers; however, the following features are generally common to most:

**Single and Dual Transducer Selection** This switch isolates the transmitter side of the circuit from the amplifier. It allows the reception of voltage generated solely by the receiver side of a dual transducer or from a transducer used as a receiver in a "pitch catch" or through transmission mode. In the case of a single transducer, the first voltage to reach the amplifier is from the pulser and transducer (in combination). This produces the signal on the left-hand side of the screen that is often referred to as the "initial pulse" or "main bang." It is obvious that the initial pulse "blinds" the display for an area under the transducer (front surface) in the test piece. Reflectors close to the test surface will not be resolved because they occur within the time of the initial pulse. Using the "dual" setting with a dual transducer essentially eliminates the "initial pulse" signal from the left-hand side of the screen, thus increasing the available test time close to the front interface (top) of the test piece.

**Frequency Selection** Transducers operate at a predetermined nominal frequency based on their thickness. The transducer oscillates at its resonant frequency, but it also produces other frequencies, some higher and some lower than the nominal center frequency. It is sometimes necessary to filter out undesirable frequencies as they can produce low-level noise. This can adversely affect (reduce) the "signal-to-noise" ratio. It is important that the signal from a reflector be visible above the background noise caused by material grain and other factors such as instrument circuit noise. To this end, some ultrasonic instrumentation is designed so that individual frequencies are user-selectable. Other instrumentation is designed so as to accept a range of frequencies that are not user-selectable. These are classified as having either narrow band receivers or broad band receivers. The narrow band receiver is usually user-selectable. In this case, the user selects the frequency nearest to that of the transducer being used. The effect is that the receiver processes only the frequency selected, within a certain "bandwidth"; e.g., a 5-MHz selector may be receptive to energy from 4 to 6 MHz, depending on the design specifications. This circuitry will filter out frequencies outside this bandwidth. Another name for this type of circuitry is called a "band-pass filter." There are other types of filters. Those that allow frequencies higher than a certain value to be processed are called "high-pass filters." Conversely, filters that blank out frequencies above a certain value are called "low-pass filters." When conducting a test on very grainy material, low-frequency energy is used to help overcome the grain noise. In this case, it is advantageous to use a low-pass filter to reject the scattered higher-frequency energy. This usually helps to increase the signal-to-noise ratio and provide superior data.

### Gates

Electronic gates are used to produce some action based on a signal being present in the gate. A gate is a device that is inserted into the time base at a user-selected location, as shown in Fig. 7-30. It is usually seen as an extra electronic line on the time base. A "positive gate" will enable any signal interrupting this gate to cause a voltage to be sent to a selected apparatus, for example, an audible or visible alarm. Conversely, if used in a negative direction, the absence of a signal in the gate can cause a similar action to take place. For instance, the back-echo signal can be gated. If there is some diversion causing the back-echo to disappear, this occurrence could signal an alarm of some kind. The alarm may also be in the form of a paint gun used to mark a discontinuity area on the item being examined, or a pen recorder used to record the event on a strip chart or an X–Y recorder. Depending on the circuitry, the signal can cause the gate to deliver a voltage that is proportional to the signal amplitude. The gate can be used to produce a numeric display of the signal's horizontal position or percentage amplitude on the screen. The gate threshold can be selected so that it functions at a predetermined signal amplitude. The user, dependent upon the application, also determines the gate position and width.

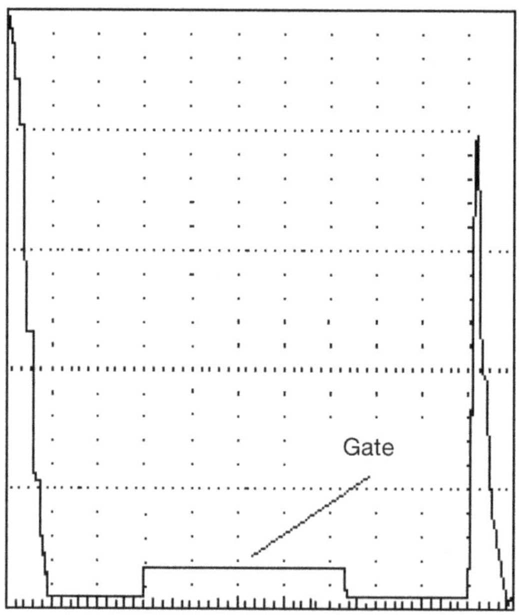


**Figure 7-30** Gate.

Modern instrumentation uses the gated signal to provide information that is used to calculate and numerically display flaw depth or, in the case of an angle beam transducer, the location distance of a reflector in front of the transducer by programmed trigonometry. Gated signals are also used to produce a "C-scan" image (see Fig. 7-34). Generally, during a manual ultrasonic examination, it is not necessary to use a flaw gate, although it can also be used to electronically mark a position or amplitude on the time base.

**Reject** This control is used to dismiss low-level "noise" on the screen. The effect of reject is visual. The principle of reject can best be described by the exercise of holding a ruler horizontally along the bottom of the screen and raising it until the noise level has been masked by the ruler. (It can be imagined that this practice would affect the instrument's vertical linearity; however, modern circuitry overcomes this problem). Reject can on rare occasions be a useful tool when conducting thickness measurements. [*Note:* The effect of reject is to electronically suppress the low-level signals on the time base. As a result of this, low amplitude signals from discontinuities may not be observed.]

### Storage Memory

Digital instrumentation usually provides the facility to store calibrations or waveforms. This is a very useful function. A multitude of calibrations can be stored for retrieval at any time. Waveforms (screen dumps) are also stored and can usually be uploaded to a computer for inclusion in subsequent printed reports.

### Displays

There are a few different ways that the ultrasonic information can be displayed. Typically, "A-scan" presentations are viewed with conventional ultrasonic flaw detection equipment. There are other displays to discuss. Figure 7-31 shows a test piece containing two reflectors with an A-scan presentation. (Note that the presentation has been rotated in the figure to show the reflectors on the screen relative to their actual positions in the test piece.)

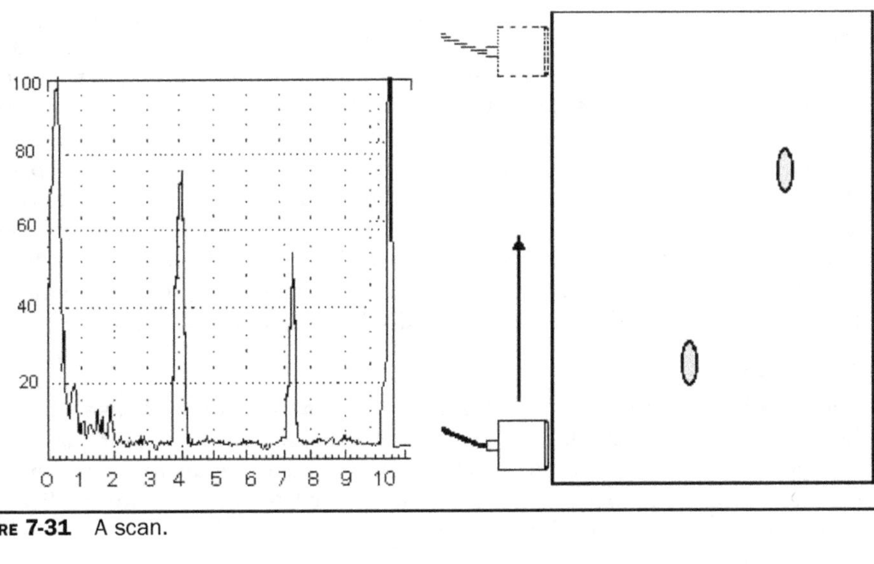


**FIGURE 7-31** A scan.

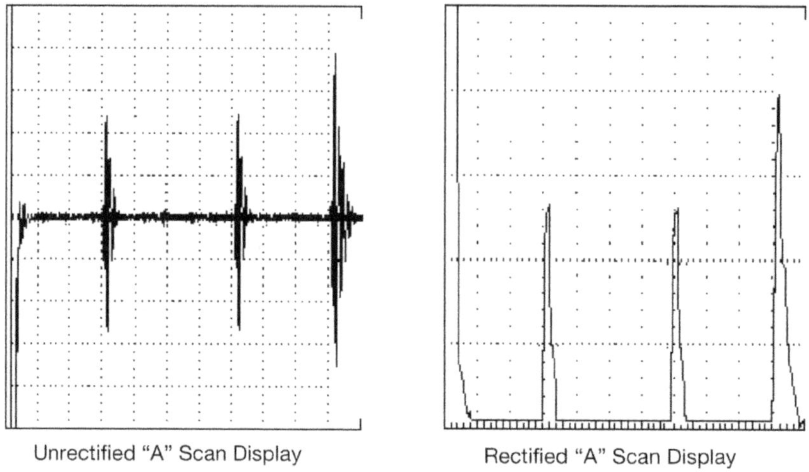


**FIGURE 7-32** A-scan displays.

A-scan displays can be "rectified" or "unrectified," as illustrated in Fig. 7-32. The unrectified trace has both positive and negative deflections. Other display options can be seen in Figs. 7-33 and 7-34.

The "B scan" provides a two-dimensional graphic display or profile of the cross section of the object being examined. See Figs. 7-33a and b. It is essentially a slice through a section of the part. The travel time of the ultrasonic wave is plotted along the vertical axis and the linear position of the transducer along the horizontal axis. It results in a display that allows visualization of the near and far surface and displays internal reflectors in relative depth. It is an effective way to show internal changes in the far surface conditions such as would be received from corrosion or thickness variations. In addition, the image may be stored and recalled for further evaluation or confirmation.

A C scan is a "map" type of display. It is a "plan" view of the test piece (see Fig. 7-34). The first C-scan recordings were produced with external recorders that were activated by a signal entering an electronic gate set-up on the time base, generally between the top and bottom surfaces of the test piece where reflectors were expected to occur. A pen device "wrote" on the recording paper in the areas that were activated by the signal in the gate. Technology today allows the recording of these images digitally and displays them in different colors on a monitor. The transducer is moved back and forth and indexed so as to scan the entire test piece as can be noted in Fig. 7-34b. (This pattern is known as a "raster" scan.) The C scan can also be displayed electronically as shown in Fig. 7-34c.

The B-scan and C-scan techniques are well suited to applications where a permanent record is required. It is common to digitally store A-, B-, and C-scan data.

### Transducers

The transducer is the actual "front end" of the system. It is analogous to the microphone in a public address system. If the public address system is of the best quality and a poor microphone is used, the sound will be only as good as the microphone. The same principle can be applied to an ultrasonic system. The use of an inferior quality transducer on the best system can result in deficient data.

(a)

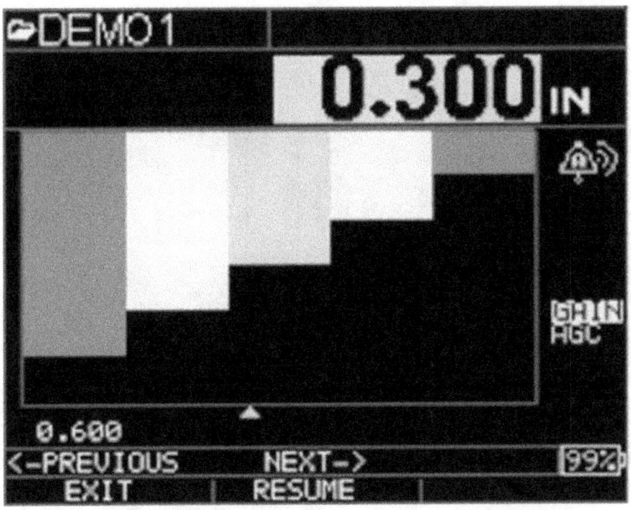


(b)

**Figure 7-33** (*a*) B-scan equipment showing the variations on thickness on the display. (*b*) The B-scan display. (*Courtesy of Danatronics*.)

Other terms used to describe the transducer are "probe," "search unit," and "test head." The word "transducer" is from the Latin "transducere," which means to lead across or transfer. The function of the transducer is to transfer electrical energy to mechanical energy and vice versa.

As far back as the year 1880, the Curie brothers, Pierre and Jacques-Paul, discovered that when sectioned in specific planes certain crystal materials would generate a voltage when distorted. This is called "piezoelectricity"—electricity due to pressure. The opposite

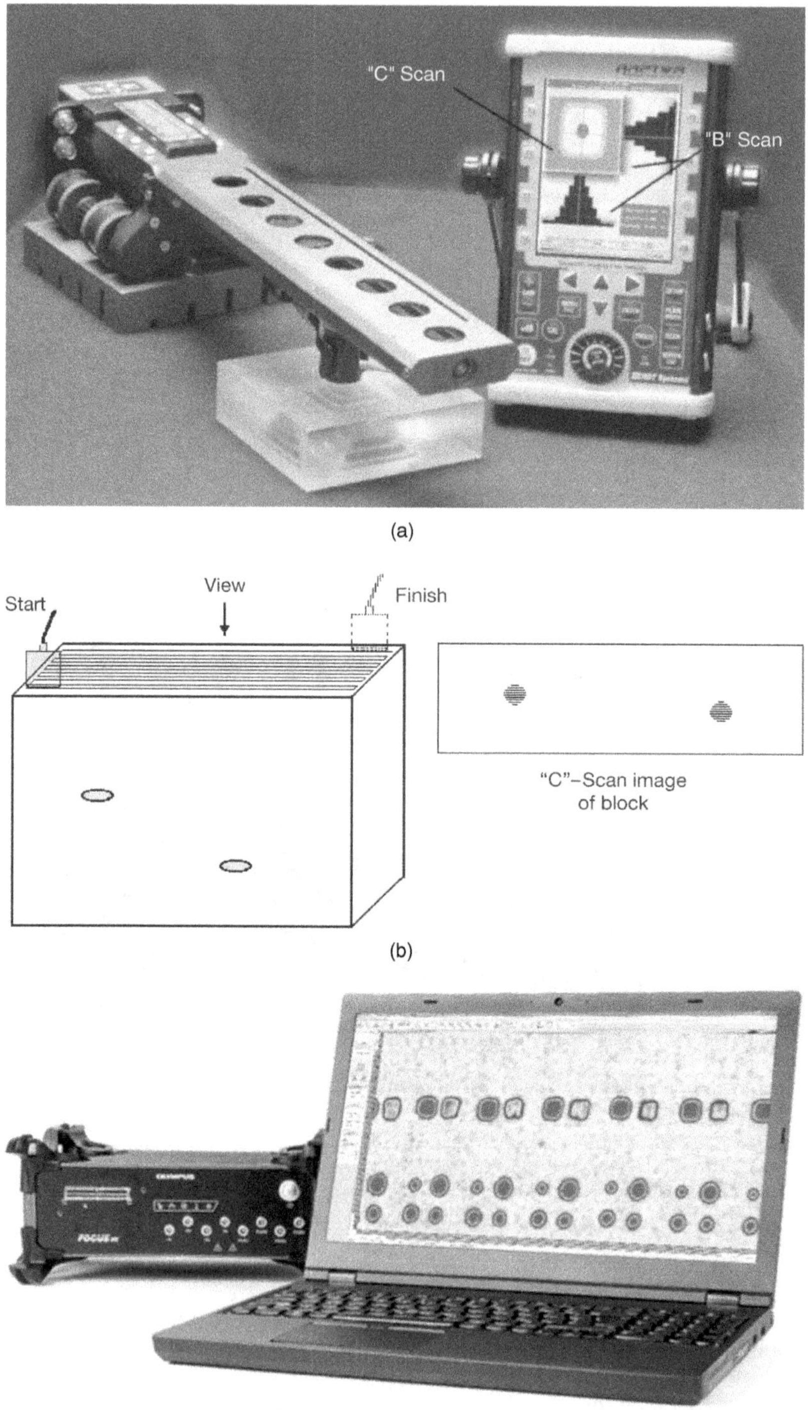


**FIGURE 7-34** C-scan display. (*Courtesy of NDT Systems.*)

effect is also valid; i.e., if a voltage is applied to the crystal material, it will distort. Lippman documented this about a year later. Quartz crystal is a prime example of this type of crystal. Other naturally occurring piezoelectric materials exist, such as tourmaline and Rochelle salt. These crystals were used in the early days of ultrasonic testing until polycrystalline ceramic materials—materials that do not exhibit piezoelectric properties in their original state—were developed to perform this function. Some of the more common polycrystalline materials used in transducers are lead zirconate titanate (PZT) and lead metaniobate (PMN). The material is mixed in the form of a slurry, poured into a mold, then dried under pressure. It is then sliced to the required thickness. This is the thickness at which the element resonates at its designed frequency. (Materials resonate according to their formulation and thickness. For example, a 5-MHz transducer element from PZT may be a different thickness than its counterpart made from PMN.) The slices are then placed on a lapping table and precision lapped to the final thickness. The next step is to coat the element with a very thin layer of conductive material, usually silver. This is sometimes electrostatically applied. At this stage, the element is not yet active. It comprises many microscopic piezoelectric elements that are randomly oriented. These have to be aligned or polarized in order for the element to be useful for the purpose of generating and receiving ultrasonic energy. This is referred to as the "poling" process. Electrodes are attached to the faces and the element is immersed in a bath of oil. The oil is heated to a temperature that is around the "Curie" temperature of the specific material also referred to as the "critical" temperature. A high polarizing voltage is applied to the element. The element is then allowed to cool while the field is active. Once cool, the voltage is eliminated and the element is now polarized. This means that the element will have + (positive) and – (negative) polarity. For longitudinal wave generation, the elements are polarized so that the element deformation is as shown in Fig. 7-35a. Shear waves can be generated when the elements are polarized to deform as shown in Fig. 7-35b.

*Note 1.* Transducers used for angle beam shear wave applications are usually longitudinal. The shear wave component is generated upon the energy traversing two different material velocities at predetermined angles (see Refraction and Mode Conversion).

*Note 2.* Heating a transducer above its Curie temperature will allow the microscopic elements to depolarize. This will remove the piezoelectric properties. When conducting high-temperature ultrasonic examinations, precautions are necessary so that the operational temperature does not approach the Curie temperature of the transducer. *Approximate* Curie temperatures of the more common element types are as noted in Table 7-3.

The primary advantage of polycrystalline ceramic transducers is their relative efficiency. With some of these materials, the efficiency can be as much as 60 to 70 times more efficient than their quartz equivalent.

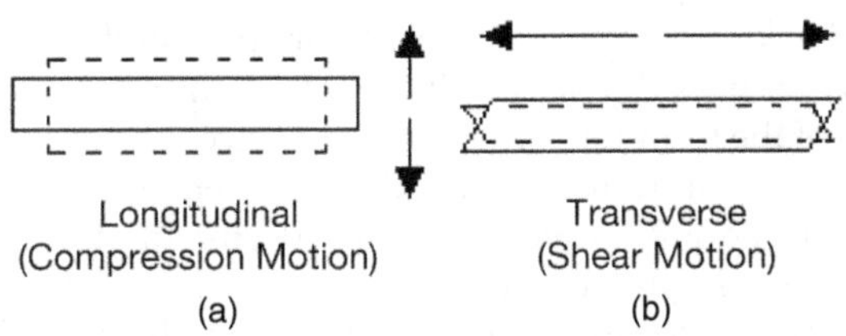


**FIGURE 7-35** Elements.

| Element Type | Curie Temperature, °C |
|---|---|
| Barium titanate | 120 |
| Lead zirconate titanate | 190–350 |
| Lead metaniobate | 400 |
| Quartz | 575 |

**Table 7-3** Approximate Curie Temperatures

Elements polarized as shown in Fig. 7-35a will function with a "piston" type motion. The element "grows" taller and thinner, and as the voltage is removed from the element, it collapses and becomes "shorter and fatter." This motion continues until it has stopped ringing. This is analogous to striking a bell with a hammer just once. The bell rings and takes a while to stop ringing or oscillating. The transducer operates in much the same manner as the bell. A voltage from the pulser strikes it and it rings. This ringing has to be arrested or "damped" quickly. Superior resolution characteristics require that the transducer exhibit the minimum number of oscillations possible. If the oscillation is not damped, signals from reflectors that are relatively close together in a test piece will appear on the screen as one combined (reflector) signal instead of two separate signals. The ringing of the element has to be shortened in order to improve resolution of reflectors, whether they are close to the surface (near-surface resolution) or close together in time (spatial resolution). Refer to the example of the bell. If the bell is struck many times and in quick succession, it is difficult to establish the number of "strikes" on the bell. If a hand is placed on the bell while it is being struck, the vibrations are quickly damped after each strike. The number of times that the bell is struck may now be individually counted. This principle applies to the transducer. This quality is accomplished by placing "damping" or "backing" material onto the rear of the element, similar to placing a hand on the bell.

Transducer technology may vary between manufacturers, each having their own zealously guarded methods of damping transducers. The bond between the backing material and the element is of primary importance, as is the acoustic impedance match between the element and the backing material. Ideally, the transducer should have a short pulse length and a high-energy output. These two features usually contradict each other (see Bandwidth). Return to the example of the bell. If there is a hand placed on the bell, it will not ring as loudly as when it is rung without the arresting hand. There is obviously a compromise between heavy damping and high-energy output. It boils down to a matter of selecting the right tool for the job to be done. Damping is accomplished by using a substance that is acoustically matched to the element material and that is also dense in structure. Usually, a tungsten and epoxy mixture are used. Another requirement of a damping material is that it attenuate the energy fast. It is extremely undesirable to have signals from energy bouncing around in the damping material. Designs of the backing material include the shaping of the backing and inclusion of scattered particles of reflective material that scatters the energy within the backing piece.

Regardless of the technique used, the transducer performance should match its useful purpose, that is, the introduction of ultrasonic energy into the part being examined. Applications that require the detection of reflectors that exist close together in the part being examined will require the use of a transducer that has a short pulse. It is preferable that the transducer is acoustically matched to its application. The transfer of energy from the transducer to the test piece or to the wedge material (in the case of angle beam units) has to be optimized (see Acoustic Impedance). The transducers used as "wedge drivers" should have a front "face," usually made from an epoxy, that has a good impedance match to the Plexiglas wedge to which it is attached. Transducers intended for use in direct contact applications, such as conventional zero degree units, have front faces that are made from material that is acoustically close to the test piece, usually aluminum or steel. The material used for this application is generally aluminum oxide. Not only is the impedance match to the element and a steel or aluminum test piece reasonably good but the material is also by nature extremely hard and therefore exceptionally wear resistant. It therefore goes without saying that the use of a wedge driver for a direct contact test is not recommended. Not only will the face wear very quickly, but from the acoustic standpoint the transfer of the energy will not be optimum.

The layer thickness of the facing material is extremely important. It should be equal to one-quarter wavelength in the facing material. This arrangement considerably enhances the emitted energy. See Fig. 7-36, an example of single element transducer construction, and Fig. 7-37 for examples of an angle beam unit. Note that the damping material is made at some angle so as to redirect the internal energy away from the element.

### *Piezocomposite Transducers*

Primarily used in the medical field, the idea of piezocomposite transducers was considered as early as the mid-1970s. The technology necessary to manufacture this type of transducer was not readily available at that time and solutions to problems such as the facility for slicing the crystal accurately in a production mode were not possible at that time. Piezocomposite transducers function in a similar manner to regular ceramic transducers. The primary difference is that the piezocomposite ceramic is made up of small-sectioned pieces of ceramic material (PZT, PMN, etc.). See Fig. 7-38.

Initially, the ceramic is sliced into squares. The small areas between the squares are filled with epoxy and the transducer is lapped to the required thickness, silvered, and polarized in a similar manner to the regular transducer elements. The difference to the user is in the performance. Because of the damping material (epoxy) around each square, the transducer exhibits superior bandwidth and therefore superior resolution. There is little or no need to apply a backing; therefore, the efficiency is greater than the conventional damped ceramic transducer. Because of the lack of backing, the height profile of these units can be reduced for accessibility to small areas. These transducers are particularly useful when testing grainy materials having increased signal-to-noise ratios.

### *PVDF (Polyvinylidene Fluoride)*

Ceramics are not the only materials that can produce the piezoelectric effect. Piezoelectric polymer sensors have been available in the nondestructive testing industry for some time. In the 1960s, it was discovered that whalebone exhibited a very weak piezoelectric effect. This inspired the search for other materials that could exhibit piezoelectric activity. Relatively high activity was found to exist in polarized polyvinylidene fluoride. The advantages of this material are:

1. PVDF has an acoustic impedance close to that of water, so that it allows efficient transfer of energy.
2. PVDF has a large frequency range and therefore a very broad bandwidth. Because of this, it exhibits very favorable spatial and near-surface resolution compared with conventional piezoelectric element material.

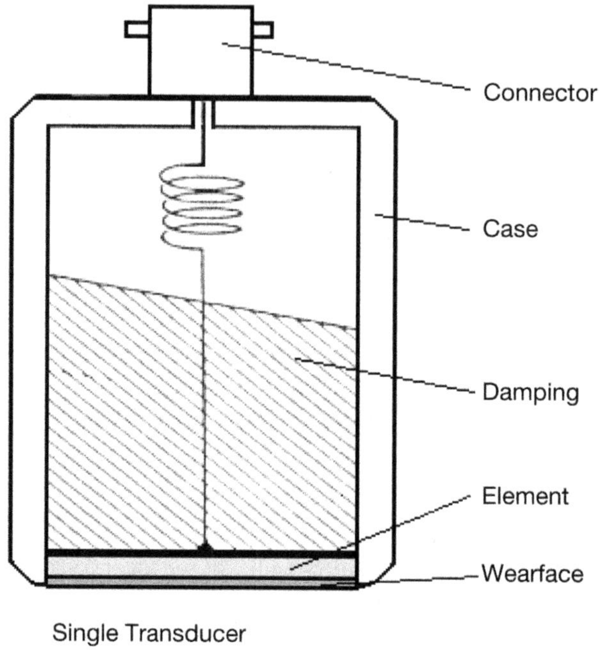


(a)

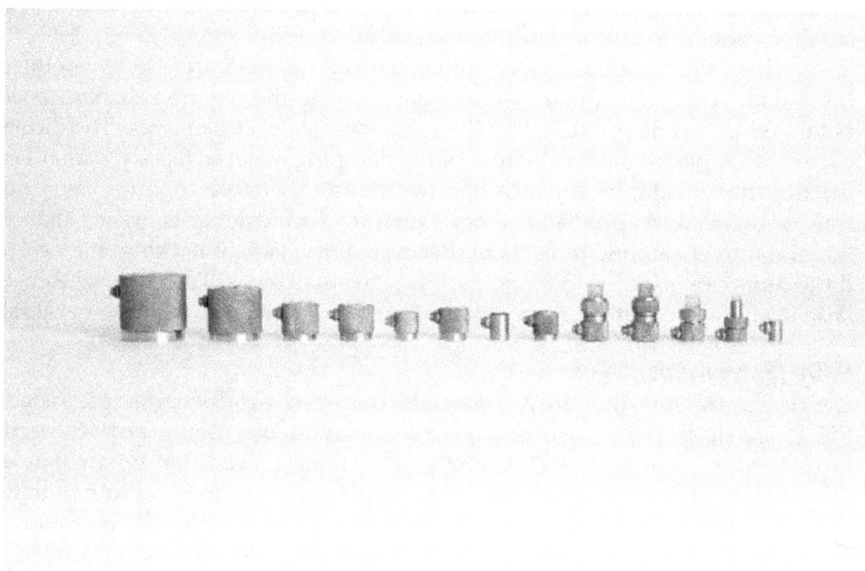

(b)

**FIGURE 7-36** (*a*) Single transducer. (*b*) Examples of compression wave (straight beam) transducers. (*Courtesy of Olympus*.)

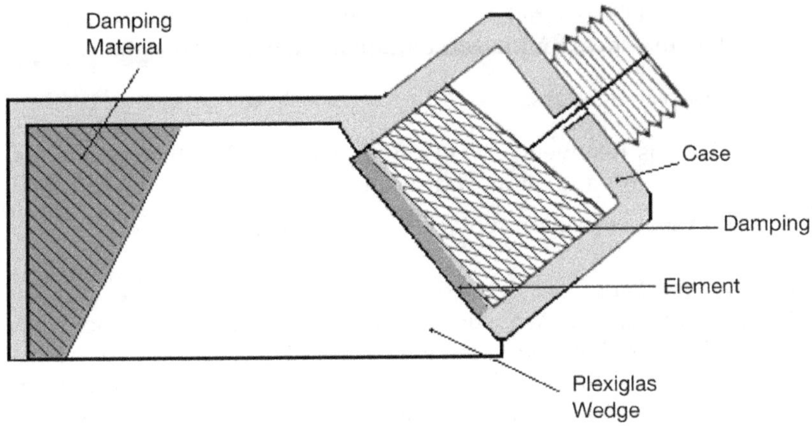


(a)

(b)

**Figure 7-37** (*a*) Angle beam transducer. (*b*) Examples of angle beam transducers. (*Courtesy of Danatronics*.)

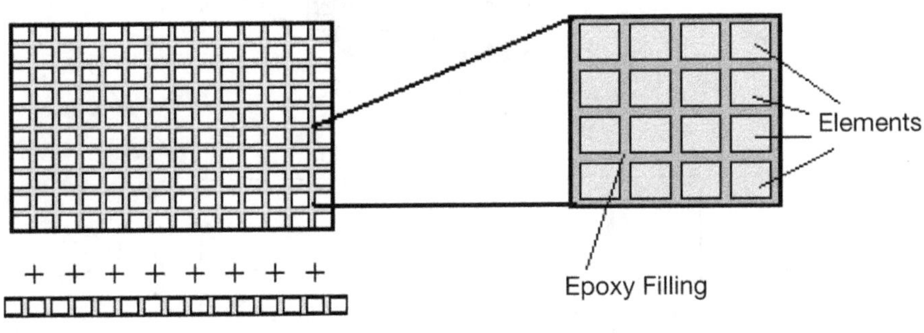


**Figure 7-38** Piezo composite element.

3. The PVDF material is flexible and can be shaped for beam focusing. It is therefore ideal for high-resolution immersion applications.

The disadvantages are:

1. PVDF is relatively low in power compared with conventional piezoelectric transducers and, in most cases, requires additional amplification.
2. It cannot be used in contact applications.

### Noncontact Methods

EMAT (ElectroMagnetic Acoustic Transducer) (see Fig. 7-39) technology is an alternative method of generating and receiving ultrasonic energy. These are transducers that are made up of coils that are placed in close proximity to the test piece. The coils produce a magnetic field that interacts with the metal, producing a deformation in the surface of the material. This deformation produces the wave of ultrasonic energy. The advantages of these units are:

1. There is no need for a couplant. An EMAT is a noncontact transducer.
2. They lend themselves to applications that normally have limitations, such as the examination of high-temperature components. Because this type of transducer depends on the induction of a field, the transducer has to work in close proximity to the work surface. The strength of the magnetic field is reduced as the distance between the transducer and the component surface increases.
3. The gap between the transducer and the work face need not be composed of air. Examination of components that have been coated with some protective layer is possible. It is the front surface of the component material that actually generates the ultrasonic energy.

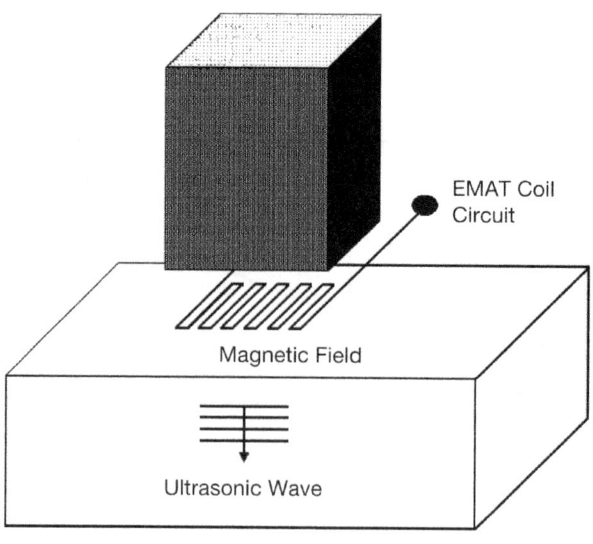


**FIGURE 7-39** EMAT transducer.

4. Focusing of the beam is also possible, as is steering the beam at various angles.
5. Horizontally polarized shear wave energy can be produced. The polarity is important in that horizontally polarized shear waves do not mode convert when striking surfaces that are parallel to the direction of polarization. This has certain advantages, particularly when examining austenitic welds and other materials with dendritic grain structure, e.g., certain cast stainless steels.

The disadvantages include:

1. Low efficiency compared with piezoelectric transducers.
2. Relatively large transducer size.
3. Producing ultrasonic energy in nonconductive material is only possible if a conductive layer is applied to the surface.

### Laser-Generated Ultrasound

The use of laser technology for the generation of ultrasonic energy has been known since the latter part of the 1970s. Acoustic propagation is accomplished by briefly heating or "ablating" the surface of the test material. This brief heating of the surface causes the generation of thermal expansion on the surface of the material, which in turn results in the formation of a wavefront that travels through the material. The technology generally employs two separate lasers for this application—one to ablate the surface and produce the wave, and a second (a laser interferometer) to detect the movement of the surface due to the disturbance from the reflected wave. The technology is commercially available and has been developed for (but not limited to) applications such as the inspection of composite materials in the aircraft industry. Reception of the energy is also possible with a conventional transducer coupled to the surface.

The advantages are:

1. When used with laser interferometry for reception, there is no need for couplant.
2. The laser can be located remote from the test surface (10 inches or more in certain instances).

### Bandwidth

Transducer bandwidth can best be described in ultrasonic terms as the spectrum or range of frequencies that occur when pulsing the transducer. To make this more meaningful for comparison purposes, the bandwidth is measured within a given amplitude range. This is usually at either the –3 or –6 dB levels from maximum amplitude (see Fig. 7-40). This is a standard method of comparing the transducer's "quality factor" or "Q" that is varied by the mechanical damping placed on the element. The higher the Q, the greater the sensitivity. The lower the Q, the greater the resolution characteristics. A short-pulsed transducer provides a broader bandwidth but less energy, therefore a "low Q." See Fig. 7-41.

### Dual Transducers

Transducers that incorporate separate elements for transmission and reception of the ultrasonic energy are referred to as "dual" transducers. Designs vary, but a typical dual transducer construction is as shown in Fig. 7-42. As can be seen, the elements are mounted on separate delay lines that are separated by an acoustically opaque

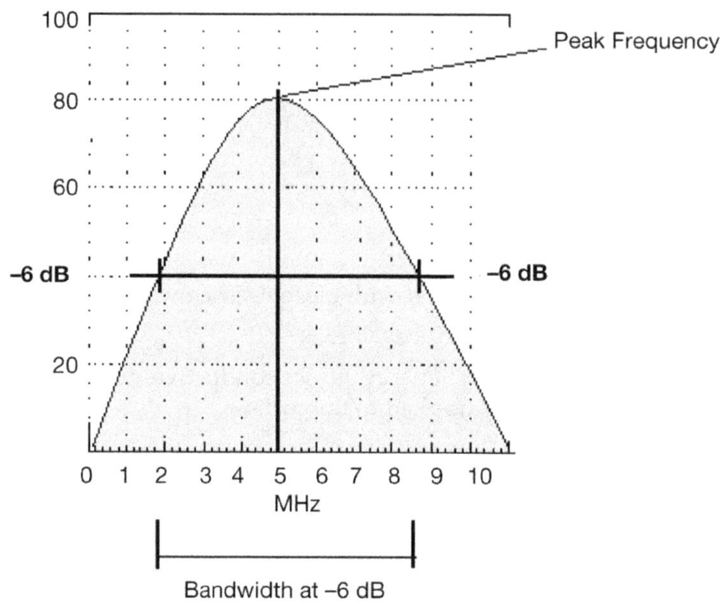


**FIGURE 7-40** Spectrum.

(nontransmitting) material such as cork. The receiver element must not be able to receive energy directly from the transmitter element. In other words, they must not indulge in "cross-talk." The primary purpose of this type of transducer is to enhance the near-surface resolution capability of the ultrasonic system. The dual transducer can be designed to resolve reflectors that are very close to the scanning surface. To this end, they are designed with a certain "roof angle" or "squint angle." This produces energy in the test piece that is refracted in the direction of the receiver side of the transducer. It is obvious that the greater the roof angle, the better the resolution close to the

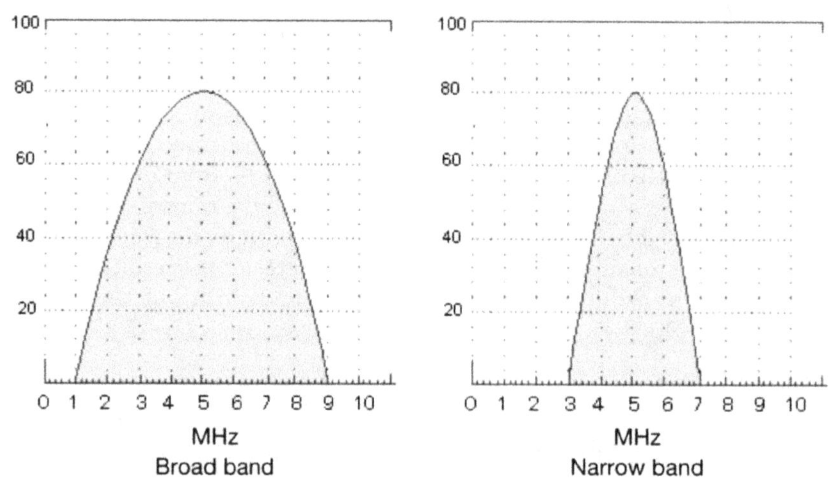


**FIGURE 7-41** Bandwidth.

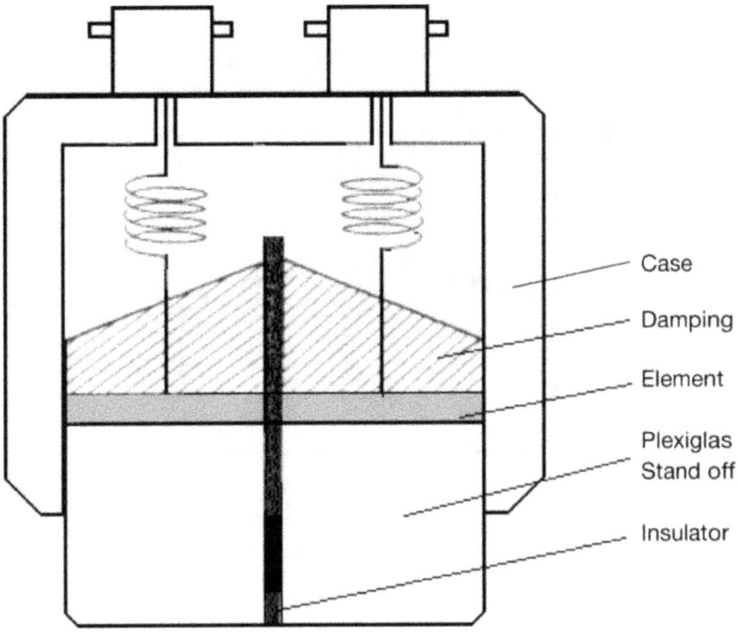


FIGURE 7-42 Dual transducer.

scanning surface. It must be realized that the detection efficiency further out in the part will suffer. Again, it is a matter of selecting the correct transducer for the application. The dual transducer can be used in angle beam applications. In this case, the same advantages are present. The sound energy intersection area can be predetermined to enhance a specific area in the test piece or component under test. It is incorrect to refer to this type of transducer as a "focused" transducer, but the technique does limit the received reflected energy to the zone where the beam from the transmitter intersects the hypothetical area of the receiver beam. Note that there is no energy from the receiver; however, because the angle of incidence is equal to the angle of reflection, and because the receiver is "aimed" at the area of the transmitted beam, planar reflectors in the beam path will reflect the energy toward the receiver. The energy that is detected by the receiver will be from the intersecting zone of the transmitter (see Fig. 7-43).

### Focused Transducers

The principles of optics apply to sonics. Focusing light is achieved with the use of lenses. Focusing sound energy is accomplished in a similar manner. Immersion testing, in particular, lends itself to the use of focused transducers (see Fig. 7-44).

There are essentially three approaches that can be used to accomplish the task of focusing the beam:

1. Shaping the actual transducer element.
2. Attaching a concave lens to the transducer face.
3. Inserting a biconvex lens into the path of the sound energy, similar to focusing the light from the sun using a magnifying glass.

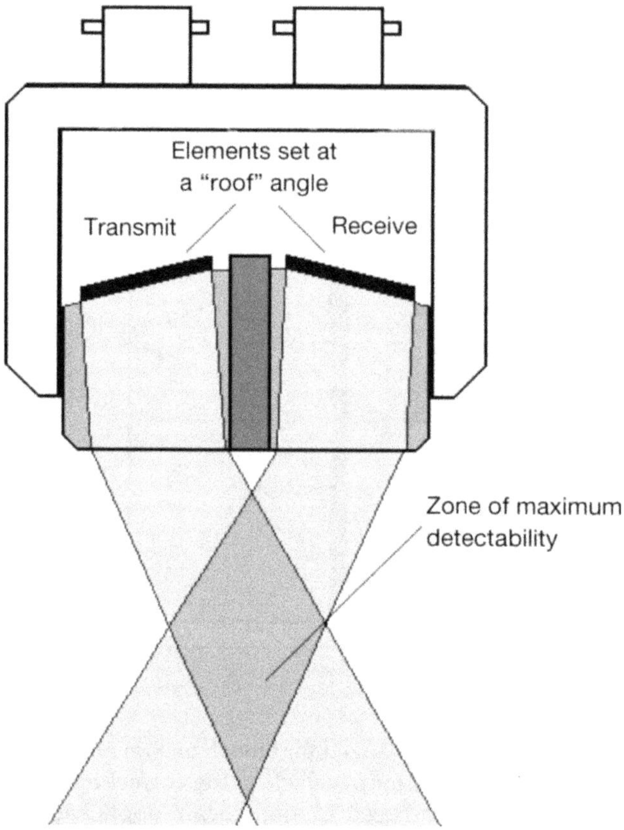


**FIGURE 7-43** Angled dual transducer.

The sound energy can be focused so that it concentrates the energy at a particular depth or area within the test piece. This improves the detection of small reflectors at a predetermined depth. This is due to the small beam dimension at the focal point relative to the size of the reflector.

Contact angle beam transducer units that focus the energy have also been produced. This is accomplished by contouring the wedge under the transducer to a specific radius, dependent upon the desired focal length (see Fig. 7-45). This radius is filled with a fluid having a velocity different from the wedge material, thus providing a lens to the energy. This will produce a refracted (converging) beam in the wedge that refracts again to a focal point after entry into the test piece.

### Phased Array Transducers

These transducers incorporate elements that are arranged in certain patterns for the purpose of dynamically focusing or steering the energy. Sequentially pulsing the elements, using a combination of elements in the array, and timing the pulses used to excite these elements, produces a beam focused at variable depths in the test material (see Fig. 7-46). Multiple wavefronts are combined to form a beam of a particular shape. Element configurations can be circular or rectangular, depending on the

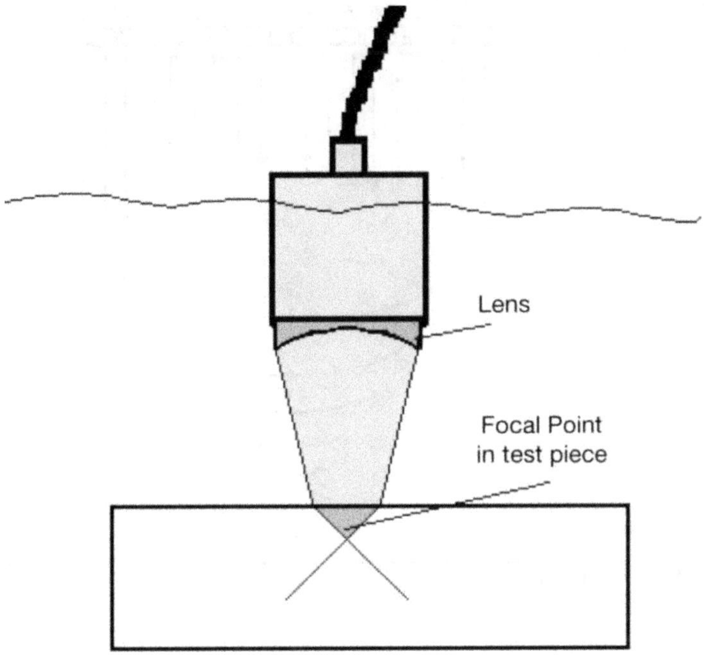


**FIGURE 7-44** Focused transducer.

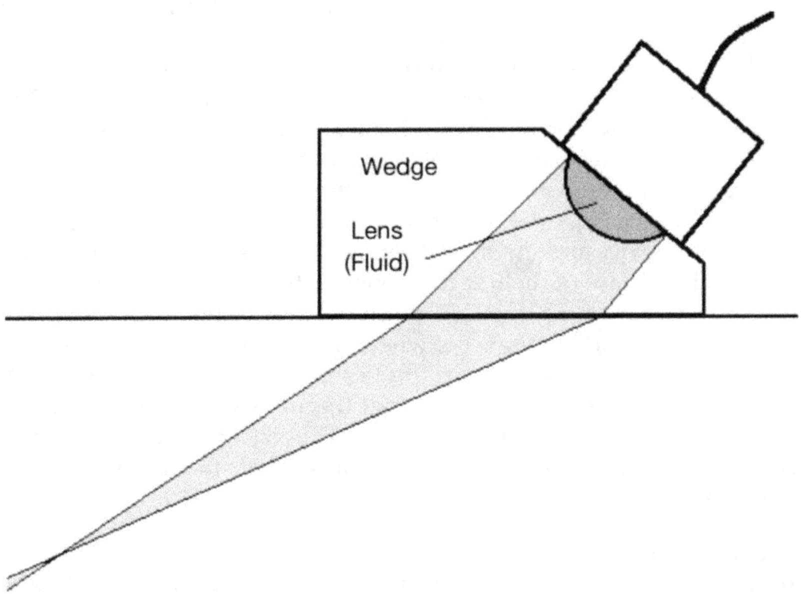


**FIGURE 7-45** Focused wedge.

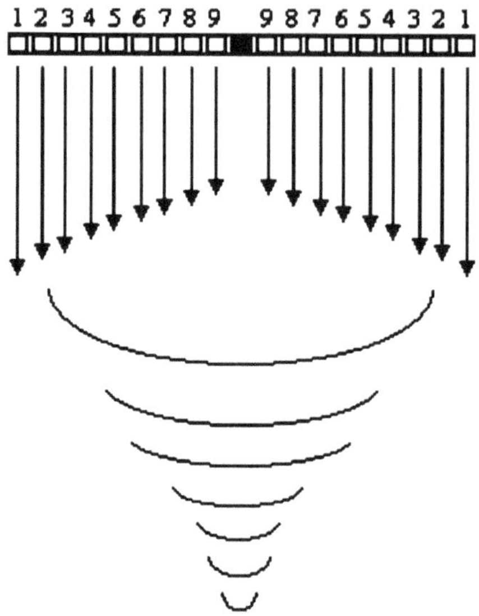


**Figure 7-46** Phased array.

desired beam shape and direction of energy propagation. Linear array units are described in Chapter 12, Phased Array Ultrasonic Testing.

### *Bubblers and Squirters*

Immersion testing is not limited to placing the test material or component into a tank of water. An immersion test can be simulated by a device called a "bubbler" or a "squirter."

A bubbler is a device wherein the transducer is located within a housing that includes a fluid reservoir. The fluid is continuously pumped into the housing. The transducer is placed slightly above the level of the holder. When the unit is placed on the test surface, or more often, the test surface is placed over the unit, the housing fills with fluid and provides a film of couplant to the front of the transducer (see Fig. 7-47).

A squirter (see Fig. 7-48) comprises a holder with a nozzle. The transducer is located inside of this device and the smooth jet of water that exits the nozzle carries the sound energy to the surface of the test component. This device is generally incorporated into automated test systems such as those for production line inspection of pipe, or inspection of aircraft components where immersion of the component is impractical. Through-transmission techniques are best suited for the squirter. In the pulse-echo mode, extraneous signals may appear at the surface of the test component due to the water striking the surface of the test part. The size of the ultrasonic beam can be controlled by varying the size of the nozzle orifice. (Note that the example in Fig. 7-48 does not have a smooth water jet.)

### *Wheel Transducers*

Immersion testing is found in many forms. The wheel transducer is another variation. Transducer(s) are located in a holder(s) within a sealed wheel that has been filled with fluid. The fluid can be water, thin oil, or other suitable materials. The transducers can be placed at strategic angles within the wheel so as to provide the desired refracted

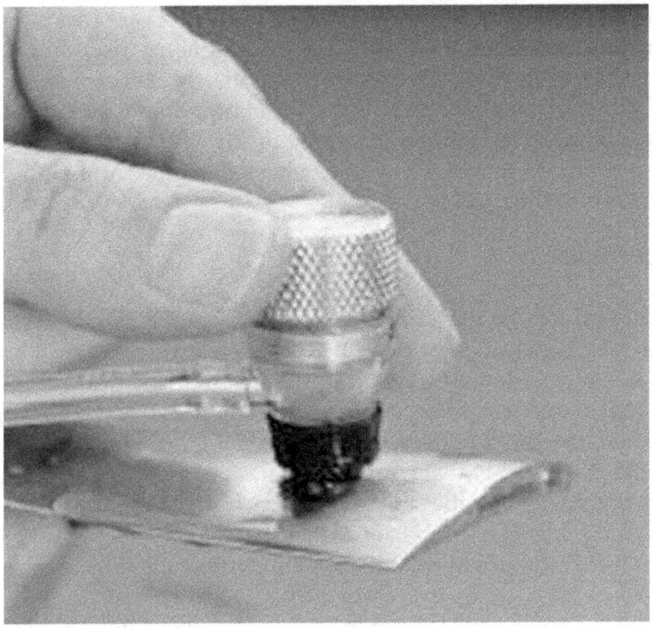

**FIGURE 7-47** Bubbler. (*Courtesy of Olympus*)

**FIGURE 7-48** Squirter system designed for through transmission testing. (*Courtesy TecScan.*)

test angles. This device is very useful for examination of moving components where surface couplant is to be limited. They are commonly used in the railroad industry to test carriage wheels and railroad track on the fly. The wheels usually require a fine spray of water to couple the tire to the test component. There are special wheel transducers that do not need couplant on the tire surface (dry coupled). The wheel tire material can be made so that it is virtually invisible to the ultrasound by matching the tire characteristics to the fluid in the wheel.

#### *Transducer Problems*

As mentioned previously, the ultrasonic test begins with the transducer. It is therefore of utmost importance that the transducer is performing as required. Problems with the transducer need to be identified prior to any test. It is advisable to document the performance of each transducer when purchased so that its performance can be monitored during its lifetime. Items such as pulse width, amplitude, and angle (if appropriate) should be quantified. Photographs of the spectrum and waveform should be acquired. These parameters should be checked on a regular basis and the performance verified. It is advisable to verify the parameters on the same system each time it is used. Variables such as pulsers, receivers, and cables can exhibit anomalies that may be erroneously attributed to the transducer alone. Possibilities of changes in performance include, but are not limited to:

1. Pulse length. This can increase for the following reasons:
   (a) The backing or damping material becomes detached from the transducer.
   (b) The wear face becomes detached from the transducer.
   (c) The element or wear-face is cracked.
   (d) The tuning coil is detached.
2. Low sensitivity. This can occur for a number of reasons such as, but not limited to:
   (a) Deterioration of or damage to the transducer element material.
   (b) Detachment of the transducer face.
   (c) Detachment of the tuning coil.
3. Beam skew (apparent) due to partial detachment of the transducer face.
4. Faulty connections. Damaged or loose wires in the transducer housing.
5. In the case of angle beam transducers, wearing down of the Plexiglas wedge may cause the refracted angle to change.

In conclusion, the transducer is a precision component in the ultrasonic system. Treated with due care, it will provide many years of service.

## 7.4 Techniques

The ways in which sound waves propagate through materials and are attenuated, reflected, or transmitted dictate the different ultrasonic methods or techniques used to detect the many types of discontinuities that can exist in materials. These techniques fall into two main categories, one called "pulse-echo," and the other called "through-transmission." Any of the techniques that may be used requires *calibration* of the ultrasonic system so that the time base can have some meaning in terms of material thickness.

This is also true for through-transmission techniques where depth information is not available, since the practitioner needs to be sure that an adequate time base is available to show the transmitted energy.

## 7.4.1 Calibration Techniques

The process of calibration needs to be just as disciplined as the inspection technique. In fact, the whole inspection relies on the calibration process. Some of the following calibration techniques are concerned with measuring and documenting the characteristics of the transducer and flaw detector.

## 7.4.2 Transducer Characteristics

### *Compression Wave Transducers*

**Resolution** The ultrasonic pulse consists of a few cycles of sound energy at the test frequency. Therefore, the pulse occupies some space in time, or distance, within the material. The amount of space occupied is called the "pulse width." Physically, it is the number of cycles in the pulse multiplied by the wavelength of that frequency in the test material, or, mathematically:

$$W = n \times \lambda$$

where:
$n$ = the number of cycles in the pulse
$\lambda$ = wavelength

Pulse width is important because while one pulse is being processed, another closely following pulse cannot be properly processed. This means that two reflectors that only have a small separation in depth will only show as a single, messy signal on the trace. Because the two signals overlap, the time difference between them cannot be measured. The ability of an ultrasonic system to discriminate between two reflectors that are close together is called "resolution."

Resolution for a given transducer can be determined in one of several ways:

1. By calculation of the pulse width and expressing that in terms of distance occupied in the test material. For example, the wavelength of a 5-MHz compression wave in steel is 0.047 in. (1.192 mm). If the transducer in question has two-and-a-half cycles in the pulse, then the pulse width for the transducer is 0.047 in. × 2.5, which comes to 0.117 in. (2.98 mm). Signals from surfaces closer together than this will begin to merge.
2. By measuring the pulse width on the calibrated trace. If the time base has been calibrated for the test material, any signal positioned clear of any other can be used to measure pulse width. Figure 7-49 shows a trace calibrated to 25 mm full scale. The signal in the middle of the trace starts at 11 mm and ends at 14 mm. The pulse width is 14 − 11 = 3 mm.
3. The resolution is checked on a suitable calibration block such as the IIW block shown in Fig. 7-50. The transducer is placed opposite the machined slot.

If the transducer is capable of resolving the distances, three distinct signals should be shown on the trace (see Fig. 7-51). The block has been designed so that the distance

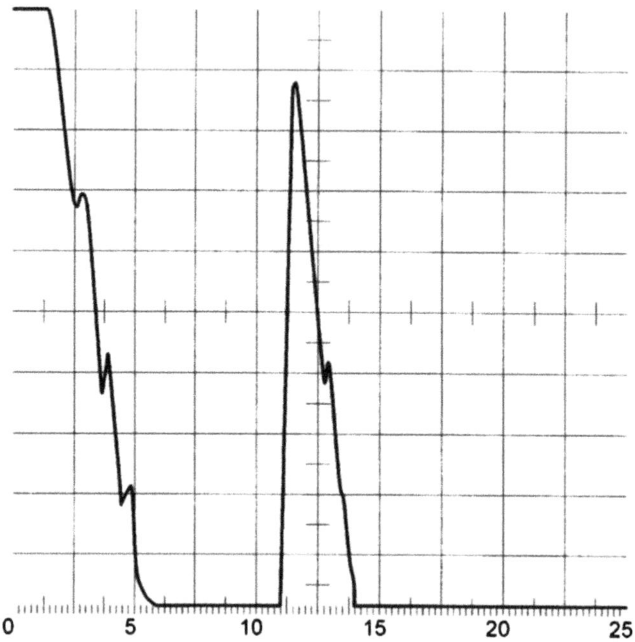


**Figure 7-49** A-scan 3.

between the first and second reflecting surfaces is 1 μsec at steel velocity. For the round trip, the time difference between the first and second signals is 2 μsec. A disadvantage of the resolution check that can be performed on the IIW V1 block is that it is confined to straight beam transducers. An alternative approach is illustrated in Fig. 7-52, which shows a pair of holes drilled on a common axis from opposite sides of the block. The two holes are of different diameters and meet in the center of the block. A sound beam, at any angle, aimed at the region where the two holes meet, will reflect from both holes. If the two echoes are resolved on the trace, the transducer can be said to have sufficient resolution for that spacing. If the holes are 0.5 in. (12 mm) and 0.25 in. (6 mm) in diameter, respectively, the time difference between the two is equivalent to 0.125 in. (3 mm). The advantage of this approach is that it is applicable to both straight beam and angle beam transducers.

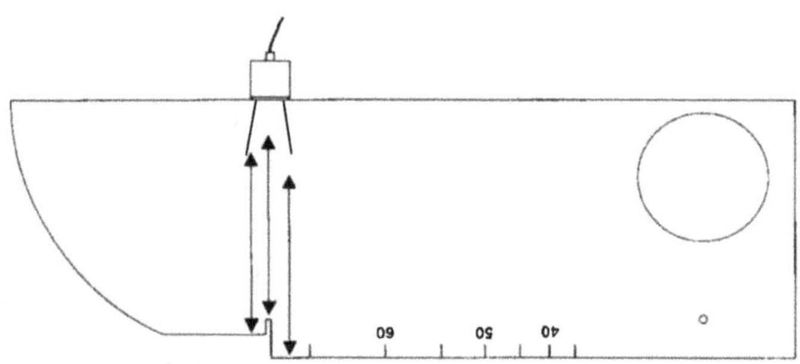


**Figure 7-50** Resolution on IIW block.

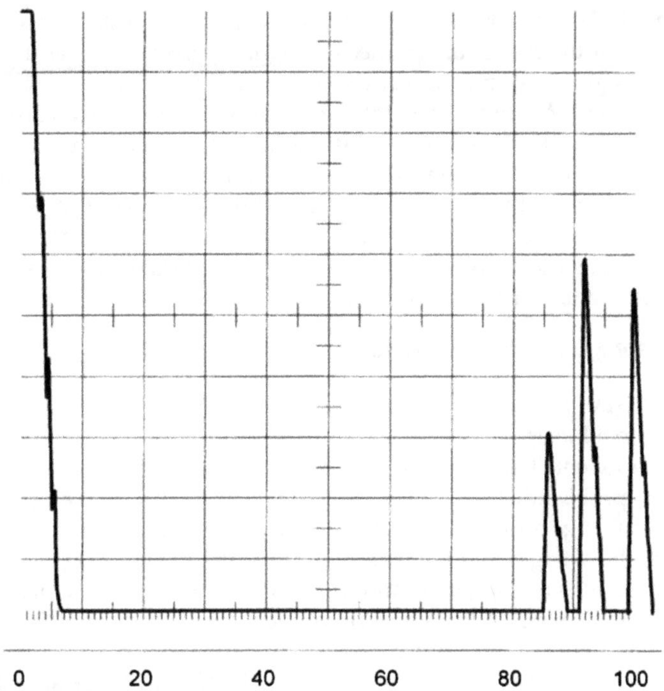


**FIGURE 7-51** A-scan resolution.

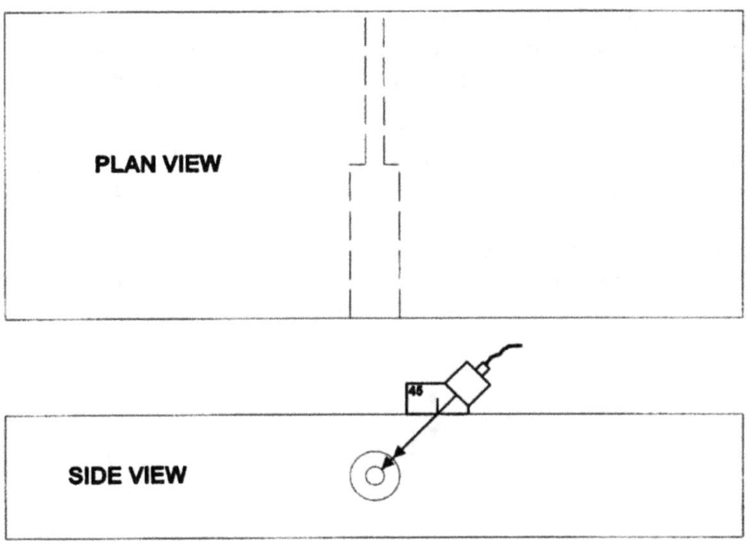


**FIGURE 7-52** Angle beam resolution.

**Dead Zone—Initial Pulse (Main Bang)** When using single crystal transducers, the initial pulse and ringing of the crystal are connected to the receiver circuit and displayed on the time base. This signal occupies the early part of the trace, and indicates the "dead zone" in the material. While the crystal is transmitting, the trace cannot display received signals clearly in this region. For this reason, it is usual to use dual crystal transducers for components less than 0.5 in. (12.5 mm) in thickness. The dead zone can be measured on a calibrated time base by noting the point at which the initial pulse dies away. It should be remembered that the initial pulse, and hence the dead zone, will increase as the gain is increased. So it is necessary to measure the dead zone at the working gain for the inspection.

### *Angle Beam Shear Wave Transducers*

**Probe Index (Exit Point)** With angle beam transducers for contact scanning, the crystal is mounted on a Plexiglas incidence wedge in order to generate the desired refracted angle in the component. The center of the sound beam emerges from the wedge at a position called the "probe index" or "beam exit point." This position is engraved on the side of the transducer wedge and is the position from which horizontal distance (HD) is measured. With use, the wedge may become worn and the actual exit point can move. The new probe index must be measured and marked if accuracy is to be preserved.

The IIW V1 block allows several angle beam calibration checks to be carried out. One of these checks is to determine the true probe index. Figure 7-53 shows a 45° shear wave angle beam transducer positioned on the IIW V1 block so that the beam center is aimed at the 100 mm radius. The intersection of the narrow slot with the scanning surface marks the center of the 100 mm radius. As the transducer is moved forward or back past the slot, the echo from the 100 mm radius will be seen to rise and fall. When it reaches maximum amplitude, the beam center is aligned with the radius center, and is perpendicular to the tangent to the radius. The slot marking the center of the radius is now aligned with the true probe index. If this differs from the marked index, a new probe index must be marked on the wedge or the discrepancy measured and recorded so that allowance can be made during subsequent operations.

**Beam Angle** The beam angle of an angle beam transducer can be measured on the V1 block once the true probe index has been established. The large Plexiglas-filled hole in the V1 block provides the target for beam angle measurement. From the center of this hole, angles have been projected to the edges of the block and the intersection engraved with the angle every five degrees from 35° to 70°.

When an angle beam transducer is scanned toward the hole, as shown in Fig. 7-54, the signal maximizes when the beam is aiming straight at the center of the hole. As this

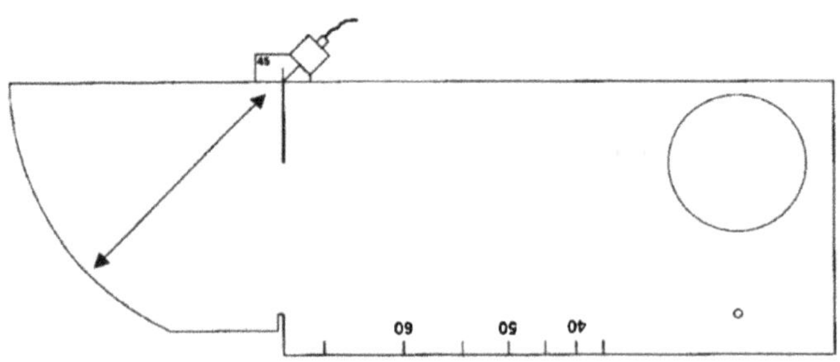

**FIGURE 7-53** Angle beam on IIW block.

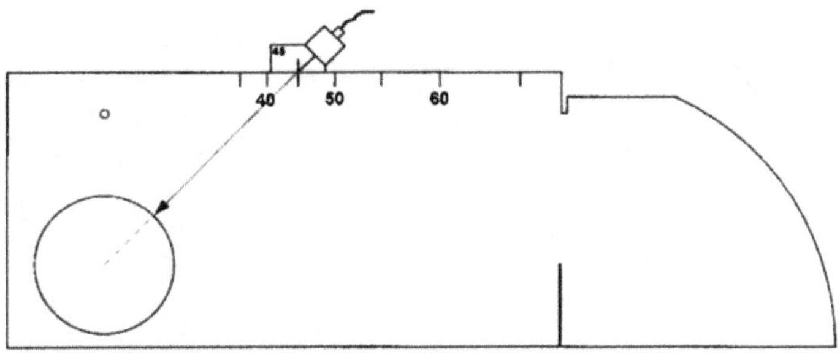


**Figure 7-54** Angle beam on IIW block 2.

occurs, the probe index is aligned with the true beam angle mark on the edge of the block. The actual beam angle should be checked against the nominal angle; any variation must be noted and the actual angle should be used when conducting discontinuity plotting and sizing techniques.

**Beam Spread Diagram** Some discontinuity sizing techniques make use of beam spread diagrams to identify the edges of a discontinuity. For this, a practical beam spread diagram must be drawn for the probe being used. To obtain an accurate beam spread diagram, the following information is required:

1. Beam angle of the probe in question
2. The beam limits to be used (–6 dB or –20 dB)
3. The exact probe index (exit point) for the probe in question

Beam angle and probe index are defined in Section 7.4.1, Calibration Techniques. The choice of 20 dB drop, or 6 dB drop will be dictated by the relevant inspection code or standard. Once the above information has been obtained, the next step is to prepare the plotting aid shown in Fig. 7-53.

The card shown in Fig. 7-55 has a horizontal scale representing the horizontal distance along the scanning surface from the probe index to a reference mark (the center line of a weld, for instance). The vertical scale represents the thickness of the component. From the intersection of these two scales, the beam path scale is drawn in at the measured beam angle for the probe (not the nominal angle marked on the probe).

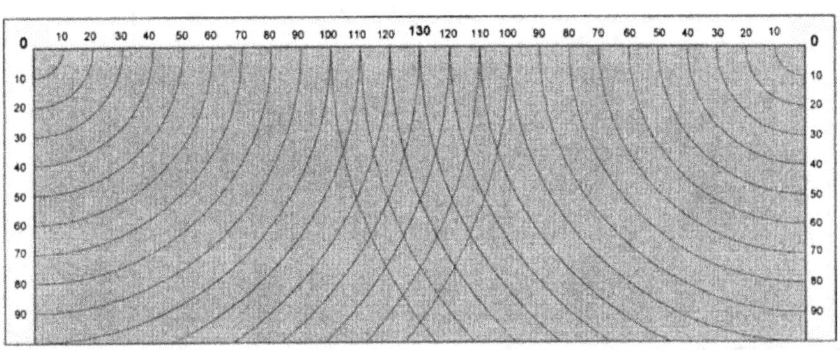


**Figure 7-55** Plot card.

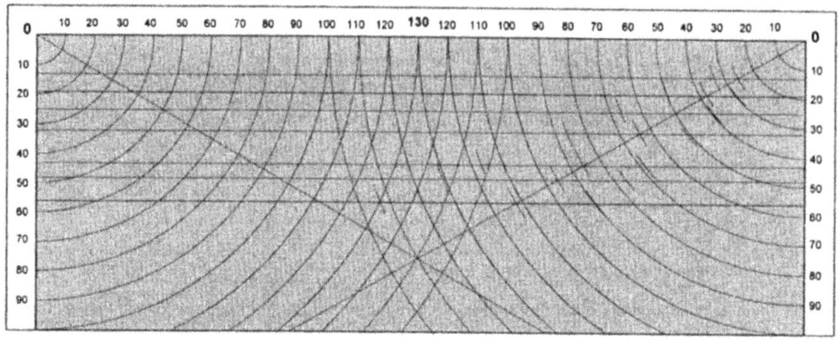


**FIGURE 7-56** Plot card.

The next step is to measure the beam spread at various depths so that a beam spread diagram can be constructed. This is done using a calibration block such as the IIW block (see Fig. 7-54). This block contains a series of side-drilled holes 1/16 in. (1.5 mm) diameter and at various depths below the surfaces of the block. On the card shown in Fig. 7-56, lines are drawn parallel to the horizontal scale at the depths of the holes that are to be used for the beam spread diagram. It is only necessary to plot the beam spread to the maximum range that will be used on the time base. In specimen thickness terms, this should be at least twice specimen thickness if discontinuities at full skip distance are to be plotted.

If, for example, the component being tested is 25 mm thick, then parallel lines would be drawn at IOW block hole depths of 13, 19, 25, 32, 43, 48, and 56 mm, as shown in the diagram. Each of the holes would then be scanned from the appropriate surface. The echo from the hole is maximized by moving the probe along the scanning surface, as shown in Fig. 7-57. When the signal reaches its maximum amplitude, the beam center is pointing at the center of the hole. The signal height is then adjusted to a convenient value, usually 80 percent of full screen.

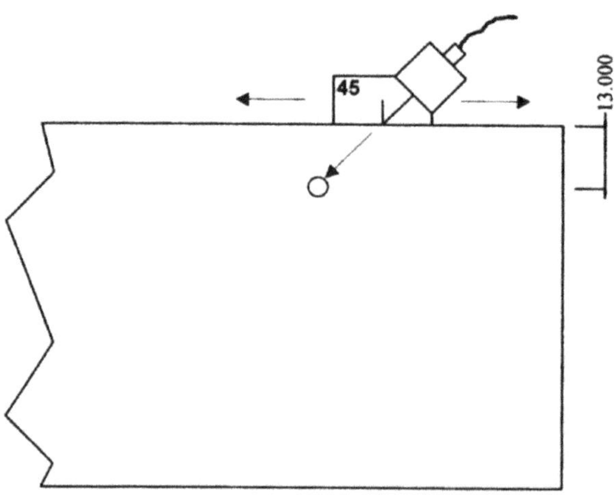


**FIGURE 7-57** Depth hole.

The gain setting in dB is noted and the gain reduced by the appropriate intensity drop (–6 dB or –20 dB). The new signal height is noted and marked on the screen with a wax pencil before returning the signal to the original (80 percent) height. The beam path length is taken from the time base and recorded in a table for that depth of hole and beam center position.

The transducer is then scanned forward until the signal amplitude has decreased to the wax pencil line (i.e., by the required intensity drop). The time base range is again noted and recorded as the beam *back-edge* position.

The last step is to move the transducer back to locate the maximum amplitude and then scan back until the signal again reaches the wax pencil line. Note this time base range and record it as the beam *front-edge* position.

The three recorded time base ranges are then transferred to the plotting card by drawing arcs centered on the card zero. The arcs cut the parallel line for the hole depth in question and the beam center arc also cuts the beam path scale. These arcs are shown in Fig. 7-56. The above steps are repeated for each of the seven holes to be used in the beam spread diagram. Once all the arcs have been plotted, the edges of the beam can be drawn as shown in Fig. 7-58 and the beam spread diagram is complete.

### 7.4.3 System Checks and Calibrations

The techniques discussed so far have concentrated on the transducer, but in the overall preparation for an inspection, the combination of transducer and instrument must be addressed. Many of the checks rely on the practitioner being certain of the position of a signal with respect to the transducer or scan surface, which, in turn, relies on time base calibration.

#### *Time Base Calibration—Compression Waves*

Time base calibration is concerned with establishing a known depth scale across the trace for the material to be inspected. This requires a calibration block made of the same material as the intended workpiece. The IIW V1 block is a calibration block for steel. It has reflecting surfaces that allow direct calibration of the time base for 25, 50, 100, and 200 mm. By using multiples of those values, other time base ranges can be established.

The objective for the time base calibration is that "zero" on the trace represents the scanning surface and "full scale" represents a known depth or thickness. Between zero

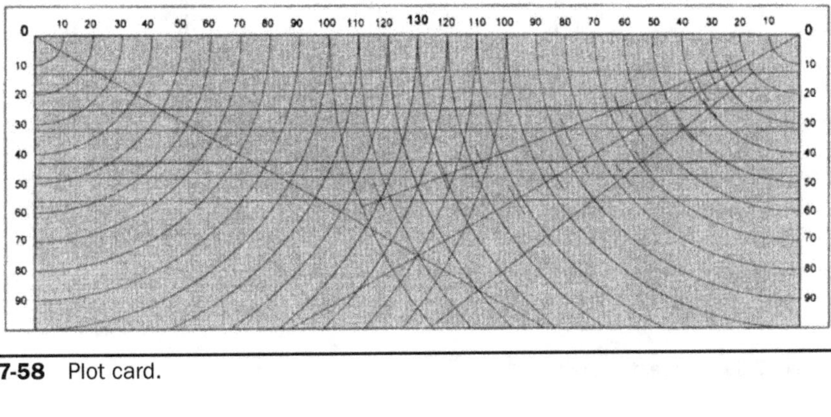


**FIGURE 7-58** Plot card.

and full scale, the graticule should represent a linear change of depth. There is a major problem facing the practitioner, because the point on the trace that represents entry of the beam into the surface is not obvious. In the case of a single crystal transducer, sound starts as the applied pulse voltage is released. The start point is therefore somewhere in the "initial pulse," where the signal is saturated.

Of course, the calibration block is of known thickness, so all that needs to be done is to identify the first echo from that thickness on the block and position it at "full scale," right? Wrong! The zero has to be correct too, or the time base will be a little too long or too short. Somehow, there has to be a pattern display with two signals that are exactly a known distance apart, and whose start points can be identified.

When looking at a multiple echo pattern as sound bounces up and down inside the material, the one certain fact is that between one back wall reflection and the next, the sound has traveled through the specimen thickness. So, for instance, the space from the beginning of the first back reflection to the beginning of the second (first multiple of the back reflection) is directly related to the wall thickness. Figure 7-59 shows a single crystal transducer positioned on the V1 block for the 25 mm range.

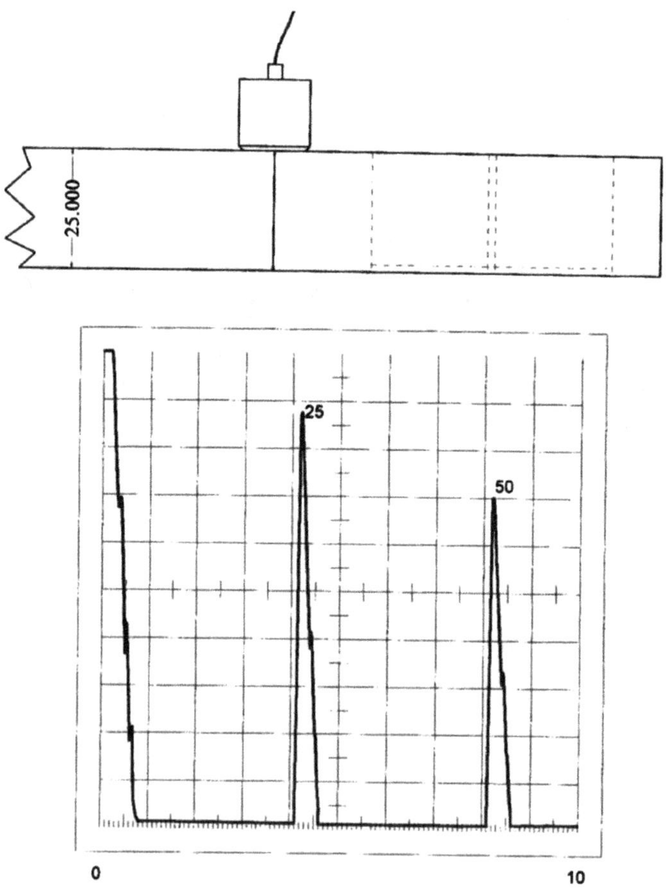


**FIGURE 7-59** Repeat signals.

To the left of the trace (time base) is the initial pulse and to the right are the back reflection and first multiple signals. The two signals are spaced apart by one round trip in the block.

To exploit this known distance, the left-hand edge of the first back reflection is moved using the "delay" control until it coincides with "zero" on the graticule. Then, using the "depth" or "range" control, the left-hand edge of the first multiple back-reflection signal is moved to coincide with "10" on the graticule. The position of the first back-reflection signal is checked and adjusted, using the "delay" control if necessary, and the first multiple signal checked and adjusted using the "Depth" or "Range" control. These adjustments are made until the left-hand edges of the two signals are correctly aligned with the graticule, as shown in Fig. 7-60.

At this stage, the time base is calibrated for a range of 25 mm, but zero is 25 mm, and ten is 50 mm (i.e., the back reflection and first multiple echoes). The requirement was that zero means scanning surface (0 mm). However, all is not lost! Lock the depth or range then use delay to place the first back reflection echo at 10, and the job is done! Zero is now 0 mm and 10 is 25 mm.

The above technique, using delay and depth (range) controls to adjust the back reflection and first multiple echoes to fit the graticule, can be used for any of the direct ranges on the V1 block. Because there is a delay due to the Plexiglas column, and because the transmitter is separated from the receiver, there is no way in which the

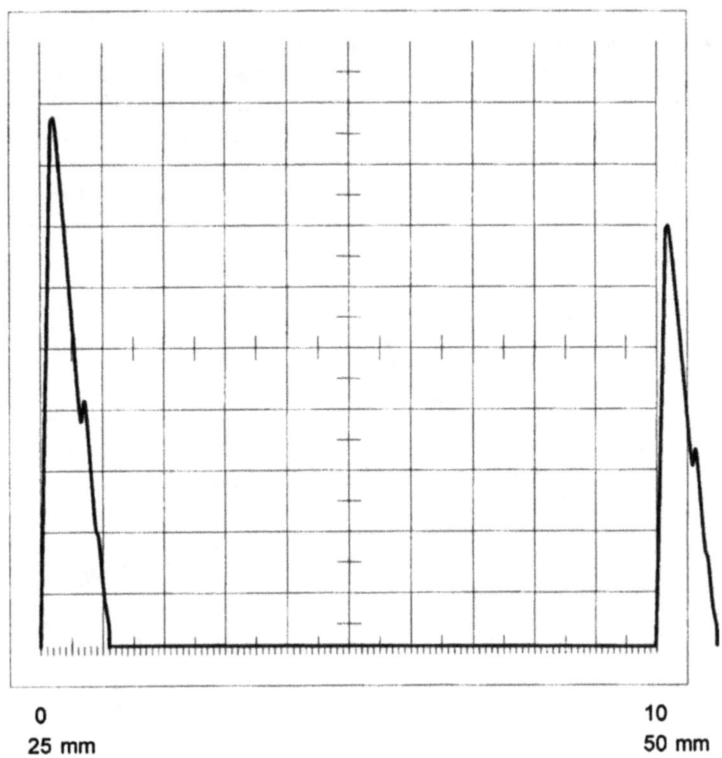


**FIGURE 7-60** Signal from 50 mm.

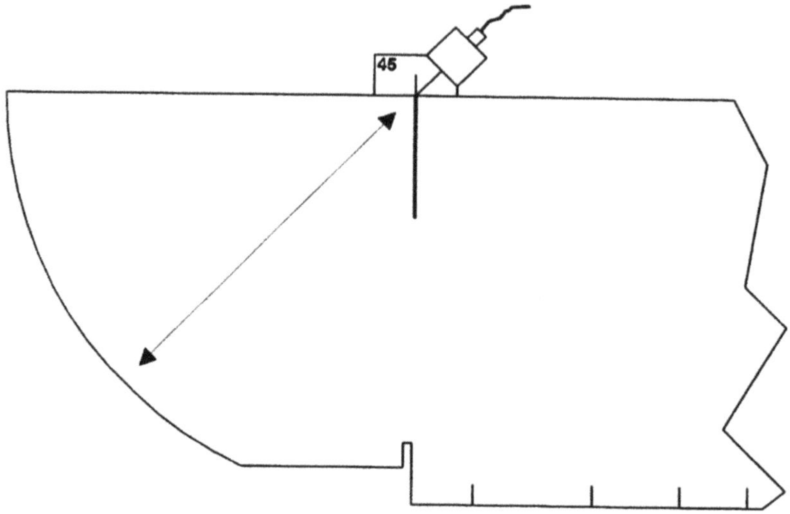


FIGURE 7-61 IIW radius.

beam entry point can be identified with a dual crystal transducer. The only way to calibrate is to use the above procedure.

### *Time Base Calibration—Shear Wave Transducers*

Calibration of the time base for a known time base range is also necessary in most angle beam inspections. This calibration can be carried out on the V1 block for 100 mm beam path range, and on the DIN 54 122 (V2) block for 25 mm and 50 mm. The procedure is similar to the one described above (see Figs. 7-61 and 7-62).

With the V1 block, the narrow slot that marks the center of the 100 mm radius also serves another purpose. It makes a corner reflector at the scanning surface to enable the

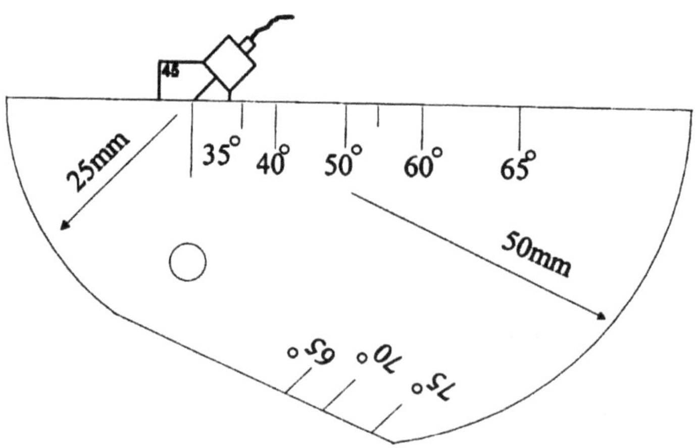


FIGURE 7-62 V2 block.

first return echo to reflect back to the radius to give a multiple echo pattern. The calibration of the time base can therefore be set from the first back echo to the second. This is achieved by using the delay control to bring the first back echo to zero on the time base. The range control is then used to position the second back echo to 10 on the time base. The time base now represents 100 mm of the material, but is set from 100 to 200 mm. To correct this setting to read 0–100 mm, the delay control is used by adjusting the control until the first back echo is at 10.

#### Calibration of Amplifier (Test Sensitivity)

Just as it was necessary to calibrate the time base in order to correctly position reflectors, so it is necessary to calibrate the amplifier to a known amount of gain. This is necessary to make sure that enough gain is used to detect critical discontinuities, that the gain is not so high that spurious indications appear, and to ensure repeatability between inspections.

The simplest form of setting gain is to choose a known target in a calibration block, and use the gain to set the amplitude of the echo from the target to a predetermined height. The target could be the back wall, a side-drilled hole, a flat-bottomed hole (FBH) in the calibration block, or even a keyway or oilway in the component itself. The main objective is that the actual target used is documented so that anyone can repeat the test.

#### Distance and Area Amplitude Blocks

Since the amplitude of an echo can be a function of the area and depth of the reflector, targets of known area and depth can be used to set test sensitivity. There are calibration blocks that are designed to perform this function. Essentially, flat-bottomed holes are used as targets. They range in size from 1/64 in. (0.4 mm) to 8/64 in. (3.2 mm) diameter in 1/64 in. steps. Sets of blocks in those sizes are available in a range of scanning depths.

Sensitivity is set by selecting an appropriate FBH size and scanning depth in a block of the same material as the component to be tested. The transducer is placed on the block and scanned to obtain the maximum echo amplitude from the FBH. The gain is adjusted to bring this echo to a predetermined height. This is known as "reference sensitivity" or "reference gain." Reporting levels are generally compared to this signal amplitude.

#### Gain as a Function of Beam Path Distance

In Section 7.2, it was shown that the intensity of sound energy decreases with distance along the beam due to attenuation. This means that simple signal amplitude references, like those just described, do not apply the same sensitivity across the entire time base. There are, however, techniques to overcome this problem.

#### Distance Amplitude Correction (DAC)

Imagine listening to the radio while sitting close to the speaker. If the distance from the speaker to the ear increases, the volume will decrease relative to the distance from the speaker to the ear. This is because the sound is being attenuated in the air. If it were necessary to hear the sound at the same level, regardless of the distance from the ear to the speaker, a remote volume control could be used and the volume (amplitude) could be gradually increased relative to the distance from the source. This would enable the sound from the speaker to be heard at the same level when near the speaker or far away (this is distance amplitude correction). By electronically maintaining the same signal amplitude to the ear as the distance from the speaker is increased or decreased, the volume to the ear has remained the same.

As in the above scenario, when the distance between the transducer and a reflector increases, the amount of energy reaching and returning from the reflector to the transducer decreases. This principle is made apparent by observing the amplitude of the reflected signal from a given size reflector placed at different distances from the transducer. The apparent amount of signal reduction will be dependent on, but not limited to, such factors as material attenuation, transducer frequency, and transducer size.

Ultrasonic systems can include circuitry that electronically increases the gain according to the distance of the reflector, taking into consideration the amount of attenuation in the test material and the other factors mentioned above. This is also known as "swept gain." It is effectively a controlled increase in gain along the time base (sweep). For example, a signal from a 1/4-in.-diameter flat reflector placed at a depth of 1 inch with its signal amplitude set at 80 percent can now electronically appear at the same amplitude on the time base if the reflector was at perhaps two, four, five, ten, or more inches, away from the transducer. This type of presentation is necessary when factors such as attenuation affect gating conditions in automated test systems. The variable of material attenuation is reduced or eliminated in this way.

A curve can also be plotted to graphically express the reduction in sound energy as it moves through the test material. This is known as a "DAC" curve (see Fig. 7-63). Certain specifications require that a DAC curve be generated for a specific examination. This is sometimes needed to determine an acceptance or rejection level based on the amount of energy that is returned to the transducer by a reflector (see note below). In this instance, a DAC curve is generated by observing and noting the amplitudes of signals from known sized reflectors, e.g., flat-bottomed holes or side-drilled holes at various depths in the test piece. The signal amplitudes are plotted on the screen as the reflectors are interrogated at the varying distances (depths) from the transducer. A line is drawn to connect the signal amplitude points along the time base.

*Note:* There are many variables to be considered when determining the size of a reflector. Estimating or measuring reflector size based solely on amplitude is not a very practical or accurate method.

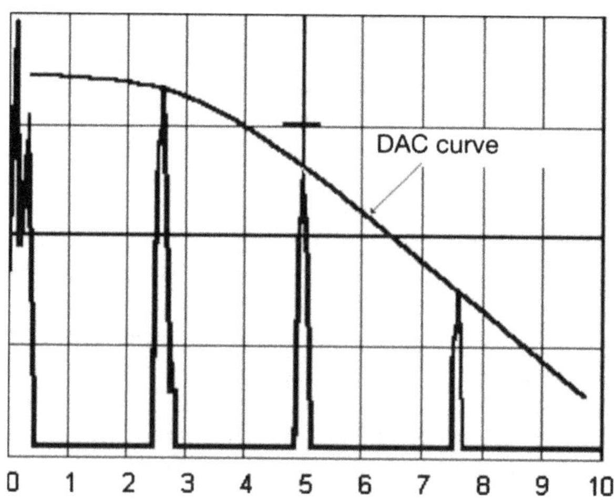


**FIGURE 7-63** DAC curve.

Reasonable size comparisons using flat-bottomed holes drilled perpendicular to the scanning surface can be made under controlled conditions. In terms of quantifying the amount of energy that returns to the transducer from a reflector, a number of variables need to be considered, e.g., the size, shape, and orientation of this reflector.

### *Transfer Corrections*

When reference amplitudes are established using calibration or reference blocks, there may be differences between the calibration block and the component that affect the sensitivity. These may be due to differences in surface condition, material thickness, or attenuation. The differences can be compensated for; this process is known as "transfer correction" (transferring from calibration block to component). The following example is for angle beam transducers, but the same process also applies to straight beam transducers.

**Transfer Correction Technique** The transducers are placed on the calibration block and the received signal from one full skip distance is maximized. The signal amplitude is set at 80 percent FSH (full screen height). The beam path length along the time base is noted.

The transducer is moved to locate the echo from the second full skip. With a wax pencil, the signal amplitude and the beam path length at its time base location are noted.

By dividing the dB difference by the beam path length, the energy losses can be calculated in dB per unit distance (inches or mm). This value should be noted.

The transducers are then placed on the part and the above procedure is repeated.

The calculation is performed as above and the dB difference per unit distance is noted.

The difference is subtracted in order to determine the amount of change that needs to be made to the reference sensitivity for the examination of the part.

By performing the above procedure, compensation for attenuation, differences in curvature, surface condition, and beam spread losses will be taken into account.

### *Distance, Gain, Size (DGS) Technique*

The DGS system was developed by Krautkramer in Germany as a method of standardizing inspection and to provide assistance with acceptance and rejection decisions. The principles are based on the known beam characteristics of each transducer. The way in which specific reflectors will respond to the beam can be predicted. DGS plots curves for two types of reflectors. The first is a total beam reflector, effectively back wall reflectors (i.e., bigger than the beam) at increasing depths. This forms one boundary of the DGS diagram. Inside the back wall curve is the second type—a series of curves for reflectors that are smaller than the beam. These reflectors are FBH targets; each curve represents one diameter of FBH at increasing depths (see Fig. 7-64).

Curves are calculated and produced for each transducer type and are available in data sheet form and as transparent accessories to clip in front of the screen. The calibration procedure uses the echo from a full back wall as the setting up target. This is adjusted using the gain control to a specific height on the CRT (40 percent FSH, for example). The next step is to refer to the "size" curve for the FBH diameter specified. Read down the "distance" line to the distance equivalent to the component thickness and note the dB difference between the "back wall" curve and the "size" curve at

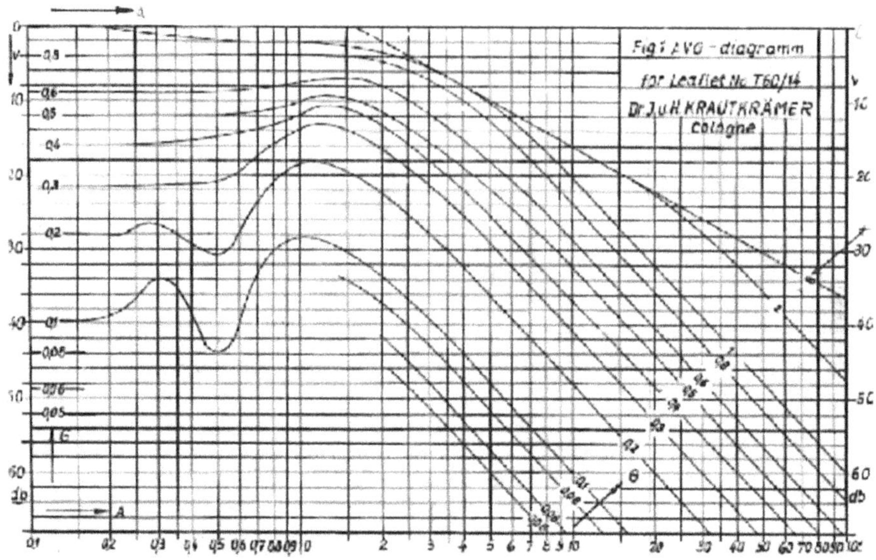


**Figure 7-64** DGS diagram. (*Courtesy of Krautkramer Branson.*)

that distance. Go back to the flaw detector and increase the gain by that dB difference. At the new gain setting, the reference FBH reflector at the back wall depth would give a 40 percent FSH signal. At any other depth, the signal from the reference FBH would follow the size curve for the depth concerned.

In angle beam shear wave applications, there is not likely to be a back wall echo. Instead, the echo from the 100 mm radius on the IIW V1 block is used. Because this is not the back wall echo from the component, a transfer correction has to be made. This entails measuring the sound intensity through the V1 block and through the component, expressing that as a dB difference, and adding that difference to the reference echo as well as adding the size difference.

The greatest attribute of the DGS system is probably the fact that it allows the determination of what would be the smallest discontinuity that could be detected with that transducer at that depth. Since an FBH reflector is a near-perfect reflector, any real discontinuity would give a weaker response. Therefore, a lot of time can be saved by *not* looking for a discontinuity that can't be found.

### System Checks

Various checks and calibrations are available for testing the entire system, i.e., transducer, cable, and flaw detector. The simplest is the check for overall system gain, using the large Plexiglas insert in the V1 block. The transducer is placed on the Plexiglas and the gain increased until 2 mm of noise (grass) appears or full gain is reached. The number of back wall echoes that appear on the screen is taken as a measure of the system gain for that transducer, frequency, and flaw detector.

Other checks need to be carried out from time to time to ensure that the performance of the system has not deteriorated. These include linearity of the amplifier (vertical linearity) and linearity of the time base (horizontal linearity).

**Vertical Linearity** It has been established that signal amplitude has some meaning and that signal amplitudes are comparable in a measurable denomination (dB). In order to compare amplitude data, it is necessary that the measuring device, or comparator, be "linear"; in this case, that the measured signal amplitude is accurate within a given tolerance. Knowing this, there is confirmation that the measured signal amplitudes displayed on the screen are meaningful for the intended purpose. This provides information for the comparison of reflectors.

The basic procedure for verification of vertical linearity is one that demonstrates the instrument's ability to maintain an amplitude ratio between two signals, throughout the instrument's "linear range," while varying the gain. This practice is referred to as "screen height linearity verification." The basic procedure is as follows:

1. Two signals from either one or two reflectors are established at 80 percent FSH and at 40 percent FSH.
2. The gain is reduced until the 80 percent signal is positioned at 70 percent FSH. At this point, the lower signal should have an amplitude of 35 percent.
3. This procedure is repeated in steps that reduce the higher signal amplitude by 10 percent. The lower signal amplitude (if the display is linear) should remain at 50 percent of the higher signal amplitude throughout the vertical range. The specification to which this evaluation is being conducted may include acceptance criteria.

*Note:* There are several standards that provide the methodology for conducting these checks:

**Amplitude Control Linearity** Verification that the gain control is linear with respect to the signals noted on the screen is also an essential requirement. This procedure is known as "amplitude control linearity verification" and is as follows:

1. A signal from a reflector is set at an amplitude of 80 percent FSH. The gain is reduced by –6 dB on the control. The signal should now have an amplitude of 40 percent FSH.
2. If the signal amplitude is set to 10 percent FSH and 20 dB of gain is inserted, the amplitude should become 100 percent FSH.
3. Other values throughout the range of interest are selected and verified.

To summarize: screen height linearity checks are conducted to verify that the display is linear throughout its useable vertical range. Amplitude control linearity checks verify that the gain control on the instrument is in fact adjusting the signal amplitude by the selected number of dB.

**Horizontal Linearity** Horizontal linearity is a measure of the instrument's ability to display time (distance) in a linear fashion across the screen. This means that in any given material, repeat signals from parallel surfaces will be represented on the screen as equidistant signals along the time base. For example, a ruler is calibrated to measure in inches or millimeters. If the ruler is not divided into some known denomination, it cannot be used as a measuring device. The same principle applies to the ultrasonic instrument. It is necessary to know that if the horizontal position of the signal on the screen indicates that a reflector is, for example, indicating 6 inches of steel thickness under the transducer, this measured distance is actually 6 inches.

With ultrasonic instruments, the ability to calibrate the circuitry (sweep/range) to display either inches, millimeters, or even microseconds is possible. The instrument screen displays a combination of the speed of sound in the test material and the distance that the sound has traveled. To do this, a "calibration standard" is necessary. This is referred to as "horizontal linearity." The instrument is calibrated and verified by using a standard having a known thickness. With the transducer placed on this standard, the instrument's controls (sweep and delay) are set to display signals in multiples of this thickness, which coincide with selected equal increments on the screen. If the signals can be aligned with these equally spaced increments, the instrument can be considered to have achieved horizontal linearity.

### 7.4.4 Calibration and Reference Blocks

#### *Basic Calibration Blocks*

The term *calibration block* is defined in British Standard BS 2704 as: "A piece of material of specified composition, heat treatment, geometric form, and surface finish, by means of which ultrasonic equipment can be assessed and calibrated for the examination of material of the same general composition." Therefore, a calibration block may be a simple step wedge in a particular material to allow the time base to be calibrated for accurate thickness measurement, or it may be a more complex block like the A.2 block described in BS 2704, which allows calibration of time base, plus calibration of probe index, angle, resolution, etc.

The term *reference block* is defined in BS2704 as: "An aid to interpretation in the form of a test piece of the same material, significant dimensions and shape as a particular object under examination, but not necessarily containing natural or artificial defects." So, for example, a section of an aircraft wing forging may be prepared as a reference block so that a technician may become familiar with the standard signal patterns from the various changes in section and more easily recognize a defect quickly when examining the component on an aircraft. More usually, the block would contain known reflectors from which the sensitivity (gain) used in the test could be set.

The BS 2704 A.2 calibration block, also known as the International Institute of Welding (IIW) block, is illustrated in Fig. 7-65. In the United States, this block is known as the IIW Type A2 block and the dimensions are in inches. The block can be used for the following assessments:

- Calibration of the time base in terms of thickness
- Checking linearity of the time base
- Assessing overall sensitivity of probe and amplifier
- Determination of the angle of refraction
- Determination of beam characteristics
- Assessment of dead zone
- Checking linearity of the amplifier
- Checking resolution
- Determination of probe index
- Finding the correct zero point

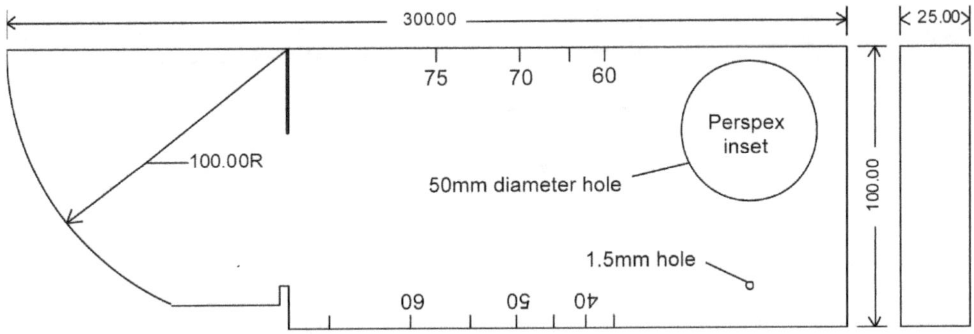


**FIGURE 7-65** IIW Type A2 calibration block.

In some instances, the Perspex inset is included. This provides the equivalent of a 2-in. steel dimension.

The A.2 block was derived from the original "Dutch block" designed by RTD Rotterdam and accepted by IIW as the IIW V1 block. In its original form, the deep slot at the center of the 100-mm radius was a scribed line and a 25-mm radius slot was positioned as shown in Fig. 7-66. This design is still used in some parts of the world, and has the advantage that shear wave calibration can be achieved for ranges other than multiples of 100 mm. In all other respects, it is the same as the A.2 block and is referred to as the IIW Type I Calibration block.

The BS 2704 A.4 Calibration block, also known as the V2 block, is a more compact form of the V1 block suitable for site use, although somewhat less versatile in its functions. Figure 7-67 illustrates the A.4 block.

The Institute of Welding (IOW) Beam Profile calibration block is designed primarily for beam profile measurement and has four 1.5-mm-diameter side-drilled holes giving eight depths from two scanning surfaces. These can be examined by direct scan for probes of various angles, and at several more ranges for each probe, using indirect scans by reflecting from the far surface. There are two series of five holes on an inclined axis to measure shear wave probe resolution and to simulate an inclined discontinuity. The block is illustrated in Fig. 7-68.

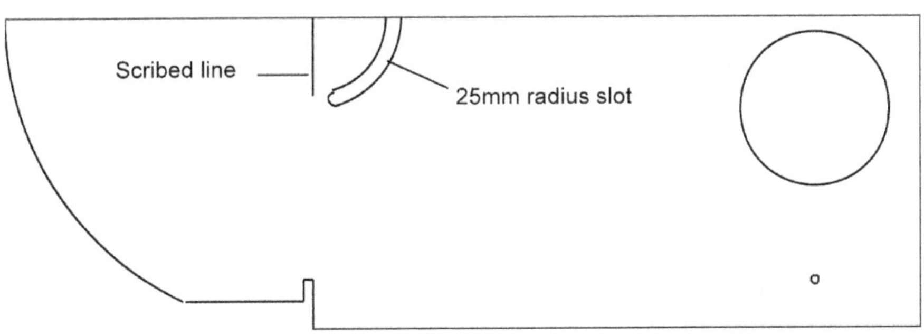


**FIGURE 7-66** IIW Type I calibration block.

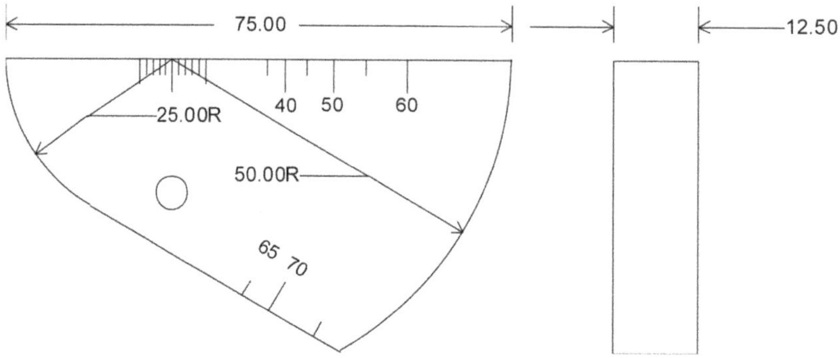


**FIGURE 7-67** A.4 (V2) Calibration block.

### *Reference Blocks*

Distance/area amplitude blocks (see Fig. 7-69) are mainly used for setting sensitivity levels and accept/reject levels for discontinuity sizes with reference to signal height. Blocks are produced in a range of scanning depths and each set of blocks contains flat-bottomed holes. In the area amplitude block set, the distance from the scan surface to the hole is the same, but the hole diameter varies in increments of 1/64 in. In the distance amplitude block set, the hole diameter is the same, but the distance from the scan surface varies with each block.

**Area Amplitude Blocks** If the signal heights from different sized flat-bottomed holes located at the same distance from the test surface are compared, they will be proportional to the reflecting areas of the holes, as shown in Fig. 7-70.

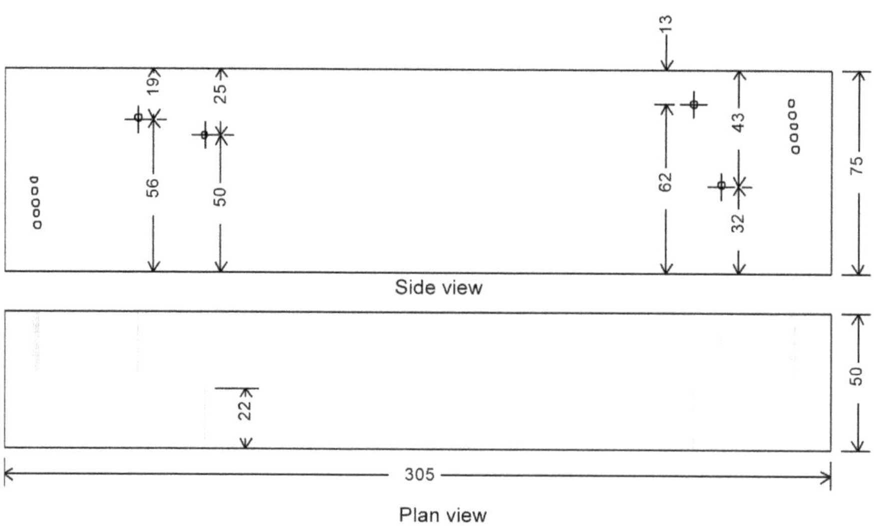


**FIGURE 7-68** IOW Beam Profile calibration block.

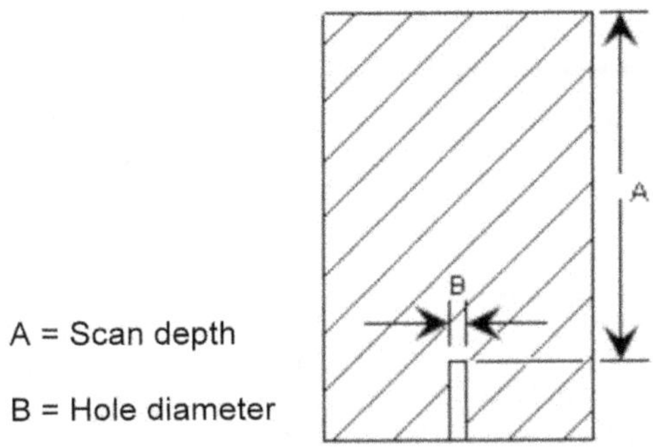


**FIGURE 7-69** Design of typical distance/area amplitude blocks.

### *Distance Amplitude Blocks*

However, if flat-bottomed holes of identical reflecting areas are located at different depths from the test surface, reflectors intercepted at greater distances will produce lower amplitude signals as shown in Fig. 7-71.

Angle beam distance/amplitude correction (DAC) reference blocks are made from the same thickness and grade of material as the workpiece. They contain an artificial flaw (a side-drilled hole). The change of signal height with changes in scanning distance (multiple skips) is noted and plotted on the display as a DAC curve so that a signal amplitude can be specified to cover all depths within the working range for reporting, acceptance, or rejection purposes. Figure 7-72 shows a typical ASME DAC block and Fig. 7-73 shows a DAC curve.

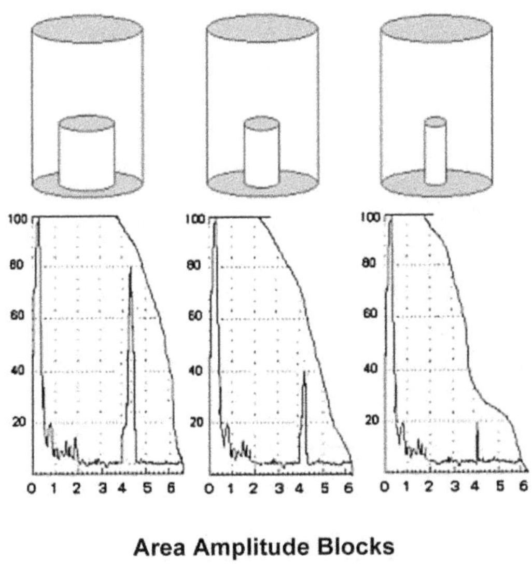


**FIGURE 7-70** Area amplitude blocks.

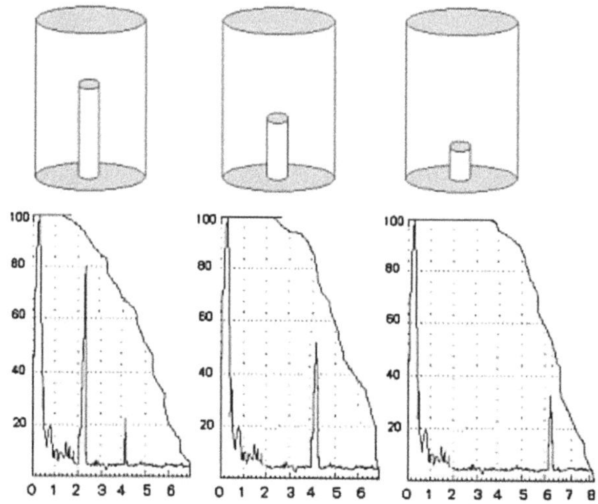


**Figure 7-71** Distance amplitude blocks.

## 7.4.5 Inspection Techniques

### *Pulse-Echo Techniques*

**Contact Scanning Using Compression Waves** Compression wave pulse-echo techniques usually employ either a single or dual crystal transducer directing ultrasonic energy perpendicular or near perpendicular to the scanning surface. These techniques are often known as "straight beam testing" techniques. There are some special techniques using compression waves at steeper angles used, for instance, in detecting cracks in

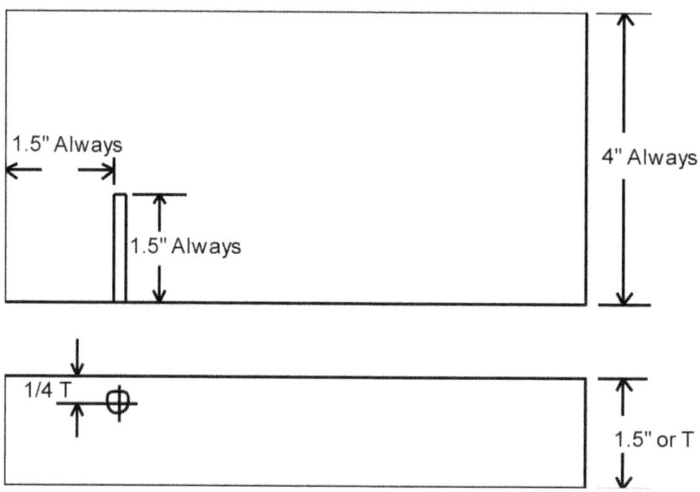


**Figure 7-72** Typical ASME DAC block.

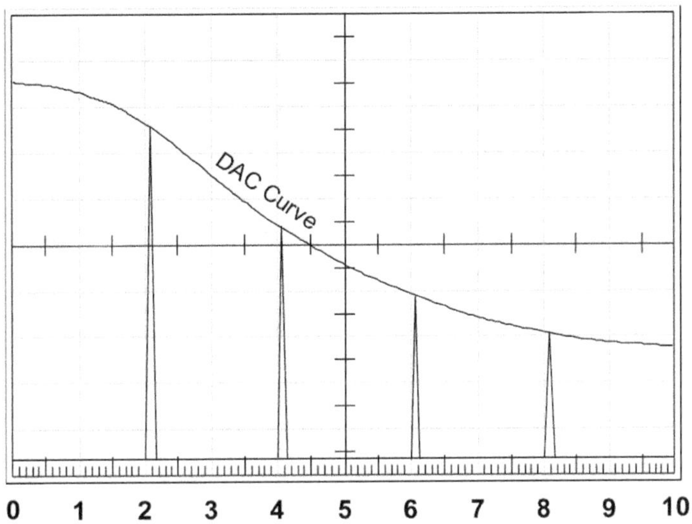


**FIGURE 7-73** Typical DAC curve.

ferrous materials under stainless steel cladding and for time of flight diffraction (TOFD) testing, but these will be covered separately.

In the standard compression wave techniques, reflections from the back wall and discontinuities are used to assess the suitability of a component for service. In order to obtain a reflection, it is necessary for the reflector to be orientated so that part of its surface is parallel to the scanning surface, in other words, normal to the beam. Laminar discontinuities and volumetric discontinuities like gas pores and nonmetallic inclusions are all suitably orientated. Discontinuities that are angled to the scanning surface either may not reflect at all or may reflect the sound away from the transducer.

Figure 7-74a shows a single crystal compression wave transducer set up for thickness measurement of a metal part, and Fig. 7-74b shows the corresponding ultrasonic A scan trace in which the time base has been calibrated for 25 mm full scale.

The initial pulse appears at zero on the left of the trace and the back reflection signal appears three-quarters along the time base, indicating a sample thickness of 18.75 mm. Thickness gauging is one of the simplest examples of compression wave testing. Notice that the initial pulse occupies almost a quarter of the time base so that 6 mm of metal path are obscured. This obstructed area is known as the "dead zone."

Figure 7-75a shows a dual element transducer set-up for thickness measurement on a sample that is 4 mm thick. Figure 7-75b shows the trace for this sample with the time base again calibrated for 25 mm. Notice that the selection of "dual" operation of the flaw detector isolates the transmitter from the receiver circuit, so there is no initial pulse and therefore, no dead zone. The first back reflection signal (also called "back wall echo" or "BWE") shows at 4 mm on the time base. Notice also that multiples of the BWE appear at 8, 12, 16, 20, and 24 mm on the time base.

One way in which the reading accuracy can be improved is to take a reading from a multiple and divide the result by the number of passes corresponding to that multiple. Take, for example, the reading at 24 mm (which is the sixth signal). Divide 24 by 6 and the answer is 4 mm. However, suppose the actual thickness was 4.15 mm.

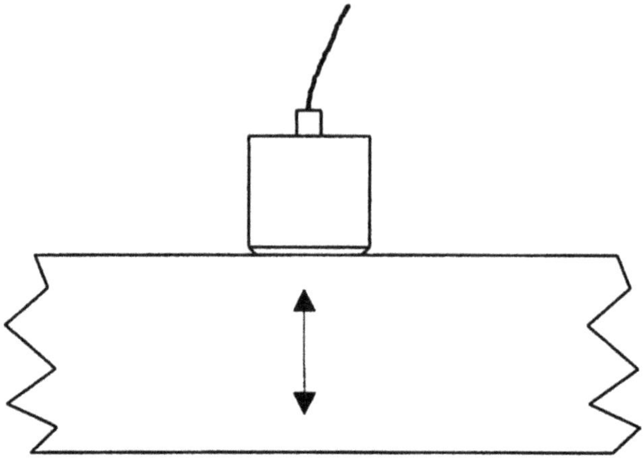

(a)

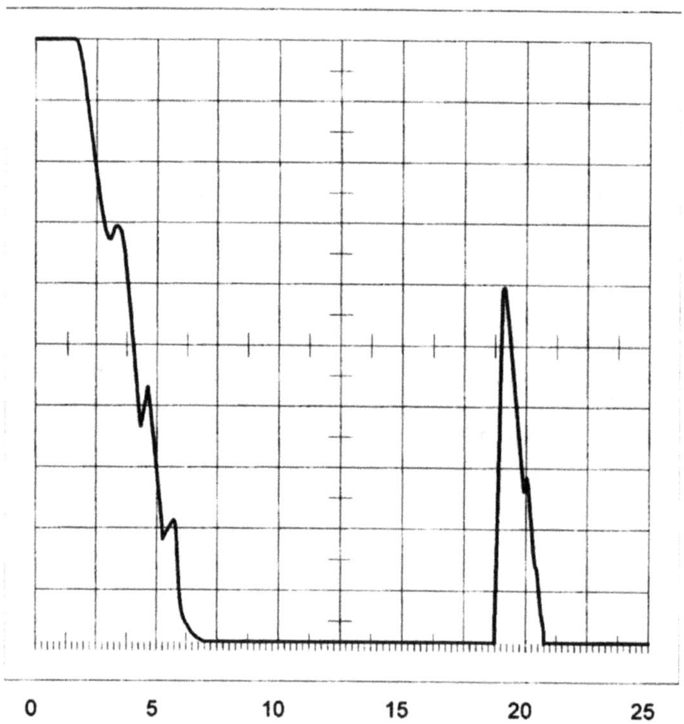


(b)

**Figure 7-74** Thickness measurement.

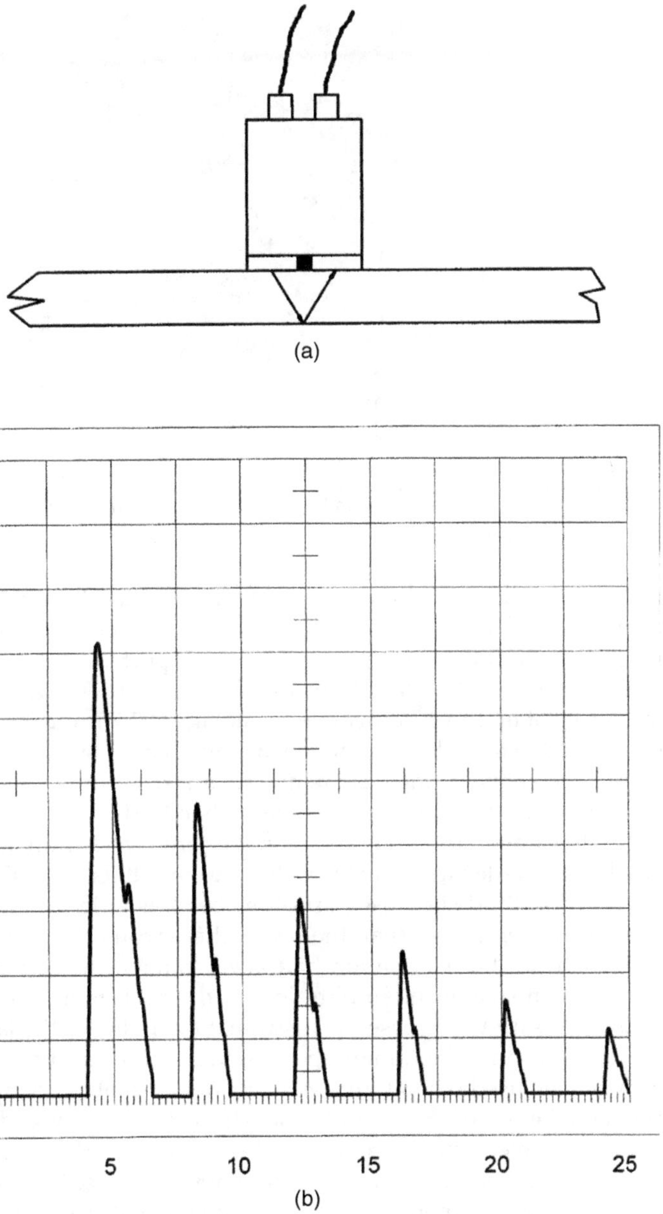


**FIGURE 7-75** (*a*) and (*b*) Thickness measurement.

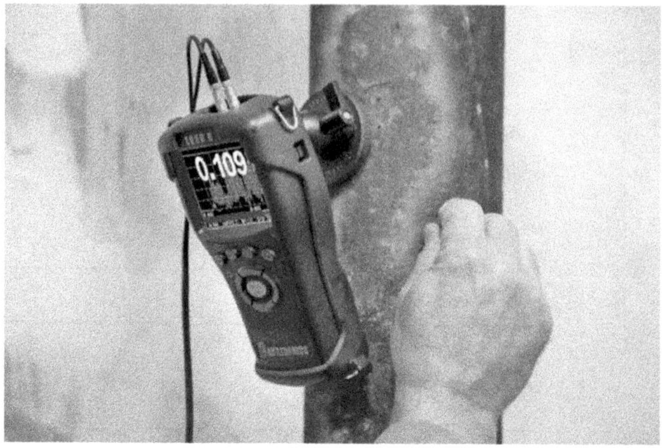


(c)

**FIGURE 7-75** (c) Measuring the pipe wall thickness with an ultrasonic thickness gage. (*Courtesy of Danatronics.*)

It would be difficult to read that accurately on the first back reflection signal, but the sixth signal would have been judged at 24.9. This number divided by 6 equals 4.15 mm. In practice, this sort of accuracy could only be expected on samples with very smooth scanning and back wall surfaces. For thickness measurement in the field on corroded surfaces, errors of up to ±0.5 mm are typical. Figure 7-75c illustrates an actual thickness gage being used to read the thickness of a plate that is corroded.

Figure 7-76a shows a single element transducer set-up to detect laminations in steel plate 20 mm thick. The lamination is smaller than the beam. Notice in Fig. 7-76b that the position of the lamination echo occurs on the screen at 11 mm below the scanning surface and the back reflection echo indicates a 20-mm thickness. The back reflection echo is reduced in amplitude because part of the beam is reflected by the lamination. If the lamination had been bigger than the beam, there would be no back reflection echo. If, on the other hand, the lamination had been smaller, the signal from the lamination would have been smaller in amplitude and the back reflection echo bigger. It might have been necessary to increase the instrument gain to see the lamination at all. In the extreme case, the lamination might be so small that it could not be detected at the test frequency or gain used. Therefore, the detection of suitably orientated discontinuities is dependent on the size of the discontinuity, the test frequency, and the gain used. Higher frequencies can detect smaller reflectors due to their shorter wavelength.

It must be remembered that attenuation of the ultrasonic beam also has an effect on detection. As the energy penetrates deeper into the material, it weakens. Eventually, the beam is too weak to allow small echoes to get back to the receiver. The higher the test frequency, the greater the attenuation and the less penetration that can be achieved. The material and its grain structure also affect attenuation. The practitioner must balance the conflicting requirements of discontinuity size to be detected, material properties, and ultrasonic beam properties in the choice of transducer and test frequency.

Figures 7-77a and b show two discontinuities unfavorably oriented to the sound beam. The inclined discontinuity in Fig. 7-77a is reflecting the energy away from the transducer, but also obscuring the back wall. The result would be no signals visible on

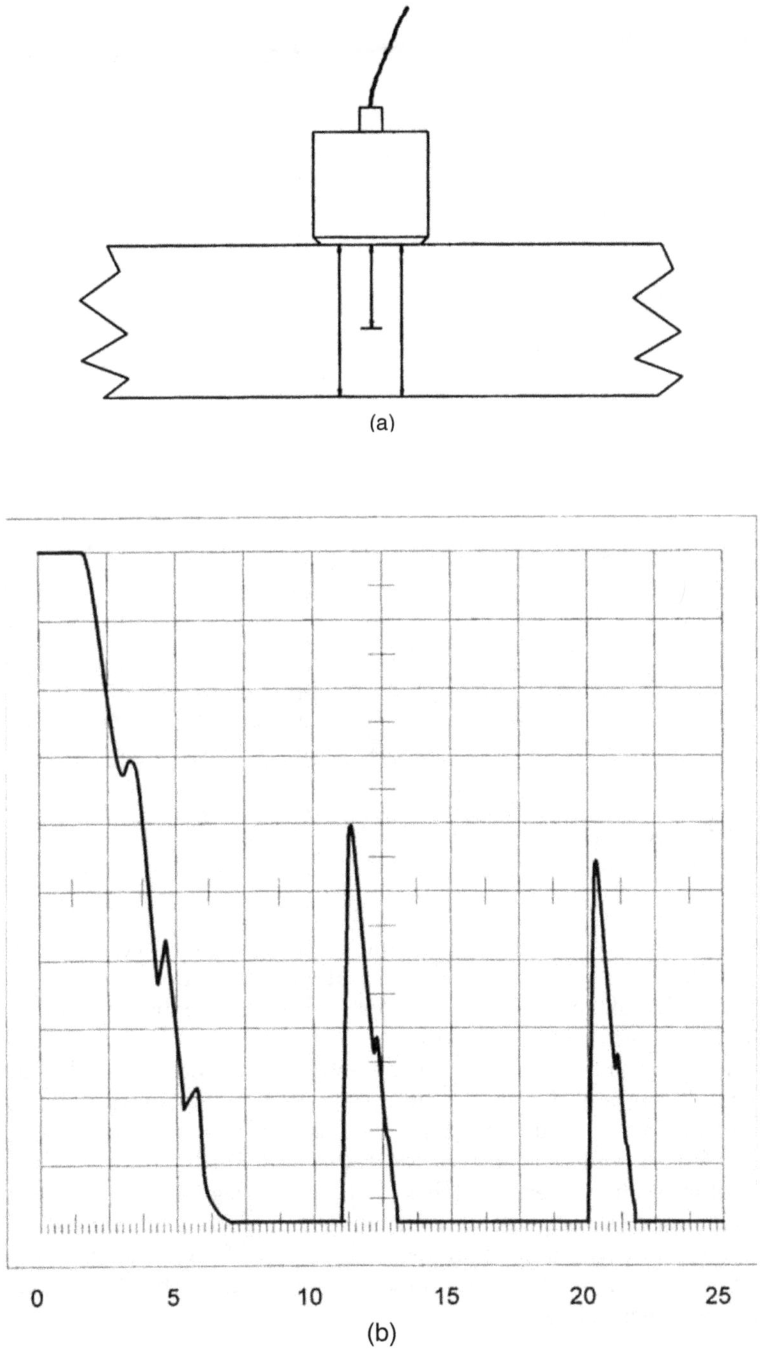


**FIGURE 7-76** Perpendicular reflector.

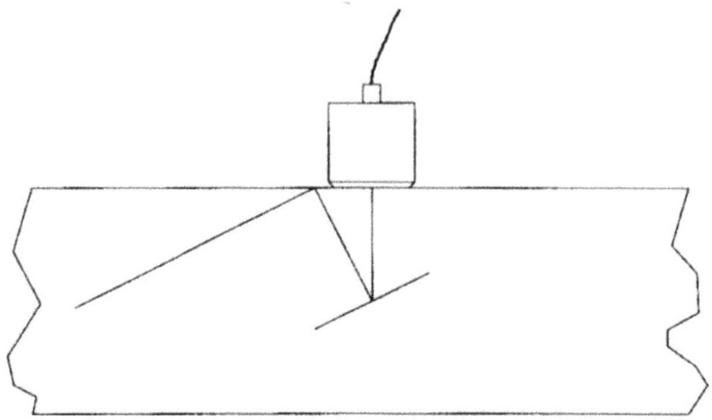
(a)

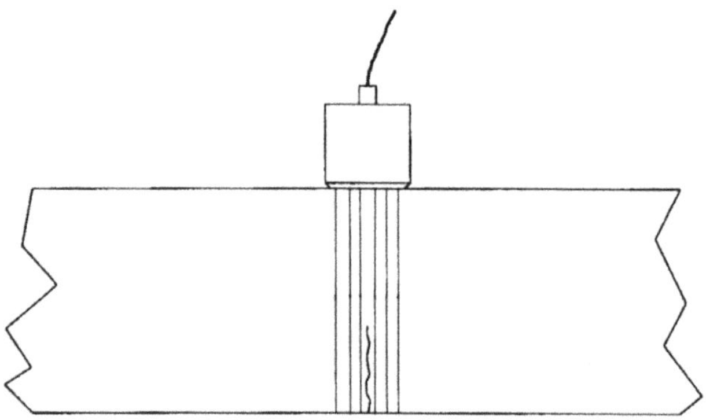
(b)

**Figure 7-77** Adverse reflectors.

the display, but there would be a reduction in the back reflection. In Fig. 7-77b, the vertical crack allows the sound to pass on either side without reflecting but would give a normal back wall echo. The possible orientation of the discontinuity must also be considered in devising a test procedure.

Finally, a test technique for the detection of laminations in thin plate is illustrated in Figs. 7-78a, b, and c. The technique is called the "multiple echo" technique for reasons that are obvious from Figs. 7-78b and c. The time base has been calibrated for 50 mm for a sample 3 mm thick. With the transducer in position 1 (sound material), the multiple echo pattern stretches to 30 mm (14 signals) as shown in Fig. 7-78b. With the transducer in position 2, over the lamination, the multiple echo pattern only stretches to 15 mm, as can be noted in Fig. 7-78c. This is because the echoes are closer together; so close, in fact, that they interfere with each other, leaving no clear time base in between echoes.

**Contact Scanning Using Angle Beam Shear Waves** If the possible orientation of any discontinuity is considered to be unfavorable to a beam perpendicular to the scanning surface, it will be necessary to tilt the beam to an appropriate angle to ensure that the beam strikes the discontinuity as near perpendicular as possible. For small angles (up

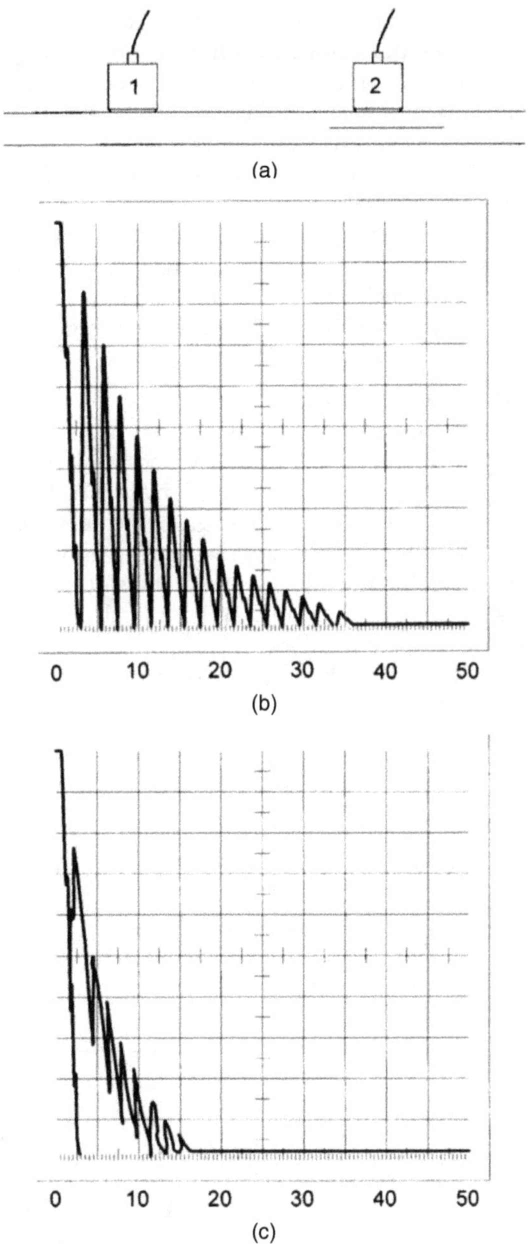


**Figure 7-78** Lamination detection.

to about 10° in the test material), compression waves may be used. However, for larger angles, mode conversion to shear wave energy makes the use of compression waves alone impossible. It therefore becomes necessary to increase the incident angle beyond the first critical angle, leaving only a shear wave in the part.

The lowest practical angle for testing with a shear wave alone is about 35° refracted shear wave angle. This does not mean that testing at angles between 10° and 35° is impossible. However, if an angle in this range needs to be used, the practitioner must consider carefully the geometry of the part. The next decision is whether to use the compression wave or the simultaneous shear wave, depending on what happens to the unwanted mode. Regular off the shelf transducers are either straight beam compression wave or shear wave angle transducers of 35° to 70°.

The common or "preferred" angles available in ultrasonics for shear wave testing are 45°, 60°, and 70°, although other angles can be made to order. The angles marked on a shear wave transducer are for steel, unless followed by an identifying letter for other materials. For instance, "45°Al" would denote a 45° shear wave angle transducer for aluminum.

### *Half Skip and Full Skip Techniques*

The simplest angle beam test is one that looks only for vertical surface-breaking discontinuities originating at the scanning surface or back wall. Figure 7-79a shows a 45° shear wave transducer positioned on a 20-mm-thick plate. The beam reflects at the bottom and then top surfaces at an angle equal to the angle of incidence of the shear wave beam, 45° in this case. Note that with no discontinuity present, no echo energy returns to the transducer.

Figure 7-79b shows the transducer aimed at a slot breaking the bottom surface. This slot will produce an echo arriving back at the transducer at a fixed time that is twice the beam path distance A–B divided by the shear wave velocity. The trace is shown at Fig. 7-79c; the time base is calibrated for 100-mm return trip time, and the beam path distance is shown as 28 mm.

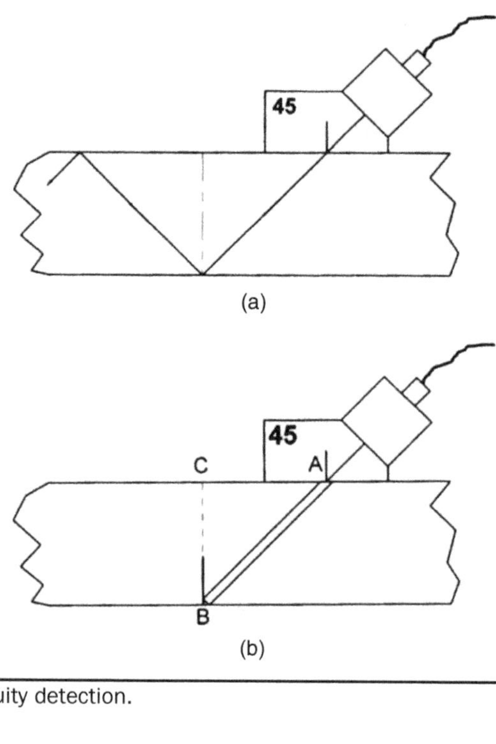


**FIGURE 7-79** Discontinuity detection.

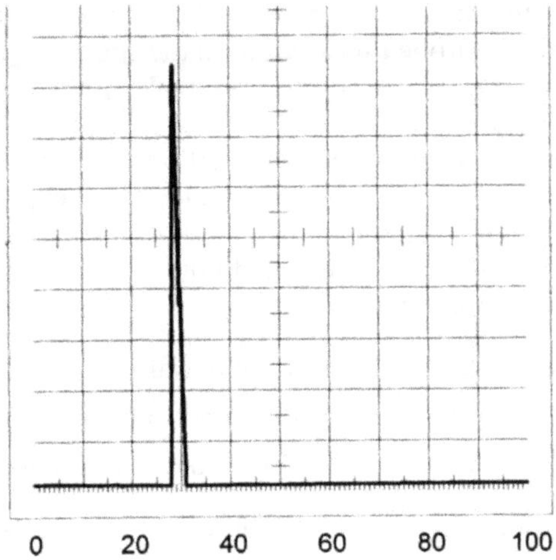


(c)

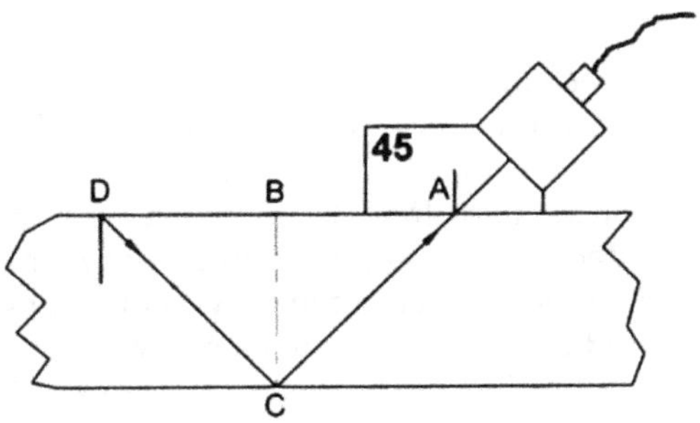


(d)

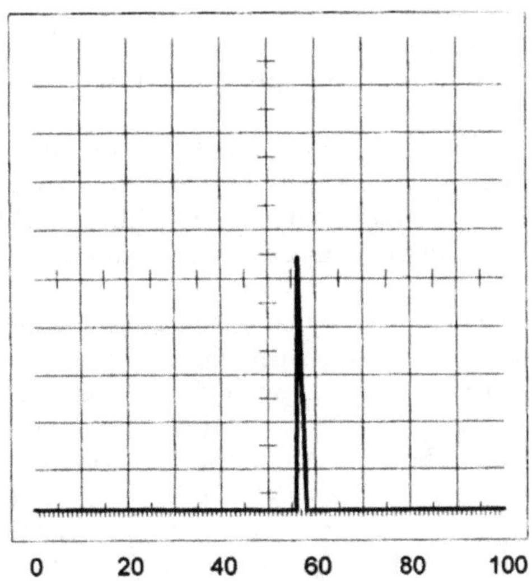


(e)

**Figure 7-79** *(Continued)*

In Fig. 7-79d, the distance A–B along the scanning surface is called the "half skip" distance and for a 45° probe this is equal to the specimen thickness. Position C is called the half skip position. Any discontinuity breaking the bottom surface of this part will produce a signal at 28 mm.

Figure 7-79d shows the transducer positioned to reflect from a top surface-breaking slot. The distance A–D is called the "full skip" distance and a full skip signal would appear at 56 mm on the time base, as shown in Fig. 7-79e. If the transducer is scanned along a 20-mm plate containing top and bottom surface-breaking discontinuities, only three signal patterns are possible:

1. No signal representing sound material
2. A signal at 28 mm representing a bottom corner reflector
3. A signal at 56 mm representing a top corner reflector

For the practitioner, interpretation of results is relatively simple since there are only two screen locations on which to concentrate. The technique is commonly used to detect fatigue cracks during the in-service inspection of critical components.

### Beam Path Distance Techniques

Of course, not all discontinuities occur at the top or bottom surfaces. In welds or castings, for example, planar and volumetric discontinuities may occur anywhere within the volume of the part. In order to detect, correctly assess, and position such discontinuities, it is necessary to determine the distance along the beam path at which the reflection occurs. This distance together with the known beam direction and angle allows the position of the discontinuity to be plotted.

Volumetric discontinuities, such as gas pores or slag inclusions in welds, are not very sensitive to beam angle. Approached from almost any direction, there is likely to be a facet of the discontinuity that will reflect back to the transducer. On the other hand, planar discontinuities, such as lack of side wall fusion in welds and angular cracks, are very sensitive to beam angle. The practitioner must be aware of the types of discontinuities that might occur during fabrication and service in a part to be inspected.

To be able to measure beam path distance, the time base must be calibrated for the shear wave velocity in the material to be inspected. This process of calibration requires suitable calibration blocks. Such calibration blocks should be capable of being used for a variety of beam angles. An example of a calibration block (IIW) is shown in Fig. 7-80.

Figures 7-81a–f illustrate how this technique can be used to locate a region of lack of side wall fusion in a weld 20 mm thick. Figure 7-81a shows the prepared edges before welding. Figure 7-81b shows the completed weld, with dotted lines to show the original prepared edges. In Fig. 7-81c, a 60° angled shear wave transducer is shown positioned to obtain a maximum echo from the side wall discontinuity. The angle has been chosen to ensure that the beam meets the original prepared face at 90° for maximum sensitivity. Figure 7-81d is a plan view of the same set-up, showing that the center line of the weld has been marked as a reference for measurements. Figure 7-81e shows the trace as the echo maximizes.

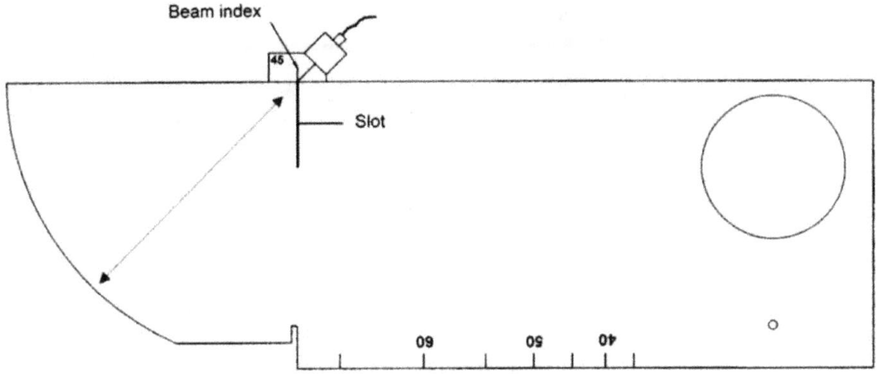


**FIGURE 7-80** Beam index.

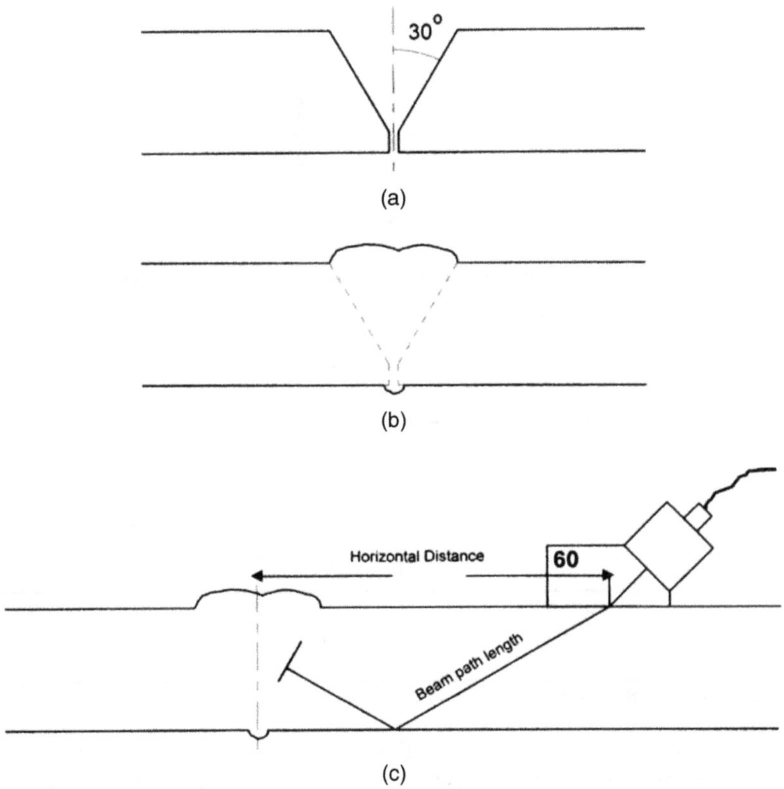


**FIGURE 7-81** Discontinuity detection in weld.

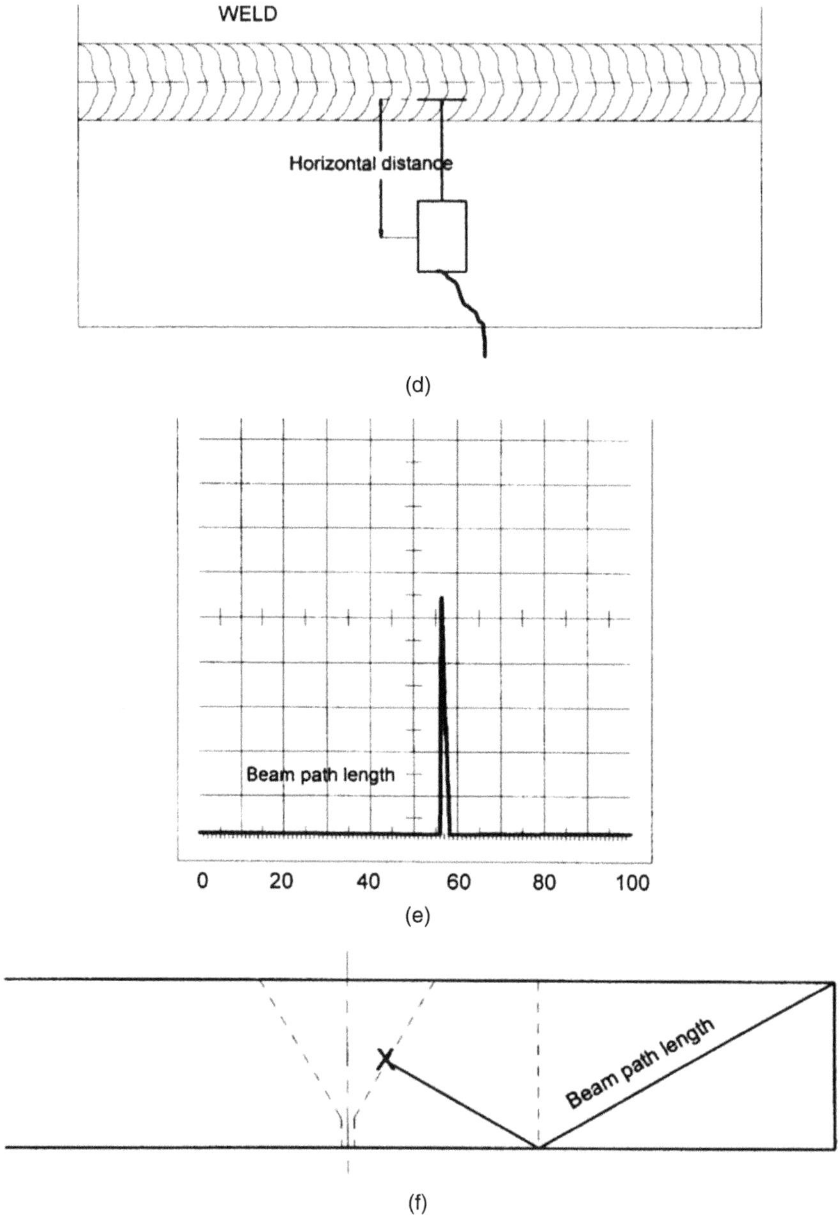


**Figure 7-81** *(Continued)*

The practitioner, once satisfied that the echo is maximized, then makes two measurements; one from the time base, the other on the scanning surface using a ruler. The first measurement is the time base range giving the beam path length to the discontinuity. The second measurement is the horizontal distance from the beam index to the weld center line. These two measurements, together with knowledge of the beam angle, are enough to accurately plot the position of the reflecting point.

Figure 7-81f shows the process for plotting this position. A scale drawing of the weld is made. The center line of the weld is drawn and another perpendicular line through the parent metal at the measured horizontal distance from the weld center line. From the intersection of this perpendicular with the scanning surface, a beam path line is drawn at 60° to the perpendicular to meet the bottom surface. A reflection line from this intersection is then drawn up toward the top surface. The position of the discontinuity is plotted by measuring along the beam path line a distance equal to the measured time base range. The position is marked on the scale drawing ("X" in Fig. 7-81f). In practice, dedicated plotting aids are used to make the process easier.

### Rod and Pipe Techniques

Figure 7-82 shows an angle beam transducer placed on the circumference of a metal bar in order to detect longitudinal discontinuities along the outer diameter (OD). The beam path length at which an echo from such a discontinuity will appear on the time base can be calculated from

$$\text{beam path} = \text{OD} \times \cos\theta$$

where:
$\theta$ = probe angle

If the time base is calibrated to a suitable range, the entire circumference can be examined by scanning from A to B in Fig. 7-82.

A similar set-up is shown in Fig. 7-83a, but this time the probe is placed on a pipe. The angle at which the beam strikes the bore of the pipe depends on the transducer angle and the ratio of the inside diameter (ID) to the OD. The beam path length A–B to the ID can be calculated from

$$\text{beam path} = \frac{t}{\cos\theta}$$

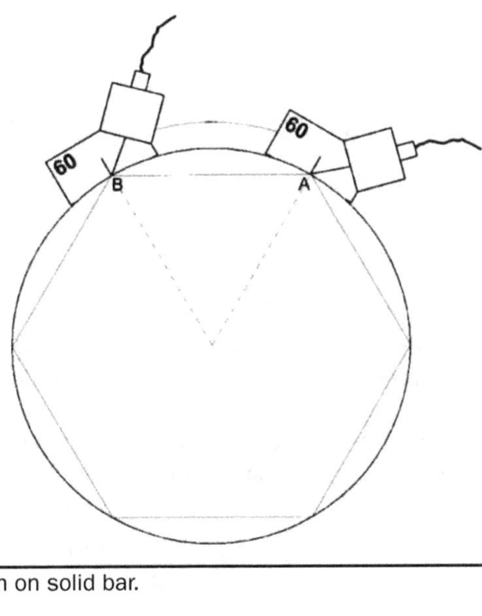


**FIGURE 7-82** Angle beam on solid bar.

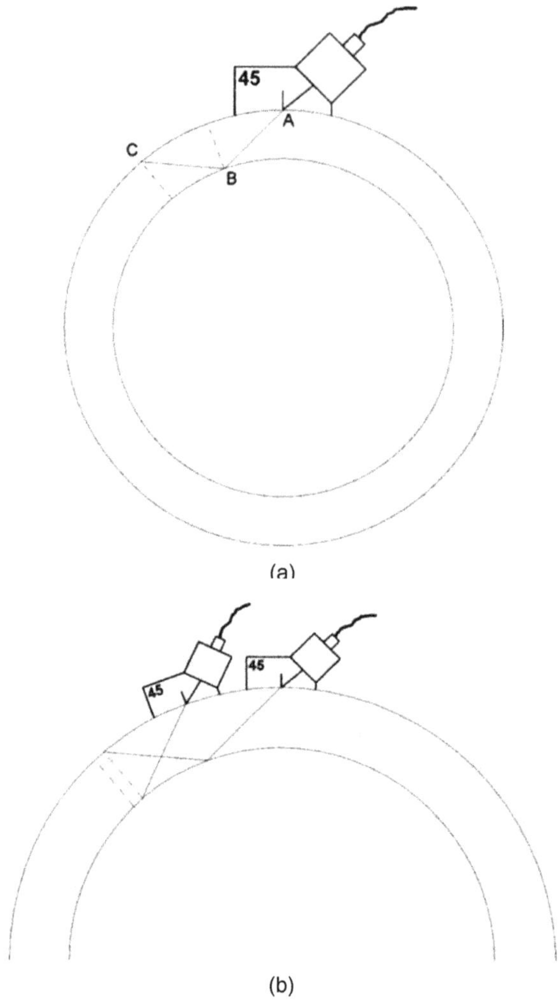


**FIGURE 7-83** Single and tandem transducers.

where:
$t$ = pipe wall thickness
$\theta$ = probe angle

When inspecting pipe for OD or ID surface breaking discontinuities it is often easier to calibrate the time base using a reference block similar to the one shown in Fig. 7-83b; the drilled hole provides both ID and OD reflecting points. If the pipe wall is very thick, the beam center never reaches the bore of the pipe. This is illustrated in Fig. 7-84.

For any given beam angle, the maximum wall thickness for a given pipe OD can be calculated from

$$t = \frac{\text{OD} \times (1 - \sin\theta)}{2}$$

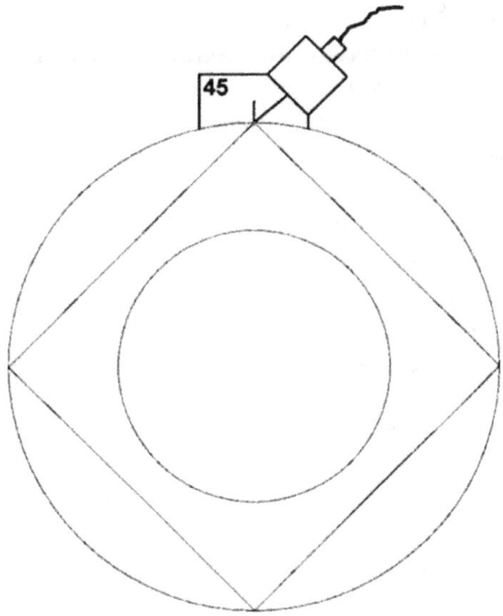


**FIGURE 7-84** Angle beam on hollow bar.

where:
$t$ = maximum wall thickness
$\theta$ = probe angle

From this equation, it is possible to derive another, which will allow the calculation of optimum probe angle to examine the bore of any given pipe. The equation becomes

$$\sin\theta = 1 - \left(\frac{2t}{\text{OD}}\right)$$

For convenience, Table 7-4 gives the maximum wall thickness that can be examined with common standard probe angles for a range of typical pipe diameters.

### *Multiple Transducer Techniques*

**Through-Transmission** Ultrasonic tests using through-transmission predate the pulse echo technique and were used in the initial experimentation procedures. By definition, through-transmission implies that two transducers are placed on opposite sides of a test

| Pipe OD, in. (mm) | Probe Angle 35°, in. (mm) | Probe Angle 45°, in. (mm) | Probe Angle 60°, in. (mm) |
|---|---|---|---|
| 4 (100) | 0.825 (21) | 0.57 (14.5) | 0.250 (6.5) |
| 6 (150) | 1.26 (32) | 0.87 (22) | 0.375 (10) |
| 8 (200) | 1.67 (42) | 1.14 (29) | 0.5 (13) |
| 10 (250) | 2.1 (53) | 1.44 (36.5) | 0.65 (16.5) |
| 12 (300) | 2.5 (64) | 1.75 (44) | 0.8 (20) |

**TABLE 7-4** Approximate Maximum Wall Thickness

piece facing each other. Whereas with dual transducers only single-sided access is necessary, through-transmission techniques require access to two sides of a component. The test material is either moved through or rotated within the sound field and scanned. It is a valuable technique when used for velocity measurements and in the characterization of material by comparison with a standard.

Certain applications lend themselves to this technique. These are usually situations in which the material has high attenuation and pulse-echo techniques are not suitable. Composite material used in the aircraft industry is one example. Other applications include production plate scanning, where banks of transducers are used with water columns (squirters) in the through-transmission mode to inspect for laminations. Other situations exist where discontinuities are close to the surface and can only be detected by observing the loss of transmitted energy.

Advantages of through-transmission are:

1. Discontinuities in highly attenuative materials may be detected by noting a reduction in the received energy.
2. The complete section can be tested. There are no initial pulse constraints.

The disadvantages are numerous. Through-transmission testing relies on energy reduction as it is "shadowed by" a material discontinuity. Energy reduction can be caused by factors other than the anomalies sought. The possibilities are:

1. Loss of couplant on either side.
2. Misalignment of transducers.
3. Change in material attenuation.
4. There is no positive indication of a reflector and therefore discontinuity depth information is not possible.
5. Variations in surface finish (roughness).

**Tandem Techniques** These are common terms for techniques using two or more transducers arranged so that reflections of the transmitted pulse are detected by a receiver (or receivers) positioned at their predicted exit points. As an example, Fig. 7-85a shows a vertical discontinuity that does not break either the top or bottom surface of a metal plate. Such a discontinuity will not reflect sound back to the transmission point. This means that a single angle beam transducer cannot be used. Instead, the arrangement of two probes, one to transmit and one to receive, is used, as can be noted in Fig. 7-85b. This is known as the "tandem" technique.

The distance A–B between the probes is dependent on the aiming point of the transmitter. The illustration shows an aiming point in the middle of the plate. If the aiming point were to be nearer the top surface, then the probes would need to be further apart. If the aiming point were to be lower, then the probes would have to be closer together.

### *Lamb Wave Techniques*

Lamb waves, like Rayleigh waves, propagate parallel to the test surface and have a particle motion that is elliptical. They occur when the thickness of the test material is only a few wavelengths at the test frequency and where the test piece is of uniform thickness. In other words, they can exist best in plate, tube, and wire.

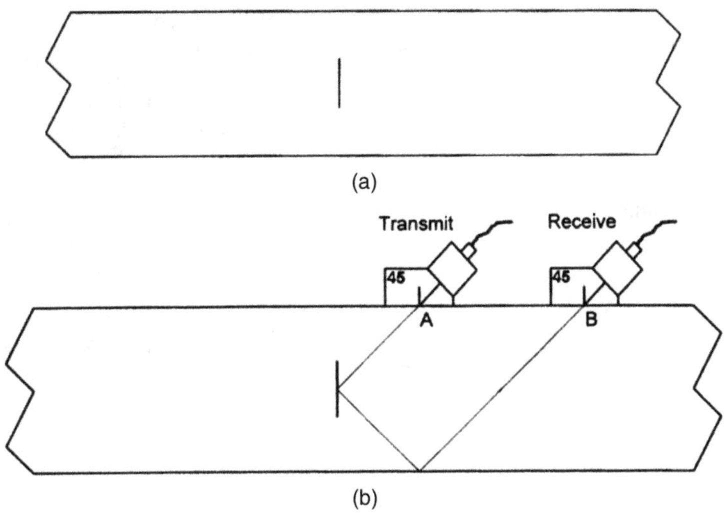


**FIGURE 7-85** Tandem transducers on plate.

The Lamb wave affects the entire thickness of the test material in such a way that it flexes. Figure 7-86 illustrates a type of Lamb wave where the crests of the wave on the near and far surfaces coincide. These are called symmetrical Lamb waves. Figure 7-87 shows another type of Lamb wave where the crest on one side coincides with a trough on the other. These are called asymmetrical Lamb waves.

These waves are generated at incident angles that depend on the test frequency and material thickness. These parameters also determine the number of modes of Lamb wave that can exist in the test material. In order to generate a Lamb wave, the velocity at which the incident compression wave in the Perspex (Plexiglas) sweeps along the interface must coincide with the velocity of the Lamb wave in the material. This is achieved by adjusting the angle of incidence $i°$. This velocity can be calculated from

$$V = \frac{V_c}{\sin i°}$$

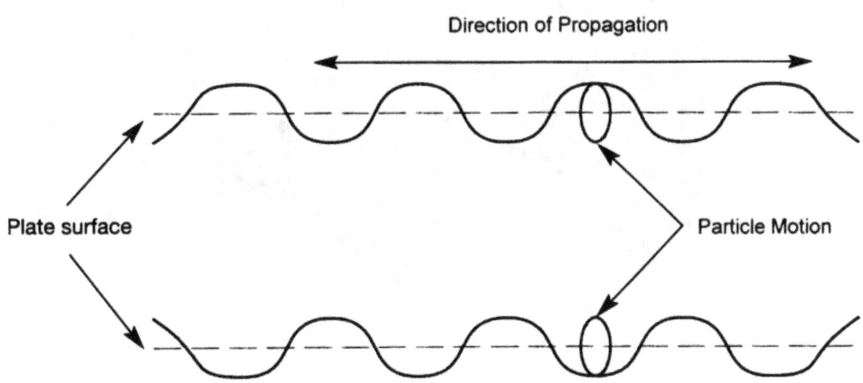


**FIGURE 7-86** Lamb wave, symmetrical.

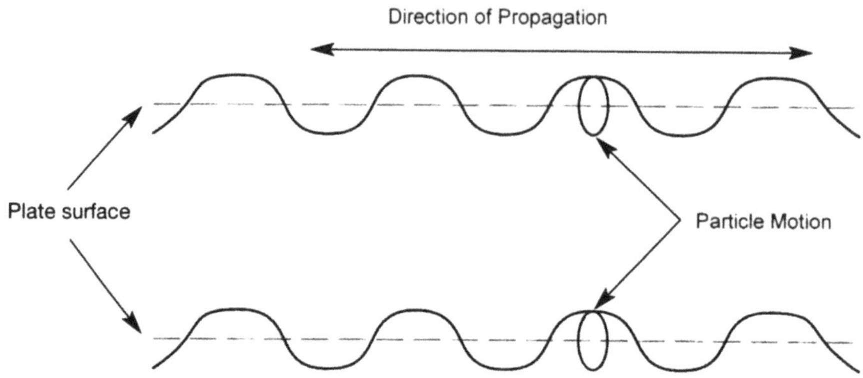


**FIGURE 7-87** Lamb wave, asymmetrical.

where:

$V$ = the velocity of the incident wavefront along the test surface

$V_c$ = the incident compression wave velocity in Perspex (Plexiglas)

$i^\circ$ = the angle of incidence in the Perspex

Figure 7-88 illustrates the above formula.

Lamb waves can travel several meters in steel, so they can be used for rapid scanning of plate, tube, and wire.

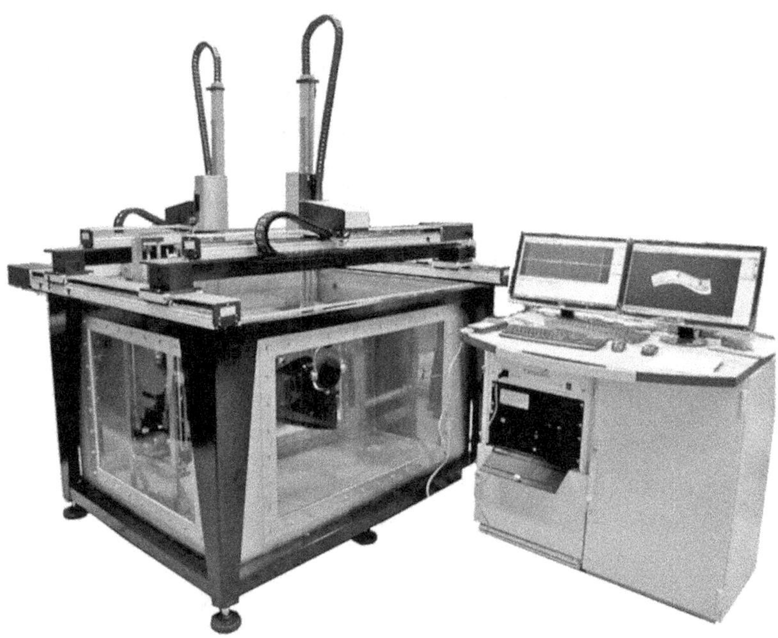

**FIGURE 7-88** Multipurpose lab scanner. (*Courtesy of TecScan*.)

### *Immersion Techniques*

Contact scanning techniques are mostly used for scanning small areas of components for in-service inspections and for scanning large components that cannot be moved to an immersion set-up. Immersion techniques are far quicker and more convenient for scanning large areas of plates, pipes, and wrought products during manufacture. The techniques lend themselves to automated scanning and recording systems (see Fig. 7-88).

There are several advantages of automated immersion testing over manual scanning techniques. These include:

1. Consistent coupling conditions and scanning capability
2. Variable beam angles and beam focusing
3. Interface gating and contour following is possible
4. Improved near-surface resolution and use of higher frequencies

Most contact scanning techniques can be used in immersion testing. In addition, the same transducer can be used for both compression wave and angle beam techniques. To generate any angle of shear wave, the transducer is simply tilted to the appropriate incident angle derived from Snell's law.

Disadvantages include:

1. Setting up is more complicated.
2. The part must be compatible with water.
3. Air bubble formation interferes with the test.

Figure 7-89 illustrates a typical immersion set-up for compression wave scanning. The water path distance (w) is important because there will be multiple echoes from the

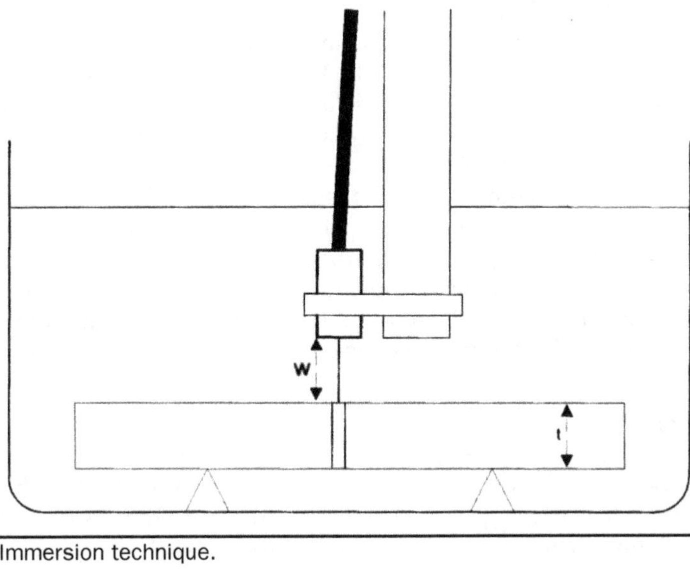


**FIGURE 7-89** Immersion technique.

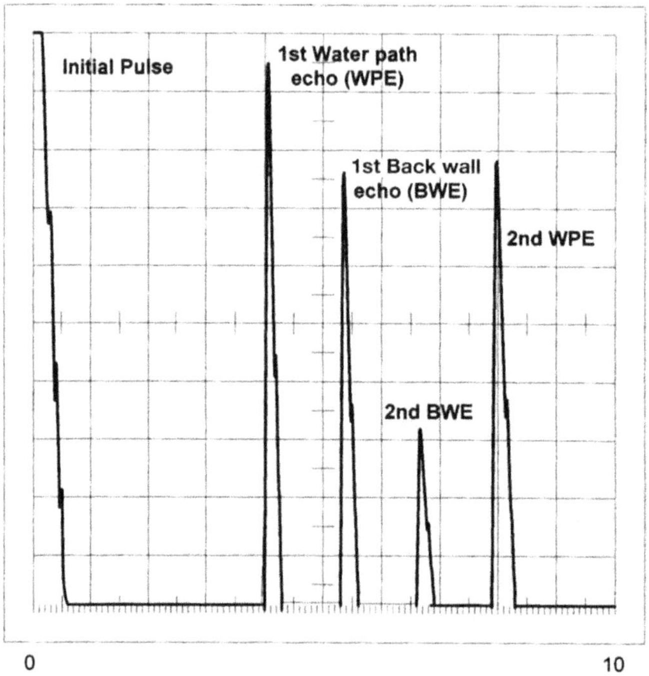


**FIGURE 7-90** A-scan of immersion technique.

top surface of the part (entry surface). The spacing between these multiple echoes must be longer than the time taken by the transmitted sound to travel the back wall and return. As a rough guide, for steel and aluminum the minimum water path should be about one-quarter of the material thickness. Since sound in water travels at about a quarter of the speed it does in those metals, the back reflection (back wall echo) will appear just before the water multiple. Figure 7-90 shows a typical trace for the above immersion set-up. In this trace, no delay has been used. By using the delay control to get rid of the water path, and calibrating the time base for a suitable range, the trace shown in Fig. 7-91 can be obtained.

This trace can be compared to that shown in Fig. 7-76b in the contact scanning techniques above. Note that the entry surface echo in Fig. 7-91 occupies less space than the initial pulse shown in Fig. 7-76b and therefore offers better near-surface resolution.

Figure 7-92 shows the same transducer tilted to an angle of incidence to generate a shear wave beam at 45° to the normal. It can be seen that other refracted angles can be generated by simply changing the incident angle.

### *Normalization*

Normalization is the name given to the process of ensuring that the transducer is actually perpendicular to the entry surface before any other calibration procedure is commenced. The name given to the probe holder in immersion testing systems is the "manipulator." The manipulator allows rotation of the probe in two directions

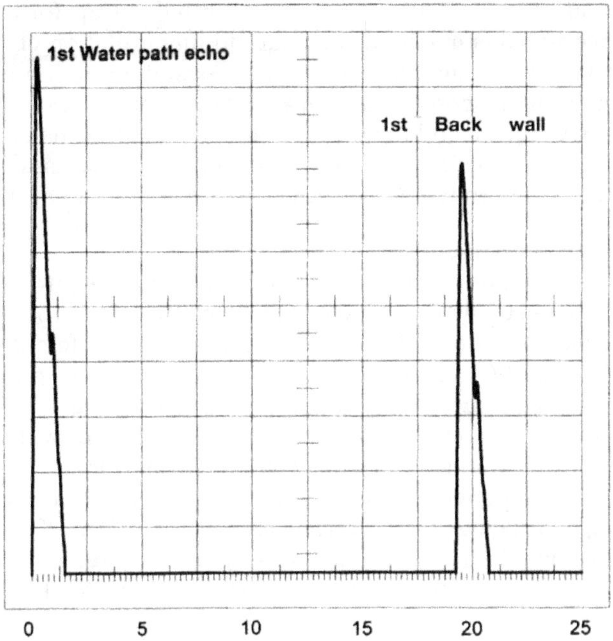


**FIGURE 7-91** Delayed immersion A-scan.

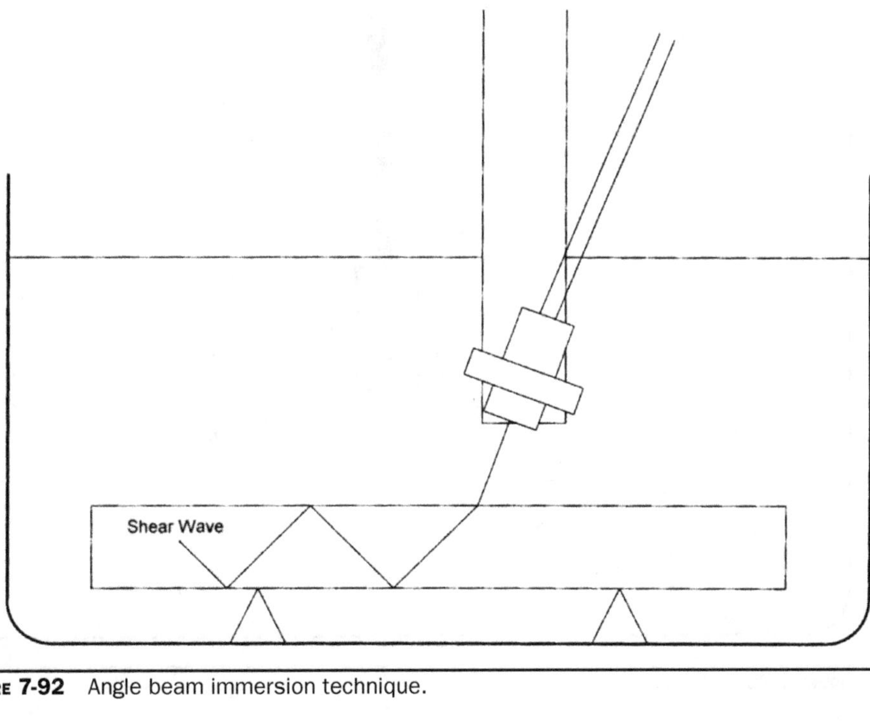


**FIGURE 7-92** Angle beam immersion technique.

perpendicular to one another. During the initial set-up for an immersion test, the entry surface signal is maximized by careful manipulation of the probe in each axis. As the entry surface echo maximizes in an axis, the rotation is stopped in that axis. When a maximum response has been achieved in both axes, the probe is perpendicular to the entry surface and the transducer has been "normalized."

The normalized position would be used for the compression wave techniques and as a starting point from which to set the incident angle for shear wave techniques.

### *Interface Triggering (Gating)*

Interface triggering or "gating" is often used when the entry surface, because of the component geometry, is not at a fixed distance from the probe face. For example, the stepped surface shown in Fig. 7-93 presents three different water path distances (surfaces A, B, and C). Interface triggering starts the sweep as the first entry signal arrives. It acts as an automatic delay control that always ensures that zero on the trace represents the entry surface of the test piece regardless of water path. This is particularly useful, for example, if a round component is being scanned with a spiral scan pattern as the component is rotated past the scanning head. If the component is not truly round, or is being rotated eccentrically, without interface triggering, the whole signal pattern would be moving left and right. Interpretation of the trace visually or automatically under those circumstances would be difficult.

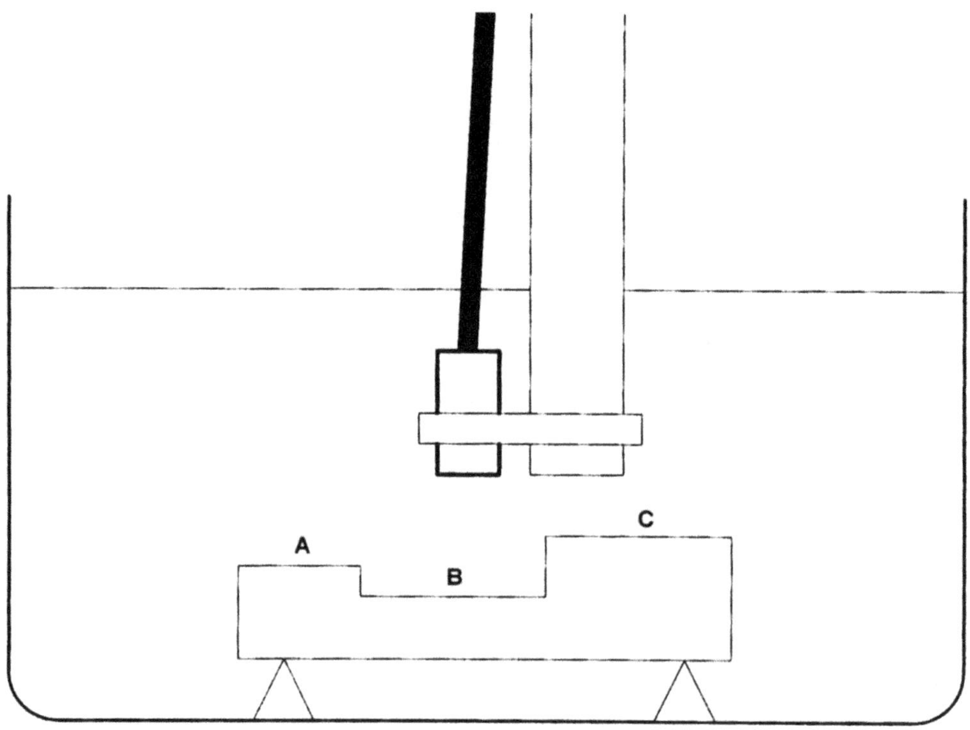


**FIGURE 7-93** Stepped surface.

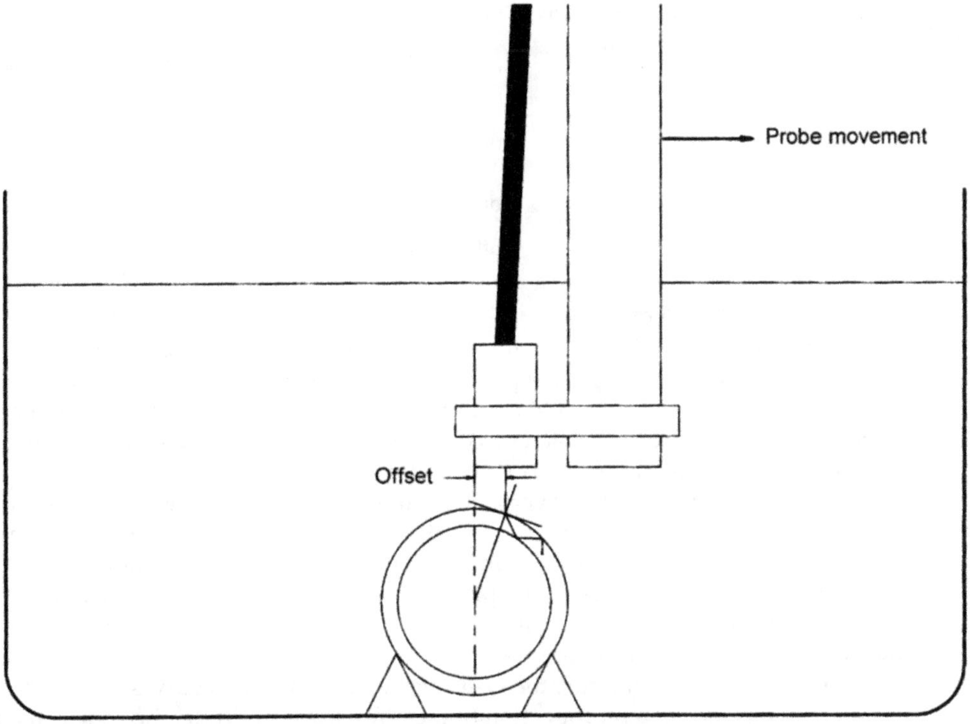


**FIGURE 7-94** Offset transducer.

### *Offset Method for Generating Shear Waves*

When carrying out immersion testing of tubular components for longitudinal discontinuities such as the one shown in Fig. 7-94, it is not necessary to tilt the probe in order to generate shear waves of a given angle. The appropriate angle of incidence can be achieved by offsetting the probe from the center line by an appropriate amount. The amount of offset for a given required incident angle can be calculated using the following equation:

$$\text{offset} = \frac{D}{2} \times \sin i^\circ$$

where:
$D$ = component OD
$i^\circ$ = required angle of incidence

**Example 10** Calculate the offset required to generate a 45° shear wave in a 50-mm-diameter steel rod. Given that the shear wave velocity in steel is 3,240 m/sec, and the velocity of sound in water is 1,480 m/sec, from Snell's law, the required angle of incidence is

$$\sin i^\circ = \frac{1{,}480 \times \sin 45^\circ}{3{,}240}$$

$$\sin i^\circ = 0.3230$$

$$i^\circ = 18.84^\circ$$

Putting the calculated angle into the formula

$$\text{offset} = \frac{D}{2} \times \sin i^\circ$$

$$\text{offset} = \frac{50}{2} \times \sin 18.84^\circ$$

$$\text{offset} = 25 \times 0.3229$$

$$\text{offset} = 8.03 \text{ mm}$$

### Discontinuity Sizing Techniques

The operating procedure for carrying out the inspection of a specific component must include instructions for setting test parameters that will ensure detection of a given minimum reflector. This can mean that discontinuities will be detected that do not render the component unfit for service. Each discontinuity detected must, therefore, be evaluated against the applicable inspection code or standard. In most cases, the acceptance criteria are set in terms of the length or vertical extent of the discontinuity. In order to correctly evaluate the results, it will be necessary to derive these critical dimensions from the ultrasonic information. Techniques for doing this are known as "sizing techniques."

The techniques available to the practitioner for assessing the severity of discontinuities fall into two categories. First, there are those procedures that quickly "screen" the echo on the trace with reference to an amplitude standard. Examples are the area/amplitude reference standards, the DAC curves, and the DGS curves. In most codes, if the signal exceeds a given proportion of the reference amplitude, the code calls for further investigation (see DAC, page 357).

The second category is those procedures that attempt to measure or plot the through-thickness dimension and length of the discontinuity. The techniques include intensity drop, maximum amplitude, and time of flight diffraction (TOFD).

**Intensity Drop Technique** The intensity drop technique uses the beam spread to determine the edges of the reflector. Figure 7-95 shows a 20-dB beam spread diagram on a

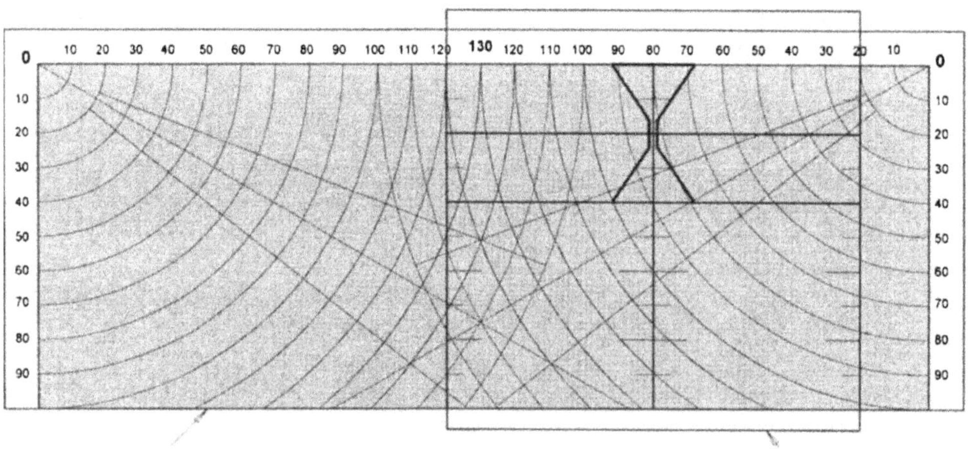

**FIGURE 7-95** Beam plot card.

card to which a transparent cursor has been fitted. The cursor has a horizontal line scribed to coincide with the horizontal scale on the card, and it also represents the scanning surface of the component. A vertical line is scribed to denote the reference point (e.g., weld center line). In this example, a weld is being examined and the weld profile has been drawn in wax pencil on the cursor. Note that a mirror image of the weld has also been drawn to allow full skip ranges to be plotted.

### *Sizing Procedure—Through-Thickness Dimension*

Figure 7-96 shows the plan view of a weld with the transducer at a position where a maximum echo height has been obtained from a discontinuity. At this point, the echo height is adjusted to the reference level (i.e., 80 percent) and the horizontal distance (HD1) measured from the probe index to the weld center line. The corresponding time base range (TB1) is also noted.

The transducer is scanned forward until the signal drops to the intensity drop wax pencil line. At this position, the bottom edge of the plotted beam will be reaching the top of the discontinuity. The new horizontal distance (HD2) and time base range (TB2) are recorded.

Last, the transducer is scanned backward, through the maximum again and back until the signal again drops to the wax pencil line at HD3. The distance and range (TB3) are recorded. This position represents the top edge of the beam just striking the bottom of the discontinuity.

There are now three sets of data for the through-thickness dimension of the reflector. These are the probe position (HD) and time base range (TB) for top, middle,

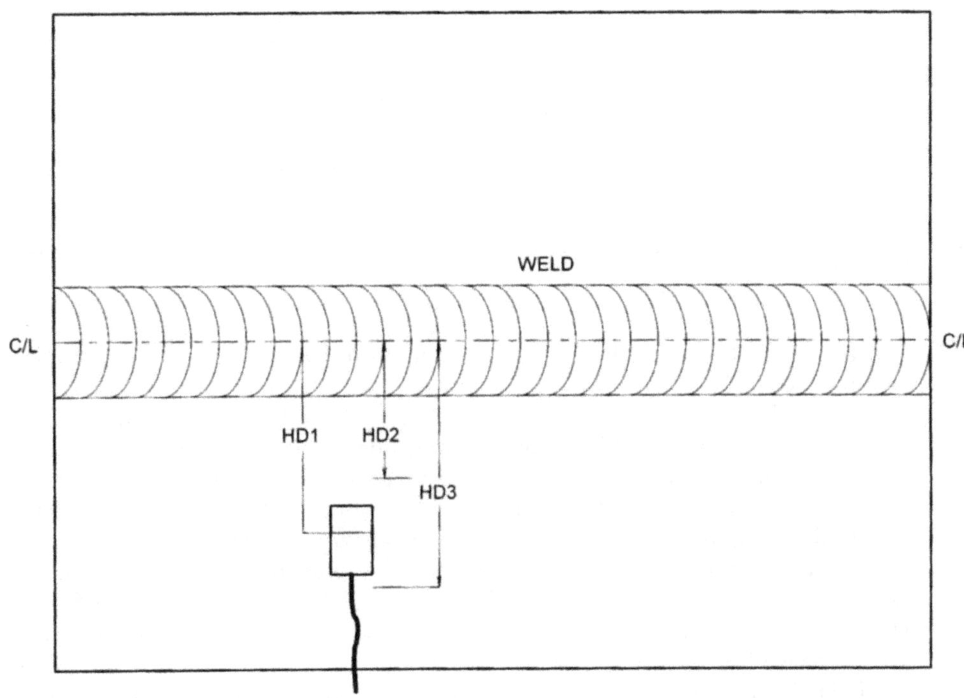


**FIGURE 7-96** Weld center line.

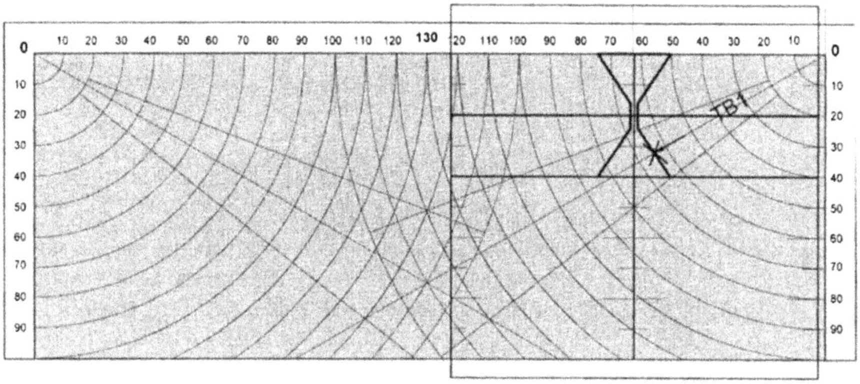


(a)

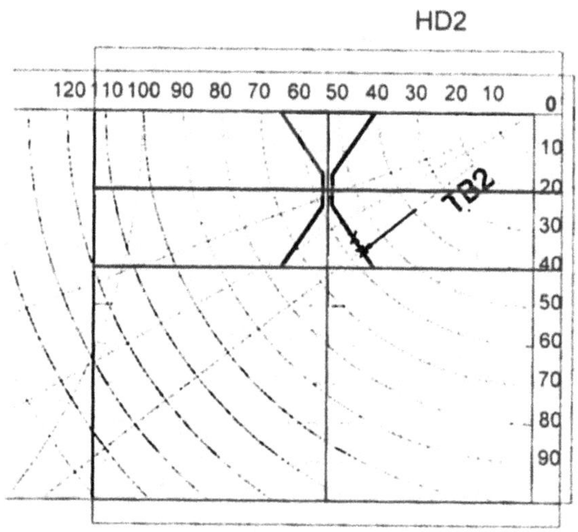


(b)

**Figure 7-97** Plotting the discontinuity.

and bottom of the discontinuity. These data are now transferred to the cursor of the plotting aid.

Figures 7-97a, b, and c show the steps for plotting the discontinuity. In Fig. 7-97a, the cursor has been set with the weld center line positioned at a distance HD1 on the card's horizontal scale. Going down the beam path scale, a cross has been marked with a grease pencil at the time base range for the middle of the discontinuity (TB1). Note that in this case the distance TB1 has put the cross in the mirror image on the cursor. This means that the reflection has come from beyond the half-skip position when the beam is on its way up to the full skip position. Because the weld profile has been sketched onto the cursor it is immediately obvious that this discontinuity is close to the weld side wall.

In Fig. 7-97b, the cursor is shown to be repositioned at HD2 and a new cross marked at range TB2. Note that this cross is marked where the bottom edge of the beam on the

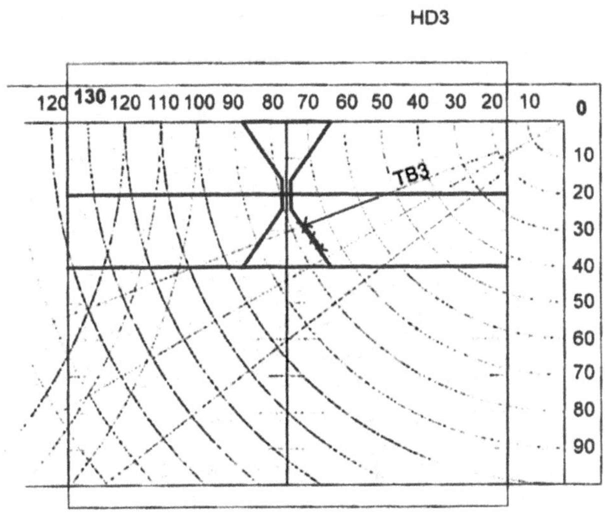


(c)

**Figure 7-97** *(Continued)*

beam spread diagram intersects with the beam path scale. The new cross marks the top of the discontinuity. Note that it also lies near to the side wall.

The bottom of the discontinuity is plotted as shown in Fig. 7-97c. The distance HD3 and range TB3 place a cross where the top edge of the beam intersects the beam path scale. Again, the cross falls close to the side wall. The three crosses can now be joined to show the extent of the discontinuity in the cross section of the weld. This can be measured and evaluated against the relevant acceptance standard.

### *Sizing Procedure—Discontinuity Length*

Using the intensity drop method to measure the length of the discontinuity is much simpler than the procedure for the through-thickness dimension. All the measurements and marks are made on the test surface. Figure 7-98 shows the transducer positions for the measurement.

Position 1 is where the transducer detects a maximum echo. The position is marked with grease pencil. The transducer is scanned sideways until the echo has dropped by 6 dB and the position marked on the surface (position 2). The transducer is then scanned through position 1 and on to position 3, where the signal has again dropped by 6 dB. Position 3 is marked on the surface. The distance "L" between positions 2 and 3 is the length of the discontinuity. Care should be taken to ensure that the –6 dB point is the "end" of the discontinuity.

### *Maximum Amplitude System*

The maximum amplitude system (MAS) uses the same plotting aid as the intensity drop system. However, with MAS only the beam center is used to plot the extent of the discontinuity. The method is particularly suitable for sizing discontinuities that are multifaceted, such as cracks or slag inclusions. Figure 7-99 shows an angle beam transducer aimed at an internal crack-like target.

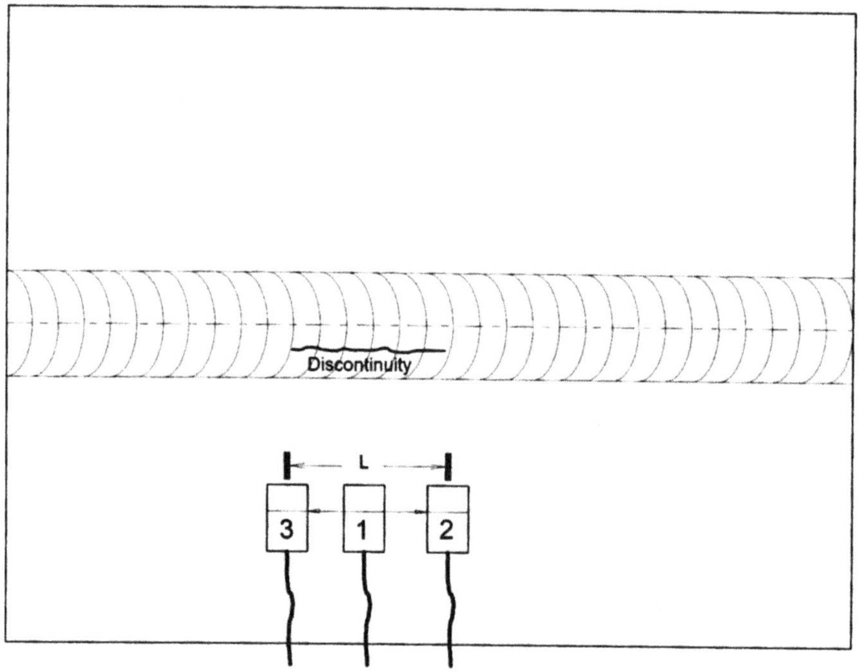


FIGURE 7-98 Plan view of discontinuity.

Across the beam, various facets of the target are oriented perpendicular to that part of the beam. The result is that several echoes get back to the receiver in a cluster over a short time period. This gives a signal pattern similar to the one shown in Fig. 7-100. The signal from each facet of the discontinuity reaches a maximum as the beam center moves onto it. If the transducer is moved, forward or back, from the position on the test surface at which the overall signal pattern is maximized, the signal pattern as a whole decays. So far everything is the same as for the intensity drop method. However, if the

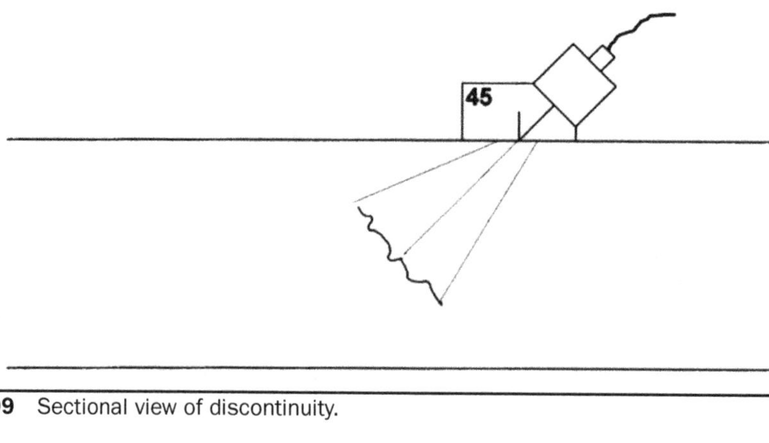


FIGURE 7-99 Sectional view of discontinuity.

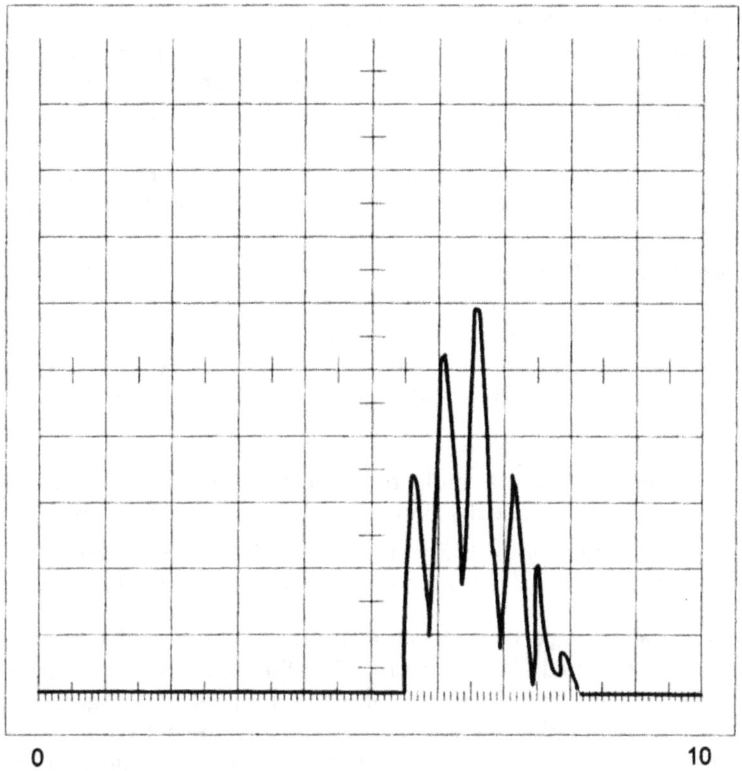


**Figure 7-100** A-scan of discontinuity.

dynamic signal pattern is studied in detail during this decay, signals from the individual facets will be seen to rise and maximize as the beam center reaches that area of the discontinuity. As the transducer is scanned right across the discontinuity region, each facet in turn will grow, maximize, and decay. Each maximum represents a point on the face of the discontinuity.

The sizing procedure starts in the same way as the procedure for intensity drop by maximizing the overall signal from the discontinuity. The horizontal distance to the weld center line and the time base range are recorded as before. As the transducer is scanned forward from this position, the signal pattern is scrutinized to detect the rise in signal of a facet echo while the main pattern is falling. As each facet echo maximizes, the HD and TB distances are recorded until the last facet has started to fall and the overall signal is lost.

The process is repeated on the backward scan from the overall maximum position until the last maximum in that direction has been recorded. The coordinates of all the recorded points can then be transferred to the cursor of the plotting aid.

The set-up for the plotter is the same as that shown in Fig. 7-95, except that only the beam path scale is used to plot the points. There will be more points to plot for each discontinuity than for the intensity drop method. Joining these points on the cursor traces the shape and plane of the discontinuity after the fashion of "connect the dots" picture books for children.

During the scanning procedure, as each maximum is being detected, the gain can be increased to bring in the weaker signals from the smallest facets. If this is done, the last maximum detected is likely to be a tip diffraction signal from the end of the discontinuity. This signal was described by Sproule to be –30 dB below the amplitude of a corner reflector at the same depth. In practice, this can be difficult to identify, but if it does appear, it very accurately pinpoints the extremes of the discontinuity.

### *Tip Diffraction Techniques*

The tip diffraction techniques follow on from the "last maximum" in the maximum amplitude technique. If the extremities of the discontinuity can be accurately plotted, it follows that the discontinuity can be "sized." Since the diffraction signal originates at a "tip," it would appear to be an ideal feature to exploit. There are two basic approaches to tip diffraction sizing:

1. Using a standard angle beam transducer to carry out an additional scan specifically to size discontinuities that have already been detected
2. Using the time of flight diffraction (TOFD) technique in which two angled compression wave transducers are used to both detect and size in one pass

**Using a Standard Angle Beam Transducer** When a beam of sound strikes the tip of a reflector, the reflected energy "diffracts" at the tip to form a spherical wave, as shown in Fig. 7-101. The spherical wave will eventually reach a large part of the component surface. This means that the diffracted energy can be detected over a much larger area than the main beam, which is very directional. The diffraction energy is greatest as the beam center encounters the tip. The energy as this point is reached is typically 30 dB less (i.e., –30 dB) than a corner reflector at the same depth.

This means that the tip diffraction signal is weak and may only be two or three times the noise level (i.e., the signal-to-noise ratio is typically around 2:1–3:1). These small signals can be difficult to identify on the screen. Nevertheless, the signal can frequently be identified, and in these cases it becomes a useful sizing tool.

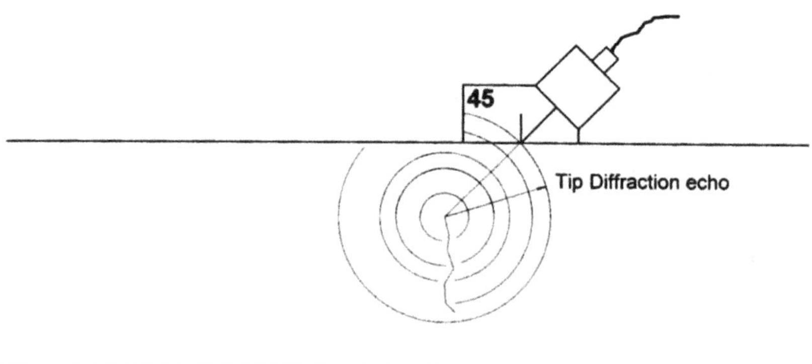


**Figure 7-101** Tip diffraction echo.

The procedure is similar to the maximum amplitude method. Once the discontinuity has been identified and positioned within the component, the transducer is arranged to obtain a corner reflection at a depth close to the discontinuity depth. The signal from the corner reflector is adjusted to a reference height (typically 30 percent FSH). The gain is then increased by 30 dB (+30 dB) and the transducer moved to the discontinuity region. As the transducer interrogates the discontinuity, the screen is studied for a signal that rises to the reference level as the main echo signal decays. As this signal reaches its maximum, the coordinates are noted and the tip position plotted as before.

It is often considered preferable to use an unrectified (RF) display to carry out tip diffraction techniques. This is because there is a phase reversal between diffraction signals originating at the top tip and those originating at the bottom tip of a discontinuity. With small discontinuities, it is sometimes possible to identify top and bottom of the discontinuity at the same time.

**Time of Flight Diffraction (TOFD) Technique** The TOFD technique, first used by M. G. Silk in 1977, uses tip diffraction to identify the top, bottom, and ends of a discontinuity in one pass. Silk chose to use an angled compression wave for the TOFD technique rather than a shear wave, for two reasons. First, the tip diffraction signal is stronger than a shear wave diffraction signal, and second, a lateral wave is produced that can be used to measure the horizontal distance between the transmitter and the receiver.

The tip diffraction signal is generated at the tip of the discontinuity—effectively a "point" source. According to Huygens, a point source produces a spherical beam. Figure 7-102 shows both the lateral wave and a diffraction beam from the tip of a reflector.

Figure 7-103 shows a typical TOFD transducer set-up on a component with a vertical discontinuity. There are four sound paths from the transmitter to the receiver. Path "A" is the lateral wave path traveling just below the surface. Path "B" is the tip diffraction path from the top of the discontinuity. Path "C" is the tip diffraction path from the bottom of the discontinuity, and path "D" is the back wall echo path.

Figure 7-104 shows a typical unrectified trace for the four signals. Note that the phase relationships A and C are in opposite phase to B and D. The important difference

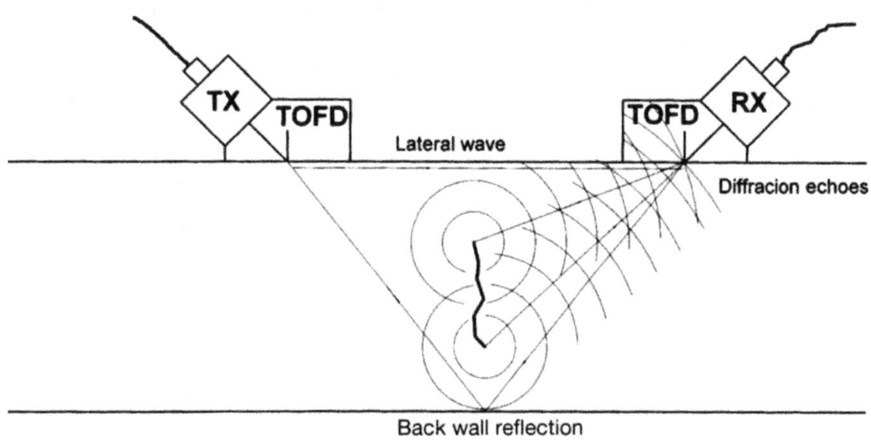


**Figure 7-102** TOFD.

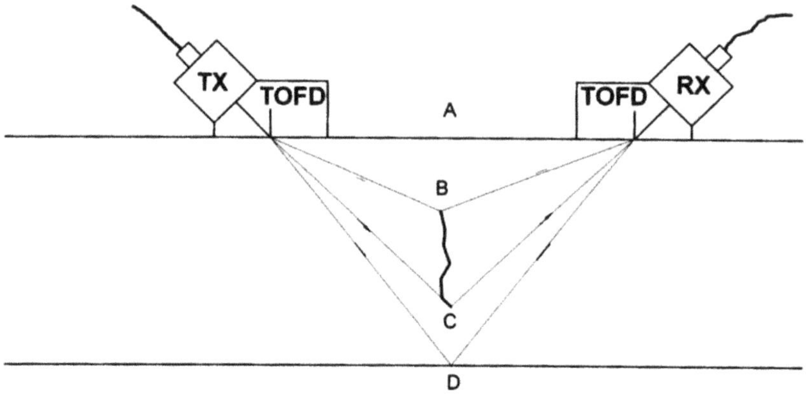


FIGURE 7-103 TOFD.

to note is between B and C—the top and bottom diffraction signals are in opposite phase. This phase difference allows the practitioner to identify those points.

Assuming that the diffracting tip is centered between the two transducers, the depth of the tip below the surface can be calculated from

$$\text{depth} = \sqrt{\left(\frac{BPL}{2}\right)^2 - \left(\frac{HD}{2}\right)^2}$$

where:
BPL = beam path length for the signal in question
HD = beam path length for the lateral wave

The distance measurements taken from the ultrasonic trace must be made from the same part of each waveform. In the example trace shown in Fig. 7-104, the largest half-cycle would be selected. For signals A and C, this is negative, and for signal B, positive. Advances in computer technology have made it possible to carry out all the calculations and for plotting to be handled automatically and stored for subsequent evaluation. The method that has been chosen to display this TOFD data presents the information in a special "B-scan" form that is easy to assimilate. The way in which the positive and negative half cycles are displayed needs explaining.

An echo arriving at the receiver is a pulse of a certain pulse width and amplitude. In conventional B-scan displays, this pulse is displayed as a bright spot whose diameter

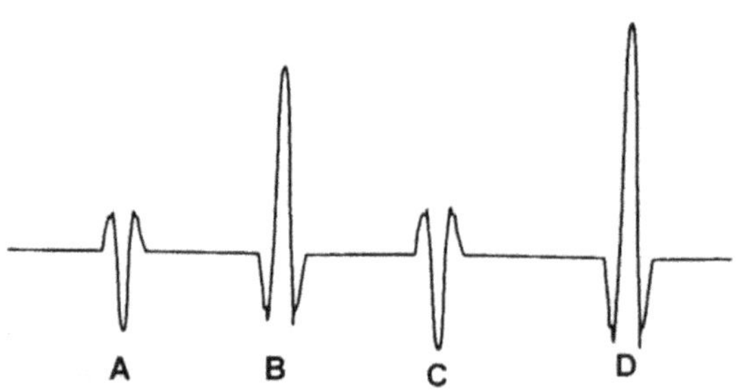


FIGURE 7-104 TOFD RF signals.

is proportional to the pulse width and whose brightness is proportional to the signal amplitude. In some ways, it is like a broad pencil tip that can be used to draw pictures in light or bold broad strokes. The pulse is really a short burst of a few cycles of alternating waveform. In the TOFD system, the waveform is depicted in grayscale, with positive half-cycles tending toward white, and negative half-cycles tending toward black (see Fig. 7-105). This allows particular half-cycles to be identified for measurement purposes, and phase changes to be recognized for determination of top or bottom echo.

Figure 7-105 shows a typical computer screen for a TOFD inspection. The image shows details of the component (in this case, a weld) as well as the TOFD B-scan image and an A-scan trace. In this image, left to right represents the component thickness, and the vertical dimension represents scan length.

The A-scan trace shown corresponds to a slice through the weld at the location indicated by the cross hairs of the cursor. The striped band on the left of the TOFD image represents the lateral wave, and the bold striped band to the right of the image represents the back wall signal. The difference in boldness is due to the different signal amplitudes. Following the horizontal cross hairs and about halfway between the lateral wave and back wall "stripes," a series of faint horseshoe-shaped stripes can be seen. These are diffraction signals from a small discontinuity. The A-scan trace shows the signal clearly.

In this example, the discontinuity has a very small dimension in the through-thickness dimension, but close study of the A-scan shows a small phase shift in the

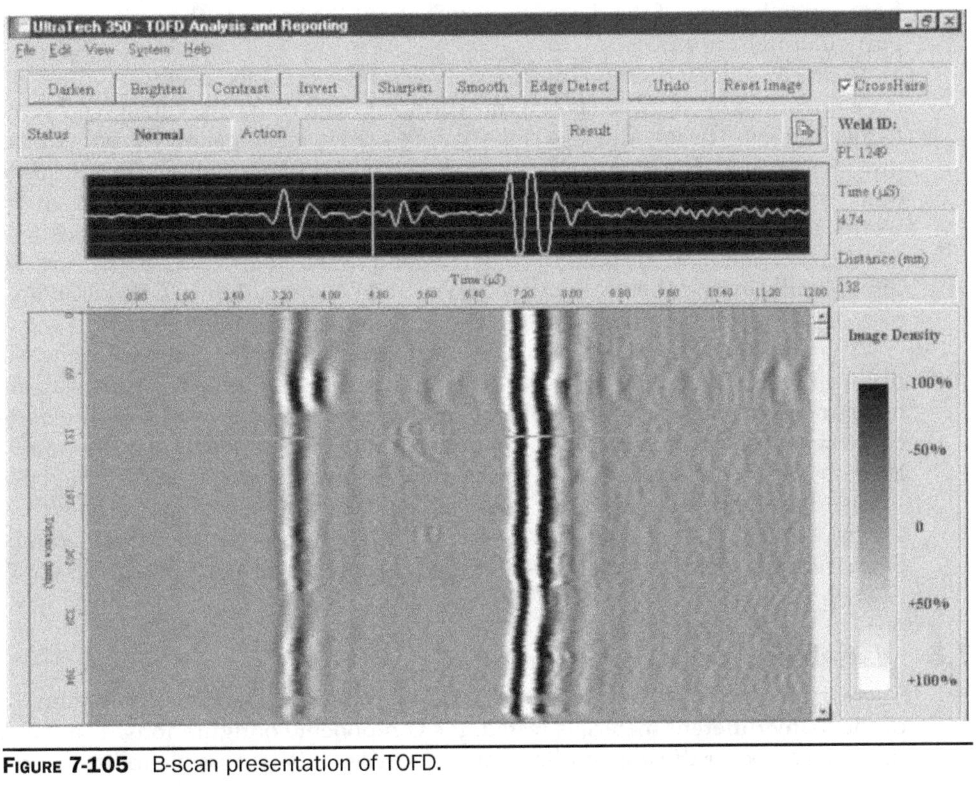


FIGURE 7-105 B-scan presentation of TOFD.

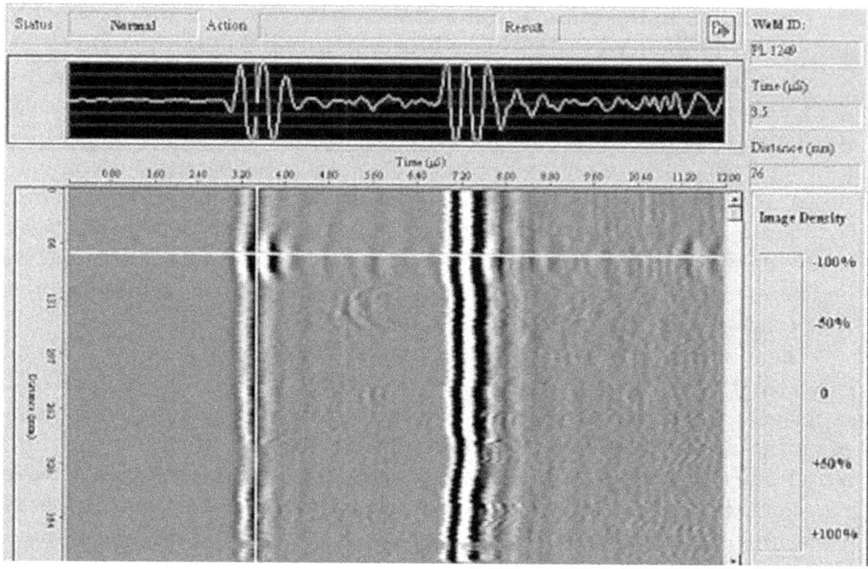


**Figure 7-106** B-scan presentation of TOFD.

last half-cycle of the discontinuity signal. This tells the practitioner that the distance from top to bottom of the discontinuity is about the same as the pulse length for this particular discontinuity.

A much bolder indication can be seen toward the top of the lateral wave line, suggesting a discontinuity at, or just below, the surface. In Fig. 7-106, the cursor has been moved to this location. The lateral wave signal can be seen to be longer and stronger than at the previous location. The fact that the wave shape stays in phase suggests that the diffraction echo, which is extending the signal, has the same phase as the lateral wave. In other words, it is a bottom tip signal. However, it is not possible in this case to see where the lateral wave ends and the bottom tip begins, and so it is not possible to say how deep the discontinuity extends below the surface. The TOFD method is limited in its ability to size near-surface discontinuities when the depth is similar to the pulse length.

The transducers used in these TOFD techniques are angled compression wave transducers (refracted longitudinal wave). The common angles used are 60° and 70°, although other angles may be used if the component thickness makes it necessary. The design and construction of the transducer is important in order to promote a good lateral wave. Previous theory has suggested that a shear wave should also exist in the component, and this is true—it does. Figure 7-107 shows a little more of the trace for the above example. On the extreme right of both the A-scan and TOFD B-scan, the shear wave can be seen. Since it arrives well after the other signals, it does not present a problem in this application.

## 7.5 Variables

There are a number of variables that can influence an ultrasonic examination. Because of the many different materials tested, the component configurations, and the anomalies that can occur during an ultrasonic inspection, it is impossible to list or predict all

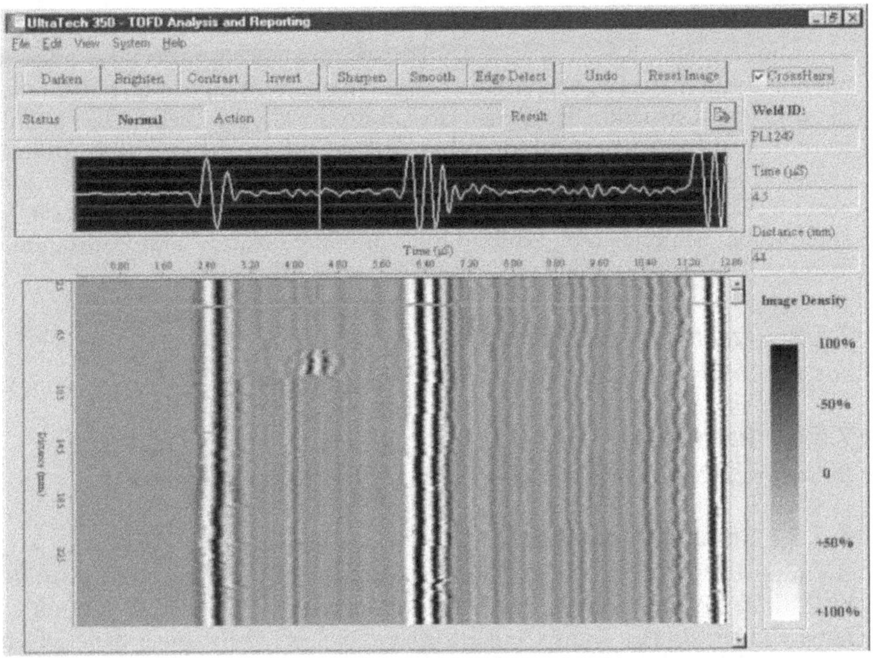


**FIGURE 7-107** B-scan presentation of TOFD showing shear wave.

of these variables. The following subsections deal with some of the more common variables that may have an effect on the results of a test.

## 7.5.1 Temperature and Velocity

Temperature is an important factor when considering ultrasonic examination. Temperature affects the velocity of sound in most materials to a greater or lesser degree depending on the material. Water, for instance, undergoes velocity changes with temperature. The velocity of water at 68°F (20°C) is approximately 1,480 meters per second, whereas at 86°F (30°C) the velocity is approximately 1,570 meters per second. At the other end of the scale, water at approximately 34°F (1°C) is approximately 1,414 meters per second. Velocity changes in metals are not as dramatic as with water, but changes do occur. When conducting examinations, consideration should be given to temperature variations between the calibration standard and the component. Some specifications require that the temperature difference of the calibration standard and the component be within 20°F (–7°C). Of greater influence than the component is the temperature in the Plexiglas wedge or delay line. Temperature variations can cause beam angle changes and/or alter the apparent delay on the time base. When conducting examinations at elevated temperatures, special high-temperature transducers and couplants are usually necessary. Calibration blocks need to be heated to a temperature similar to that of the component to be examined prior to calibration, and the transducers should be allowed to warm up to the examination temperature before using them to calibrate the system. Temperature variations will also result in dimensional changes to the part, which must be taken into consideration when calibrating and testing material.

### 7.5.2 Attenuation

In addition to the change in velocity, temperature can also affect the amount of attenuation in some materials. Apparent changes in attenuation can be indicative of changes in the material structure. Hardened steel, for example, may exhibit less attenuation than its untreated counterpart. Sudden changes that are noted in signal amplitude can yield valuable data about the component being examined (see Hydrogen Embrittlement in Section 7.7.5). Microporosity in the material may have the same effect on the apparent attenuation. Material that exhibits high attenuation may require examination using lower test frequencies due to the probable larger grain size.

### 7.5.3 Frequency and Grain Size

For details regarding this variable, see Scatter in Section 7.2.

### 7.5.4 Resolution

Pulse length can affect the resolution characteristics of the system.

### 7.5.5 Surface Conditions

An important variable is that of surface condition. The differences in surface finish can result in large variations in the results of an examination. Paint or other coatings can have similar effects. Tightly adhering coatings on the surface generally allow good transfer of the energy. Loose or flaking coatings are undesirable and should be removed prior to conducting the examination. When calibrating the equipment for reference sensitivity on critical applications, it is essential to evaluate the component for any energy losses due to surface condition and apparent attenuation variations. The procedure for this is fairly simple and is performed by using two transducers in a "pitch–catch" technique. See Transfer Correction Technique in Section 7.4 for details of this procedure.

### 7.5.6 Diameter Changes

Changes in the diameter of the test surface can result in changes in test sensitivity. The effective transducer size is limited to its contact area. It is highly desirable to perform the system calibration on a surface with a diameter similar to the one being scanned on the component under test. Compensation for diameter changes should be made by either adding or subtracting gain after calibration, prior to examination of the component.

### 7.5.7 Contact Pressure and Couplant

The amount of couplant used and the contact pressure on the transducer can create differences in signal amplitude. Too little couplant will leave the surface dry and therefore create an air boundary between the transducer and the component surface. Excessive pressure can squeeze the couplant from under the transducer.

### 7.5.8 Dendritic Structures

Dendrites are branch-like grains that exist in certain metal structures and can cause problems, particularly in stainless steel welds. These dendrites form in the direction of heat dissipation during the welding process. A single grain can grow from one weld pass to the next, leaving elongated grains that can effectively redirect the sound energy.

Indications that are plotted to originate from, e.g., the fusion line may actually originate from the root of the weld if the energy has been redirected. Special procedures including the use of refracted compression wave transducers may be necessary to reduce this effect.

### 7.5.9 Gain

The use of excessive gain can exaggerate otherwise insignificant indications. This emphasizes the need for precise calibration.

### 7.5.10 Other Factors

Other factors such as transducer frequency, diameter, and angle can affect the examination results. It is necessary to follow a qualified procedure when carrying out an examination, particularly when repeatability is an issue. Ultrasonic examination requires careful consideration of all the variables. The practitioner needs to be mindful of these variables and others that may present themselves. Attentiveness, awareness, and the ability to recognize anomalies are important.

## 7.6 Evaluation of Test Results

In order that the terminology used in this section be understood, the flowchart in Fig. 7-108 should be considered.

When NDT is specified, the following information must be provided:

1. The component description
2. The test method
3. The specification for the test

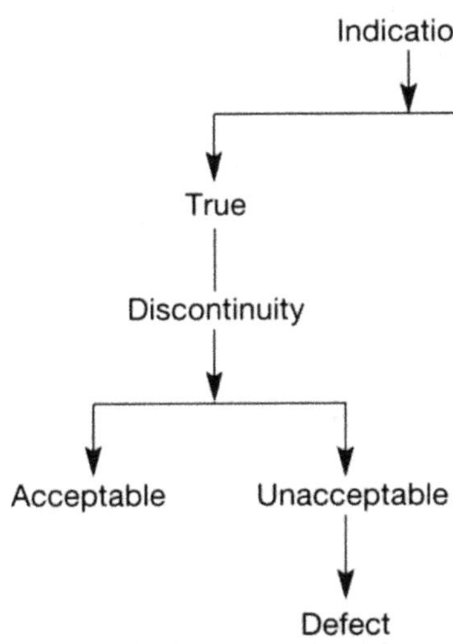


**FIGURE 7-108** Procedure flowchart.

The specification for the test is most important. If there is no agreed upon definition regarding the type, size, and quantity of discontinuities, meaningful accept/reject determination will not be possible. The specification used should consider the component's "fitness for purpose." This means that the component should not have any discontinuities that would lead to its failure. This determination should be made by the designer of the specific component, who will use "fracture mechanics" to make this judgment. Evaluation of any indication should be made in consideration of the code or specification referenced in the purchase order or contract.

### 7.6.1 Reporting

Unless a report is generated at the conclusion of an examination, there will be no record of it having been conducted. Examination repeatability is important, especially where the condition of a component is to be monitored periodically. Test reports should be developed that require certain information. A situation may arise where specific test reports are unavailable and the examiner may have to improvise.

Always ensure that as a minimum the following details are recorded clearly and concisely.

### 7.6.2 Identification

Date, time, and place of examination

Examiner's name and certification level

Component examined and its serial number if applicable

Procedure, specification, or standard to which the test was performed. Note any revision number, changes, or deviations from the procedure.

#### *Equipment*

Instrument used, including serial number

Transducers used, including frequency, serial number, and angles

Calibration standards or reference block(s)

Couplant used—include batch number where appropriate

Calibration

Time base (range and units per division)

Calibration sensitivity and DAC curve (if used)

Scanning sensitivity

Reference level for recording

Examination details

Area scanned

Limitations and interferences

Percentage complete coverage

#### *Results*

Indications noted—percent DAC (as appropriate), location of indications

Classification of discontinuities (if required)

Scale drawing or plot of indications

Comments—surface condition, temperature, etc.

*Note:* Reports should be concise and accurate. The person reading the report may not be very familiar with UT. A drawing or sketch is easier to follow than a lengthy description.

## 7.7 Applications

Product forms commonly subjected to ultrasonic inspection can be wrought, cast, welded, composite, or other materials. There are two main applications for nondestructive testing (NDT) in general. These are production-type tests of either raw materials and/or new components, and inspections that take place at a time when the component is in service. There is virtually an infinite list of areas where ultrasonic testing can be applied. It is, however, prudent to remember that the shape and orientation of a discontinuity does not always lend itself to ultrasonic inspection.

### 7.7.1 Economic Factors

The application of almost any method of nondestructive testing is dependent on factors such as the nature of the material, accessibility to critical areas, component geometry, discontinuities that are sought, and cost factors. From an economic standpoint, the cost of dismantling a system for NDT plus the cost of the NDT itself has to be considered against the cost of replacement of the component or system. The amount and method of inspection necessary to ensure an acceptable product or system as well as the probability of discontinuity detection has to be evaluated. For example, it is certainly possible to scan a complete tank floor and map the corrosion ultrasonically; however, the cost of performing this may exceed the cost of replacing the tank floor. In this case, there are other less time-consuming and therefore more economical means of accomplishing this task; in this instance, magnetic flux leakage.

A general description of some applications for UT follows.

### 7.7.2 Thickness Measurement

Probably the most common UT application is the measurement of component thickness. The "one side only access" condition for ultrasonic thickness measurement is very desirable. Thickness measurement of tubes and pipes using inside diameter (ID) access can be conducted with either contact transducers or immersion-type systems. The contact systems usually incorporate a device in which the transducer(s) is spring loaded against the tube wall. In the case of immersion systems, either static or revolving transducer(s) scan the tube wall for thinning or corrosion. The data output is either digital or in the form of a "B scan" profile. Thickness measurements are either tabulated or can be determined from the "B scan" type data. Hand held thickness gauges are usually in abundance at many plant maintenance departments. They are relatively simple to use, very portable, relatively accurate, and also generally inexpensive.

When considering the use of an ultrasonic thickness gage, certain prerequisites should be considered in advance to provide the best and most accurate readings. These include material type, thickness range to be measured, accuracy/tolerances required, part temperature, surface conditions, and couplant to be employed.

Materials with coarse grain structure or high attenuation may require the use of special gages designed to provide better sound transmission through the material. If the part being measured has a coating, the type and thickness of the coating must also be considered. Paint, for example, will introduce an error since the velocity of steel is approximately 2.5 times higher than the velocity of paint. Therefore, if the thickness of the paint is 0.010 in. (0.25 mm), it can result in a 0.025 in. (0.625 mm) higher reading error. Some UT thickness gages will have a feature called "echo to echo." This feature measures the thickness between two successive multiples therefore ignoring the paint coating.

Other factors to consider in advance include the format to be used for recording the results of the thickness measurements. Manual recording can be very time consuming and introduces the possibility of error. Whenever there will be significant readings to be taken, data logging should be considered. This feature can save and send reading directly to a computer. Printout can highlight low (or high) readings and provide trending data for the future comparison.

Surface accessibility and condition are other factors to be evaluated up front. The cost is also something to be considered. UT gages can currently be purchased for as little as several hundred dollars and go up from there depending on the features and manufacturer. Like purchasing any other electronic instrument, the reputation of the manufacturer should be considered. Gages with more features typically will cost more, so careful consideration prior to purchase is essential.

As with other NDT methods, training and certification of the personnel using the gages is highly recommended and, in some cases, required. Conventional UT Level I and II courses will generally also include training for thickness using the UT gages. However, specialized courses designed for thickness testing only are usually available, or the course can be conducted in-plant if the number of students warrants the cost. UT certification, as with other NDT methods, will consist of written and practical (hands-on) tests. This certification should be considered as "limited" for thickness only.

### 7.7.3 Discontinuity Detection

A discontinuity is an imperfection that may render the component in which it exists unfit for service. It is the time of the formation of the discontinuity that determines the inspection category into which it falls, i.e., whether it is production or service related. Components that are presented for inspection prior to being placed in service are said to be subjected to preservice inspection (PSI). Those components or systems that are to be inspected during service or during plant "shut down season" are said to be subjected to in-service inspection (ISI).

When considering PSI-type discontinuities, cognizant personnel should bear in mind that certain acceptable discontinuities can mask those that may occur in service. An example of this would be a lamination in a pipe near the weld joint. This may cause an "acoustic shadow" that will reflect the sound energy away from a weld discontinuity. Design personnel should be aware of the inspection requirements of their designed components and should also be knowledgeable of the technology of NDT to the extent that it may influence their design philosophies.

### 7.7.4 Elevated Temperature

In situations where the component is "on line" and shutting down the system is costly, ultrasonic testing can be extremely cost effective; for instance, in cases where a pressure vessel is working to temperature and pressure. Transducers can be designed for either thickness measurement or discontinuity detection at elevated temperatures.

### 7.7.5 Hydrogen Embrittlement

Hydrogen embrittlement occurs in steel during the solidification process. This is usually due to moisture entrapment and manifests itself as extremely small ruptures close to the grain boundaries. These discontinuities, by nature, are extremely small and usually cannot be detected by most other nondestructive methods. With ultrasonics, the effect of hydrogen embrittlement is to scatter the sound energy. This appears as increased attenuation. The ultrasonic methodology includes the comparison and measurement of apparent attenuation differences in the component. Attenuation is calculated in dB per unit length and measurements are made at the suspected sites. These measurements are compared with a standard section of material that does not contain discontinuities. This process should be conducted with care. Attenuation measurements are made by observing signal amplitude; therefore, transducer contact is critical. Variations in contact or contact area can cause changes in signal amplitude.

### 7.7.6 Bond Testing

Because there is reflection at an interface, and since the amount of reflection varies with the material on either side of the interface, bond integrity can be assessed with ultrasonic techniques. In its simplest form, the increase in signal height at a bond line will usually indicate a lack of bonding. This is due to a change in the acoustic impedance between the two (or more) layers. A change in impedance also causes a change in the phase of the signal. Special equipment is available that is designed specifically to detect these changes. Multiple bond layers can be interrogated from one surface. An example of this is multiple layer "skin" on aircraft bodies.

### 7.7.7 Inaccessible Components

Components that are inaccessible may be inspected remotely using ultrasonic techniques by attaching transducers to rods or manipulators that can be inserted into an opening so that they reach the component to be examined. Examples of these conditions could be high-radiation areas or the rotor shaft within the gearbox of a helicopter, where the component cannot be seen without dismantling the gearbox.

### 7.7.8 Fluid Level Measurement

An ideal method of measuring the level of a fluid in a vessel, pipe, or container is with ultrasonics. Sound passing through the liquid will be transmitted to the opposite side of the container, and the level of the fluid can be established by monitoring the presence or absence of a signal from the opposite side of the vessel. (The opposite side must of course be parallel to the scanning side.) Alternatively, the height of fluid can be monitored by placing a transducer on the bottom of the container and monitoring the signal from the fluid to air interface.

### 7.7.9 Stress Analyses

The monitoring or measurement of component stress, either residual or induced, can be accomplished with ultrasonic techniques. This is because the sound velocity changes in the material with mechanical strain. Because these changes are relatively small, instrumentation designed for this purpose incorporates extremely accurate velocity measurement timing devices. Stress measurement can be made either along the surface of a component with precision surface wave transducers, or through a section, usually with the use of transducers that oscillate in the "*Y*" axis and produce shear wave energy in the material. Because shear wave energy is polarized, the direction of stress can be determined by rotating the transducer and observing velocity changes in anisotropic material.

### 7.7.10 Ultrasonic Extensometers

High-technology fasteners that have precision tensioning specifications can be tensioned using ultrasonics to measure this parameter. A transducer is sometimes located within a socket wrench and the extension of the bolt being torqued is measured. This is a more accurate tension-measuring device than a conventional torque wrench, as it is not affected by thread friction.

### 7.7.11 Liquid Flow Rate

Usually used in medical applications, this technique is used industrially when liquid flow rates in pipes must be monitored. This technique usually incorporates a through-transmission configuration where the sound frequency is monitored. Variations in the flow velocity produce changes in the frequency of the sound due to the "Doppler effect."

### 7.7.12 Other Applications

New applications involving ultrasonic technology evolve almost on a daily basis. With computer technology, the application horizon broadens and is often limited only by the imagination.

## 7.8 Advantages and Limitations

There is no nondestructive test method that is a panacea. Each method has its advantages and limitations. It is a matter of selecting the test method that offers the most effective approach to solving the examination problem. When determining whether ultrasonics is the most appropriate test method, consideration should be given to the following:

1. Part and geometry to be examined
2. Material type
3. Material thickness
4. Material process—cast, wrought, etc.
5. Type of discontinuities to be detected
6. Minimum discontinuity size to be detected
7. Location of the discontinuities—surface-breaking or internal
8. Orientation of discontinuities (very important when selecting a test technique)
9. Accessibility to areas of interest

10. Surface conditions
11. Type of examination record required

Ultrasonic inspection is ideal for locating small, tight discontinuities assuming the following:

1. The sound energy can be projected at some angle that will respond favorably to the orientation of the reflector.
2. The relationship between the size of the discontinuity and the material's grain structure allows for an acceptable signal-to-noise ratio.
3. The surface condition is suitable for scanning. A poor scanning surface will not only require a more viscous couplant but possibly the use of a lower test frequency. This may not provide the necessary resolution for the test.

The advantages of ultrasonic examination are as follows:

1. Inspection can be accomplished from one surface
2. Small discontinuities can be detected
3. Considerable control over test variables
4. Varieties of techniques are available using diverse wave modes
5. High-temperature examination is possible with the correct equipment
6. Examination of thick or long parts
7. Inspection of buried parts, e.g., shafts in captivated bearing houses
8. Accurate sizing techniques for surface-breaking and internal discontinuities is possible
9. Discontinuity depth information
10. Surface and subsurface discontinuities can be detected
11. High speed scanning is possible with electronic signal gating and alarm system
12. "Go/No-Go" testing of production components
13. Test repeatability
14. Equipment is light and portable
15. Area evacuation of personnel is not necessary
16. Special licenses are not required as with radiation sources
17. Minimum number of consumables

Some of the limitations of ultrasonic examination are as follows:

1. Discontinuities that are oriented parallel with the beam energy will usually not be detected. Orientation of the discontinuity (reflector) is the most important factor in detecting discontinuities.
2. Discontinuities that are similar to or smaller than the material's grain structure may not be detected.
3. Thin sections may present resolution problems or require the implementation of special techniques.

4. Uneven scanning surfaces can reduce the effectiveness of the test.
5. Signals can be misinterpreted. This includes spurious signals from mode conversion or beam redirection, etc.
6. In general, this method requires a high level of skill and training.
7. Permanent record of the examination results is not typical. The records are limited to physical documentation rather than an actual reproduction of the test, e.g., as is possible with radiography.

## 7.9 Glossary of Terms

**A-Scan** A method of data presentation on an ultrasonic display utilizing a horizontal baseline, that indicates distance, and a vertical deflection from the baseline, that indicates amplitude.

**A-Scan Presentation** A method of data presentation utilizing a horizontal baseline to indicate distance, or time, and a vertical deflection from the baseline to indicate amplitude.

**Amplitude** The vertical height of a signal, usually base to peak, when indicated by an A-scan presentation.

**Angle Beam** A term used to describe an angle of incidence or refraction other than normal to the surface of the test object, as in angle beam examination, angle beam search unit, angle beam longitudinal waves, and angle beam shear waves.

**Area Amplitude Response Curve** A curve showing the relationship between different areas of reflection in a material and their respective amplitudes of ultrasonic response.

**Attenuation** A factor that describes the decrease in ultrasound intensity or pressure with distance. Normally expressed in decibels per unit length.

**B-Scan Presentation** A means of ultrasonic data presentation that displays a cross section of the specimen, indicating the approximate length (as detected per scan) of reflectors and their relative positions.

**Back Echo** See Back Reflection.

**Back Reflection** An indication, observed on the display screen of a UT instrument, that represents the reflection from the back surface of a reference block or test specimen.

**Back Surface** The surface of a reference block or specimen that is opposite the entrant surface.

**Beam Spread** A divergence of the ultrasonic beam as it travels through a medium.

**Bubbler** A device using a liquid stream to couple a transducer to the test piece.

**C-Scan** An ultrasonic data presentation that provides a plan view of the test object and discontinuities.

**Collimator** A device for controlling the size and direction of the ultrasonic beam.

**Contact Testing** A technique in which the transducer contacts directly with the test part through a thin layer of couplant.

**Couplant** A substance, usually a liquid, used between the transducer unit and the test surface to permit or improve transmission of ultrasonic energy.

**Critical Angle** The incident angle of the ultrasonic beam beyond which a specific refracted wave no longer exists.

**DAC** Distance amplitude correction. Electronic change of amplification to provide equal amplitude from equal reflectors at different depths. Also known as swept gain, time corrected gain, time variable gain, etc.

**DAC Curve** A curve (usually drawn on the screen) derived from equal reflectors at different depths.

**Damping, Search Unit** Limiting the duration of a signal from a search unit subject to a pulsed input by electrically or mechanically decreasing the amplitude of successive cycles.

**dB Control** A control that adjusts the amplitude of the display signal in decibel (dB) units.

**Dead Zone** The distance in the material from the surface of the test specimen to the depth at which a reflector can first be resolved under specified conditions. It is determined by the characteristics of the search unit, the ultrasonic instrumentation, and the test object.

**Decibel (dB)** Logarithmic expression of a ratio of two amplitudes or intensities. (UT) dB = 20 log10 (amplitude ratio).

**Delay Line** A column of material such as Plexiglas that is attached to the front of a transducer. It behaves similarly to a water path and allows the initial pulse to be shifted off the screen. This often improves "near surface resolution."

**Delay Sweep** An A-scan or B-scan presentation in which an initial part of the time scale is not displayed.

**Discontinuity** A lack of continuity or cohesion; an intentional or unintentional interruption in the physical structure or configuration of a material or component.

**Distance Amplitude, Compensation (Electronic)** The compensation or change in receiver amplification necessary to provide equal amplitude on the display of an ultra-sonic instrument for reflectors of equal area that are located at different depths in the material.

**Distance Amplitude, Response Curve** See DAC. A curve showing the relationship between the different distances and the amplitudes of an ultrasonic response from targets of equal size in an ultrasonic transmitting medium.

**Distance Linearity Range** The range of horizontal deflection in which a constant relationship exists between the incremental horizontal displacement of vertical indications on the A-scan presentation and the incremental time required for reflected sound to pass through a known length in a uniform transmission medium.

**Doppler Effect** The change in frequency of a sound wave due to movement of the reflector. Movement toward or away from the sound will result in a change in frequency (e.g., the tone of a train whistle changing as the train passes).

**Dual Search Unit** A search unit containing two elements, one a transmitter, the other a receiver.

**Dynamic Range** The ratio of maximum to minimum reflective areas that can be distinguished on the display at a constant gain setting.

**Entrant Surface** The surface of the material through which the ultrasonic waves are initially transmitted.

**Far Field** The zone of the beam (beginning at the Y0 point) where equal reflectors give exponentially decreasing amplitudes with increasing distance.

**Flaw** A discontinuity in a material or component that is unintentional.

**Flaw Characterization** The process of quantifying the size, shape, orientation, location, growth, or other properties of a flaw based on NDT response.

**Frequency (Examination)** The number of cycles per second (Hz).

**Frequency, Pulse Repetition** The number of times per second that a search unit is excited by the pulser to produce a pulse for ultrasonic imaging. This is also called pulse repetition rate or pulse repetition frequency (PRF).

**Gate** An electronic means of selecting a segment of the time range for monitoring, triggering an alarm, or further processing.

**Immersion Testing** An ultrasonic examination technique in which the search unit and the test part are submerged (at least locally) in a fluid, usually water.

**Impedance, Acoustic** A mathematical quantity used in computation of reflection characteristics at boundaries. It is the product of wave velocity and material density ($\rho c$).

**Indication** A response or evidence of a response disclosed through an NDT that requires further evaluation to determine its full and true significance.

**Initial Pulse** The response of the ultrasonic system display to the transmitter pulse (sometimes called "main bang").

**Lamb Wave** A specific mode of propagation in which the two parallel boundary surfaces of the material under examination (such as a thin plate or wall of a tube) establish the mode of propagation. The Lamb wave can be generated only at particular values of frequency, angle of incidence, and material thickness. The velocity of the wave is dependent on the mode of propagation and the product of the material thickness and the examination frequency.

**Linearity, Amplitude** A measure of the proportionality of the amplitude of the signal input to the receiver and the amplitude of the signal appearing on the display of the ultrasonic instrument or on an auxiliary display.

**Linearity, Time or Distance** A measure of the proportionality of the signals appearing on the time or distance axis of the display and the input signals to the receiver from a calibrated time generator or from multiple echoes from a plate or material of known thickness.

**Longitudinal Wave** A wave in which the particle motion of the material is essentially in the same direction as the wave propagation (also called compressional wave).

**Metal Path** See Sound Path.

**Mode** The type of ultrasonic wave propagating in the material as characterized by the particle motion (e.g., longitudinal, transverse, etc.).

**Mode Conversion** Phenomenon by which an ultrasonic wave that is propagating in one mode refracts at an interface to form ultrasonic wave(s) of other modes.

**Multiple Back Reflections** Successive signals from the back surface of the material under examination.

**Near Field** The region of the ultrasonic beam adjacent to the transducer having complex beam profiles and intensity variations. Also known as the Fresnel zone.

**Noise** Any undesired signal (electrical or acoustic) that tends to interfere with the interpretation or processing of the desired signals.

**Normal Incidence (also see Straight Beam)** A condition in which the axis of the ultrasonic beam is perpendicular to the entrant surface of the part being examined.

**Penetration Depth** The maximum depth in a material from which usable ultrasonic information can be obtained and measured.

**Probe** See Search Unit.

**Pulse-echo Technique** An examination method in which the presence and position of a reflector are indicated by the echo amplitude and time.

**Pulse Length** A measure of the duration of a signal as expressed in time or number of cycles.

**Range** The maximum distance that is presented on a display.

**Rayleigh Wave** An ultrasonic surface wave in which the particle motion is elliptical and the effective penetration is approximately one wavelength.

**Reference Block** A block of material that includes reflectors. It is used both as a measurement scale and as a means of providing an ultrasonic reflection of known characteristics.

**Reflector** An interface at which an ultrasonic beam encounters a change in acoustic impedance and at which at least part of the sound is reflected.

**Reject, Suppression** A control for minimizing or eliminating low-amplitude signals (electrical or material noise) so that true signals are emphasized.

**Relevant Indication** An indication caused by a discontinuity that requires evaluation.

**Scanning** The movement of a transducer relative to the test part in order to examine a volume of the material.

**Search Unit** An electroacoustic device used to transmit and/or receive ultrasonic energy. The device generally comprises a piezoelectric element, backing, wearface, and/or wedge. Sometimes known as a "probe" or "transducer."

**Sensitivity** A measure of the smallest reflector that produces a discernible signal on the display of an ultrasonic system.

**Shear Wave** Wave motion in which the particle motion is perpendicular to the direction of propagation (transverse wave).

**Skip Distance** In angle beam testing, the distance along the test surface from sound entrant point to the point at which the sound returns to the same surface. It can be considered the top surface distance of a complete "vee" path of sound in the test material.

**Sound Path** The path of the sound energy from the time that it leaves the transducer and reflects back to the transducer.

**Transducer** A piezoelectric element used to produce ultrasonic vibrations.

**Through-Transmission Technique** A technique in which ultrasonic waves are transmitted by one search unit and received by another at the opposite surface of the material being examined.

**Vee Path** The angle beam path in materials starting at the search-unit examination surface, through the material to the reflecting surface, continuing to the examination surface in front of the search unit, and reflecting back along the same path to the search unit. The path is usually shaped like the letter V.

**Water Path** The distance from the transducer to the test surface in immersion or water column testing.

**Wedge** In angle beam examination by the contact method, a device used to direct ultrasonic energy into the material at an angle.

**Wheel Search Unit** An ultrasonic device incorporating one or more transducers mounted inside a liquid-filled flexible tire. The beam is coupled to the test surface through the rolling contact area of the tire. Also known as a "wheel probe" or "roller search unit."

## References

J. C. Drury, *Ultrasonic Flaw Detection for Technicians,* 3rd ed-, Silverwing Limited. Swansea SA5 8JF, Wales UK.

*Nondestructive Evaluation and Quality Control,* ASM International Metals Handbook, 9th ed, 1989.

# CHAPTER 8
# Eddy Current Testing

## 8.1 History and Development

Eddy current testing is one of the oldest nondestructive testing (NDT) methods. However, it wasn't until the last few decades of the 20th century that the eddy current method started to reach its true potential in the marketplace. One reason for this is that general-purpose, user-friendly eddy current instruments are a relatively recent phenomenon. Whereas portable ultrasonic instruments offering considerable versatility have been available since the 1960s, comparable eddy current portables only became available in the 1980s. In addition, it is only recently that eddy current theory became widely understood by NDT professionals. The early 1980s, in particular, produced excellent explanatory material that made eddy current theory understandable to persons without advanced technical backgrounds. Modern microprocessor-based instruments, plus the availability of high-quality operator training, ensure the continued growth of this versatile, high-performance NDT method.

### 8.1.1 Significant Discoveries about Electromagnetism

Development of the eddy current method was based on certain discoveries made during the early-19th century about the relationship between electricity and magnetism. In fact, the relevant electromagnetic principles were discovered in the same sequence in which they occur during an eddy current test.

In 1820, Hans Christian Oersted, a Dane, discovered electromagnetism—the fact that an electrical current flowing through a conductor causes a magnetic field to develop around that conductor. Oersted discovered electromagnetism accidentally. While demonstrating that heat is developed when an electric current passes through a wire, Oersted observed that the needle of a magnetic compass deflected perpendicular to the wire while the current was passing through it. Electromagnetism is the principle on which eddy current coils operate. Whereas Oersted was using direct current developed from a battery voltage when he discovered electromagnetism, an eddy current instrument employs alternating electric current flowing through the test coil in order to develop an alternating magnetic field around the coil.

In 1831, an Englishman, Michael Faraday, discovered electromagnetic induction—the fact that relative motion between a magnetic field and a conductor induces a voltage in that conductor, causing an electric current to flow. Consequently, when the alternating magnetic field of an eddy current instrument's coil is brought in contact with a conducting test object, a voltage is developed, causing a current to flow in the test object.

Thus, electromagnetic induction is considered to be the operating principle of eddy current testing. Joseph Henry also independently discovered electromagnetic induction in the United States at about the same time. In fact, the unit of measure for induction is named after him.

In 1834, Heinrich Lenz stated the principle that defines how the properties of the test object are communicated back to the test system. Lenz's law states that the direction of current flow in the test object will be such that its magnetic field will oppose the magnetic field that caused the current flow in the test object. This means that, in practice, the eddy currents communicate with the test coil by developing a secondary flux that cancels a portion of the coil's flux equivalent to the magnitude and phase of the flux developed by the eddy currents.

The theory describing the chain of events of an eddy current test may thus be fully described by the discoveries of Oersted, Faraday, Henry, and Lenz. The existence of eddy currents themselves, however, was not discovered until 1864. They were discovered by James Maxwell, who is famous for stating the defining equations of electromagnetic theory. The first use of eddy currents for nondestructive testing occurred in 1879 when D. E. Hughes used these principles to conduct metallurgical sorting tests.

### 8.1.2 Modern Eddy Current Testing

The development of the eddy current method progressed slowly until the late 1940s, when Dr. Friedreich Foerster founded the Institut Dr. Foerster, which made great strides in developing and marketing practical eddy current test instruments. By the late 1960s, the Institute had developed a product line covering virtually every application of the eddy current test method and worked with American manufacturers to firmly establish the method in the United States. Two major contributions of Foerster were the development of impedance plane display, which greatly aided in communication of test information to the practitioner, and formulation of the Law of Similarity, which enables the practitioner to duplicate the same eddy current performance under a variety of test situations.

The next major contribution to the advancement of the method, multi-frequency testing, was also developed by an equipment manufacturer, Intercontrolle of France, in 1974. Driving the test coil at multiple frequencies helps to overcome what has traditionally been the major limitation of the eddy current method, the fact that the various conditions to which the method is sensitive can vector into a single displayed signal that is difficult to interpret. Originally developed to suppress the display of undesired test variables, multi-frequency testing can also optimize an eddy current test for normally conflicting performance variables such as sensitivity and penetration as well as aid in identifying the nature of a particular test response. Multi-frequency testing is a very significant innovation that has markedly advanced the state of the art.

The development of microprocessor-based eddy current instruments since the mid-1980s has also enhanced the potential and user-friendliness of the method. It has improved recording capability, provided sophisticated post-inspection signal analysis, and has allowed automatic mixing of multi-frequency signals. Modern microprocessor-based eddy current instruments offer a breadth of useful features virtually unimaginable in the days of analog equipment. Manufacturers such as Zetek, Hocking, Foerster, Nortec, ETC, and Magnetic Analysis have been important contributors. GE Inspection Technologies and Olympus have now entered the field as full-service NDT equipment manufacturers.

In addition to mainstream eddy current testing, more specialized techniques are employed for certain applications. These include magnetic flux leakage, remote field eddy current, AC field measurement, and modulation analysis inspection. In classifying nondestructive test methods for the purpose of qualifying and certifying test personnel, the American Society for Nondestructive Testing (ASNT) classifies AC field measurement, eddy current, and remote field under the umbrella of the electromagnetic testing method (ET). Magnetic flux leakage has been classified as a separate method.

### 8.1.3 Material Variables Detectable by Eddy Currents

During more than a century of development as a test method, eddy current testing has found application due to its sensitivity to the following variables:

- Conductivity variations
- Detection of discontinuities
- Spacing between test coil and test material (*lift-off* distance)
- Material thickness
- Thickness of plating or cladding on a base metal
- Spacing between conductive layers
- Permeability variations

Eddy current testing is suitable for inspection of the surface and just beneath the surface of conductive materials, volumetric inspection of thin conductive materials, and lift-off measurement to determine thickness of nonconductive materials adhering to or resting on the surface of conductive materials.

### 8.1.4 Major Application Areas

The versatility of the eddy current method has resulted in broad applications usage. However, the major application areas include the following:

- In-service inspection of tubing at nuclear and fossil fuel power utilities, at chemical and petrochemical plants, on nuclear submarines, and in air conditioning systems
- Inspection of aerospace structures and engines
- Production testing of tubing, pipe, wire, rod, and bar stock

## 8.2 Theory and Principles

Eddy current theory is based on the principles of electricity and magnetism, particularly the inductive properties of alternating current. The discussion begins with a review of some basic principles.

### 8.2.1 Electricity

All matter is made up of atoms, the atom being the smallest unit of any element that retains the properties of that element. The center of an atom, the nucleus, has a positive electrical charge. Orbiting the nucleus and rotating on their own axes are negatively

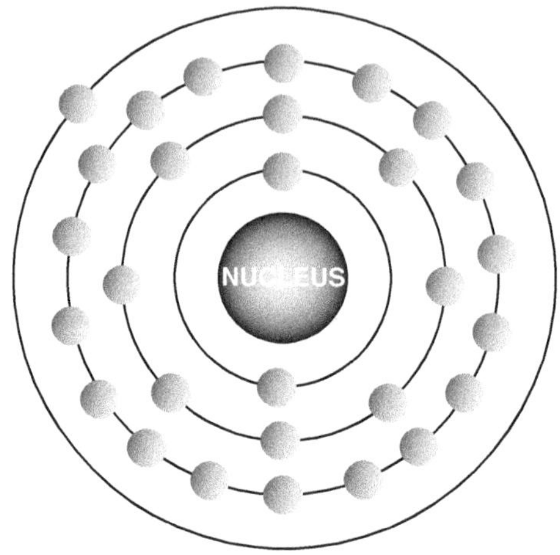


**Figure 8-1** Copper atom.

charged particles called electrons. As shown in the illustration of the copper atom (Fig. 8-1), orbits of electrons around the nucleus resemble the orbits of planets around the sun in that there can be several orbits, called "shells." However, atomic structure differs from the solar system in that a given shell can contain multiple electrons.

From the perspective of eddy current testing, one is concerned specifically with the outer shell of a material's atoms, because the number of electrons in the outer shell determines whether the material will conduct electricity. The outer shell can contain a maximum of eight electrons, and when the outer shell contains as many as seven or eight electrons, the material will not conduct electricity and is called an insulator. However, materials whose atoms have only one, two, or three electrons in the outer shell can conduct electricity and are, in fact, called conductors. Materials whose outer shells contain an intermediate number of electrons are called semiconductors and, although important in the design of computer circuitry, are not significant here.

If undisturbed by outside forces, a conductor's electrons will repeatedly orbit the nucleus. However, when voltage [also called electromotive force (EMF) or potential] is applied to a conductor, its electrons will advance from one atom to the next. That is, there is a flow of electrical charges called current or electricity. Voltage causes electrons to flow because it can attract and repel them; that is, voltage applies polarity to electrons. A battery is an example of a voltage source. Electrons, being negatively charged, will be attracted to a battery's positive terminal and repelled by its negative terminal. As shown in the illustration of a flashlight circuit (Fig. 8-2), electrons flow through the bulb's filament from the negative to the positive terminal of the battery.

Although a conductor's atoms will permit current flow when voltage is applied, there is always some opposition to flow, due to the attraction of electrons to their atoms. This opposition varies among the atoms of different materials. Willingness of a test specimen to allow current flow is a key point in eddy current testing, detailed in the following definitions:

- *Conductivity* is the relative ability of a material's atoms to conduct electricity.
- *Resistivity* is the opposition of a material's atoms to the flow of electricity; it is the inverse of conductivity.

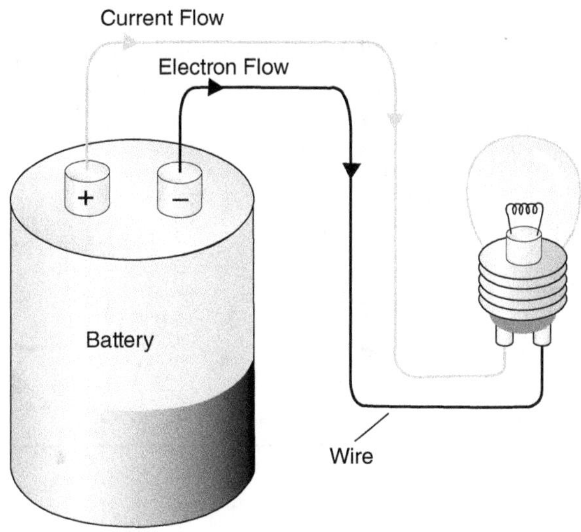


**Figure 8-2** Flashlight circuit.

- *Conductance* is the ability of a particular component to conduct electricity. Conductance depends on a component's conductivity, length, and cross section.
- *Resistance* is the inverse of conductance. It is the opposition that a particular component offers to the flow of electricity. Like conductance, it depends on a component's conductivity, length, and cross section.

Conductivity is the material property of most interest to us in eddy current testing, whereas resistance is an important element in the display of test information. Material conductivities are compared on a scale called the International Annealed Copper Standard (IACS). Pure unalloyed annealed copper at 20°C is the base value on this scale, with a value of 100 percent. Other materials are assigned a percentage depending on their ability to conduct electricity relative to copper.

Having now identified resistance, as well as voltage and current, these terms can be tied together, showing their units, using the most basic formula of electricity, Ohm's law:

$$I = \frac{V}{R}$$

where:
$I$ = current in amperes
$V$ = voltage in volts
$R$ = resistance in ohms

Thus, current flow increases when voltage increases and current flow decreases when resistance increases. There are two types of current: direct current (dc), which flows in only one direction; and alternating current (ac), which continually reverses direction.

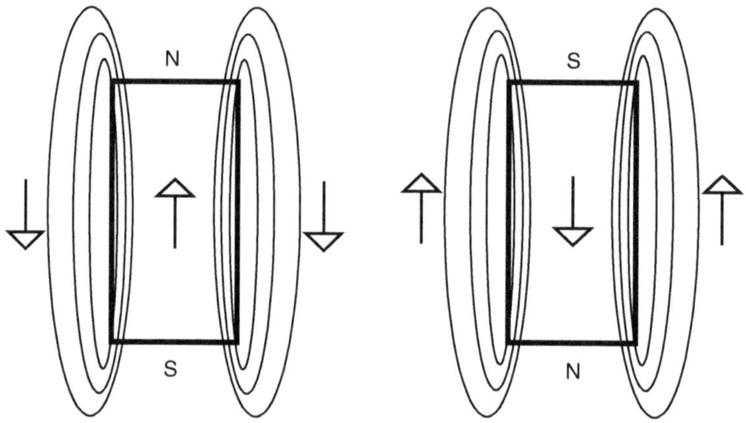


**Figure 8-3** Magnetic polarity.

### 8.2.2 Magnetism

Magnetism is a mechanical force of attraction or repulsion that one material can exert upon another. The opposite ends of a magnet exhibit opposing behavior called *polarity*. Thus, the ends of a magnet are called poles—one north and one south.

A magnet has a force field that can be visualized as a number of closed loops that flow through the magnet, travel around the outside of the magnet, and then reenter the magnet at the other end (Fig. 8-3). These magnetic loops are called *lines of force* or *flux lines*. The word "flux" literally means "flow" and relates to the fact that the lines of force flow from the north to the south pole around the outside of a magnet, and from the south to the north pole within the magnet. Units of measurement and definitions for magnetism include the following:

*Magnetic flux* ($\phi$ or phi) is the entire set of a magnet's flowing lines of force.

*Flux density* (B) is the number of flux lines per unit area, perpendicular to the direction of flow.

The *maxwell* (Mx) is one magnetic field line or line of force.

The *weber* (Wb) is $1 \times 10^8$ lines or maxwells.

The *gauss* (G) is one line of force per square centimeter.

The *tesla* (T) is one weber per square meter.

Field intensity depends on flux density. Flux density is greatest within the core of a magnet and at the poles. Flux density decreases with distance from the magnet according to the inverse square law (Fig. 8-4); that is, flux density is inversely proportional to the square of the distance from the poles of the magnet.

When like poles of two magnets are brought together, the magnets push apart as their force fields repel each other. When unlike poles of two magnets are brought together, the magnets attract as the two force fields attempt to combine.

There are two types of magnets: permanent magnets and electromagnets. Permanent magnets are physical materials, having a property called ferromagnetism, which means that they can become magnetized when their *domains* have become aligned (Fig. 8-5).

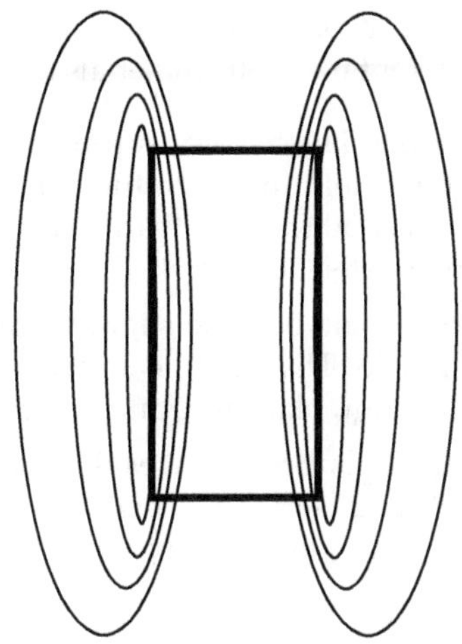

**FIGURE 8-4** Flux density distribution.

Domains are miniature magnets consisting of groups of atoms or molecules present within a material's individual grains.

Permanent magnets were discovered in ancient times and are often produced in bar and horseshoe shapes. The fact that they retain a magnetic field without activation by electrical current is what distinguishes them from electromagnets. *Permeability* is the measure of a material's ability to be magnetized; that is, a material's ability to concentrate magnetic flux. The more flux density obtained from a material by a given quantity of applied magnetizing force, the greater the permeability of that material.

Materials may be classified as *magnetic* or *nonmagnetic*. Although there are three types of magnetic materials—*ferromagnetic*, *paramagnetic*, and *diamagnetic*—the term "magnetic"

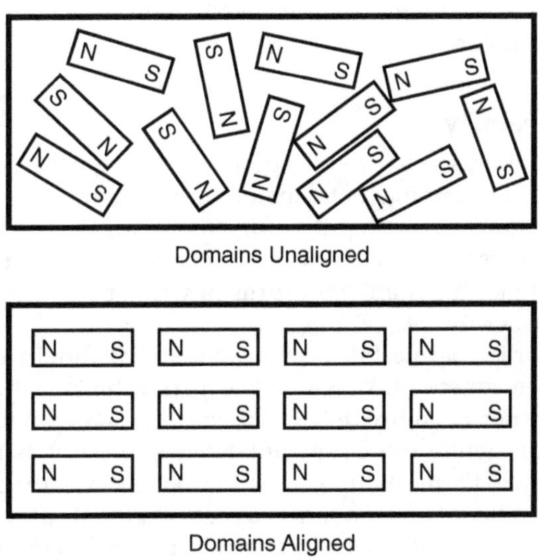


**FIGURE 8-5** Magnetic domains.

usually refers to *ferromagnetic* materials, which have much higher permeability than the other two types. The different magnetic materials may be characterized as follows:

1. Ferromagnetic materials become strongly magnetized in the same direction as the magnetizing field in which they are placed. Their permeability ranges from approximately 50 to more than 100,000. Examples are iron, carbon steel, 400-series stainless steel, and nickel.
2. Paramagnetic materials become slightly magnetized in the same direction as the magnetizing field. Their permeability is slightly more than 1. Examples are aluminum, chromium, platinum, and oxygen gas.
3. Diamagnetic materials become weakly magnetized in the opposite direction from the magnetizing field. Their permeability is slightly less than 1. Examples are copper, gold, silver, and hydrogen.

### 8.2.3 Electromagnetism

Electromagnetism is the phenomenon whereby the passage of electrons through a conductor causes a magnetic field to develop concentrically around the conductor, perpendicular to its axis. A concentrated magnetic field, similar to that obtained from a bar magnet, can be obtained by winding a conductor into a coil. A coil functioning as an electromagnet is called a solenoid, although an ideal solenoid has a length much greater than its diameter. A solenoid concentrates a magnetic field inside the coil, with opposing poles at each end of the coil, and flux lines completely encircling the loops of the coil. If direct current is applied to the coil, its magnetic field will flow in only one direction and it can perform the same work of attracting ferromagnetic materials as a permanent magnet. In addition, the coil can be wound around a ferromagnetic core for increased field strength.

If the current is alternating, the electromagnetic field will likewise alternate and the coil will exhibit a quality called inductance, $L$, whose unit is the henry. Inductance is the ability of a conductor to induce voltage in itself or in a neighboring conductor when the current varies.

### 8.2.4 Permeability

Electromagnets are used to produce permanent magnets. A conducting wire or cable is wound around the ferromagnetic material to be magnetized. Direct current is passed through the conductor, causing it to function as an electromagnet. When the resulting magnetic field enters the material to be magnetized, the material's domains become aligned. The higher the number of turns of wire or cable and the stronger the applied current, the greater the *magnetizing force.*

As stated earlier, permeability defines a material's ability to be magnetized, its ability to concentrate magnetic flux. Numerical permeability values for different materials, termed relative permeability ($\mu_r$) are stated in comparison to the permeability of air or a vacuum. Permeability can be quantitatively expressed as the ratio of flux density to magnetizing force. Permeability can be a problem in eddy current testing because the relative permeability of a given ferromagnetic material can vary during testing, causing permeability noise signals that can override the eddy current signals being sought. A *hysteresis loop* (Fig. 8-6) is a plot of a material's flux density (B) variations as magnetizing

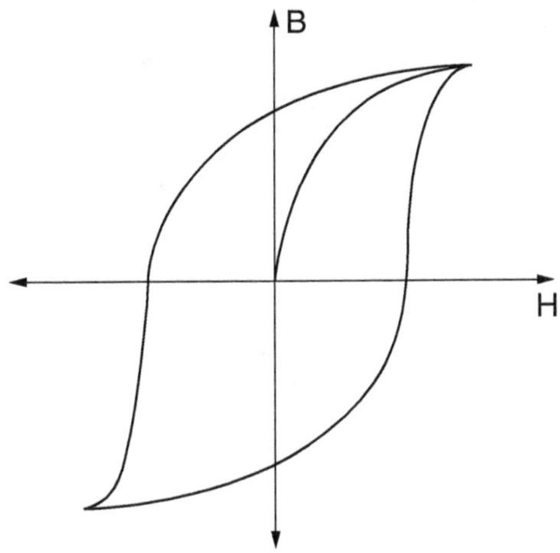


**FIGURE 8-6** Hysteresis loop.

force (H) is varied. By magnetically saturating the test material, permeability becomes constant and eddy current testing can proceed without interference from permeability variations. *Saturation* (Fig. 8-7) occurs at that point on the loop where further increases in magnetizing force do not cause significant increases in flux density.

At the completion of testing, the material will retain a certain amount of *residual magnetism* (Fig. 8-8), the amount of flux density remaining in the material after the magnetizing force has been reduced to zero. The residual magnetism must be eliminated by demagnetization, to prevent problems such as the material attracting ferromagnetic debris.

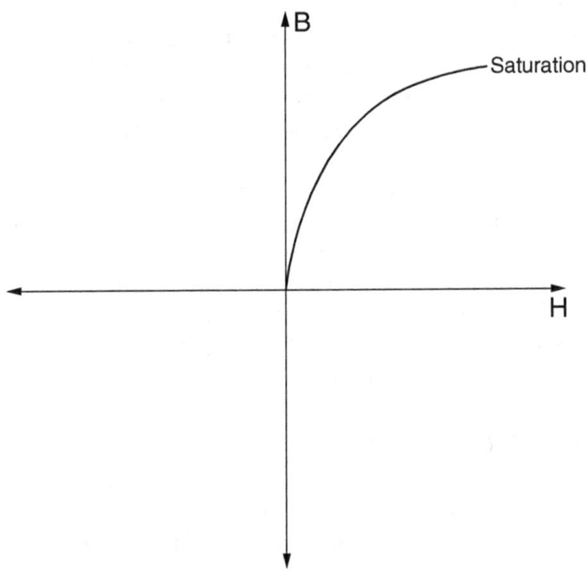


**FIGURE 8-7** Magnetic saturation.

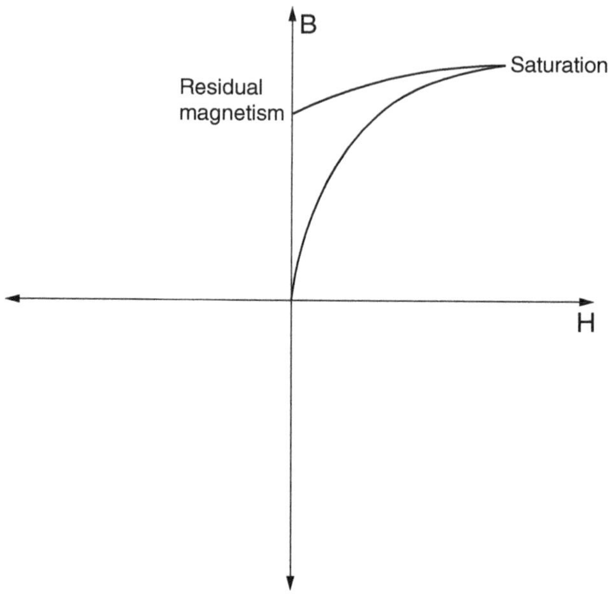


**FIGURE 8-8** Residual magnetism.

## 8.3 Alternating Current Principles

### 8.3.1 Sinusoidal Variation

Alternating current flows in a cyclical manner, behavior that is accurately illustrated by a sine curve (Fig. 8-9).

The current begins its cycle at zero amplitude and, as time elapses, rises to a peak in one direction, falls back to zero, rises to a peak in the opposite direction, and falls back to zero again to complete the cycle. The end of one cycle is the starting point of the next cycle. One complete 360° cycle is called a *sinusoid*. Activity exhibiting the behavior of a sinusoid is termed sinusoidal.

The 360 degree points of a sinusoid correspond to the 360 degree points of a complete circle. Thus, both a sinusoid and a circle express one complete cycle. However, the sinusoid adds dimensions of amplitude and polarity to the cycle concept. Recall that the distance from the center of a circle to the edge is its radius. The portion of a circle's arc that corresponds to the length of its radius is called a *radian*, which occupies an arc of 57.3°. Note that 360° divided by 57.3° equals 6.28 or $2\pi$, an important constant in alternating current calculations.

### 8.3.2 Electromagnetic Induction

Prior to discussing eddy currents, the induction process will first be modeled using a pair of coils as primary and secondary elements of a mutual induction circuit. Sinusoids are included to illustrate time and polarity relationships. The symmetry of two coils aids in explanation of the polarity relationships that occur in mutual induction processes such as eddy current tests. This illustration models an eddy current test in its

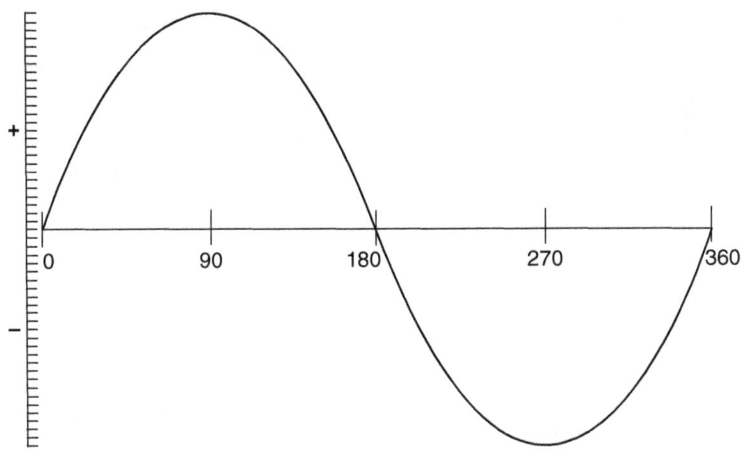


**Figure 8-9** Sinusoid.

most basic configuration, where a single coil is the primary circuit and the test material is the secondary. Initially, the primary coil will be examined without the influence of a secondary. The primary coil will be multi-turn to represent a typical eddy current test coil. The secondary, when introduced, will be a single-turn coil, which validly represents any eddy current test specimen. It must be remembered that the secondary coil in this discussion represents strictly the test specimen, *not* the second coil of various multi-coil configurations that will be described later in this chapter.

### The Induction Process

1. An alternating current generator applies alternating voltage to a coil circuit (Fig. 8-10a). A portion of this voltage, $V_R$, is applied across the resistance of the coil wire. The $V_R$ amplitude rises from zero amplitude at zero degrees (Fig. 8-10b). The "R" subscript in $V_R$ identifies this voltage, sometimes called "resistance voltage," as a force needed to move current through the resistance of the coil wire. $V_R$ is a component of the circuit's total voltage, $V_T$, which will be explained later.
2. $V_R$ causes a current, $I_P$, to flow through the coil (Fig. 8-10c), in phase with $V_R$ (Fig. 8-10d). The "P" subscript in $I_P$ identifies the coil current as the primary current.
3. *Electromagnetism* occurs. The alternating current flowing through the coil causes an alternating magnetic field, $\Phi_P$, the primary flux, to develop around the coil (Fig. 8-10e), in phase with $V_R$ and $I_P$ (Fig. 8-10f).
4. *Self-induction* occurs. Since the coil is standing in the field of its own varying flux, Faraday's law applies and electromagnetic induction is imposed on the coil wire. That is, $\Phi_P$ induces an additional voltage, $V_L$, often called "back voltage," into the coil (Fig. 8-10g). The "L" subscript identifies $V_L$ as an "induced voltage." This voltage is separate from the $V_R$ voltage that caused $I_P$ to flow. According to Faraday's law, the quantity of induced voltage is proportional to the rate of flux variation. Since $\Phi_P$ is varying the most through the 0°, 180°, and 360° points, and not varying through the 90° and 180° points,

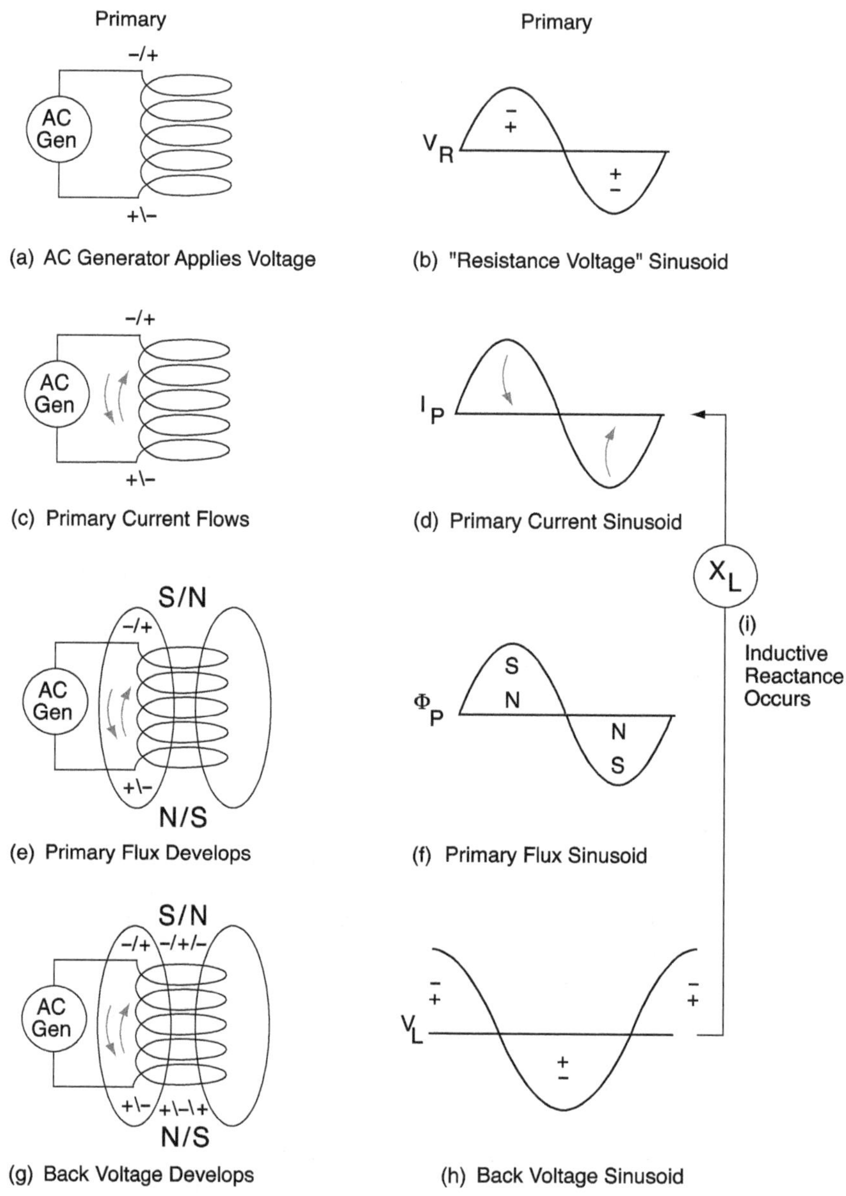


**FIGURE 8-10** Self-induction process. (*a*) AC generator applies voltage. (*b*) "Resistance voltage" sinusoid. (*c*) Primary current flows. (*d*) Primary current sinusoid. (*e*) Primary flux develops. (*f*) Primary flux sinusoid. (*g*) Back voltage develops. (*h*) Back voltage sinusoid. (*i*) Inductive reactance occurs.

the back voltage is induced 90° out of phase (Fig. 8-10h) with the coil current and flux.

5. *Inductive reactance* occurs. Since the back voltage is 90° out of phase with the coil current, it will oppose changes in the coil current (Fig. 8-10i). In that

amplitude change is the very nature of alternating current flow, opposition to change in ac is effectively opposition to flow of ac. This opposition, called *inductive reactance*, $X_L$, is distinct from resistance; $R$. Resistance simply opposes flow of current and can occur in either a DC or AC circuit. Inductive reactance, which, strictly speaking, opposes *change* of current flow, can occur only in an AC circuit.

Inductive reactance depends on coil design and test frequency to the extent that, as more flux lines cut across more coil turns per unit time, inductive reactance increases.

The variables influencing inductive reactance are detailed in the following equations:

$$X_L = 2\pi f L$$

$$L = \mu_r \frac{N^2 \times A}{l} \times 1.26 \times 10^{-6}$$

where:
$X_L$ = inductive reactance
$f$ = test frequency
$L$ = coil inductance
$\mu_r$ = relative permeability of the coil core
$N$ = number of turns
$A$ = cross-sectional area
$l$ = coil length

***Note:*** The permeability of air, $1.26 \times 10^{-6}$, must be multiplied by an appropriate value for $\mu_r$ which would be 1 (one) for an air core coil or some higher value in the case of a coil with a ferromagnetic core.

6. If a secondary circuit is placed in proximity to the primary, mutual induction will occur and $\Phi_P$ will induce a voltage into the secondary circuit (Fig. 8-11a). This voltage is appropriately called secondary voltage, $V_S$, and is 180° out of phase (Fig. 8-11b) with the inducing primary flux.

7. $V_S$ causes a current, $I_S$, to flow in the secondary circuit (Fig. 8-11c), with the same phase (Fig. 8-11d) as $V_S$. In an actual eddy current test, where the secondary circuit is the test specimen, $I_S$ will be the eddy currents.

8. With current now flowing in the secondary circuit, electromagnetism will again occur and a secondary flux, $\Phi_S$, will develop (Fig. 8-11e), with its phase (Fig. 8-11f) determined by the secondary current.

9. The 180° phase difference between primary and secondary activity indicated in Figs. 8-11b, d, and f is the effect of Lenz's law, which states that the induction process in the secondary circuit causes a phase reversal, which results in the secondary flux being opposite in polarity to the primary flux. Due to this state of opposition, the secondary flux will cancel a portion of the primary flux, resulting in an overall decrease in net flux for the two coils. The reduced amplitude of net flux (dashed line in Fig. 8-11g) results in a reduced rate of net flux variation. Less flux variation results in reduced back voltage (Figs. 8-11h and 8-11i), which results in a reduction of inductive reactance (Fig. 8-11j).

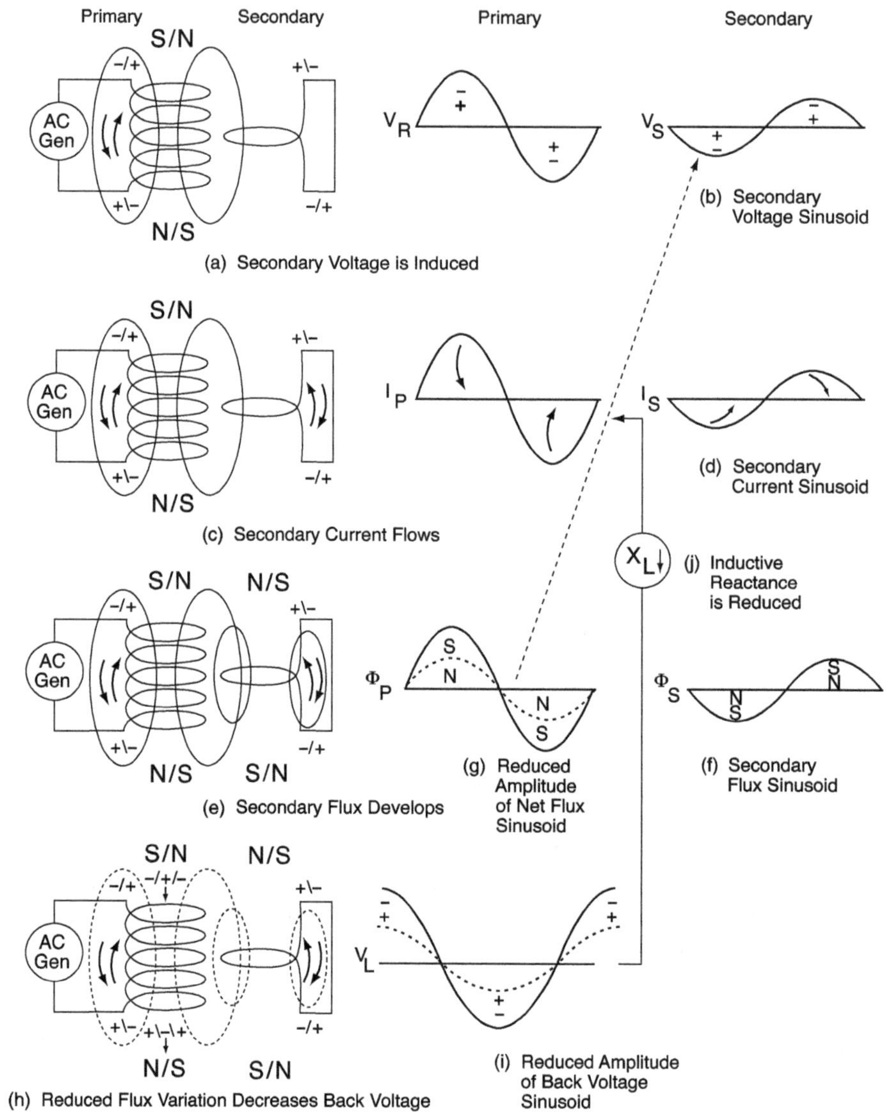


**FIGURE 8-11** Mutual induction process. (*a*) Secondary voltage is induced. (*b*) Secondary voltage sinusoid. (*c*) Secondary current flows. (*d*) Secondary current sinusoid. (*e*) Secondary flux develops. (*f*) Secondary flux sinusoid. (*g*) Reduced amplitude of net flux sinusoid. (*h*) Reduced flux variation decreases back voltage. (*i*) Reduced amplitude of back voltage sinusoid. (*j*) Inductive reactance is reduced.

Observe that the induction process occurs in a certain order: voltage drives a current, which develops an electromagnetic field, which then induces a voltage to again initiate the process. During an eddy current test, a primary circuit (the test coil) induces eddy currents into a secondary circuit (the test material). Any factors that affect current flow in the secondary circuit, such as primary/secondary coupling or conductance variations, will affect the amplitude of both $V_L$ and of the inductive reactance in the primary circuit.

Variations in the test material change not only the test coil's inductive reactance but also a quantity called effective resistance. Although the resistance of the coil wire itself does not change, the eddy currents in the test material encounter friction as they circulate, thus dissipating a portion of their energy as heat. That is, the secondary circuit acts as a load on the primary circuit, with electrical energy converting to thermal energy. This energy loss in the circuit is counteracted by an increase in $V_R$ to keep the coil current constant. Thus, both $V_R$ and $V_L$ vary with change of test material properties during an eddy current test.

## 8.3.3 Signal Output

### *Voltage Plane*

As stated previously, there is a 90° phase difference between $V_R$ and $V_L$. These two voltages can be vectorially added to produce a quantity called $V_T$, which is the total voltage in the primary circuit, the output of the instrument's alternating current generator.

Figure 8-12 illustrates a voltage plane diagram with $V_R$ and $V_L$ values plotted as the base and elevation of a right triangle, and $V_T$ as the hypotenuse. Thus the Pythagorean theorem

$$c^2 = a^2 + b^2$$

may be expressed using voltage values

$$V_T^2 = V_R^2 + V_L^2$$

and restated to solve for $V_T$ by vector addition:

$$V_T = \sqrt{V_R^2 + V_L^2}$$

The basic information available from an eddy current test is the magnitude of $V_T$ and its phase relative to $I_P$. As shown in the sinusoids of Figs. 8-10b and 8-10d, $I_P$ is in phase with $V_R$. $I_P$ can therefore be placed along with $V_L$ on the horizontal axis of the voltage plane. Although test output could be shown as a sinusoid, indicating variation in magnitude and phase of $V_T$, a display showing just the tip of the vector arrow as a dot, called "vector point display," provides all necessary information in a simple

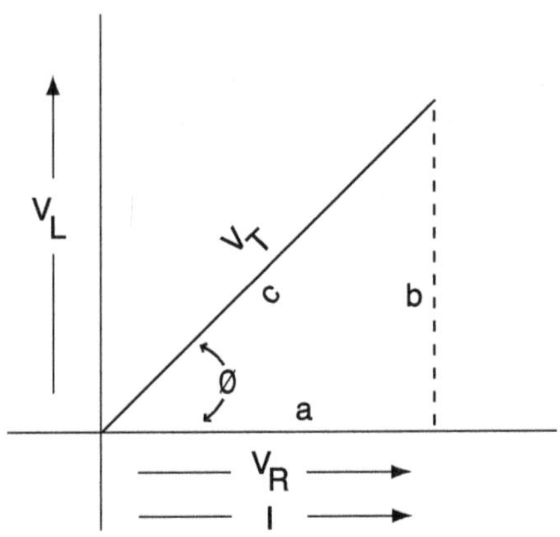


**Figure 8-12** Voltage plane.

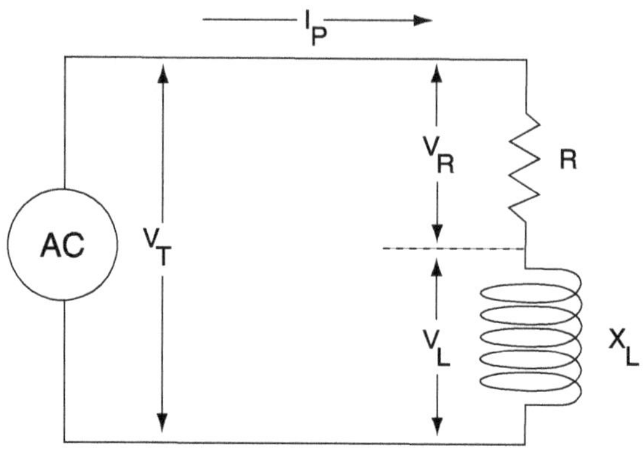


**FIGURE 8-13** Coil circuit driven by alternating voltage.

manner and lends itself especially well to eddy current signal analysis. When the magnitude and phase of $V_T$ are plotted on the voltage plane, the tip of the $V_T$ vector indicates the magnitudes of $V_R$ and $V_L$. Figure 8-13 summarizes how the various voltage and impedance components fit into a coil circuit driven by alternating voltage, with the resistive and reactive properties separately identified.

### *Impedance Plane*

Just as $V_R$ and $V_L$ can be combined into $V_T$, the combined effects of $R$ and $X_L$ on the alternating current in the coil can be expressed as a quantity called impedance. Specifically, *impedance amplitude* (Z) is the magnitude of the vector sum of inductive reactance and resistance, and is the coil's total opposition to current flow. That is

$$Z = \sqrt{R^2 + X_L^2}$$

*The impedance phase angle* ($\phi$), the proportional relationship between inductive reactance and resistance, can be calculated from:

$$\phi = \arctan \frac{X_L}{R}$$

Impedance amplitude and phase angle are illustrated in Fig. 8-14.

It is important to understand that there can be only one current in the circuit, flowing through both the resistance and inductive reactance, and influenced by both $V_R$ and $V_L$. Ohm's law then shows how a voltage plane can be converted into a corresponding impedance plane:

$$R = \frac{V_R}{I}$$

$$X_L = \frac{V_L}{I}$$

$$Z = \frac{V_T}{I}$$

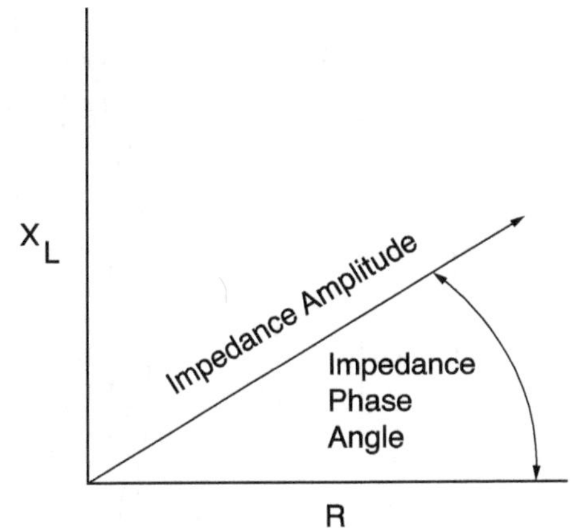


**FIGURE 8-14** Impedance amplitude and phase angle.

Although eddy current signal variations represent voltage variations as well as impedance variations, the impedance plane is the convention for expressing eddy current signal variations. That is, voltage variations are used to represent impedance variations in the coil. However, before exploring the different patterns of signal variation, it is necessary to examine how eddy currents behave in the test material in order to produce such patterns.

## 8.4 Eddy Currents

When a test specimen is brought into proximity to the alternating flux field of an eddy current coil, coil flux causes electrons in the specimen to circulate in a swirling eddy-like pattern; hence the term "eddy currents." Eddy current behavior depends on the properties of both the flux and the specimen itself.

### 8.4.1 Eddy Current Flow Characteristics

Eddy currents have a number of flow characteristics that affect their test performance:

1. They flow only in closed, concentric loops. Their flow paths are circular when unimpeded by the intrusion of material boundaries or discontinuities. Also, the flow paths are parallel to the turns of the bobbin-type coil shown in Fig. 8-15 and perpendicular to the axis of the coil's flux field.
2. The orientation of the coil to the test material therefore determines the orientation of the eddy current flow pattern in the test material. Orientation of the coil to the test material can be controlled and varied for optimum results by selection of the proper coil configuration. Several options are shown in Section 8.5.2 of this chapter.
3. Discontinuities are detectable by the eddy current method in proportion to the degree to which they disturb the flow pattern. Thus, a discontinuity is least detectable when its longest dimension is parallel to eddy current flow

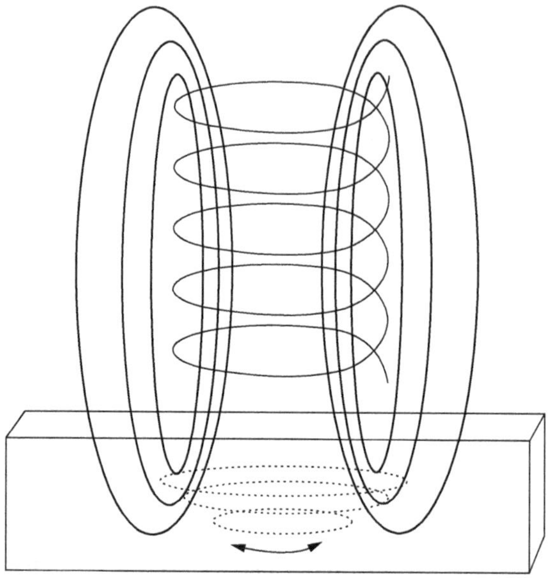

**Figure 8-15** Bobbin-type coil's flux and eddy currents.

paths (see Fig. 8-16a) and most detectable when the longest dimension is perpendicular to the flow paths (see Fig. 8-16b). Discontinuities with smaller volumes may not be detectable when oriented parallel to the flow paths. Ensuring that discontinuities of all likely orientations are detectable is an important part of test coil design and selection. Moreover, eddy currents always follow the path of least resistance around nonconducting obstacles, flowing under long, shallow discontinuities and flowing around short, deep discontinuities.

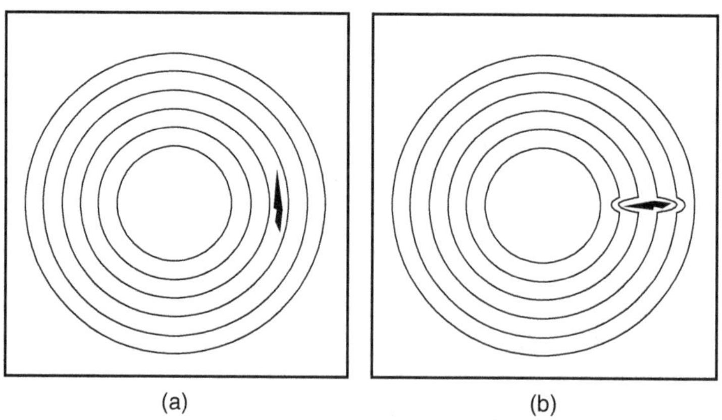


**Figure 8-16** Discontinuities in eddy current flow patterns. (*a*) Discontinuity parallel to flow paths. (*b*) Discontinuity perpendicular to flow paths.

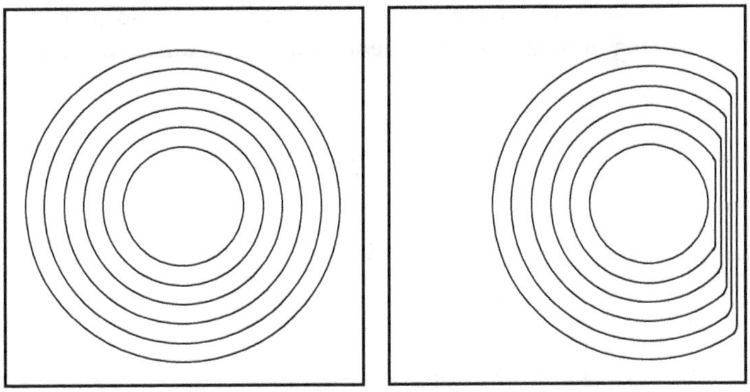


**Figure 8-17** Effect of material boundaries. (*a*) Eddy currents undisturbed by material boundaries. (*b*) Eddy currents compressed by material boundaries.

4. Eddy currents behave like compressible fluids. Although the flow paths are circular as long as the eddy currents are undisturbed by non-conducting material boundaries and discontinuities (see Fig. 8-17a), the flow paths will distort and compress to accommodate intrusions into their flow (see Fig. 8-17b).
5. Since an alternating flux field develops eddy currents, their flow in the test material likewise alternates clockwise and counterclockwise. The frequency of alternation of the eddy currents depends on the frequency of alternation of the flux field.
6. Eddy current density varies in the test material as follows:
   a) Eddy currents exhibit a *skin effect*. That is, current density is maximum at the material surface and decreases exponentially with depth. Thus, in thicker materials, eddy current testing operates only on the outer "skin" of the test material and test sensitivity decreases rapidly with depth. Volumetric tests are possible only in thin specimens.

      Skin depth ($\delta$), also called standard depth of penetration, is defined as the depth at which eddy current density (the portion of electrons active at a particular depth as compared to the material surface) has decreased to $1/e$, where "e" is the so-called natural logarithm, the number 2.71828, a device representing the natural rate of decay for many phenomena. Eddy current density for one, two, and three skin depths calculates as:

$$1\delta = \frac{1}{e} = \frac{1}{2.71828} = 0.368 = 36.8\%$$

$$2\delta = \frac{1}{e^2} = \frac{1}{7.38906} = 0.135 = 13.5\%$$

$$3\delta = \frac{1}{e^3} = \frac{1}{20.08554} = 0.0498 = 5.0\%$$

Beyond three skin depths eddy current density is too small to provide a displayable signal. (See effective depth of penetration in paragraph 6c below.)

Standard depth of penetration in English and metric units for a particular material and test frequency can be calculated as follows:

$$\delta(\text{in.}) = 1.98\sqrt{\frac{\rho}{f \times \mu_r}}$$

$$\delta(\text{mm}) = 25\sqrt{\frac{\rho}{f \times \mu_r}}$$

where:
$\delta$ = standard depth of penetration
$\rho$ = resistivity
$f$ = frequency
$\mu_r$ = relative permeability

b) The skin depth formula is truly valid only in the case of infinitely thick test material and large coils. However, at a material thickness of at least five skin depths, where eddy current density is only 0.0067 percent of surface density, the effect of restriction in thickness is so slight that the material may be considered to be "effectively infinite," rendering the formula accurate enough for practical purposes, providing that coil size is adequate. Conversely, if coil size is adequate but material thickness is restricted, current density at the opposite surface will exceed the calculated values. This is a fortunate circumstance, enhancing the possibility of volumetric inspection on thin-walled specimens.

c) The extent of a coil's flux field varies with coil diameter such that effective eddy current penetration is approximately limited to the diameter of the coil. Consequently, if the coil is too small, current density at a particular depth will be less than that indicated by the skin-depth equation.

   *Effective depth of penetration* is defined as the depth at which eddy current density decreases to 5 percent of surface density. This is the minimum eddy current density necessary to develop sufficient secondary flux to change coil impedance by a displayable amount. Thus, effective depth will equal three skin depths only where coil diameter is at least as great as three skin depths. Coil diameter, however, is a double-edged sword: As diameter increases, sensitivity to small defects decreases.

7. Eddy currents exhibit a linear phase lag with depth. To visualize the phase lag phenomenon, one may imagine a tall glass of water filled with small ice cubes all the way to the bottom of the glass. If one dips a teaspoon a short distance into the glass and begins to stir, the ice cubes on the surface circulate first and those at greater depths go into motion progressively later as the energy introduced by the spoon proceeds toward the bottom of the glass. In similar fashion, as depth increases, eddy current activity is progressively delayed. Phase lag in the test material proceeds at the rate of one radian (57.3°) per standard depth of penetration. The phase lag signal indicates discontinuity depth and material thickness in eddy current testing.

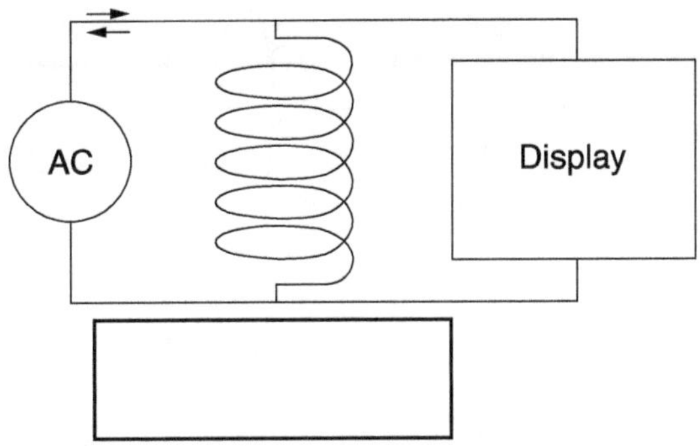


**FIGURE 8-18** Voltage causes current to flow.

### 8.4.2 Eddy Current Test Sequence

The induction process was previously modeled using a pair of coils as primary and secondary circuits. The eddy current test process is now summarized with an actual test specimen as the secondary circuit.

1. The test instrument's AC generator applies an alternating voltage of a certain frequency to the test coil, causing an alternating current to flow through the coil (Fig. 8-18).
2. The current in the coil develops a primary magnetic field around the coil (Fig. 8-19). The primary magnetic field initiates the following induction processes:
   (a) The coil's flux induces a back voltage into the coil, causing inductive reactance (Fig. 8-20).
   (b) The coil's flux induces a voltage into the test material, causing eddy currents to circulate (Fig. 8-21).
3. The eddy currents generate a secondary magnetic field, which reacts with the primary field that the coil is generating (Fig. 8-22).

   Any changes in the flow of eddy currents will cause changes in the magnetic field that the eddy currents return to the test coil. Any changes in this magnetic

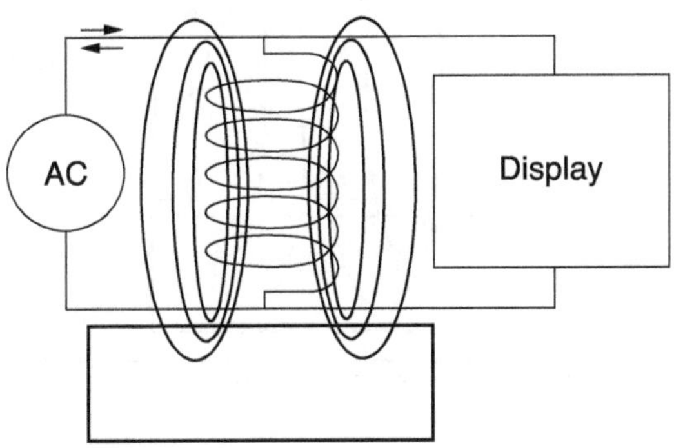


**FIGURE 8-19** Electromagnetism.

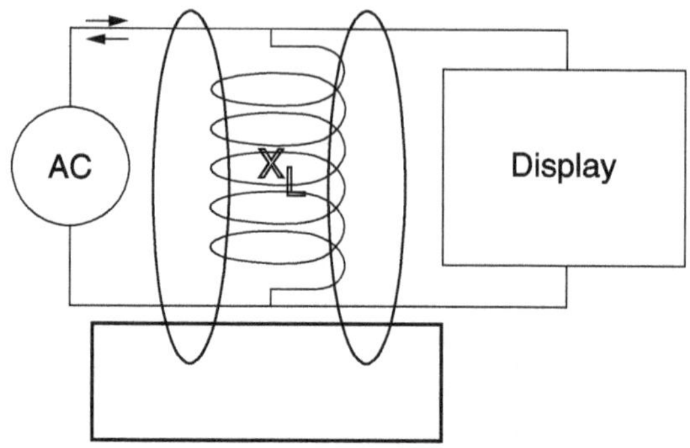


**Figure 8-20** Varying flux induces back voltage, which causes inductive reactance.

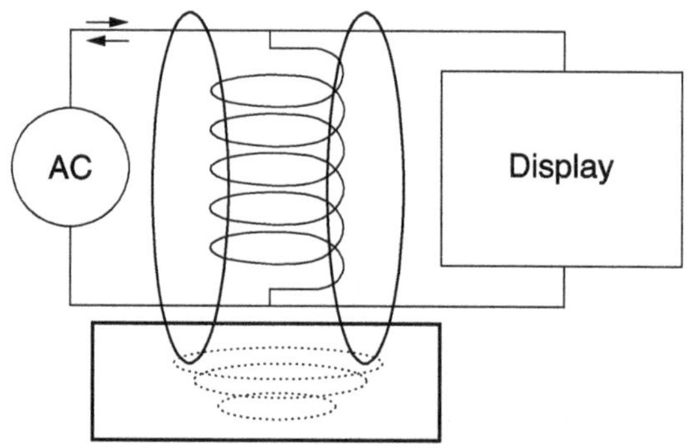


**Figure 8-21** Eddy currents in test material.

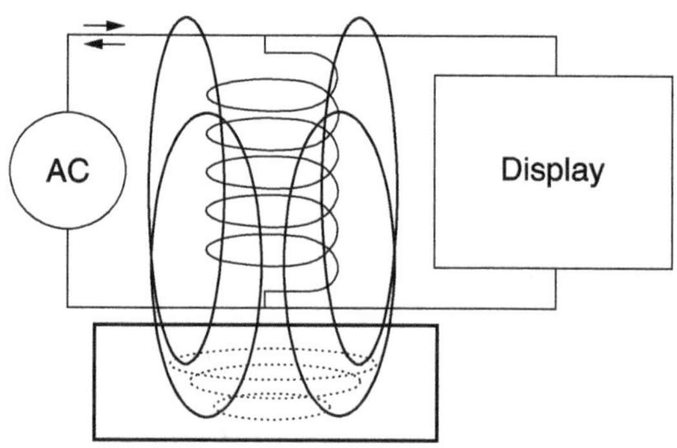


**Figure 8-22** Secondary flux interacts with primary flux.

field will cause changes in the inductive reactance and effective resistance of the coil, resulting in changes in current flow through the coil.

4. Finally, any changes in current flow through the coil will produce a change in the impedance indication on the instrument's display.

### 8.4.3 Test Performance

Test performance criteria do not seem to be as formally defined for eddy current testing as for ultrasonic testing and radiography. However, the same performance-related terminology can be usefully employed as follows:

- *Sensitivity:* The minimum size of discontinuity that can be displayed from a given material depth. Surface sensitivity is especially important with the eddy current method.
- *Penetration:* The maximum depth from which a useful signal can be displayed for a particular application.
- *Resolution:* The degree to which separation between signals can be displayed.

Test performance depends primarily on material conductivity, permeability, test frequency, coil design, and lift-off. Test frequency and coil design are readily selectable, and are therefore the primary controls over test performance. Guidelines for realizing optimum results for a specific application are given in Section 8.6. The following paragraphs summarize how major test variables affect performance.

1. *Conductivity*. The greater the conductivity of the test material, the greater the sensitivity to surface discontinuities, but the less the penetration of eddy currents into the material. Initially, this reduced penetration may seem contradictory, but is actually quite logical. As the coil's flux field expands, voltage is induced first on the surface and then at increasing depths in the test material. In high-conductivity materials, a considerable eddy current flow and thus a strong secondary flux are developed at the surface. This results in a substantial cancellation of primary flux. Because the primary flux has been greatly weakened, less primary flux is available to develop eddy currents at greater depth.
2. *Permeability*. This variable applies only to ferromagnetic materials. As material permeability increases, noise signals resulting from permeability variations increasingly mask eddy current signal variations. This effect becomes more pronounced with increased depth. Permeability thus limits effective penetration of eddy currents. The problem can be eliminated by magnetically saturating the material. However, the opportunity to saturate is limited by coil/test material geometry.
3. *Frequency*. Eddy current testing is performed within a frequency range of approximately 50 Hz to 10 MHz, although most applications are performed well within the extremes of that range. As test frequency is increased, sensitivity to surface discontinuities increases, permitting increasingly smaller surface discontinuities to be detected. As frequency is decreased, eddy current penetration into the material increases. In addition, as frequency is decreased, the speed of coil motion must be decreased in order to obtain full coverage.

   The test frequency for obtaining adequate penetration in a given material can be estimated using the skin depth equation or by using a penetration chart

| | Conductivity | | | 36.8% Depth of Penetration | | | | | |
|---|---|---|---|---|---|---|---|---|---|
| Metal | % IACS | Resistivity | Permeability | 1 kHz | 4 kHz | 16 kHz | 64 kHz | 256 kHz | 1 MHz |
| Copper | 100 | 1.7 | 1 | 0.082 | 0.041 | 0.021 | 0.010 | 0.005 | 0.0026 |
| 6061 T-6 | 42 | 4.1 | 1 | 0.126 | 0.063 | 0.032 | 0.016 | 0.008 | 0.004 |
| 7075 T-6 | 32 | 5.3 | 1 | 0.144 | 0.072 | 0.036 | 0.018 | 0.009 | 0.0046 |
| Magnesium | 37 | 4.6 | 1 | 0.134 | 0.067 | 0.034 | 0.017 | 0.008 | 0.0042 |
| Lead | 7.8 | 22 | 1 | 0.292 | 0.146 | 0.073 | 0.37 | 0.018 | 0.0092 |
| Uranium | 6.0 | 29 | 1 | 0.334 | 0.167 | 0.084 | 0.042 | 0.021 | 0.0106 |
| Zirconium | 3.4 | 70 | 1.02 | 0.516 | 0.258 | 0.129 | 0.065 | 0.032 | 0.0164 |
| Steel | 2.9 | 60 | 750 | 0.019 | 0.0095 | 0.0048 | 0.0024 | 0.0012 | 0.0006 |
| Cast steel | 10.7 | 16 | 175 | 0.018 | 0.0089 | 0.0044 | 0.0022 | 0.0011 | 0.0006 |

**TABLE 8-1** Typical Depths of Penetration

plotted from the skin depth equation for various conductive materials. Table 8-1 shows skin depths obtained at various frequencies for a selection of materials. However, because of the number of variables affecting eddy current behavior, this frequency should only be used as a starting point. The optimum frequency is best determined by experimentation.

4. *Coil Design.* Penetration and sensitivity are affected by conflicting requirements for coil geometry. Sensitivity to small surface discontinuities requires that the eddy current field be sufficiently compact so that it will be adequately distorted by the discontinuity. Conversely, penetration requires that the eddy current field extend to the required depth in the test specimen. The rules of thumb are that eddy current penetration is limited to a depth equivalent to coil diameter while sufficient sensitivity requires that coil diameter be limited to the minimum length of discontinuity to be detected.
5. *Lift-off.* Since flux density decreases exponentially with distance from the test coil, the amount of lift-off, or separation between the coil and test specimen, has a significant impact on sensitivity. The closer the coupling between coil and test specimen, the denser the eddy current field that can be developed, and thus the more sensitive the test to any material variable. Conversely, close coupling increases sensitivity to lift-off noise due to causes such as probe wobble.

## 8.5 Test Equipment

Basic eddy current hardware includes instruments, coils and coil fixtures, coil/test specimen transport equipment, recording devices, and reference standards. Test instruments can be either general purpose or designed for a specific application. Coils are usually designed for a particular category of application.

### 8.5.1 Eddy Current Instruments

A broad variety of eddy current instruments is available for use, from simple to complex. Although these instruments vary greatly in applications flexibility as well as size, most of them operate on similar principles.

In addition to a power supply, all eddy current instruments require at least three circuit elements: AC generator, coil circuit, and processing/display circuitry. The level of flexibility designed into each of these elements generally determines how eddy current instruments differ from each other.

- *AC generators* provide the voltage that drives the coil. They can operate at a single fixed frequency, provide a selection of switchable frequencies, be continuously variable, or even provide multiple frequencies simultaneously. In some instruments, there is adjustment for amplitude of the voltage applied to the coil.
- *Coil circuits* range from designs intended to work with only a single specific coil, a limited range of specified coils, or with virtually any coil configuration available.
- *Displays* can range from single LED and meter readouts to multi-frequency presentations on multicolor display screens.

### Dedicated Instruments

Dedicated instruments are designed for a specific application and are usually able to perform that application more efficiently than general-purpose instruments. Examples of dedicated instruments are crack detectors, coating thickness gauges, and conductivity meters. Conductivity meters, for example, can give direct readout of conductivity in IACS values. In addition, some crack detection instruments provide lift-off suppression to prevent noise signals caused by variations in coil to test material spacing. For example, the unit shown in Fig. 8-23 can be set to compensate for both lift-off variations and

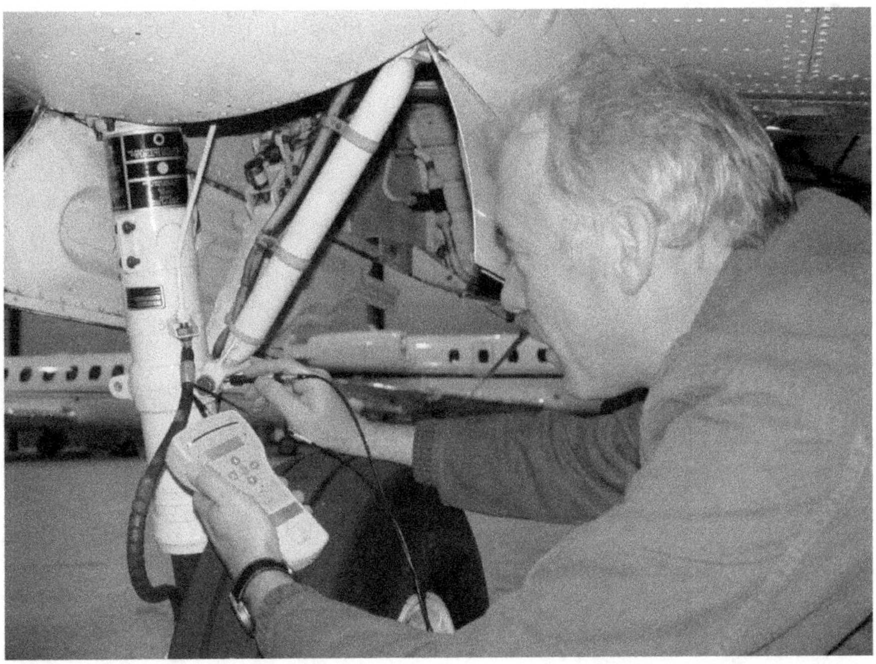

**FIGURE 8-23** Portable eddy current meter, capable of numerous crack detection applications. (*Courtesy of Foerster.*)

probe wobble. When there is sufficient work in a given application to justify investment in a single-purpose instrument, it is likely to be the best choice. However, one must be careful using eddy current meter-type instruments. Because they do not provide the quantity of information available from an impedance plane display, meter-type instruments can mislead less-qualified users. Since meter instruments can display only upscale or downscale deflections, they must be operated so that only one material variable is displayed. However, with impedance plane display instruments, each type of material condition deflects the display dot in a characteristic manner, facilitating separation of variables and interpretation of signals.

### Standard Impedance Plane Display Instruments

The AC generator of a standard impedance plane display instrument drives the test coil at only one frequency, which is usually selectable from a wide range of frequencies. These general-purpose instruments can perform an extensive variety of eddy current applications. The ability to view actual impedance plane signals provides the knowledgeable user a great deal of valuable information. Some newer impedance plane instruments have flat displays, offering enhanced portability, such as the equipment shown in Fig. 8-24. However, test-system-type instruments (Fig. 8-25) do not provide portability, but may be expected to operate 24 hours a day to accommodate continuous production at tubing, pipe, rod, or wire mills.

Impedance plane display instruments show variation of both inductive reactance and resistance during testing. Control functions of impedance plane instruments can include, but are not limited to, the following:

- *Frequency:* Adjusts the frequency at which the AC generator drives the test coil
- *Gain (Sensitivity, dB):* Adjusts amplification of the bridge output signal for display (see Mode of Operation)

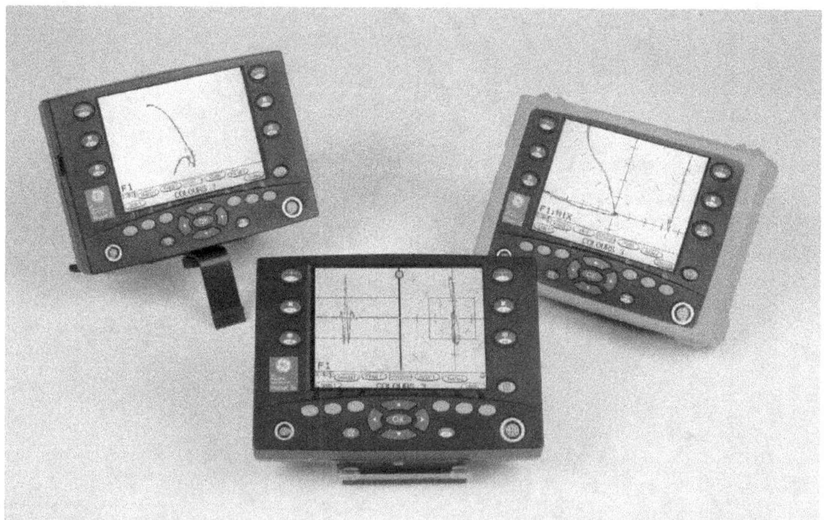

**FIGURE 8-24** Portable eddy current instruments with various display options, including impedance plane. (*Courtesy of GE Inspection Technologies.*)

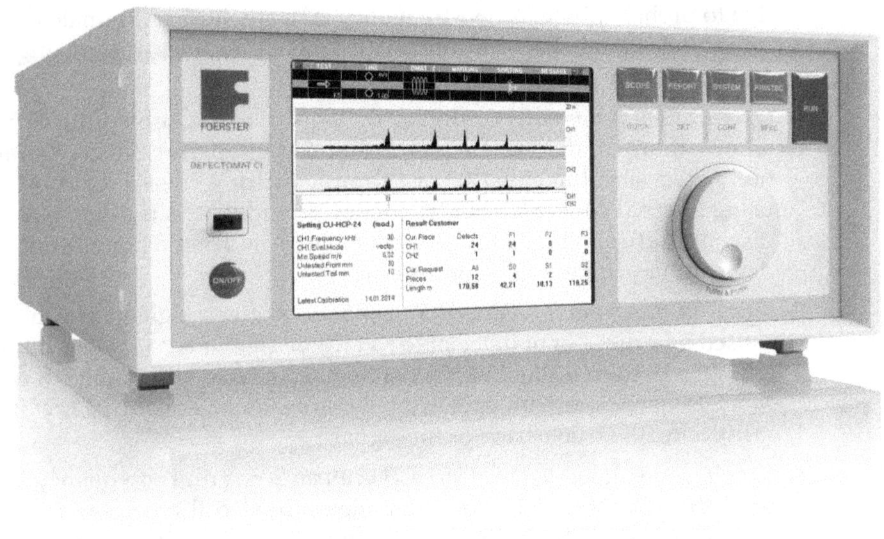


**FIGURE 8-25** Production line system instrument. (*Courtesy of Foerster.*)

- *Horizontal/Vertical Dot Position:* Adjusts dot position on the display
- *Phase Rotation:* Rotates the direction of dot deflection
- *Balance (Null, Zero):* Adjusts impedance to be identical on both sides of the bridge
- *Erase (Clear):* Erases the display
- *Gate:* Sensitizes some portion of the display to trigger an alarm
- *Filters:* Prevent display of signal above and/or below a certain frequency range
- *Probe Drive:* Adjusts voltage amplitude applied to the test coil
- *Horizontal and Vertical Display Amplification:* Allows one axis of the display to be expanded relative to the other for signal enhancement

#### *Multi-Frequency Instruments*

Development of multi-frequency instruments was one of the most significant advances in the evolution of eddy current testing hardware. These instruments practically eliminate what had been one of the most severe limitations of the method, the fact that signals caused by different material variables can vector into a combined signal that becomes difficult to interpret. In addition, they offer potential for substantial enhancement of performance. Driving the test coil at more than one frequency, multi-frequency instruments can not only display the test activity at each frequency separately but can also show a so-called "mixed output" of different frequency signals subtracted from each other. These capabilities result in the following four advantages:

1. *Suppression of Undesired Variables.* The ability to subtract signals from each other and display the difference as mixed output permits elimination of undesired signals on the display. This feature is the reason why multi-frequency instruments were originally developed: to suppress signals from steel supports during inspection of non-ferromagnetic tubes, as well as reduce lift-off noise

due to probe wobble. A two-frequency instrument can eliminate one source of unwanted signal. Each additional frequency enables the mixing out of an additional type of signal.

2. *Optimization of Normally Contradictory Test Variables.* Use of multiple frequencies allows more than one frequency-dependent performance variable to be optimized simultaneously. For example, during in-service tube inspection using internal coils, a higher frequency provides sensitivity to inner diameter discontinuities, while a lower frequency provides the penetration needed to detect outer diameter discontinuities.
3. *Signal Identification by Pattern Recognition.* A given signal deflection could be caused by a number of detectable conditions. However, each condition exhibits a unique pattern of behavior when viewed over a wide range of frequencies. Multi-frequency instruments display this behavior, enhancing the likelihood of identifying the true nature of the signal.
4. *Simultaneous Absolute/Differential Operation.* Some multi-frequency instruments have the advantage of allowing a single dual coil assembly to be operated simultaneously in both absolute and differential mode (see Mode of Operation, under Section 8.5.2), cutting in half the required testing time when the inspection must be performed using both of these techniques.

Two types of multi-frequency instruments are available: multiplexed and multichannel. Multiplexed equipment operates at only one frequency at a given instant, rapidly switching among the available frequencies. Thus, the test is not being performed simultaneously at all frequencies, although the display gives the illusion that this is the case. Multichannel equipment is the equivalent of having more than one eddy current instrument sharing a single display screen. Early multi-frequency instruments required that signal mixing be performed manually by the technician. More recent designs, such as the unit shown in Fig. 8-26, perform the mixing automatically.

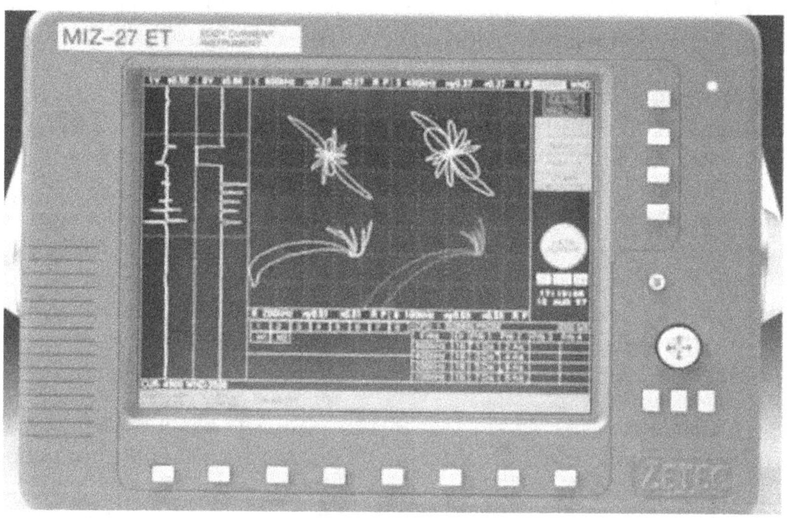


**FIGURE 8-26** Multi-frequency instrument. (*Courtesy of Zetec.*)

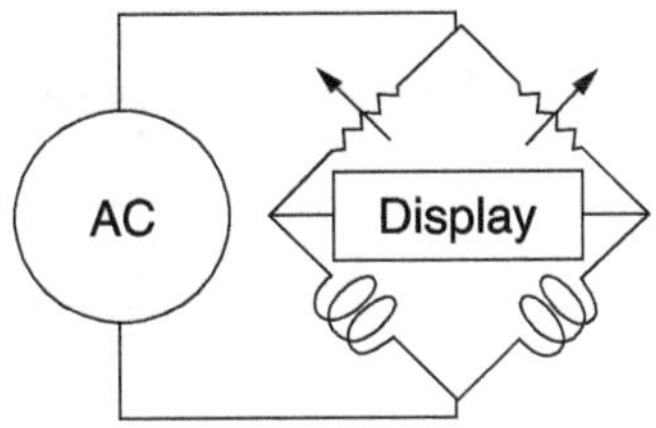


FIGURE 8-27 Bridge circuit.

## 8.5.2 Test Coils

Eddy current techniques are often classified according to the mode of operation and basic configuration of the test coil assembly. *Mode of operation* determines how the instrument interfaces with the test specimen, such as whether it is comparing coil input from the test specimen to a reference coil (absolute operation) or whether it is comparing coil input from two adjacent portions of the test specimen to each other (differential operation). *Basic configuration* determines how coils are physically packaged to "fit" the test object; that is, whether the coil approaches a portion of the test surface in a probe-like fashion (surface coil), whether it fully encircles the outer circumference of the test object (encircling coil), or whether it passes through the inside of tubular product (internal coil). Coil design, as well as magnitude and frequency of the applied current, all affect the electromagnetic field developed by the coil.

### *Mode of Operation*

With most eddy current instruments, the coil assembly is connected to the instrument via a bridge circuit, as illustrated in Fig. 8-27. Bridges are capable of detecting very small impedance variations. At the start of the test, the instrument operator balances the bridge to provide a reference signal. During testing, the display provides a readout of bridge imbalance caused by interaction of the coil with the test material.

*Absolute* coil configurations (Fig. 8-28) place a single coil on the test material and employ a second coil, called a *balance load,* remote from the test material to balance the bridge. Absolute coils detect any condition that affects eddy current flow. Although this

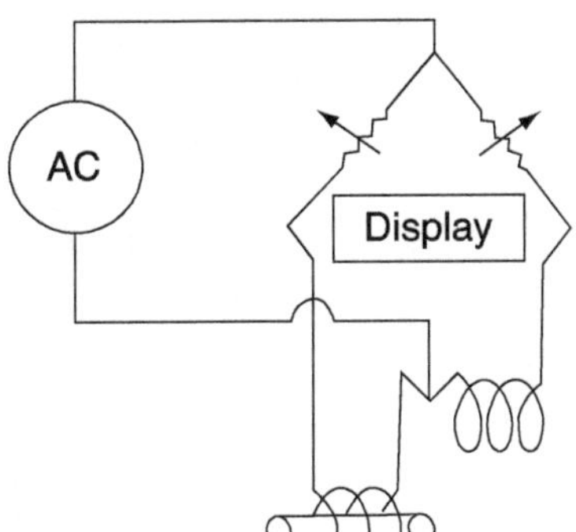


FIGURE 8-28 Absolute coil configuration.

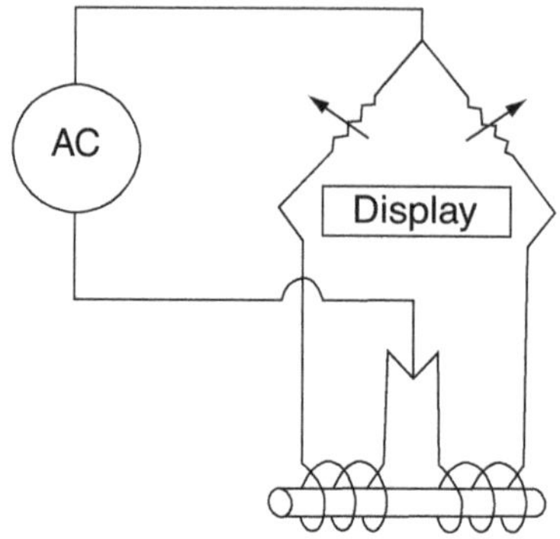


**Figure 8-29** Differential coil configuration.

means that they are capable of detecting any type of condition to which the eddy current method is sensitive, it also means that they are sensitive to potentially unwanted signals such as lift-off and material temperature variations.

*Differential* coil configurations (Fig. 8-29) use a matched pair of coils to perform a comparison. Both coils are coupled to the test material, with one portion of the test material being compared to another. Conditions sensed by both coils are not detected, whereas conditions sensed by only one coil are detected. This has the advantage of suppressing temperature and lift-off variations. Suppression of lift-off helps small discontinuities to be distinguished from lift-off noise. The downside to differential coils is that they provide no signal when a defect condition is simultaneously detected by both coils. Thus, differential coils will only display the ends of long discontinuities; they are not sensitive to gradual discontinuity variations and could ignore a long discontinuity entirely if its ends are very narrow. Differential coil signals are also difficult to interpret: The displayed signal represents the *difference* between two coils' impedances, rather than the impedance of a single coil's interaction with the test material.

*External Reference* coil configurations (Fig. 8-30) combine features of both the absolute and differential modes, placing one coil in contact with the test material and the other coil coupled to a reference standard. This technique provides an indication whenever the test material differs from the standard.

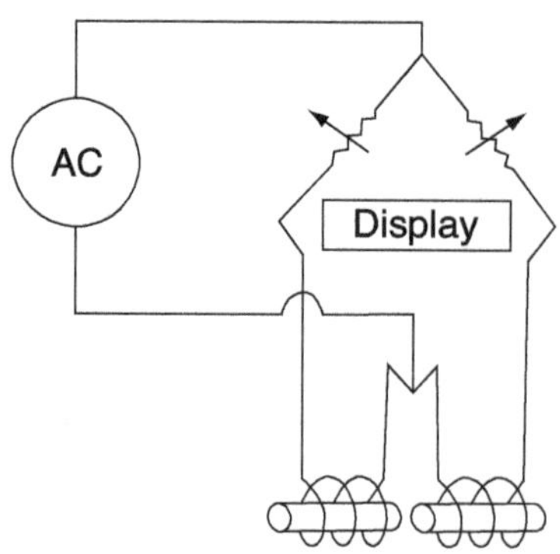


**Figure 8-30** External reference coil configuration.

### Basic Configurations

**Surface Coils** Surface coils are usually designed to be hand-held and are encased in probe-type housings for scanning material surfaces. Surface coils are available in different shapes and sizes to meet different application needs. There is vastly more variety in surface coil design than with encircling and internal coils. Some of them are astonishingly small, wound with wire finer than human hair. Some surface coils can perform a variety of applications, whereas others have been configured to fit a specific size and shape of test specimen. For example, surface probes have been fitted with guides to enable tracing the coil along the edge of turbine blades.

Most surface coils are "bobbin wound" like a spool of thread (Fig. 8-31a) and are designed so the axis of the coil is perpendicular to the surface of the test specimen. Such coils are sensitive to surface cracks and discontinuities that are oriented perpendicular to the test surface; they are generally insensitive to planar subsurface discontinuities. Planar discontinuities can be detected using so-called "horseshoe" or "gap" probes (Fig. 8-31b). These probes employ a pair of coils wound on each end of a U-shaped ferrite form so that the flux field flows from one pole of the "horseshoe" to the other and therefore parallel to the test surface. The eddy current field thus flows perpendicular to the test surface, providing sensitivity to planar discontinuities.

Wide surface coils permit rapid scanning and deeper penetration but are less effective at pinpointing the location of small discontinuities. They are often selected for conductivity testing because they tend to average out localized conductivity variations along material surfaces. Conversely, narrow coils are preferred for detecting and pinpointing the location of small surface discontinuities. Because of their smaller diameter electromagnetic fields, narrow coils are less susceptible to edge effect. Surface coils are made in numerous configurations to meet specific application needs.

Typical surface coil geometries include the following:

*Pencil Probe (Fig. 8-32a):* Shaped, as its name implies, to be held between the fingers and drawn across the test specimen.

*90° Probe (Fig. 8-32b):* Similar in function to a pencil probe, except that the coil is at a right angle to the probe housing for use where access is limited

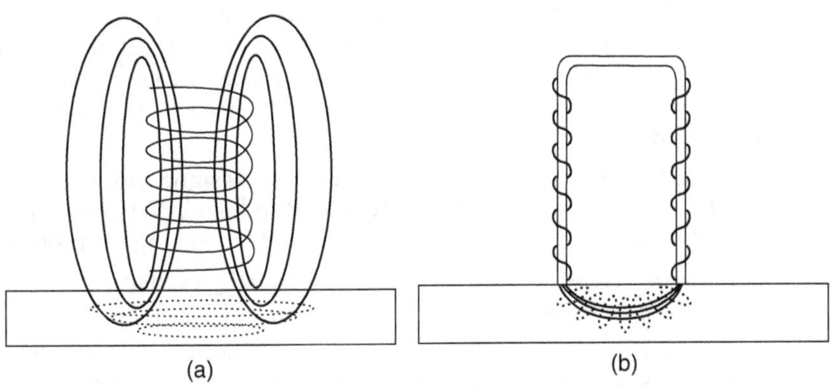


**FIGURE 8-31** Surface coils: (*a*) bobbin-wound, (*b*) horseshoe probe.

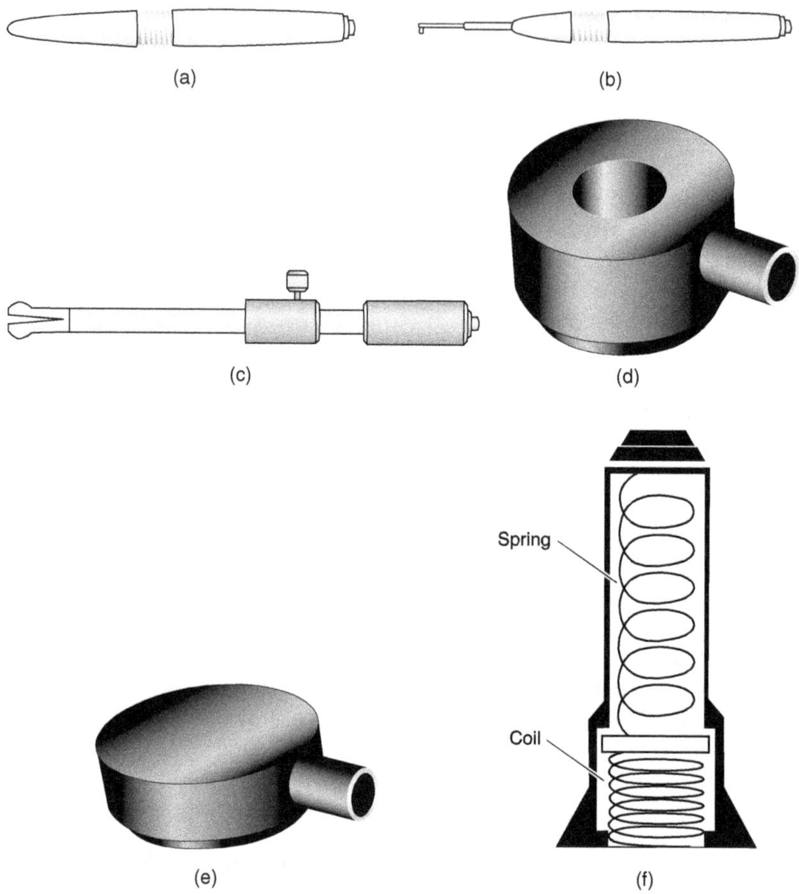


**Figure 8-32** Typical surface coil configurations: (*a*) pencil probe, (*b*) 90° probe, (*c*) bolt hole probe, (*d*) fastener (doughnut) probe, (*e*) pancake probe, (*f*) spring-loaded surface coil.

*Bolt Hole Probe (Fig. 8-32c):* Designed to fit inside bolt holes with the coil axis perpendicular to the wall of the hole. Manual bolt hole probes are often fitted with retainers so that they can be rotated at a certain depth in the hole and may be fork-shaped to ensure a snug fit. Motorized bolt hole probes are also available. Their output is generally shown on a time base display, allowing the user to determine circumferential position of discontinuities.

*Fastener (Doughnut) Probe (Fig. 8-32d):* A probe designed to fit above the fastener (rivet) holes of an aircraft fuselage. It is used to inspect for cracks around the hole and can be fitted with a clear plastic sight to aid in aligning the probe with the hole.

*Pancake Probe (Fig. 8-32e):* A low-profile coil generally used for scanning surfaces that have little or no curvature.

*Spring-Loaded Surface Probe (Fig. 8-32f):* The coil is mounted like a piston in a cylinder, spring-loaded so that it retracts into an outer housing when pressed against the test surface, thereby minimizing lift-off noise due to probe wobble.

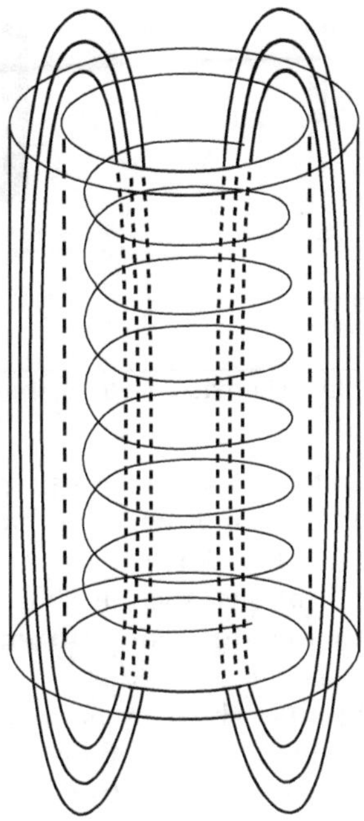

**Figure 8-33** Shielded coil.

Surface coils can be further classified according to the following functional features:

*Shielded coils* are encased within a cylinder of ferrite, a nonconductive, ferromagnetic material. As shown in Fig. 8-33, shielding contains the coil's flux field to prevent interaction with test material boundaries. However, since shielding only operates in the lateral direction, it does not impair penetration.

The *cross-axis* coil assembly consists of a pair of adjacent coils interacting with the test material, with the coil axes oriented 90° to each other. Thus, there is sensitivity to defects of all orientations. Cross-axis coils can be placed in a side-by-side configuration as shown in Fig. 8-34a, where one coil generates eddy currents parallel to the test surface

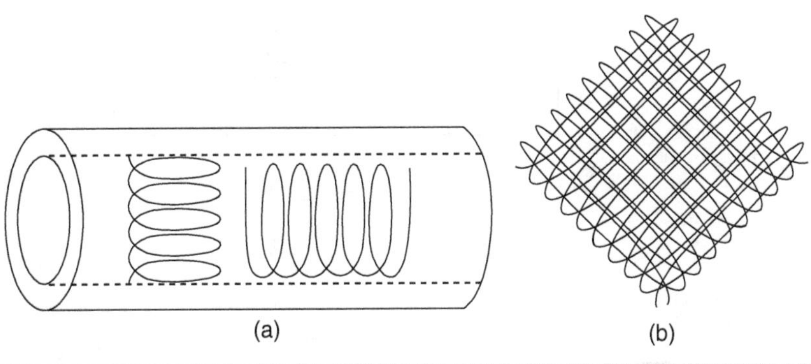



**Figure 8-34** Cross-axis coils: (*a*) side by side, (*b*) X-wound.

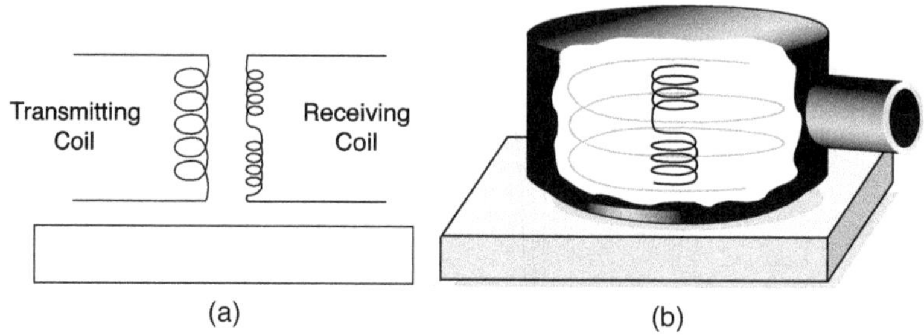


**FIGURE 8-35** Reflection coil: (*a*) schematic diagram, (*b*) actual configuration.

while the other coil generates eddy currents perpendicular to the test surface. Cross-axis coils can also be wound as a unit with alternate layers wound at 90° angles to each other (x-wound), as shown in Fig. 8-34b.

*Transmit–receive* configurations, such as reflection coils and through-transmission coils, use one coil assembly to induce eddy currents into the test material and a second coil assembly to sense the secondary field.

The *reflection coil* technique, shown schematically in Fig. 8-35a, employs two coil assemblies in a single housing, positioned on the same side of the test object. A large single outer coil functions as a transmitter surrounding a pair of smaller, stacked inner coils that form a receiver circuit (Fig. 8-35b). The term "reflection" coil results from the fact that the inner coils form a matched pair, wound in opposition. The outer coil induces eddy currents into the test specimen. Secondary flux developed by the eddy currents then induces voltage into the inner coil closest to the test specimen, causing an imbalance in the two-coil receiver circuit, thus providing a signal. Reflection coils have the advantage of being insensitive to temperature drift. They perform well at low frequencies and can function over a broad frequency range.

The *through-transmission* technique (Fig. 8-36), positions transmitting and receiving coil assemblies on opposing sides of the test object. It provides a valuable performance

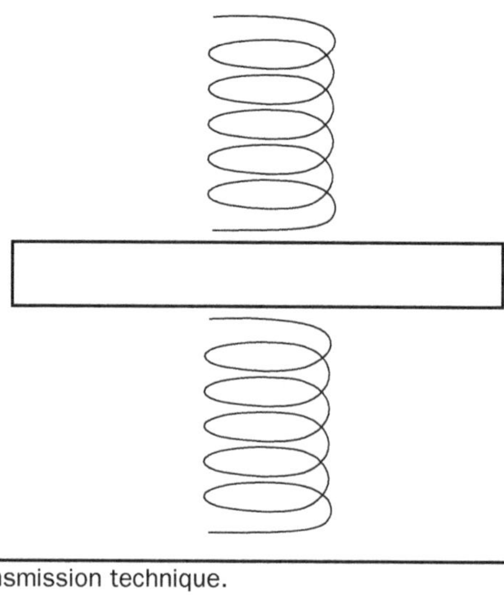

**FIGURE 8-36** Through-transmission technique.

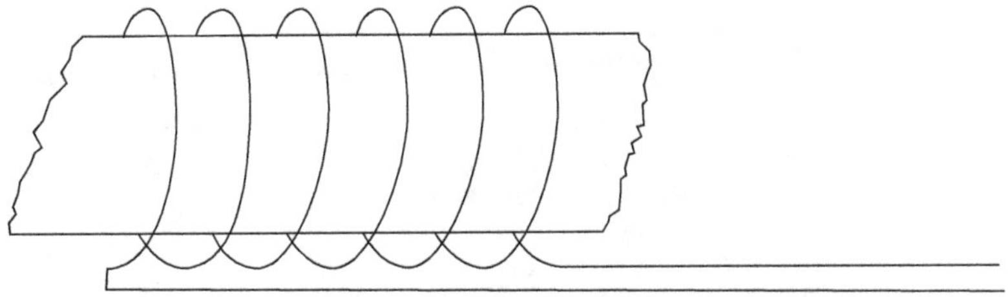

**Figure 8-37** Encircling coil.

advantage in that discontinuities can be detected at greater depths. However, this is offset by the fact that discontinuity depth cannot be displayed because the technique does not provide phase information.

**Hall Detector** Another transmit–receive technique employs a coil to provide primary flux for generating eddy currents, but uses a solid-state device called a Hall detector as a receiver. In 1879, E. H. Hall discovered that a small voltage is developed across a current-carrying conductor when an inducing magnetic flux is perpendicular to the direction of current flow. Although the voltage was not significant using typical conductors, certain semiconductors develop voltage of magnitude suitable for eddy current test purposes. An ordinary receiver coil operates according to Faraday's law, with induced voltage proportional to the time rate of change of the inducing flux, and therefore depends on the effects of frequency as well as amplitude, with the result that output decreases as frequency decreases. However, a Hall detector responds only to the instantaneous magnitude of the inducing flux, offering good performance at low frequencies. In addition, a Hall detector can be physically quite small.

In addition to the widely known coil types described above, a number of advanced designs have been developed in recent years. A notable contributor in this area is Atomic Energy of Canada, Limited, whose staff has developed a reputation for developing state-of the-art coils for power utility applications.

**Encircling Coils** Encircling coils (Fig. 8-37) completely surround the test material and are normally used for production testing of rod, wire, pipe, tube, and bar stock. Material tested with encircling coils should be centered in the coils by means of guides, so that the entire circumference will be tested with equal sensitivity. Production line encircling coils experience more heavy-duty usage than any other type of eddy current coil. Some of them may operate continuously, with product moving through them as fast as 5000 feet per minute. A heavy-duty, production line encircling coil assembly with interchangeable coil size modules is shown in Fig. 8-38a. Light-duty, handheld encircling coil pairs that can be fastened together for differential use or separated for absolute or external reference use are shown in Fig. 8-38b.

Because of the "center effect," eddy currents oppose and therefore cancel themselves at the center of solid cylindrical materials tested with encircling coils. Thus, discontinuities located at the center of rods and bar stock cannot be detected with encircling coils. Since encircling coils inspect the entire circumference of the test object, they cannot pinpoint the exact location of a discontinuity along the circumference.

(a)

(b)

**Figure 8-38** Typical encircling coil configurations: (*a*) production line, (*b*) handheld.

So-called *spinning coils* (Fig. 8-39), which are actually surface coils that revolve around cylindrical test material, are employed when identification of circumferential location is required in encircling coil applications. Since spinning coils couple to only a limited segment of test material circumference, they are not subject to the center effect. However, spinning coils inspect with a spiral pattern, so their material coverage depends on coil rotation speed versus material transport speed. Care must be taken to ensure adequate coverage.

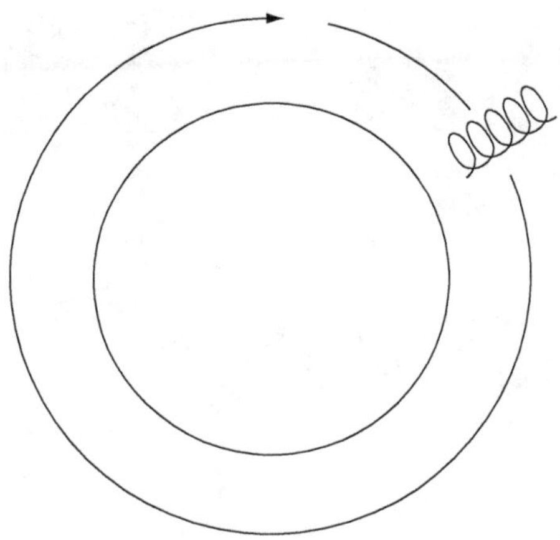

**Figure 8-39** Spinning coil.

**Internal Coils** Internal coils (Fig. 8-40) pass through the core of hollow product and are normally employed for in-service inspection of pipes and tubes. Like encircling coils, standard bobbin-wound internal coils inspect the entire circumference of the test object at one time, but cannot pinpoint the exact location of a discontinuity along the circumference. Again, special designs are available to pinpoint circumferential location of discontinuities. Both manual and automatic means are used to propel internal coils down the length of a long tube. Flexible "u-bend" coil assemblies are available for navigating extreme curvature of tubing. A selection of typical internal coils is shown in Fig. 8-41. Inspection of heat exchanger tubing, Fig. 8-42 is the most common application of internal coils.

### 8.5.3 Reference Standards

Test calibration or standardization is the process of adjusting the instrument display to represent a known reference standard so that the test can be a comparison between the test material and the reference standard. The validity of the test thus depends upon the validity of the reference standard. Moreover, the test system should be checked at regular intervals against the reference standard to ensure that it is operating properly and is still set up correctly for the test being performed. If a variation in instrument performance or setup is discovered, all materials tested since the last verification of proper performance and setup should be retested.

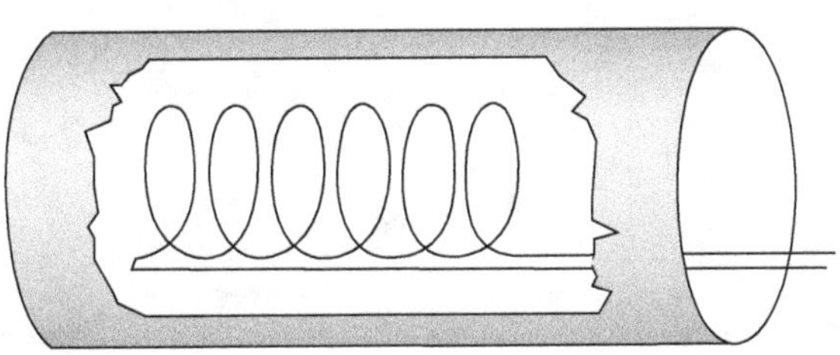

**Figure 8-40** Internal coil.

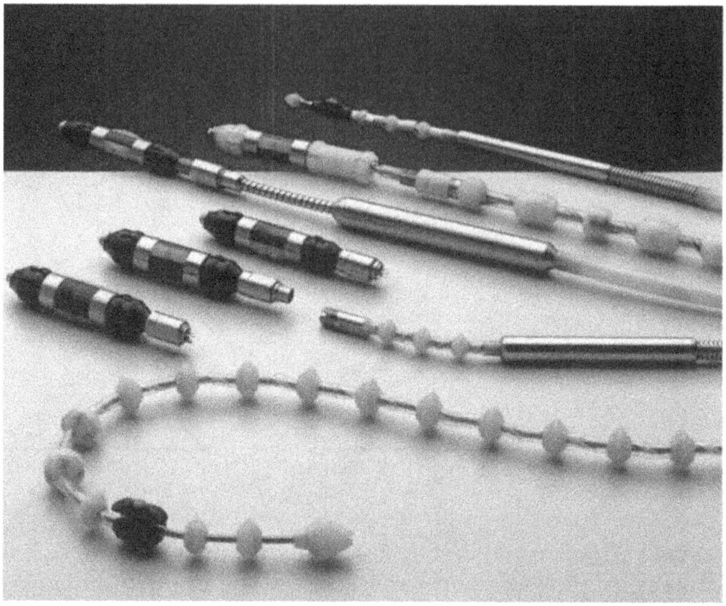

**FIGURE 8-41** Typical internal coil configurations.

**FIGURE 8-42** Inspection of heat exchanger tubing.

Since there is an infinite variety of discontinuity conditions, it is neither possible nor practical to have a set of reference standards so complete as to replicate every possible condition that could be detected during a test. Testing is therefore not a matter of matching each test signal with an identical reference signal. Instead, one obtains practical reference standards that contain a manageable number of representative discontinuity conditions. Signals that vary from these must then be interpreted through techniques such as impedance plane analysis.

The following rules apply to the selection and fabrication of standards:

1. The test standard should be of the same material, with the same wall thickness and configuration, and receive the same processing as the material to be tested.
2. The artificial discontinuities in the standard should model the natural discontinuities expected in the test material. For example:
   (a) Drilled holes can simulate pits.
   (b) EDM notches or saw cuts can simulate cracks.
   (c) Thickness reduction in tubing can simulate wear.
   (d) Heat treatment can simulate a conductivity change.
3. Artificial discontinuities in test standards should be sufficiently separated so that their signals will not interfere with each other.
4. Standards should contain no discontinuities other than those intended to produce reference signals.

Figure 8-43 shows tubing standards, as well as standards used to calibrate eddy current instruments for detection and evaluation of cracks in material surfaces and inside bolt holes. To augment standards containing artificial discontinuities, it is helpful to build a library of natural discontinuities by obtaining actual specimens removed from service or rejected during manufacture. These are invaluable in sharpening one's interpretive skills.

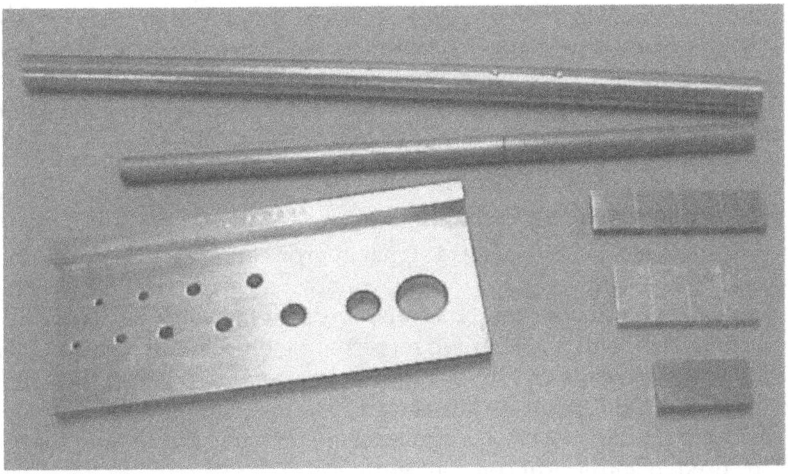

**FIGURE 8-43** Reference standards.

## 8.6 Eddy Current Applications and Signal Display

### 8.6.1 Eddy Current Display Media

Eddy current testing employs a variety of display media, including:

- Graphic displays, such as cathode ray tubes, liquid crystal screens, and electroluminescent screens
- Analog and digital meter displays
- Simple "go/no-go" displays such as light emitting diode illumination to indicate a deviation of impedance from a reference signal (e.g., detection of a surface crack)
- Recording devices, such as strip charts, magnetic tape, and flash media that enable post-inspection analysis of test data

Graphic display screens are the prevalent media for obtaining detailed, real-time test information. These displays include:

- A sinusoid showing amplitude and phase variations of total voltage across the coil
- An ellipse that tilts and varies in shape as the test material varies in geometric properties and conductivity
- Deflections of a time base sweep
- Impedance plane display
- X–Y output of impedance plane display

The sinusoid and ellipse presentations are essentially obsolete, having been replaced by impedance plane displays; but test systems using these outputs may still be in use. The time base is used to show discontinuity location over a defined range of test coil movement, such as the 360° sweep of a coil rotating in a bolt hole; however, time base display is limited in that it indicates signal amplitude but not phase information. Impedance plane display shows complete eddy current signal information and is generally the most useful for signal analysis. The X–Y output of the impedance plane can be sent to various recording and storage media, and is also available on the display screen of some instruments.

### 8.6.2 Impedance Plane Display

The impedance plane is the key to signal interpretation as well as to obtaining optimum test results. As stated earlier, the impedance plane is a graph of test coil impedance variations with inductive reactance displayed on the vertical axis and resistance on the horizontal. The point on the graph representing the test coil when it is remote from any conductive material is an important display reference position. It is called the "coil in air" or simply "air" point for surface and internal coils and the "empty coil" point for encircling coils (Fig. 8-44a). For the remainder of this discussion, the term "air point" will be used, regardless of coil configuration.

As the test coil is brought into proximity with a test specimen, the display dot makes a trajectory from the air point to a point representing the impedance of the test

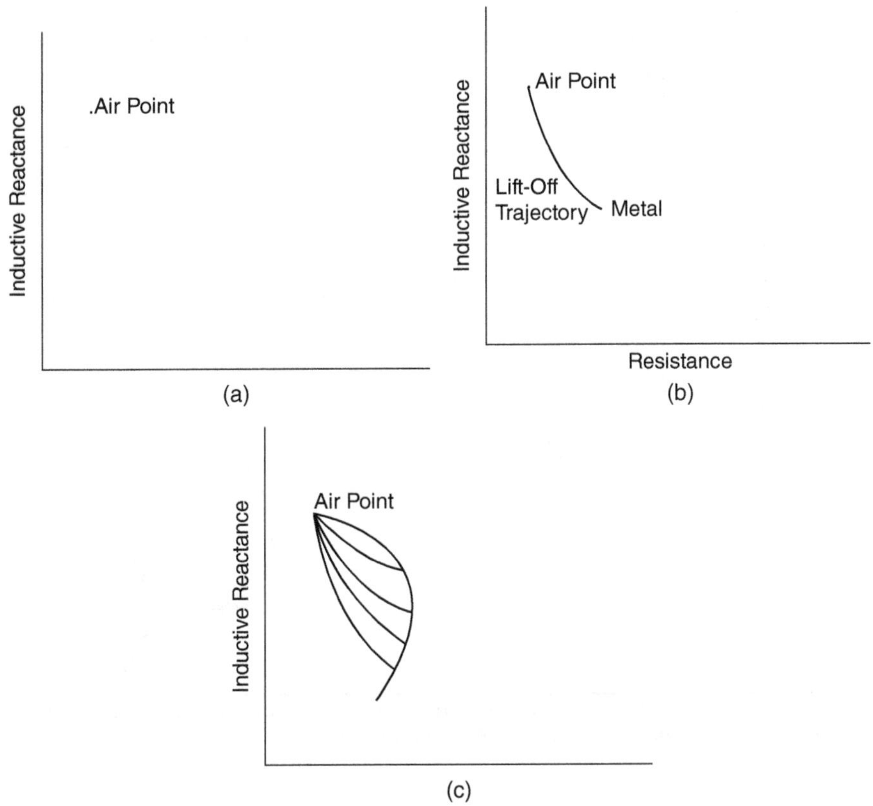


**FIGURE 8-44** Impedance plane: (*a*) air point, (*b*) lift-off trajectory, (*c*) conductivity curve.

specimen (Fig. 8-44b). Each condition that eddy current testing can detect is characterized by a unique position or pattern on the impedance plane, with test variables generally arranged along curves, such as the conductivity curve shown in Fig. 8-44c, that are obtained by joining the end-points of the series of lift-off curves. The curves may properly be termed "loci" in that they trace a set of points that represent the range of some variable to which eddy current testing is sensitive.

The impedance plane can be manipulated to advantage by altering parameters such as test frequency and coil diameter. The best presentation is usually a compromise, in that sensitivity, penetration, and resolution are all affected by changes in frequency and coil specifications. By studying impedance graphs and their manipulation, the practitioner can anticipate and optimize test signals for a given application.

### Normalization of the Impedance Plane

The sensitivity of the impedance plane to alteration of test parameters does present one problem: Changes in these parameters, since they change coil impedance, also change the position and scale of the curves that indicate test material variations. Thus, every change in frequency and/or coil specifications results in a new set of impedance curves. The problem can be illustrated by the effect of frequency variation

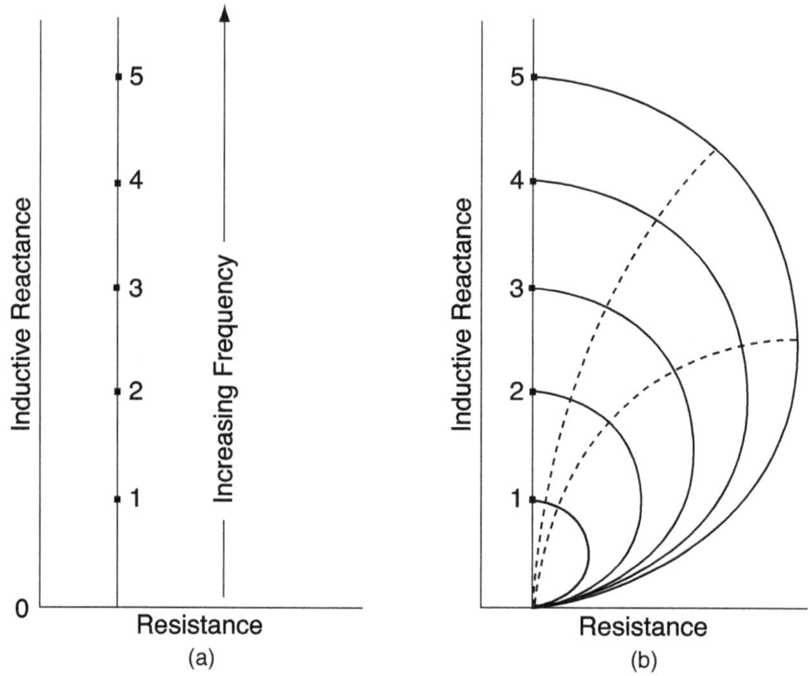


**FIGURE 8-45** Impedance plane before normalization. (*a*) Movement of air point with change in frequency. (*b*) Variation of conductance with frequency.

on inspection of very thin-walled tube with an encircling coil. Figure 8-45a shows that, without the tube inserted, changes in frequency cause changes in inductive reactance values for the coil. That is, five different frequencies result in five different positions for the air point. Figure 8-45b shows curves for variation in conductance from each air point to infinity, with the dashed lines showing the effect of inserting tubes of different conductance into the coil. As frequency is increased, a given tube's impedance takes a more clockwise position on a larger curve. However, the proliferation of curves can be solved with normalization. The impedance plane is normalized by converting the vertical axis from a scale of inductive reactance of the test coil in ohms to a scale of inductive reactance of the test coil compared to the inductive reactance of the coil's own air point; this is done by dividing actual coil inductive reactance by inductive reactance at the air point. This has the effect of mathematically pulling all of the curves shown in Fig. 8-45b into a single curve. Additionally, the horizontal axis needs to be normalized; this is done by subtracting variations in coil wire resistance from total resistance (one is interested in displaying test specimen variations, not coil variations) and dividing the remainder by the inductive reactance at the air point. Thus, the normalized impedance plane would be labeled $\omega L/\omega L_0$ on the vertical axis and $R/\omega L_0$ on the horizontal.

Most of the remaining illustrations will show a normalized impedance plane. However, in some cases it will be more convenient to illustrate the subject matter using a non-normalized impedance plane, with the assumption that frequency and coil properties are fixed.

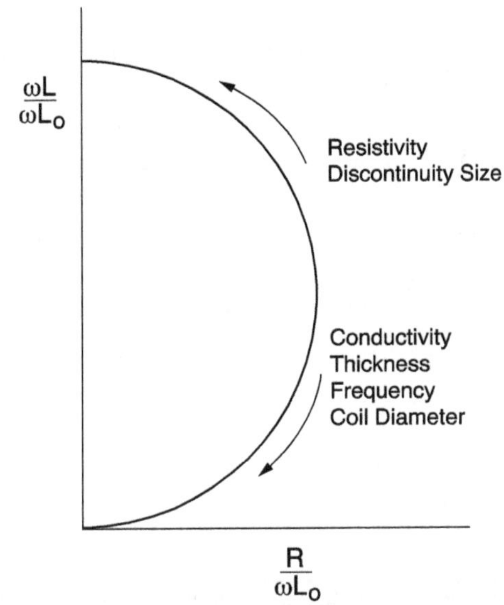


**FIGURE 8-46** Effect of test variables on impedance curve operating point.

### *Effect of Phase Lag on the Conductance Curve*

Figure 8-45b showed that impedance traces a semicircular path when conductance in a very thin-walled tube is varied. The semicircular pattern results when the specimen wall is so thin that phase lag is insignificant. Factors that increase current flow result in clockwise movement of the impedance operating point along the semicircle, while those that restrict current flow result in counterclockwise movement of the operating point, as seen in Fig. 8-46.

As specimen wall thickness is increased, display of test variables becomes more complicated, with the semicircle evolving into a comma shape and dimensional variations appearing as appendages to the comma curve. The shaded area in Fig. 8-47a shows how the semicircular impedance curve changes from a semicircle to a comma shape as a very thin-walled tube inside an encircling coil transforms into a solid bar. The semicircular curve is also valid for very thin-walled tubes inspected by an internal coil. In the case of a solid bar inside a long encircling coil, a "perfect comma" is achieved, with the lower end of the curve intersecting the origin of the graph at a 45° angle. The comma-shaped curve that evolves due to increased wall thickness is normally called a conductivity curve, although variations in frequency and coil diameter also cause movement along this curve on the normalized impedance plane. Internal coils testing thicker tubes, as well as surface coil applications, produce a more flattened impedance curve, similar to that shown in Fig. 8-47b.

### *Lift-Off Curves*

When a test coil is remote from any conductive material, impedance is at a position of high inductive reactance and low resistance. The high value for the air point on the inductive reactance scale occurs because there is no secondary flux available to reduce primary flux; the low value on the resistance scale occurs because the only resistance detected is that of the coil wire.

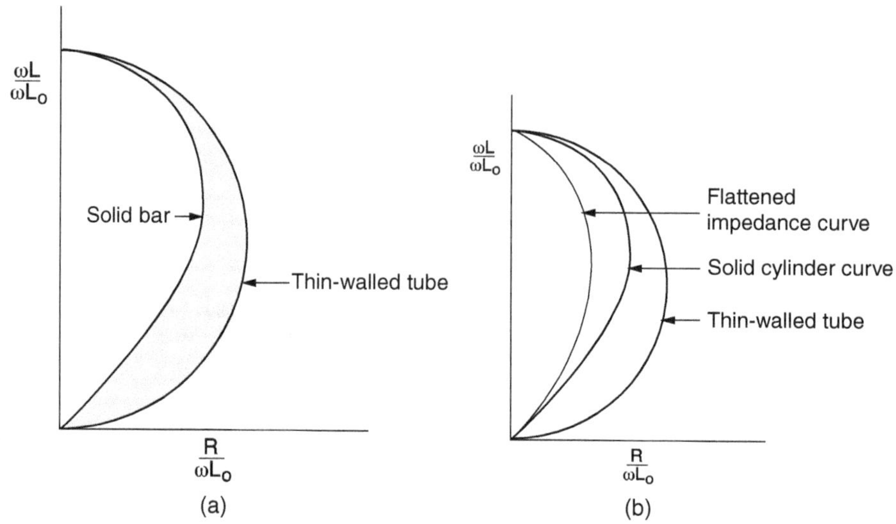


**FIGURE 8-47** Variation of impedance curve shape with specimen configuration. (*a*) Transition of impedance curve from semicircle to comma shape. (*b*) Impedance curves for internal coil testing thicker tubes and for surface coils.

As the coil approaches a conductive and/or ferromagnetic object, the impedance of the coil changes and the display dot moves. If the coil approaches a non-ferromagnetic conductive specimen, secondary flux cancels a portion of primary flux, resulting in a decrease in inductive reactance, as shown in Fig. 8-48a. The more conductive the test material, the greater the cancellation of primary flux and the further downward the display dot moves. Simultaneously, the specimen acts as a resistive load on the coil and the impedance point advances along the resistance scale. However, if the coil approaches a specimen that is both ferromagnetic and conductive, the specimen's flux adds to the coil's flux and the impedance point moves up both the inductive reactance and resistance scales, as shown in Fig. 8-48b. If the material is ferromagnetic and nonconductive

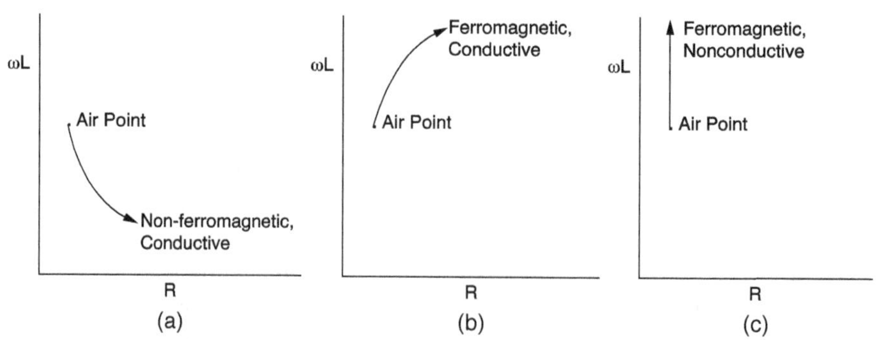


**FIGURE 8-48** Lift-off curves. (*a*) Non-ferromagnetic conductive specimen. (*b*) Ferromagnetic, conductive specimen. (*c*) Ferromagnetic nonconductive specimen.

(such as ferrite), the impedance point advances up the inductive reactance axis only, Fig. 8-48c, with no movement along the resistance axis. This effect is useful in orienting phase rotation to achieve a vertical deflection for inductive reactance variation. Regardless of direction, the display dot trajectory is the vector sum of movement along both axes and is called a lift-off curve.

Just as the spacing between a surface coil and the test material is called "lift-off," the spacing between either an internal coil or encircling coil and concentrically positioned test material is called "fill factor." As stated previously, lift-off is useful for measuring the thickness of nonconductive coatings on metals. Moreover, it can be used to measure the thickness of any nonconductive material that is simply resting on a conductive surface. Fill factor can be used to measure diameter of bars and rods placed inside encircling coils. Sensitivity to lift-off and fill factor depends on flux density and thus decreases as coil to test material distance increases. The decrease in sensitivity is, of course, nonlinear because flux density decreases according to the inverse square law.

The downside of sensitivity to lift-off and fill factor variations is that inadvertent movement of the coil relative to the test material will cause noise signals that can obscure the signals for which the test is being performed. Prevention of noise signals due to lift-off and fill factor variations is an important element in most eddy current inspections. Special techniques, fixtures, and instrument circuit designs have been developed to suppress these signals. When performing lift-off and fill factor applications, it is best to operate at high test frequencies. This reduces penetration into the test material, thus lessening the effects of material variables on test results.

### *Edge Effect*

Just as increasing the spacing between a surface coil and conductive material causes a liftoff curve, moving the coil toward the edge of the material will cause an edge effect trace (Fig. 8-49). This results from compression of the field and then reduction of current density as the coil is moved off the edge of the test specimen.

If the eddy current field simultaneously intercepts a discontinuity while approaching the edge, the two conditions will produce a combined response, rather than separate edge and discontinuity indications. Thus, the discontinuity may not be detected. The problem can be eliminated by scanning the coil parallel to the material edge at a

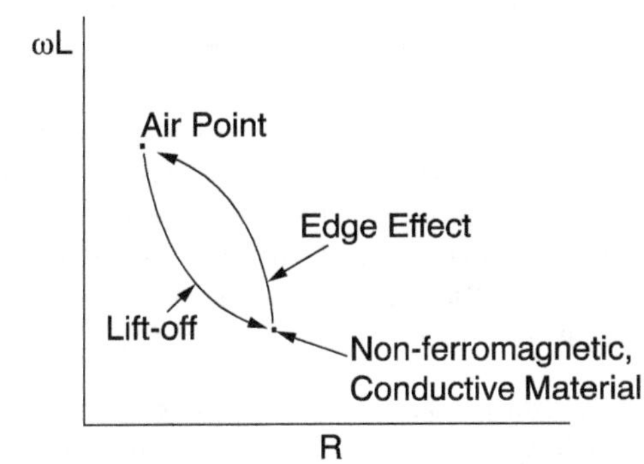


**FIGURE 8-49** Edge effect versus lift-off.

constant distance from the edge to first establish a uniform edge effect indication. Interception of a discontinuity will then cause an additional response. Edge effect is intensified by the wider eddy current fields developed by large-diameter coils and lower test frequencies. Use of smaller-diameter coils reduces edge effect; use of shielded coils virtually eliminates it.

### Surface Coil Impedance Curves for Material and Performance Variables

Although there are some differences among the signals obtained from surface coils, internal coils, and encircling coils, surface coils are ideal for demonstration purposes. This is because they are easily manipulated, are employed for a wide variety of applications, and retain good coupling to the test specimen when their diameter is varied (in contrast to internal and encircling coils, whose diameter must depend on the diameter of the test specimen). The following sections show a variety of material and performance variables using handheld surface coils.

### Conductivity Variation

Figure 8-50a shows the so-called conductivity curve, the locus of the end points of the lift-off curves for all non-ferromagnetic, conductive materials. In order for materials to be positioned on this curve, their thickness must be effectively infinite relative to electromagnetic penetration. The counterclockwise extreme of the conductivity curve represents zero conductivity; the clockwise extreme of the curve represents infinite conductivity. Varying test frequency shifts the impedance points for materials of different conductivities along the conductivity curve on the normalized impedance plane. At higher frequency (Fig. 8-50b), the impedance points for the various conductivities advance clockwise along the curve, with lower conductivity materials spreading apart while higher conductivity materials compress at the bottom end of the curve. Conversely, at lower frequency (Fig. 8-50c), the impedance points move counterclockwise, with higher conductivity materials spreading apart while the lower conductivity materials become compressed at the top end of the curve. Thus, higher frequencies provide greater separation for conductivity tests on lower conductivity materials, while lower frequencies provide greater separation for conductivity tests for high-conductivity materials.

Frequency adjustment also helps separate the lift-off and conductivity variables during conductivity tests, as shown in Fig. 8-51. At low frequencies, lift-off curves for low-conductivity materials are almost parallel to the conductivity curve. As frequency is increased, the operating point advances clockwise along the conductivity curve, increasing the angle between the lift-off curve and conductivity curve. Optimum separation is achieved near the so-called "knee" of the conductivity curve. Advancing beyond the knee will lengthen the lift-off curve, increasing sensitivity to lift-off noise caused by probe wobble.

From the standpoint of eddy current testing, conductivity is a material's most important variable. A material must be conductive in order for it to be tested by eddy currents. In addition, conductivity affects test performance. There are two groups of factors that can cause a material's conductivity to vary: those that result in useful applications and those that can sabotage the test process. Factors affecting conductivity variations that can result in test applications include the following:

1. *Variations in chemical composition.* The various metallic elements and alloys can be sorted as long as none of the materials have overlapping conductivities. Certain ranges of copper alloys have overlapping conductivities, for example, and therefore could not be sorted.

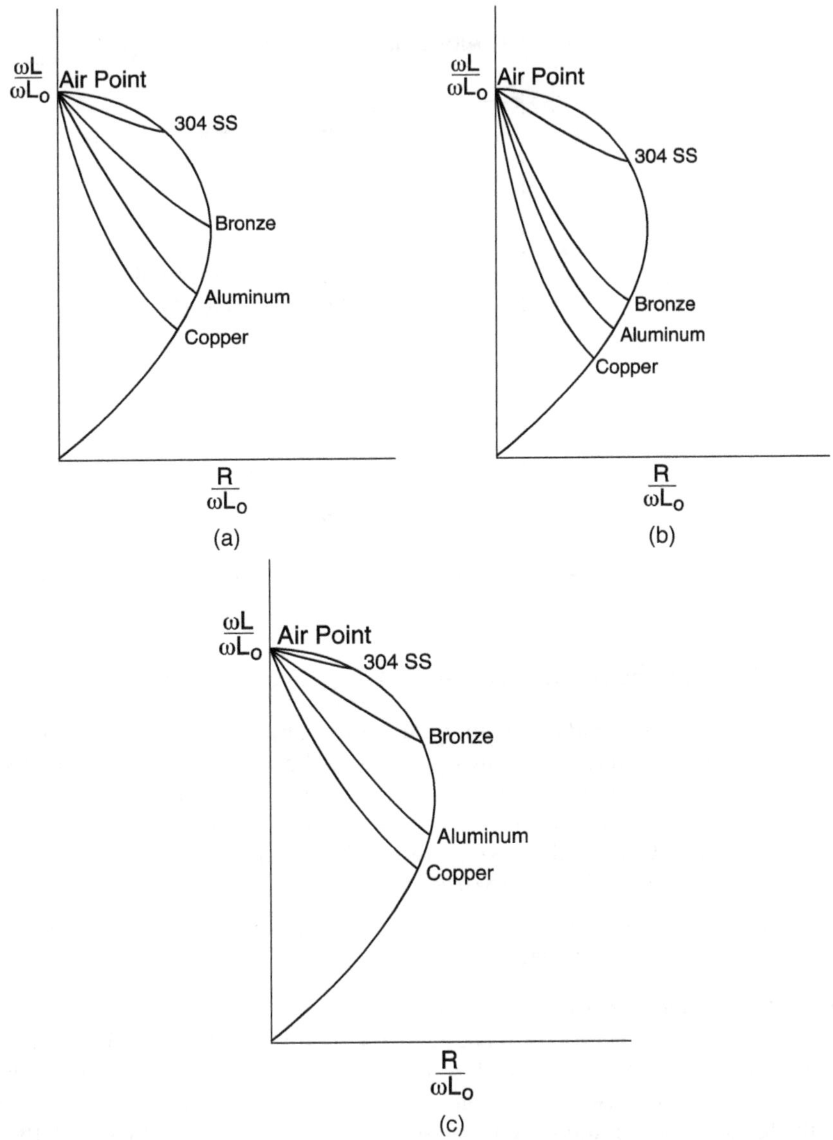


**FIGURE 8-50** Conductivity curve: (*a*) medium frequency, (*b*) high frequency, (*c*) low frequency.

2. *Mechanical processing.* Cold working affects lattice structure, causing minor conductivity changes.
3. *Thermal processing.* Heat treatment causes hardness changes that are detectable as conductivity variations. Because of this, eddy current testing can be used to some extent as a process control for heat treatment.
4. *Variations in thickness of plating or cladding* are a combination of both conductivity and dimensional variables.

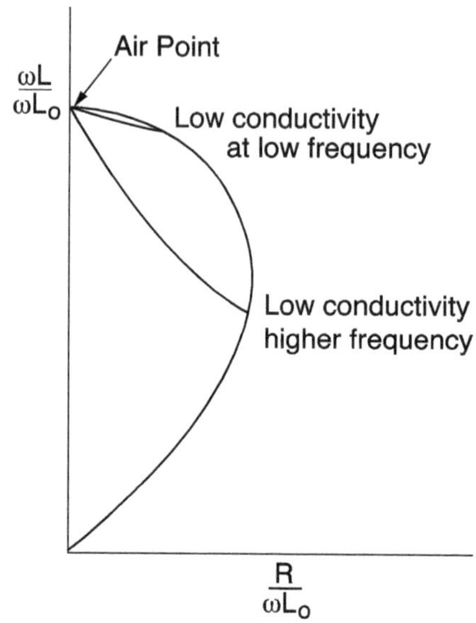


**FIGURE 8-51** Effect of frequency on separating signals.

Factors affecting conductivity that can negatively impact test performance include:

1. *Material temperature.* As material temperature increases, conductivity decreases. Variations in temperature can be caused by environment, materials processing, and eddy currents themselves. Care must be taken that material temperature does not vary during testing and that reference standards are the same temperature as the test material. Temperature variation during testing can cause the display dot to drift away from the balance point.
2. *Unrelieved residual stress* causes unpredictable conductivity variations. Thus, it is an undesirable variable.

### *Ferromagnetic Materials*

It follows that the effects of ferromagnetic test materials on the impedance of the coil will be similar to those obtained when a ferromagnetic core is used to enhance the flux density of an eddy current surface coil. The coil's alternating electromagnetic field causes the domains of the ferromagnetic material to rotate so that its poles alternate along with the coil, the two fields combining into a field of increased flux density. This increase in primary coil flux density (IP) causes an increase in back voltage, which results in an increase in inductive reactance. The impedance point therefore moves upward on the impedance plane.

The difference between a coil with a ferromagnetic core versus bringing the coil into contact with ferromagnetic test material is that since the core is a permanent part of the coil assembly, it causes the air point to be positioned higher on the inductive reactance axis. Conversely, proximity between the coil and ferromagnetic test material can be varied, with the result that the air point remains lower and inductive reactance increases only as the coil is brought closer to the test material. Moreover, the coil can be brought into proximity to ferromagnetic test materials of various conductivities. Thus, there is theoretically a conductivity curve for every permeability value.

| Discontinuity Depth | Current Density | Phase Lag In Material | Phase Lag on Display |
|---|---|---|---|
| Surface | 100.0% | 0.0° | 0.0° |
| $1\delta$ | 36.8% | 57.3° | 114.6° |
| $2\delta$ | 13.5% | 114.6° | 229.2° |
| $3\delta$ | 5.0% | 171.9° | 343.8° |

**TABLE 8-2** Eddy Current Density and Phase Lag for Subsurface Discontinuities

### *Discontinuity Signal Display*

One of the most important applications of eddy current testing is detection and evaluation of cracks and other discontinuities. The impedance plane responds to discontinuities depending on the density and phase lag of interrupted eddy current circulation. Displayed signal amplitude depends on the total quantity of deflected electrons, whose density is decreasing exponentially with increased depth. Displayed phase lag depends on the weighted average of the phase lags of all deflected electrons, whose phase lag is increasing linearly with increased depth. In fact, since a discontinuity of any size will possess some degree of thickness, it will interrupt eddy currents over a *range* of current densities and phase lags, depending on the dimensions, shape, and orientation of the discontinuity.

Table 8-2 shows eddy current density and phase lag at one, two, and three standard depths of penetration. The table indicates that displayed phase lag represents a round trip of lag time from the material surface to the discontinuity and back again to the surface. To obtain meaningful depth information, the skin depth formula must first be employed so that actual depth values can be assigned to 18, 28, and 38.

### *Subsurface Discontinuities*

Assuming that material thickness exceeds five skin depths and that a large-diameter coil is used, theory states that the display of a series of fixed-size voids (see Fig. 8-52a) would perform as indicated in Table 8-2 as their depth from the surface varies and produce the display shown in Fig. 8-52b.

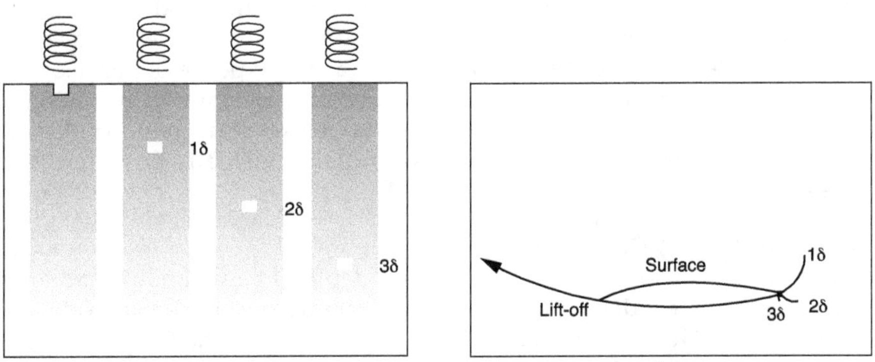


(a) (b)

**FIGURE 8-52** Subsurface discontinuities: (*a*) variation in void depth from surface, (*b*) signal display for voids of different depths.

The lift-off signal represents the test material surface and serves as the 0° reference. Since a so-called "surface void" must necessarily penetrate the surface, it will deflect eddy currents with a phase lag greater than 0°, with its signal exhibiting some rotation clockwise from the lift-off signal. Similarly, subsurface voids cannot exist only at depths of 16, 26, and 36, as indicated in Table 8-2; the voids must have some finite thickness and extend a measurable amount above and/or below the stated depths, with a consequent variation in displayed phase lag.

Changing the frequency affects penetration, sensitivity, and resolution. Figure 8-53 compares the display of a series of nine subsurface voids at 300, 50, and 1 kHz. At 300 kHz (Fig. 8-53a), penetration is just adequate to display the top three subsurface voids well-separated at approximately 18, 28, and 38. As frequency is decreased, the signals rotate counterclockwise on the display. Thus, at 50 kHz (Fig. 8-53b), deeper voids are displayed as penetration increases. Finally, at 1 kHz (8.53c), the nine voids are now only a small portion of one skin depth, with their signals virtually superimposed and lower in amplitude. Penetration has been enhanced at the cost of resolution and sensitivity.

### Surface-Breaking Discontinuities

The behavior of surface-breaking discontinuities is more complicated. The key to understanding the difference in the display pattern between a series of subsurface discontinuities of increasing depths and a series of surface-breaking discontinuities of increasing depth is, of course, to consider the effect of each on skin effect and phase lag. The subsurface series represents a void of a *fixed volume* whose position is being shifted down to locations of decreased current density and increased phase lag; the surface series represents a void whose *volume is expanding* into locations of decreased current density and increased phase lag. Thus, in that the surface void disturbs fewer additional electrons each time its depth increases by a fixed amount, its signal amplitude increases by progressively smaller amounts. This likewise causes the weighted average of deflected electrons to similarly increase by smaller amounts, resulting in displayed signal phase correspondingly increasing by progressively smaller amounts.

As with subsurface discontinuities, changes in frequency affect penetration, sensitivity, and resolution. Figure 8-54 compares the behavior of the surface-breaking voids in a 304 stainless steel plate at 100, 10, and 1 kHz. At 100 kHz (Fig. 8-54a), the eddy currents are concentrated on the surface, giving good resolution of the shallowest void, but there is insufficient penetration to resolve the deeper voids. As frequency decreases, penetration is gained at the expense of surface sensitivity. At 10 kHz (Fig. 8-54b), there is sufficient density distribution to obtain adequately resolved signals from the shallowest void as well as the deepest. At 1 kHz (Fig. 8-54c), decreased surface sensitivity and phase rotation virtually obscure the signal from the shallowest void.

### Shape of the Eddy Current Field

Terms such as "skin depth" and "effective depth of penetration" tend to imply that a decrease in current density and phase lag occur only in the downward direction in the test material. However, the coil's flux, which develops the eddy currents, extends in all directions away from the coil. Therefore, the eddy current field is necessarily somewhat ring-shaped, as shown by the field of the surface coil in Fig. 8-55a, with current density and phase lag varying laterally as well as downward, indicated by

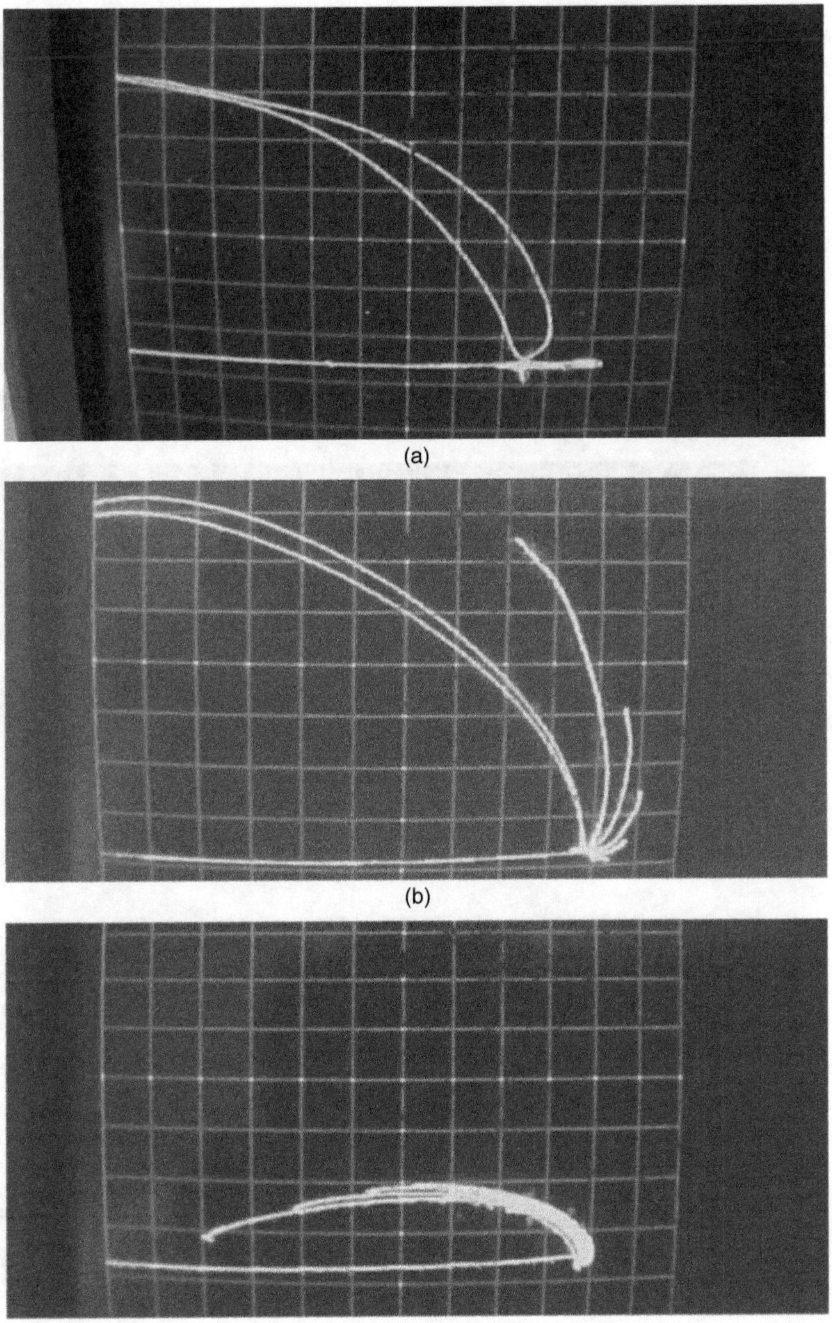

(a) (b) (c)

**FIGURE 8-53** Subsurface voids at different frequencies: (*a*) 300 kHz, (*b*) 50 kHz, (*c*) 1 kHz.

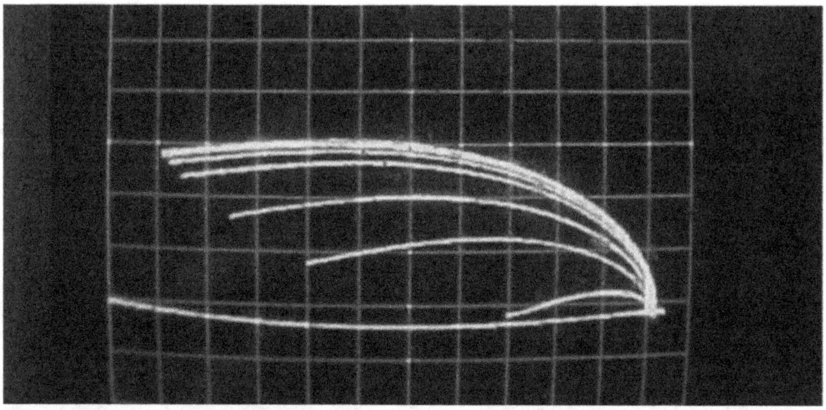

(a)

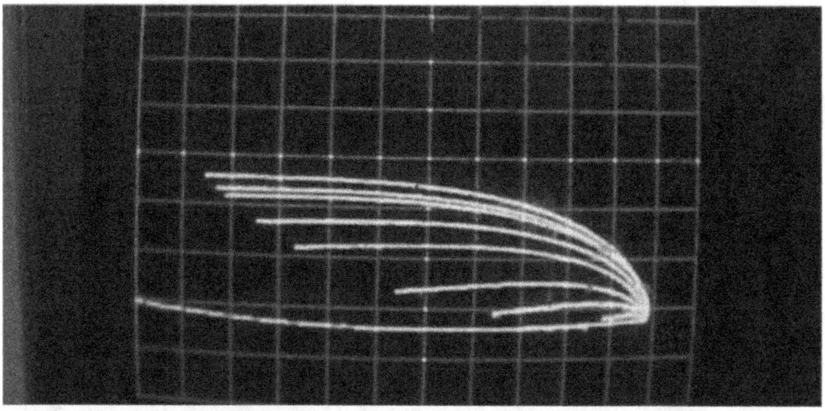

(b)

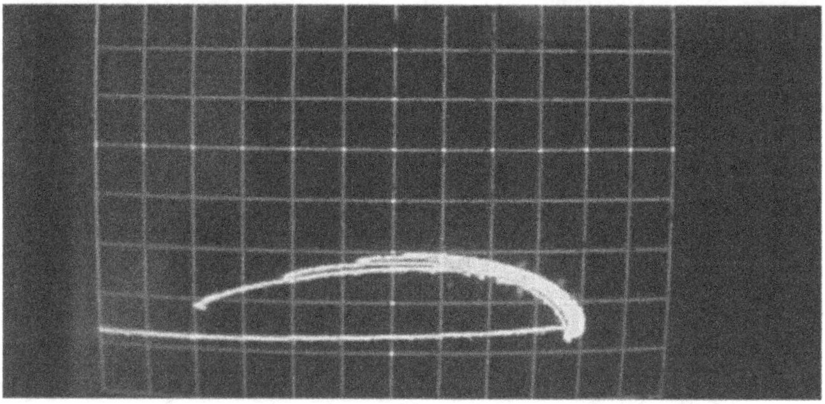

(c)

**Figure 8-54** Surface-breaking discontinuities: (*a*) 100 kHz, (*b*) 10 kHz, (*c*) 1 kHz.

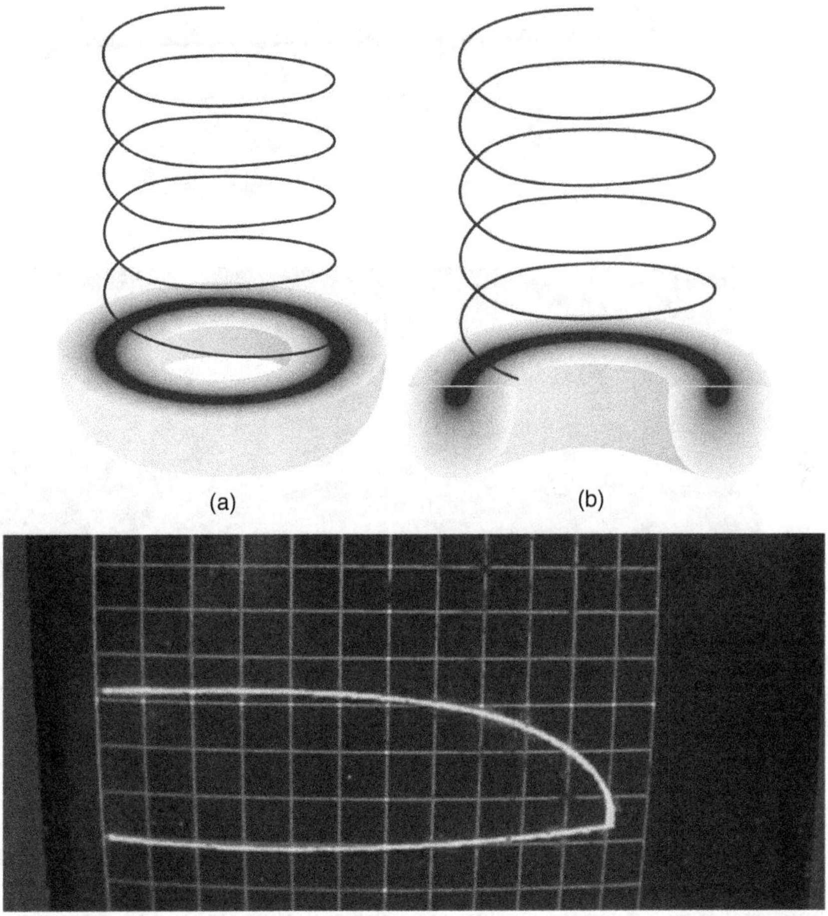

(a) (b)

(c)

**FIGURE 8-55** Effect of field shape on signal display. (*a*) Eddy current field shape for surface coil. (*b*) Cross section of eddy current field. (c) Signal for coil approaching surface-breaking discontinuity.

progression from black through increasingly lighter shades of gray in the cross-sectional view of Fig. 8-55b. These field properties become obvious when one examines the signal produced by a surface-breaking discontinuity (Fig. 8-55c). Close examination shows that the trace initially proceeds vertically, then curves increasingly toward the left, indicating an initially high phase lag that then decreases as the coil approaches the discontinuity.

### *Display of Discontinuity Orientation*

Figure 8-56a shows a portion of an aluminum bar with machined notches at angles decreasing in 10° increments from 90° to 10°. Figure 8-56b shows how phase lag can be used to indicate the varying orientation of these surface-breaking discontinuities.

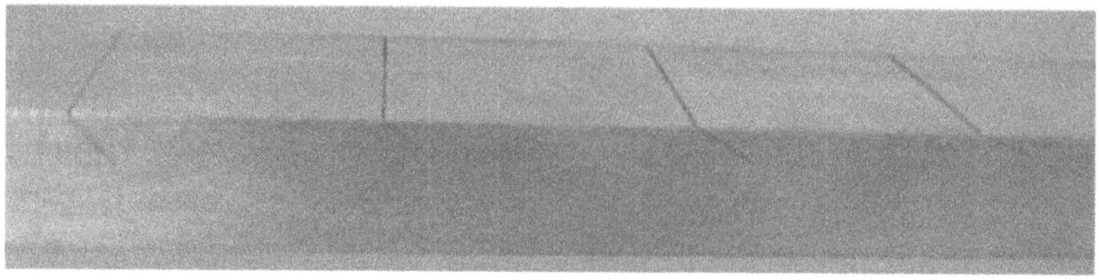

(a)

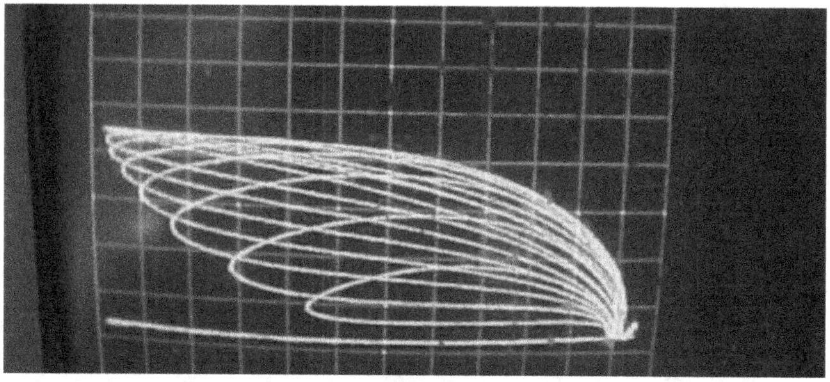

(b)

**Figure 8-56** Display of varying discontinuity orientations. (*a*) Bar with discontinuities of varying orientations. (*b*) Signal variation as orientation is varied.

As discontinuity orientation tilts away from the perpendicular, an increasingly open series of loops is traced on the display, indicating a progressively less rapid variation of phase lag. It is also possible to state the direction in which the discontinuity is leaning. If the signal trace advances clockwise as the coil approaches the discontinuity, then the field is being interrupted in an area of lower phase lag, signifying that the discontinuity is being intercepted where it breaks the surface. However, if the trace advances counter-clockwise, greater phase lag is initially indicated, the result of the discontinuity being intercepted deeper into the material.

### Enhancing Signal Display

In addition to the instrument gain control, which adjusts amplification of bridge output, most impedance plane display instruments provide capability to adjust the ratio of amplification for horizontal versus vertical display of test signals. This feature can be especially useful for improving signal resolution. Figure 8-57a shows a display of surface-breaking notches of increasing depth, with the inevitable reduction of phase separation as notch depth increases. In Fig. 8-57b, the vertical amplification has been increased at the expense of the horizontal, stretching apart the signals for improved resolution. The enhancement is accompanied by a corresponding amount of signal distortion, as indicated by the increased curvature of the lift-off trace. The user should be careful to return the horizontal/vertical display amplification to normal settings after completing the application, to preclude undesired distortion during subsequent tests.

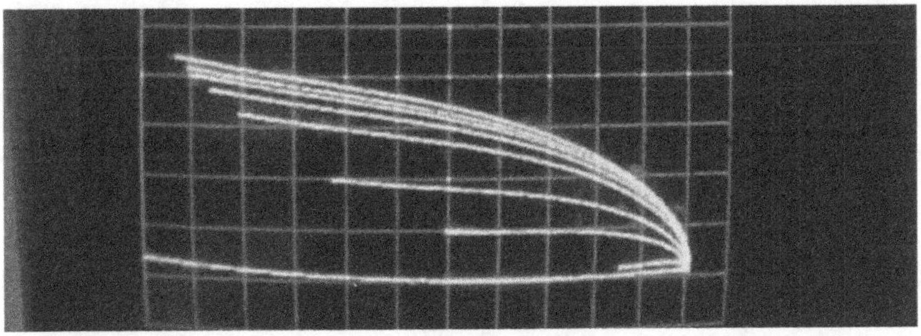

(a)

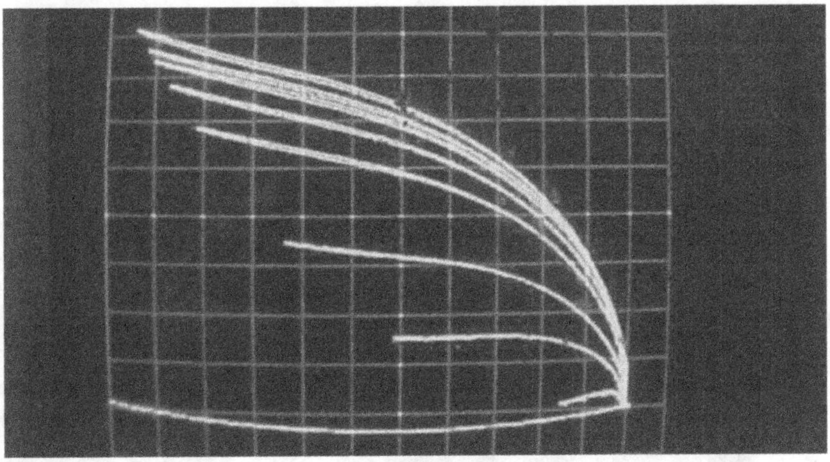

(b)

**Figure 8-57** Signal enhancement using display amplification. (*a*) Surface-breaking notches without enhancement. (*b*) Surface-breaking notches with enhancement.

### *Thickness Variation*

Thickness variations exhibit the same display behavior as subsurface discontinuities, except that they represent an infinite-size void whose depth is increasing. The phase rotation pattern is the same, but the signal amplitude is greater. Figure 8-58a shows the lift-off curves for a series of aluminum samples of increasing thickness. When the end points of the lift-off curves are joined, a spiral curve is produced. Such a spiral would be traced if the coil were drawn along the length of a wedge covering the same thickness range.

Frequency variation also follows the same pattern as for subsurface discontinuities. Decreasing frequency improves penetration and rotates signals from greater depth in a counterclockwise direction. Increasing frequency reduces penetration, but improves resolution of signals from thinner materials by rotating the signals clockwise. These effects of frequency variation are illustrated using a series of eleven aluminum steps ranging in thickness from 0.012 to 0.100 in. In Fig. 8-58b, at a frequency of 10 kHz, the thinner steps are well separated, but penetration is only sufficient to resolve the first nine steps. When frequency is reduced to 2 kHz (Fig. 8-58c), penetration is increased to

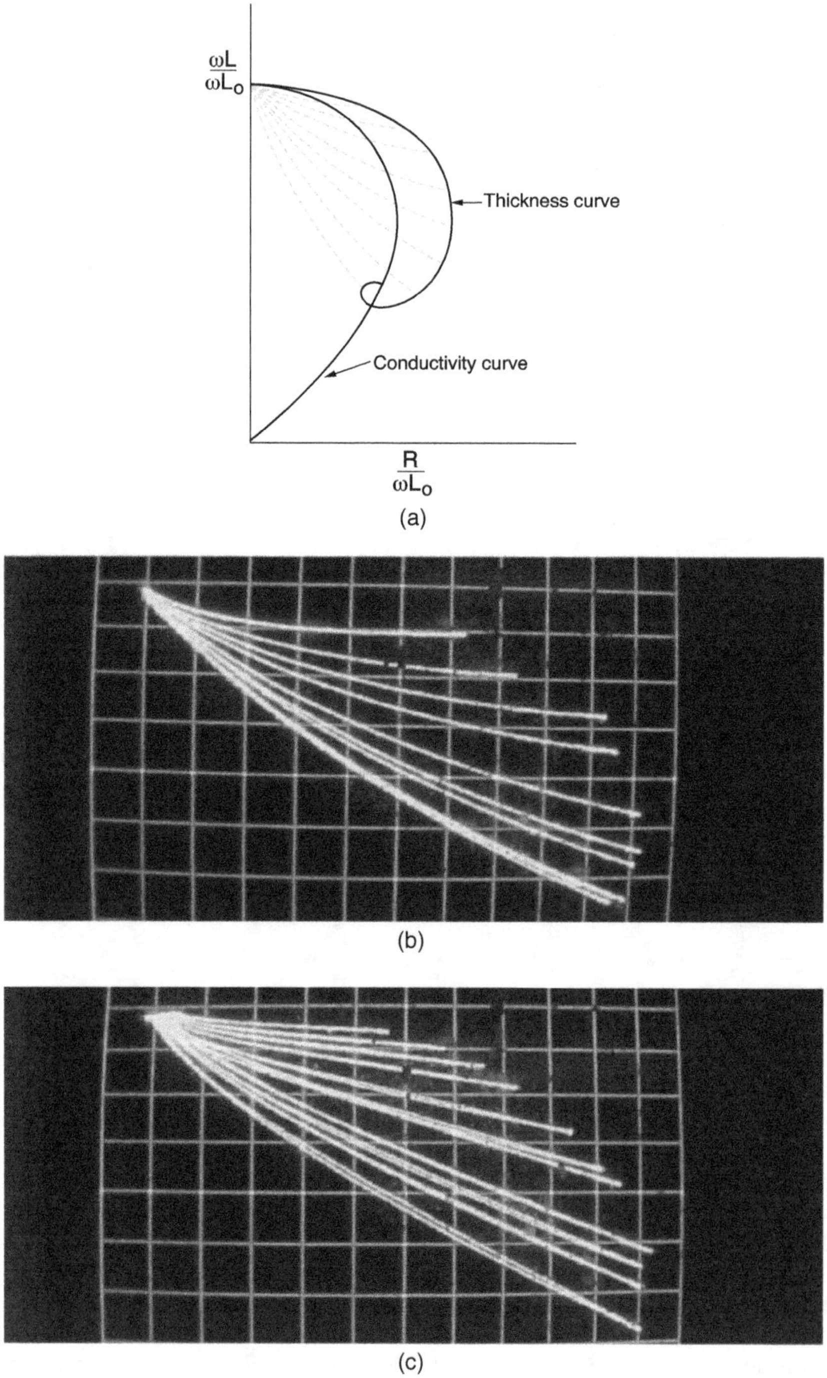


**Figure 8-58** Thickness testing. (*a*) Thickness spiral. (*b*) Thickness signal display at 10 kHz. (*c*) Thickness signal display at 2 kHz.

the point where all 11 steps are resolved, but the reduction in frequency has reduced displayed separation of the thinner steps.

### Plating and Cladding

Variation in thickness of one conductive non-ferromagnetic material bonded over another is shown in Fig. 8-59. Figure 8-59a shows how a wedge of high-conductivity material (copper) bonded over low-conductivity material (304 stainless steel) would be displayed. As thickness increases, a spiral advances from the 304 stainless position on the conductivity curve and reaches the copper position on the conductivity curve when the thickness of the copper becomes effectively infinite and the effects of the 304 stainless can no longer be detected. In effect, a thickness curve for the copper has been grafted onto the conductivity curve, with 1δ, 2δ, and 3δ appearing at their usual positions of phase rotation. Figure 8-59b shows that low-conductivity material bonded over high-conductivity causes the thickness spiral to be inverted.

The display for non-ferromagnetic (304 stainless steel) material bonded over ferromagnetic (carbon steel) is shown in Fig. 8-60. As thickness increases, a curve proceeds from the impedance plane position for carbon steel to the position for 304 stainless on the conductivity curve for non-ferromagnetic materials. The illustration shows the display for a wedge of 304 stainless positioned on carbon steel, as well as lift-off curves for various thicknesses of 304 stainless bonded over carbon steel.

### Spacing between Conducting Materials

The coil's flux can penetrate multiple layers of conducting material. Therefore, it is possible to use eddy currents to measure thickness of adhesives or air gaps between two metal samples, providing that the combination of conductivity, frequency, and coil diameter provides adequate penetration. Figure 8-61 shows the spacing curve traced for a wedge-shaped air space between two layers of aluminum. As the coil is scanned along

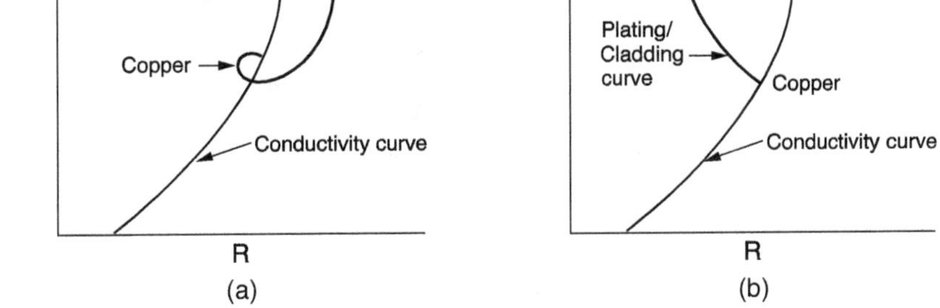


**FIGURE 8-59** Thickness of non-ferromagnetic plating/cladding over non-ferromagnetic base material. (*a*) High conductivity bonded over low conductivity. (*b*) Low conductivity bonded over high conductivity.

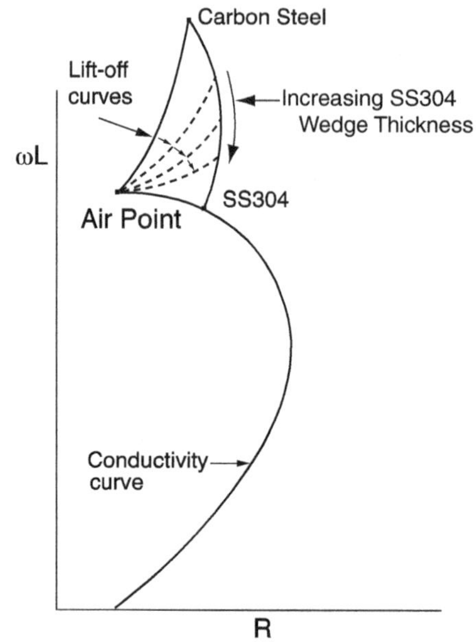


**FIGURE 8-60** Thickness of non-ferromagnetic plating/cladding over ferromagnetic base material.

the upper metal surface in the direction of increasing gap thickness, a curve is traced from the point of infinite aluminum thickness on the conductivity curve to the point on the aluminum thickness curve that represents the thickness of the upper metal surface. Frequency must be selected so that the position for the thickness of the upper metal layer is far enough up the thickness spiral for a usefully long spacing curve to be traced.

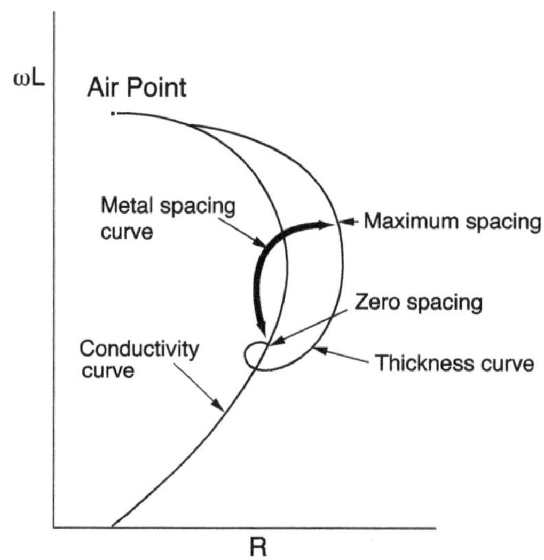


**FIGURE 8-61** Metal spacing.

### Differential Surface Coil Display

For purposes of convenience and simplicity, all display illustrations in this section have been those of surface coils operating in the absolute mode. Figure 8-62 compares display of absolute (Fig. 8-62a) versus differential (Fig. 8-62b) surface coil indications for surface-breaking voids of increasing depth. The differential coil pair was oriented so that one coil followed the other as the discontinuities were intercepted. However, if the probe were rotated 90°, so that the coils jointly intercept each discontinuity, the signals would almost completely cancel each other. Thus, while differential coils have the advantage of suppressing unwanted signals such as lift-off noise, the user must be well aware of their characteristics.

### Encircling and Internal Coil Display

Figure 8-63 shows signal traces from a series of external circumferential grooves on the same tubing standard intercepted by both encircling (Fig. 8-63a) and internal (Fig. 8-63b) differential coils. The grooves are detected as surface-breaking by the encircling coil and subsurface by the internal coil.

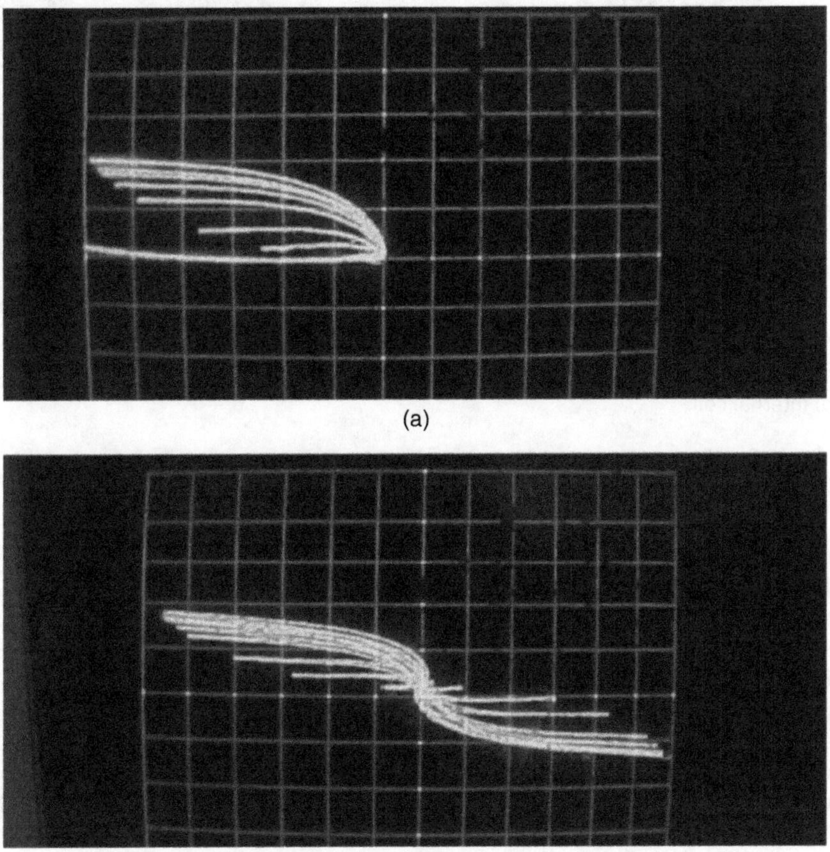

(a)

(b)

**FIGURE 8-62** Absolute versus differential coil systems. (*a*) Absolute. (*b*) Differential.

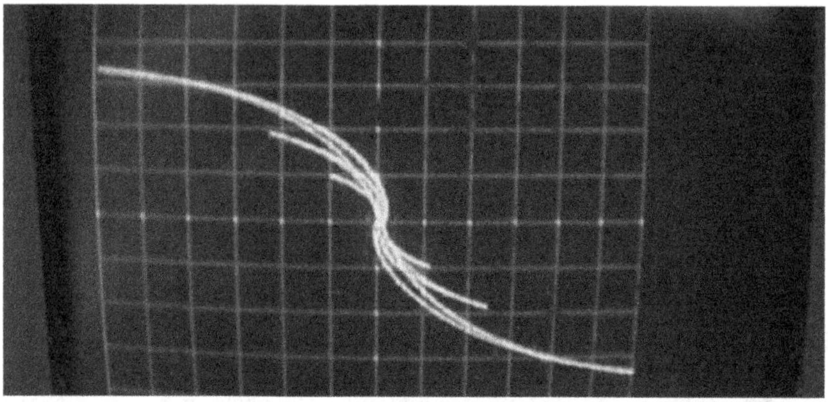

(a)

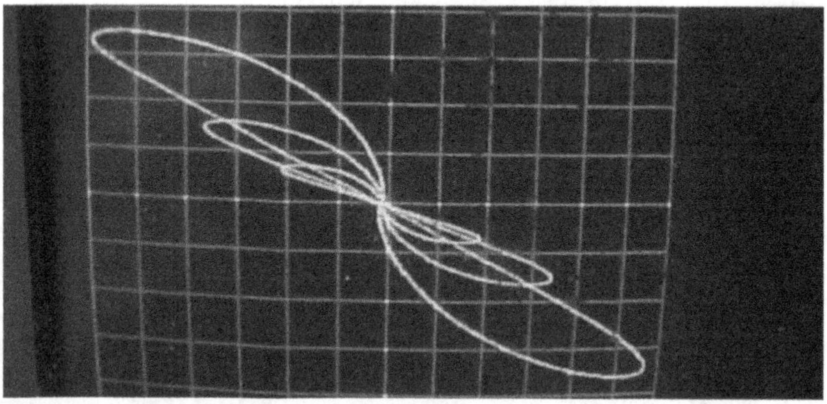

(b)

**Figure 8-63** Display of signals for encircling and internal differential coils. (*a*) Encircling coil. (*b*) Internal coil.

### Law of Similarity

In the discussion of conductivity, it was shown that positioning along the comma-shaped impedance curve advances clockwise as conductivity increases. It was also shown that individual conductivity positions shift along the curve as test frequency is varied. In the case of surface coils, for example, similar shifts also occur with changes in permeability and coil diameter. Thus, a given position along the impedance curve depends on a combination of factors. If positioning along the impedance curve can vary with conductivity, frequency, permeability, and coil diameter, the question arises as to what a given curve position actually represents. The answer is that a given position along the curve represents a specific distribution of eddy currents in the test material and that conductivity, frequency, permeability, and coil diameter all affect that distribution. Since distribution of eddy currents affects the major performance criteria—sensitivity, penetration, and resolution—position along the impedance curve largely determines test performance.

Various systems have been devised for assigning reference values for different positions along the impedance curve. The most well-known system is probably the one defined by Foerster for encircling coils enclosing a long bar, the $f/f_g$ ratio. The $f$ value represents the frequency at which the test instrument is driving the coil. The $f_g$ value is a mathematically derived reference value known as the *limit frequency,* calculated as follows:

$$f_g = \frac{5.066\,\rho}{\mu_r D^2}$$

where:
$f_g$ = limit frequency (kHz)
$\rho$ = resistivity in microohm-centimeters
$\mu_r$ = relative permeability (dimensionless)
$D$ = bar diameter

Ascending $f/f_g$ values are plotted clockwise along the impedance curve, as shown in Fig. 8-64. Each position represents a specific eddy current distribution in the test material and, hence, a specific combination of sensitivity, penetration, and resolution. This leads to the concept of the similarity law, which states, in effect, that if test conditions are manipulated to obtain the same $f/f_g$ ratio in two different test objects, performance will be the same for both. For example, a given signal display for an identical discontinuity in both a low-conductivity and a high-conductivity material can be obtained by manipulating test frequency to obtain the same $f/f_g$ ratio for each of them.

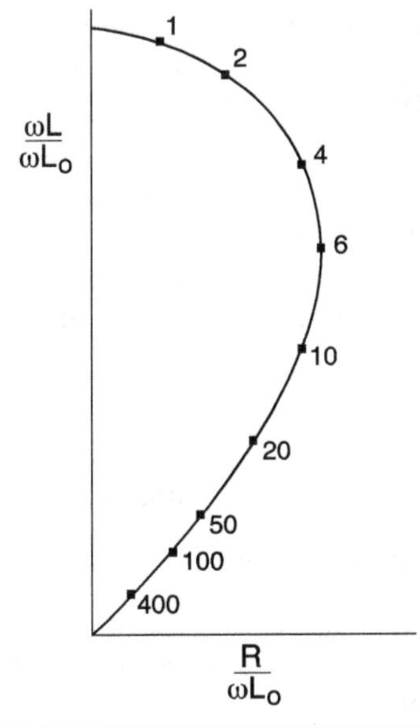


**FIGURE 8-64** Impedance curve for *f/fg* variations with encircling coil inspecting long bar.

## 8.7 Advantages and Limitations

One of the major advantages of the eddy current method is also one of its most severe limitations. That is, the eddy current method is sensitive to many variables including material conductivity and thickness, size of surface and subsurface discontinuities, thickness of plating or cladding on base metals, spacing between conductive layers, spacing between test coil and test material (lift-off), and permeability variations. However, the major limitation of the eddy current method is that response to these variables is vectorially additive, with the result that when more than one variable is detected by the test coil, all variables combine into a single response that may be difficult to resolve into its separate components. Ability to suppress and render variables separately identifiable is an important element of the user's knowledge. The techniques used to overcome this problem range from relatively simple to complicated and potentially expensive, such as the use of multi-frequency instruments.

Additional advantages of the eddy current method include the following:

1. Equipment available for field use has become increasingly portable and lightweight. In addition, many instruments are microprocessor-based, which permits test setups to be saved to memory and test results stored for archiving and analysis.
2. The method is nondestructive. No couplants, powders, or other physical substances need to be applied to the test material. The only link required between the probe and test material is a magnetic field.
3. Test results are usually instantaneous. As soon as the test coil responds to the test specimen, a qualified user can interpret the results. However, data on a large quantity of test material can be acquired so rapidly, as in the case of in-service tube testing, that it becomes more practical to initially record the data and review it later at a more reasonable pace. Moreover, the use of multi-frequency equipment can greatly complicate the amount of data acquired at a given instant and render it overwhelming unless it is recorded and later reviewed as separate elements.
4. Eddy current testing is ideal for "go/no-go" inspections. Audible and visual alarms, triggered by threshold and box gates, as well as gates that can be set to almost any desired shape, are available to automate testing.
5. It is not necessary for the coil assembly to touch the test material. This permits high speed production testing to be done without friction, thus preventing wear of test coils.
6. Eddy current testing is safe; there is no danger from radiation or other such hazards.
7. Material preparation is usually unnecessary and cleanup is not required.

Additional limitations of eddy current testing include the following:

1. The test material must be electrically conductive. However, it is possible to measure the thickness of nonconductive coatings on conducting materials using the lift-off technique.
2. It is difficult to assess subsurface conditions in ferromagnetic materials. Consequently, testing of ferromagnetic materials is limited to detecting surface

discontinuities only, unless the material has been magnetically saturated. Magnetic saturation is limited to the testing of geometries that can accommodate saturation coils, primarily encircling coil applications. It is possible to perform inner-diameter inspection of tubing as long as the magnetic field is not so strong as to cause the probe to lock onto the tubing. In addition, magnetically saturated test objects may have to be demagnetized after testing is completed, so that they will not attract ferromagnetic debris.

3. Even on non-ferromagnetic materials, the eddy current method has limited penetration, which varies with the material conductivity and test frequency. As a rule, penetration is limited to fractions of an inch in most materials.
4. Inspection speed may have to be limited as a function of test frequency.
5. Much of the eddy current theory is complicated, presenting a challenge to practitioners requiring mastery of the method.

## 8.8 Other Electromagnetic Test Techniques

In addition to the conventional eddy current testing techniques described above, electromagnetic test techniques and equipment have been developed to solve specific applications. Although these techniques cannot be as broadly applied as conventional eddy current testing, they often provide the best solution to a given problem.

### 8.8.1 Flux Leakage Testing

It was stated earlier that eddy current testing cannot detect subsurface discontinuities in ferromagnetic test material unless the material is magnetically saturated. Saturation, a common technique using encircling coils, is possible but sometimes inconvenient using internal coils, and is generally impractical using surface coils. Flux leakage testing is used at steel mills and fulfills an important need in the in-service inspection of ferromagnetic tubing.

When a ferromagnetic object that is free of discontinuities is magnetized, an unbroken flux field flows through the object, forming a path between its poles. Flux density decreases toward the outer material surfaces, and a detector of flux leakage would pick up a very weak signal from the surface. However, if the surface is interrupted by a discontinuity, poles are formed at opposite sides of the resulting gap, and flux lines flow externally from the north to south pole. The flux leakage lines are detected by a test probe passing over the discontinuity and a signal is displayed. Disturbance of the material's original flux lines also extends beneath the discontinuity, and if the material wall is thin enough, a discontinuity on one wall surface may be detected from the opposite wall. Therefore, it is possible to use internal-type flux leakage probes to detect both inner- and outer-wall discontinuities on tubing.

### 8.8.2 Remote Field Eddy Current Testing

Remote field eddy current testing is another technique useful for inspection of ferromagnetic tubing. Compared to impedance-type eddy current techniques where detection is limited to surface discontinuities in ferromagnetic materials unless saturation is employed, volumetric inspection of ferromagnetic tubing is possible using remote field testing. This technique responds to material conditions that may not be detectable with flux leakage testing, but this advantage is counterbalanced by possible distortion of the received signal. Although the technique can be used on either ferromagnetic or nonferromagnetic tubing using internal coils, it is not as effective overall as standard eddy current testing, so it is primarily used to obtain penetration in ferromagnetic materials.

Remote field eddy current testing is essentially a geometric variation of conventional send–receive eddy current test techniques, employing a bobbin-wound sending coil that induces eddy currents in the tube wall and a sensor consisting of a single coil or an array. Test frequency is selected such that skin depth is at least as great as tube wall thickness. As usual, eddy current skin effect operates laterally as well as downward from the coil, with eddy currents flowing for some distance in the lateral direction. The eddy currents' secondary flux cancels the coil's primary flux on the inner tube wall near the coil. As distance from the coil increases, this cancellation effect decreases. In the so-called remote field, laterally removed from the sending coil, the sensor can detect phase and amplitude information from the secondary flux developed by the eddy currents. Standard impedance plane display eddy current instruments can be used for remote field inspection. However, the drive voltage for the sending coil must be amplified substantially above the levels normally used for conventional impedance-type eddy current inspection.

### 8.8.3 Modulation Analysis

Modulation analysis is a production system technique whereby the test material is moved past the coil assembly at a constant high velocity. Impedance amplitude variations are displayed on a strip chart recorder. However, displayed information is selective in that filtration blocks frequencies that contain undesired information from the test material. For example, a high-pass filter would prevent display of the gradual impedance changes characteristic of wall thickness variations, whereas a low-pass filter would block more abrupt impedance variations. Because the amplitude-only display makes it difficult to interpret the type of material variable that causes test signals, material transport speed and filter frequency must be well controlled so that only relevant information is displayed.

### 8.8.4 Pulsed Eddy Current Testing

Pulsed eddy current (PEC) is an electromagnetic nondestructive testing (NDT) screening tool applicable to conductive and ferromagnetic metallic structures such as steel pipes, pressure vessels walls, and the like. PEC is best suited for the detection and estimation of wall loss in situations where the metal surface is not directly accessible, or when the surface condition would prevent the use of conventional techniques such as ultrasonic testing (UT). For instance, PEC is particularly adapted for the detection of corrosion under insulation (CUI) or corrosion under fireproofing (CUF), and for the assessment of corrosion under composite wraps or surface-forming scabs.

In contrast to conventional eddy current testing techniques, PEC employs square magnetic pulses to generate eddy currents, rather than sinusoidally modulated magnetic fields. The low frequency content of PEC pulses provides penetration into the metal and information about the entire thickness of the part under test. For this reason, PEC is sensitive to both near-side and far-side discontinuities. Also, the pulse lengths can be adjusted to enable the inspection of wall with thickness exceeding the capability of most competing techniques: some PEC implementations can penetrate up to 4 in. of steel.

PEC is still considered by many as a relatively new NDT technique. Compared to other eddy current testing techniques, this point of view is somewhat justified since the technology is still being actively developed and improved. The concept of using pulsed eddy currents to probe metallic components dates back to the early 1950s, when D. L. Waidelich first proposed to employ pulsed magnetic fields to measure the thickness of the corrosion-resistant jacket encasing the fission rods used as fuel elements in nuclear reactors.[1] However, it took almost 30 years for the emergence of the first industrial uses of PEC: in the 1980s, the first industrial PEC systems were developed by oil and gas companies to test the integrity of insulated pipelines. The scientific and technical advancements made in those years paved the way for further technical developments, and more extended trials and validations of the technology were conducted in the 1990s. New applications emerged: PEC has notably been used for material characterization[2] and, in the aircraft industry, for rivet corrosion detection and crack detection in wings with multilayered structure.[3,4] However, industrial and commercial applications of PEC remained relatively limited. Over the last few years, PEC has seen a revival as new, better performing and easier-to-use system appear on the market. The latest noteworthy advancement is the appearance of the first commercially available PEC array probes, which significantly improve the inspection productivity of this technology compared to conventional single-sensor PEC probes.

### *Principle of Operation*

In its simplest expression, PEC consists of pulsing magnetic energy into a conductive component and then stopping the pulse to listen to how fast eddy currents decay in the component. Data acquisition is therefore discrete rather than continuous. A PEC acquisition has three distinct operating phases: emission, cutoff, and reception (see Fig. 8-65) each of which will be explained in greater detail in the following sections.

**Emission Phase** A coiled wire is placed at some distance (referred to as liftoff; usually the coating and/or insulation thickness) from the conductive component under test where it fires a magnetic pulse. The magnetic field lines loop from the probe and into the component where eddy currents initially form to resist the change in magnetic field.

It remains active long enough for the magnetic field to penetrate and stabilize in the full thickness of the surface under test and for the eddy currents to die down.

This phase can last from a few milliseconds to hundreds of milliseconds or more, depending on the material's electromagnetic properties and thickness. For example, a 0.5-in. (12.7-mm) carbon steel plate requires a pulse duration of a few tens of milliseconds whereas a 1-in. (25.4-mm) wall of the same material could require more than a hundred milliseconds.

**Cutoff Phase** During this phase, the magnetic pulse emitted from the probe is abruptly shut down by quickly interrupting the coil excitation current. To oppose this change in the magnetic field, eddy currents appear in the metal mass, which induce a response

[1]D. L. Waidelich, J. A. DeShong, and W. J. McGonnagle, A Pulsed Eddy Current Technique for Measuring Clad Thickness. Metallurgy Program 12.1.2, United States, 1955.

[2]X. Chen and Y. Lei, Electrical Conductivity Measurement of Ferromagnetic Metallic Materials Using Pulsed Eddy Current Method, *NDT E Int.*, 2015, 75: 33–38.

[3]R. A. Smith and G. R. Hugo, Transient Eddy-Current NDE for Aging Aircraft—Capabilities and Limitations, *Insight: Non-Destructive Testing and Condition Monitoring*, 2001, 43(1): 14–25.

[4]Y. He, F. Luo, M. Pan, and F. Weng, Pulsed Eddy Current Technique for Discontinuity Detection in Aircraft Riveted Structures, *NDT E Int.*, 2010, 43(2): 176–181.

magnetic field. Ideally, the coil shutdown should happen instantaneously. In reality, the emitter coil has a nonzero inductance and it requires some time to discharge. During that spillover period, the decay of the field from the emitter coil can mask the response magnetic field coming from the sample, creating a blind zone and preventing the measurement of very thin features. Probe emitter coils are designed to achieve the best possible balance of magnetic field intensity and rapid shutdown response.

**Reception Phase** During this phase, the eddy current-induced magnetic field emitted back from the component decays over time. It is detected by a receiving magnetic sensor, which generates a voltage signal, displayed as an A-scan. The A-scan shape and decay rate relate directly to the thickness and magnetic properties of the component under test. The length of the reception phase usually mirrors that of the emission phase, lasting longer for thicker components.

In most advanced PEC systems, the cutoff and reception phases are well separated, so that no magnetic field is emitted by the probe during the reception phase. This provides a significant signal-to-noise advantage and simplifies the signal analysis compared to systems with nonzero magnetic emission during the reception phase. This also contrasts with conventional eddy current methods, where the emitter is always active.

The A-scan acquired during the reception phase is usually displayed using either a log-log or a log-lin scale. PEC signals have two distinct stages.

In the early stage, the signal varies as an inverse power law (Eq. 8-1) and appears as a straight line in the log-log scale. During this stage, the signal is rich in high-frequency content from the component's surface and from the weather jacket (or cladding).

$$V \propto t^{-\alpha} \tag{8-1}$$

The power law parameter "$-\alpha$," which rules the decay of the signal in this stage, depends chiefly on the material's electromagnetic properties. Its value will vary noticeably when moving from carbon steel to cast iron but very little within the range of an inspection on a single component. The negative sign represents the decrease in amplitude over time.

In the late stage, PEC signals decay exponentially (Eq. 8-2) plotted as a slope using a log-lin scale. The eddy currents have diffused through the whole bulk, hitting the back wall. The signal is rich in low-frequency content, which bears information on the whole bulk and its thickness.

$$V \propto e^{-t/\tau} \tag{8-2}$$

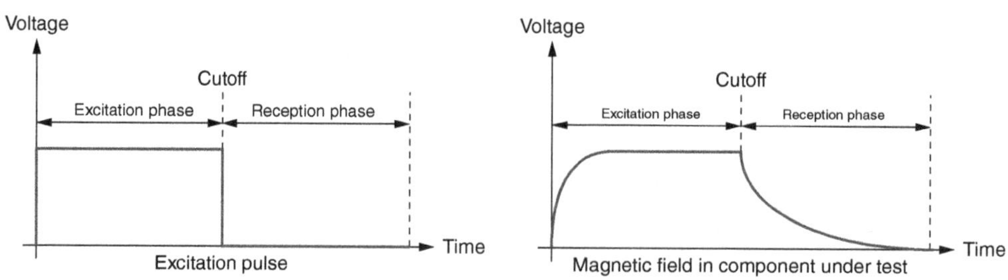


**Figure 8-65** PEC excitation pulse in relation to the magnetic field inside the component under test for the three operating phases. (*Courtesy of Eddyfi Technologies.*)

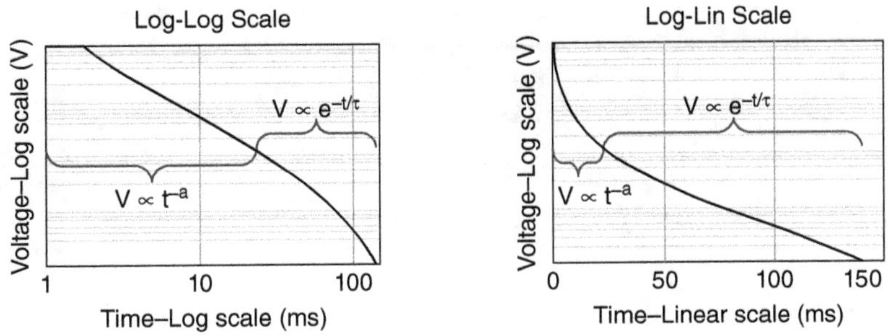


**FIGURE 8-66** A-scans of PEC signal decay according to different scales. (*Courtesy of Eddyfi Technologies.*)

The time constant "$-\tau$" describes how fast the signal decays in the late stage. It is directly linked to the thickness of the component being probed as well as its electromagnetic properties.

Between the two stages is a gradual transition, commonly referred to as the signal's bending point. The position of the bending point and the slope of the signal with a log-lin scale (corresponding to the time constant of exponential decay) varies with thickness, as shown in Fig. 8-66. Historically, the position of the bending point has been used to measure variations in wall thickness but given the difficulty to pinpoint a precise bending point in what is essentially a gradual transition, this method has often proved unreliable. More modern systems analyze the shape of the whole A-scan to provide a thickness measurement, relative to a known thickness on the same component. As seen in Fig. 8-67, a wall thickness thinning appears in the log-log A-scan as an earlier drop-off and on the log-lin A-scan as a steeper slope.

Conventional eddy current testing measures the mutual impedance of the probe-sample system and as such is very dependent on small variations in the length of the airgap between the two, also known as liftoff, which directly affects the overall magnetic reluctance. Since PEC measures a rate of decay, it is much more resilient to liftoff variations.

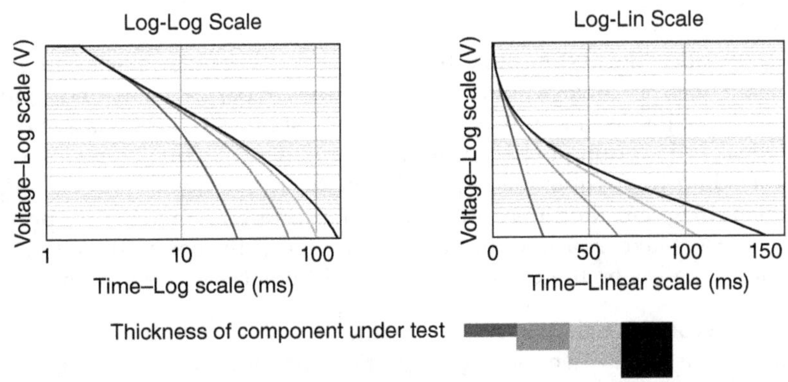


**FIGURE 8-67** PEC signal decay varies as a function of wall thickness shown in log-log and log-lin scales. Magnetic field decay inside the component varies as the probe passes over a discontinuity, as shown in the corresponding A-scan. (*Courtesy of Eddyfi Technologies.*)

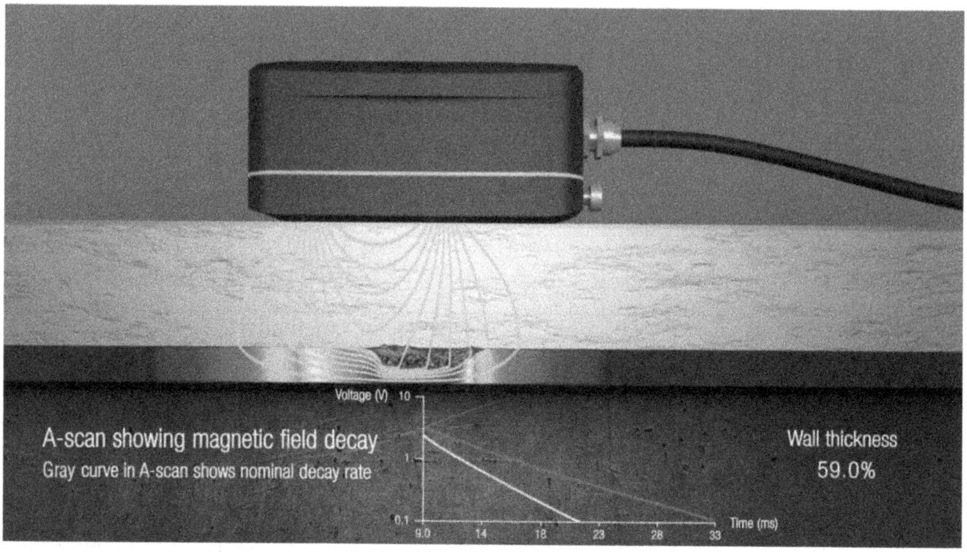


**FIGURE 8-68** A-scan showing magnetic field decay. (*Courtesy of Eddyfi Technologies.*)

**Calibration and Measurement** The pulsed eddy current technique can provide a thickness measurement, relative to a known thickness on the same component. This was stated in the previous section but bears repeating given that it is one of the most important concepts tied to the technique. This is because the PEC technique is cross-sensitive to wall thickness, permeability and conductivity variations all at once, hence it cannot automatically discriminate between the effects of these three parameters. For this reason, PEC requires a calibration on a point deemed "nominal" to which every other measurement is compared (see Fig. 8-68). It is assumed that, within a component, the material's electromagnetic properties are fairly constant. Considering this, the instrument calibration factors out the permeability and conductivity when estimating the wall thickness. These estimations lead to very good results so long as the previous assumption remains true.

In practice, the assumption that the electromagnetic properties of a component are constant does not always hold. Local variations in permeability and conductivity affect the shape of the A-scan in ways that are very similar to the signature of wall thinning. This makes discerning between these variations and actual corrosion sometimes difficult, leading to possible false calls. Temperature gradients, poor manufacturing, and remnant magnetization may all lead to local property changes. To prevent false calls, trained users take into account their knowledge of the component, the shape, and profile of the feature on the C-scan and subtle differences in the A-scan.

**Magnetic Footprint and Undersizing** Due to its resilience to high liftoff and liftoff variations, PEC is most often used to inspect components at a wide range of liftoffs, going from a few millimeters to tens of centimeters. For this reason, the concept of footprint is integral to the technique. During the emission phase, the PEC probe projects a magnetic footprint onto the component under test. The affected area is energized by the pulse and will emit back the eddy current induced magnetic field during the reception phase. This means that the signal measured doesn't come from a single point directly under the point, but rather from a small area. Because the magnetic field projected by the probe onto the component under test is cone-shaped, the further away the probe is from the component (greater liftoff), the larger is the energized zone. This footprint is instrumental in establishing the sizing and detection capabilities of a probe for different discontinuity types and shapes.

As a first-order approximation, one can assume that PEC provides a volumetric measurement, meaning that the response is given by the average wall thickness of the energized area inside the probe's magnetic footprint. When discontinuities are smaller than this averaging area, their depth is underestimated. This is something referred to as *undersizing* (Fig. 8-69). In other words, when a PEC probe goes over a discontinuity that is smaller than its averaging area, the inspection signal is a combination of the nominal thickness of the component under test and the discontinuity. In this situation, the thinnest region (the discontinuity) can be somewhat averaged out by the thicker wall surrounding it, which prevents the PEC system from accurately measuring the wall thickness of small indications. There are, however, a number of ways to compensate this effect that rely on a more detailed analysis of the A-scan around the transition period, analysis of the resulting C-scan, or a combination of the two. Advanced analysis tools can allow for a better estimation of remaining wall thickness even for smaller discontinuities.

**Equipment and Accessories** Basic equipment for Pulsed Eddy Current inspections usually consists of a handheld probe and a portable acquisition instrument. Some instruments have a touchscreen and onboard computer to control the data acquisition while others additionally require a laptop.

*Probe* Pulsed eddy current probes come in various shapes and sizes. This is mainly due to the design trade-off between signal intensity and the duration of the dead zone stemming from the emitter coil inductance. To function at high liftoff, a stronger signal is necessary. This can be achieved either by increasing the inductance of the coil and/or circulating a large current through it. Both approaches have limitations, and an optimal mixture of the two is usually found for a range of liftoff and wall thickness. High inductance increases the signal, but the emitting coil takes longer to shut down, resulting in a dead zone which obfuscates the early signal and can limit measurements of thin components or very deep discontinuities. On the other hand, increasing the emission current

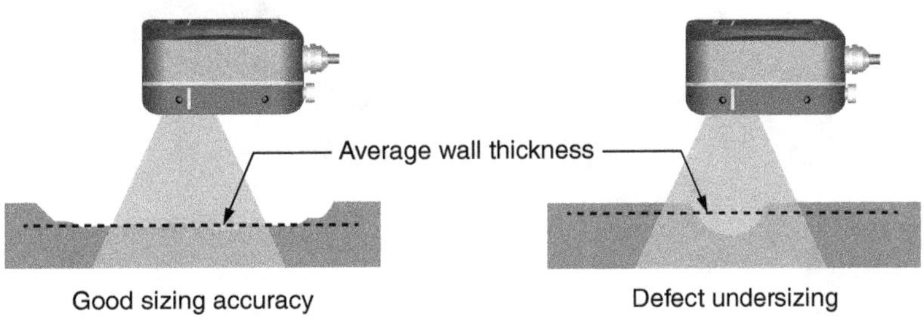


**FIGURE 8-69** Sizing accuracy depends on the discontinuity size relative to the probe's magnetic footprint. (*Courtesy of Eddyfi Technologies.*)

requires to lower the electrical resistance of the coil, achieved either by decreasing the number of turns or increasing the volume (and weight) of the coil: two unattractive options. In addition, high current creates issues with energy dissipation and electronics.

At its simplest expression, a PEC probe usually consists of an emission coil and a magnetic sensor, which is often a second coil. Many technological improvements have been made over the years in the shape of the coil(s) and the inclusion of advanced magnetic materials to improve the signal-to-noise ratio, resilience to adverse conditions and versatility.

For these reasons, manufacturers offer a range of standard probes and users must use the best one depending on liftoff and wall thickness of the component. Standard probes are usually handheld and have a few buttons to operate the data acquisition remotely.

Historically, PEC acquisition has consisted of taking discrete data points on a grid to fill the inspection C-scan. In recent years, improvement in probe design, control electronics, and signal analysis has allowed dynamic acquisition. In this acquisition mode, the user moves the probe at constant speed along a line and data acquisition is spatially encoded. Scanning speed depends directly on the wall thickness of the inspected component and can range from very fast on thin components to impractically slow on thicker ones. For example, on a 5-mm wall component, scanning speed can reach several centimeters per seconds but go down to a few millimeters per second on 25-mm wall components.

Another recent improvement in the way data acquisition is carried out with PEC is the introduction of array probes. Productivity has long been a limiting factor in the widespread adoption of the technique, especially in cases where other techniques were viable. Array probes consisting of several elements firing simultaneously allow to cut down inspection time considerably without compromising data quality. However, array probes are, by definition, bulkier and their usage is limited to components where obstruction is limited.

In addition, there are many different types of speciality probes designed for a slew of non-standard applications. This includes underwater probes to be used by divers to inspect submerged pipelines, offshore drilling platform risers or any other undersea component. Flat, elongated probes (Fig. 8-70) are also available from most manufacturers

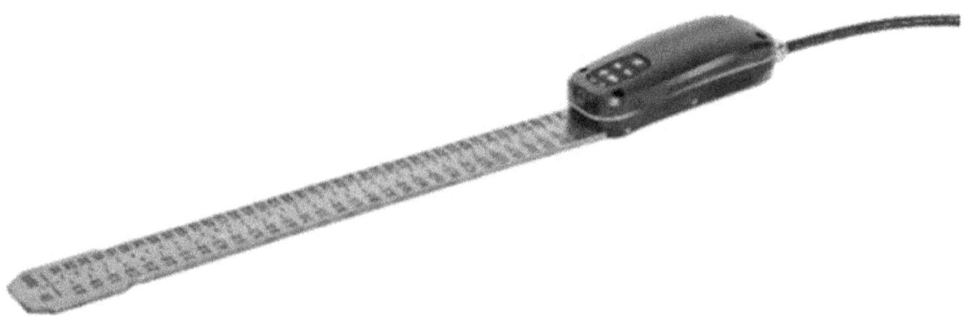

**FIGURE 8-70** Elongated PEC probe for storage tank annular ring inspection. The PEC sensor is on the tip of the probe blade and PEC pulses are emitted upward. (*Courtesy of Eddyfi Technologies*.)

and are used to inspect the annular ring of storage tanks, where few other techniques can get reliable results.

*Instrument* The instrument is the unit controlling all three phases of the acquisition cycle: emission, cutoff, and reception. The instrument holds the control electronics and the pulser. It is controlled by a computer, either housed in the same box and equipped with a touchscreen or via a connected heavy-duty laptop. Data analysis is possible either on the instrument embedded screen or on a separate computer.

Most advanced PEC systems are portable, highly rugged, and battery operated to allow easy deployment in the field. See Fig. 8-71. Probes are connected to the instrument via a cable of a few meter length. Extension cords up to a hundred meters are generally available but may cause a slight drop-off in pulse strength and signal quality at extreme range.

**Advantages and Applications** PEC is an effective screening tool most commonly used on ferromagnetic materials such as carbon steel, but its working principles allow it to be used on any conductive material, including cast iron, aluminum, etc. PEC measures the decay rate of the eddy currents–generated magnetic field, which is directly related to the component's relative thickness. For this reason, it is best suited to detect corrosion or other forms of wall loss.

Furthermore, PEC can operate at high liftoff through any nonconductive insulation including wet or dry mineral wool, calcium silicate, cellular glass, foams, perlite,

**FIGURE 8-71** Example of a commercial PEC system. (*Courtesy of Eddyfi Technologies.*)

polyethylene, concrete, etc. Depending on the probe used, inspections can typically be performed in contact with the part under test or up to liftoffs of 305 mm (12 in.). Remember, however, that the greater the liftoff, the larger the probe's footprint and the larger discontinuities need to be in order to be detected. Hence, PEC cannot detect very small indications (like pitting) or cracks.

Accordingly, the detection of corrosion under insulation (CUI) and corrosion under fireproofing (CUF) are among the most common applications that rely on PEC. By allowing the screening of wall loss through the insulation, asset owners can minimize the cost associated with the unwrapping, cleaning and visual inspection of insulated components. PEC inspections can therefore be performed on in-service components, allowing the assessment of the installation integrity year-round, leading to a better planification and allocation of always limited shutdown inspection time. Inspection through insulation is particularly important for cryogenic components: removal of insulation during operation may cause condensation and frost buildup on the cold metal surface, potentially promoting corrosion.

Other common applications for PEC include the detection of flow-accelerated corrosion (FAC), tank floor annular ring inspection, and estimation of remaining ligament under corrosion scabbing. PEC is not affected by water, so it can be used in marine environments for splash zones and underwater inspections.

**Limitations** Several factors can negatively influence the quality of data acquisition. Selecting the right probe that offers the optimal balance of signal and resolution for the inspected component is always the foremost concern. Signal-to-noise ratio is an important factor to consider and can be negatively affected by the presence of external noise. This can take many forms, the most important one being electromagnetic (EM) noise emitted from powerlines, nearby motors or other high-voltage sources in the vicinity. PEC probes are essentially large antennae aiming to collect the faint signal emitted from the component during the reception phase. While most PEC systems offer some form of EM noise mitigation, it remains a challenge in some situations. Mechanical vibrations of the component, instability of the probe during the reception phase or any kind of rapid probe movement can also lead to a degradation of the signal. Poor quality data can lead to false calls or missed calls, but this problem can be identified and mitigated by analyzing the shape of the A-scans.

It is common for insulated pipes to be wrapped in weather jackets (also called cladding). These are typically aluminum, stainless steel, or galvanized steel, depending on the industry and region of the world. Because weather jackets are conductive, they interfere with the component signal to various degrees during a PEC inspection. Aluminum and stainless steel are non-ferromagnetic, so their impact is marginal.

Galvanized steel (GS) weather jacket adversely affects PEC inspection in many ways. First, as the material is ferromagnetic, it screens part of the magnetic field generated by the PEC probe during the emission phase. Hence, only a fraction of the magnetic field emitted by the probe reaches the component. The intensity of detected PEC signal in the reception phase is also reduced, negatively affecting the measurement of the signal-to-noise ratio. To give a sense of the importance of this effect: with just 0.5 mm of GS, the field magnitude reaching the component surface may be reduced by up to 60 percent. This is illustrated in Fig. 8-72, which is a simulation of the magnetic flux density B distribution in on a typical component covered with GS jacket.

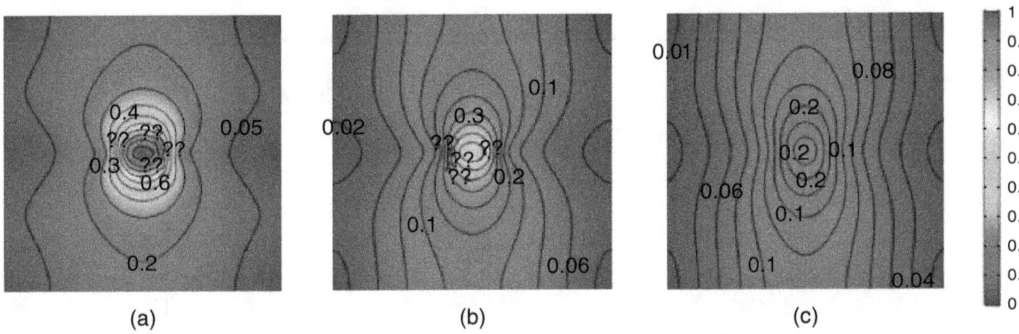


**FIGURE 8-72** Normalized magnitude of the magnetic flux B in the carbon steel plate. (*a*) No GS jacket. (*b*) 0.5 mm GS. (*c*) 1 mm GS. The same color scale is used in all images. (*Courtesy of Eddyfi Technologies.*)

In addition, as the GS weather jacket is conductive, eddy currents are generated within the jacket during the reception phase. The decay of these eddy currents generates a magnetic field that dominates the beginning of the A-scan. Fortunately, the decay of the GS contribution is relatively fast compared to the signal coming from the usually much thicker plate under test. Figure 8-73 shows an example of a typical A-scan measured on a typical steel plate: the GS contribution may partially mask the signature of some types of flaws, particularly discontinuities smaller than the probe's footprint.

Another detrimental effect of GS is that it enlarges the magnetic footprint of the probe on the carbon steel plate, reducing the ability of the technique to detect small discontinuities. Finally, the GS may vibrate during the PEC emission and reception phases, further degrading the measurement SNR.

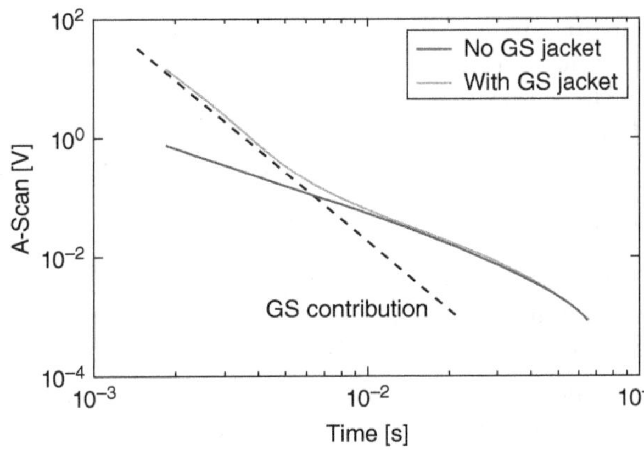


**FIGURE 8-73** Typical A-scans measured with and without GS jacket. (*Courtesy of Eddyfi Technologies.*)

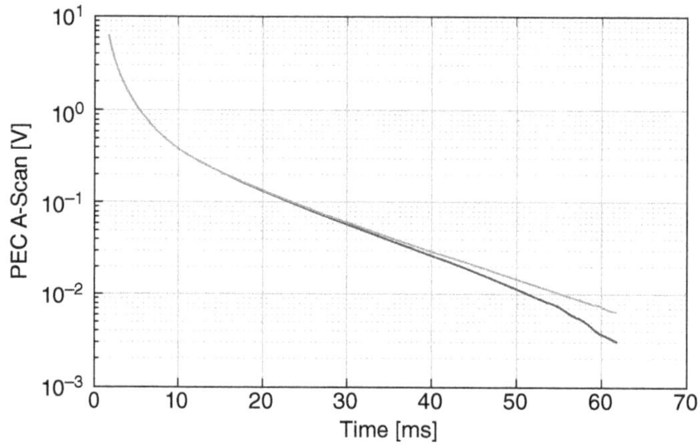


**FIGURE 8-74** Typical A-scan distortion from overspeed (black curve) compared to static A-scan (upper curve). (*Courtesy of Eddyfi Technologies.*)

Fortunately, recent advances brought to the market dramatically improved PEC performances on GS cladding.[5] Ultimately, whether an inspection can be practically performed over galvanized steel cladding depends on the component, the insulation, and the cladding thicknesses.

While dynamic scanning drastically improves productivity, it also brings additional sources of potential signal disruption: a speed limitation must be respected to capture good quality A-scans. The speed limitation stems from two phenomena: first, during the emission phase the magnetic field must have the time to penetrate the wall thickness while the probe remains approximately at the same position all along the emission pulse. That is why the acceptable dynamic speed decreases rapidly with increasing wall thickness. During the reception phase, the probes moves away from the region that was energized during the emission phase. If the dynamic acquisition speed is excessive, this movement causes a distortion of the detected A-scan, as shown in Fig. 8-74, and may lead to erroneous PEC measurements.

PEC signals may be affected by liftoff variations and weather jacket overlaps, but most systems on the market now effectively compensate for these effects, so they can be disregarded because they do not affect the quality of inspections.

Using advanced software algorithms, nozzles, flanges, and insulation support rings, which can also interfere with PEC signals, are now relatively simple to filter out.

**PEC Inspection Workflow** A typical PEC inspection workflow could be:

- The operator performing the inspection will first usually draw a grid on the component under test at the recommended resolution (depending on the probe's footprint and the desired resolution).
- The operator would then calibrate the PEC probe on an area of the component under test known to be nominal. Because the shape and decay rate of the inspection, A-scans relate directly to the thickness and magnetic properties of

[5]C. Dalpé et al., Pulsed Eddy Currents: Overcoming Adverse Effects of Galvanized Steel Weather Jacket, Cofrend 2017, Strasbourg.

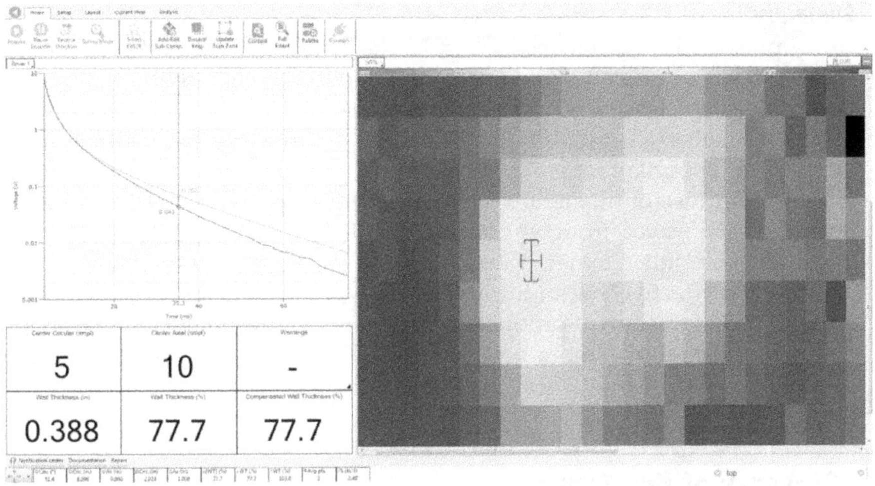


**FIGURE 8-75** A-scan and C-scan from insulated pipe (50 mm/2 in. liftoff, 13 mm/0.5 in. nominal wall thickness) showing localized natural corrosion. (*Courtesy of Eddyfi Technologies.*)

the surface under test. PEC is usually calibrated this way instead of using a calibration standard. This has the effect of making the material under test constant, enabling the extraction of a relative thickness variation between two points (nominal and corroded).

- The operator would finally acquire data, building an inspection C-scan. With older systems, inspectors could only do this in grid mode, where they had to press a button for each point on the grid. More modern systems allow this to be performed as a dynamic scan, greatly speeding up the process as less gridding is necessary and it can acquire more data points in less time. Recently, PEC array probes have become available, speeding up even more the process of scanning large sections of insulated vessels or pipes. See Figs. 8-75 and 8-76.

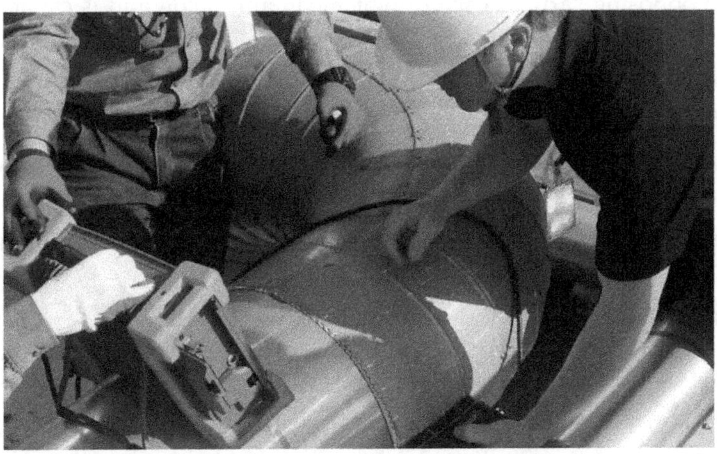

**FIGURE 8-76** NDT technicians performing PEC measurements on an in-service insulated pipe. (*Courtesy of Eddyfi Technologies.*)

***Summary***

Today, pulsed eddy current technology has become one of the preferred methods for asset owners to screen insulated and fireproofed pipes, vessels, sphere legs, and other carbon steel or cast iron structures. Various types of material loss can be detected without having to take the inspected structures out of service or stripping them of their insulation. This provides an important save in inspection cost and time, because PEC screening allows a more focused use of quantitative measurement inspection methods. The use of pulses instead of sine waves, from which pulsed eddy current draws its name, makes the technique uniquely suited to detect wall thickness variations at high liftoff.

PEC can effectively scan through most types of nonconductive insulation, marine growth, and corrosion by-products, and recent improvements to the technology enable it to scan through the more prevalent cladding types worldwide, while minimizing traditionally detrimental effects on the technology.

## 8.9 Glossary of Key Terms

**Absolute Coil Technique** An eddy current test in which one arm of the instrument's bridge is connected to a "test coil" that is interfacing with the test material and the opposing arm of the bridge is connected to an electrically similar coil, called a "balance load" or "dummy load," that is remote from the test material.

**Alternating Current** A current whose direction of flow is periodically reversing.

**Atom** The smallest unit of matter; hence, the smallest unit within an electrical conductor. The atom consists of a positively charged nucleus surrounded by one or more "shells" or "orbits" of negatively charged electrons.

**Back Voltage** Voltage induced into a conductor by means of the changing electromagnetic field of a varying current passing through that conductor.

**Bridge Circuit** Formally known as a "Wheatstone Bridge," this design is commonly used as the test coil input circuit for eddy current instruments. In this application, it consists of two opposing "arms," each of which contains a coil and is capable of sensing very small impedance differences between the two arms.

**Coil in Air** A surface coil that is not in proximity to and is therefore unaffected by conducting material.

**Conductance** The ability of a particular component to conduct electricity. Conductance depends on a component's conductivity, length, and cross section.

**Conductivity** The opposite of resistance, that is, the relative "willingness" of a material to allow the flow of current.

**Conductor** A material whose atoms contain very few (e.g., one or two) electrons in their outer shell. When atoms contain only a few electrons in the outer shell, it is easy to cause electrons to move from one atom to the next.

**Cross-Axis Coil** A test coil design in which the coil assembly consists of two coils wound 90° to each other in order to make the coil assembly less directional in sensitivity.

**Current** The flow of electrical charges, measured in amperes.

**Differential Coil Technique** An eddy current test in which two opposing arms of the instrument's bridge are each connected to a separate test coil that is interfacing with the test material. Also called the differential self-comparison technique.

**Direct Current** The flow of electricity in only one direction.

**Domains** Miniature magnets consisting of groups of atoms or molecules present within a material's individual grains.

**Eddy Currents** Circulating electrical currents induced in an isolated conductor by an alternating magnetic field.

**Effective Depth of Penetration** The maximum material depth from which a displayable eddy current signal can be obtained, arbitrarily defined as the depth at which eddy current density has decreased to 5 percent of the surface eddy current density.

**Electricity** The flow of electrons through a conductor from one atom to the next.

**Electromagnet** A magnet operating on the principle of electromagnetism, usually consisting of a solenoid coil containing a ferromagnetic core. Electromagnets are distinguished from permanent magnets by the fact that they require electricity in order to operate.

**Electromagnetic Induction** Relative motion between a magnetic field and a conductor causes a voltage to be induced in that conductor.

**Electromagnetism** The phenomenon whereby passage of electrons through a conductor causes a magnetic field to develop concentrically around the conductor, perpendicular to it.

**Electron** A negatively charged particle of electrical energy that orbits the atom.

**Empty Coil** An encircling coil which is not in proximity to and is therefore unaffected by conducting material.

**Encircling Coil** An eddy current test coil designed so that the test material can pass through its interior wall. It is used primarily for production testing of long, continuous product such as pipe, tube, rod, wire, and bar stock.

**External Reference Coil** A type of absolute coil bridge connection where the bridge's reference coil is sensing a reference standard.

**Feed-Through Coil** See Encircling Coil.

**Ferromagnetism** A property whereby materials can become magnetized when their domains have become aligned.

**Fill Factor** The calculated coupling effectiveness between the inner surface of an encircling coil and the outer surface of a specimen enclosed within it or between the outer diameter of an internal coil and the inner surface of a specimen surrounding it.

**Flux** See Magnetic Flux.

**Flux Density** The number of flux lines per unit area perpendicular to the direction of current flow.

**Gap Probe** See Horseshoe Probe.

**Gauss** The unit of measure for flux density. One gauss equals one line of force passing through an area of one square centimeter.

**Horseshoe Probe** A surface probe configured with the coil wrapped around a horseshoe-shaped permanent magnet, to provide an eddy current field circulating perpendicular to the test surface.

**I.D. (Inner-diameter) Coil** See Internal Coil.

**Impedance** The vector sum of inductive reactance and resistance.

**Inductance** The property of an electrical component, such as an eddy current coil, whereby the component's alternating electromagnetic field induces voltage into the component itself or into a nearby secondary conducting component such as a test specimen.

**Inductive Reactance** Opposition that induced voltage offers to the alternation of alternating current.

**Internal Coil** An eddy current test coil designed to pass through the interior walls of tubes and pipes. It is used primarily for in-service inspection of tubing.

**International Annealed Copper Standard (IACS)** A scale used to compare material conductivities with pure, unalloyed, annealed copper at 20°F, as the base value at 100 percent conductivity. Other materials are assigned percentage values relative to copper.

**Lenz's Law** The phenomenon whereby, when a primary circuit induces voltage into a secondary circuit; the direction of current flow in the secondary will be such that the polarity of the secondary flux will be opposite to the primary flux.

**Lift-Off** The variation in impedance as the distance between a probe coil and a conductor is varied. Lift-off can be a positive effect because it can be used to measure the thickness of nonconductive substances located between the coil and a conductor; but it is often a negative effect because its strong signals can mask the weaker indications of variables that an eddy current examination is intended to measure.

**Locus (plural: loci)** A mathematical term indicating a set of points that satisfy a given set of conditions. For eddy current purposes, a set of points that trace a line or curve representing the range of a displayed variable such as conductivity or lift-off.

**Magnetic Field** A force field radiating from permanent magnets and electromagnets, which decreases in strength inversely as the square of the distance from the poles of the magnet.

**Magnetic Flux** The lines of force that make up a magnetic field. In the case of an eddy current test coil, which is an electromagnet, the lines of force flow out of one end (pole) of the coil, around the outside of the coil, and back into the other end (pole) of the coil.

**Magnetism** The mechanical force of attraction or repulsion that one material can exert upon another.

**Normalization** Confirmation to a standard. The impedance plane is normalized to preclude the need to develop separate impedance graphs for each test coil that is used. The vertical axis is normalized by dividing actual inductive reactance of the test coil by inductive reactance of the coil when remote from conductive material. The horizontal axis is normalized by subtracting test coil resistance from total resistance and then dividing the remainder by inductive reactance of the coil when remote from conductive material.

**Nucleus** The central portion of the atom, containing protons, which carry a positive charge and thus balance the negative charge of the orbiting electrons, as well as neutrons, which carry no charge.

**Ohm's Law** A formula defining the relationship of current, voltage, and resistance, as follows:

$$\text{current} = \frac{\text{voltage}}{\text{resistance}}$$

**Permanent Magnets** Materials that retain their magnetism after they are removed from a magnetic field.

**Permeability** The measure of a material's ability to be magnetized, that is, a material's ability to concentrate magnetic flux.

**Polarity** The phenomenon whereby opposite ends of a magnet exhibit opposing forces. This is due to the fact that the lines of force flow from the north to the south pole around the outside of a magnet, and from the south to the north pole within the magnet. This causes like poles to repel and unlike poles to attract.

**Poles** Nomenclature for the opposite ends of a magnet, one north and one south.

**Probe Coil** An eddy current test coil, configured in a probe-type housing, designed to be placed in contact with a surface of the test material.

**Reflection Coil** A coil assembly in which a large outer coil induces primary flux into the test specimen and a pair of stacked inner coils, wound in opposition, sense the secondary flux. The term "reflection" derives from the fact that the inner coils, being wound in opposition, function as electrical mirror images of each other.

**Resistance** Opposition to the flow of current. Its unit of measure is the ohm.

**Resistivity** The opposition of a material's atoms to the flow of electricity. It is the inverse of conductivity.

**Self-Induction** Induction of voltage into a conductor, such as a test coil, resulting from the coil's own magnetic field expanding and collapsing against the turns of the coil.

**Shielded Coil** An eddy current surface coil whose turns are surrounded by a cylinder of ferromagnetic material that concentrates the coil's flux field to suppress lateral extension of the field.

**Sinusoid** One complete 360° cycle of a sine curve.

**Skin Depth** See Standard Depth of Penetration.

**Skin Effect** Eddy current density is maximum at the material surface and decreases exponentially with depth. Thus, in thicker materials, eddy current testing operates only on the outer "skin" of the test material. Test sensitivity decreases rapidly with depth, and volumetric tests are possible only on thin specimens.

**Solenoid** A coil of conducting wire whose length substantially exceeds its width.

**Standard Depth of Penetration** The material depth at which eddy current density has decreased to 1/e (36.8 percent) of surface eddy current density.

**Surface Coil** An eddy current test coil, often configured in a probe-type housing, designed to be placed in contact with a surface of the test material.

**Through-Transmission** A technique whereby test coils are positioned on opposite sides of the test material, with one coil transmitting and the other coil receiving. This technique increases penetration of eddy currents, but does not provide display of phase information.

**Voltage** The polarity applied to electrons, causing current to flow through a conductor. The unit of measure is the volt.

## Suggestions for Further Reading


API Recommended Practice 583, Corrosion Under Insulation and Fireproofing, 2nd ed., American Petroleum Institute.

ASM International Handbook Committee. Metals Handbook, Ninth Edition, Volume 17, *Nondestructive Evaluation and Quality Control*. ASM International, Metals Park, Ohio, 1989.

Cecco, V. S., S. P. Sullivan, J. R. Carter, and L. S. Obrutsky. *Innovations in Eddy Current Testing*. Engineering Technologies Division, Nondestructive Testing Development Branch, Chalk River Laboratories, Chalk River, Ontario, 1995.

Cecco, V. S., G. Van Drunen, and F. L. Sharp. *Advanced Manual For Eddy Current Method*. Canadian General Standards Board, Ottawa, Canada, 1986.

*Eddy Current Classroom Training Handbook*, 2nd. ed. American Society for Nondestructive Testing, Inc., Columbus, Ohio, 1979.

Grob, B. *Basic Electronics*, 4th ed. Gregg Division, McGraw-Hill, New York, 1977.

Hagemaier, D. J. *Fundamentals of Eddy Current Testing*. American Society for Nondestructive Testing, Inc., Columbus, Ohio, 1990.

Hull, B. and V. John. *Nondestructive Testing*. Macmillan Education Ltd., London, England, 1988.

ISO 20669:2017, Non-destructive testing—Pulsed eddy current testing of ferromagnetic metallic components, ISO international Standard, March 2017.

Libby, H. L. *Introduction to Electromagnetic Nondestructive Test Methods*. Robert E. Krieger Publishing Company, Huntington, New York, 1979.

Lovejoy, D. *Magnetic Particle Inspection*. Chapman & Hall, London, England, 1993.

McGonnagle, W. J. *Nondestructive Testing*, 2nd. ed. Gordon and Breach, New York, 1961.

McMaster, R. C. *Nondestructive Testing Handbook*, First Edition, American Society for Nondestructive Testing, Inc., Columbus, Ohio, 1959.

McMaster, R. C., P. McIntire, and M. L. Mester. *Nondestructive Testing Handbook*, 2nd. ed. American Society for Nondestructive Testing, Inc., Columbus, Ohio, 1986.

Mix, P. E. *Introduction to Nondestructive Testing*. Wiley, New York, 1987.

# CHAPTER 9
# Thermal Infrared Testing

## 9.1 History and Development

Humans have always been able to detect infrared radiation. The nerve endings in human skin can respond to temperature differences of as small as 0.0162°F (0.009°C). Although extremely sensitive, nerve endings are, however, poorly designed for thermal nondestructive e valuation (TNDE). Even if humans had the thermal capabilities of a pit viper, which can find its warm-blooded prey in the dark, it is most probable that better detection tools would still be needed for TNDE. Thus, as inventive beings, we have turned to mechanical and electronic devices to allow us to become hypersensitive to heat. These devices, many of which produce thermal images, have proved invaluable for thermal evaluations and inspections in countless applications.

Sir William Herschel, who in 1800 was experimenting with light, is generally credited with the beginnings of thermography. Using a prism to break sunlight into its various colors, he measured the temperature of each color with very sensitive thermometers. Much to his surprise, the temperature increased when he moved out beyond red light into an area he came to term the "dark heat." This is the region of the electromagnetic spectrum referred to as infrared and recognized as one of the forms of electromagnetic radiation that, when absorbed, causes a material to increase in temperature.

Twenty years later, Seebeck discovered the thermoelectric effect, that quickly lead to the invention of the thermocouple by Nobili in 1829. This simple contact device is based on the premise that there is an EMF (electromotive force) or voltage that occurs when two dissimilar metals come in contact, and that this response changes in a predictable manner with a change in temperature. Melloni soon refined the thermocouple into a thermopile (a series arrangement of thermocouples) and focused thermal radiation on it in such a way that he could detect a person 30 ft away. A similar device, called a bolometer, was invented 40 years later. Rather than measuring a voltage difference, a bolometer measured a change in electrical resistance related to temperature. In 1880 Longley and Abbot used a bolometer to detect a cow over 1,000 ft away!

Herschel's son, Sir John, using a device called an evaporograph, produced the first infrared image, crude as it was, in 1840. The thermal image was caused by the differential evaporation of a thin film of oil. The image was viewed by light reflecting off the oil film. During World War I, Case became the first to experiment with photo-conducting detectors (thallium sulfide) that produced signals not as a result of being heated, but by their direct interaction with photons. The result was a faster, more sensitive detector.

During World War II, the technology began to expand and resulted in a number of military applications and developments. The discovery by German scientists that cooling the detector increased performance was instrumental in the rapid expansion of the infrared technology.

It was not until the 1960s that infrared thermal imaging began to be used for nonmilitary applications. Although systems were cumbersome, slow to acquire data, and had relatively poor resolution, they proved useful for the inspection of electrical systems. Continued advances in the 1970s, again driven mainly by the military, produced the first portable systems usable for thermal nondestructive testing (TNDT).

These systems, utilizing a cooled, scanned detector, proved both rugged and reliable. However, the quality of the image, although often adequate, was poor by today's standards. Still, by the end of the decade, infrared was being widely used in mainstream industry, for building inspections, and for a variety of medical applications. It became practical to calibrate systems and produce fully radiometric images, meaning that radiometric temperatures could be measured throughout the image. Adequate detector cooling, which had previously been accomplished through cryogenics—using either compressed or liquefied gases—was now done much more conveniently using thermoelectric coolers.

Less expensive tube-based pyro electric vidicon (PEV) imaging systems, basically heat-sensitive television cameras, were also developed and produced. Although not radiometric, PEVs were lightweight, portable, and could operate without cooling. The fact that they were nonradiometric, however, limited their use and, by the mid-1990s, the demand for detector tubes had dropped to the point they were no longer being made.

In the late 1980s, a new technology, the focal plane array (FPA), was released from the military into the commercial marketplace and development has progressed rapidly ever since, displacing nearly all other technologies. The FPA employs a large array of thermally sensitive semiconductor detectors, similar to those in charge coupled device (CCD) visual cameras. This was a significant improvement over the single-element, scanned detector and the result was a dramatic increase in image quality and spatial resolution. Arrays of $160 \times 120$, $320 \times 240$, and $640 \times 480$ elements are now commonly used, and for specialized applications, arrays are available with densities up to $1{,}000 \times 1{,}000$.

Development of the FPA technology has exploded since the 1990s. Both long- and mid-wave sensing systems are now available in fully radiometric versions, as are systems with data capture rates over 1,000 frames per second and sensitivities of 0.09°F (0.05°C or 0.05 K) or less. The cost of radiometric systems has dropped significantly, and the quality has increased dramatically. The digital storage of data has allowed 12- and 14-bit digital files; the result is access to far more image and radiometric detail than ever imagined. Figure 9-1, taken with a typical FPA system, shows the clarity of detail available and its appeal for specialized work like this study of moisture intrusion into the insulation on a liquid $CO_2$ storage tank.

The most significant advances, however, have been a reduction in price, dramatic improvements in image quality and advanced, handheld designs that are much easier than ever to use. The least costly imaging systems, though limited in their capabilities, are selling for about $1,000 USD. Thermographers have a variety of choices for systems fully capable of nearly all applications related to building diagnostics and industrial maintenance for under $10,000. Specialized systems have also been developed for NDE and medical thermography at a much lower price point than in the past.

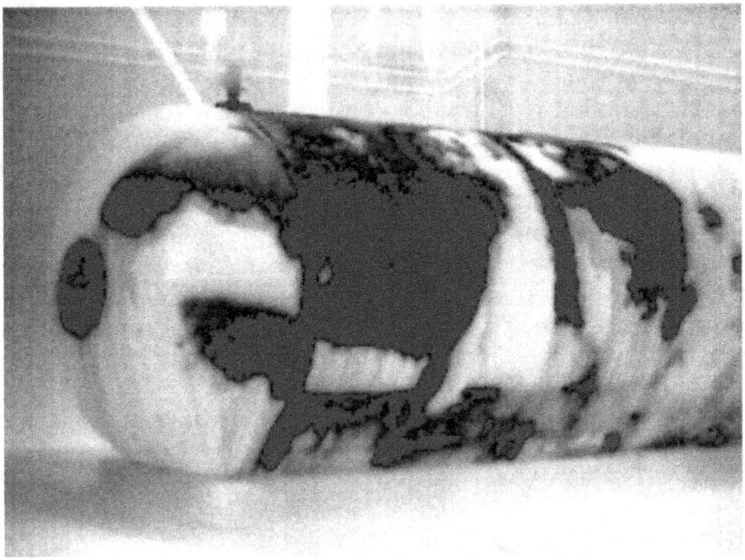

**FIGURE 9-1** The dark areas (colder) indicate places where moisture has intruded into the insulation on a liquid $CO_2$ storage tank reducing its insulating value.

Not only are modern imaging systems more thermally sensitive, the advances in image processing has resulted in image quality far beyond what was imagined possible even two decades ago. The size of many imaging systems has shrunk to less than 1 lb (0.45 kg), and designs are ergonomically excellent with software interfaces that are remarkably intuitive. The practical effect of these changes means that many more people are now using the technology successfully.

Concurrently, the use of processing software has grown tremendously. Nearly all systems commercially available today offer software programs that facilitate analysis and report writing. Reports can be stored digitally and sent electronically over the Internet.

With so many advances being made so rapidly, it is difficult to imagine what is next. Certainly, we will see continued development of new detector systems with a promise of higher data capture speeds and greater sensitivities. "Tunable" sensors capable of capturing and processing images at several wavelengths will also become commercially available. Many industry experts believe the greatest area of growth will be in the development of automated systems in which the infrared instrument, combined with a data processing computer, will capture and analyze data and provide control feedback to a process or system being monitored; these applications, however, are often more complex than they may appear on the surface and their development, to date, has been slow except in a few applications. Systems are currently in use for certain processes in the cement, metals, and chemical industries and have significantly reduced failures and costs, while dramatically increasing quality and profits.

As equipment use has increased, so has the need to qualify inspection personnel and standardize inspection methodologies. In the late 1980s, members of the American Society for Nondestructive Testing (ASNT) met to discuss a strategy for having infrared adopted as a standard test method. ASNT, a volunteer professional organization, is responsible for developing personnel qualification programs. Certification, which under ASNT is granted by the employer, is based on the practitioner having had the required training and experience, as well as passing written and practical examinations.

In 1992, infrared testing was officially adopted by ASNT and thermographers could for the first time be certified to a widely used, professionally recognized standard. Since that time, a number of large companies in the United States have begun certifying their thermographers in compliance with ASNT recommendations. In 2011 the Residential Energy Service Network (RESNET), an organization charged with implementing energy efficiency in the housing market in the United States, developed their own certification. Though based broadly on ASNT, the RESNET certification and an accompanying inspection standard both are part of an important effort to meet the specialized needs in this market sector.

Additionally "central certification" schemes have been developed in many countries around the world to align with standards developed under the auspices of the International Standards Organization (ISO). Rather than being granted by the employer, these schemes call for certification to be provided by a government agency or its designate. Unfortunately the implementation of ISO-compliant certification has not been uniformly adopted in all countries. This has led to some confusion when thermographers travel between countries, especially those in Europe.

The development of inspection standards over the past two decades has also been an interesting process. The American Society of Heating, Refrigeration, and Air Conditioning Engineers (ASHRAE) developed a building inspection standard, which was later modified and adopted by the International Standards Organization (ISO). The American Society for Testing and Materials (ASTM) also developed several standards used for determining the performance of infrared systems, as well as standards for roof moisture and bridge deck inspections. Concurrently, a number of relevant ASTM standards were developed for radiation thermometry, which also has application to infrared thermography. As previously mentioned, the RESNET standard is based on both ASNT and ASTM standards related to buildings. The National Fire Protection Association (NFPA) has developed a standard, NFPA70-B—the maintenance of electrical systems, which incorporates the use of thermography. Even though inspection standards continue to be developed, there are many applications at this time that lack inspection standards. Without standards, inspection results too often lack repeatability, which typically results in misleading, perhaps even dangerous, analysis of data.

The technology has become readily accessible as the value of imaging products has improved and their cost reduced. Companies investing in developing solid programs, including inspection procedures and qualified personnel, have a distinct advantage because they are able to take advantage to uniformly high-quality, long-term benefits provided by infrared.

## 9.2 Theory and Principles

Even though today's infrared equipment is quite easy to use, it is still essential to understand the basics of the technology. Its value, therefore, depends on the thermographers ability to interpret the data, which requires a practical understanding of the basics of thermography, radiometry, image optimization, heat transfer, and operational characteristics of the equipment being inspected.

### 9.2.1 Thermal Energy

Energy can be changed from one form to another. For instance, a car engine converts the chemical energy of gasoline to thermal energy. That, in turn, produces mechanical energy, as well as electrical energy for lights or ignition, and energy for the defroster or air conditioner. During these conversions, although the energy becomes more difficult to harness, none of it is lost. This is the First Law of Thermodynamics. A by-product of nearly all energy conversions is heat or thermal energy.

When there is a temperature difference between two objects, or when an object is changing temperature, the net flow of energy is transferred from warmer areas to cooler areas (unless work is done to reverse that) until thermal equilibrium is reached. This is the Second Law of Thermodynamics.

Heat can be transferred by any of three modes: conduction, convection, or radiation. Heat transfer by conduction occurs primarily in solids, although also to some extent in fluids, as warmer molecules transfer their energy directly to cooler, adjacent ones. Convection takes place in fluids and involves the transfer of energy caused with the mass movement of molecules. Radiation is the transfer of energy between objects by electromagnetic radiation. Because radiation needs no transfer medium, it can take place even in a vacuum, the sun being the most obvious example.

Transfer of heat can be described as either steady-state or transient. In the steady-state condition, the rate of heat transfer is constant and in the same direction over time. A fully warmed-up machine under constant load transfers heat at a steady-state rate to its surroundings. In reality, there is no such thing as absolute steady-state heat transfer as there are always small transient fluctuations. A more accurate term is quasi-steady-state heat transfer. When heat transfer and temperatures are constantly and significantly changing with time, heat flow is said to be transient. A machine warming up or cooling down is a good example. Because thermographers are often concerned with the transfer of heat and with the resulting temperatures, it is vital to understand what type of heat flow is occurring in a given situation.

Heat energy has been typically measured in British thermal units (Btu) or calories (c). A Btu is defined as the amount of energy needed to raise the temperature of 1 lb of water 1°F. A calorie is the amount of heat needed to raise the temperature of 1 g of water 1°C. One wooden kitchen match, burned entirely, gives off approximately 1 Btu or 252 calories of heat.

The units to describe the energy content of food are also termed "calories." These calories are actually kilocalories (Kcal or C) and are equal to 1,000 calories (c). Various devices, such as a calorimeter or a heat flow meter, can measure the flow of heat. These devices are not commonly used, however, except by scientists in research laboratories.

Temperature is a measure of the relative "hotness" of a material *compared to some known reference*. Thus, a cup of coffee may be "too hot" to comfortably drink or a winter's day too cold for the clothing being worn. In all instances a comparison is either specified or implied. Ultimately we can use an absolute measurement scale (Kelvin or Rankine) to compare anything to absolute zero, or no heat.

There are many ways to measure temperature. The most common is to use our sense of touch. We also use comparisons of various material properties, including expansion (liquid and bimetal thermometers), a change in electrical voltage (thermocouple), and a change in electrical resistance (bolometers). Infrared radiometers *infer* a temperature measurement from detected infrared radiation. This radiometric measurement, to be accurate, must be corrected for several variables.

Regardless of how heat is transferred, thermographers must understand that materials also change temperatures at different rates due to their thermal capacitance. Some materials, like water, require a great deal of energy to warm up or cool down, while others, like air, change temperature with little energy required. The thermal capacitance or specific heat and density of a material describes this relationship between energy and temperature; this is not necessarily time-based, a common misunderstanding, because the rate of change depends primarily on the mode of transfer as well as the amount of energy needed. A great deal of heat is transferred, for instance, when changing the temperature of a swimming pool of water compared to a room of the same volume filled with air. We cannot, however, infer that one will change more or less rapidly unless we know how the energy is being supplied or removed. Without an understanding of these concepts and values, thermographers will not be able to properly interpret their findings, especially with regard to transient heat flow situations. Although potentially confusing, understanding these properties can be used to our advantage. Finding liquid levels in tanks, for example, is possible because of the differences between the thermal capacitance of the air and the liquid.

### 9.2.2 Latent Heat

As materials change from one state (solid, liquid, or gas) to another, heat is released or absorbed. When a solid changes state to a liquid, energy is absorbed in order to break the bonds that hold it as a solid. The same thing is true as a liquid becomes a gas; heat must be added to break the bonds. As gases condense into liquids, and as liquids freeze into solids, the energy used to maintain these high-energy states is no longer needed and is released.

This energy, which can be quite substantial, is called latent energy because it does not necessarily result in the material changing temperature. Water, for instance, can exist as either a liquid or a solid at 32°F (0°C) yet to change from ice to water or vice versa clearly involves a substantial amount of energy, either added or released. The impact of this energy being released or absorbed during state change often affects thermographers because it is pulled from or added to the surrounding materials. A wet surface or a rain-soaked exterior wall will not warm up until it is dry, as the energy is used to evaporate water rather than heat the surface; this may mask over any subsurface thermal anomalies. On the positive side, state changes may enable thermographers to see remarkable phenomena, such as whether or not solvents have been applied evenly to a surface, as indicated by the temperatures of the materials involved.

### 9.2.3 Conduction

Conduction is the transfer of thermal energy from one molecule or atom directly to another adjacent molecule or atom with which it is in contact. This contact may be the result of physical bonding, as in solids, or a momentary collision, as in fluids. Fourier's law describes how much heat is transferred by conduction:

$$Q = k/L \cdot \Delta T \cdot A$$

where:

$Q$ = heat transferred
$k$ = thermal conductivity
$L$ = thickness of materials
$A$ = area normal to flow
$\Delta T$ = temperature difference

The thermal conductivity ($k$) is the quantity of energy that is transferred, in this case through 1 ft$^2$ of a material, which is 1 in. thick, during 1 hr when there is a 1°F temperature difference across it. The metric equivalent is W/(m × °C) and assumes a thickness of 1 m.

Materials with high thermal conductivities, such as metals, are efficient conductors of heat. We use this characteristic to our advantage by making such things as cooking pans and heat sinks from metal. Differences in conductivity are the basis for many thermographic applications, especially the evaluation of flaws in composite materials or the location of insulation damage in various applications.

Materials with low thermal conductivity values, such as wool, fiberglass batting, and expanded plastic foams, do not conduct energy very efficiently and are called insulators. Their insulating value is due primarily to the fact that they trap small pockets of air, an inefficient conductor.

The term R-value, or thermal resistance, is a measure of the resistance to conductive heat transfer. It is defined, as the inverse of conductivity, or $1/k$. R-value is a term that is generally used when describing insulating materials.

Table 9-1 lists the thermal conductivities and R-values of some common materials.

Another important material property is thermal diffusivity. Thermal diffusivity is the rate at which heat moves throughout the volume of a material after a temperature change is induced somewhere in the material. Diffusivity is determined by the ratio of the material's thermal conductivity to its thermal capacitance. Differences in diffusivity and consequent heat transfer are the basis for many active thermography applications in TNDT.

## 9.2.4 Convection

Heat is transferred in fluids, either gases or liquids, by convection. During this process, energy is transferred by conduction from one molecule to another and by the subsequent mixing or the mass movement of molecules.

In natural convection, this mixing or diffusing of molecules is driven by the warmer (less dense) molecules' tendency to rise and be replaced by more dense, cooler molecules.

Cool cream settling to the bottom of a cup of hot tea is a good example of natural convection. Forced convection is the result of fluid movement caused by external forces such as wind or moving air from a fan. Natural convection, in which gravity and density

| Material | $k$ (Btu × in./ft$^2$ × hr × °F)* | $R$ (ft$^2$ × hr × °F/Btu × in.) |
|---|---|---|
| Extruded polystyrene (1 in.) | 0.20 | 5.0 |
| Fiberglass batts (1 in.) | 0.32 | 3.125 |
| Soft wood (average) | 1.0 | 1.0 |
| Brick (average) | 10.0 | 0.1 |
| Concrete (average) | 13.0 | 0.07 |
| Steel | 314.0 | 0.003 |
| Cast iron | 331.0 | 0.0036 |
| Copper | 2724.0 | 0.00036 |

*Metric units for $k$ are W/m × °C.

**Table 9-1** Thermal Conductivities and R-values of Common Materials

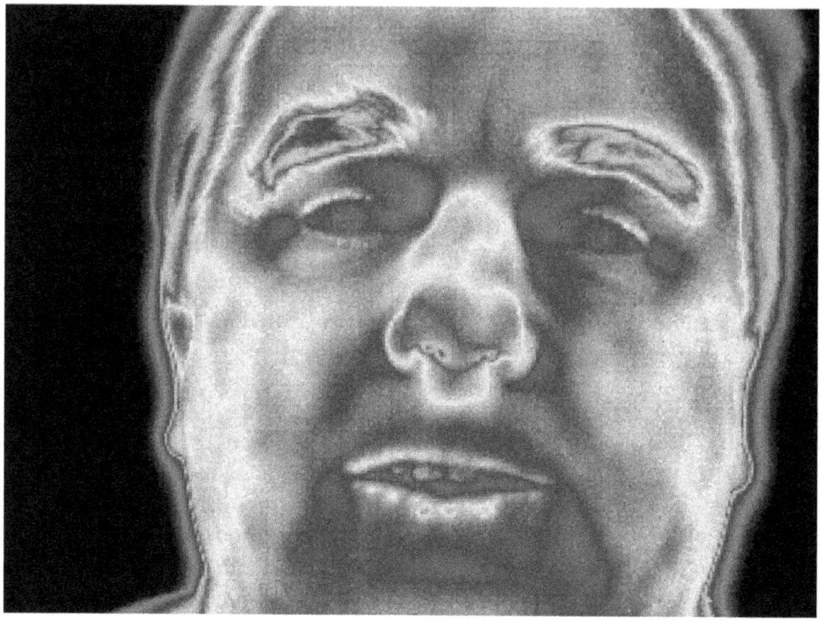

**FIGURE 9-2** The human face provides an interesting thermal signature. Eyebrows and hair are cooler because they are detached from the circulation system; the teeth and lips are cooler due to evaporative cooling; the nose is cooler because of the extensive convection that takes place both internally and externally; the area around the tear ducts is warmer because of the proximity of arteries to the skin's surface, and the cheeks are mottled due to the presence of layers of insulating fat.

define the movement of the molecules, is quickly overcome by these external forces. Figure 9-2 shows the typical, yet dramatic, pattern associated, in large part, with the cooling effect of forced internal and natural external convection on a person's nose as well. The mouth has been cooled by evaporation, and the eyebrows are closer to the cooler ambient temperature rather than body temperature.

Newton's law of cooling describes the relationship between the various factors that influence convection:

$$Q = h \times A \times \Delta T$$

where:

$Q$ = heat energy
$h$ = coefficient of convective heat transfer
$A$ = area
$\Delta T$ = temperature difference

The coefficient of convective heat transfer is often determined experimentally or by estimation from other test data for the surfaces and fluids involved. The exact value depends on a variety of factors, of which the most important are velocity, orientation, surface condition, geometry, and fluid viscosity.

Changes in $h$ can be significant due merely to a change in orientation to the flow. The topside of a horizontal surface can transfer over 50 percent more heat by natural convection than the underside of the same surface.

In both natural and forced convection, a thin layer of relatively still, fluid molecules adheres to the transfer surface. This boundary layer, or film coefficient, varies in thickness depending on several factors, the most important being the velocity of the fluid moving over the surface. The boundary layer has a measurable thermal resistance to conductive heat transfer. The thicker the layer, the greater the resistance. This, in turn, affects the convective transfer as well. At slow velocities, these boundary layers can build up significantly. At higher velocities, the thickness of this layer and its insulating effect are both diminished.

Thermographers must always be concerned with convection because it often affects temperatures by quickly moving large quantities of energy. As forced convection, such as the wind, increases, heat transfer also increases and can have a significant impact on the temperature of a heated or cooled surface. Regardless of velocity, air at ambient temperature has no cooling or warming effect on surfaces that are at the same ambient temperature. Thermographers inspect a variety of components where an increase in temperature over ambient is an indication of a potential problem and forced convection, most commonly the wind for outdoor applications, is capable of masking these indications by cooling the surface below the threshold of detection.

### 9.2.5 Radiation

In addition to heat being transferred by conduction and convection it is also be transferred by radiation. Thermal infrared radiation is a form of electromagnetic energy similar to light, radio waves, and X-rays. All forms of electromagnetic radiation travel at the speed of light, 186,000 m/sec ($3 \times 10^8$ m/sec). All forms of electromagnetic radiation travel in a straight line and are considered to behave as both waves and massless particles called photons; they differ only in their wavelength and energy. Infrared radiation that is detected with thermal imaging systems has wavelengths between approximately 2 and 15 microns (µm). Transfer of electromagnetic radiation requires no medium and, thus, can travel through a vacuum, as demonstrated by the sun's warming effect from a distance of 93 million miles of space.

All objects above absolute zero radiate infrared radiation. The amount and the exact wavelengths radiated depend primarily on the temperature of the object. It is this phenomenon that allows us to see radiant surfaces with infrared imaging systems.

Due to atmospheric absorption, significant transmission through air occurs in only two "windows" or wavebands: the mid-wave (2–6 µm) and long-wave (8–15 µm) bands. Both can be used for many thermal applications. With some applications, one waveband may offer a distinct advantage or make certain applications feasible. These situations will be addressed in subsequent sections.

The amount of energy emitted by a surface depends on several factors, as shown by the Stefan–Boltzmann formula:

$$Q = \sigma \times \varepsilon \times T^4 \text{ absolute}$$

where:

$Q$ = heat transferred by radiation

$\sigma$ = the Stefan–Boltzmann constant ($0.1714 \times 10^{-8}$ Btu/hr $\times$ ft$^2 \times R^4$) ($5.6704 \times 10^{-8}$ W/m$^2 \times K^4$)

$\varepsilon$ = the emissivity value of the surface

$T$ = the absolute temperature of the surface

When electromagnetic radiation interacts with a surface several events may occur. Thermal radiation may be reflected by the surface, just like light by a mirror. It can be absorbed by the surface, like our skin in the sunshine, in which case it often causes a change in the temperature of the surface. In some cases, the radiation can also be transmitted through the surface; light passing through a window or X-rays passing through skin are both good examples. The sum of these three components—the energy reflected, absorbed, and transmitted—must equal the total amount of radiant energy involved.

This relationship, known as the conservation of energy, is stated as follows:

$$R + A + T = 1$$

where:
$R$ = reflected energy
$A$ = absorbed energy
$T$ = transmitted energy

Radiation is never perfectly transmitted, absorbed, or reflected by a material. In the real world, as radiation intercepts a material, two or three of these phenomena are always occurring simultaneously. For example, one can see through a window (transmission) and also see reflections in the window at the same time. It is also known that glass absorbs a small portion of the radiation because the sun can cause it to warm up. For a typical glass window, 92 percent of the light radiation is transmitted, 6 percent is reflected, and 2 percent is absorbed. One hundred percent of the radiation incident on the glass is thus accounted for.

Infrared radiation, like light and other forms of electromagnetic radiation, also behaves in this way. Energy that has been absorbed is also re-emitted based on the temperature of the material. So, if we can quantify the energy emitted, we can determine the temperature. When a surface is viewed with a thermal imager it senses all the radiation coming from the surface, a combination of emitted, reflected, and transmitted radiation. Neither the transmitted nor reflected radiation provides any information about the temperature of the surface. While corrections can be made to account for transmitted or reflected energy, the reality of doing so in many instances makes getting an acceptable degree of measurement accuracy very challenging.

The combined radiation from a surface seen by the infrared system is called its radiosity. The job of the thermographer is to distinguish the emitted component from the others so that more about the target temperature can be understood.

Only a few materials transmit infrared radiation very efficiently. The lens material of the thermal camera is obviously one. Transmissive materials can be used as thermal windows, allowing viewing into enclosures. The atmosphere is also fairly transparent, at least in the two wavebands mentioned previously. In the rest of the thermal spectrum, water vapor, carbon dioxide and ozone, among others, absorb the bulk of the thermal radiation. As can be seen from Fig. 9-3, radiation is transmitted quite readily in both the mid-wave (2–6 μm) and long-wave (8–14 μm) bands. Infrared systems have been optimized to one of these bands or the other. Broadband systems are also available and have some response in both wavebands.

A transmission curve for glass shows us that glass is somewhat transparent in the mid-wave band but opaque in the long-wave band. It is surprising to look at a window with a long-wave thermal imager and not be able to see anything on the other side! Many thin plastic films are transparent in varying degrees to infrared radiation. A thin

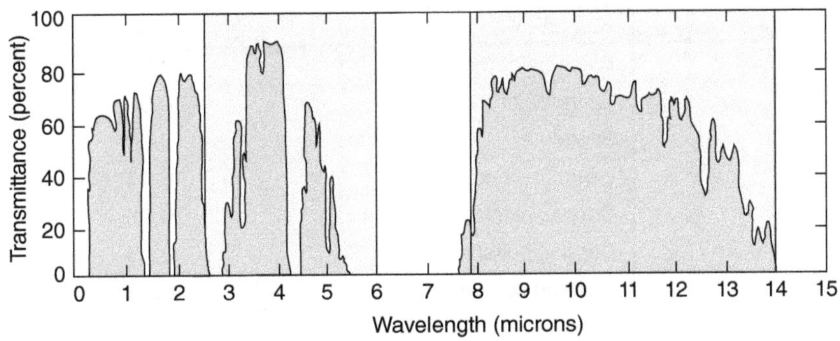


**FIGURE 9-3** Electromagnetic radiation passing through the earth's atmosphere is transmitted quite readily in both the mid-wave (2–6 μm) and long-wave (8–14 μm) bands. Other bands, where energy is absorbed by the atmosphere, show little or no transmission.

plastic bag may be useful as a camera lens protector in wet weather or dirty environments. Be aware, however, that all thin plastic films are not the same! While they may look similar, it is important to test them for transparency and measure the degree of their thermal attenuation. Depending on the exact atomic makeup of the plastic, they may also absorb strongly in very narrow, specific wavebands. This unique characteristic allows us to measure the temperature of a thin plastic film by viewing through a filter that limits the radiation to those areas where absorption (and re-emission) occurs.

The vast majority of materials are opaque and *not* transparent to thermal radiation. This simplifies the task of looking at them thermally by leaving one less variable to deal with. This means that the only radiation we detect is that which is reflected and that which is absorbed and re-emitted by the surface ($R + A = 1$).

If $R = 1$, the surface would be a perfect reflector. Although there are no such materials, the reflectivity of many polished shiny metals approaches this value. They are, in essence, heat mirrors. Kirchhoff's law says that for opaque surfaces the radiant energy that is absorbed is also re-emitted, or $A = E$. By substitution, it is concluded that the energy detected from an opaque surface is either reflected or emitted ($R + E = 1$). Only the emitted energy provides information about the temperature of the surface.

In other words, an efficient reflector is an inefficient emitter, and vice versa. For thermographers, this simple inverse relationship between reflectivity and emissivity forms the basis for interpretation of nearly all of that is seen. Highly emissive objects reveal a great deal about their temperature; reflective surfaces do not. In fact, commonly, very reflective surfaces, which emit very little of their own thermal energy, typically hide their true thermal nature by also reflecting their thermal surroundings.

If $E = 1$, all energy is absorbed and re-emitted. Such an object, which exists only in theory, is called a blackbody. Human skin, regardless of color, has an emissivity of 0.98, and is nearly a perfect blackbody. In fact many painted surfaces, regardless of their color, approach being thermally "black" because they absorb and re-emit most of the incident energy and reflect very little of it.

Emissivity is a surface characteristic of a material that indicates the relative efficiency with which it emits radiation. Emissivity is defined as the ratio of thermal energy emitted by a real surface to that emitted by a blackbody of the same temperature. Emissivity is a value between zero and one. Most nonmetals have emissivities above 0.8.

| Human skin | 0.98 |
|---|---|
| Black paint (flat) | 0.90 |
| White paint (flat) | 0.90 |
| Paper | 0.90 |
| Lead, oxidized | 0.40 |
| Copper, oxidized to black | 0.65 |
| Copper, polished | 0.05 |
| Aluminum, polished | 0.10 |

*Values will vary with exact surface type and wavelength.

**TABLE 9-2** Emissivity Values*

Shiny metals, on the other hand, typically have emissivities below 0.2. Materials that are not blackbodies—in other words everything—are called real bodies. Real bodies always emit less radiation than a blackbody at the same temperature. Exactly how much less depends on their emissivity.

Several factors can affect what the emissivity of a material is. Besides the material type, emissivity can also vary with surface condition, temperature, and wavelength. The emittance of an object can also vary with the angle of view.

It is not difficult to characterize the emissivity of most materials that are not shiny metals. Many of them have already been characterized, and their values can be found in tables such as Table 9-2. These values should be used only as a guide. Because the exact emissivity of a material may vary from these values, competent thermographers also need to understand how to determine and measure the actual value.

It is interesting to note that cracks, gaps, and holes absorb thermal energy at a higher rate than the surfaces in which they are found. The same is true for visible light. The pupil of your eye is black because it is a cavity, and the light that enters it is absorbed within that cavity. When all light is absorbed by a surface, we say it is "black." The emissivity of a cavity will approach 0.98 when it is seven times deeper than it is wide and will appear "black" regardless of whether it is visible or thermal radiation.

From an expanded statement of the Stefan–Boltzmann law, the impact that reflection has on solving the temperature measurement problem for opaque materials can be seen:

$$Q = \sigma \times \varepsilon \times T^4 + \boldsymbol{[\sigma \times (1 - \varepsilon) \times T^{4\,\text{background}}]}$$

The second part of the equation (in boldface) represents that portion of the radiation that comes from the reflected energy. When using a radiometric system to make a measurement, it is important to characterize and account for the influence of the reflected background temperature.

Consider these two possible scenarios:

- When the object being viewed has a very low emissivity (i.e., is highly reflective), the temperature of the reflected background becomes quite significant.
- When the background is at a temperature that is extremely warmer than the object being viewed, the influence of the background becomes more pronounced.

It becomes clear that repeatable, accurate radiometric measurements can be made only when emissivities are high. This is a fundamental limitation within which all thermographers work. Generally, it is not recommended to make temperature measurements of surfaces with emissivities below approximately 0.60, in other words all shiny metals, except under tightly controlled laboratory conditions. However, with a strong understanding of how heat is transferred and a working knowledge of radiation, the value of infrared thermography as a noncontact temperature measurement tool for nondestructive evaluation is unsurpassed.

## 9.3 Equipment and Accessories

Infrared systems can generally be classified as either thermal imaging systems or point temperature measuring systems. Thermal imaging systems can be further subdivided into quantitative or radiometric (measuring temperatures) and qualitative or nonradiometric systems (thermal images only). Very few nonradiometric thermal imaging systems are used nowadays. Hybrid systems are much more common, systems that measure the radiometric temperature only at a small center spot in the image; when the image is processed in the manufacturer's proprietary processing software, however, all points in the image can be corrected and measured accurately. Fully radiometric systems are also available allowing temperature measurement of the entire thermal scene in the thermal camera.

Quantitative thermography involves the use of calibrated instruments. By using correction factors, accurate radiometric temperatures can be determined from the thermal image. For the greatest degree of accuracy and repeatability, calibration is best made for the specific temperature range(s) required. Accuracy is often improved by calibrating to several narrow ranges rather than one broad one.

Non-imaging radiometric measuring systems, called point or spot radiometers, are typically lower-cost instruments, which, although they do not give an image, can be valuable for many types of NDT. For limited areas and specific applications, and *when used properly*, spot radiometers can provide accurate, real-time radiometric measurement across a wide range of temperatures. Typically, higher-cost spot radiometers have emissivity correction capabilities but not all have background correction. The distance to spot size ratio can vary, from 1:1 on lower cost systems to 100:1 or greater on higher cost systems. Spot radiometers also come in a variety of wavebands to match the specific needs of materials to be measured, surface temperatures and imaging environment.

Without an image, it can be difficult to determine the exact area the spot radiometer is measuring, possibly resulting in erroneous data. Some models include a laser-aiming device to assist with this task. It should be carefully noted, however, that the area actually being measured is always larger than the small spot illuminated by the laser resulting in the temperature of a larger area being averaged into the single measurement. Some models are available with a projected laser beam that also outlines the exact measurement area, alleviating any misunderstanding.

The waveband response of most spot system detectors is typically longwave for a wide variety of general applications. Mid-wave systems can be filtered to detect narrower wavebands for specialized applications. This is particularly useful for such applications as measuring temperatures of thin-film plastic or glass, especially during manufacturing. Spot systems can also be fixed or mounted to constantly monitor process temperatures unattended. The data is typically fed back to a computer or control system. Such thermal data can be correlated not only to temperature, but, with the right design and software, also to approximate thickness, moisture content, material integrity, material type, or parts presence detection. When combined with visual automated systems, a powerful information system is available.

Even though spot radiometers are less expensive than thermal imaging systems, they should not be thought of as a substitute for an imaging system. Both types of systems may, however, have their place in many NDT programs.

A variation of the spot radiometer is the line scanner. It is also very useful for many fixed-mounted NDT inspections. This instrument employs a single detector or small linear array of detectors over which the scene is then scanned. If the line scanner views a moving object, such as a rolling drum, an image can be created. Line scanners are particularly useful for viewing equipment that is moving at a constant rate because the image is built up one line at a time as the product passes the detector. Two common applications are viewing the refractory in rotating limekilns or determining the cross-roll moisture profile on the moving sheet in a paper machine. Line scanners, which can produce very high-quality images with excellent radiometrics, are typically less expensive than full imaging systems and lend themselves well to these types of automated inspections.

It is useful to have a general understanding of how imaging systems work. The purpose of the imager is to detect the infrared radiation coming from (or through in the case of non-opaque targets) the object being inspected, commonly called the target. The radiation coming from the target is focused by the thermally transparent optics onto a detector, which is sensitive to this "heat" radiation. A response from the detector produces a signal, usually a voltage or resistance change, which in turn is read by the processing electronics of the system and to create a thermal image on a display screen, typically an LCD screen or viewfinder. In this seemingly simple process, we are able, without contact, to view the thermal image, or thermogram, that corresponds to the thermal energy coming off the surface of the target. With correction for emissivity and background temperatures can also be inferred.

The first infrared imaging systems used a technique called "scanning" to create a full screen image using a single, small detector. A series of rotating or oscillating mirrors or prisms inside the scanner direct the infrared radiation from a small part of the scene being viewed onto the detector. The detector response results in the display of a portion of the field of view. Very quickly, the mirror or prism moves to scan another part of the scene with the resultant display. This process of making a series of single measurements happens thousands of times per second, so that a complete image is built up and viewed in what appears to be "real time," i.e., at 30 frames per second or greater. For some types of detectors to respond quickly, accurately, and in a repeatable fashion they must be cooled to extremely low temperatures. Systems with the highest-quality images use liquid nitrogen ($LN_2$) at −321°F (−196°C). The nitrogen is simply poured into an insulated container, called a "Dewar," that surrounds the detector or is fed continuously from a reservoir. Obviously, great care must be used in handling cryogenics. Because of these safety problems and the problems of dealing with cryogenic substances in the field, cryogenically cooled systems are seldom used anymore except in laboratory situations.

Focal plane array systems use a matrix of detectors to sense the infrared radiation. These typically have spatial resolutions of 160 × 120, 320 × 240, or 640 × 480 pixels. Resolutions of 1,000 × 1,000 pixels are now widely used for specialized applications such as aerial thermography. Although early FPA systems had to be cooled and were only sensitive in the mid-wave band, uncooled, long-wave sensing systems are now the norm.

There are several types of FPA systems based on characteristics of the detector. Typically detectors are classified as either thermal or photonic. Photonic or photon detectors come in both long-wave and mid-wave configurations. These systems have cooled detectors, which actually count the number of photons received in a specific time period called the integration time. Photon detecting systems typically have very reliable radiometrics and excellent imagery. Thermal detectors are typically all long-wave and do not have to be cooled, although they must be temperature stabilized. Thermal detectors actually change temperature as a result of the incoming thermal radiation. The most typical thermal detectors use microbolometer technology, measuring the change in electrical resistance relative to its temperature. Very reliable radiometrics and excellent imagery is now possible from thermal imagers using microbolometer detectors.

Infrared imaging systems vary widely in price from $1,000 for simple low-resolution units to over $80,000 for sophisticated specialized systems. Very capable handheld systems are now available in the $3,000 to $10,000 range that meets a wide range of imaging requirements for general condition monitoring, building diagnostics, and many TNDE applications.

Infrared imaging systems are generally made up of several common components, including lens, detector, processing electronics, controls, display, data storage, data processing and report generation software, and filter.

### 9.3.1 Lens

Lenses serve to focus the incoming infrared radiation on the detector. Most infrared lens materials are actually opaque to visible light. The most common lens material used today is germanium (Ge), but lenses are also made of a variety of thermally refractive materials such as silicon (Si), and zinc selenide (ZnSe) based on specific system requirements.

Lenses can be of different focal lengths. A normal lens (with a field of view from 16° to 25°, approximately) is useful for most applications. Where inspection space is limited or a wide view required, the use of a wide-angle lens is recommended. Over longer distances, such as when viewing a disconnect switch in a substation, a telephoto lens may be warranted. Lens selection also impacts spatial and measurement resolution.

Coatings used on some lenses may contain small amounts of radioactive material (thorium). When using the system in a radioactive area, make sure to test the system for its baseline characteristics prior to entering radioactive environments.

### 9.3.2 Detector

The radiation is focused on the detector, where it produces a measurable response. Materials commonly used for detectors are vanadium oxide (VOx), amorphous silicon (ASi), platinum silicide (PtSi), mercury cadmium telluride (HgCdTe), and indium antimonide (InSb).

### 9.3.3 Processing Electronics

The response from the detector is processed electronically to produce a thermal image, a radiometric temperature measurement, or both. Typically, radiometric systems also feature some very powerful analysis functions in the instrument itself, such as spot or area measurement and isotherm display. Again, these features often make fieldwork simpler and more effective.

### 9.3.4 Controls

Various controls on the system allow for adjustments to control the input of infrared radiation or the output of data. These typically include adjustments to range and span, thermal level, polarity, emissivity, and other temperature measurement functions.

### 9.3.5 Display

The processed data is output to a display, typically a liquid crystal display (LCD) screen. Most instruments can display the thermal image in either grayscale or a variety of color paletes.

### 9.3.6 Data Storage

Data is stored either as a still digital image or as a digital video file stored on a PC card or internal flash memory. Digital voice data and supporting text data can also be stored with an image on some systems as well. Storage capacity is commonly up to 1,000 images or more with accompanying voice data.

Many models of imager now also incorporate visual cameras and capture not only a thermal image but also a corresponding visual image. These can be manipulated in post-processing to appear side-by-side or overlaid in various "fusion" modes, providing an extremely useful means of documentation.

It is also possible on advanced systems to output digital data directly to a computer. This is the basis of most systems used for nondestructive evaluation of composite materials, for instance, as well as for many process monitoring and security applications.

### 9.3.7 Data Processing and Report Generation Software

Most of the software that is available today is both powerful and very easy to use. Digital images are imported into the computer and may be displayed in any of several grayscale or color palettes. Adjustments can be made to all radiometric parameters, such as emissivity, background temperature, span, and level. Analysis functions may include spot, area, isotherms, and line thermal measurement, as well as the ability, with a reference, to take distance measurements off the image. Analysis can extend beyond the image by displaying the numerical data in a spreadsheet or in various standard graphical forms such as a histogram.

When an image has been analyzed and processed, notation labels can be added and the image inserted into a report template. From there, the completed report can be sent to a printer, stored electronically or sent via the Internet.

### 9.3.8 Filters

Some thermographic applications depend upon the use of specialized filters to obtain a useful image or measurement. Before using filters, it is important to know the exact spectral response of the imaging system, as determined by the detector and the lens material. Exact responses within the long- or mid-wave bands can vary from system to system. It is also important to understand how the selected filter interacts with the detector's response.

There are three generic filter designations:

1. High-pass filters, which allow only shorter wavelengths to pass
2. Low-pass filters, which allow only longer wavelengths to pass
3. Band-pass filters, which allow only a designated band of wavelengths to pass

Among the most commonly used infrared filters are:

1. Flame filters. These suppress the peak flame radiation by excluding all radiation except that in the 3.9 μm band (for mid-wave or broadband systems) or, less commonly, 10.8 μm band (for long-wave systems). This enables viewing through flames to hot objects beyond.
2. Glass filters:
   a) For mid-wave systems one type of filter cuts off wavelengths below 4.8 μm. This allows the system to see the surface of the glass rather than through the glass.
   b) Another glass filter allows a mid-wave system to see primarily through glass by limiting radiation to a 2.35-μm band-pass.
3. Filters for viewing thin plastic films come in various band passes, depending on the exact plastic that is being inspected. Many thin plastic films are very transmissive in both long and short wavelengths. In all cases, the filter limits the radiation detected to that narrow band at which the plastic is opaque (nontransparent). The result is that the surface of the plastic can be seen and measured thermally. For polyethylene, a 3.45-μm band-pass filter is used, whereas for polyester a 7.9-μm band-pass filter is used.

To maintain radiometric accuracy, these systems must be calibrated for each filter used. Some systems automatically load the correct calibration curves when the filter is installed. Some systems detect in the full 2 to 12 μm broadband and can be used in either mid-wave or in long-wave mode simply by switching a filter into place and referencing the proper calibration.

Infrared imaging systems have become much smaller over the past few years. The need to push a cart full of accessory equipment has become a thing of the past! With digital data storage digital video, and on-board visual, audio, and text annotation, the packages have become very robust. We can expect new developments in the near future that tie the imager to the World Wide Web in ways that were unimaginable even a few years ago.

Temperature calibration can only be done by the manufacturer. Most manufacturers suggest calibration be done on an annual basis but the logic is flawed. What is actually important is to verify calibration. Verification should be done on a frequency based on the customer's needs. For example, fever detection standards call for continuous verification by inclusion of a black body reference in the field of view at all times. The verification of calibration of systems used for inspection of a home, on the other hand, are usually done much less often, typically monthly or annually because, except in a few cases, temperature measurement is typically not part of the scope of work.

Verification of calibration is easily accomplished by comparing the radiometric reading to the temperature of a stable, known, high-emissivity source such as a calibrated blackbody reference, under controlled conditions. Records of verification should be kept in case calibration is ever questioned.

A side benefit of implementing a valid verification procedure is a cost savings of both time and money. Verification at three temperature set points can be accomplished for an imaging system in about an hour. The cost of sending the same instrument back to the manufacturer for calibration can be over a thousand dollars and involve several weeks of time. Of concern, however, is the fact there is really no assurance the imager is still in calibration when it is shipped back. The only way to ensure a camera is in calibration is to verify the calibration against a known source at the time of inspection!

When accurate radiometric measurements are required, additional data may need to be collected at the time the image is captured. Without this data, results may be misinterpreted or incorrect. The following environmental data may be important to collect:

- Ambient air temperature
- Emissivity and background temperature
- Relative humidity
- Distance to the object being inspected
- Information about wind or other convection
- Impact of the sun
- Specific operational conditions of the target

It is necessary to measure or estimate the emissivity of the object. In many instances, especially where safety allows, it may also be important to increase the emissivity of the surface by some sort of modification to even make measurements possible. These measures typically include adding paint, tape, oil or some other high-emissivity material to the surface.

## 9.4 Techniques

Thermographers have developed a number of very useful techniques for expanding the use of the technology. These techniques can consistently result in higher quality data that is often simpler to gather than in the past. While some development time is often required when applying these techniques, the potential returns usually far outweigh the investment.

Some techniques involve specialized tools or equipment, for instance, lasers used to heat the sample, while others are simply better ways to enhance the flow of heat into or out of a sample, such as wetting the surface of a tank to induce evaporative cooling and thus make the liquid level more clearly visible.

The basic strategy used in many thermal applications is quite simply termed "comparative thermography." It is no more complex than comparing similar samples or components under similar conditions. When the comparative technique is used appropriately and correctly, the differences between the two (or more) samples will often be indicative of their condition. For this method to be effective, the thermographer must eliminate all but one variable from the comparison. Too often, this simple but essential requirement is not achieved due to the complex circumstances of a test, changing or varied ambient conditions, or the poor work habits of the thermographer. As a result, data can be inconclusive or misleading.

As an example, when inspecting three-phase electrical systems, it is very useful to compare the three phases. When the loads are even, which is often the case, all phases will appear thermally similar. However, when the loads are uneven, the phase carrying more load will typically appear warmer. Unless loads are measured and understood, a thermal image of such a situation has little or no value. Misdiagnosis in such a situation can, at best, result in an embarrassing loss of credibility or, at worst, cause the loss of valuable equipment.

Prior to inspecting a component or system, it is critical to determine what the normal or desired condition is. What is the "ground truth" or baseline? In some cases it

**Figure 9-4** The lighter areas of this heated building show increased heat loss. While such indications are normal for windows and doors, the irregular areas on the walls suggest insulation, probably a material blown into the wall cavity after construction, was poorly placed or is otherwise missing.

may be easier to determine what the abnormal condition will look like. For flaw detection in aerospace composites, this determination can involve costly theoretical modeling as well as construction and extensive testing of sample test pieces. Understanding a baseline signature may also be simply intuitive. For example, Fig. 9-4 shows an outside view of a heated building. The light areas in the thermal image indicate the location of areas of greatest heat transfer including windows and doors as well as wall sections that are poorly insulated.

In comparative thermography, it is useful to know as much as possible about the sample being tested, such as its construction, basic operations, known failure mechanisms, direction of heat transfer, or history. Because this knowledge is often not readily available, the thermographer must become adept at asking clear, simple questions and, even more important, listening carefully to the answers. Many thermographers fail at either one or both of these tasks, and the work quality suffers. Communications skills are as important as technical skills, especially when working with unfamiliar equipment or materials.

A variation of comparative thermography is thermal mapping. Here, temperature distributions over large areas of a sample are compared with other, previously saved images. Baseline thermal mapping is used extensively for inspecting mechanical equipment where the thermal patterns may be complex and the signatures of failure often develop slowly over time. Figure 9-5 shows the heating caused by friction as a sheave was found to be slipping on the shaft of this belt driven motor/fan unit.

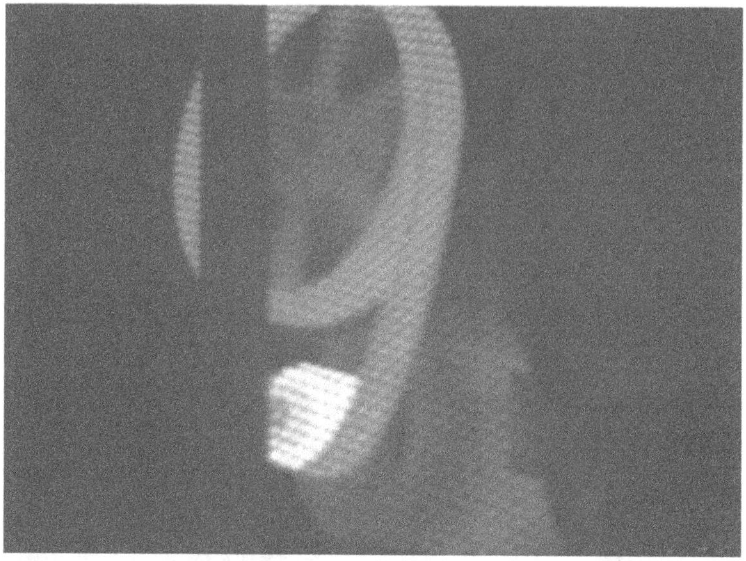

**Figure 9-5** A by-product of mechanical friction is heat. The lighter areas in this thermal image clearly show a sheave or pulley not adequately attached to the drive shaft and, as a result, it slips as the belt rotates resulting in abnormal heating.

Again, it is critical to eliminate, or at least understand, all the variables involved with any temperature related event. Because the maintenance of many mechanical systems has a strong time-based component, techniques that trend the changes over time, such as vibration monitoring, are already widely used and accepted. While these thermal mapping techniques can reveal a very accurate and useful temperature trend of past performance, it is important to remember that trending only implies, rather than predicts, the future.

Thermal maps can be made of a material sample and compared to other known references, such as the failure point of a material or the temperature at which contact injury could occur. Thermal mapping has been used with remarkable success in this regard in the automotive industry to validate the temperatures of such areas as the sheet metal floor pan or a heated seat. Typically, this temperature data has been gathered via thermocouples, an expensive, time-consuming process that sometimes fails to identify the high-temperature problem areas. Thermal mapping, as shown in the heated seat image of Fig. 9-6, not only verifies that the thermocouple is in the proper location; it also provides a wealth of information about each and every location on the area. These thermal maps can be compared to alarm limits, such as a threshold temperature for the fabric or the temperature at which a human could be burned.

Thermal mapping has also been used extensively in the electronics industry to look at printed circuit boards (PCBs). Printed circuit boards can be inspected during design, manufacturing, or repair. Typically, the inspection is conducted from the time a cold board is energized until it warms up to steady-state operating temperatures. Image subtraction, another valuable mapping technique, is often used in this application. Specialized image processing software is used to subtract the thermal image of the board being inspected, pixel by pixel, from a thermal image of a normal or "golden"

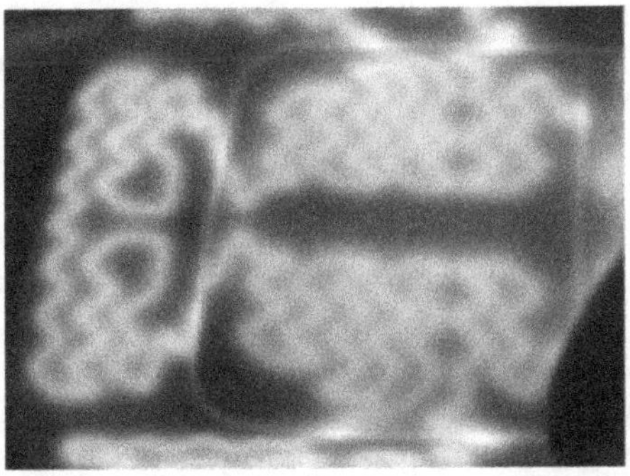

**Figure 9-6** The electric heating elements in an automobile seat are clearly visible in the lighter portions of this thermal image. Such imagery is just one of many ways automotive design engineers use thermography to better understand comfort, quality and safety issues.

board. The difference between the two images represents anomalous components or situations.

Manipulation of images using other software routines provides a wide range of techniques for improving thermal sensitivity, spatial resolution, and image contrast. Standard image manipulation routines, such as multiplication, offsetting, and edge enhancement, can be used to beneficially change the mathematical values of an individual or group of pixels. Some experimentation is required to get the best results, but with practice, remarkable detail can often be revealed in otherwise uncooperative images.

These techniques can also be pushed beyond the pictorial qualities of the image. The numerical values of each of the image pixels can be used in a spreadsheet or histogram function for analysis. Because the human eye is limited in what it can distinguish, fewer than 32 of the 256 levels of intensity used in a typical image, these numerical analysis routines allow for a much greater depth of analysis. It is possible to distinguish tens of thousands of discreet thermal levels with the latest infrared systems.

Image averaging, which can be done in the field using many imaging systems, is a particularly powerful routine used to dramatically increase thermal sensitivity. During this routine, a specified number of images taken over time are averaged together into a single image. Although real-time viewing is sacrificed, it is possible to dramatically reduce noise and note very small temperature differences with this technique.

One additional software-based comparative thermography technique utilizes either a saturation palette or isotherm function to highlight alarm limits. Data points that are above or below the set point in the image will automatically be portrayed as a different color.

Isotherms can be used in a similar manner to highlight areas either above or below an alarm or indicate areas of similar apparent temperature. Isotherms in thermal images are analogous to elevation contours on a geographical map. They are often used to show an area on the side of a boiler, for instance, that corresponds to degraded refractory insulation. If the boiler operating conditions remain the same, during the next inspection these same isotherm settings will reveal any changes in the damaged area. This information allows for the planning of the material and labor that will be necessary during a repair shutdown.

Several manufacturers are marketing various "indicators," notably to show insulated or uninsulated areas or areas below the dew point. These are nothing more than isotherms set to temperatures that may indicate the appropriate condition. A word of warning, however, that these are not an adequate substitute for reason! An "indicator" by itself is accurate only when properly used within the limitations of the tool.

Another interesting variation on comparative thermography techniques is termed "the null method." The temperature of a material (a fiber leaving a spinneret, for example) is viewed against a blackbody surface of the exact same temperature in the foreground. As it becomes thicker or thinner than specifications allow, the fiber changes temperature and becomes visible against the blackbody. Machine logic can be developed to allow the device to control the fiber-making process.

Many techniques used by thermographers depend upon seeing thermal patterns caused by differences in the rates of conduction, capacitance, or thermal diffusivity (a combination of the first two properties). The differences between several materials, or between normal and flawed sections within a given material, become thermally obvious when these techniques are employed. For instance, when inspecting a graphite epoxy honeycomb composite, heat is supplied by a high-powered flash lamp producing a pulse of high intensity light. The surface absorbs much of the light, converting it to heat, which begins to flow into the material. Where the honeycomb is properly bonded to the graphite epoxy face sheet, conductivity is high and the heat continues to be transferred unimpeded. Around disbonded areas, the flow of heat slows down. Unable to move forward at the same rate as the bonded area, the heat remains in the vicinity of disbond and causes it to heat up. Using these "pulse-active" techniques, a disbond will appear warm when viewed from the heated side.

The use of pulse-active thermography has grown exponentially in the past decade, especially in the aerospace industry. High-speed data collection and improved spatial resolution enables us to now view the transfer of heat in and out of a material. Thus, it is possible to view a sequence of thermal images—snapshots of heat transfer in space and time—that show us not only the location of a flaw but its relative size and depth. Typically, the length of time required until a flaw can be observed is a function of the square of the depth of the flaw. This relationship implies that detectable flaws will be relatively shallow and that contrast will weaken at greater depths. Typically, the radius of the smallest detectable flaw will be at least one to two times larger than its depth.

Inclusions in a composite material will typically have a different thermal capacitance than that of an unflawed sample. They are thus thermally obvious when the sample is made to change temperature. Active thermography techniques are at this time being refined and promise to become more powerful in the near future.

Flash lamps are the prime heating source for these new active thermographic techniques, but many other heat sources are employed in other applications of active thermography. Heat lamps and hot air guns have been used with great success on materials that are not highly diffusive. Microscopic and hairline cracks can be located by injecting heat into the sample across the face of the sample. As it reaches a crack, which has a higher resistance, the transfer of heat is reduced. Similar results can be gained by subjecting the sample to cold; for instance, inserting one end in a cold water bath, which causes heat to flow toward the bath.

On a larger scale, imaging damage in laminated or composite marine hulls is now common. The hull often consists of inner and outer shells with the space between injected with plastic foam. The foam bonds to the shells to form a very strong laminate. Unfortunately, impact damage to such structures seriously weakens them but is not readily visible to the eye. If, however, the boat is heated from the inside while in dry dock, the structural anomalies become thermally obvious when viewed from the exterior. Delaminations, which resist the transfer of heat, will appear cooler, whereas voids, lacking foam entirely, will usually appear warmer. Heat can also be applied from the exterior using heat lamps or guns.

The heat of the sun is used as the heating source for many applications including roof moisture inspections, delamination of building facades, fluid levels in outdoor tanks, and the analysis of delaminations in bridge decks and runways. Again, patterns resulting from differential heat transfer through the composite material allow detection of the defects.

## 9.5 Variables

At this time, most applications of infrared thermography require a qualified person to interpret the data. This is largely due to the many variables that are often difficult to understand and sometimes impossible to quantify. For the sake of discussion, in addition to the operator of the instrument, variables can be grouped simply into three categories. These relate to (1) the target, (2) ambient conditions of the system in which the target is operating, as well as changes that may affect the temperature in the future, and (3) the instrument itself.

Target variables include emissivity, spectral characteristics, temperature, heat transfer relationships in and around the target, thermal capacitance, and diffusivity characteristics. Of these, emissivity is often the most important. Unless it is very low (below 0.6, approximately), emissivity can be easily characterized and corrected for. As indicated previously, measurements of surfaces with emissivities lower than this are subject to unacceptable error. In an effort to avoid problems regarding absolute measurement of low-emissivity surfaces, many thermographers mistakenly measure the temperature difference between two similar surfaces using an emissivity value of 1.0. The results are nearly always wrong due to the fact that the radiational relationship between two surfaces with different emissivities is exponential rather than linear.

Some targets, such as glass, thin plastic films, and gases, have transmissivities that can vary significantly with wavelength. An understanding of the spectral characteristics of the target, and their relationship to actual wavelengths detected by the infrared instrument is thus necessary. The target temperature may also influence emissivity. Stainless steel, for example, has an emissivity of 0.25 at ambient temperatures but at 600°F (316°C) it is 0.4. The point to remember is to measure or determine emissivity using the conditions that most closely resemble those you will encounter at the actual target.

The target's thermal diffusivity affects the area and intensity of the thermal patterns as well as, when considered along with the heat transfer relationships, the speed of the change. Aluminum has a very high thermal diffusivity, which means that heat moves through the volume very quickly making location of any change in temperature difficult to pinpoint.

Among possible variables in ambient conditions are wind speed and direction, solar absorption, radiational cooling, precipitation, surface effects of evaporative cooling, ambient air temperature, background temperature, distance to object, relative humidity, and the presence of emitting/absorbing gases.

The impact of wind-driven convection, in particular, can be profound. Although precise corrections cannot easily be generalized, a commonly used rule of thumb suggests that a 10-mph (16-kph) wind may reduce the temperature difference ($\Delta T$) from a heated target to an ambient target by approximately one-half; a 15-mph (24-kph) wind may reduce this $\Delta T$ by as much as two-thirds. Be aware that these rules of thumb are based on a very simple analysis; a real-life situation will be much more complex and difficult to predict. Temperature data for outdoor electrical inspections are therefore often rendered meaningless without corresponding local wind speed data.

Precipitation will usually result in the surface being cooled due to evaporation. However, as freezing occurs—for example, of water absorbed into a masonry surface or of molten plastic hardening—latent heat of fusion is released, causing the materials to warm, often with confusing consequences.

Ambient air temperature can add to or subtract from target temperature. An abnormally hot electrical connection, for example, can easily be 100°F (56°C) warmer in the summer than in the winter if no other variables other than ambient air temperature change. Given the positive relationship between increased temperature and the probability of failure, it is not unusual for the rate of failure to increase during the summer. Even if the temperature was measured accurately during a winter inspection, if summer increases, and consequent degradation, are not anticipated, poor decisions about repair priority will probably be made.

Changes in the background or reflected temperature can be significant. These can affect the actual surface temperature and, especially for low-emissivity targets, the apparent or measured radiometric temperature. Whenever possible, it is best to have a thermally uniform background free of extremes. If gases with strong spectral emittance/absorptance characteristics, such as $CO_2$ or water vapor, are present, radiational transfer—and thus both actual temperature as well as radiometric measurements—will be attenuated. Mid-wave signals in particular are susceptible to attenuation by the atmosphere and, as a result, correction for relative humidity and distance to object are recommended.

The primary thermal variable relating to the target is whether transient or steady-state heat transfer conditions exist. If heat transfer is transient, knowing the position in the thermal cycle, as well as the rate of temperature change, is critical to interpreting the thermal image. Regardless of whether the heat transfer is steady-state or transient, it is important to know the thermal condition of the target at the time of viewing with respect to the extreme thermal situation that the target may experience (or has experienced) at a different time. Heat transfer from a heated building, for example, is usually considered steady-state during the late evening and early morning hours, but the effects of the sun can cause the direction of net heat transfer to reverse, even in very cold weather. During these transient situations, thermal patterns can be very confusing. An advantage of active thermography—based entirely on variations of induced transient heat flow—is that it is typically controlled, making thermal images much easier to understand.

There are other variables related to the infrared instrument, including the precise waveband detected, thermal sensitivity, detectivity, rate of data acquisition, dynamic range, field of view, spatial and measurement resolution, and system calibration. Mid-wave systems are particularly sensitive to problems with solar glint or excessive reflection from the sun which is rich in mid-wave energy. For outdoor work long-wave systems are generally preferred. Long-wave systems, on the other hand, are generally more susceptible to error when used to measure temperatures of surfaces with low emissivities.

Some applications require detectors with greater thermal sensitivity. InSb detectors, for example, are favored over less sensitive PtSi detectors for some R&D applications.

The detectivity ($D^*$) of a detector is a measure of its response relative to wavelength. Depending on whether or not this response coincides with the spectral characteristics of the target, one detector may be more useful than another.

Most instruments operate at 30 or 60 Hz in order to be compatible with standard video frame rates. Using high-speed computers to acquire and store the data, some photonic detectors such as InSb allow for acquisition at rates in excess of 500 Hz.

The dynamic range of the acquired data is another important variable. Most of today's focal plane array systems can store 14 bits of data for each image. The result is that a great deal of thermal data is available for analysis beyond the 8 bits that can be viewed as an image at any one, given moment.

Using a thermal imager with appropriate spatial and measurement resolution is essential. Imagine having to measure a kilometer using a meter stick, or a meter using a car odometer. Each infrared instrument has particular characteristics of spatial and measurement resolution determined by the detector and array size and the lens used, and the relationship among these. On most instruments, the lens can be changed. This results in a change of the field of view (the area viewed by the instrument), as well as a change in both spatial and measurement resolution. Spatial resolution—the smallest detail seen—is greater than the measurement resolution, which is the smallest detail that can be measured. It is not uncommon for an inexperienced thermographer to either miss small anomalies or, because of a failure to understand the differences between spatial and measurement resolution, to measure them inaccurately. Knowing the specification for the instrument and lens being used, and working within those limits, is essential to accurate work.

FPA detectors are particularly subject to thermal drift and non-uniform response over the array. Most manufacturers provide for some sort of non-uniformity correction (NUC), typically performed automatically upon start up. In critical inspections, however, the NUC is performed manually immediately prior to the inspection and in some cases several times during the inspection period.

Because they are temperature-stabilized rather than cooled, micro-bolometers, in particular, can be influenced by changes in ambient temperature. Where temperature measurement needs are critical, allowing them to thermally stabilize for up to 20 min may be warranted or whenever ambient conditions change significantly.

Simple field calibration checks are essential for day-to-day accuracy of a thermographers work. These checks can be as simple as checking the temperature of a human tear duct at instrument startup and shutdown. For more critical work, the inclusion of a thermocoupled surface or a calibrated reference blackbody in the field of view may be warranted. Good practice suggests the calibration of an instrument be checked against a calibrated reference blackbody on a periodic basis. The exact frequency of verification depends on the requirements of the customer or the application. Calibration and verification records should be kept on file.

Of course, the thermographers themselves are probably the greatest variable! Stated simply, the question is "is the thermographer qualified?" Qualification is based on appropriate training, experience, and testing. Especially with the industry-wide confusion over the use of the term "certification," variability in personnel qualification is dangerously wide. Success for many applications also requires the operator to have additional related skills or experience. Regardless of the application, the thermographer must have mastered the basic communications skills of asking questions and listening, without undue "filtering" of the answers.

Today's instruments are easy to use—literally pressing one button can result in a remarkable image—but many operators fail to learn about the many other variables that are not optimized when that single button is pressed. Much of thermography is in fact, simply comparative or qualitative work. By comparing the target to another that is similar, it may be possible to minimize or eliminate the effect of one or more of these variables. Clearly, training and experience are fundamental to the process. As an example, it is not uncommon when inspecting a three-phase electrical system to find one phase being warmer than the other two. If the load is balanced across all three phases, this heating would be considered abnormal, probably associated with an overload condition, an undersized conductor, or possibly heating from a nearby source. But under some conditions, an unbalanced loading situation, resulting in an identical thermal pattern, may be acceptable or even normal. The skill and experience of the thermographer are essential to tip the scales in favor of the correct interpretation.

Quantitative thermography requires an even greater understanding of the variables impacting radiometric measurement, as well as a grasp of its limitations. It is vital to determine what margin of error is acceptable before beginning an inspection, and to work carefully to stay within those bounds. Too many thermographers believe that all data output from their instrument is accurate, or that the impact of all variables can easily be understood and measured. This is simply not true!

Considering the variables that have been discussed, it becomes obvious that there is considerable potential for error in both qualitative and quantitative thermography. Only careful consideration of all variables by a qualified and skilled thermographer can result in good data and interpretation of that data, and even then it is important to understand—and work within—the margin of error remaining.

## 9.6 Evaluation of Test Results and Reporting

Much of thermography is still highly dependent on the skill of the thermographer to conduct the inspection correctly, understand the limitations of the test, record all relevant data, and interpret the results properly. As we have seen in the previous section, the number of variables is staggering. Inspection personnel, as a consequence, must be adequately qualified for the task at hand.

It is generally recommended that a Level I qualified thermographer collect data but only if working under the supervision of a Level II thermographer. The use of written inspection procedures developed with the support of a Level III thermographer is also critical to success. To write procedures or validate data, it is often necessary to set up other tests. It may be necessary to make emissivity measurements, for instance, of typical components before an inspection methodology can be developed. For NDT of composite materials, for another example, it may be necessary to develop costly test pieces to develop test procedures as well as define the limits of detectability.

Procedures should detail the knowledge and skills required to conduct the inspection. For instance, to conduct an electrical inspection, many companies require the thermographer to have a basic working knowledge of electrical systems together with both electrical safety and CPR training. The procedures should also clearly spell out the weather and system conditions necessary for a successful inspection. Any limitations should be clearly spelled out, such as a caution for using mid-wave sensing systems outside in the sunshine.

Evaluation of data is also the function of the Level II thermographer. Sometimes additional, supplemental testing using other test methods, as well as specific engineering or Level III support, may be required. On a limited basis, it may be possible for Level I personnel to make "pass/fail" assessments based on clearly written procedures. For instance, when conducting an inspection of boiler refractory, a Level I thermographer might mark refractory as "failed" when the surface temperature exceeds a value set forth in the procedures.

If personnel are qualified and good procedures are followed, evaluation of the data can proceed. A primary challenge is to determine the reliability of measurement, especially for radiometric values. Generally, radiometric measurements are not recommended when emissivity falls below 0.6. Even when qualitatively evaluating low-emissivity materials, one must proceed with caution. System variables, such as load and wind, must also be carefully accounted for, even if it cannot be done precisely.

Even after the data is correctly evaluated, the results must be clearly communicated to others in a report. Part of this process often requires educating the person who is reading the report. If they do not understand the potential problems associated with radiometric temperature measurements or heat transfer, for instance, serious misinterpretation of the results is a possibility. Generally, the problem for the thermographer is to have the customer understand the strengths and limitations of the technology or the findings. To what lengths the thermographer should go in the process varies, but is probably directly proportional to his or her investment in having the report make a difference in the inspection process.

Reports can take many forms. As a minimum, the report should include the thermographers name; the instrument model and serial number(s); and relevant ambient conditions, such as wind speed and direction and ambient temperature. System conditions, such as load, component identification and location information, component emissivity, instrument parameter settings, and, in most instances, a thermal image and matching visual image, should also be included.

These details should be displayed in a way that does not clutter the report, but instead supports the presentation of essential information in an easily understood fashion. The best reports are intuitive in nature with a natural flow of data to support the thermal and visual images.

## 9.7 Applications

Many of the applications for thermography are not considered mainstream NDE/NDT. Instead, they have come from other areas, such as industrial predictive maintenance, building science and forensic investigation. Regardless of terminology, applications of thermography are nondestructive in nature, allowing the effective evaluation of both materials and systems.

The use of thermography in more traditional NDE/NDT has recently become more widely accepted. Growth has been concentrated in two industries in particular, electronics and aerospace materials, but new applications are being developed in many other industries as well.

### 9.7.1 Electronics Industry

In the electronics industry, thermography has become a powerful tool for design and manufacturing of integrated circuit boards (ICBs) as well as, in some cases, their repair. As designers have been pushed to make smaller boards, component density on the boards has increased and so have the problems associated with the heat that they produce.

During the design stages, infrared is used to observe the distribution of this heat and its impact on components. The boards are next inspected in situ to evaluate the impact of any further heat transfer conditions that might affect their performance.

The emissivities of many of the components on an ICB vary widely and many are quite low. Two methods have been used successfully to reduce the difficulties of inspection. Boards are first "heat soaked" in an oven; the imager is then used to "back calculate" the emissivity from these known temperatures and reflected background temperatures. An emissivity map of the ICB is created from this analysis, which is used to plot accurate radiometric readings of the board at operating temperature. The second method involves the application of a relatively high-emissivity, conformal coating to the ICB, thereby allowing direct thermal viewing and analysis.

At the manufacturing stage, infrared can be used to inspect the actual ICB being produced. The board being inspected is energized in a test stand. Its thermal signature is compared to that of a "golden board," meaning an ICB that performs within specifications, both during transient warm-up as well as after it has achieved steady-state operating temperature. Using computerized processing, one image is superimposed on the other, and an image subtraction routine performed. Any anomalous areas are immediately pinpointed. The same technique is used during the repair of more costly boards. The result has been a dramatic reduction in repair costs and an increase in quality assurance.

### 9.7.2 Aerospace Applications

Nowhere has the use of active thermography grown as it has in the aerospace industry. As more and more composite materials are used, the need to monitor quality of both new and existing stock has also increased. Techniques have evolved that range from simple handheld evaluation to automated inspections using sophisticated computer-based vision systems.

The basic premise of aerospace thermography is that properly laminated or bonded material will have relatively uniform thermal characteristics, including conductivity, capacitance, and diffusivity. Disbonds, delaminations, and inclusions of foreign substances can be located and characterized based on their anomalous thermal signatures. Typically, the component being inspected is heated from one side and viewed from either the same side or the opposite side, depending on the specific requirements. Of paramount importance is that the application of energy should be relatively uniform across the area to be inspected. Uniform heating results in heat transfer into the material. Active thermography can be accomplished with a variety of heat sources, such as hot air guns, heat lamps, or capacitor driven xenon flash lamps.

Several of today's active systems use a carefully designed system of high-powered xenon flash lamps as a heat source and a computer to capture the thermal data. Flash-thermography, as it is often called, is now widely used for production and repair of both commercial and military aerospace components. During flash thermography, the pulse of light energy is absorbed by the surface of the component. The heat conducts into the surface and diffuses through the part at a uniform and predictable rate until it encounters a discontinuity. Because a delamination or disbond usually has a lower rate of conductivity, the heat over the disbond cools slower than the surrounding area, thus indicating the location, and relative size of the discontinuity.

Data, captured at rates of up to 500 frames per second (60 to 100 frames per second being typical), reveals the required detail. Because the entire process takes only a few seconds to complete, the area covered by a single flash pulse—approximately two to four square feet—can result in a large area being inspected in a short time. Advanced techniques, such as TSR®, in which the raw thermal data is reconstructed into mathematical functions, allow images with less noise and thus find anomalies that were before undetectable. This method also has the ability to determine the depth of anomalies.

While much of this work is being performed by hand by qualified technicians, many systems now utilize robots that can move the flash in a preprogrammed route over the entire aerospace structure. The data is then processed and analyzed by the computer system. Of course, considerable experimentation is required to set up such an operation, but, especially for testing of larger components, these automated systems are tremendous time and cost savers.

Moisture or fluid intrusion into honeycomb structures can also be detected. Successful inspections have been conducted right after an aircraft has landed. Because water has a high thermal capacitance, it remains cool (from the high altitude) long after the rest of the structure has reached warmer temperatures. Active thermography can also be used to locate moisture. The area of the material with no water intrusion heats uniformly, whereas the areas with water trapped in the cells appear cooler due to the greater thermal capacitance of the water. Among several other techniques being used successfully are lock-in thermography and vibro-thermography. With lock-in thermography a series of thermal pulses is applied to the surface of the material being inspected. The frequency of the thermal pulses is adjusted based on the materials thermal properties and the depth of the thermal anomaly. This method is useful for locating a wide range of subsurface defects but is especially useful for finding deeper anomalies than other methods. Vibro-thermography utilizes high energy acoustic waves usually induced by devices such as ultrasonic welders. The mechanical waves move throughout the part without heating it up, but if they intercept tight cracks or close fitting delaminations frictional heating will occur resulting an surface temperature indications that are seen by the infrared camera.

### 9.7.3 Electrical Inspections

Probably the most widely used application for thermography is inspecting the integrity of electrical systems. It has the distinct advantages of being both fast and noncontact. While heat is not a perfect indicator of all problems in electrical systems, heat produced by abnormally high electrical resistance almost always precedes an electrical failure. Much of electrical thermography work is qualitative, comparing the thermal signature of the similar components. This is quite simple with three-phase electrical systems—the heart of all utility, commercial, and industrial installations—where the phases should almost always appear similar. One or more components at a different temperature (usually warmer) suggest a problem is at hand.

The technique for inspecting electrical systems is quite straightforward. The component is viewed directly while energized. Particular attention is paid to any connection or point of electrical contact. These areas are most susceptible to abnormal high electrical resistance and associated heat-related failure. It is possible to inspect a good-sized power substation, containing thousands of points of contact—*each* a potential problem—in less than an hour. Figure 9-7 shows several classic patterns associated with a loose or corroded connection points in the fuse system of a distribution panel. Electrical current imbalances among phases are also readily seen. Often, these are considered normal, such as in a lighting circuit, while in other parts of the electrical system they can

**FIGURE 9-7** Heat is generated as electricity flows through any conductor but when resistance is abnormally great, excessive heat—often quite damaging—develops. The lighter colored areas in the thermal image of this group of fuses show abnormal heating at several points of electrical contact.

result in very costly failures, such as when a phase imbalance is caused by an internal fault in a large motor.

Although the application is straightforward and widely used, it is often used ineffectively. A good electrical system's thermographer must contend with several problems related to the component, the infrared instrument, and the interpretation of the data. Many electrical components are bright metals, highly conductive electrically but also having extremely low emissivities. This, unfortunately, means even the best of thermographers will have difficulty accurately measuring temperatures. The low-emissivity nature of the components also means that they must be quite warm before they radiate enough energy to even be detected. Often, by the time you see a high-resistance problem, some damage has already been done. Although it can represent a significant investment, an increasing number of plants are placing high-emissivity targets on these critical components in order to increase the probability of early fault detection and improve the reliability of measurement.

Some thermographers advocate measuring nearby surfaces that have higher emissivities—the electrical insulation, for instance. While this can be useful, care must be taken in doing so. Most components have such a high thermal conductivity and diffusivity that a large thermal gradient can exist even over a small distance. This means that the measured temperatures at a nearby high-emissivity site may be much less than those at the site where high-resistance heating is actually occurring.

Temperature also varies widely as the electrical load on the system changes. The heat output from a high-resistance connection is predictable and can be correlated with changing loads; unfortunately, the *temperature* change of the connection is much less predictable. All that can be known is that as load increases, the temperature of a connection will increase at a rate that is greater than linear and less than exponential. The National Fire Protection Association (NFPA) standard 70-B, Maintenance of Electrical Systems, recommends inspections be done with a 40 percent minimum load and, when possible, at the highest normal load. Even seemingly insignificant problems should be noted when inspecting systems that are temporarily under a light load.

Too many thermographers still try to inspect electrical equipment without opening the enclosures to directly view the components. Except where extenuating circumstances make a direct view impossible or dangerous, components *must* be viewed directly. Opening enclosures must be done with great care, however, because the thermographer is often exposed to the potential for extreme danger, especially from a high energy arc flash explosion. These explosions, which can be triggered simply by opening an enclosure door, can reach temperatures in excess of 15,000°F in less than a half-second.

Every effort should be made to understand and minimize these dangers of inspecting electrical systems. NFPA 70-E, which is accepted as law throughout most of the United States, details the steps necessary to protect the thermographer. These include, among others, education and personnel protective equipment (PPE). PPE may include clothing designed to withstand an arc flash, as well as protection for the eyes from the high-energy ultraviolet radiation given off in such an explosion.

Some equipment may be so dangerous to access that other measures are advised. Infrared transparent windows are now being installed in some instances to allow viewing into an enclosure without opening it. Careful placement is required to ensure that all components can be seen. It may also be possible to use other types of inspection equipment that detects airborne ultrasound. Even micro-arcing, which a failing connection produces, yields a detectable ultrasound signature that can be heard through a crack or small hole in the enclosure.

Where enclosures cannot be opened easily, such as on an overhead-enclosed bus, the thermal gradient between the problem and the viewed surface is typically very large. A rise as small as 5°F (2°C) surface signature on an enclosed bus often indicates that the metal bus bars a few inches away may have already begun to melt! Great care must be used when interpreting the data from inspections of unopened enclosures.

For outdoor inspections, wind is a variable that can quickly reduce the value of thermography unless results are interpreted carefully. Convective cooling by the wind (or inside plants by any forced air currents) will quickly reduce the temperature of a high-resistance connection. What appears to be a thermally insignificant problem during these windy conditions can, in fact, become very significant when the wind speed drops. Many thermographers try to compensate for this by measuring a temperature difference from one phase to another, with the mistaken notion that the effects of the cooling will be neutralized. Such is not the case!

When making delta measurements between like components when subjected to cooling by wind or air currents, one can fail to recognize that the normal phase, operating at only a few degrees above ambient, can never be cooled below ambient temperature. Thus, in the wind, the delta measurement will also seem deceptively small. The results of outdoor inspections when wind speeds are greater than approximately 10 mph (16 kph) should be viewed only with a great deal of caution. Even "minor" temperature differences may represent extremely serious problems. It is likely that many problems will not even be warm enough to be detectable, much less accurately measured. To be of any value at all, a measurement or estimate of wind speed *at the component location* must always accompany temperature measurements. The use of an accurate air velocity measurement tool or, if that is not readily available, another means of estimating wind speed, such as Beaufort's Wind Scale, is recommended.

Two other difficulties crop up during outdoor inspections. The sun can cause heating of any thermally absorbent surface. Dark surfaces in particular can be heated to the point where they are difficult to view with any reasonable chance of understanding what is happening. Usually, inspecting during the morning hours or on overcast days is sufficient to minimize difficulties. More problematic is using mid-wave sensing systems outside in sunshine. These systems are susceptible to seeing solar "glint," or reflection of the sun's high quantity of mid-wave thermal radiation. For extensive outdoor work during sunny periods, long-wave sensing systems will give superior results. Mid-wave systems should be used only on a limited basis or, if loads and other conditions allow, on overcast days or at night.

Although many believe otherwise, the job of acquiring reliable thermographic data regarding electrical systems is not easy. Even with good data in hand, too many people use the data incorrectly when prioritizing the seriousness of their findings. Temperature is often not a reliable indicator of the criticality or severity of a problem. The difficulties are further exacerbated given the limitations of making radiometric measurement on low-emissivity electrical components.

Unfortunately this reality does not prevent many thermographers, or people purchasing thermographic inspections services, from believing that the hotter a finding is, the more serious the problem. Or, conversely and even more dangerously, they believe that if a finding is not very hot, it is not a problem. Nothing could be less true! The problem shown in Fig. 9-8 is a fault inside on the bushing connection of a pole-mounted transformer tank; the temperatures seen on the outside of the tank are quite cool compared to the temperature inside at the actual point of high-resistance heating.

The effects of convection on heated surfaces have been previously described. While conducting electrical inspections inside a plant, it is not unusual for warm air to be convected up and over components, causing them to be warmed more than might be expected. High resistance in a load side (lower) connection of a breaker, for instance, will result in heating, but convection from it may also cause the line side (upper) connection to appear warm.

When inspecting three-phase electrical systems outside, it is also important to remember that wind (at ambient air temperature) moving over a normal component

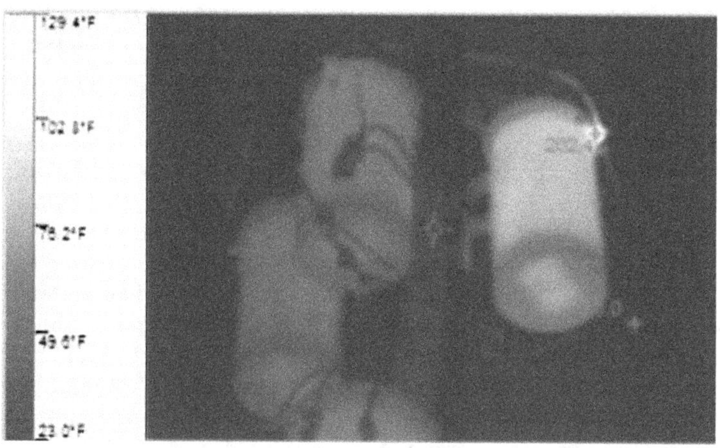

**FIGURE 9-8** The small lighter area at the top of the right phase of this bank of pole-mounted transformer tanks indicates a high-resistance electrical fault has occurred internally. It is likely the heat is being generated at the bushing connection that is submerged in insulating oil in the tank. The temperatures seen on the outside of the tank are quite cool compared to the temperature inside at the actual point of high-resistance heating.

(also at or near ambient temperature) results in little cooling effect. However, the same wind blowing over a high-resistance hot connection will dramatically cool it. The result is that when the wind is blowing, the hot problem connection may not be visible or, if it is, it may not appear to be very warm.

The seriousness of many findings will not be made obvious by temperature alone. It must be recognized that any components having low emissivity or a high thermal gradient located inside an enclosure subject to convective cooling, or under light load, will probably not appear very warm. In fact, in most cases, they could be extremely unstable. Further, little is understood about the actual failure mechanisms for the diverse collection of electrical components that can be inspected.

Some generalizations are possible, such as the fact that lower melting points will cause aluminum materials to fail before copper alloys. Spring tension, an essential characteristic of many switch and contact components, begins to fail after the material has been at approximately 100°C for a month. The blocked cooling fins (darker) shown in Fig. 9-9 also indicate a very serious problem. Insulating oils in transformers break down rapidly when subjected to overheating, and this problem will result in overheating when loads are at peak and ambient temperatures are high.

Rather than basing prioritization strictly on temperature, a more useful approach is to look at all the parameters involved and consider how they interact with each other. This can be done either with a simple approach or a more complex analysis. On the simple end, a weighted prioritization matrix can be established using any parameters deemed significant. These might include, for instance, safety, loading condition, reliability of the radiometric measurement, and a history of similar components. Certain parameters

**FIGURE 9-9** Problems are not always indicated by heat. The radiators on this large power transformer are designed to remove damaging heat from the oil-filled device and the cooler (darker) tubes indicate oil is not circulating properly. This abnormal heating can lead to the breakdown of the oil and the failure of the transformer. This can happen very rapidly during already warm summer peak periods when an outage can be very costly.

| | Greater Chance | | | | |
|---|---|---|---|---|---|
| | 1 | 2 | 3 | x | = |
| Will a failure injure people? | | | * | 2 | |
| Is the component critical? | | | | | |
| Will loads increase? | | | | | |
| Will the wind decrease? | | | | | |
| Is the measurement unreliable? | | | | | |
| If it is reliable, is *AT* great? | | | | | |
| | | | Total score | | |

*Immediate action required.

**Table 9-3** Priority of Parameters

can be given added value or "weight" as a total score is figured. In Table 9-3, the issue of personnel safety is given a weight of twice the other factors. A simple system like this, if used consistently, can become "smarter" with time if a feedback loop is incorporated. Although this approach is particularly useful for electrical components, it can be used for a myriad of mechanical equipment as well. Other factors, values, and weights can be used, depending on the needs of the thermographer.

A more complex analysis may be required in some instances. This might involve full engineering studies and incorporate data from other sources or test methods. A volatile gas analysis, for instance, provides excellent indicators of various internal faults in oil-filled equipment such as transformers.

Because the consequences of not doing thermal electrical inspections are so great—production outages, decreased product quality, and the potential for injury or death—care should be taken to structure and implement a program that will achieve optimum results. Companies that do this are able to virtually eliminate unscheduled downtime due to electrical failures.

### 9.7.4 Mechanical Inspections

Mechanical inspections encompass a diverse variety of equipment. The technology has proven to be an invaluable tool for looking at motors, rotating equipment, steam traps, refractory, and tank product levels, among others. Most of these applications are qualitative in nature, often comparing the current thermal image to a previous one and understanding the cause and extent of any changes. In fact, conducting baseline inspections of new or rebuilt equipment has proven to be of particular value. As with electrical equipment, it is beneficial to establish inspection routes and periodic frequencies based on needs and resources.

Approximately 90 percent of the electricity used in the United States passes through electric motors. Although there are a number of excellent means to monitor the electrical and mechanical condition of motors, infrared has proven particularly useful as a fast screening device. High resistance, abnormal current flow, or excessive friction produces heat. If problems are indicated by one of these abnormal thermal signatures, other testing is often implemented to determine the exact nature of the problem.

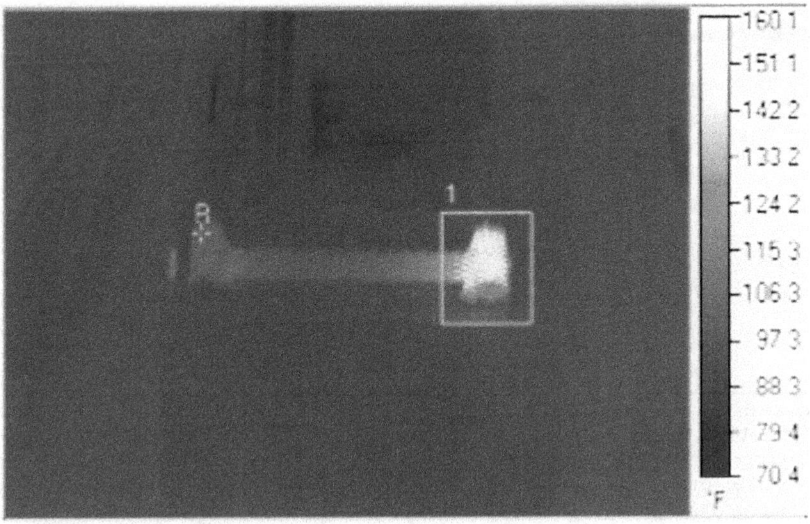


**Figure 9-10** Excessive friction in the pillow block bearing of the shaft of a large fan unit is overheating. After such a problem is located, the root cause can be determined and corrected and repairs made.

Thus, the abnormal heating of a misaligned motor coupling usually precedes a measurable vibration signature. Left unresolved, the problem will be compounded as it wears on the motor bearing itself. Many problems that are related to the motor's electrical system also have characteristic thermal signatures. An internal short caused by a breakdown of insulation on motor windings will result in an increased temperature internally, as well as on the motor casing.

Heating of bearing surfaces caused by abnormal friction can provide an excellent indicator of the location and nature of many problems for all types of rotating equipment. The line shaft, pillow block bearing in Fig. 9-10 shows a typical pattern of an overheating bearing, in this case due to lack of lubrication. As often as not, the problem stems from too much lubrication rather than a lack of it! Especially for low-speed equipment, such as trolleys on overhead conveyors, where vibration signatures may not be reliable, thermography has proved a boon for finding friction and alignment issues.

### 9.7.5 Refractory Inspections

In a variety of applications where high-temperature refractory is used, such as in boilers, rotating kilns, and torpedo cars, infrared can be used to monitor outside skin temperatures. With this information, engineers can validate the condition of the insulation or back-calculate the thickness of remaining refractory. It is also possible to identify any hot spots related to refractory failure. Typically, refractory is inspected regularly with a frequency determined by age and the estimated consequence of failure. Figure 9-11 clearly shows a problem with the refractory in this high-temperature rotary kiln. Rotary kilns benefit from monthly inspections, whereas torpedo cars may require weekly inspections. Because they are under less stress, boilers are often inspected on an annual basis prior to shut down. All refractory should also receive a baseline inspection at startup so that trends can be established from consequent inspections.

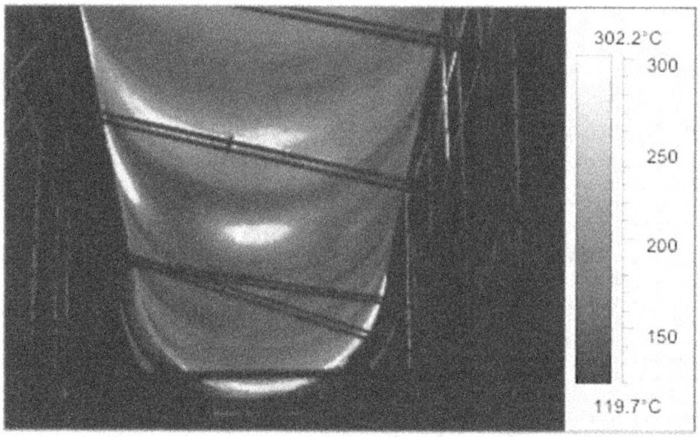


**FIGURE 9-11** Rotating kilns and furnaces are used in many industries and applications. Containing the heat is essential to both efficiency and safety. The thermal images show extensive areas (lighter) where heat loss is abnormally great due to the wear and breakdown of the internal refractor insulation. These "hot spots" can become warm enough to damage the exterior structural shell, a not-so-uncommon and very costly consequence.

### 9.7.6 Steam Traps and Lines

Steam traps are highly susceptible to a "stuck open" failure mode, which can often be detected using infrared testing. Although airborne ultrasound is one of the primary methods of investigation, infrared has the advantage of being fast and able to be used even at a distance. A trap that is functioning properly will typically have a detectable temperature difference between the steam and condensate sides. This difference will temporarily disappear as it cycles. If both sides of the trap are warm, particularly if they are at steam temperatures, the indication is good that there has been a failure.

The inspection of steam lines using thermography is also valuable. Where insulation has failed on aboveground lines, a thermal signature is usually a clear indicator of the location of the problem. This is particularly true outdoors where water has saturated the insulation. Because there are more variables, buried underground lines are more difficult to inspect. Using a map of the system, it is possible to walk a steam line with a good handheld infrared system and locate both steam and condensate leaks. Signatures are generally clearer at night, but variations in ground moisture, bury depth, and soil type must be considered.

Heat tracing (both steam and electric) on process lines can also be inspected, except when insulation levels are very high, but this is rarely the case. More problematic is the use of aluminum or stainless steel sheet metal coverings over the insulation. These make inspections practically impossible due to the low emissivity (and consequent high reflectivity) of the coverings.

Many processes depend on maintaining internal temperature at a precise level. Operating temperatures may fall below these thresholds as a consequence of either the insulation or the tracing failing. In the confectionery and petrochemical industries, in particular, this can lead to blockage of a line as the transported product freezes. When blockages do occur, thermography can be used to help locate them while there is still a thermal difference between liquid and solid.

### 9.7.7 Tank Level Verification

One of the most surprising, and usually easy, applications for thermography is locating or confirming product levels in tanks and silos. Although many tanks have some sort of

**FIGURE 9-12** Thermography has proven invaluable for locating levels in tanks or validating data from the existing level indication equipment. When the tank is in a thermal transition, the various elements change temperature are different rates delineating the interfaces between. For tanks that are out-of-doors the transition between day and night is nearly always sufficient to drive the process.

instrumentation, it often does not work or readings must be independently confirmed. As with line blockages, the differing thermal capacities of solids, liquids, and gases and the mode by which heat is transferred mean that they will change temperatures at different rates during a transient heat transfer cycle. Gases change most quickly. In fact, in large outdoor tanks the sun can cause a detectable thermal change in a matter of minutes. The levels in the left and center tanks in Fig. 9-12 are clearly visible. The right tank is completely full. In the late evening or early morning (before sunrise), the pattern would be reversed but the levels would still be visible.

Solid sediments typically change temperature next, as heat is conducted in or out of the material next to the tank wall. The liquid and sludge levels in a pair of tanks are both evident in Fig. 9-13. Next, any floating material, such as waxes, can usually be distinguished from the liquid. The liquid, because it may have a higher thermal capacitance and is also subject to convective heat transfer, typically takes the longest to change temperature. For tanks that are indoors, a normal diurnal swing in ambient air temperature is often enough to reveal the levels. Even in air-conditioned spaces, small ambient fluctuations may be enough. The materials inside many tanks are heated or cooled, and thus provide their own thermodynamic, driving force.

Uninsulated tanks and silos are quick to reveal their levels. Where insulation is present, the image may take longer to appear or may require some enhancement. Levels can be enhanced using simple active techniques like applying heat or by wetting the surface to induce evaporative cooling. Simply spraying water briefly on a tank is often enough to reveal several of the levels.

Where insulation is covered with an unpainted metal covering, difficulties may arise due to the low emissivity of the surface. Even in this case, however, a high-emissivity vertical stripe, such as paint or a removable tape, can be applied. The levels will be visible on the stripe.

There are numerous other applications for inspecting mechanical equipment thermographically. It is not possible to describe all of them, but it is useful to point out several generalizations. A good baseline inspection is critical. This should be done both

FIGURE 9-13 These two tanks also show interfaces between both the air and liquid and the liquid and solids. The thermal transition necessary to drive the differences in temperature may be provided by normal changes in the ambient environment or they may need to be created artificially by increasing or reducing the temperature.

at the transition during startup and also during the cool-down period. Be aware that all components of a machine will not go thorough transitional swings at the same rate. Other images should be made when the machine is operating at steady-state temperatures. Again, it is important to understand the impact of variations in conduction and convection and how they affect the surface patterns being seen. Subsequent images taken later on a periodic basis should be compared to this baseline data, noting any changes that may indicate problems. Careful data collection and record keeping is required for these methods to be effective.

### 9.7.8 Roof Moisture Inspections

The use of infrared roof moisture inspections has grown tremendously over the past decade as a cost-effective maintenance tool. There are literally tens of thousands of acres of flat commercial and industrial roofs throughout the world having a replacement value in the billions of dollars. For a number of reasons related to design, installation, and maintenance, most develop leaks within a year or two. While the damage caused by the actual leak may be substantial, the hidden long-term damage is usually far more costly.

From top to bottom, a typical cross section of a low-sloped roof is composed of the membrane, insulation, and a structural deck (see Fig. 9-14). Water that enters a roofing system becomes trapped in the roof system, especially in the insulation, resulting in degradation of the system and its premature failure. By locating and replacing the wet roof insulation, subsurface moisture is eliminated, and the life of a roof can be greatly extended beyond the current average. Because water has such a high thermal capacitance, it is

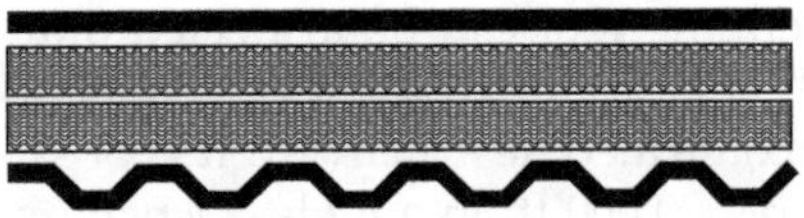


**FIGURE 9-14** The cross section of most low-slope roofs consists of three layers, a structural deck (bottom), thermal insulation (middle) and a water-proof membrane (top). The exact types of materials used can vary widely depending on a variety of factors.

slow to change temperature during a transient heat transfer cycle. During the daytime, the sun warms a roof. The dry insulation heats faster and to a higher temperature than does the wet insulation. In theory, it would be possible to locate the wet insulation during the daytime, but the sun typically heats the surface of the roof enough to mask these differences. After sunset, however, especially if the night is clear, the roof will begin to cool rapidly. Excessive wind may result in uneven cooling and confusing thermal patterns. At some point in the evening, usually an hour or two after sunset, the dry insulation will become cooler than the wet. It is possible, as this "inspection window" opens, to locate areas of wet insulation by their warmer thermal signature and characteristic patterns. Figure 9-15 shows a typical pattern associated with moisture intrusion into absorbent insulation; water has probably intruded into the system from around the vent system seen in front of the image; the roof is also being marked with spray paint by the person at the left of the image.

**FIGURE 9-15** The lighter rectangular patterns on this built-up roof are indications of moisture intrusion into the insulation layer. Two people working together, the thermographer and the assistant who is marking the roof with spray paint, ensure a higher degree of safety and efficiency.

Once patterns are seen, the roof can be inspected quite quickly, marking all wet areas with paint directly on the roof surface. If necessary, the actual presence of moisture in a wet area can be confirmed destructively. It is possible, on a fairly uncluttered roof, to inspect upwards of 500,000 $ft^2$ in a night. Depending on a variety of variables the inspection window may remain open quite long into the evening, until it is closed by wind, decreasing temperature differences or the formation of dew on the surface.

Typically, radiometric temperatures are not recorded during a roof moisture inspection. A video recording or a series of individual images of all anomalies should be made.

Exactly what pattern will be seen and when depends in large part on conditions and the type of roof insulation. Absorbent insulations, such as fiberglass, wood fiber, and perlite, yield clear signatures. These types of insulation are typically used in built-up roofs. A "board edge" pattern, with its characteristic right angles, results because each board of insulation tends to saturate with water before spilling over onto the next one.

On the other hand, nonabsorbent foam-board insulation, which is often used in single-ply roof systems, is extremely difficult to inspect due to the fact that little water is trapped compared to other roof types, although still more than enough to cause degradation. The roofing industry recommends installing an absorbent layer of insulation on top of foam board, but this is often ignored. The unfortunate consequence is that infrared has limited value as an inspection tool on many foam-board roofs. Many single-ply roofs are also ballasted with pavers or a heavy layer of stone. These, too, can render infrared of little or no value. Clearly, the potential for inspection should be considered at the time of design!

Patterns are also influenced by weather. The roof surface must be dry or else evaporative cooling will prevent it from heating up adequately. If heavy cloud cover forms after sunset, the roof will probably not cool quickly enough for the inspection window to open. Wind in excess of approximately 10 mph will quickly cool a roof and erase all thermal signatures of value. During cold weather, heat transfer from the building through the roof section can produce detectable thermal patterns in the wet insulation, once again resulting in these sections appearing warm. Very good results can usually be achieved with ambient air temperatures as low as approximately 40°F (4.4°C).

Rooftop conditions themselves also help determine patterns. A roof section that is in the shadow of a higher wall, for instance, will not be heated to the same degree as the unshaded areas of the roof. A west-facing wall, on the other hand, will reradiate its energy to the roof long into the night, preventing it from cooling adequately. Extra gravel or flashing material will stay warmer, sometimes masking over the wet insulation beneath. Where the roof has been previously repaired, differences in the type of insulation may cause variations in the thermal signature. Understanding these influences and the result they have on the thermal patterns is relatively easy for a qualified thermographer.

Roofs should be inspected shortly after installation to establish a baseline signature. Another inspection would be warranted after any potentially damaging incident, such as a heavy hailstorm or hurricane. When leaks occur, as they are bound to, a quick follow-up infrared inspection will help locate the leak and indicate the extent of the wet insulation needing replacement. It is often possible to inspect a well-designed and maintained roof as infrequently as every 3 to 5 years.

Great care must be used during roof inspections to prevent accidents and unnecessary injuries. Look for safe access to the roof, and notify all local security officials of the inspection. Never work on the roof alone. Thermographers are most vulnerable because the brightness of their instrument displays prevents their eyes from adjusting to the low light conditions found on most roofs. It is critical to become familiar with, and adhere to, any relevant company or government regulations. Especially when people are within 10 ft of the roof edge, it is important to be protected from an accidental fall by an appropriate barrier or fall protection device. Falling over the edge is not the only danger. Even stumbling over a small change in elevation can result in serious injury and a damaged infrared instrument.

A walking rooftop inspection is obviously very labor-intensive, relatively slow, and highly weather-dependent. A great deal of information can be obtained very quickly by performing an aerial infrared roof moisture inspection. Using either a helicopter or fixed-wing aircraft, the instrument can be flown at altitudes between 500 and 1500 ft from the roof, depending on local air traffic requirements and the spatial resolution of the test instrument. Obviously it is important to determine what the size of the smallest detectable area must be. If the conditions are right, data for millions of square feet can be acquired in an evening. Of course, bad weather or roof conditions can result in a costly cancellation of the mission.

Great care must be taken to identify the target roofs. A daytime flyover to conduct a visual pre-inspection is highly recommended. Most professionals videotape their survey and include some type of on-screen geographical positioning information. Once the data has been collected, analysis and reporting will require additional office time.

Although infrared has some limitations as an inspection tool (mainly that its use is weather-dependent), it is fast, thorough, and reliable. Unlike other methods, it also allows for the inspection of every square foot of roof. On roofs with absorbent insulation, the results of an infrared inspection are extremely accurate, measuring areas of wet insulation within several inches.

The potential savings from infrared roof moisture surveys are astounding. It is not unusual to extend the life of a roof by 50 percent or more. The lifecycle cost drops significantly even when the life is increased by a year or two. As waste disposal costs continue to climb, especially for the hazardous materials that many roofs are constructed of, there is a strong incentive to repair and keep a roof in place. Even when the useful life of the membrane has been reached, it may be possible to install a new membrane over the existing insulation, again at great savings.

### 9.7.9 Building Energy Surveys

During the "energy crisis" of the late 1970s, thermography was used extensively for inspecting buildings. The technology lends itself to locating both conduction (insulation) and convection (air leakage) problems as well as moisture intrusion and delamination of facades. When inspecting a building, it is vital to understand exactly how it is constructed. This is often not the same as looking at the drawings for the building. In fact, the thermographers job is often to find performance flaws in a design or discrepancies between the design and its execution. A simple destructive examination often proves a valuable supplement for the thermographer.

Inspections are also simplest if accomplished during steady-state heat transfer situations. Generally, an 18°F (10°C) temperature difference from inside to outside is needed for conduction problems to express themselves. Figure 9-16 shows a thermal pattern from the exterior of a heated building in the winter; the irregular light areas are where the insulation was missing due to poor quality installation of blown-in cellulose. When the building surfaces are heated directly by the sun, however, a reversal of heat transfer can occur resulting in confusing patterns. These effects can be noticeable for up to 6 hours

**FIGURE 9-16** Buildings are insulated to control heat transfer, in this case to contain heat inside the building in a cold climate. The lighter areas are clear indications of poor placement of blown-in insulation.

after the sun has left a building facade. For this reason, it may be difficult to conduct inspections during sunny days in warm or cold weather.

With optimum conditions, however, missing, damaged, or ineffective insulation, as well as the location of framing, becomes obvious. Patterns vary with the type of insulation and the exact conditions. The conductive inspection is best done from inside the building in order to minimize the effects of wind and sun, and because the inside wall is typically better connected thermally to the outside. If possible, an outside inspection is also conducted to provide views of large sections of facades.

Any conditioned space is influenced by pressure differences caused by natural or forced convection. These result in air movement, which will often leave a characteristic thermal signature. Excessive air leakage can be identified with thermography, even when temperature differences are only a few degrees. During the heating season, signatures will be seen as cold streaks along the interior building surfaces or warm "blooms" on the outside of the building where heated air is exfiltrating. Some air movement may be evident inside the walls, even interior or insulated exterior walls. By artificially inducing a pressure difference on the building using the HVAC system or a device called a blower door fan, air leakage patterns can be enhanced and to some extent quantified.

Although the inspections of residential-scale buildings are quite straightforward, those of large commercial units can be more complicated, although the returns on the investment, which can be huge, usually warrant a thorough inspection. When possible, such buildings should be inspected during construction as each floor is closed in, insulated, and the relevant finish installed. This allows for design or construction problems to be identified and corrected before the entire building is completed and occupied.

Reducing excessive energy consumption is important, but, generally, a well-planned inspection will also increase occupant comfort, which may also lead to reduced energy use. Several case studies have documented not only an overall reduction in energy use, but more importantly, a reduction in peak use or the size of the system required to condition the space. Other issues may be even more important, such as minimizing unwanted moisture condensation and consequent mold blooms, eliminating the undue buildup of ice on the roof, checking air circulation in conditioned spaces, and preventing the plumbing from freezing.

## 9.8 Advantages and Limitations

Previous discussion has revealed many of the advantages and limitations of infrared thermography. In some applications, thermography has proven itself as an inspection technique that is highly effective and very easy to use, whereas other applications may require sophisticated analysis.

Probably the two advantages thermography has over many other inspection techniques are that it is fast and it can create a thermal image. For many applications, an experienced and trained thermographer can make a determination of condition almost immediately upon viewing the thermal image with the right equipment, and good thermographers can create an excellent thermal image in seconds.

Today's equipment is extremely capable of doing amazing things. Radiometers are capable of accurate non-contact temperature measurement. Thermal imaging systems can resolve temperature differences of less than 0.07°F (0.04°C). The spatial resolution of the latest imaging systems provides such detail that many thermographers are often foregoing capturing a corresponding visual image. In-camera processing allows sophisticated field analysis in a package that weighs less than a pound.

As has been seen, thermography is a very versatile inspection method. The applications where thermography is effective are numerous. The aerospace industry is now augmenting, and in some cases supplanting, ultrasonics and radiography with thermography for determining the location of subsurface flaws and inclusions in state-of-the-art composites. Using similar techniques, a roofing contractor can locate areas of wet roof insulation for replacement. From identifying production problems in printed circuit boards, to finding loose connections in electrical systems, to the thermal mapping of complex industrial machines, thermography's list of applications is large and diverse. Where heat is a by-product of a process, or where an object undergoes a thermal cycle, thermography may very well have the ability to provide information about the operation and/or the internal integrity of a component.

The limitations of the technology can be summed up rather briefly. Only the surface of an object can be seen thermally. The thermal pattern is the result of either subsurface differential heat transfer or heat reflecting off the surface. As discussed, some material surfaces are so thermally reflective that they require preparation with a high-emissive coating. If the reflective surface cannot be made more emissive, the subsurface condition may not be detectable, or the internal temperature will have to be increased until the surface temperature is sufficient to overcome the limited capability to emit. The thermal image must be interpreted. This requires knowledge of the application along with training and experience in thermography.

## 9.9 Glossary

**Absolute Temperature Scale** Temperature scales that are measured from absolute zero.

**Absolute Zero** The point on the Kelvin and Rankin temperature scales that indicates zero. Commonly known as the temperature at which no molecular activity occurs.

**Background** The radiating objects that are reflected by a surface to the infrared instrument, usually from "behind" the instrument.

**Beaufort's Wind Scale** A simple scale developed by a sea captain named Beaufort that allows one to estimate wind speed based on visual indications, such as the characteristics of the waves or leaves on a tree. It is useful to thermographers who need to estimate wind speed at a specific location.

**Blackbody** A surface that absorbs and reradiates all energy incident upon it. Perfect blackbodies do not exist, but surfaces that are close to blackbodies do exist and, if traceable to a standard source, can be used to check the calibration of a system.

**Boundary Layer** A thin layer of fluid that builds up next to a surface during convection. It reduces heat transfer in proportion to its thickness.

**British Thermal Unit (Btu)** A unit of energy defined as the amount of heat required to raise the temperature of a pound of water one degree Fahrenheit at sea level (at standard pressure). A Btu is equal to approximately 1055.06 joules.

**Calorie** Commonly referred to as the amount of heat needed to raise the temperature of one gram of water one degree Celsius. The modern definition is the amount energy equal to about 4.2 joules. Its symbol is "c" or cal.

**Cavity Radiator** A hole, crack, scratch, or cavity that will have a higher emissivity that the surrounding surface because reflectivity is reduced. A cavity seven times deeper than wide will have an emissivity approaching 0.98.

**CCD** Charge coupled device; one of several types of electronic readout devices used in focal plane array systems.

**Coefficient of Thermal Conductivity** See *Thermal Conductivity*.

**Conduction** Heat transfer from warmer (more energetic) to cooler (less energetic) areas in a substance due to the interaction of atoms and molecules. This is the only way heat is transferred in solids. Heat transfer by conduction is also present in fluids (liquids and gasses) when atoms or molecules of different energy levels come in contact with each other.

**Conductor** Loosely defined as a material that conducts heat well, usually in comparison with materials that don't conduct well (insulators). Most metals are good conductors of heat.

**Convection** The movement of fluids in response to a temperature difference.

**Density** The mass of a substance per unit volume. Units are pounds per cubic foot.

**Detectivity (D*)** Sensitivity of an infrared detector.

**Electromagnetic Radiation** Vibrating electrical and magnetic fields in the form of waves that travel at the speed of light.

**Electromagnetic Spectrum** Electromagnetic radiation at all possible wavelengths from gamma rays to radio waves.

**Emissivity** A property of a material that describes its ability to radiate energy by comparing it to a blackbody (a perfect radiator) at the same temperature. Emissivity values range from zero to one.

**Film Coefficient** See *Boundary Layer*.

**First Law of Thermodynamics** Energy in a closed system is constant; it can't be created or destroyed.

**Forced Convection** Convection caused by wind, fans, pumps, stirring or some other added force.

**Foreground** The radiating objects that are "in front of" the camera and usually behind the target.

**Fourier's Law** The rate equation that describes conductive heat transfer, where energy equals thermal conductivity × area × temperature difference.

**FOV** Field of view; a measure of the angular view for a given system and lens combination, usually measured in degrees.

**Heat** Also known as thermal energy; energy transferred from regions of higher temperature to areas of lower temperature when a material changes temperature.

**Heat of Fusion** The latent heat released as a material changes from a liquid state to a solid state or absorbed as it changes from solid to liquid.

**IFOV** (instantaneous field of view) A measure of the smallest area that can be seen by the system at any one instant, i.e., spatial resolution.

**$IFOV_{meas}$** (instantaneous field of view—measurement) A measure of the smallest area that can be measured by the system at any one instant. It is a measurement, not a specification of spatial resolution.

**InSb** ("Ins-bee" or indium antimonide) A photon detector material with excellent performance in the mid-wave band.

**Insulator, Insulation** Loosely defined as a material that restricts the flow of heat, especially in comparison with materials that conduct well (conductors).

**Isotherm** A software function that outlines areas of apparent similar temperature or radiosity in the image.

**Kilocalories** One thousand calories. Commonly used for expressing the energy value of foods. The symbol is Kcal or C.

**Kirchhoff's Law** For an opaque object, radiant energy absorbed equals radiant energy emitted.

**Latent Energy** Energy used to make or break the bonds in the phase of a material.

**Latent Heat of Fusion** The energy that is used to create or break the bonds in the solid phase of a material.

**Latent Heat of Vaporization** The energy that is used to create or break the bonds in the gaseous phase of a material.

**Long-Wave** Radiation with wavelengths between 8 and 15 μm.

**Mid-Wave** Radiation with wavelengths between 2 to 6 μm.

**Natural Convection** Convection occurring only due to changes in fluid density.

**Newton's Law of Cooling** The rate of heat transfer for a cooling object is proportional to the temperature difference between the object and its surroundings.

**Opaque** Nontransparent; $T = 0$.

**Phase Change** The process that matter goes through when it changes from a solid to a liquid to a gas.

**Qualitative Thermography** Thermal imaging using nonradiometric equipment or images to compare the radiation coming from various targets without making radiometric measurements of temperature.

**Quantitative Thermography** Thermal imaging using radiometric equipment (radiometers) or radiometric images to make radiometric measurements of the target temperature.

**Quasi-Steady-State Heat Flow** A thermal condition that is assumed to be steady-state for the purpose of analysis.

**R-Value** The measure of a material's thermal resistance. It is defined as the inverse of thermal conductivity.

**Radiation** The transfer of energy by electromagnetic waves or radiation.

**Radiometric** An image or system that is calibrated to infer temperature measurements from the detected infrared radiation.

**Radiosity** All radiation coming from a surface, including that which is emitted, reflected, or transmitted.

**Resistance** The measure of a material's ability to resist the flow of energy by conduction. Its value is the reciprocal of its conductivity.

**Second Law of Thermodynamics** Heat cannot flow from a cooler object to a warmer one unless additional work or energy is added. Also stated as "heat cannot be totally changed into mechanical work."

**Slit Response Function (SRF)** A test used to determine the smallest object that can be seen or measured by a system.

**Solar Glint** A phenomenon usually associated with mid-wave sensing infrared systems whereby reflections of the sun off any shiny surface are very predominant.

**Span** The set of temperature values that can be measured within a preset range. Thermal "contrast."

**Spatial Resolution** A specification, usually in milliradians (mRad), of the smallest size object that can be seen by the system.

**Specific Heat** The amount of heat required to raise a unit mass of a given substance by a unit temperature.

**Spot Radiometer** Non-imaging radiometric device that outputs a temperature or other radiometric measurement; also called infrared thermometer.

**Spot Size** The size of an area that can be measured at a given distance by a radiometric system.

**State Change** See *Phase Change*.

**Steady-State Heat Transfer or Flow** A hypothetical thermal condition in which temperature difference across a material or system is unchanging.

**Stefan-Boltzmann Constant** $5.7 \times 10^{-8}$ W/m$^2\cdot$K$^4$

**Stefan-Boltzmann Law** Total energy radiated by a blackbody surface is proportional to its absolute temperature to the fourth power.

**Temperature** The relative measure of hotness or coldness of a material or substance.

**Thermal Background** The radiation sources "behind" the system that are reflected to the detector.

**Thermal Capacitance** The ability of a material to store thermal energy. It is defined as the amount to heat energy (in joules) required to raise the temperature of 1 kg of material 1°. It is arrived at by multiplying a material's specific heat times its density.

**Thermal Conductivity** The symbol for thermal conductivity is "k." It is the measure of materials' ability to conduct thermal energy. It is defined as the rate at which (watts) flows through a material of unit area and thickness, with a temperature gradient (Kelvin), over a unit of time. In SI units it is W/m$^2\cdot$K.

**Thermal Resistance** The inverse of thermal conductivity. It is the measure of a material's ability to resist the flow of thermal energy.

**Thermodynamics** The study of energy, how it changes, and how it relates to the states of matter.

**Thermograph** A visual picture of thermal data; a thermal image.

**Thermographer** A person who is qualified to use thermography equipment.

**Thermography** From the root words for "heat pictures."

**Transient Transfer or Flow** A thermal condition in which the heat transfer through a material or system is changing over time.

## Bibliography and References

ASTM Standards (misc.), available from ASTM, 100 Barr Harbor Drive, West Conshohocken, PA 19428-2959; phone 610-832-9500/fax 610-832-9555.

DeWitt, D. P. and G. D. Nutter. *Theory and Practice of Radiation Thermometry.* Wiley, New York, 1988.

Guyer, E. C. (Ed.). *Handbook of Applied Thermal Design.* McGraw-Hill, New York, 1989.

Incropera, F. P. and D. P. DeWitt. *Fundamentals of Heat and Mass Transfer,* 3rd ed. Wiley, New York, 1990.

*ISO 6781 Thermal Insulation, Qualitative Detection of Thermal Irregularities in Building Envelopes, Infrared Method.* American National Standards Institute, Washington, DC

*Manual for Thermographic Analysis of Building Enclosures (149-GP-2MP).* Canadian General Standards Board, Ottawa, Canada, K1A 1G6.

Lloyd, J. M. *Thermal Imaging Systems.* Plenum Press, New York, 1982.

Maldague, X. P. V. *Nondestructive Evaluation of Materials by Infrared Thermography.* Springer-Verlag, London, 1993.

*NFPA 70-B, Recommended Practice for Electrical Equipment Maintenance,* and *NFPA 70E, Standard for Electrical Safety Requirements for Employee Workplaces.* National Fire Protection Association, 1995. NFPA, PO Box 9101, Quincy, MA 02269; 800-344-3555.

Nondestructive Testing Handbook Volume 3, Infrared and Thermal Testing. ASNT, Colombus, OH, 2001.

*Proceedings of Thermosense,* 1980–1999, SPIE, Bellingham, WA.

*SNT-TC-1A-1996, Guidelines for the Qualification and Certification of Nondestructive Testing Personnel.* American Society for Nondestructive Testing (ASNT), 1711 Arlingate Lane, P.O. Box 28518, Columbus, OH 43228-0518; 800-222-2768.

Stanley, R. K. and P. O. Moore. *Nondestructive Testing Handbook, Volume nine: Special Nondestructive Testing Methods.* ASNT, Columbus, OH, 1995.

Wolfe, W. L. and G. J. Zissis (Eds.). *The Infrared Handbook,* ERIM, Ann Arbor, MI, 1993.

# CHAPTER 10
# Acoustic Emission Testing

## 10.1 History

The phenomenon that we now call acoustic emission (AE) was not studied extensively until the 1960s, but the actual history did not start there. Early man recognized that the detection of unusual sounds could be a useful tool in defense of a predator or enemy. In the production of domestic objects, such as clay pots, the sound of cracking from the kiln indicated that the pot had cracked and was unusable. And who could ignore the sound of ice breaking as a heavy load is transported over it or the sound of a tree branch before it gives way? Later, in the metal age, tinsmiths were very familiar with "tin cry," which is the sound that tin makes when it is deformed. The first scientific research into the detection of sound from materials under stress is credited to Joseph Kaiser who, as a graduate student in the 1950s, discovered the irreversibility of the sound and its strain rate dependence. Research flourished in the 1960s and was greatly aided by the establishment of the Acoustic Emission Working Group. By the close of the 1960s, several entrepreneurs were producing primitive sensors and instrumentation. During the 1970s the technology accelerated as instrumentation improved, useful applications were developed, and training courses became available. It was also during this period that industry leaders began to participate in code-writing organizations, such as ASTM and ASME, to standardize terms, formalize definitions, and produce test procedures. The name "acoustic emission" was officially chosen over other possible candidates such as "microseismic emission" and "stress wave emission." During the 1980s, AE matured into a valuable tool for NDE.

Acoustic emission testing (AE) has become a recognized nondestructive test (NDT) method, commonly used to detect and locate faults in mechanically loaded structures and components. AE could provide comprehensive information on the origination of a discontinuity (flaw) in a stressed component and also provide information pertaining to the development of this flaw as the component is subjected to continuous or repetitive stress.

## 10.2 Overview of Acoustic Emission

Discontinuities release various forms of energy as a material is subjected to mechanical loading or stress. The acoustic energy travels away from the discontinuity in the form of stress waves. These waves, or mechanical oscillations, are detected with the use of sensors (transducers) that in turn convert the energy into a voltage. This voltage is electronically amplified and with the use of timing circuits is further processed as AE signal data. Analysis of the collected data comprises the characterization of the received voltage (signals) according to their source location, voltage intensity, frequency content, etc.

The AE source, being a short impulse, produces a wave carrying energy of all frequencies from very low to some very high, on the order of 1,000 kHz and higher. Experience has shown that high sensitivity to these emissions is most useful when captured using contact sensors between 100 and 500 kHz.

Some of the lower-frequency emissions (approximately 50 Hz to 15 kHz), if they are loud enough, can be heard with the unaided ear. This confirms the idea that the energy of emissions is spread over a very wide frequency range, and the theory that AE comprises frequencies all the way down to zero is evidenced by the largest acoustic emissions of all, earthquakes, which shake buildings a hundred miles away, at frequencies of a few hertz and less.

The amount of acoustic emission energy released, and the amplitude of the resulting wave, depends on the size and the speed of the source event. A strong event produces a greater signal than will a weak event. One popular theory is that the amount of emission energy is proportional to the area of the new surface that is created while the emission amplitude is proportional to the crack velocity. Therefore, a sudden, discrete, crack event will produce a greater signal than will a slow, creeping advance of the crack tip over the same distance. One useful example of this is the observation that an increment of fatigue crack growth will produce much more emission than an increment of ductile tear.

The major difference between the AE method of NDT and the other NDT methods is that it is passive, whereas the others are active. With ultrasonic, radiographic, or the other NDT methods, the source of information is derived by creating some effect in or on the material by external application of energy or compounds. Conversely, AE relies on energy that is initiated from within the component or material under test. Detectable flaws in metal include crack growth due to metal fatigue, hydrogen embrittlement, stress corrosion, and creep. Detectable flaws in composite materials include fiber fracture, delamination, and matrix cracking. Other sources including high-pressure leaks, partial discharge, etc. can also be detected and isolated.

The AE process is illustrated in Fig. 10-1. It begins with forces acting on a body; the resulting stress is the stimulus that causes deformation and with it, acoustic emission. The stress acts on the material and produces local plastic deformation, which is the

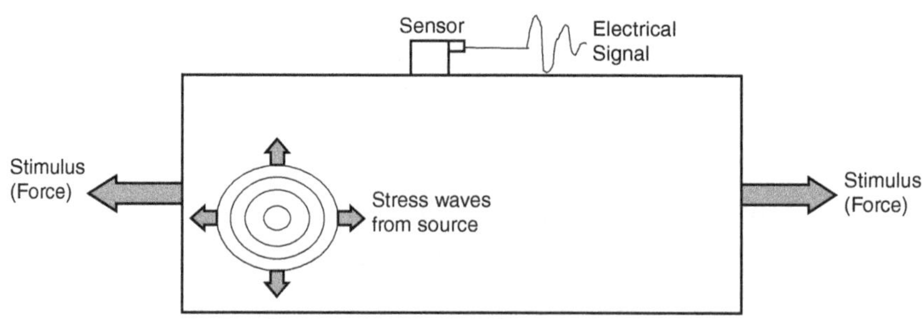


**FIGURE 10-1** Schematic of the acoustic emission process.

breakdown of the material at specific places. This material breakdown produces acoustic emission: an elastic stress wave that travels outward from the source, moving through the body until it arrives at a remote sensor. In response to mechanical displacement, the sensor produces an electrical signal, which is passed to electronic equipment for further processing.

## 10.3 Sources of Acoustic Emission

There are three categories of AE sources: primary, secondary, and noise. Primary sources are generally those which could be of structural significance. There are numerous primary sources, which are best understood if broken down into subcategories of material type and loading situation.

### 10.3.1 AE from Metals as a Direct Result of Mechanical Loading

There are two general situations here that relate to loading and condition of the metal being loaded. In metals that have no flaws (no stress concentrations), the source of emission is from microstructural sources such as dislocations. In metals that have defects, the emission comes from the growth of the defects as well as from microstructural sources.

#### *Microstructural Sources*

As previously mentioned, the AE process begins with stress. Stress is a familiar concept to engineers and designers. It can be described as an internal force field in a structure that adjusts to balance the externally imposed forces (load). Depending on the load's directional properties, stress may be described as tensile, compressive, bending, shear, or torsion. Stress is measured as load divided by the cross-sectional area on which the load is applied, i.e., PSI (pounds per square inch).

Stress can be thought of as a three-dimensional field having different components in different directions at each point in a structure. In response to stress, the structure of the material changes in shape. This change in shape is called "strain." The material deforms elastically and if the stress is high enough, it will deform plastically as well. "Plastic" in this context means "permanent." On the atomic scale, plastic deformation involves the sliding of atomic planes over one another, through the agency of atomic-scale irregularities known as dislocations. The movement of dislocations is the microscopic mechanism that underlies the gross changes in shape that are recognized as yielding, buckling, denting, etc. Acoustic emission that results from the movement of dislocations, or "strain," has been extensively studied with special laboratory techniques.

A typical representation of the stress/strain relationship is known as the "stress strain curve" (see Fig. 10-2). At low load (low stress), the stress and strain form a linear relationship known as the elastic range. In the elastic range, the material will return to its original shape when load is removed. If load is increased above the "yield stress," the result will be plastic (permanent) deformation. This is called the "plastic" (or "work hardening") range. The second curve shown in Fig. 10-2 is the AE behavior that results when a material (with no defects) is simply loaded to failure. Note that AE is very low in the beginning of the elastic range and that it peaks near the yield stress. Note also that there is little AE in the plastic range until just prior to failure when the metal begins to tear. The AE that does occur during a single load to failure is relatively low in amplitude and somewhat controlled by the strain rate.

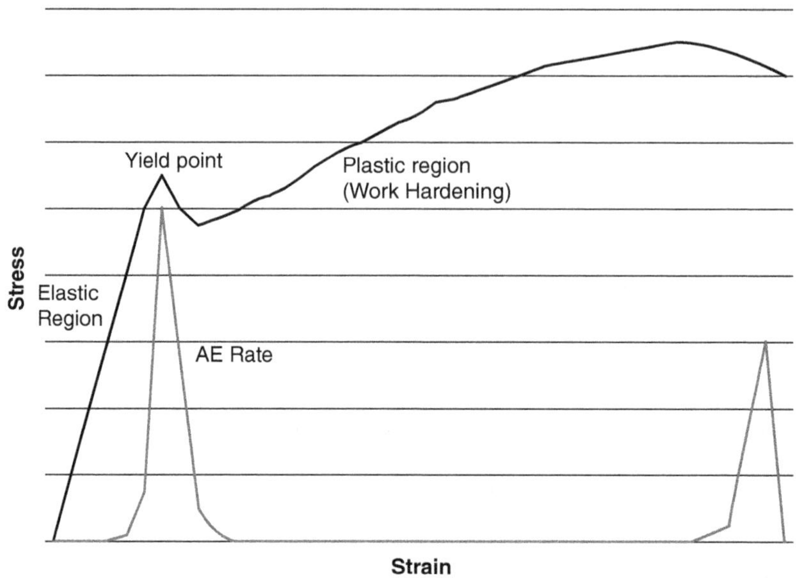


**FIGURE 10-2** Stress versus strain curve with AE rate.

On a microscopic scale inside a piece of steel, the constituents most likely to break are specks of sulfide, oxide, and carbide, and other nonmetallic materials. The smallest of these items are the carbide "precipitates" scattered within the metal grains, for example, microscopic plates of iron carbide, only a few hundred atoms thick, distributed in "pearlite colonies." These precipitates play a big part in governing the steel's mechanical properties. On a larger scale, there are nonmetallic "inclusions" lying between the metal grains, such as manganese sulfide "stringers" formed during the rolling of the steel plate and slag inclusions introduced during welding. There may also be nonmetallic corrosion products intimately connected to the metal surface. All these nonmetallic components are less ductile than the metallic matrices in which they are embedded. As a result they break more easily when the metal is strained. The breaking of these nonmetallic components is the main source of the acoustic emission observed when defect-free metals are deformed.

When a material deforms and emits energy, the deformation tends to relieve the high local stresses. Often the load is transferred to some other part of the structure. This has a stabilizing effect. The elastic and plastic components of the strain occur immediately after the load is applied. The stress/strain field inside the material is quickly redistributed such that all the forces are balanced.

If the structure is unloaded and then reloaded to the same level, the regions that deformed the first time will tend to be stable the second time. Thus, the emission sources will tend not to reemit the second time around, unless the load exceeds the previous maximum. This irreversibility is known as the Kaiser Effect.

### *Macrostructural Sources*

Larger-scale sources, such as cracks, originate at stress concentrations. Stress concentrations are regions within a material where the stress is raised by local geometry. Stress concentrations exist at weld details, changes in section, and structural discontinuities. The stress concentrations at weld details are the reason why fatigue cracks initiate at these locations. When metal is cracked, there is a different kind of acoustic emission source and this source is the most important one in nondestructive testing. The occurrence of a crack instantly creates a new surface. This is a major threat to a structure's integrity and is also the best recognized source of high amplitude acoustic emission. Detection of emission from growing cracks has been the primary goal in many applications of AE technology. When a surface-breaking crack grows, the structure opens up in response to the applied forces. This is far more serious than the opening of an inclusion, which would tend to have no more than a local effect. Cracks tend to produce higher amplitude signals that are more readily detectable.

The tip of a crack grows in size as a result of work hardening and repeated loading. A crack begins to grow when the tip has experienced several load cycles which have exceeded the (local) yield stress and the area in front of the crack tip has become work hardened (embrittled). This nucleation process will allow a crack to form and create new crack surfaces. The crack will grow just far enough to be blunted by metal which has not been work hardened sufficiently. More load cycles will be necessary to work harden the new crack tip so that propagation can continue. The process of work hardening followed by propagation is known as subcritical fatigue crack propagation. This process is illustrated in Fig. 10-3.

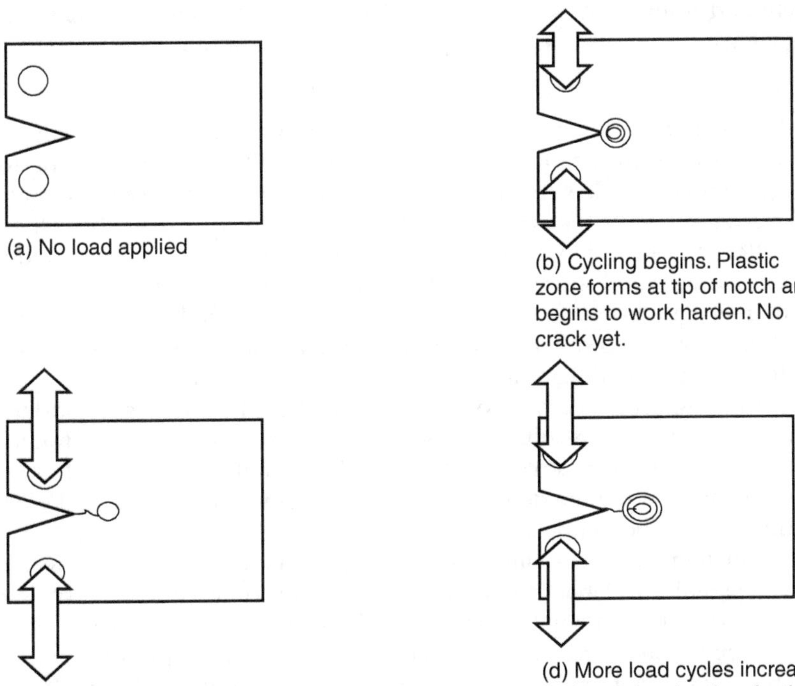

(a) No load applied

(b) Cycling begins. Plastic zone forms at tip of notch and begins to work harden. No crack yet.

(c) Crack propagates into work hardened plastic zone and is arrested.

(d) More load cycles increase the size of and work harden the plastic zone allowing the crack to propagate again

**FIGURE 10-3** AE from fatigue crack propagation.

The emission from growing fatigue cracks has been attributed to two sources. First, there may be emissive particles, typically nonmetallic inclusions, in the stress-concentrated region (plastic zone) near the crack tip. As the crack advances toward these particles, the local stress levels rise, and their breaking will produce burst emission. The other source is the movement of the crack tip itself. Crack tip movement is typically taking place in a mixed mode: some of the new surface is created by dislocation activity and some of it is created by small-scale cleavage, a sudden separation of the material in a region of local weakness. Crack tip movement by dislocation activity is typically not detectable, but cleavage is an abrupt and relatively gross mechanism that produces plenty of AE energy.

When material in a component deforms in response to any type of loading, the deformation tends to relieve and smooth out the local stresses. This means that after crack has advanced, the elastic energy stored in the stress field will have been reduced. The energy released from the stress field is used to create new deformations that will warm the material and produce the acoustic emission. Stated another way, the source of the acoustic emission energy is the energy stored in the elastic stress field produced by the loading of the structure. Acoustic emission is produced at the source, as a short pulse of elastic and kinetic energy that travels through the material as an elastic stress wave.

## 10.3.2 AE from Nonmetals as a Direct Result of Mechanical Loading

Nonmetals that have been studied with AE include composites, ceramics, and polymers. Composites, because they are an important structural material, have been studied extensively and found to be extremely copious emitters. Composite material testing represents one of the largest practical applications of AE testing.

### *Composites*

Composites are materials that are made up of two or more distinct materials. There are many common examples ranging from fiberglass (FRP) to ultra-high-performance metal matrix composites. In general one component, which is often a fiber, provides the bulk of the strength while the other is a matrix such as epoxy, used to hold the fibers in position. Often, all of the constituents of a composite material are brittle and the fracture of each one is a good source of AE. A crack in the matrix material will only propagate until it encounters a fiber and is arrested. A cracked fiber will result in a redistribution of stress to other fibers. Among the numerous failure mechanisms are fiber/matrix debonding, delamination, and fracture of the fiber.

Matrix cracking occurs because of the different strength and elastic constants of the materials. Matrix cracking results in emission that is relatively low in amplitude. This source of AE can occur at relatively low load and does not produce load hold emission unless accompanied by time-dependent fiber elongation near failure. Delamination between the layers of the composite layup is often the result of weakness due to fabrication flaws. Delamination is a matrix failure but many are occurring over a wide area, so the AE amplitude is significant. Fiber fracture is of major concern since load-carrying capacity may be diminished. It is among the highest amplitude sources found in any material.

Since the emission sources tend to produce different AE amplitudes' it is noteworthy that an early presentation of this data was in the form of a "tri-modal" amplitude distribution. As noted in Fig. 10-4, the progression of damage in the composite sample begins with matrix crazing (lowest amplitude) then delamination (higher amplitude) and finally fiber fracture (highest amplitude). The ability to determine what sort of

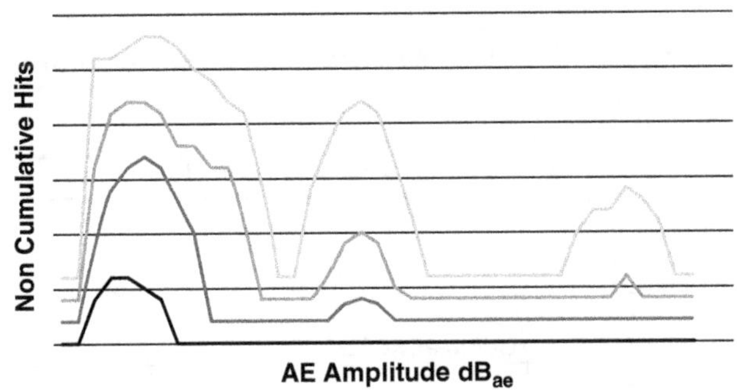


**FIGURE 10-4** Tri-modal amplitude distribution from FRP.

failure mechanism was taking place at a particular load has been of great use to those designing structures from composites.

A second very illuminating presentation of AE data from composites is known as the "knee in the curve." As seen in Fig. 10-5, there is very little emission at low load up to the appearance of the knee. With load above the knee, the amount of AE is significantly greater. The knee in the AE versus load curve has become very useful as a material parameter. It can be used to characterize a particular layup of a composite material in the same way as the material parameter, yield stress, is used to characterize a particular alloy of a metal. Service loads that exceed the knee value will damage the structure and shorten its service life.

Another useful observation shown in Fig. 10-6 is the presence of AE data during load hold. When a composite material is loaded, emission is produced as expected. But under certain circumstances, the emission continues at constant load. Some of the

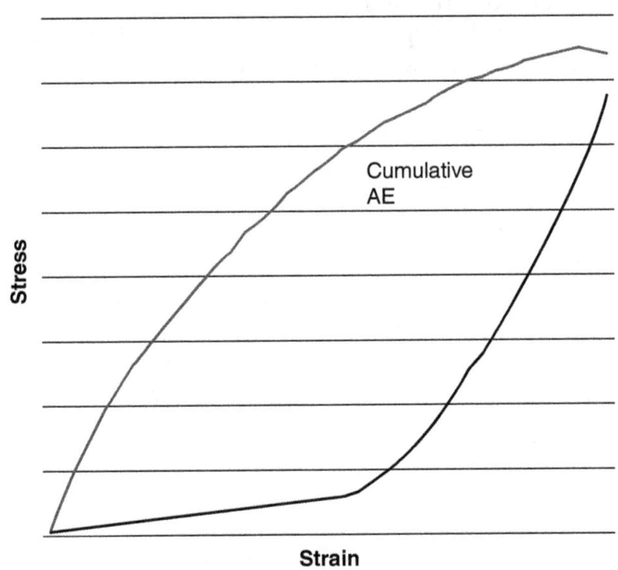


**FIGURE 10-5** Knee in the AE curve from FRP.

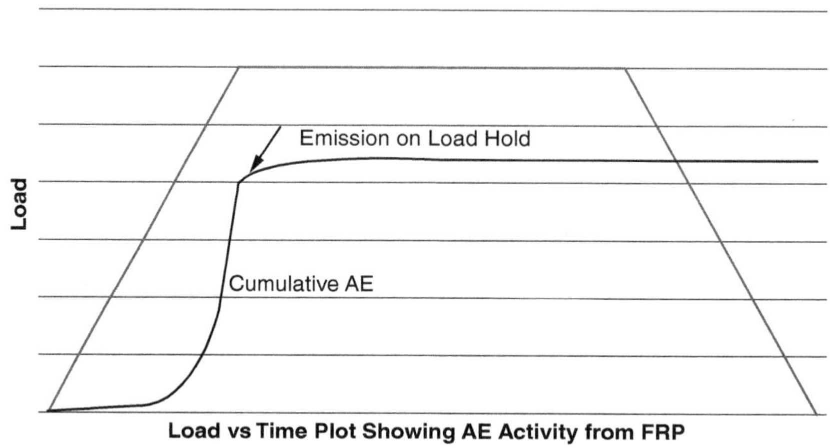


**FIGURE 10-6** Emission on hold from FRP.

deformation is immediate but some of it is delayed. In FRP the reason for the delay is often cited as a visco-elastic nature of glass whereby the redistribution of stress takes place over a period of time as the material comes to equilibrium. Delayed deformation of nonmetallic materials is quite familiar. In time, plastics creep and stretch, and wooden beams sag. Acoustic emission is a very sensitive indicator and will often reveal time-dependent behavior that would otherwise go unnoticed.

Another characteristic of structurally significant defects in composites is that they tend to emit on a second loading to the same level. If a second loading is carefully monitored, one often sees a small amount of emission before the previous maximum load; not nearly as much as the first time, but not zero either. This emission can be an important indicator of structural instability.

As can be observed in Fig. 10-7, emission is plotted directly against load. In this scenario, the load is raised, lowered, raised again to a higher level, lowered, and finally

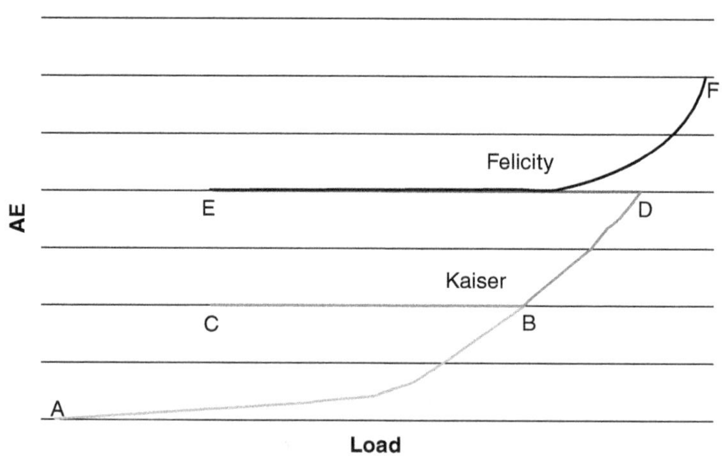


**FIGURE 10-7** AE activity versus load illustrating Kaiser effect and Felicity effect.

raised to a higher level still. Emission is generated during the first load rise (AB), but as the load is lowered (BC) and raised again (CB), there is no more emission until the previous load maximum is exceeded. Emission continues as the load is raised further (BD), and stops as the load is lowered for the second time (DE). On raising the load for the last time, a different emission pattern is observed: the emission starts up before the previous maximum load is attained (F). Emission continues as the load is increased (FG). The behavior observed at point B (no emission until previous maximum load is exceeded) is known as the Kaiser effect. The behavior observed at point F (emission at a load below the previous maximum) is known as the Felicity effect. Insignificant flaws tend to show the Kaiser effect while structurally significant flaws tend to show the Felicity effect.

#### *Ceramics*

Fracture in nonmetals such as ceramic materials differs from that of metals in that ceramic materials are more brittle in nature. They do not possess the plasticity and deformation characteristics of metals. Under load, they tend to produce micro-cracks rather than dislocation slip in the structure. When the stress, applied to a ceramic, exceeds the strength of the interatomic forces, the result is large-scale crack initiation/propagation in the process of brittle fracture. This type of fracture produces very high amplitude AE signals but provides very little warning of impending failure.

#### *Polymers*

Some polymers (plastics) are brittle in nature and deform by crazing. This form of progressive brittle fracture produces copious amounts of AE. Other polymers are very ductile and soft.

#### *Wood*

Wood is a form of natural composite and produces AE in a fashion similar to FRP.

#### *Rock*

Stone is brittle by nature and produces AE in a fashion similar to ceramics.

### 10.3.3 Secondary Sources

Emission sources that are not directly related to structural failure are often called secondary sources. They may in fact be very useful in determining structural integrity and used or ignored in any particular test procedure.

#### *Crack Face Rubbing*

As well as causing large-amplitude AE as they progress, cracks produce small-amplitude AE waves from material deformation at the crack tip. Emission can also be produced from the rubbing or "fretting" of crack surfaces as they open and close and grind in response to changing loads (Fig. 10-8). Corrosion products forming on the crack surfaces can enhance this emission, which make the crack even more emissive.

#### *Leaks*

Leaks of gas or liquid from a pressure boundary will generate AE. A leak may or may not be of significance to the primary objective of the test; however, the presence of emission from a leak can disrupt the ability to detect data from more critical sources such as cracks. In many instances, it is necessary to halt a test in order to repair leaks before continuing to higher loads.

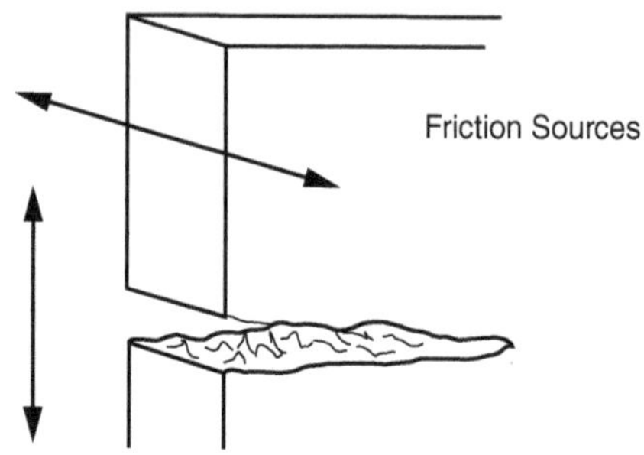


**Figure 10-8** Crack face rubbing can produce secondary emission.

#### *Partial Discharge*

Partial discharge is the result of failure in the insulation between two conductors when the voltage is high enough to produce ionization. Partial discharge occurs in voids in the insulation or on the surface of the insulation. In large power transformers, electric arcing associated with the discharge vaporizes the dielectric fluid in the area of the arc and creates cavitations. As the cavitation bubble collapses, an AE signal is launched into the fluid. Sources of partial discharge can be located and scheduled for repair.

#### *Phase Transformation*

Martensitic phase transformation is a diffusionless shear phase transformation in steel (and other metals). When steel is hardened, by rapid cooling from an elevated temperature, the martensite phase is nucleated in the high-temperature austenite phase, and an interface sweeps along converting austenite to martensite. The shearing mechanism results in large deformations that produce AE. AE is commonly used to detect phase transformation during welding.

### 10.3.4 Noise

Noise in AE testing means any unwanted signal and is a major issue in the use of acoustic emission techniques. Unwanted signals may be spurious bursts from the material under test or from environmental sources. High continuous noise levels can overwhelm low-amplitude AE sources, and burst-type signals can be confused with relevant data. Elimination of noise is a primary objective of the AE practitioner.

#### *Pressurization System Noise*

There are numerous noise sources associated with the pressurization system ranging from fluid flow tones to noisy pumps and motors. Some such as flow noise are continuous in nature and can be reduced by lowering the pressurization rate. Others, such as leaks at pressure boundaries, must be eliminated before a test can proceed.

#### *Physical Contact*

Any mechanical contact with the test object may cause a slip-stick (frictional) mechanism resulting in fretting or rubbing. This type of contact produces a burst-type AE signal. Impact from airborne objects such as windblown sand, or even rain, also produces a burst- type signal. Burst-type signals are often confused with relevant data.

### EMI and RFI

EMI (electromagnetic interference) and RFI (radio frequency interference) are generated by numerous sources including radio transmitters, welding machines, fluorescent lighting, and electric motor switches. These sources may be high in amplitude and within the bandwidth of the AE instrumentation. EMI/RFI is one of the most difficult sources of background noise to eliminate because the source is seldom obvious.

## 10.4 The Signal-Shaping Chain

The signal-shaping chain is shown in Fig. 10-9. It has four links: the source, the propagation of the wave, the sensor, and the signal-conditioning electronics. Each link has a controlling influence on the size and shape of the measured signal. The final signal is drastically different in shape from the original motion at the source.

An important consideration in discussing the signal-shaping chain is the frequency content. All signals can be analyzed as to their sine wave frequency components. This is the field of Fourier analysis, one of the most powerful tools in the science and technology of signal processing. The term "frequency" refers to the repetition rate of an oscillation within a given time period, i.e., the number of cycles per second. Anyone who has turned the knob to tune in a car radio has been selecting the frequency that he or she wants to receive. Each radio station broadcasts at a particular frequency. By turning to a station, the receiver becomes sensitive to the desired frequency. An acoustic emission source, however, is not like a radio station radiating just one frequency. It is more like a lightning bolt! When listening to a radio during a thunderstorm, the radio will pick up the lightning discharge anywhere on the AM band. This is known as a "broadband" signal, in contrast with the "narrow band" radio station. Like lightning, the impulsive acoustic emission source radiates energy at many frequencies.

From the fact that the source is broadband and radiating a variety of frequencies, it follows that there is a choice of frequencies to use for sensing the wave and

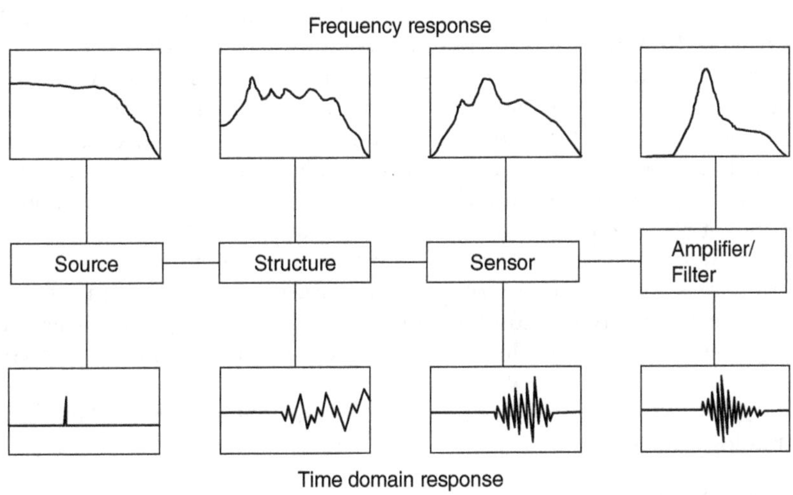


**FIGURE 10-9** Signal-shaping chain.

measuring the resulting signal. This is the rationale for selecting the sensor and electronic filtering.

The frequency response of AE sensors can be either broadband or narrow band. Broadband sensors offer higher fidelity, a more faithful rendering of the actual motion of the metal surface at the sensor location; yet, in most practical AE testing, narrow band sensors are preferred. There are several reasons for this. Resonant sensors are generally more sensitive and much less expensive than the broadband types. They have the advantage of operating in a known and well-established frequency band, which can be chosen to optimize system performance when faced with wave attenuation and background noise. Broadband sensors have the potential to deliver extra information, through the use of advanced signal-processing techniques, but, unfortunately, the best of this information is at the low end of the spectrum, just where the noise problems are worst. Thus, high-frequency resonant sensors are recommended for most practical structural monitoring.

Once the resonant frequency sensor has been selected, the frequency bandpass of the amplifier/filter combination in the AE instrument is normally set to match it. Thus, the last two links of the signal-shaping chain have been established. Now the signal measurements will be altered only by changes in the first two links. The same kind of source event at different locations can produce different signals; and different kinds of source events at the same location can also result in different signals. To obtain meaningful information from the AE signals, it is necessary to have made a suitable choice of sensor type and frequency filter, as well as to have a working understanding of source behavior and wave propagation effects.

### 10.4.1 Propagation of the Wave

The AE wave can be visualized by recalling the ripples produced when a stone is dropped into a pond. The ripples spread out from the source and eventually reach the banks, reflecting in a complicated pattern and ultimately dying away. Reacting to these ripples, a floating cork near the bank of the pond will bob up and down for several seconds in a complicated rhythm, even though the impact of the stone was over in a fraction of a second. The same principle applies to the propagation of AE waves. The short pulse from the source is only the beginning of the AE process. The motion at the sensor is quite unlike the original shock at the source. This is the part played by wave propagation in the signal-shaping chain. The principal difference between the ripples on a pond and the AE waves in a structure is that the AE process occurs many times faster. The typical AE source motion is finished in a few millionths of a second. The wave takes perhaps a thousandth of a second to reach the sensor, and it takes, on the order of, a hundredth of a second for the motion to die away.

Another difference between the pond and the structure is that wave propagation in a structure is generally more complex. First, the structure has many surfaces that repeatedly reflect the wave. Second, solids (unlike liquids) support shear forces as well as compressional forces. This leads to the existence of several different wave types (modes) that can all be excited simultaneously.

Aspects of the wave propagation process that are particularly important in AE technology areas are:

- Attenuation—the loss of amplitude as the wave travels outward from the source. This is important when detecting waves from distant sources.

- Wave velocity—the speed with which the disturbance travels through the structure is important for some source location techniques. In addition, signal-shaping effects have a profound influence on the measured values of the detected AE signal.

### Attenuation

As the acoustic wave travels through the structure, its amplitude decreases. This effect is known as attenuation and is illustrated (for a steel plate) in Fig. 10-10. In this figure, amplitude is plotted on the *Y*-axis using the $dB_{ae}$ decibel scale. On this scale, each increment of 20 decibels is a tenfold increase in the signal's peak voltage. The graph shows that the signal's peak voltage 9 feet from the source is about one-thirtieth of the voltage very close to the source. The $dB_{ae}$ scale is universally used and very convenient because, being logarithmic, it condenses the very wide range of AE signal amplitudes. $dB_{ae}$ is a logarithmic measure of acoustic emission signal amplitude referenced to 1 microvolt out of the sensor element.

Attenuation is due to several factors. In most structures, the important ones are geometric spreading, scattering at structural boundaries, and absorption. When the source is only a few inches from the sensor, as in local monitoring of weld details, the geometric spreading effects are the most important. These effects are therefore especially important for crack detectability. At distances greater than a couple of feet, energy absorption and structural scattering are the most influential properties. The understanding of these effects is also important for identifying extraneous noise.

Geometric spreading effects are fundamental to wave propagation. Basically, the sound wave is "trying" to spread throughout the volume of the structure. Near to the source, the change in the static stress field and the wave motion are at their maximum.

In theory, for a structure that is large in all dimensions (infinite half-space), a stress wave would simply spread in a spherical fashion and continue to attenuate. In realistic structures, the boundaries force the stress wave to remain in a confined space such that attenuation from beam spreading is limited. In a small, well-defined structure such as a rod, the attenuation from geometric beam spreading is minimal and stress waves may travel great distances. The data shown in Fig. 10-10 illustrate this relationship (approximately). The curve falls steeply at short distances and more gradually at larger distances. The 30 percent figure applies to a circular wave front spreading out in a

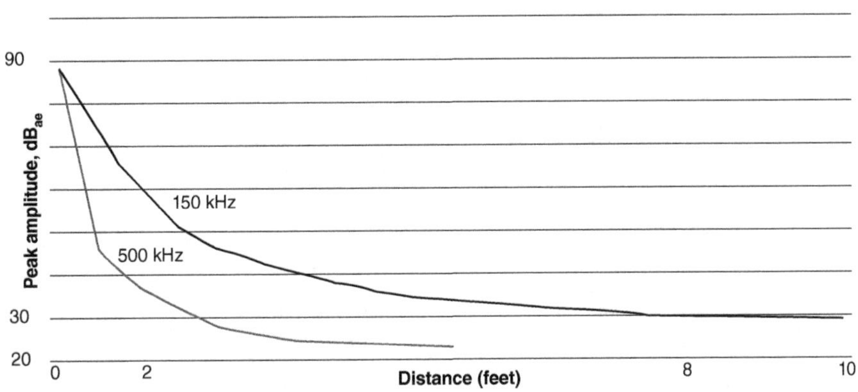


**FIGURE 10-10** Signal attenuation amplitude ($dB_{ae}$).

plate-like structure such as a girder web. In a solid medium like concrete, the wave spreads out in all three dimensions and it loses amplitude more rapidly—50 percent for each doubling of distance. Conversely, in a rod, the wave is channeled and cannot spread, so attenuation is relatively low.

The second major cause of attenuation is reflection (scatter) at structural boundaries and geometrical discontinuities. Whenever a wave meets a discontinuity, some of the wave energy is reflected. Discontinuities also produce mode conversions. These effects are especially important in the complex geometry of many structures, where there may be changes in direction, connections, stiffeners, and other boundaries along the acoustic path from source to sensor.

The third cause of attenuation is absorption. Here the elastic and kinetic energies in the wave are absorbed and converted into heat by the material through which the wave is passing. Steel absorbs very little at the frequencies used for AE testing. Nonmetallics, in general, and paint, in particular, tend to absorb energy more than steel. Absorption is greater at higher frequencies due to the shorter wavelengths at these frequencies. The distance/amplitude reduction mechanism for absorption is different from the mechanism of beam spread/amplitude reduction. In the case of distance/amplitude, there is a constant number of decibels absorption per foot from the source. This is the same at all distances from the source. For example, if there is a 6-dB reduction after the energy has traveled 1 foot, there will be an additional 6-dB reduction after the energy has traveled an additional foot, and so on. Thus, whereas energy losses due to beam spreading near the source are high, absorption also accounts for large losses of energy as the wave fronts move further away from the source.

Attenuation measurements are easily made with a simulated AE source. The most widely used simulated AE source is the breaking of a pencil lead pressed against a structural member, as illustrated in Fig. 10-11. As the lead is pressed against the structural member, the applied force produces a local deformation that is suddenly relieved when the lead breaks. With good technique, the resulting stress wave is amazingly reproducible.

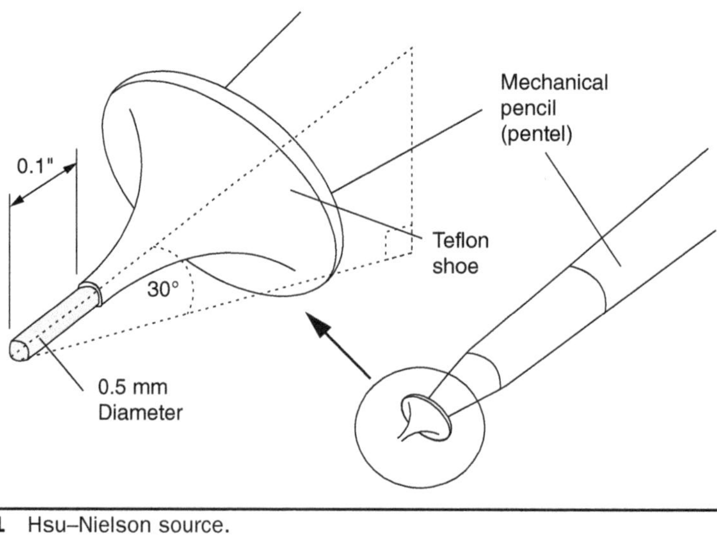


**FIGURE 10-11** Hsu–Nielson source.

The Hsu–Nielsen pencil lead break technique is a convenient, inexpensive aid that has been enormously valuable in practical AE testing. The breaking of the lead creates a very short-duration, localized impulse that is quite similar to a natural acoustic emission source, such as a crack. Furthermore, the amplitude from a 0.3-mm lead-break source is well within the range of typical crack signal. The larger signal from a 0.5-mm lead break may be appropriate in some instances. The Hsu–Nielsen pencil lead-break technique has become accepted as a simulated AE source for sensor/system performance verification, attenuation studies, determination of sensor spacing, etc.

The usual procedure for developing a graph such as shown in Fig. 10-10 is to break lead several times at each of several different distances from a sensor, and then to record the amplitude for each break. The amplitudes for each distance are averaged and, subsequently, the average amplitudes are plotted against distance.

The attenuation curve is an important aid in determining the sensor placements for the specific application. In many acoustic emission applications, the goal of the inspection is to monitor the entire structure. In this case, it is important that all parts of the structure are within detection range of at least one sensor. In global monitoring tests of this kind, the test procedure typically specifies how the attenuation curve can be used to determine acceptable maximum sensor spacing.

In some tests, the recommended strategy is limited-area monitoring focusing on a known crack at a known position. Here the attenuation curve has different applications. It shows how sensitivity to the crack compares with the sensitivity to distant noise sources and whether distant noise sources will reach the sensor or whether they will be attenuated before reaching it. Also, it shows that sensors close to the crack must be placed at similar distances in order to get similar responses.

### *Wave Velocity*

Source location calculations are based on the difference in time of arrival of the wave at two or more sensors. These arrival times depend on the velocity with which the waves travel through the material. Understanding and predicting wave velocities is an important aspect of the science of physical acoustics.

The type of waves that are launched at the AE source are bulk waves consisting of both longitudinal and shear components. Bulk waves are nondispersive (each mode travels at a characteristic velocity). At the surface, part of the wave energy is converted to other wave modes including Lamb waves (wall thickness less than one wavelength) and Rayleigh waves (wall thickness greater than one wavelength). Lamb waves are dispersive, and their velocity is a function of frequency and material thickness. Bulk waves are very useful for very small test objects but attenuate quickly, so it is generally Lamb and Rayleigh waves that are predominant in AE testing of structures.

Of great practical importance for testing large structures are Lamb waves. This wave mode is named after Horace Lamb, an English acoustician who in the 1920s developed the mathematical theory of sine waves propagating in finite plates. This kind of theory seeks to describe wave propagation in terms of wave modes—patterns of oscillatory motion that can propagate in a stable way, maintaining their shape as they travel. In plates, Lamb identified two families of wave modes, and he developed equations that describe their velocities of propagation. In the first family, the motion is symmetrical about the median plane of the plate, and in the second family it is asymmetrical. The parent members of these families are called the $s_0$ and $a_0$ modes, respectively. The form of the motion for these modes is shown in Fig. 10-12.

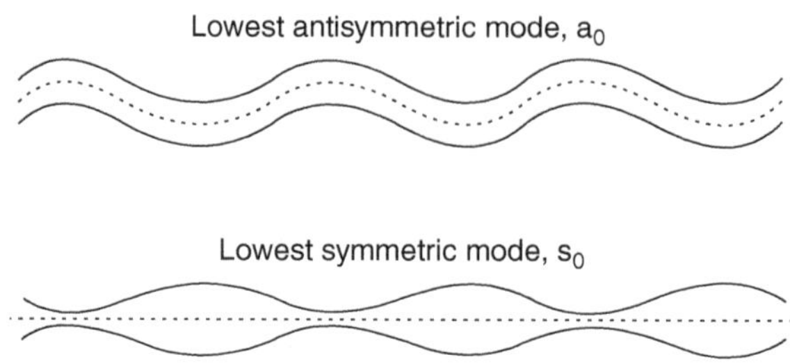


**Figure 10-12** Basic Lamb wave modes in plates.

The $s_0$ mode, often called the "extensional mode," is a rippling movement in which the plate is alternately stretching and compressing in the direction of the wave motion. Coupled with this in-plane movement, the plate surfaces are "breathing" in and out symmetrically as the ripples run through the plate. To produce this kind of wave motion, the exciting force needs to be directed parallel to the plate. A sudden release of in-plane tension will also tend to produce this kind of wave motion.

The $a_0$ mode, often called the "flexural mode," is a mode in which the body of the plate bends with the two surfaces moving in the same direction. Most of the motion is transverse to the plate; there is relatively little motion in the plane of the plate. Exciting forces perpendicular to the plate produce this kind of wave motion. In addition to this, forces that are parallel to the plate but offset from the center line can produce this wave mode.

The underlying theory of elastic wave modes is the general mathematical theory of "eigenfunctions." According to this theory, wave modes travel independently and do not interfere with one another. Several wave modes can be traveling at the same time in the same material; in this case, the motion at any point in the material is simply the sum of the motions of the various modes. The wave modes are like the raw ingredients of a recipe, which can be combined in various proportions to produce many different flavors.

As can be noted from Fig. 10-13, the two Lamb wave modes travel at different velocities. Specifically, at 100 kHz in a 0.65-in.-thick steel plate, the $s_0$ mode travels at approximately 4.7 km/s (180,000 in./s). The $a_0$ mode travels somewhat slower at

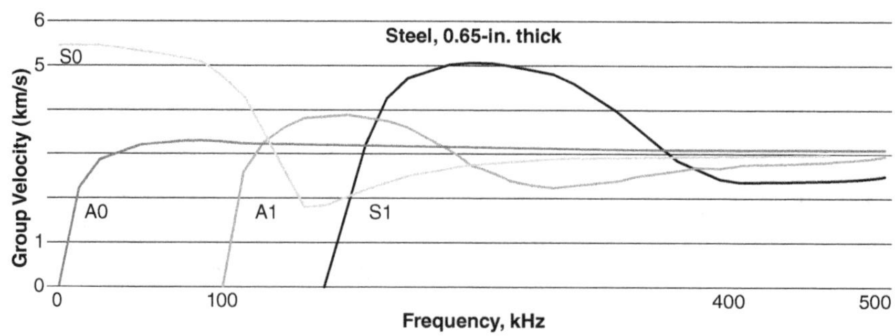


**Figure 10-13** Dispersion curves for 0.65-in.-thick steel plate.

approximately 3.3 km/s (140,000 in./s). Thus, the $s_0$ mode will reach the sensor before the slower-moving, $a_0$ mode. When setting up AE source location systems, the technician usually has to enter into the computer the sensor positions and also the velocity of sound that will be used in the source location calculations. Sensor positions will usually be specified in inches, and in this case the velocity must be entered, correspondingly, in inches per second.

The $a_0$ Lamb wave mode (flexural mode) usually produces a higher amplitude wave than the $s_0$ mode, and in typical structures it also travels slower and therefore arrives at the sensor later. Other members of the two families of Lamb waves (called $a_1$, $s_1$, etc.) can travel faster than either the $a_0$ or the $s_0$ modes, but their amplitudes tend to be low and they are relatively unimportant.

### *Signal-Shaping Effects*

Wave propagation is a tremendously important stage in the signal-shaping chain. It is wave propagation that largely determines the signal's size, shape, and shape-dependent signal features, such as rise time and duration. Because several successful interpretive techniques are based on signal shape, it is important to have some understanding of these effects.

Figure 10-14 shows a typical record of an AE waveform recorded from a structure. Note the timescale on the X-axis, running from –245 μs to 1.39 ms (1,390 μs), a total span of 1,635 μs. Thus, the wave has occurred and decayed in about a thousandth of a second. In this time, there are several hundred positive and negative oscillations of the signal voltage. Some of the individual oscillations can be recognized as they stand out from the background. The dominant frequency during these oscillations is the resonant frequency of the sensor.

The waveform envelope (overall shape) shows a fast rise and a much slower decay. This is typical of AE signals. The waveform at the start of the time base is from direct waves from the source; the later part of the time base comprises waves that have been reflected back and forth many times before arriving at the sensor. The rising part of the time base is determined by strong reflections from nearby surfaces. The falling part of the time base is shaped by more remote reflectors and by the acoustic damping in the structure.

## 10.4.2 Sensing and Measurement

Secondary to wave propagation, the next essential stage in the signal-shaping chain is the sensor. For sensing AE waves, piezoelectric crystals are used. The Greek word

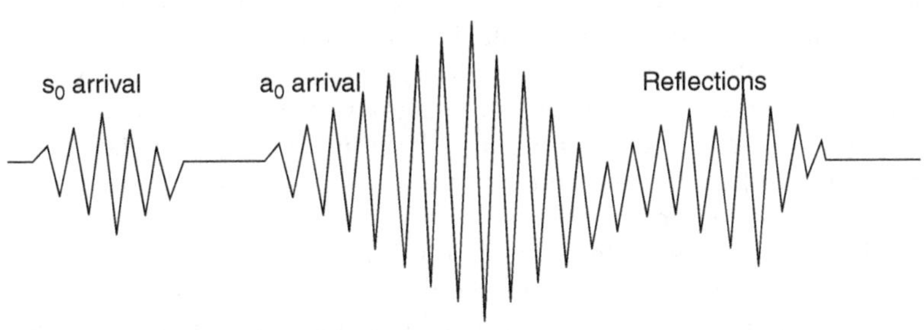


**FIGURE 10-14** Typical AE waveform in a structure.

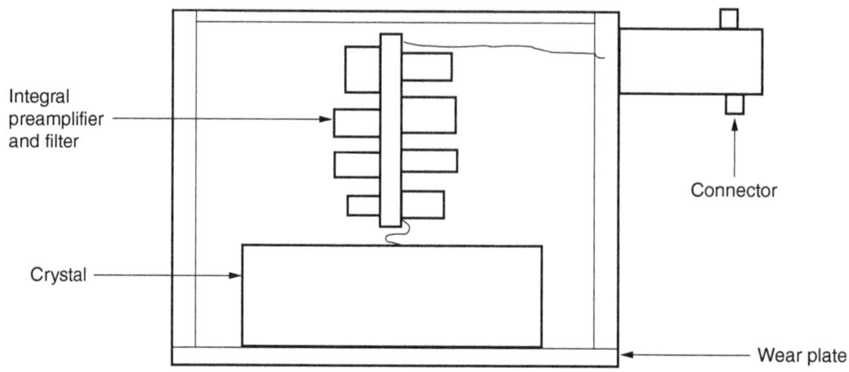


**FIGURE 10-15** AE sensor schematic.

*piezein* means "to squeeze." Piezoelectric materials generate an electrical voltage and a corresponding separation of charge when they are squeezed (deformed). In the AE sensor, the deformation is produced by motion; it is the elastic response of the piezoelectric crystal when it is struck by the incoming stress wave. The piezoelectric element is mounted inside the sensor enclosure, as illustrated in Fig. 10-15 along with an integral preamplifier. The electric voltage is generated by the element material itself then amplified and filtered. The element does not need an external power supply, but the preamplifier does.

### *Sensor Performance and Calibration*

When struck by a sudden impulse, a piezoelectric element will vibrate like a bell, "ringing" at its resonant frequency. In fact, there may be many resonant frequencies all excited together. If shaken by a vibratory motion, a piezoelectric element will produce a corresponding oscillating voltage at the same frequency as the motion. The element has a linear response; if the input motion is doubled, the output voltage will also double.

The ratio (output voltage amplitude)/(input motion amplitude) is a measure of the sensitivity of the sensor. It has been established that this sensitivity depends strongly on the frequency of the motion (i.e., the number of oscillations per second). The sensitivity of the element is greatest at its resonant frequency.

Sensor calibration curves show how sensor sensitivity varies with frequency. An example is shown in Fig. 10-16. The shapes of these curves are characteristic of the sensor type and vary widely in their sensitivity and frequency response. For a given sensor type, the calibration curves of individual sensors should be closely matched.

With calibration curves such as that shown in Fig. 10-16, the amplitude of the input motion may be specified in units of displacement, velocity (meters per second), or pressure (microbars). In all cases, it refers to the amplitude of a sinusoidal oscillation at the frequency indicated on the $X$-axis.

Calibration curves often have their $Y$-axis (sensitivity) scaled in decibels (dB). Decibels are a relative measure: Each decibel corresponds to an increase of 12.2 percent. Compounding upward, with each decibel being a 12.2 percent increase, it is easy to show that 6 dB is a doubling and 20 dB is a tenfold increase in sensitivity.

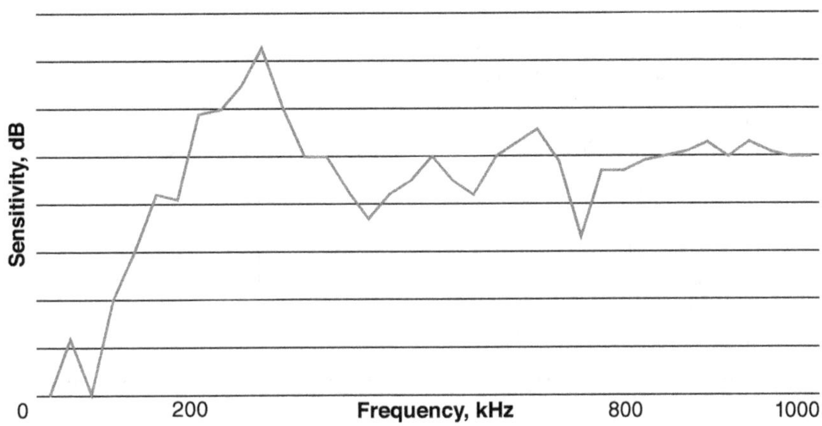


**Figure 10-16** Typical sensor calibration curve.

The sensitivity curves on sensor calibration certificates have their $Y$-axis labeled in decibels relative to a stated reference level, such as 1 volt per meter/second. That is, 0 dB means a sensitivity of 1 volt per meter/second. Mathematically, the formula is:

$$dB = 20 \log (S/S_{ref})$$

where $S$ is the sensitivity and $S_{ref}$ the reference level, both stated in volts per meter/second.

The sensitivity of the sensor depends not only on frequency, but also on the direction of the motion. Unlike accelerometers, which are carefully designed to measure only the component of motion parallel to their axis, AE sensors respond to motion in any direction. Therefore, to understand fully the significance of a sensor calibration curve it is necessary to know about the type and direction of the wave that is being used for the calibration.

In the design of AE sensors for nondestructive testing, the manufacturer's priorities are high sensitivity, well-defined and consistent frequency response, robust performance in the working environment, and immunity to unwanted noise. Unwanted noise can also include electromagnetic interference originating from radio stations, navigation systems, etc. Immunity from this kind of noise was greatly improved with the development of the integral preamp sensor. This type of sensor has a preamplifier built into the housing along with the piezoelectric element. For field testing, it has considerable advantages over earlier sensor types, which required separate preamplifiers to be mounted within a few feet of the sensor.

The sensor must be in good acoustic contact with the structure under test so that it will detect the motion of the AE wave and deliver a strong signal. Coupling and mounting techniques are very important. An acoustic couplant in the form of an adhesive, a viscous liquid, or grease is applied to the sensor face and the sensor is pressed against the structure. The surface on which the sensor is mounted must be smooth and clean; some preparation may be necessary. The sensor must be securely held in place with adhesive, magnetic hold-down, or other means. After mounting, the system performance is verified by simulating an AE signal and checking the system response.

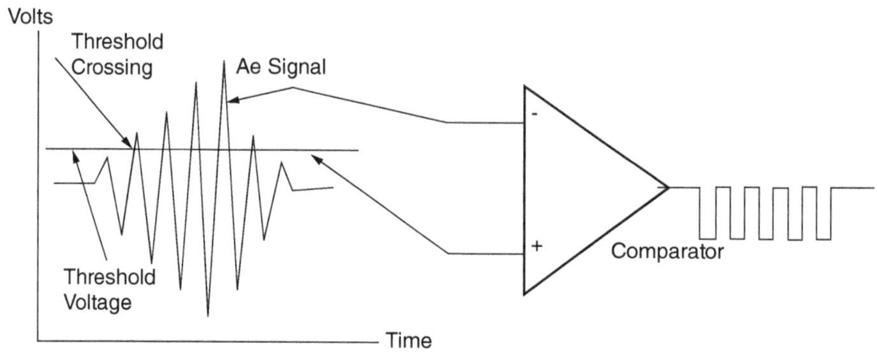


**FIGURE 10-17** Comparator circuit.

### *Signal Conditioning, Detection, and the Hit Process*

The signal produced by the sensor is amplified and filtered, detected and measured. Amplifiers boost the signal voltage to bring it to a level that is optimum for the measurement circuitry. Along with several stages of amplification, frequency filters are incorporated into the AE instrument. These filters define the frequency range to be used and attenuate low-frequency background noise. These processes of amplification and filtering are called "signal conditioning." They "clean" the signal and prepare it for the detection and measurement process (evaluation).

After conditioning, the signal is sent to the detection circuit. This is an electronic comparator, a simple circuit that compares the amplified signal with an operator-defined threshold voltage. The principle is illustrated in Fig. 10-17.

Whenever the signal voltage rises above the threshold voltage, the comparator generates a digital pulse. The first pulse produced by the comparator marks the start of the "hit." This pulse is used to trigger the signal measurement process. As the signal continues to oscillate above and below the threshold level, the comparator generates additional pulses. While this is happening, electronic circuits are actively measuring several key features of the signal. In time, the amplitude of the signals reduces to a point where there are no more threshold crossings. After the predetermined time (the "hit definition time") has passed without any further pulses from the comparator, the system determines that the hit has ended. The hit detection process is then brought to a close. Control circuitry terminates the measurement process and passes the results to a microprocessor. Finally, the measurement circuitry is reset and rearmed ready for the next hit.

### *Signal Features and Their Measurement*

Measurements are made of the relevant features of the hit such as the amplitude, duration, signal energy, and counts. These features and their relationship to the detection threshold are shown in Fig. 10-18. This diagram shows a typical AE signal as a voltage–time curve and illustrates the measured features of the hit.

"Amplitude," in AE testing, refers to the largest voltage present in the signal waveform. It is one of the most important measures of signal height. It is fundamental

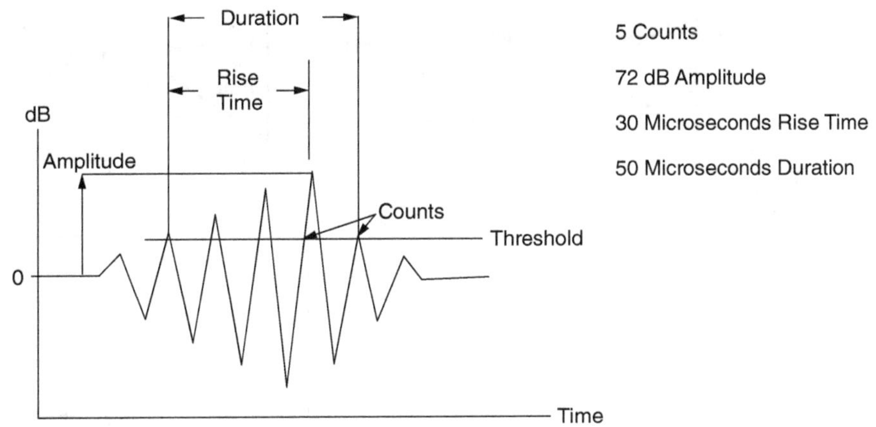


**Figure 10-18** Measured hit features from the signal waveform.

because for a signal to be detected, its amplitude must exceed a predetermined threshold. Amplitude is usually measured in $dB_{ae}$, a decibel scale running from 0 to 100 $dB_{ae}$ is defined as an amplitude of 1 microvolt at the preamplifier input.

"Duration" is the length of time from the first threshold crossing to the last, measured in microseconds (millionths of a second). The relationship between duration and amplitude tells the user about the signal's shape: whether it is a short sharp "click," or a long drawn-out "scrape."

"Signal energy" is the area under the voltage–time envelope, i.e., the area under the dotted line shown in Fig. 10-18. This is another important measure of signal size and is a widely used measure of AE activity. When a structure produces many emissions in response to loading, the energies of the individual signals can be added to produce a total amplitude. Of all the techniques that have been used to describe emission quantity in a single number, this has been the most successful.

"Counts" are the comparator output pulses corresponding to the threshold crossings. A single hit may provide only a few counts or it might furnish hundreds of counts, depending on the size and shape of the signal. For the electronics designer, this is the easiest measurement to make, and in the early years of AE, "counts" were the most common way to describe and report AE quantities. During the 1980s, energy replaced counts as the preferred measure of AE activity. However, counts are still useful for data interpretation; used in conjunction with amplitude or duration, they can give valuable information on signal shape.

Several other signal features may be measurable, depending on the equipment available, but those noted above are the most widely used. A block diagram for typical instrumentation used for making these measurements is shown in Fig. 10-19.

As well as measuring the features of the individual signals, the AE instrument usually measures also the times at which they are detected and the environmental variables that may be causing the activity. The broader aspect of instrument architecture is discussed in the next section.

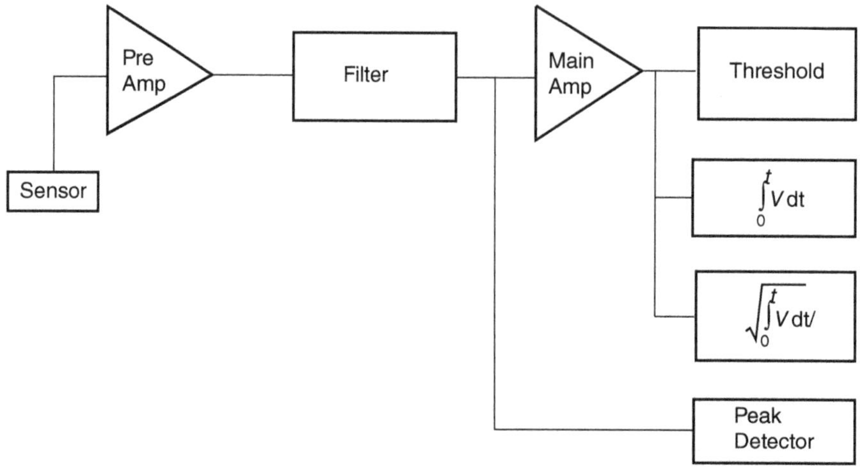


**Figure 10-19** Signal measurement block diagram.

## 10.5 Instrumentation Architecture

The most common overall design for AE instruments is the so-called "hit-based" architecture. Typical AE activity consists of a series of distinct signals. These signals occur at irregular time intervals and have widely varying shapes and sizes. Hit-based architecture is designed for the efficient measurement and recording of this kind of activity.

In hit-based architecture, the system remains dormant until a signal is detected by the threshold circuitry. The signal is measured and a microprocessor stores a data record containing the time of detection and the results of the measurements made. After this, the measurement circuitry is reset, ready for the next hit. Thus, the information passed to the computer consists of a series of "hit data sets." Each hit data set takes up typically 30 bytes of storage space.

The test data builds up in computer memory, or is converted into displays or written to disk or some other medium according to the system design. Environmental data such as pressure, force, displacement, temperature, etc. may also be measured at the same time, and the results may be written along with the signal data.

From these data received, the computer can generate many different kinds of displays, both during data acquisition and also at any time afterward by replay of data files written to a disk or other storage medium. Representative architecture for a hit-based system is shown in Fig. 10-20. This diagram shows a channel with amplification, filtering, detection, measurement of signal features, and the transfer of data and timing information to data storage and processing circuitry.

A major advantage of hit-based architecture is that it delivers a very detailed description of the emission while making very economical use of data storage space. The storing of hit data sets optimizes the use of storage media and gives the greatest scope for data interpretation and evaluation. The data may be retrieved weeks or even years later for evaluation. The test record is available in the same form as the day it was stored. The permanence of the data record and its ready availability for reanalysis is one of the true advantages, not only of this particular architecture but also of AE testing as a whole.

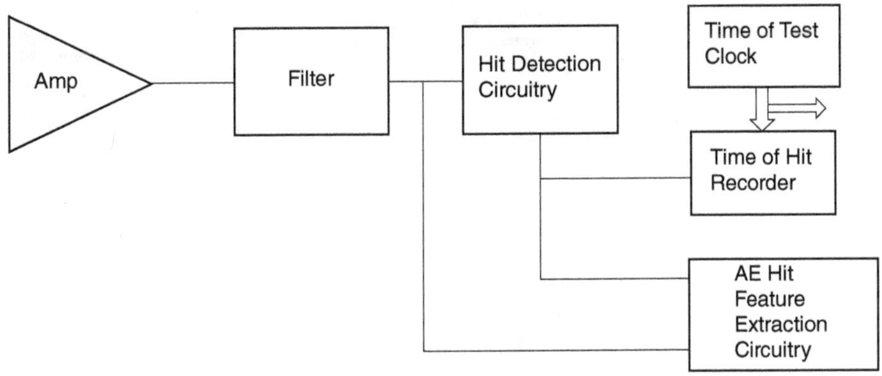


**FIGURE 10-20** Instrument functional diagram.

Hit-based architecture is aimed at comprehensive data acquisition and versatile analysis capability. There are other kinds of architecture with different goals, as well as variations within the broad class of hit-based architecture. Some instruments sacrifice performance in favor of portability. Some systems sacrifice versatility in favor of automatic operation optimized for specific applications. There are simple systems that accumulate AE energy or counts in hardware counters and output a corresponding voltage for plotting on a chart-type recorder. There are dedicated industrial instruments in which the computer prints a predefined set of graphs showing various aspects of the AE data but does not produce a permanent record of every hit. Instruments exist with special alarms or interpretive software. Both research laboratories and commercial vendors have implemented many different AE instrumentation ideas through the years. Out of several hundred diverse AE instrument designs that have been constructed, only a few have been successful enough to warrant production in substantial quantities. A state-of-the-art AE instrument is shown in Fig. 10-21. This is a complete digital 52 channel AE system.

## 10.5.1 Source Location

### *AE Hits and Events*

AE instrumentation can include many channels. A single AE *event* is produced when hits are detected from several channels. These hits will occur in very quick succession, typically all falling within a few hundred microseconds as the wavefront ripples through the sensor array. By comparing the arrival times of hits at different sensors, we can calculate the location of the source event. This capability is one of the most useful features of AE testing. When AE is used to inspect a large structure with just a few widely spaced sensors, source location techniques can be used to pinpoint the exact regions that should be inspected later, using other NDT methods to determine the exact position, size, and severity of the defect.

When AE is used to determine if a known flaw is active, source location techniques can be critical in enabling the user to discriminate successfully between the known flaw and extraneous noise sources. Figure 10-22 shows more details of the basic concept.

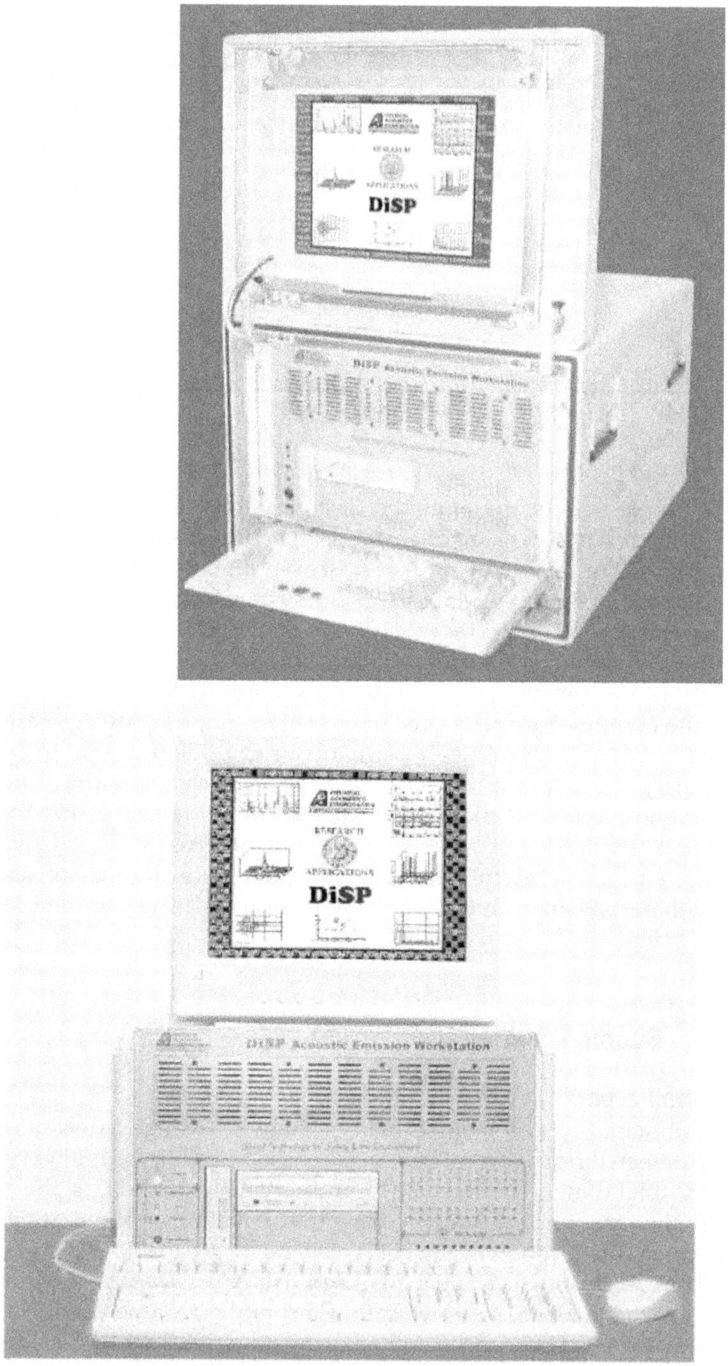


**Figure 10-21** Typical AE system.

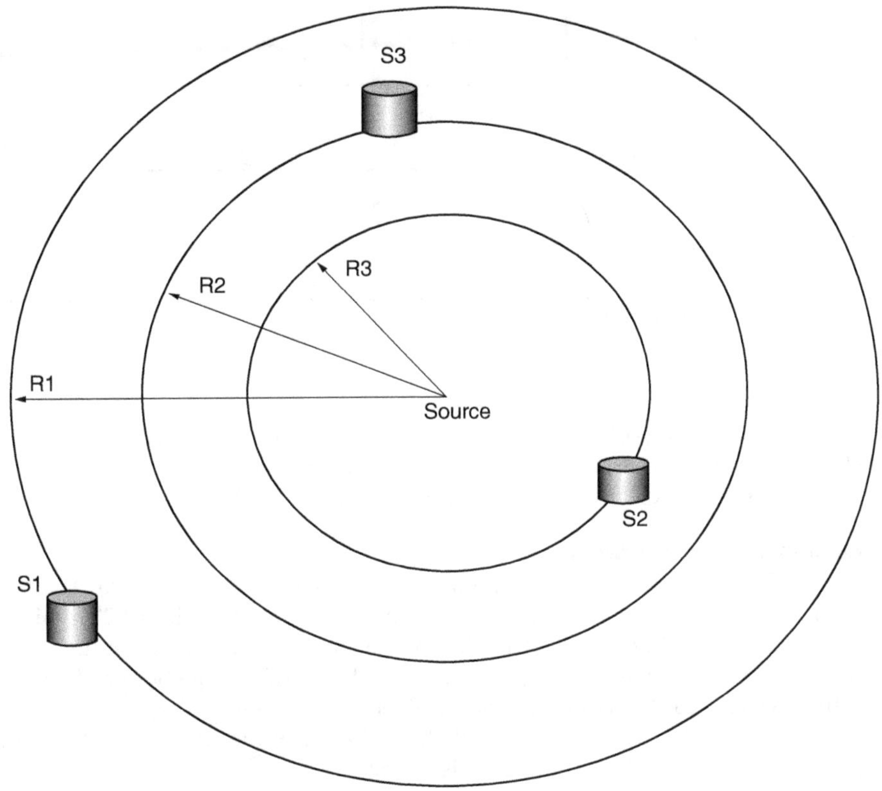


**Figure 10-22** Basic principle of AE source location.

As the wave spreads out from the source, it reaches first the nearest sensor, S2, then the farther sensors, S3 and S1. As the resulting signals cross the detection thresholds on each channel, "hits" are produced on the corresponding channels in the measurement instrument. The first hit, on sensor S2, occurs at time $r_2/c$ after the source event, where $r$ is the distance from source to sensor and $c$ is the wave velocity. The second and third hits occur at times $r_3/c$ and $r_1/c$, respectively. Note that the instrument has no direct awareness of the time at which the event occurred. It can measure only the times of the resulting hits; but from the differences between those hit times, it can calculate the location.

There are a variety of practical techniques derived from this general concept. The intention might be to focus on one or more predefined areas of interest, to discriminate against one or more predefined noise sources, or to set about the test without preconceptions and take an approach using a location-type overview of all sources on the entire structure. There are techniques based on computation from measured time differences and there are also simplified techniques that only consider the sequence of the hits, without even measuring the actual time differences. There are also techniques based on two, three, four, or more sensors with computations oriented to linear, planar, solid, and complex structural geometries.

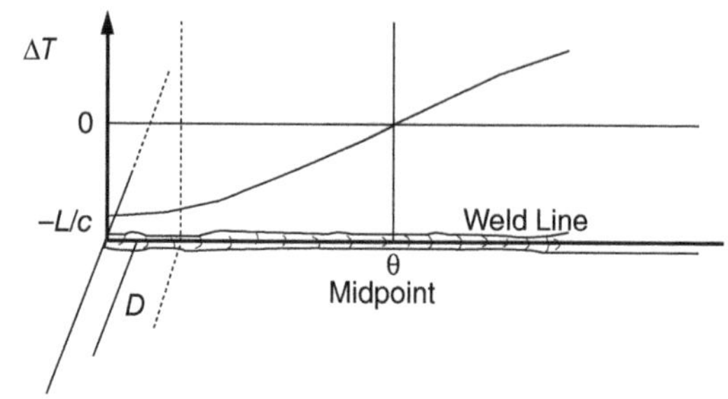


**Figure 10-23** Principle of linear location.

With AE instruments, techniques are executed sometimes in hardware, sometimes in acquisition-time software or posttest software, and sometimes according to the user's choice. Three major approaches—computed source location, zone location, and guard techniques—are explained in the sections below.

### *Computed Source Location*

Two of the most widely used location techniques are known as linear location and planar location. The principle of linear location is illustrated in Fig. 10-23, which shows the time of arrival difference as a function of the source position. When the source is at the midpoint, the wave arrives at sensors S1 and S2 simultaneously, and the time of arrival difference is zero. $\Delta T$ ("delta T") is the algebraic symbol that stands for the time difference between hits. The greater the distance between the source and the midpoint, the greater the delta T. The relationship is linear:

$$\Delta T = \frac{2X}{c}$$

where $X$ is the distance of the source from the midpoint, and $c$ is the velocity of sound. This relationship holds so long as the source is between the sensors. However, if the source is beyond one of the sensors, the delta T has a constant value of $L/c$, where $L$ is the distance between sensors. Thus, a source between the sensors can be pinpointed, but not a source beyond a sensor. Note also in the above equation that an accurate calculation of the source location requires both an accurate measurement of the delta T, and an accurate knowledge of the wave velocity $c$. If the delta T differs from true because of some accident of wave propagation, the calculated location will be in error.

If it is desired to use two sensors to locate sources along a weld in a plate but the sensors are offset a distance $D$ from the weld line, a situation results as is depicted in Fig. 10-24. The relationship between source position and delta T is quite similar to the one shown in Fig. 10-23 but the straight lines have become curves because the sensors are offset from the weld line.

In this case, it is possible, in principle, to pinpoint a location on the weld line beyond the sensors as well as between them. The greater slope of the curve in the region between the sensors indicates that the technique will be most accurate and reliable in the central region. Beyond the sensors, a small change (or error) in the measured delta T produces a large change (or error) in the calculated source position. In other words, the technique is accurate for locating sources along a straight line connecting the sensors. But accuracy is degraded when the source is anywhere else.

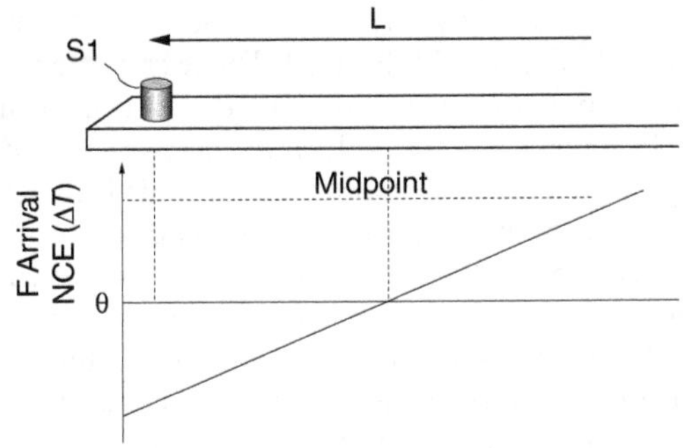


**Figure 10-24** Source location with two sensors offset from a weld line.

The examples in Figs. 10-23 and 10-24 illustrate how, with two sensors (one delta T), it is possible to locate the X position of a source along a line. With three sensors (two delta T's), it is possible to locate the X and Y positions of a source on a plane. This is the situation illustrated in Fig. 10-22. There are many mathematical approaches to source location on a plane, some using three sensors and some using four sensors.

Just as with linear location, the best accuracy for planar location is achieved when the source is well within the sensor array, rather than outside of it. In fact, with only three sensors on a plane, the source location mathematics is ambiguous over most of the plane. For many points outside the triangle, it is possible to find matching points inside the triangle that will produce the same delta T's. The resulting regions of ambiguous location are shown in Fig. 10-25 for an array of three sensors in an equilateral triangle.

Location tends to be ambiguous when the first arrival occurs long before either of the other arrivals. When a three-sensor array is used for monitoring a specific area of interest, it is clearly important to place the area of interest in the area of unique location, not in the area of ambiguous location. Alternatively, the problem of ambiguous location can be resolved by the use of an extra delta T from a fourth sensor.

There are many mathematical techniques for accomplishing source location on a plane. Some use three sensors, and some use four. Several exact analytical solutions

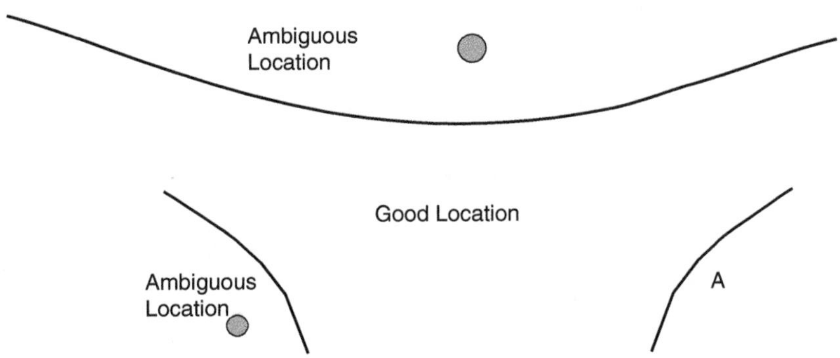


**Figure 10-25** Good location and ambiguous location with a three-sensor array.

have been described, some slow and clumsy, some fast and sophisticated. Approximation solutions have also been widely used. These can offer a good combination of speed, accuracy, and robustness when faced with the option of inferior data. These include iterative approaches, least-squares solutions, and linear and quadratic approximations.

The accuracy of the computed source location is limited more by wave propagation factors than by the mathematical approaches. The conventional approaches all assume that the wave travels directly from source to sensor with a well-defined velocity. In practice, reflections, multiple wave modes, and other propagation effects that produce uncertainties in the effective velocity, as well as scatter in the experimental data, can upset this assumption. The result is that computed source location using conventional approaches is very accurate in some situations but prone to serious error in others. Structural geometry and operating frequency govern whether conventional source location techniques will work well or not.

With computed source location, several hits are needed or there can be no computation. Smaller events that only hit one sensor are liable to be ignored. This can be dangerous, especially in wide-area monitoring. Overreliance on computed source location has sometimes led users to ignore important data and to draw the wrong conclusions about structural integrity. Therefore, computed source location must be used wisely. The sensor spacing must be properly adapted to the wave attenuation on the particular structure being monitored, and to the desired sensitivity.

### Zone Location and Guard Techniques

A simpler type of source location is known as zone location. Here, conclusions are drawn from the hit sequence alone, without actually measuring the delta T's. This development began with the idea of guard sensors, which is illustrated in Fig. 10-26. The technique was conceived in the early 1960s as a way to record data from a limited area of interest, while rejecting noise from outside.

A "data" sensor D is placed on the area of interest surrounded by several guard sensors G. AE waves from the area of interest will hit the data sensor before hitting any of the guards. Waves from outside will hit at least one of the guards before hitting the data sensor. Based on this, it is easy to reject the outside noise. The concept can be implemented in several different ways. Sometimes this process is conditioned by hardware circuitry in the instrument or by software between the detecting and the recording of the hits. In these cases, the noise signals are not recorded. Sometimes this process is performed by posttest analysis on systems that record all hits, including hits on guards.

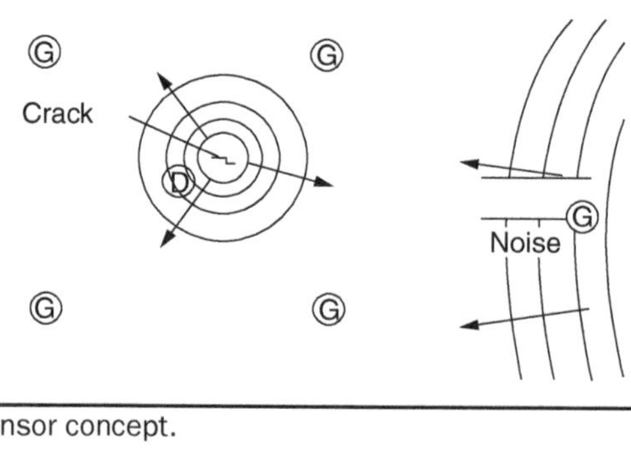


**FIGURE 10-26** Guard sensor concept.

Posttest processing is safer, but the files are longer and more work is necessary during the analysis of the data. As a variant on the guard sensor technique, guard sensors can be placed on known noise sources and data sensors can be placed around the rest of the inspection area.

The idea of first-hit zone location was a further development of the guard concept, conceived in 1972. This is a technique for wide-area monitoring. Many sensors monitor the inspection area, and a determination is made as to which channel detects the wave first. If the wave is traveling uniformly, the first-hit sensor must be closer to the source than any other sensor. By paying attention to the first-hit sensor, good information is gained about the location of the source.

With first-hit zone location, the inspection area can be divided into zones, each zone being centered on one sensor. If a particular sensor is the first one to be hit, it can be concluded that the source lies somewhere in that sensor's zone.

Figure 10-27 shows the pattern of zones for a simple layout of sensors on a plane. The zone boundaries are straight lines that evenly divide the spaces between pairs of sensors. In geometric terms, they are the perpendicular bisectors of the lines joining neighboring sensors. Figure 10-28 shows the expansion of this concept into the three-dimensional structure.

Data from structural tests is often displayed by channel, e.g., a graph of hits versus channel or a graph of a particular channel's activity. The practical value of first-hit zone location is that it allows the user to remove second-hit and third-hit information from the data, so that each channel shows only the emission that actually originated within its zone. This substantially sharpens and cleans up the data.

As with the guard technique, the first-hit zone location technique can be implemented by hardware circuitry, by software at the time of data acquisition, or by software during posttest analysis. Details of the implementation differ depending on the specific instrument design.

Zone location can be extended to take into account the second and third hits, if there are any. With the knowledge of the channel that received the second hit, the source location can be narrowed down to a small segment of the primary zone. With the knowledge of the channel that received the third hit, the source location can be narrowed down still further. However, these techniques are seldom implemented.

A benefit of the zone location approach is that only a single hit is needed for the event to be admitted into the analysis process. In the preceding discussion of computed source location, it was pointed out that several hits are needed to give the delta T's

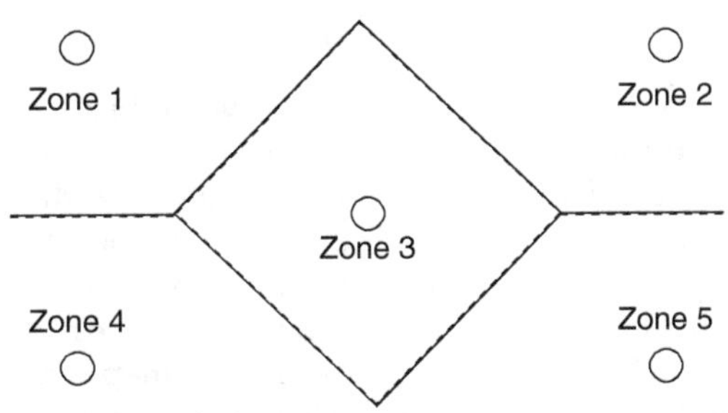


**FIGURE 10-27** First-hit zone location.

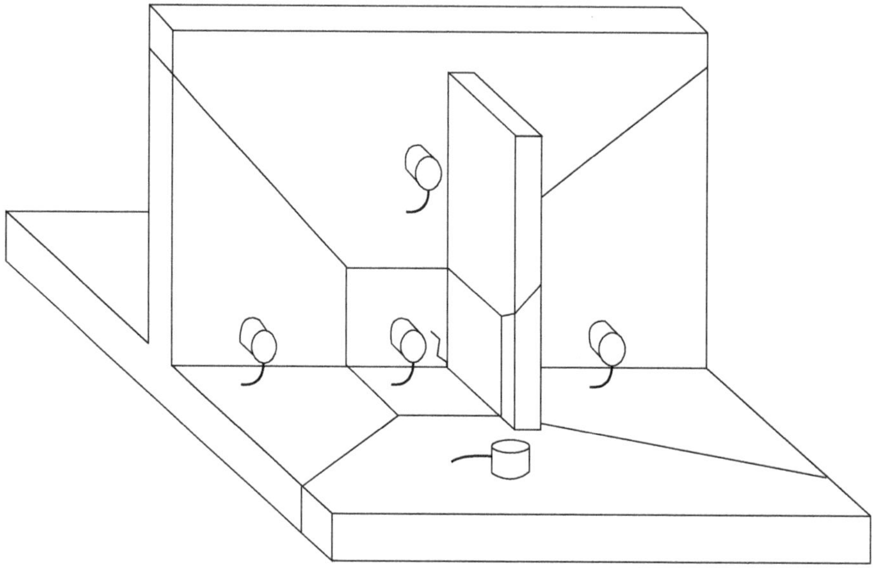

FIGURE 10-28 Zone location on a structure.

required for computation. Compared with this, zone location has, in effect, a higher sensitivity and none of the detected events are ignored.

## 10.6 Data Displays

Becoming familiar with the AE data displays and learning how to read them is a very important part of the AE user's training. Even those who are familiar with other NDT methods may find AE displays strange at first because they are basically graphs of numerical data, rather than the visual images and displays that are the working tools of some other NDT methods.

AE data displays can be grouped into four categories:

1. Location displays
2. Activity displays
3. Intensity displays
4. Data quality (cross-plot) displays

Location displays are widely used. However, other kinds of displays are very valuable for background information on-site, for investigative work, and for data analysis, especially in the more difficult situations. Therefore, with training and practice, the AE practitioner becomes familiar with all the display types. Location displays show where the emission is coming from. When computed location techniques are being used, the typical location display is a map of the inspection area. An example is shown in Fig. 10-29. The user defines the *X–Y* coordinate frame for the map as the software is set up. The sensor numbers are shown in the appropriate places on the map. When planar source location is being used, the location of each event is plotted as a dot on the map. If a source

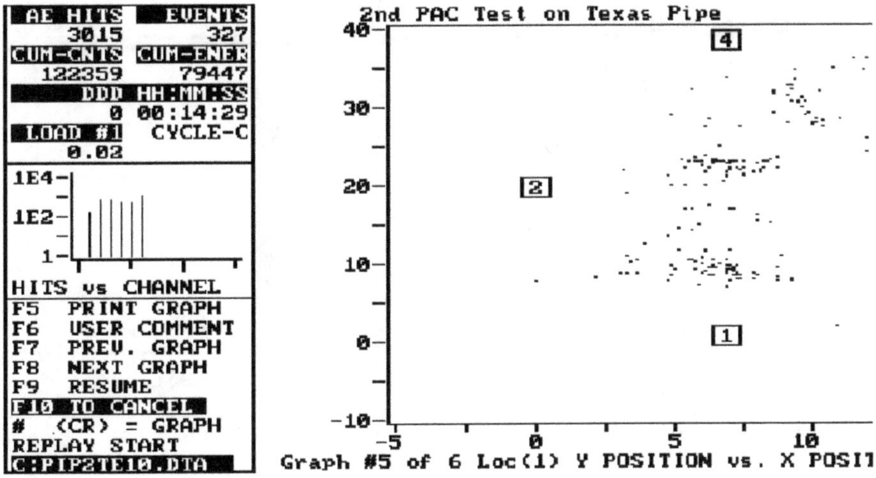


**FIGURE 10-29** A map showing a typical location display of emissions.

emits repeatedly, the dots form a cluster, as shown in Fig. 10-29. The user's eye is drawn naturally to the clusters on the screen that correspond to the most emissive sources.

When linear location is being used, the display is a histogram with the span between the sensors laid out along the *X*-axis. This is illustrated in Fig. 10-30. The *X*-axis is divided into a number of segments or "bins," typically 100. Each located event is assigned to the bin that corresponds to its location. The number of events in each bin is indicated by the height of the histogram bar. When viewing this display, the user's eye is drawn to the highest peak or the highest concentration of indications. This is where the majority of the emission activity originates.

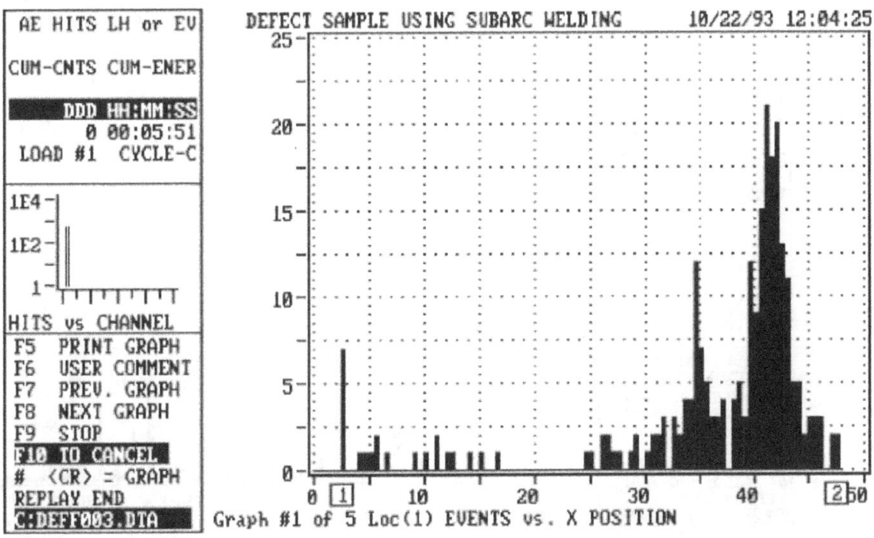


**FIGURE 10-30** Linear location display.

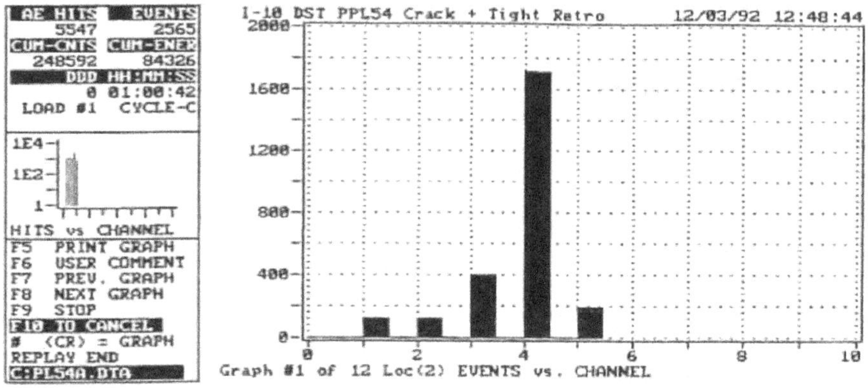


**FIGURE 10-31** Graphical display of acoustic events versus channel.

A third kind of location display is a histogram of events as related to channels. This is used for first-hit zone location, where each event is assigned to the channel that detected it first, and later hits on other channels are disregarded. This type of display is illustrated in Fig. 10-31. The height of the bar for each of the five channels shows the number of events detected by that channel first. Channel 4 shows the most activity, with 1,700 events hitting it first in 1 hour of monitoring. In this case, the crack was between channels 1 and 2; channels 3, 4, and 5 were guard sensors used to screen out extraneous noise.

There are variations on the display shown in Fig. 10-31. For example, the total AE signal energy from each zone could be plotted on the *Y*-axis instead of the total number of events. This would provide greater prominence to the zones that produced high-energy hits. The second major class of displays is activity displays. These include displays that have time on the *X*-axis (see Fig. 10-32). The displays show when the AE activity occurred and how much activity there was. As in Fig. 10-26, the *X*-axis is divided into bins. In this example, there are 100 bins from left to right, so each bin

**10.31**

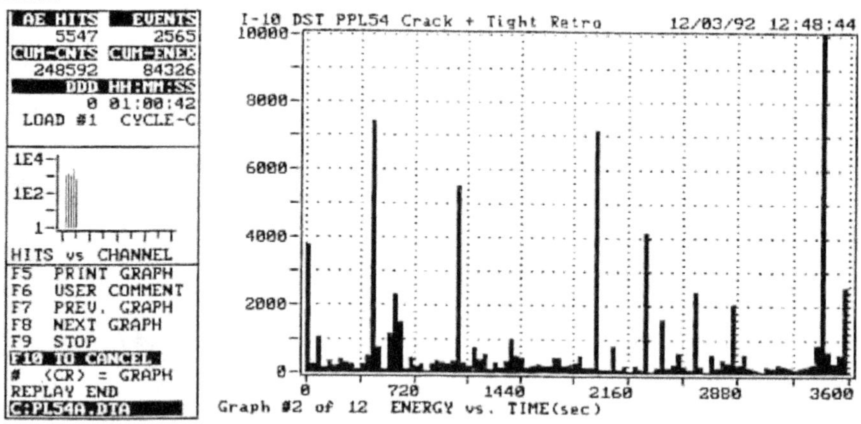


**FIGURE 10-32** Activity display: AE rate versus time.

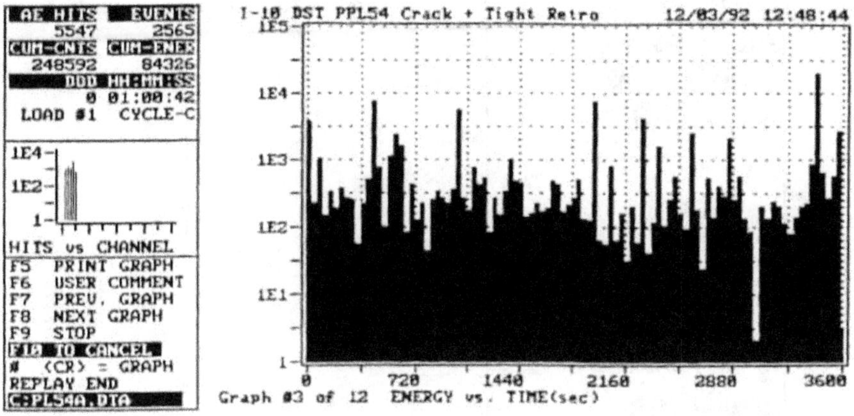


**FIGURE 10-33** Activity versus time, logarithmic scale.

represents a time slice of 36 seconds. On the *Y*-axis, the height of each histogram bar shows how much emission occurred during that time slice. As shown, there typically are detected hits, located events, or total energy shown. In older reports the *Y*-axis often showed AE counts, i.e., threshold crossing counts, as shown in Fig. 10-14.

The *Y*-axis scaling can be linear, as in Fig. 10-32, or logarithmic, as in Fig. 10-33. Looking at the linear plot, a few time slices with very high activity stand out prominently from a low-level background.

In the logarithmic plot, the scaling reduces the visual impact of these very active time slices, but provides more information about the fluctuating lower-level activity in between. In the logarithmic *Y*-axis scaling of Fig. 10-33, 1E1 stands for 101 which is 10, 1E2 stands for 102, which is 100, 1E3 stands for 103, which is 1,000, and so on. 1E0 stands for 1. Logarithmic axes can cover a large span. A 24-hour logarithmic plot would allow comparison in the same glance of large amounts of AE produced in "rush-hour traffic" and a very small amount produced in the "middle of the night." With linear axes, very small amounts may not even show above the baseline.

In becoming familiar with logarithmic plots, it is important to learn how to read the scale and how to interpret the display. In the lower part of the display, the patterns of rising and falling activity may be informative, but it must always be remembered that the actual level is low. In the upper part of the display, the level of activity is drastically greater and even a small increase in the height of the histogram bar means thousands of additional emissions detected.

Another form of activity display is shown in Fig. 10-34. Instead of showing the amount of activity in each time slice, this display shows a running total of all the activity detected since the start of the test. This is the best display for measuring the average emission rate as well as for observing the total emission quantity.

The third major class of displays is "intensity displays." These provide statistical information about the size of the detected signals. In many cases, large signals are likely to be more important than small signals. Furthermore, a given total amount of AE signal energy might be originating from a few large signals or from many small signals; it could be important to know about these origins. These are the issues that are addressed by intensity displays.

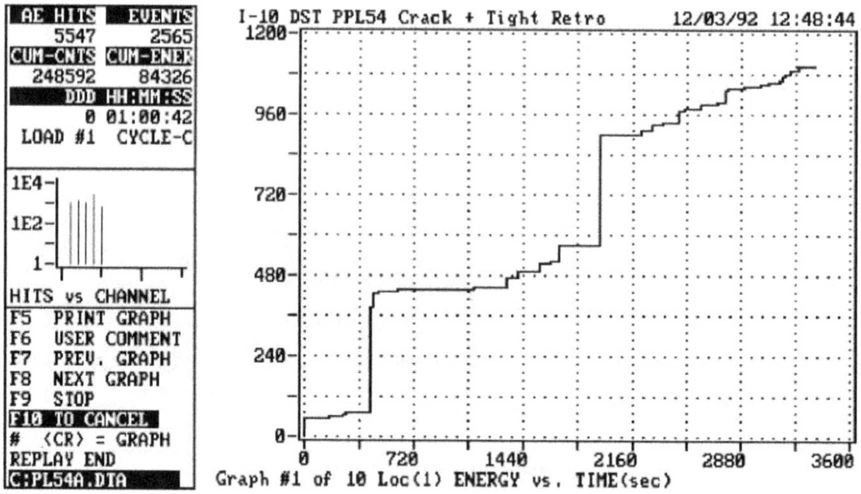


**Figure 10-34** Cumulative activity display.

The best-known intensity display is the "amplitude distribution" plot, which shows how many of the hits were large and how many were small. This display exists in two forms. The "differential" form, illustrated in Fig. 10-35, is a histogram with amplitude on the *X*-axis. The *X*-axis is divided into a hundred 1-dB slices. The heights of the bars show how many hits were recorded at each 1-dB level. Typically, there are many low-amplitude signals and fewer high-amplitude signals. Naturally, there are no measured signals with amplitudes less than the detection threshold, so the display is blank to the left of the threshold dB value.

The cumulative form of the amplitude distribution display is a line graph, presenting the number of hits that were greater than the *X*-axis amplitude. This form is derived from the differential display by scanning it from right to left and accumulating the

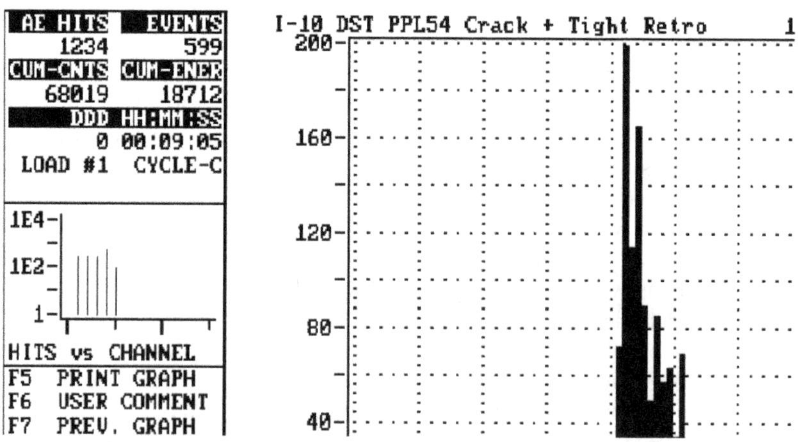


**Figure 10-35** Differential amplitude distribution.

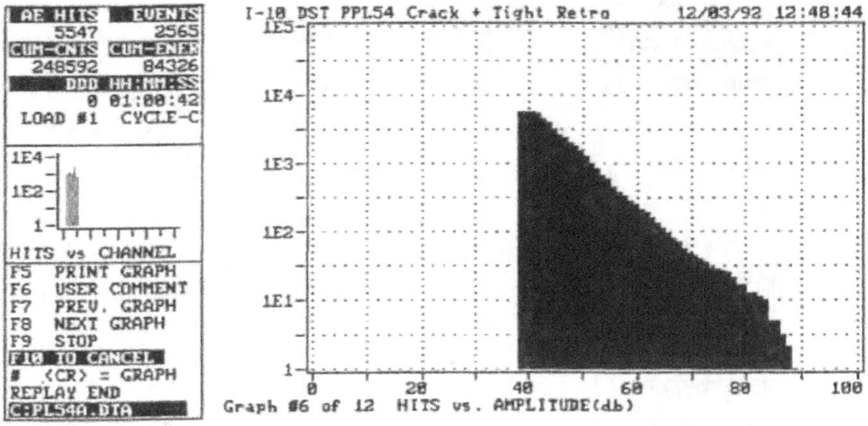


**FIGURE 10-36** Cumulative amplitude distribution.

contents of the bins. The cumulative display derived from Fig. 10-35 is shown in Fig. 10-36. Both differential and cumulative forms are often displayed with a logarithmic scale on the *Y*-axis. This makes it easier to assess the high-amplitude activity even when the low-amplitude hits are much more numerous.

The shapes of the amplitude distributions often give valuable clues about the source mechanisms that are producing them. For example, growing cracks often give straight lines when a logarithmic scale is used on the *Y*-axis. If the plot is far from linear, it is probably not a growing crack. Figure 10-36 is a typical example of a straight-line amplitude distribution. Another type of intensity display is the so-called "energy count," illustrated in Fig. 10-37. This shows whether the total signal energy is coming mostly from low-energy events or mostly from high-energy events. The *X*-axis shows the energy

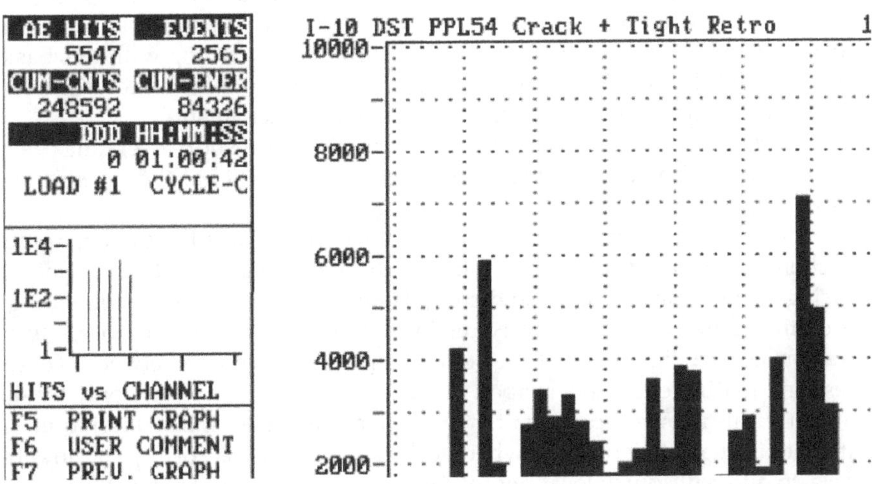


**FIGURE 10-37** Energy account.

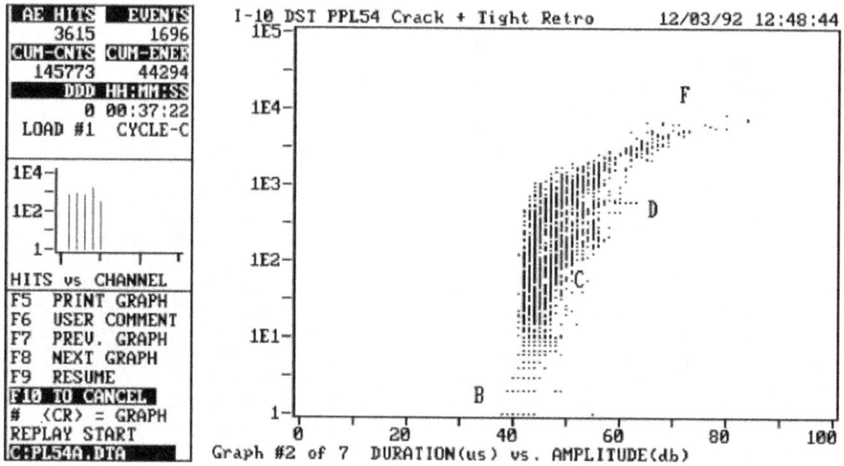


**FIGURE 10-38** Duration/amplitude cross-plot.

of the individual hits, on a logarithmic scale from 1 to 100,000, divided into 50 bins. The height of each bar shows the total energy from all the hits that fall into that bin.

Different emission source mechanisms cause amplitude distributions and energy accounts to show as different shapes. These displays can help to identify the mechanisms that are working, and can help to tell the difference between flaw-related AE and extraneous noise. The fourth and last major class of AE displays is cross-plot displays. These are used during interpretation to assess data quality. The term "cross-plot" implies a plot in which each hit gives one point on the display, showing the cross-relationship between two measured signal features. The best-established cross-plots are "duration versus amplitude," "counts versus amplitude," and "counts versus duration." A duration/amplitude cross-plot is illustrated in Fig. 10-38.

Larger signals typically have higher amplitudes and also more counts (see Fig. 10-14). Therefore, the data in the counts/amplitude cross-plot tend to fall into a diagonal band running from lower left to upper right. Along with this basic tendency, there are subtle variations that can provide additional information about the shape of the signal. Shape, in turn, depends on the source mechanism.

Short, sharp sources, such as crack growth, cause short-rise, quick-decay signals, whereas long, drawn-out source processes, such as frictional sliding, show as slow-rise, slow-decay signals. This is indicated in Fig. 10-39.

A crack-growth signal and a friction signal might have the same amplitude, but the friction signal would have a much longer duration. This provides a technique for recognizing signals from friction and distinguishing them from signals originating from crack growth. Figure 10-38 shows three distinct bands of data. There is a large, broad band at the top, a short, tight band at the bottom, and a third lying in between. These bands correspond to different signal shapes. Signal shape is influenced not only by source mechanism but also by wave propagation, so the analysis of this kind of data is quite demanding. It is not immediately obvious whether the different bands correspond to different channels, to different source locations, or to different source mechanisms. Various techniques of advanced data analysis and signal processing can be used to address these issues.

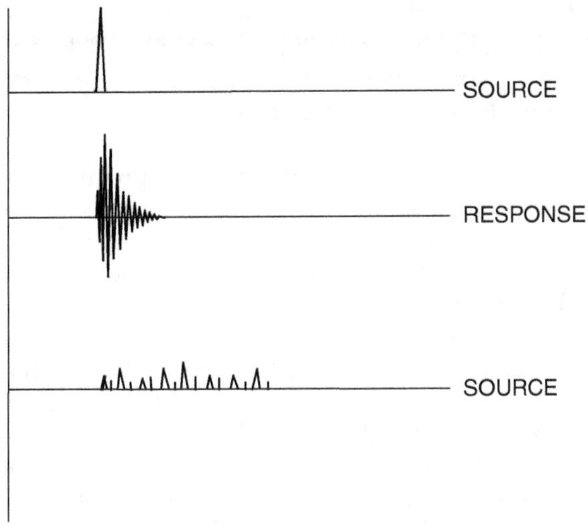


**FIGURE 10-39** Signal shapes depend on source mechanism.

## 10.7 Interpretation and Evaluation

In nondestructive testing, data interpretation and evaluation proceed as shown in Fig. 10-40. ASTM has developed the following definitions:

*Indication.* Response or evidence of a response in a nondestructive test.

*Interpretation.* The determination of whether indications are relevant, nonrelevant, or false.

*Evaluation.* The determination of the significance of relevant indications.

*False* (referring to an indication). Obtained through improper technique or processing.

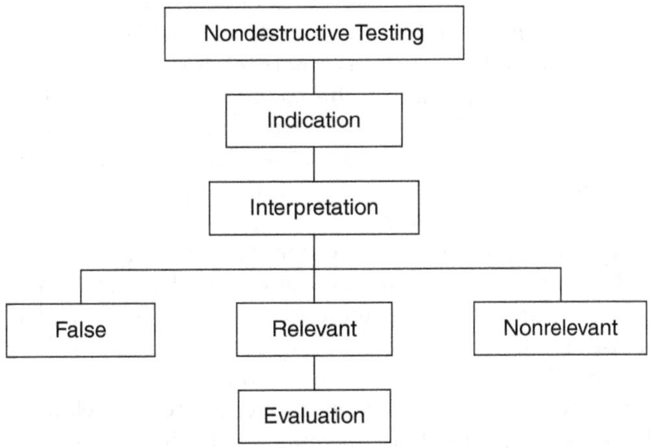


**FIGURE 10-40** Relationship between interpretation and evaluation.

In AE testing, the most basic kind of indication is simply a hit (the data record produced after the signal has crossed the threshold). Relevant indications are hits produced by the crack. Nonrelevant indications include hits produced by sources outside the inspection area. False indications include hits due to "echoes," which can occur if the instrument is badly set up.

In AE as in all other NDT methods, interpretation occurs before evaluation. Sometimes, interpretation is performed explicitly and deliberately, as a well-defined step in a documented data analysis process. In other instances, interpretation is made more implicitly and seems little more than a common sense effort to extract the most out of the available data.

When conducted explicitly, the first task in data analysis is to identify any nonrelevant and false indications in the data file, and then to filter them out. A new data file is written that contains only the relevant data, and running this filtered file through an evaluation program performs the evaluation. This process is used in several AE applications where the techniques have been standardized and a clear-cut procedure has been written. A good example is the AE testing of railroad tank cars. In this and other applications, the duration–amplitude cross-plot shown in Fig. 10-38 is a key tool in the interpretation process. The use of this cross-plot can be included as part of a well-defined procedure, and technicians using it have a common understanding and basis for discussion, refinement, and ongoing development.

Evaluation procedures have been codified for several major AE applications. The best way to arrive at documentable evaluation procedures is to obtain a broad and representative range of field experience, following up the AE tests with other NDT inspections to characterize the flaw conditions as thoroughly as possible. By documenting conditions and results with the use of a database, data can be accumulated for use in future analyses. This has been done for several leading AE applications, such as pressure vessels and tanks, railroad tank cars, jumbo tube trailers, and aircraft fuselages.

Evaluation sometimes hinges on located clusters of events, sometimes on total emission quantities, and sometimes on emission in specific parts of the test, such as high-stress load holds. Often, the data is evaluated in accordance with several applicable acceptance criteria. Comparison of AE data with specific acceptance criteria serves as a screening process. If the structure meets the acceptance criteria, it is deemed acceptable. If not, the procedure specifies either another stage of AE data analysis or verification of the AE data with other nondestructive testing methods.

In less standardized applications, interpretation and evaluation are often conducted more implicitly. In interpretation, the technician uses common sense observations to remove false indications (noise), or to focus on the parts of the test that are determined to contain the most meaningful information. The final evaluation tends to emerge from a personal blending of data analysis with the experience, common sense observation, and technical intuition. These can be subjective processes and the degree of success depends very much on the skills of the individual technician. These skills are difficult to teach and communicate, and the processes actually used by an individual will change over the years as interests and preferences shift and as new ideas are considered. This style of operation is best seen as a necessary prelude to the development of standardized procedures that can be used widely and reliably.

Evaluation criteria are only meaningful when they are related to a specific instrumentation set-up and a specific load stimulus. In testing pressure vessels, tanks, etc., the load stimulus can be numerically specified, controlled, and measured by the technician.

## 10.8 Advantages and Limitations of Acoustic Emission Testing

In contrast with most other NDT methods, in AE testing, the discontinuity itself is the releaser of energy, making its own signal (in response to stress). AE testing detects *movement* (other methods detect *geometric discontinuities*).

Advantages of AE testing:

AE can be used in all stages of testing including:

- Preservice (proof) testing
- In-service (requalification) testing
- On-line monitoring of components and systems
- Leak detection and location
- In-process weld monitoring
- Mechanical property testing and characterization
- Material anisotropy is good
- (Less) geometry sensitive
- Less intrusive
- Global monitoring
- Real-time evaluation
- Remote scanning
- Performance/price ratio

Limitations of AE testing:

- Repeatability: Acoustic emission is stress unique and each loading is different.
- Attenuation: The structure under test will attenuate the acoustic stress wave.
- History: Tests are best performed if the loading history of a structure is known.
- Noise: Acoustic emission can be subject to extraneous noise.

The advantages and disadvantages of AE testing can be summarized as in the following table.

| Acoustic emission | Most other NDT methods |
|---|---|
| Discontinuity growth/movement | Discontinuity presence |
| Stress, damage-related | Shape-related |
| Material anisotropy is good | Material anisotropy is bad |
| (Less) geometry sensitive | (More) geometry sensitive |
| Each loading is unique | Inspections are readily repeated |
| Less intrusive | More intrusive |
| Global monitoring | Local scanning |
| Principal limitations: attenuation, history dependence, and noise | Principal limitations: access, geometry, and dependence on discontinuity orientation and proximity to surface |

## 10.9 Glossary of Acoustic Emission Terms

This glossary is presented in two forms. First, selected key terms are listed in logical order to show how standard terminology is used to describe the AE process from source to data storage. Second, a more complete glossary is presented in alphabetical order. Definitions for items marked with an asterisk are drawn from ASTM E 1316, and in some cases, reflect minor changes.

### Key Terms in Logical Order

**Acoustic Emission (AE)** Elastic waves generated by the rapid release of energy from sources within a material.

**Event (AE)** A local material change giving rise to acoustic emission.

**Sensor** A device containing a transducing element that turns AE wave motion into an electrical voltage.

**Signal** The electrical signal coming from the transducing element and passing through the subsequent signal conditioning equipment (amplifiers, frequency filters).

**Source** The physical origin of one or more AE events.

**Channel** A single AE sensor and the related equipment components for transmitting, conditioning, detecting, and measuring the signals that come from it.

**Detection** Recognition of the presence of a signal (typically accomplished by the signal crossing a detection threshold).

**Hit** The process of detecting and measuring an AE signal on a channel.

**Signal Features** Measurable characteristics of the AE signal, such as amplitude, AE signal energy, duration, counts, and rise time.

**Hit Data Set** The set of numbers representing signal features and other information, stored as a result of a hit.

**Event Data Set** The set of numbers used to describe an event, pursuant to data processing that recognizes that a single event can produce more than one hit.

### Terms in Alphabetical Order

**Acoustic Emission (AE)** Elastic waves generated by the rapid release of energy from sources within a material.

**Activation (AE)** The onset of AE due to the application of a stimulus such as force, pressure, heat, etc.

**Activity (AE)** A measure of emission quantity, usually the cumulative energy count, event count, ring-down count, or the rates of change of these quantities.

**Amplitude (AE)** The largest voltage peak in the AE signal waveform; customarily expressed in decibels relative to 1 microvolt at the preamplifier input ($dB_{ae}$).

**Amplitude Distribution** A display of the number of AE signals at (or greater than) a particular amplitude, plotted as a function of amplitude.

**Attenuation** Loss of amplitude with distance as the wave travels through the test structure.

**Burst Emission** A qualitative description of the discrete signal related to an individual emission event occurring within the material.

**Channel (AE)** A single AE sensor and the related equipment components for transmitting, conditioning, detecting, and measuring the signals that come from it.

**Continuous Emission** A qualitative description of the sustained signal level produced by rapidly occurring acoustic emission events.

**Counts** The number of times the AE signal crosses the detection threshold. Also known as "ring-down counts" and "threshold crossing counts."

**$dB_{ae}$** A unit of measurement for AE signal amplitude $A$, defined by $A(dB_{ae}) = 20 \log V_p$, where $V_p$ is the peak signal voltage in microvolts referred to the preamplifier input.

**Detection (AE)** Recognition of the presence of a signal (typically accomplished by the signal crossing a detection threshold).

**Event (AE)** A local material change giving rise to acoustic emission.

**Event Data Set (AE)** The set of numbers used to describe an event, pursuant to data processing, that recognizes that a single event can produce more than one hit.

**Event Description** A digital (numerical) description of an event, comprising one or more signal descriptions and/or information extracted from them or calculated from them.

**Event Energy (AE)** The total elastic energy (in the wave) released by an acoustic emission event.

**Felicity Effect** The presence of AE at stress levels below the maximum previously experienced. (The reverse of the Kaiser effect.)

**Frequency** For an oscillating signal or process, the number of cycles occurring in unit time.

**Guard Sensors** Sensors whose primary function is the elimination of extraneous noise based on arrival time differences.

**Hit (AE)** The detection and measurement of an AE signal on a channel.

**Hit Data Set** The set of numbers representing signal features and other information, stored as a result of a hit.

**Intensity (AE)** A measure of the size of the emission signals detected, such as the average amplitude, average AE energy, or average counts.

**Kaiser Effect** The absence of detectable acoustic emission at a fixed sensitivity level, until previously applied stress levels are exceeded.

**kHz** Kilohertz, an SI unit of frequency, 1,000 cycles per second.

**Location** Relating to the use of multiple AE sensors for determining the relative positions of acoustic emission sources.

**Noise** Nonrelevant indications; signals produced by causes other than AE, or by AE sources that are not relevant to the purpose of the test.

**Parametric Inputs** Environmental variables (e.g., load, pressure, temperature) that can be measured and stored as part of the AE signal description.

**Rise Time** The time from an AE signal first threshold crossing to its peak.

**Sensor (AE)** A device containing a transducing element that turns AE wave motion into an electrical voltage.

**Signal (AE)** The electrical signal coming from the transducing element and passing through the subsequent signal conditioning equipment (amplifiers, frequency filters).

**Signal Description** The result of the hit process: a digital (numerical) description of an AE signal and/or its environmental context.

**Signal Features** Measurable characteristics of the AE signal, such as amplitude, AE energy, duration, counts, and rise time, that can be stored as part of the AE signal description.

**Signal Strength (AE)** The strength of the absolute value of a detected AE signal. Also known as "relative energy," "MARSE," and "signal strength."

**Source (AE)** The physical origin of one or more AE events.

**Source Energy (AE)** The total energy (of all forms) dissipated by the source process.

**Zone (Primary)** The area surrounding a sensor from which AE can be detected and from which AE will strike the sensor before striking any other sensors.

## References

Miller, R. K. (Ed.), *Nondestructive Testing Handbook,* 2nd ed., Vol. 5, "Acoustic Emission Testing." American Society for Nondestructive Testing, Columbus, OH, 1987.

Pollock, A. "Inspecting Bridges with Acoustic Emission," Guidelines prepared for the U.S. Department of Transportation and Federal Highway Administration (FHWA), June 1995. Technical Report No. TR-103-12 6/95. Physical Acoustics Corp., Princeton Junction, NJ 08550.

# CHAPTER 11
# Digital Radiography

Before the subject of digital radiography can be discussed, the history of filmless radiography should be considered. In the late 1950s and 1960s, several companies, manufacturers of radiation generating equipment, as well as gamma sources, had a need for developing filmless imaging technology, based on particular applications that rendered analog X-ray film imaging technology cost prohibitive, impractical, or simply capacity limited. Fluoroscopic (real-time) inspection systems were developed in response to these needs.

With all radiographic systems, one must bear in mind the three basic elements of all radiographic image:

- Contrast resolution
- Spatial resolution
- Signal-to-noise ratio (SNR)

Early fluoroscopic systems consisted of a radiation source (X-ray or gamma-ray) that produced an image on a phosphor screen. Images were viewed by directly observing the screen itself or with the use of a folding mirror configuration as illustrated in Figs. 11-1 and 11-2, due to radiation safety concerns.

These systems were widely used as process control devices and in some cases for final inspection. Typically they lacked the spatial resolution and SNR requirements of film-based techniques. These systems, however, did satisfy specific design criteria requirements and were application-sensitive.

As this technology developed, closed-circuit television (CCTV) systems were integrated into fluoroscopic screen systems, allowing higher radiation energies to be used and enabling inspection personnel to review the resultant images remotely. These systems still lacked film quality as there was a loss of spatial resolution through the video chain, and the SNR decreased due to the "real-time" multiple frames per second format of the video chain. See Fig. 11-3. However, these systems did satisfy selected blueprint requirements, were application-specific, and were widely used in many industries.

In the late 1970s and throughout the 1980s, real-time X-ray systems became a very popular solution for many RT inspection tasks. The primary imager was the X-ray image intensifier (see Fig. 11-4), which served as a conversion device converting the X-ray image using a rare earth phosphor encased in a vacuum. These devices were circular tube configurations employing input phosphors of varying diameters, which electronically focused the image onto an output phosphor of a significantly smaller

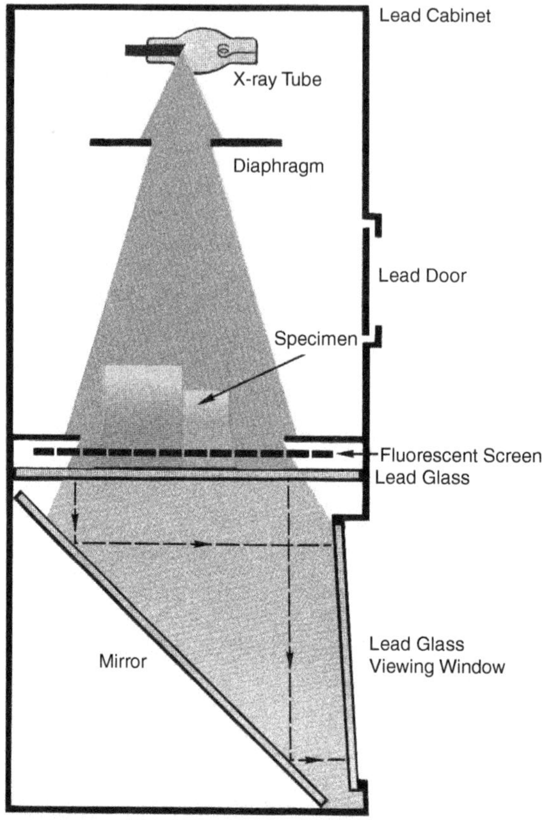


**FIGURE 11-1** Direct view fluoroscopic system.

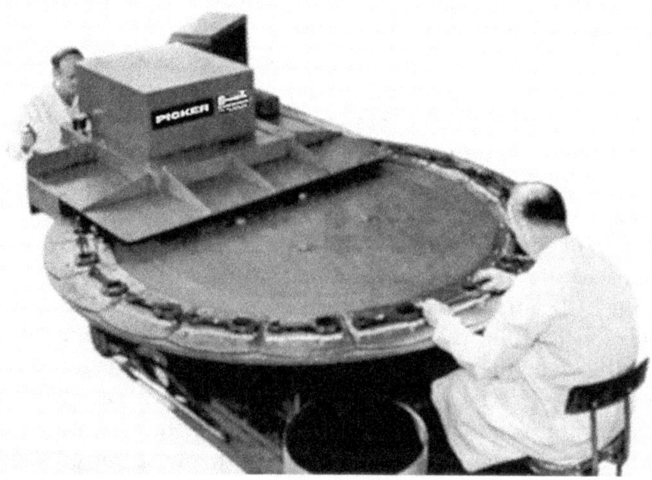


**FIGURE 11-2** Early production direct view fluoroscopic system. (*Courtesy of Picker Corporation.*)

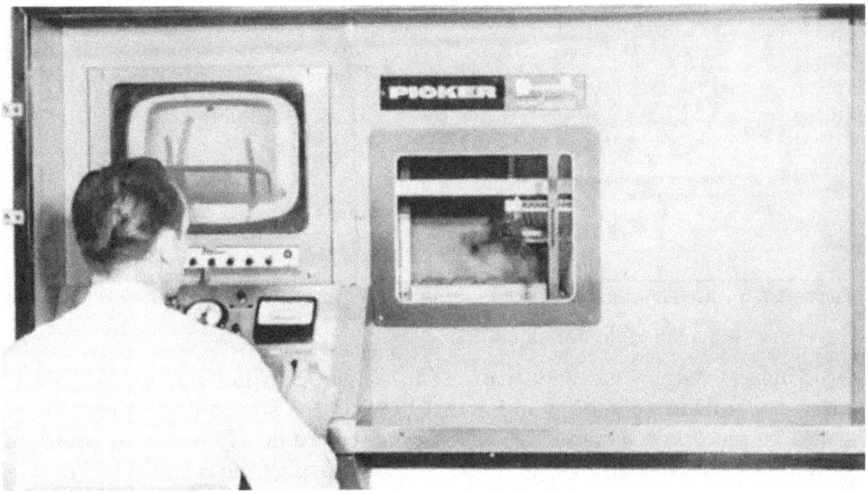


**FIGURE 11-3** CCTV fluoroscopic system. (*Courtesy of Picker Corporation.*)

diameter, resulting in gain factors. Typically a nine-inch input phosphor projecting on to a one-inch output phosphor could result in a gain of 10,000 units of light intensity.

One particular advantage of the X-ray image intensifier was the increase in the units of light intensity, its finer phosphor structure, and improved spatial resolution. The output phosphor could be easily coupled to various closed-circuit video chains. Limitations were again loss of spatial resolution through the video chain and limitations of the dynamic range due to video characteristics.

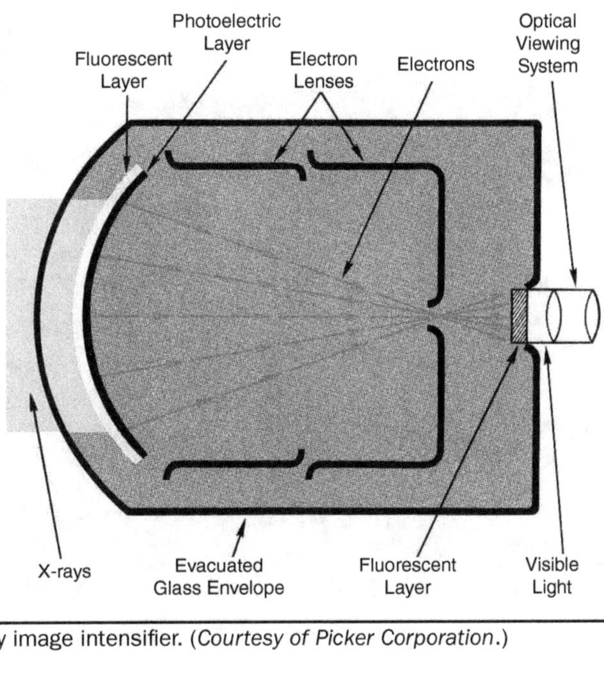


**FIGURE 11-4** X-ray image intensifier. (*Courtesy of Picker Corporation.*)

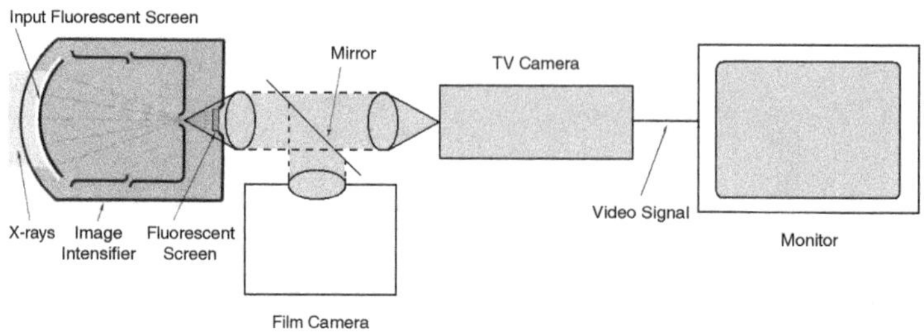


**Figure 11-5** Image intensifier coupled to a video system. (*Courtesy of Picker Corporation.*)

These systems were widely used in many applications. Inspection personnel could view dynamic images, enabling many views of the specimen in a real-time mode. Parts could be placed on a conveyor, or manipulated remotely. Viewing could be performed remotely as illustrated in Figs. 11-5 and 11-6. Suspect parts could be culled. Parts could be examined quickly and at high volumes.

It should also be noted that closed-circuit video systems were improving and also used with scintillator screen systems.

The real-time fluoroscopic systems had a significant role in further developing the filmless radiography efforts. However, they did lack true film quality with respect to the images obtained. Contrast and resolution were acceptable in many cases; however, spatial resolution was approximately half of that of analog film and the SNR was marginal.

To improve the spatial resolution of these systems, microfocus X-ray systems were developed using what was close to a point source X-ray tube. Some of these tubes employed vacuum X-ray tube technology as well as an "open" tube design, which could be constantly evacuated while in use, and also be capable of being

**Figure 11-6** Real-time X-ray system. (*Courtesy of TFI Corporation.*)

**FIGURE 11-7** Microfocus system tube head. (*Courtesy of X-Tek Corporation.*)

dismantled to facilitate the replacement of targets and filaments (see Fig. 11-7). Descriptions of this type of X-ray tube technology were labeled as "micro" or "mini" focus tubes depending on their effective focal spot sizes, which were typically 100 microns down to 10 microns.

The resultant images could be superior to films with regard to spatial resolution; however, the field of view had to be sacrificed as shown in Fig. 11-8. In many cases the field of view was reduced to between one and five inches to achieve the minimum requirements of the film quality image. The limited field of view could be compensated for by employing remotely controlled manipulating devices, some robotically controlled to assure 100 percent volumetric coverage of the specimen being examined.

Having overcome the spatial resolution issue, the SNR had to be improved. In the mid-1980s, systems were developed to capture the multiple frame image format of the CCTV systems used and to integrate those frames, improving the SNR (see Fig. 11-9).

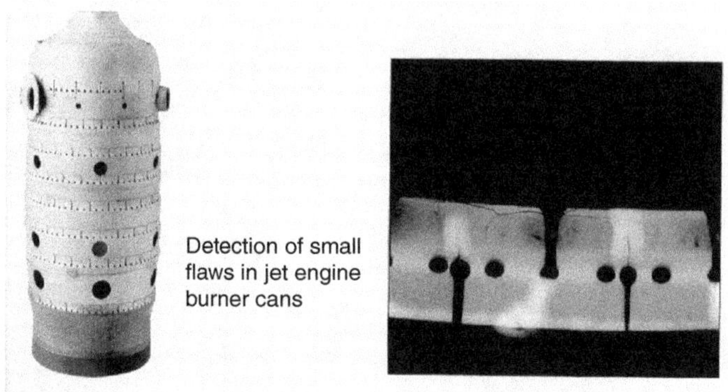


**Figure 11-8** Microfocus image. (*Courtesy of IRT Corporation.*)

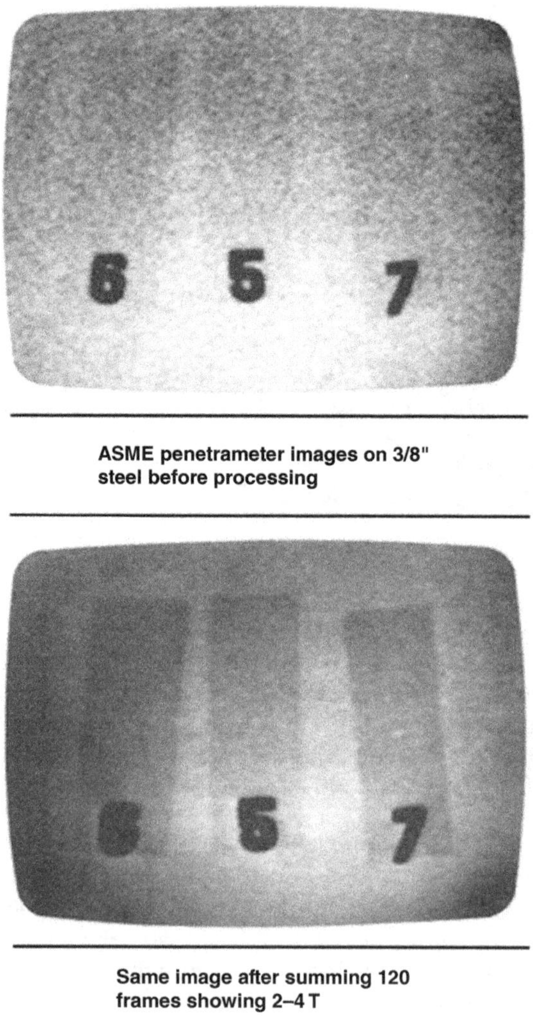


**Figure 11-9** Improved SNR by frame integration. (*Courtesy of Quantex Corporation.*)

Frames could be sequentially integrated, emulating the integration of information performed by film over time. Frames could also be recursively averaged, employing refreshed frames while replacing previously averaged frames, enabling the specimen to be moved and inspected dynamically. The only provision was that the speed of movement could not exceed the speed of the process. Naturally, the greater the degree of averaging (number of frames involved), the slower the manipulation of the product.

This was the beginning of the digital radiography age.

## 11.1 Digital Radiography Evolves

To completely comprehend digital radiography, the industrial radiographer must be familiar with the basic concepts of digital images, terminologies, file formats, and file storage.

The basic elements of the radiographic image are once more addressed here.

### 11.1.1 Spatial Resolution

Digital images are a matrix of points of light normally referred to as pixels. The pixels are arranged in a raster matrix to form the image. The pixel is the smallest addressable image element and the smallest that can be controlled. Every pixel in the image has its own gray-scale value, and $X$–$Y$ coordinates. The number of pixels in any given image is defined as the *spatial resolution* of the image. The number can be derived by simply multiplying the number of pixels per row by the number of rows in the image itself. Other methods of determining the spatial resolution of the image are the number of pixels per linear length or given areas. See Fig. 11-10.

### 11.1.2 Contrast Resolution

Contrast resolution is established by the "bit depth" of the image itself. The combination of spatial resolution and bit depth determines the size of the image file. Bit depth determines the number of the shades of gray in the image, which in turn establish the

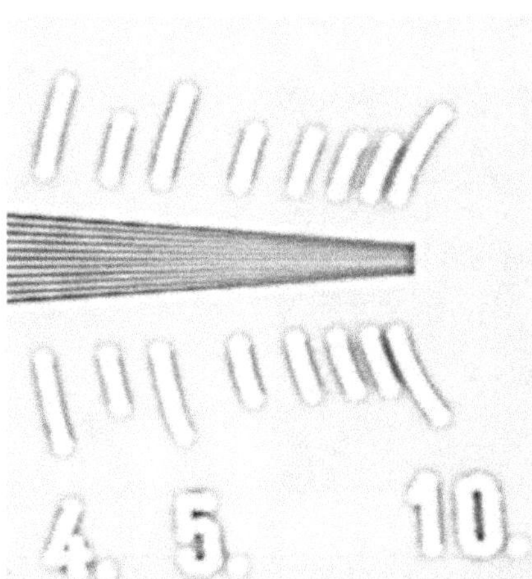


**Figure 11-10** Line pair gage (spatial resolution). (*Courtesy of Fujifilm Corporation.*)

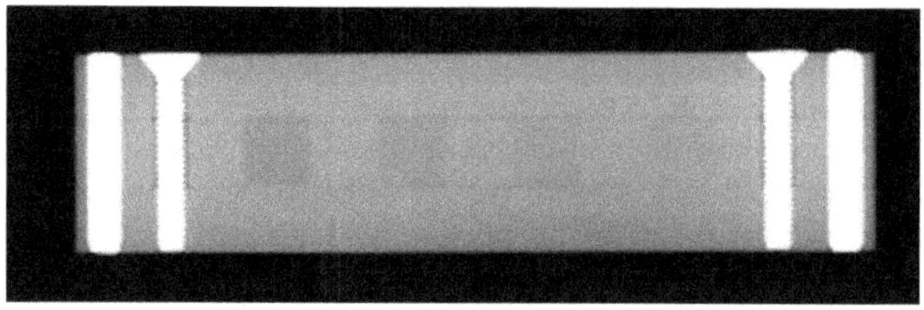

**FIGURE 11-11** Contrast resolution gage. (*Courtesy of Fujifilm Corporation.*)

contrast resolution. It should be noted that industrial radiographic films are classified by contrast resolution. Contrast is an equally important element as is the spatial resolution of the digital image. See Fig. 11-11.

### 11.1.3 Signal-to-Noise Ratio

As this is typically determined by the Hurter and Driffield sensitometric curve (H & D) (density of the analog film image), methods have been established to determine the SNR of the digital radiographic image. This method is an important tool utilized in the viewing software of the image viewer. See Fig. 11-12.

### 11.1.4 File Format

The recommended file format for digital radiographic images is DICONDE (Digital Communication Nondestructive Examination). Other file formats are available such as BMP, JPEG, TIFF, PNG, GIF, and JPEG 2000. DICONDE is the preferred format as it is an iteration of DICOM, the preeminent file format used in medical radiographic image file formats; it has distinct advantages, such as larger text headers enabling the storage of key information such as exposure parameters, exposure geometries, and image demographics, and it is written in a fashion to allow expedient image storage and retrieval. Employing the DICONDE format also allows the exchange of image files between different digital radiographic systems, as well as some alternative digital images other than radiographic images. DICONDE also assures the end user that the method of image storage and display is backward compatible.

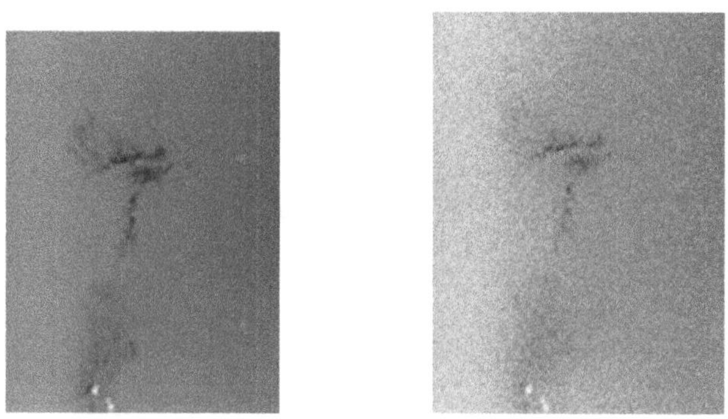

(a) (b)

**FIGURE 11-12** Good SNR and poor SNR. (*Courtesy of Fujifilm Corporation.*)

Digital radiographic imaging systems in use are primarily computed radiography (CR) systems and digital detector array (DDA) systems. DDA systems are also referred to as "DR" systems.

Both of these technologies were originally developed exclusively for medical applications. CR has been in use medically since the early 1980s. DDA systems have been in use since the late 1990s. Both technologies have significantly reduced the consumption of medical X-ray film.

Computed radiography was the first technology to be used for industrial radiographic examination. Because of its rectangular format and ease of use, CR systems were widely accepted as a technology capable of replacing many fluoroscopic-type systems. Flat-panel DDA systems were first used in the late 1990s. Both systems were application sensitive. Both systems were not true film replacement systems for the more demanding applications, which were code controlled and had to meet the three basic image elements of film quality imagery.

Since the introduction to the NDT community, both technologies have vastly improved with regard to spatial resolution and SNR. Development of faster computer processing speeds, finer-grained conversion screens, and more efficient image generation, transfer, and storage have led to many digital radiography systems capable of being utilized as true "film replacement" technology.

### 11.1.5 Computed Radiography (CR)

Introduced to the NDT community in 1990, CR had already been in use by the medical community since the mid-1980s. CR was the first truly digital radiographic technology.

Computed radiography is the oldest technology in the true digital acquisition modality. It employs "storage phosphor" technology. The key elements of the CR system are the imaging plate (storage phosphor), reader unit, which serves to extract the image from the phosphor, the imaging software, and the review station, utilizing viewing software capable of applying various enhancements and providing many tools or aids to the interpreter.

Imaging plates are used in the same fashion as X-ray film as illustrated in Figs. 11-13 and 11-14. They can be used with a variety of cassettes and lead screens. Imaging plates are flexible and are available in a variety of speeds, sizes, and shapes. Imaging plates can be erased and reused. The number of cycles depends on the radiation energies, the environment of the inspection area, the handling of the imaging plate, and the CR system reader itself.

The reusable imaging plate (IP) replaces film as a sensor. It is sensitive to X-ray, gamma ray, and high-voltage radiation. When exposed, a latent image is produced in the IP. The latent image luminesces when the IP is scanned and excited by the reader's laser, which is then accumulated by a collection device. At that time a digital file is created containing the image, and all related demographics. See Fig. 11-15.

The digital file is exported to a computer workstation loaded with viewing software coupled to a high-resolution monitor where the image can be reviewed, enhanced, categorized, and stored. Various storage methods can be used ranging from resident, external, and network storage.

To be considered as true film replacement, CR systems have been classified in ASTM E2446. However, it is possible that a particular imaging plate coupled to a working CR system can meet all imaging plate classifications. Considerations determining

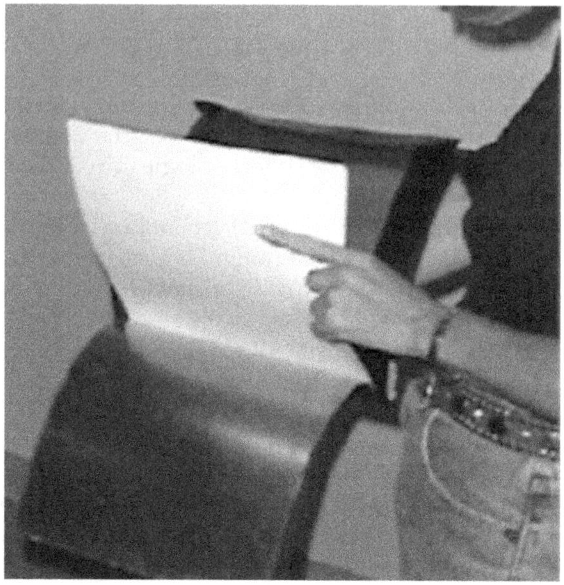

**FIGURE 11-13** Example of an imaging plate. (*Courtesy of Fujifilm Corporation.*)

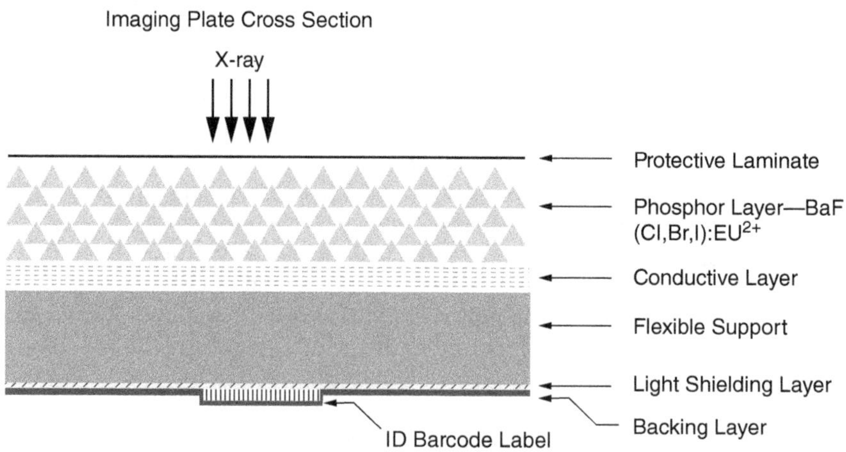


**FIGURE 11-14** Composition of the imaging plate. (*Courtesy of Fujifilm Corporation.*)

the classification of CR systems are SNR; spatial resolution, which is based on laser spot size along a given raster line; and the subsequent sampling along that given line. This is typically expressed by the number of pixels per millimeter. The accuracy of the translational movement, as well as the contrast resolution established by bit depth, are also key factors.

CR reader units can be produced in a variety of sizes ranging from floor models to tabletop models of various sizes and weights. See Fig. 11-16. CR use results in minimal change to the established radiographic process. CR systems are available

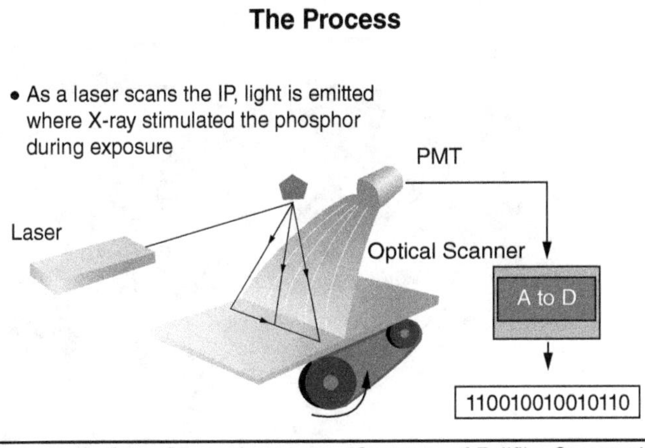


**FIGURE 11-15** Illustration of the CR read process. (*Courtesy of Fujifilm Corporation.*)

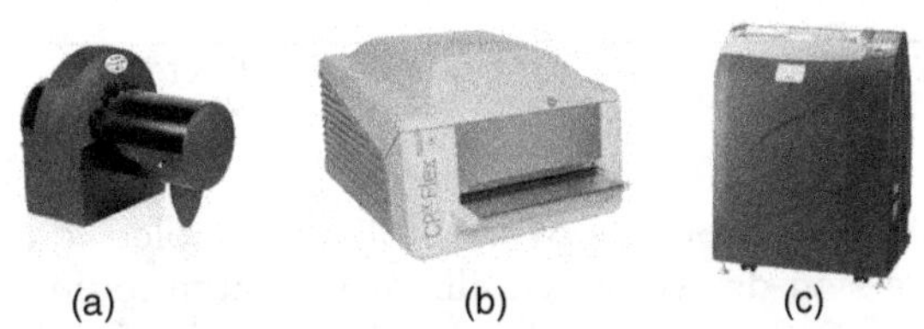


**FIGURE 11-16** CR reader units. [*Courtesy of (a) Durr Corporation, (b) General Electric Inspection Technologies, (c) Fujifilm Corporation.*]

with selectable pixel sizes ranging from 12.5 to 200 microns, which would fulfill almost any RT application.

It should be noted that another term is used when employing any conversion device, whether it be storage phosphors or fluorescent conversion screens. Quantum detection efficiency (QDE) is the ability to absorb the radiation as it passes through the specimen. In many cases, the radiation passes through the specimen without contributing to image formation. This QDE factor or capture efficiency has been a consideration with all filmless technologies. The thinner the capture media, the less the efficiency. However, the thinner the media, the greater the spatial resolution. As media thickness increases, capture efficiency improves; however, the spatial resolution suffers due to the internal scatter of the media itself.

The digital detector array (DDA, often referred to as DR) was introduced employing direct conversion screens coupled to amorphous panels as illustrated in Fig. 11-17. ASTM E2597 classifies all electronic devices capable of converting penetrating radiation into discreet signals, which are subsequently digitized and exported to a digital viewer software package installed in a computer workstation. In many cases, a scintillating material or fluorescent screen is utilized as the conversion media. There is also DDA technology that does not use a scintillator or conversion device, instead using discreet elements, directly converting radiation energies to electronic signals.

This technology includes but is not limited to linear diode arrays (line scanners), amorphous panel detection systems (selenium and silicon), and complementary metal-oxide semiconductor (CMOS) systems.

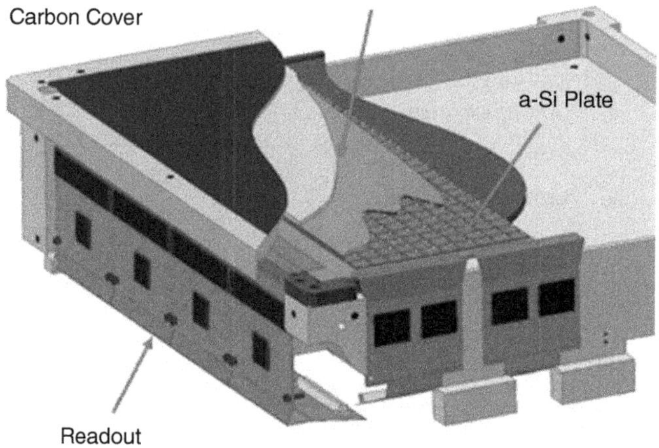


**FIGURE 11-17** Amorphous panel. (*Courtesy of General Electric Inspection Technologies.*)

The "DDA" panel consists of a scintillator coupled to an amorphous array. Typically the conversion device or scintillator is cesium iodide, which was used in most X-ray image intensifiers. The cesium iodide absorbs the X-ray photons and converts them to light, which is then sensed by the amorphous silicon panel, a matrix of sensors absorbing the light, converting it to electronic signals. Each photodiode in the amorphous panel represents a pixel in the image file. As the panel is charged and discharged, near real-time rates can be achieved, resulting in dynamic images.

A distinct advantage of the panel as compared to the X-ray image intensifier, besides being a truly digital imager, is that the entire area of the panel can be used to display images. Both CR and DDA fields of view are useable areas for display and interpretation, while the X-ray image intensifier is graded by center resolution. Being a circular "bulb" type of imager, images distort from the center to the outer edges. An example of this type of distortion was commonly known as the "pin cushion" effect.

The CMOS detector is also used in photographic and imaging sensor applications. CMOS detectors also use a conversion screen or scintillator, coupled to an array or matrix of detectors. Advantages of CMOS are that the detector arrays can be manufactured in various configurations, can be designed for high resolution (50 microns or smaller), and are not as susceptible to the effects of scattered radiation.

Linear diode array (LDA) systems commonly known as line scanners were primarily designed for security applications. The LDA is also less affected by scattered radiation, and is capable of being configured with various radii for curved specimen applications. The LDA is also an efficient conversion device requiring less radiation dose to produce usable images. LDA technology is also used in conveyor transport systems, for high-volume inspection applications.

Application dependency is the governing factor in the selection of any filmless technology. There are advantages and disadvantages to each one. Selection of the correct inspection technology for any given RT task is critical.

### 11.1.6 Selecting Digital Radiography Technologies

Computed radiography (CR) can be used as a direct film replacement. The imaging plates can be fabricated in various sizes, require no power, and are adaptable to rugged environments. Imaging plates are low-cost replacement imagers and can be used for high-radiation energy applications. Imaging plates are available in various speed configurations, which in many cases require significantly less radiation dose as compared to film.

In applications where the smaller pixel size is selected, image file sizes must be considered. A typical 14-in. by 17-in. imaging plate read at 50 micron resolution would produce a file size of approximately 130 Mbytes. However, advancements in computer processing capabilities have facilitated handling and storage of files of that size.

CR is a two-step process, requiring the placement of a cassette to image the product. Considerations must also be given to handing of the imaging plates, reader unit maintenance, and service.

DDA systems are an excellent replacement technology for the previous generations of analog imagers such as X-ray image intensifiers and fluoroscopic screen systems, as well as direct film replacement for specific applications. DDA systems can be coupled to X-ray system controller devices, eliminating the need to manually select X-ray energies. DDA systems can average collected data, integrating that data to increase SNR. DDA systems are a one-step process, producing images on completion of the radiation exposure, without any intermediate steps.

DDA systems are limited in size and field of view. They are rigid and cannot be bent or curved (with the exception of LDA systems). DDA systems are limited regarding radiation dose and must be tethered to power and data cords.

Pixel size availability ranges from 200 microns to approximately 100 microns. In some medical applications, 50 micron DDA systems are available, with limited field of view (input size of panel). In many cases, DDA systems require specialty "micro" or "mini" focus X-ray tube-heads to perform geometric enlargement techniques to achieve the required spatial resolution. Using geometric enlargement techniques limits the effective field of view of the imager, requiring that the product be remotely manipulated to adequately provide 100 percent volumetric coverage of the specimen. In some cases, the product can be manually positioned about the DDA panel; however, this defeats the concept of the one-step process.

Consideration must be given to RT applications that utilize all the area of the film package to examine quantities of smaller parts (e.g., some aircraft turbine blades). Imaging plates can be used as a true film replacement without any geometric enlargement techniques, enabling the radiographer to obtain full utilization of the area of the imaging plate, while the DDA panel would require the use of smaller fields of view, limiting the number of parts, volume, and in many cases still requiring some type of remotely controlled part manipulation.

The DDA is not as portable as the CR imaging plate and is more sensitive to radiation dose damage. Replacement cost of the DDA panel can be very expensive.

Figure 11-18 describes the salient characteristics and technological differences with respect to CR, DR or DDA, and analog X-ray film.

### 11.1.7 Digital Radiography Systems: Image Displays

The purpose of the display is to convey the information obtained by the imager to the radiographic inspector. Spatial resolution of the display, as well as the environment of

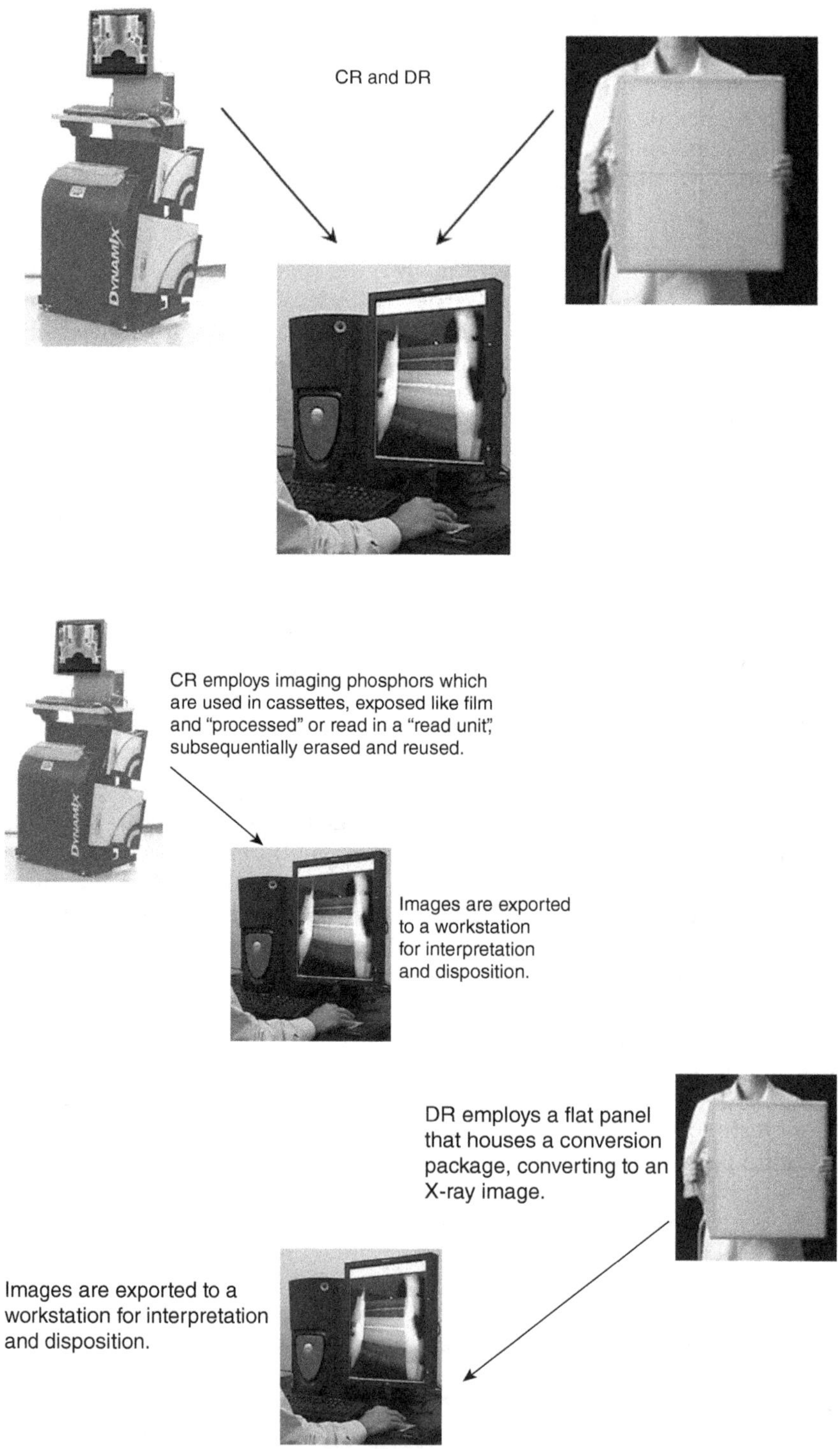


**FIGURE 11-18** Comparison of digital and analog systems. (*Courtesy of Fred Morro.*)

CR has always consisted of four major elements

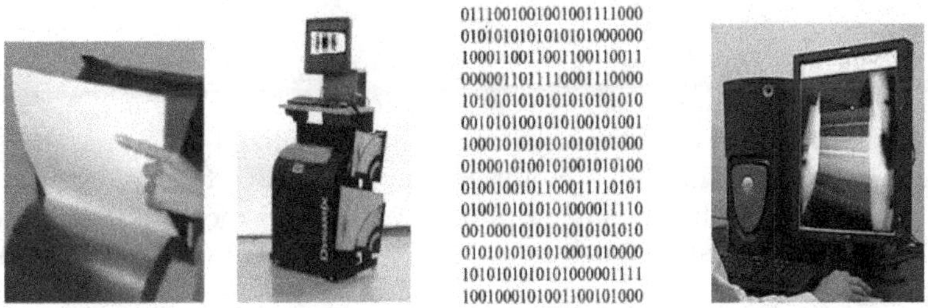

| 1 | 2 | 3 | 4 |
|---|---|---|---|
| Imaging Plates "IP's" | Reader Unit "RU" | Software | Workstation |

DR has always consisted of three major elements

| 1 | 2 | 3 |
|---|---|---|
| Imaging Plates "IP's" | Software | Workstation |

| Feature | CR | DR | Film |
|---|---|---|---|
| Resolution | Variable 50–100 m | Fixed 100 m* | Fixed-Speed |
| Steps | 2 Step | 1 Step | 2 Step |
| Flexibility | Yes | No | Yes |
| Lead Screens | Yes | No | Yes |
| Special Shapes | Yes | No | Yes |
| Longer Than 17 in. | Yes | No | Yes |
| Variable Contrast | Yes | Yes | No |
| Variable Gray Scale | Yes | Yes | Maybe** |
| Software Features | Yes | Yes | No |
| Support Multiple Cells | Yes | No | Yes |
| Special Focal Spot | No | At Times | No |
| Custom System | No | At Times | No |
| Artifact Suppression | No | Yes | No |
| Burn in 1–10 *** | 2 | 7 | 0 |
| Pixel Issue | No | Yes | No |
| Tethered | No | Yes | No |
| Benign | Yes | No | Yes |
| Portable | Yes | No | Yes |
| Real Time | No | Yes | No |
| Delicate 1–10 *** | 4 | 10 | 4 |
| SNR | Good | Good | Very Good |
| Acceptance | Yes | Yes | Yes |
| Available | Yes | Yes | Yes |
| High Energy | Yes | Yes **** | Yes |
| Cycles | Many | Many | 1 |

*Some smaller panels have better resolution
**Some analog films have less contrast
***Lower values are less susceptible, higher values are worst case
****May require special enclosures to protect electronics from higher energies

**Figure 11-18** *(Continued)*

the inspection area (especially ambient lighting), is important. Assurance of display calibration and stability are key considerations.

Early digital radiography systems initially utilized video monitors to display the resultant images. The video monitors employed CRT technology and were commercially available with inherent resolution of 1 to 1.5 megapixels. Two-megapixel CRT systems became available before the entire technology was converted to the flat-panel display.

In the mid-1990s, flat-panel displays replaced CRT technology. The flat-panel display was significantly lighter in weight and exhibited superior resolution capabilities.

The display is a critical element of the digital radiography system. The most resolute imager available would be limited by the display capabilities of the systems.

Most display panels are gray scale (monochromatic). Key elements of all displays are resolution and bit-depth. There are currently LCD displays, which are the most prevalent, and LED displays. Most displays exhibit an 8-bit bit-depth. Some displays are available with 10- and 12-bit bit-depth. Higher bit-depth exhibits smoother transitions of the gray scales in the view; however, the eight-bit display has proven to be more than adequate. Display resolution is the more significant characteristic of any of the displays currently used in digital radiography applications. The higher the display resolution, the better the image sharpness of the specimen itself as well as any inherent flaws.

Quality assurance considerations include the use of the Society of Motion Picture and Television Engineers (SMPTE) pattern to test the display contrast, spatial resolution, and linearity. There are also tests for brightness and geometric distortion. There are several approved methods for quality assurance tests for the display.

### 11.1.8 Digital Radiography Systems

Once the image file has been exported to the system workstation, interpretation, and product disposition can be performed.

Typically the system workstation consists of a high-capacity CPU, high-resolution display, local storage capability, and extensively configured viewing software. See Fig. 11-19.

The viewing software should be DICONDE compliant.

## 11.2 Software

Software is an equally critical element, if not the most critical element in any digital radiography system. Software enables the digital system to emulate analog films. This is the main intent of digital radiography.

There are various tasks assigned to the software package(s) that are associated with any digital radiography system.

1. Operator-to-machine interface is the basic controller of the system. This particular application can serve as the system controller and is available in several versions ranging from a basic manual activation, to completely automated versions. The system controller can be configured to enter image demographics, select spatial resolution settings, establish erasure dwell times, preapplying specific algorithms, determining the destination of the image file, as well as a variety of other pertinent functions. All DR systems offer software packages to perform several of the described functions.

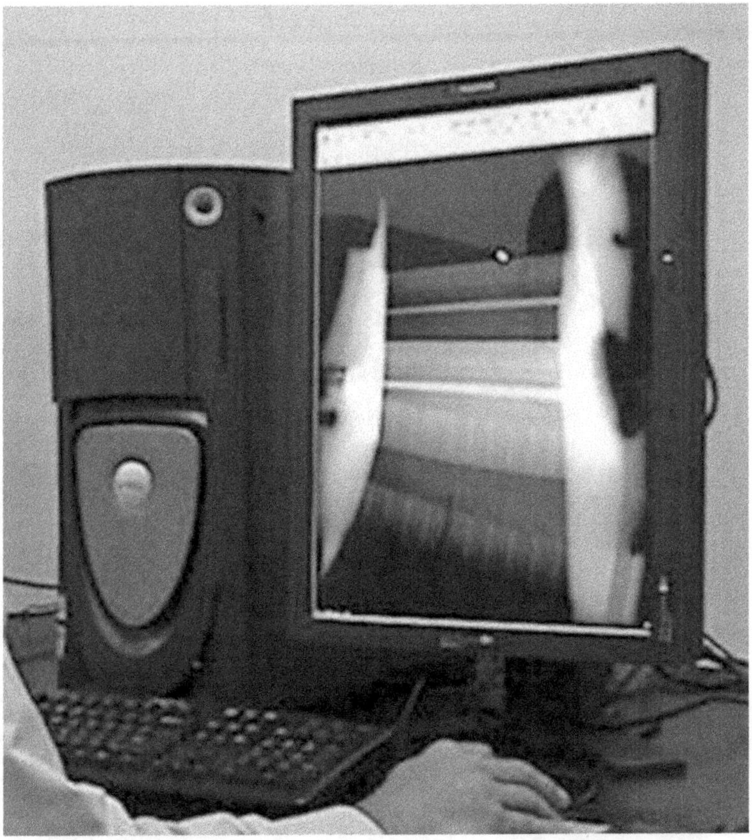

**Figure 11-19** Digital radiography workstation. (*Courtesy of Fujifilm Corporation.*)

2. Image formation software is resident to all digital radiography systems. The software generates an image to be viewed on the system viewer/workstation. Specific software algorithms are applied to the image file that generate various values for contrast levels, gray scale, various linear and nonlinear filters, dynamic range adjustments, application of various linear and nonlinear curves treating the gray-scale values in a variety of fashions to emulate true, and in some cases, improved film quality images.

   There are several algorithms available and used to enable the digital RT image to approach, and achieve analog film quality. Many apply mathematical calculations such as summing sequential frames, integrating the information with respect to "real time and near real time" applications.

   Other algorithms filter gray scales by thresholding; histogram-equalizing gray scales; sharpening edges; mathematically treating gray scales by summing, subtracting, multiplying, and averaging gray scales. Of particular importance is the gray-scale histogram equalization of the image file, which essentially histograms the gray scales, and selectively equalizes the values resulting in improved contrast; some software packages can perform this task automatically. See Fig. 11-20.

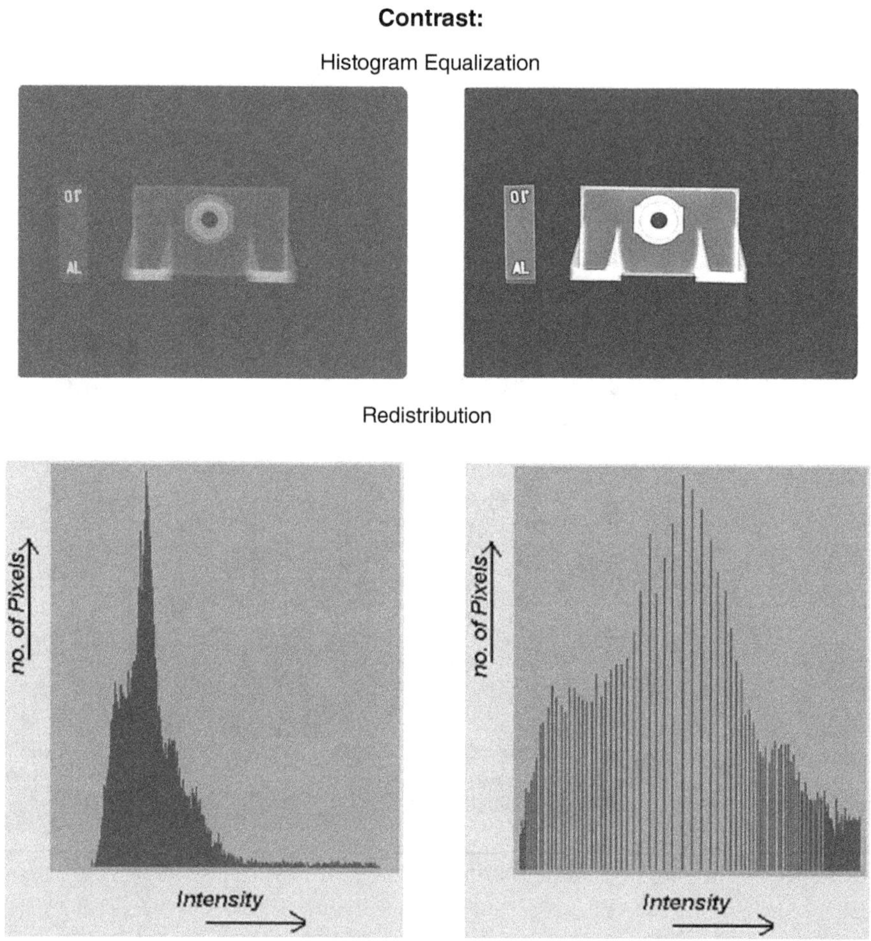


**FIGURE 11-20** Examples of gray-scale histogram equalization. (*Courtesy of Fujifilm Corporation.*)

3. Imaging "Tools" are available to adjust the displayed image, add commentary, determine image quality, and perform various export and conversion functions. Several codes and standards, including ASTM and various military and aerospace standards, have been developed that describe the specific tools required to determine the quality assurance of the displayed image itself as well as specific tools to assist the interpreter of the digital image with the disposition of the items being inspected.
4. Inventory control software is a part of digital radiography systems. DICOM and DICONDE file formats do have specific requirements for inventory control, ranging from "search" functions by date, part identification, image interpreter, and storage location.

## 11.3 Tools

Because the images are in a digital format, various tools can be employed to assist the interpreter to perform a variety of functions. Some of the tools are required while other tools can be used at the discretion of the Cognizant Engineering Organization and the responsible RT Level III. The following are examples of the tools that are available on several digital radiography systems. Some are identified as those required by various standards, and some of the tools described are important "assists" to the interpreter. It should be noted that requirements vary from industry to industry, and the newly developed standards address these requirements.

### 11.3.1 Digital Radiography Software Tools

In many standards some of these tools are an absolute requirement. Others are an assist to the image interpreter, enabling some functions to be performed that can be considered the "digital equivalent" to physical tools used during the interpretation process, as well as some new approaches. Some of the more commonly available tools are explored here.

#### *Window/Level Tool*

This particular tool enables the interpreter to adjust the gray scales of the image employing two different activities. Window adjustment is a contrast adjustment. Images can be made to display changes in the relationship between light and dark areas of a displayed image to assist the interpreter as illustrated in Figs. 11-22 and 11-23. Level is a change in the appearance of gray-scale values lightening or darkening specific areas of interest.

The concept of windowing is a newer concept, as analog film has a fixed contrast value. Leveling is the digital equivalent of adjusting the transmitted light through film typical in a film illuminator. These two functions are the most commonly used tools on all digital imaging modalities. Because of the greater dynamic range (ability to display a wider range of gray scales), leveling can in many cases eliminate the practice of employing multiloaded film cassettes (see Fig. 11-21).

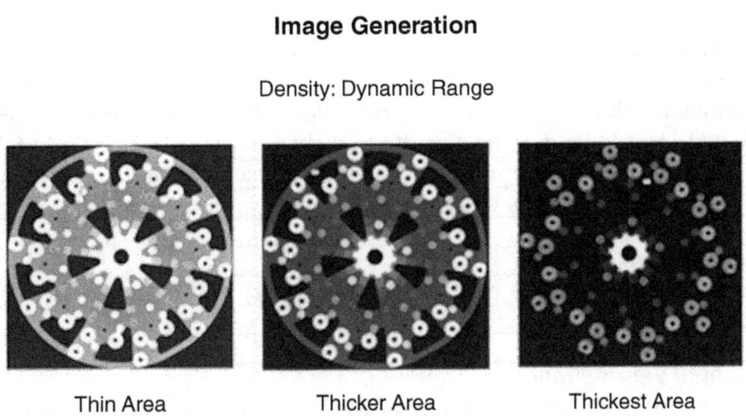


**Figure 11-21** Gray-scale leveling. (*Courtesy of Fujifilm Corporation.*)

**Figure 11-22** Low-contrast image. (*Courtesy of Fujifilm Corporation.*)

**Figure 11-23** High-contrast image. (*Courtesy of Fujifilm Corporation.*)

### *Magnification Tool*

This tool is the digital equivalent of the eye loupe or magnifying glass. Entire scene displays can be variably magnified to the pixel level of the image file. Most full-screen magnification tools also employ a roam or scroll feature as part of the tool to ensure the entire scene can be examined in a magnified mode. ROI (region of interest) magnification tools employ a user selectable size of a specific area of interest with a fixed or variable magnification factor, to emulate the common eye loupe used to interpret specific areas of interest of the image. Some ROI tools have the ability to window and level in the ROI, as well as displaying various measurement grids and scales as are currently available with eye loupes. The magnification tool is very commonly used (see Fig. 11-24).

### *Split-screen Tools*

The split screen enables more than one view of a product while comparing the quality of various products on one monitor display. Many software packages enable four or more windows to be viewed simultaneously. Picture-in-picture also falls into this category of tools. The latest practice of this tool enables the interpreter to view digital reference radiographic images to assess and disposition indications (see Fig. 11-25).

### *Thumbnail View Tools*

Thumbnail presentations of the images selected by the interpreter for review can be displayed. This can be seen in Fig. 11-26.

### *Sequential Viewing Tool*

The sequential viewing tool allows the interpreter to sequentially index the views selected for interpretation, which is especially convenient when viewing a study with many views.

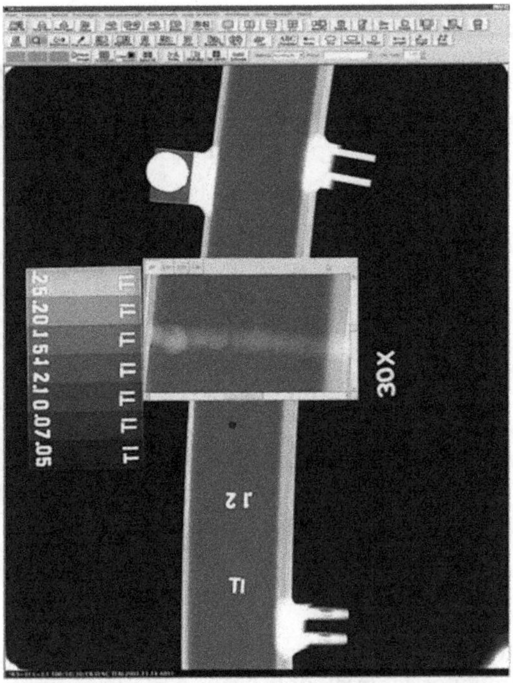

**FIGURE 11-24** ROI magnifier. (*Courtesy of Fujifilm Corporation.*)

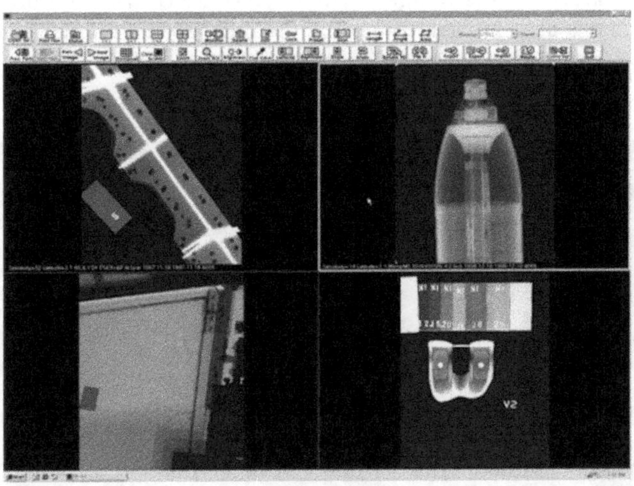

**FIGURE 11-25** Split screen images. (*Courtesy of Fujifilm Corporation.*)

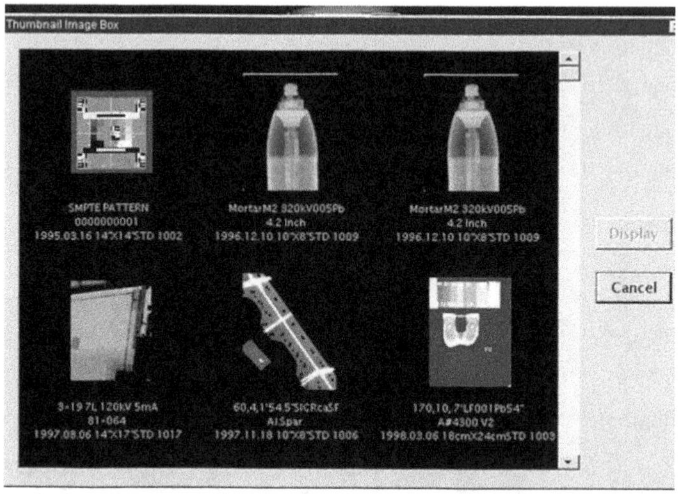


**Figure 11-26** Example of thumbnail views. *(Courtesy of Fujifilm Corporation.)*

### *Measurement Tools*

Measurement tools encompass a variety of functions ranging from linear measurement to the measurement of contrast, gray scale, and quantification of data (see Fig. 11-27).

Physical measurement can be performed by linear, angular, area, and volumetric functions in a given area of interest. Linear functions can be used to measure discontinuity length, automatically analyze wall thickness characteristics in vessels or tubing, clearance tolerance in mechanical assemblies, as well as a variety of other critical tolerances in various products. Thickness estimation tools can be used to determine a thickness of a given area of the item being examined. The volumetric and angular measurements can also be employed to accomplish these tasks. All measurement data can be displayed on the image in the form of an overlay, without affecting the image data which has been superimposed by the measurement data. The data can be saved as part of the image file.

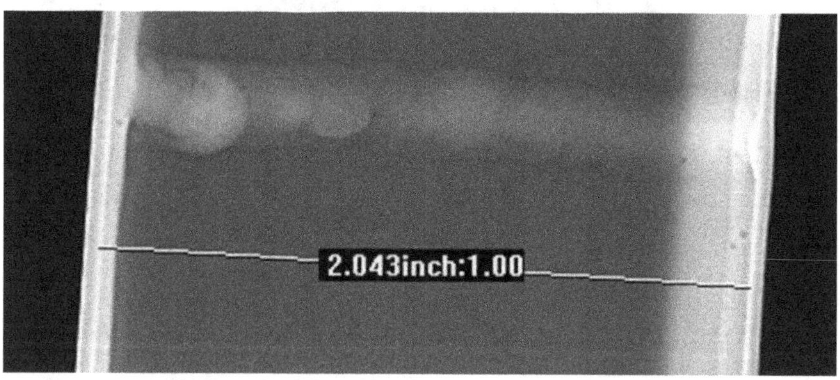


**Figure 11-27** Example of linear measurement tool. *(Courtesy of Fujifilm Corporation.)*

Geometric magnification tools can compensate for any geometric enlargement to obtain greater accuracy of measurements. A geometric enlargement factor can be applied to adjust any measurements.

Gray-scale and quantification tools can be employed to determine individual or averaged clusters of pixel values in the bit-depth range of the image itself. Profiles of the gray scales can be generated to display the actual graphical representation of a given linear series of lines of the different values of the pixels. For instance, a line can be drawn through an MTF gage or of a plaque penetrameter "T" hole to obtain a profile of the gray scales along the line to confirm the conspicuity of the hole or the spatial resolution of the gage. See Fig. 11-28.

Histogram measurement tools can also be employed to obtain quantitative results of the pixel values along a given line or area of interest.

As gray-scale leveling can be performed, changing the perceived gray value displayed after leveling, a measurement of the "digital driving level" can be obtained. This can be used to map the displayed gray scale after leveling to make comparisons between two or more images, and to perform a check of the particular characteristics of a display monitor.

All graphic representations should be superimposed on the image, not destroying the information being covered by the representations.

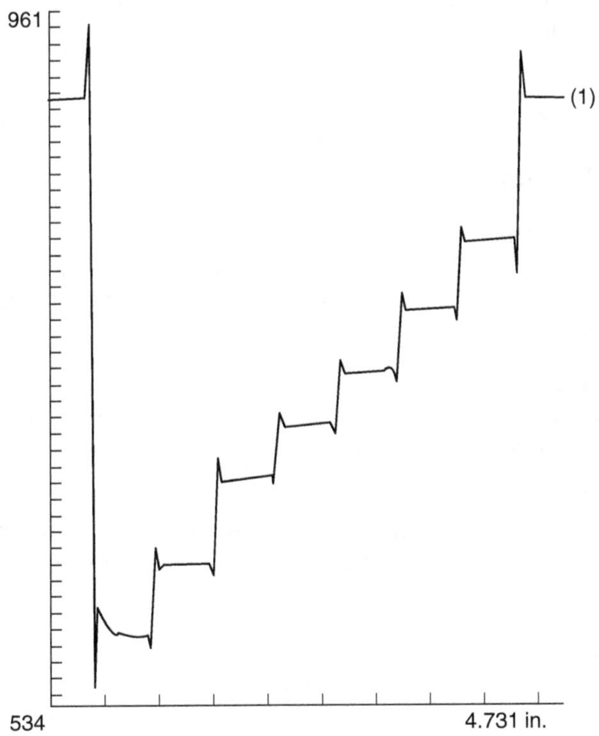


**FIGURE 11-28** Example of gray-scale profile of stepwedge. (*Courtesy of Fujifilm Corporation.*)

*Note:* It should be noted that adjustments to the gray-scale level or the contrast window do not alter the original image file. These adjustments are overlays to the file itself. Permanent change is possible but does require an administrative password to perform that type of alteration. The pixel value tool, as well as profiles and histograms are not only convenience tools but in many cases are required tools with respect to many standards.

#### Annotation Tools

Annotation tools allow commentary to be typed on the image along with call-outs such as arrows, boxes, and circles. This must be the type of tool which serves as an overlay, not destroying the information under the annotations.

#### Editing Tools

Editing tools are password protected to allow the editing of any demographic information which might have been erroneously applied to an image file.

#### Dose Measurement Tool

The dose measurement tool is used to determine the actual radiation dose impinging on a receptor available with some systems to assist with technique development and in some cases to determine the dwell time of the erasure process with regard to imaging plates.

#### Image Orientation Tool

The image orientation tool serves to allow inspection personnel to flip the orientation left to right or top to bottom.

#### File Format Conversion Tool

This tool enables the inspector to change file formats from DICONDE to other commonly used formats such as TIFF, JPEG, and BMP in order to enable certain files to be electronically transmitted.

#### Document Attachment Tool

In many cases, it is desirable to attach a variety of documents to an image file. These can be inspection reports, digital photographs confirming surface indication correlation, and other pertinent information.

#### Import and Export Tools

Import and export tools enable the interpreter to import images from other records or digital radiographic systems as well as exporting image files from the system in use.

#### Printing Tool

There are instances when a hard copy of the image is desired. Printing can be either a DICONDE print or a POSTSCIPT print on a variety of printing media including laser transparencies, emulating radiographic films.

#### Preset Tools

The preset tool is a feature which enables an image to be modified or enhanced with preset values.

### Pseudo Color Tool

Color synthesis can be applied in certain applications to particularly call out key features of an image, assisting the interpreter.

### Statistics Tool

The statistics tool (see Fig. 11-29) enables the interpreter to determine key values in a given area of interest. It can be used to determine the particular characteristics of the image, or area of the image being interpreted. These statistical values include but are not limited to:

- Minimum pixel value
- Maximum pixel value
- Average pixel value
- Standard deviation of those values
- Bit-depth of the ROI
- The number of pixels in the ROI

These values are primarily used to determine the SNR of the digital radiographic image and in many instances are required tools in inspection standards.

### Storage Tools

These are of particular importance for the storage and archiving image files to various media, servers, networks, and other databases. Various tools are available to accomplish the storage task.

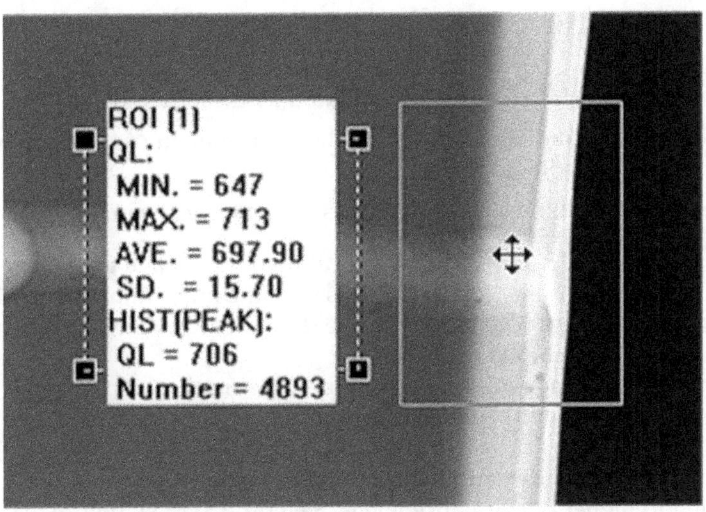


**FIGURE 11-29** Example of statistics tool. (*Courtesy of Fujifilm Corporation.*)

## 11.4 Digital Radiography Systems

### 11.4.1 Image Quality Assurance

The quality checks when employing analog film techniques are relatively expedient and easily executed [appropriate film density and applicable IQI(s)]. When employing digital imaging technology, an entirely new set of paradigms exist, most of which require specific software algorithms. These algorithms are classified as required tools, which serve to validate the quality of the obtained image.

The radiographic IQI, typically the plaque and wire penetrameter, is still a prescribed method of image quality assurance for both analog and digital images; however, in addition, new requirements have been established with regard to digital image quality.

Typically, contrast can be assured by employing the various IQIs used in routine radiographic techniques. All digital systems are capable of achieving and exceeding the specified contrast levels required by various standards.

Spatial resolution can be assured by the applicable wire diameter or plaque penetrameter "T" hole required with respect to the actual image quality requirements; however, other IQI devices such as converging line pairs and duplex wire gauges have been added to several standards to determine the performance of the system itself, as well as the quality of the resultant images. These tests are used to determine not only the classification of imaging plates but also the resultant resolution of selected images as affected by the radiographic technique itself and the effects of scatter. The control of the effects of scatter is of prime consideration when any digital radiography system is utilized.

Signal-to-noise ratio (SNR) is determined by a combination of gray-scale value and statistical data. As there is no film density to perform this function, SNR is determined by establishing the relationship of the spatial resolution as it compares the level of signal as compared to the inherent noise in the image. Note that the sources of noise can range from:

- Inherent electronic noise generated by the system itself
- Scattered radiation caused by the environment of the radiation facility and the selected radiographic technique
- The crystalline structure of the scintillator or the imaging plate

The relationship of the spatial resolution of the image to the level of signal as compared to the noise is like comparing this with the minimum acceptable film density of a radiograph (SNR). It also considers the digital equivalent of film graininess by determining pixel size (spatial resolution).

There are also activities being explored to employ minimum and maximum pixel values and equivalent penetrameter sensitivity (EPS) values to determine the SNR of digital RT images.

These activities are very fluid as the technology of digital radiography is constantly evolving to higher performance levels. Several ASTM standards have been developed to address these characteristics, as well as independently developed standards and requirements by various manufacturers and organizations, specifically addressing the requirements of their specific products.

As these systems are electronic devices subject to various environmental, electrical, and mechanical conditions, system checks for determining the stability of the imaging system have been developed.

ASTM has approved several standards in addition to Standard Guides and Standard Practices. Standards have been created and approved addressing stability, spatial resolution, system performance, imaging plate classification, system classification, and documents for manufacturers of systems as to the minimum image-quality requirements. As digital radiography is a technology with constant new developments, it is advised to routinely review all ASTM standards for any revised requirements. ASTM also has a glossary defining all the newly introduced terms with regard to employing digital radiography systems. Several standards have been developed by user groups and individual users defining image-quality requirements with respect to the three basic characteristics of image quality.

Assuring stability is another prime consideration. Physical hardware standards have been developed by ASTM committees, as well as the U.S. federal groups and military users. In particular, a "phantom" for checking CR performance can be used.

The phantom is a single device embedded with the various hardware enabling the user to determine several characteristics (see Fig. 11-30).

Another method of determining image quality is the employment of the representative quality indicator (RQI), where standard IQIs are either impractical, unavailable, or additional assurances are required. The RQI can be a physical hardware standard with actual unacceptable conditions, ranging from cast products to weldments, to mechanical assemblies. The RQI can be used as an IQI and also can be used as a stability assurance device.

ASTM digital reference images have also been developed to be used in place of the analog film reference images. To date, reference images have been converted to digital formats to be used with the developed digital imaging devices to examine cast products. Reference images can be downloaded from a storage device and inserted as a separate image as picture-in-a-picture or split-screen presentations (see Figs. 11-31 and 11-32).

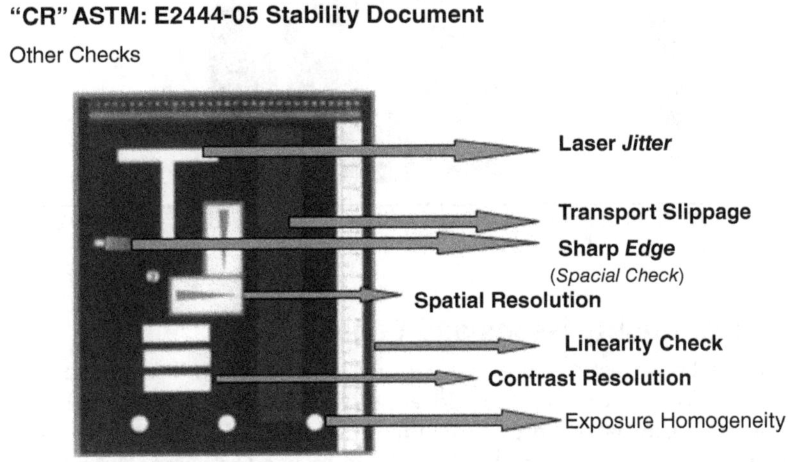


**Figure 11-30** Phantom. (*Courtesy of ASTM.*)

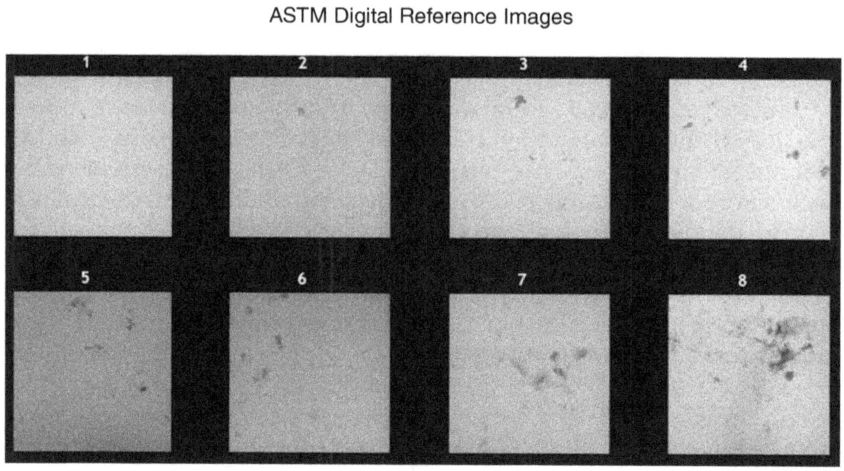


**Figure 11-31** Digital reference image. (*Courtesy of ASTM.*)

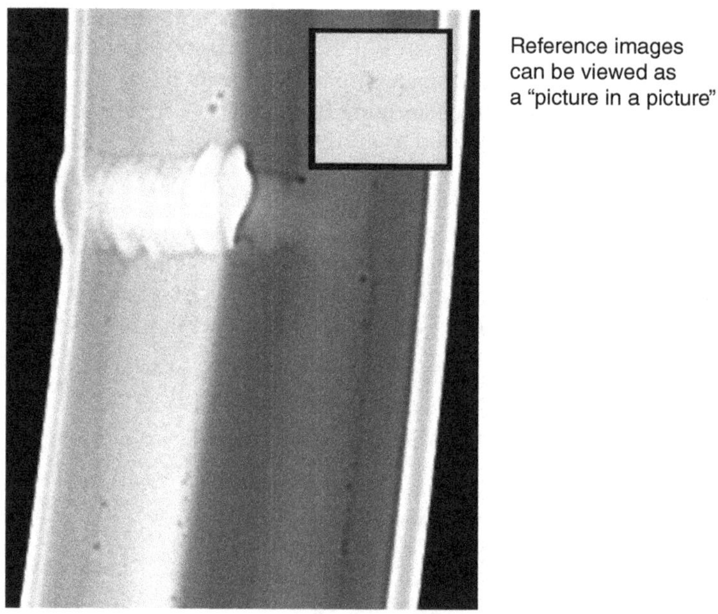


**Figure 11-32** Digital reference image used to interpret product. (*Courtesy of Fujifilm Corporation and ASTM.*)

## 11.5 Image Quality Issues and Concerns

As with all new technologies, new issues with regard to image quality and consistency have developed. New tests to address those concerns are constantly being developed.

Analog film issues are easily identified and addressed. Checks to address those concerns were developed and have been in use for years. With the new technology, there has been a significant paradigm shift.

### 11.5.1 Digital Image Issues

***CR: Image Plate Ghosting***

Retained images from previous exposures have to be controlled. Imaging plates are susceptible to damage when over-exposed and can be permanently damaged.

Procedures have been developed to address, control, and minimize this effect.

***CR: Mechanical Damage***

As imaging plates are used for many exposures, the cumulative effect of any scratches, nicks, and dents are a concern. Care must be applied when handling imaging plates. What was once a single event with film (e.g., scratches) could result in permanent damage in the case of an imaging plate. The cumulative effects of improper handing could result in the disqualification of a particular imaging plate.

***CR: Dust on the Optical Subsystem***

Dust and harsher environments can result in unacceptable image quality by introducing various image artifacts and possible damage to the imaging plate reader mechanism itself.

***CR: Accidental Imaging Plate Erasure***

Exposure to ambient lighting before the reading process can erase the imaging plate if the ambient lighting wavelengths are capable of erasing an imaging plate.

***DDA: Bad Pixels***

Bad pixels, which can be classified as "dead," "over-responding," "under-responding," and "noisy," are concerns. This condition can be a single pixel or clusters or groups of pixels, which can be measured and quantified. In some cases where the acceptance standards of the product involved will allow for this condition, controls can be established to determine the extent and severity of the bad pixel condition. In other cases, the issue of the "bad pixel" in the area of interest can be classified as the equivalent of an analog film artifact and may require additional exposures to offset the location of the area in question. The issue of interpolating or cloning adjacent pixels to make the image more appealing is still under debate and will be left up to the cognizant engineering organizations and Level III radiographer to determine the tolerances to be applied to this particular condition.

***DDA: Environmental Effects***

Many DDA devices must be packaged to reduce the effects of temperature and humidity variations.

***DDA: Electrical Safety***

DDA devices are powered and must be packaged to meet particular National Electrical Manufacturers Association (NEMA) requirements when used in hazardous environments.

***DDA: Spatial Resolution***

In many cases, the use of DDA technology requires the use of smaller X-ray tube focal spots to employ geometric magnification projections to meet the spatial resolution requirements of many analog film based applications as well as the spatial resolution of many CR systems.

#### CR and DDA: Scattered Radiation

Both technologies are extremely sensitive to the effects of scattered radiation. In many cases filters at the tube head window of an X-ray system must be employed as well as various metallic filters and screens in the imaging package itself. The same concerns exist with gamma and high-voltage systems, which require specialized packaging of the imaging devices.

## 11.6 Evolving Technology

Many new imaging techniques have been developed because of the fast-evolving digital imaging technology.

### 11.6.1 Industrial Computed Tomography (CT) Systems

Industrial computed tomography is a fast-growing technology employing various high-resolution industrial X-ray systems. The development of CT includes applications such as metrology, failure analysis, discontinuity characterization, process control, and computer-aided design.

Industrial CT systems can generate both two-dimensional (2D) and three-dimensional (3D) images. A particular advantage of CT images is the ability to view cross sections (slices) of products at various depths. Slicing can be of great value in determining the depth of particular conditions, especially in the casting industry. Color synthesis and binary slicing can be performed; this also offers the ability to examine volumetric data. The basic principle of CT technology is the ability to penetrate an object at various energy levels then reconstructing the image based on the transmission of radiation through the object. Specimen rotation allows for the high-speed collection of many views of the specimen, which, when reconstructed using very sophisticated algorithms, results in 2D or 3D displays of volumetric and multioriented views of the object being examined as illustrated in Figs. 11-33, 11-34, and 11-35.

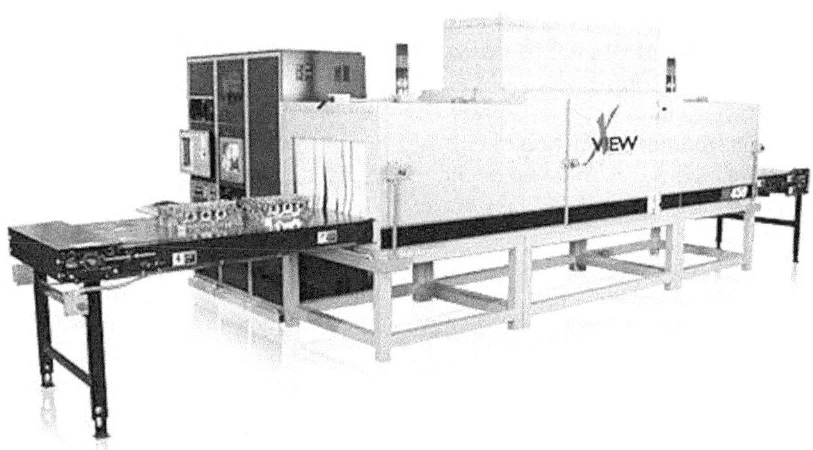

**FIGURE 11-33** Industrial CT scanner. (*Courtesy of Northstar Imaging.*)

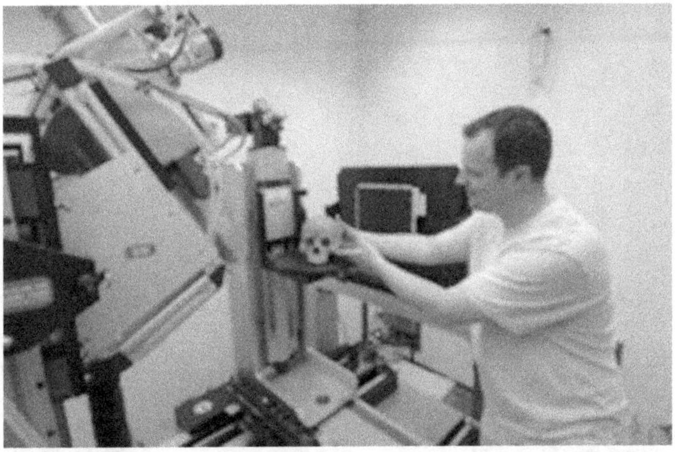

**FIGURE 11-34** Forensic application of industrial CT. (*Courtesy of eva.mpg.com.*)

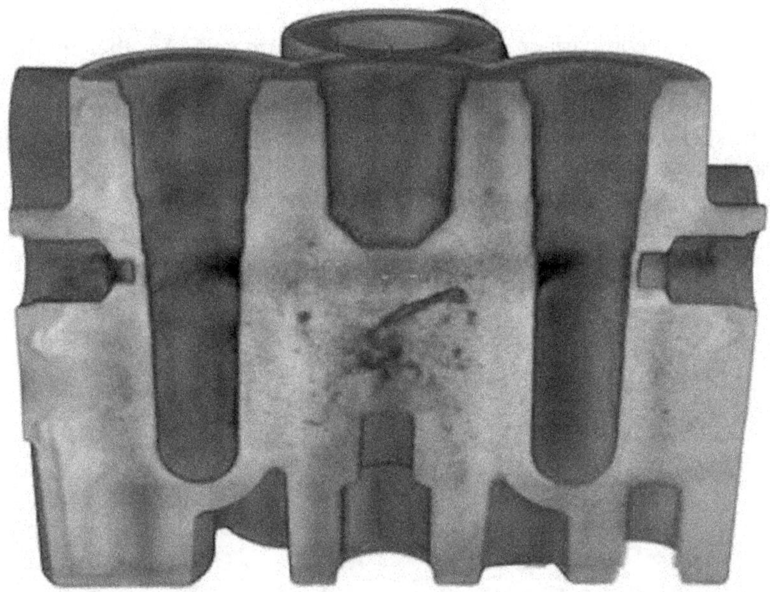

**FIGURE 11-35** CT image of a casting. (*Courtesy of Northstar Imaging.*)

### 11.6.2 Automated Defect Recognition (ADR)

ADR has been employed as an inspection technology for quantitative analysis of various objects. The radiographic image analysis of various mechanical assemblies has been successfully executed. Qualitative analysis efforts are being researched to automatically recognize various unacceptable conditions employing digital signals obtained from various digital radiographic systems. Commonly used algorithms are, for example, the use of converting image data to binary slices, gray-scale distribution analysis, and

FIGURE 11-36 ADR turnkey system. (*Courtesy of TFI Corporation.*)

direct image comparison to known standard images. Another method being investigated is the use of the gray scale or "pixel value differential" as compared to background, without performing a binary slice. This is again a search for variances in image contrast exceeding the maximum allowable level. The quantity of pixels, grouping, and location of the excessive contrast pixels are also a consideration.

To date, progress has been achieved in medical research to assist diagnosticians in locating areas of interest. However, both medical and material sciences are at high liability risk for a completely automated imaging analysis technology. See Figs. 11-36 and 11-37.

*Note:* ADR is a new evolving technology and requires significant additional research before it can evolve into an acceptable image interpretation technology. Considerations are product liability and electronic stability.

FIGURE 11-37 Digital signal processing used in ADR applications. (*Courtesy of TFI Corporation.*)

## 11.7 Glossary

Digital radiography is a technology which is revolutionary to the NDT community. Born of significant efforts and investments by the medical community, it has enabled today's industrial radiographer to not only generate image quality equal to and in some cases superior to analog film technology, but has allowed for the use of advanced tools for improved image analysis, and electronic storage and communications. New technology requires the development of new definitions. ASTM standards for NDT has a glossary section (E1316), which describes many of the new terms used with respect to digital radiography.

Some of the most commonly used are defined here.

**Amorphous Silicon (a-Si) X-ray Detector, *n*** An amorphous silicon (a-Si) X-ray detector consists of a glass substrate with a matrix of photodiodes fabricated from amorphous silicon and switches arranged in rows and columns upon it; the photodiodes are activated by light photons emitted from a scintillator, which is activated by X-rays and is usually in close contact with the diode matrix.

**Bad Pixel** A pixel identified with a performance outside of the specification range for a pixel of a DDA.

**Basic Spatial Resolution (SRb)** Indicates the smallest geometrical detail that can be resolved. It is similar to the effective pixel size/pitch and corresponds to 1/2 of the measured unsharpness.

**Contrast-to-Noise Ratio (CNR)** Quotient of the difference in the mean values of the intensity (signal) in an area in the object subtracted from the mean value of the intensity of the background, and standard deviation of the intensity (noise). The CNR depends on the radiation dose and quality, thickness/attenuation of the object, and the DDA system properties.

**Digital Image** An image composed of discrete pixels, each of which is characterized by a digitally represented luminance level.

**Digital Image Acquisition System** A system of electronic components which, by either directly detecting radiation or converting analog radiation detection information, creates an image of the spatial radiation intensity map composed of an array of discrete digital intensity values (see Pixel).

**Digital Image Enhancement** Any operation used for the purpose of enhancing some aspect of the original image.

**Digital Image Processing System** A system that uses algorithms to process digital image data.

**Ghosting** Residual signal or image from a prior exposure in a current image. Signal or image can be negative or positive and may affect interpretation of the image.

**Latent Image** A condition produced and persisting in the image receptor by exposure to radiation and able to be converted into a visible image by processing.

**Line Pair Test Pattern** A pattern of one or more pairs of objects with high-contrast lines of equal width and equal spacing. The pattern is used with an imaging device to measure spatial resolution.

**Line Pairs per Millimeter** A measure of the spatial resolution of an image conversion device. A line pair test pattern consisting of one or more pairs of equal width, high-contrast lines and spaces is utilized to determine the maximum density of lines and spaces that can be successfully imaged. The value is expressed in line pairs per millimeter.

**Photostimulable Luminescence** The physical phenomenon of phosphors absorbing incident ionizing radiation, storing the energy in quasi-stable states, and emitting luminescent radiation proportional to the absorbed energy when stimulated by radiation of a different wavelength.

**Photostimulable Luminescent Phosphor** A phosphor capable of storing a latent radiological image, which upon laser stimulation will generate luminescence proportional to the radiation intensity.

**Pixel** The smallest addressable element in an electronic image.

**Pixel, Display Size** The dimensions of the smallest picture element comprising the displayed image, given in terms of the imaged object's dimensions being represented by the element.

**Pixel Size** The length and width of a pixel.

**Representative Quality Indicator (RQI)** An actual part or similar part of comparable geometry and attenuation characteristics to that of the test part(s), which has known or measurable features, or both, representing the facets of nonconformance for which the test part is to be examined.

**Scintillators and Scintillating Crystals** A detector that converts ionizing radiation to light. This consists of a glass substrate with a matrix of photodiodes fabricated from amorphous material.

**Signal-to-Noise Ratio (SNR)** Quotient of mean value of the linearized signal intensity and standard deviation of the noise (intensity distribution) at this signal intensity. The SNR depends on the radiation dose and the CR system properties.

## References

Agfa-Gevaert. "Industrial Radiography." A Technical Manual, Teterboro, NJ.

ASTM International. "Section 3, Volume 03. Nondestructive Testing." *ASTM Annual Book of Standards*, Conshohocken, PA, 2011.

Eastman Kodak Company. *Radiography in Modern Industry*, 4th ed., Rochester, NY, 1980.

Fujifilm Holdings Corporation: Computed Radiography: "Technical Reviews No. 2, 3, and 4" "Imaging Plate Technology" 26/30. Fujifilm Holding Company NDT Systems. North Kingstown, RI, 1995.

General Electric Inspection Technologies. "Digital X-Ray Training," Module I-VI, Lewistown, PA, 2011.

Morro, Frederick A. "Computed Radiography Training," 2010 ed., Fujifilm Electronic Materials, North Kingstown, RI, 2010.

Picker X-Ray Corporation: "Picker X-Ray Corporation Special Products." Publication 52-21, Cleveland, OH, 1969.

# CHAPTER 12

# Phased Array Ultrasonic Testing

## 12.1 History and Development

In recent years, ultrasonic phased array systems have seen increasing use in industrial nondestructive testing to provide new levels of information and visualization in common ultrasonic nondestructive testing (NDT) applications that include weld inspection (as seen in Fig. 12-1), bond testing, thickness profiling, and in-service crack detection. Many people are familiar with the well-established medical application of ultrasonic imaging, in which high-frequency sound waves are used to create highly detailed cross-sectional pictures of fetal development and internal organs. Medical sonograms are commonly made with specialized multi-element transducers known as phased arrays and their accompanying hardware and software. Industrial phased array instruments utilize similar technology that has been adapted for the inspection of engineering materials rather than the human body. The diversity of materials, along with necessity to inspect with both compressional and shear wave propagation modes, makes implementation for industrial applications somewhat more challenging.

It is important to note at the start that phased array imaging represents an evolution from conventional ultrasonic testing; thus all of the basic principles of ultrasonic wave theory as well as almost all of the standard terminology and the test practices outlined in Chapter 7 apply. That important information is not repeated here. For that reason, it is strongly recommended that readers who are new to ultrasonic NDT review Chapter 7 before beginning this section of the book.

During their first several decades, commercial instruments for ultrasonic NDT relied entirely on single-element transducers that used one piezoelectric crystal to generate and receive sound waves, dual element transducers that had separate transmitting and receiving crystals, and pitch/catch or through transmission systems that used a pair of single-element transducers in tandem. These approaches are still used by the many types of conventional ultrasonic instruments designed for industrial flaw detection and thickness gaging applications; however, more advanced instruments using phased arrays are steadily becoming more important in the ultrasonic NDT field, especially in weld inspection.

Phased array testing is grounded in the principle of constructive and destructive interaction of waves, which was first demonstrated by English scientist Thomas Young in 1801 in a notable experiment that utilized two-point sources of light to create

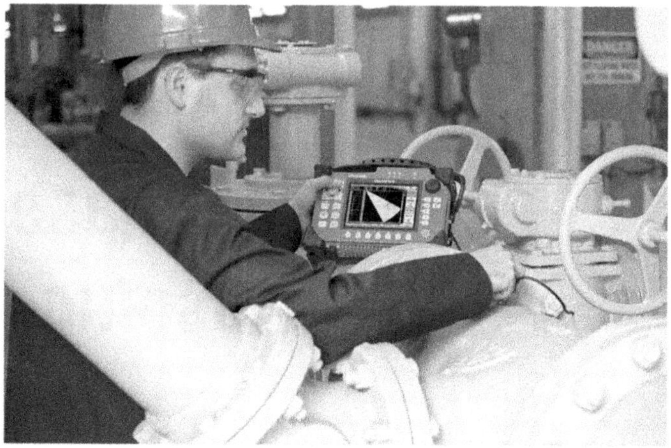

**FIGURE 12-1** Weld inspection with contemporary phased array instrument.

interference patterns. Waves whose oscillations combine in phase reinforce each other, while waves that combine out of phase will cancel each other, as seen in Fig. 12-2.

The phasing effect was first used experimentally to steer radio waves as early as 1905, and by the mid-20th century it was being employed in radar systems. Ultrasonic phased arrays using multiple transducer elements were first developed in the late 1960s, and the first linear arrays similar in concept to today's phased array transducers were announced in 1971. Initially, the use of ultrasonic phased array systems was largely confined to the medical field, aided by the fact that the predictable composition and structure of the human body make instrument design and image interpretation relatively straightforward. Industrial NDT applications, on the other hand, represent a much greater challenge because of the widely varying acoustic properties of metals, composites, ceramics, plastics, and fiberglass, as well as the enormous variety of thicknesses and geometries encountered across the scope of industrial testing.

The first industrial phased array systems, commercially introduced in the 1980s, were extremely large and heavy, and required that data be transferred to a computer for processing and image generation. These systems were typically used for in-service power generation inspections. In large part, this technology was advanced by the nuclear

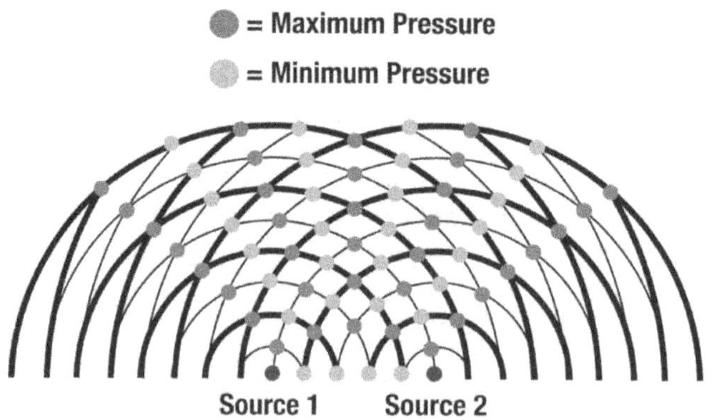


**FIGURE 12-2** Two-point source interference pattern.

power industry, where critical assessment needs encouraged the use of cutting-edge technology for improving inspection speed, resolution, documentation, and probability of detection. Other early applications included crack detection in large forged shafts and low-pressure turbine components.

Portable, battery-powered phased array instruments for industrial use appeared in the 1990s. Analog designs had required a large amount of power and space to create the multichannel configurations necessary for beam steering, but the transition into the digital world and the rapid development of inexpensive embedded microprocessors enabled more rapid development of a new generation of smaller and more powerful instruments. Techniques for the construction of multi-element phased array probes also improved, including the development of composite piezoelectric elements and advanced micromachining. In addition, the availability of low-power electronic components, better power-saving methods, and industry-wide use surface mount board design led to miniaturization of this advanced technology. This resulted in phased array tools which allowed electronic setup, data processing, display, and analysis all within a portable package, and so the doors were opened to more widespread use across the industrial NDT sector. This, in turn, encouraged the specification of standard phased array probes for common applications.

As noted above, conventional ultrasonic transducers for NDT use typically consist of either a single active element that both generates and receives high-frequency sound waves, or two paired elements, one for transmitting and one for receiving. Phased array probes, on the other hand, typically consist of a transducer assembly with from 16 to as many as 256 small individual elements that can each be pulsed independently. These may be arranged in a strip (linear array), a ring (annular array), a circular matrix (circular array), or a more complex shape. As is the case with conventional transducers, phased array probes may be designed for direct contact use, as part of an angle beam assembly with a plastic wedge, or for immersion use with sound coupling through a water path. Transducer frequencies are most commonly in the range from 2 to 10 MHz. A phased array system will also include a sophisticated computer-based instrument that is capable of driving the multi-element probe, receiving and digitizing the returning echoes, and plotting that echo information in various standard formats. Unlike conventional flaw detectors, phased array systems can sweep a sound beam through a range of refracted angles or along a linear path, or dynamically focus at a number of different depths, thus increasing both flexibility and capability in inspection setups.

Portable phased array instruments as seen in Fig. 12-3 are now in wide use in the ultrasonic testing industry. As the use of phased array systems in critical flaw detection applications increases, it is now necessary to adapt and update industry standards, test

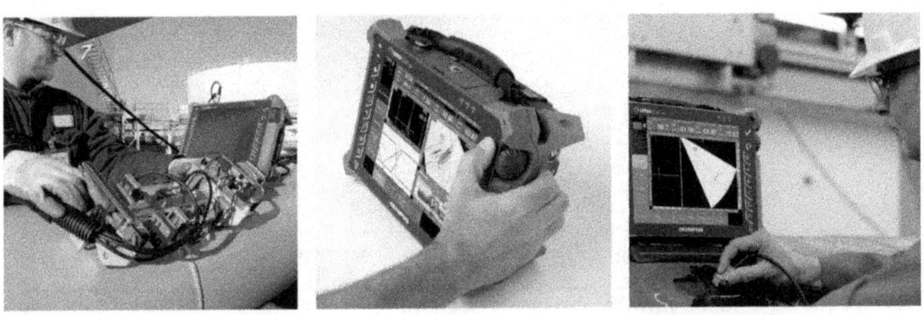

**FIGURE 12-3** Portable phased array instrument used for manual and automated inspections.

procedures, and inspection codes to accommodate this new generation of technology, as well as to train and certify operators in its use. As portable phased array instruments continue to evolve, that will be the next challenge.

## 12.2 Theory and Principles of Phased Array Testing

### 12.2.1 Phased Array Transducers

Conventional longitudinal wave ultrasonic transducers designed for flaw detection and thickness gaging operate as a piston source of high-frequency mechanical vibrations, or sound waves. As a voltage is applied, the piezoelectric transducer element (which is often informally called a crystal) deforms by compressing in the direction perpendicular to its face. When the voltage is removed, less than a microsecond later, the element springs back, generating the pulse of mechanical energy that when coupled into a test piece comprises an ultrasonic wave. Similarly, if the element is compressed by the pressure of an arriving ultrasonic wave, it will generate a voltage across its faces. Thus, a single piezoelectric element can act as both a transmitter and a receiver of ultrasonic pulses.

An array is simply an organized arrangement of multiple quantities of an object. The simplest form of an ultrasonic array for NDT would be a series of single-element transducers arranged in such a way as to increase the inspection coverage area and/or the speed of a particular inspection. While phased array transducers come in a wide range of sizes, shapes, frequencies, and number of elements, what they all have in common is a piezoelectric element that has been divided into a number of segments, effectively creating a number of separate transducers within a single housing.

Contemporary phased array transducers for industrial NDT applications are typically constructed around piezocomposite materials, which are made up of many tiny, thin rods of piezoelectric ceramic embedded in a polymer matrix. While composite transducers can be more challenging to manufacture, they typically offer a 10- to 30-dB sensitivity advantage over piezoceramic transducers of otherwise similar design. Segmented metal plating is used to divide the composite strip into a number of electrically separate elements that can be pulsed individually. This segmented element is then incorporated into a transducer assembly that includes a protective matching layer, a backing, cable connections, and an overall housing (see Fig. 12-4).

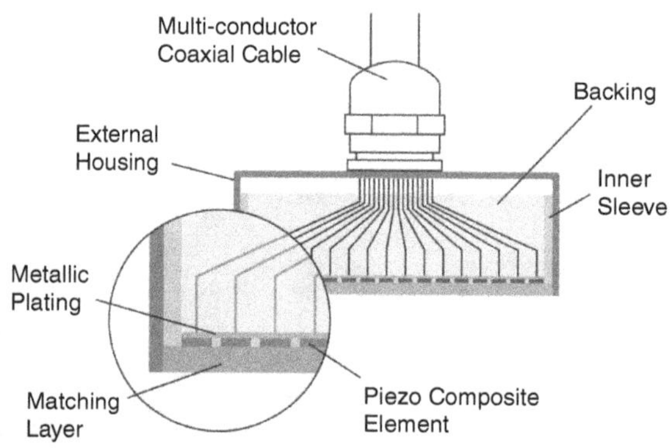


**Figure 12-4** Phased array probe cross section.

Phased array transducers are categorized according to the following basic functional parameters:

**Type:** Most phased array transducers are angle beam type, designed for use with either a plastic angle wedge or a straight plastic shoe (also known as a zero degree wedge) or delay line. Direct contact and immersion transducers are also available.

**Frequency:** Most commonly used phased array transducers fall within the frequency range between 2 and 10 MHz, which is also the most common range for conventional ultrasonic flaw detection. Lower and higher frequency probes are also available. As with conventional transducers, penetration increases with lower frequency, while resolution and focal sharpness increase with higher frequency.

**Number of elements:** Phased array transducers most commonly have from 16 to 128 elements, with some having as many as 256. A larger number of elements increases focusing and steering capability, and can increase area coverage as well, but it also increases both probe and instrumentation costs and complexity. Each of these elements is individually pulsed to create the desired wavefront. The length dimension of an array probe is often referred to as the active or steering direction.

**Size of elements:** As the width of each element gets smaller, beam steering capability increases, but large area coverage will require more elements at higher cost. The dimensional parameters of a phased array probe are customarily defined as in Fig. 12-5. This information is among the parameters used by instrument software to generate the desired beam shape. If it is not entered automatically by probe recognition software, then it must be entered by the user during initial setup.

## 12.2.2 Phased Array Wedges

In addition to the array transducer itself, phased array probe assemblies also usually include a plastic wedge. Wedges are used in both shear wave and longitudinal wave

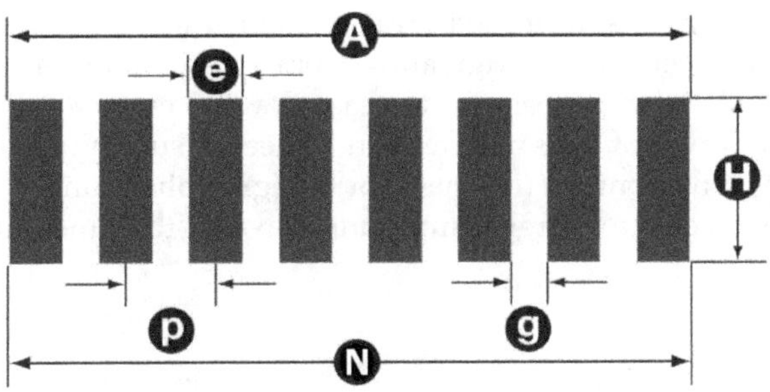


**N** = total number of elements in the array
**A** = total aperture in the steering or active direction
**H** = element height or elevation. Since this dimension is fixed it is often referred to as the passive plane
**p** = pitch, or center to center distance between two successive elements
**e** = width of an individual element
**g** = spacing between active elements

**FIGURE 12-5** Dimensional parameters of a phased array transducer.

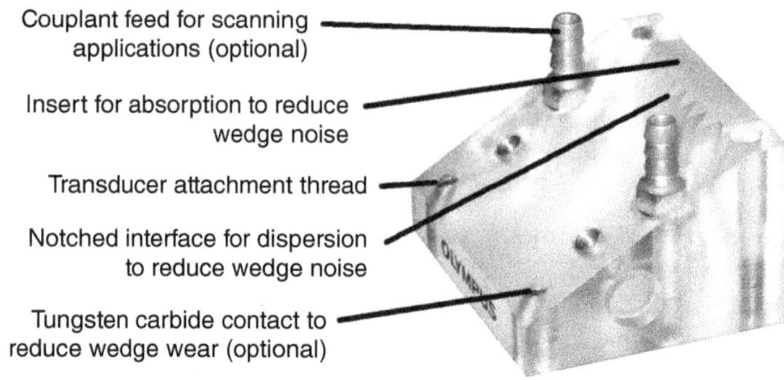


**Figure 12-6** Typical Rexolite wedge for 16-element probe.

applications, including straight beam linear scans. These wedges perform essentially the same function in phased array systems as in conventional single-element flaw detection, coupling sound energy from the transducer to the test piece in such a way that it mode converts and/or refracts at a desired angle in accordance with Snell's law. While phased array systems utilize electronic beam steering to create beams at multiple angles from a single wedge, the mechanical refraction effect and the wedge/test piece boundary is also part of the beam generation process. Shear wave wedges for phased array testing look very similar to those used with conventional transducers and, like conventional wedges, they come in many sizes and styles. Some of them incorporate couplant feed holes for scanning applications. A typical phased array wedge is seen in Fig. 12-6.

Zero-degree wedges are basically flat plastic blocks that are used for coupling sound energy and for protecting the transducer face from scratches or abrasion in straight beam linear scans, and also for low-angle longitudinal wave-angled scans. An example is seen in Fig. 12-7.

### 12.2.3 Beam Forming through Phased Pulsing

As the name suggests, a phased array system utilizes the wave physics principle of phasing. Whenever waves originating from two or more sources interact with each other, there will be an increase or decrease in wave energy at the point of combination. When elastic waves of the same frequency meet in such a way that their displacements are precisely synchronized (in phase, or 0-degree phase angle), the wave energies will add together to create a larger amplitude wave. If they meet in such a way that their

**Figure 12-7** Typical zero-degree wedge.

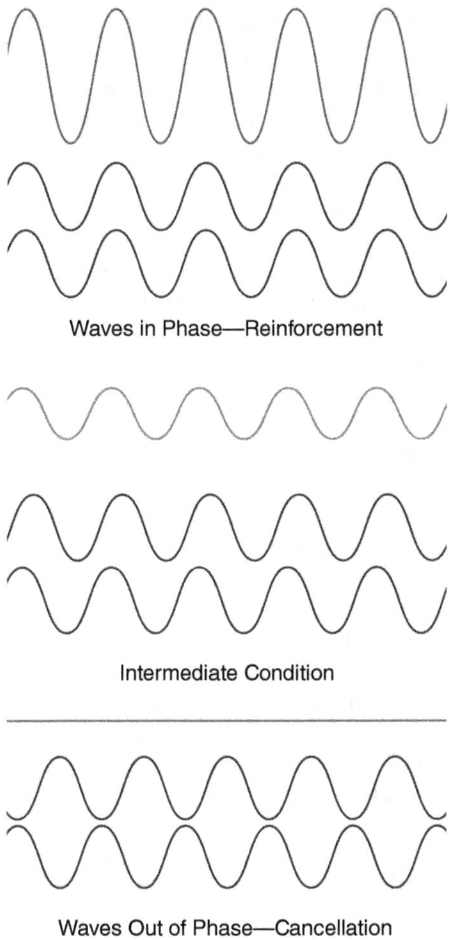


**FIGURE 12-8** Phasing effects: reinforcement and cancellation.

displacements are exactly opposite (180 degrees out of phase), then the wave energies will cancel each other. At phase angles between 0 and 180 degrees, there will be a range of intermediate stages between full addition and full cancellation. These effects are shown conceptually in Fig. 12-8. By varying the timing of the individual waves from a large number of sources, it is possible to use these effects to both steer and focus the resulting combined wavefront. This is the essential principle behind phased array ultrasonics.

In conventional transducers, these constructive and destructive interference effects occur naturally via Huygens' principle, creating the near field and far field zones and the various pressure gradients found there. Additionally, a conventional angle beam transducer uses a single element to launch a wave in a wedge. Points on this wavefront experience different delay intervals when passing through the wedge due to the wedge's tapered shape. However, these are mechanical delays, as opposed to the electronic delays that can be further employed in phased array testing.

### 12.2.4 Beam Angle Generation with Phased Array Probes

With phased array ultrasonics, these predictable reinforcement and cancellation effects are used to shape and steer the ultrasonic beam. Pulsing individual elements or groups of elements with different delays creates a series of point source waves that will combine into a single wavefront traveling at a selected angle. This electronic effect includes the mechanical delay generated by a conventional wedge, and it also includes steering by changing the pattern of pulses across the length of the array. Through constructive interference, the amplitude of this combined wave can be considerably greater than the amplitude of any one of the individual waves that produce it. Similarly, variable delays are applied to the echoes received by each element of the array to sum the responses in such a way as to represent a single angular and/or focal component of the total beam. In addition to altering the direction of the primary wavefront, this combination of individual beam components allows beam focusing at any point in the near field, as seen in Fig. 12-9. The time delay calculated to generate a particular wavefront is often referred to as a focal law.

Elements are usually pulsed in groups of 4 to 32 in order to improve the effective sensitivity by increasing the beam aperture. Larger apertures reduce unwanted beam spreading and enable sharper focusing.

Phased array instruments generate the required element pulsing pattern based on information that has been entered by the user. Software known as a focal law calculator establishes specific delay times for firing each group of elements in order to generate the desired beam shape through wave interaction, taking into account probe and wedge characteristics as well as the geometry and acoustic properties of the test material. The programmed pulsing sequence selected by the instrument's operating software then launches a number of individual wavefronts in the test material. These wavefronts, in turn, combine constructively and destructively into a single primary wavefront that travels through the test material and reflects off cracks, discontinuities, back walls, and other material boundaries like any conventional ultrasonic wave. The beam can be dynamically steered through various angles, apertures, focal distances, or focal spot sizes in such a way that a single probe assembly is capable of examining the test material across a range of different perspectives. This beam steering happens very quickly,

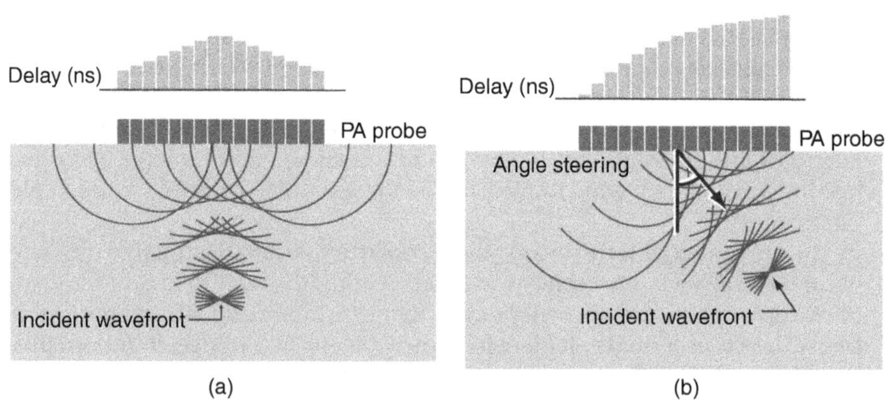


**FIGURE 12-9** Beam focusing principle for (a) normal and (b) angled incidences.

so that a scan from multiple angles or with multiple aperture positions or focal depths can be performed in a very small fraction of a second.

The returning echoes are received by the various elements or groups of elements, time-shifted as necessary to compensate for varying wedge delays, and then summed. Unlike a conventional single-element transducer, which effectively merges the effects of all beam components that strike its face, a phased array transducer can spatially sort the returning wavefront according to the arrival time and amplitude at each element. When processed by instrument software, each returned focal law represents the reflection from a particular angular component of the beam, a particular point along a linear path, and/or a reflection from a particular focal depth. The echo information can then be displayed in any of several standard formats as described in Section 12.2.9.

## 12.2.5 Beam Focusing with Phased Array Probes

Sound beams can be focused like light rays, creating an hourglass-shaped beam that tapers to a minimum diameter at a focal point and then expands once past that focal point. The beam from unfocused transducers diverges, becoming wide with distance, but within its near field a transducer can be focused to create a beam that converges rather than diverges. Narrowing the beam diameter to a focal point increases sound energy per unit area within the focal zone, thus increasing sensitivity to small reflectors. Conventional transducers usually accomplish this with a refractive acoustic lens, while as noted previously, phased arrays do it electronically by means of phased pulsing and the resulting beam-shaping effects. The depth at which the beam from a phased array focuses can be varied by changing the pulse delays. As with conventional transducers, the near field length in a given material defines the maximum depth at which a sound beam can be focused. A beam cannot be focused beyond the end of the near field in the test material (see Fig. 12-10).

A focused transducer's effective sensitivity is affected by the beam diameter at the point of interest. The smaller the beam diameter, the greater will be the amount of energy that will be reflected by a small flaw. Additionally, the small beam diameter at the focal point can improve lateral resolution. In ultrasonic NDT, beam diameter is commonly calculated based on the point at which the

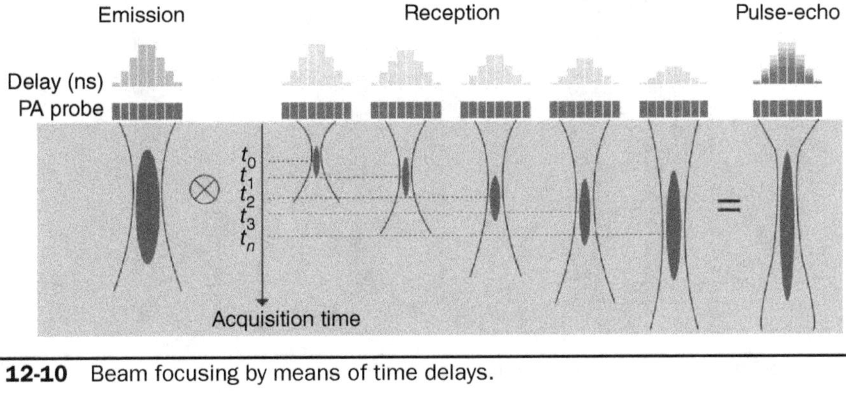


**FIGURE 12-10** Beam focusing by means of time delays.

energy drops to 50 percent of maximum, known as the –6 dB beam diameter. This can be calculated for a focused transducer at the focal point as follows:

$$-6 \text{ dB beam diameter} = 1.02Fc/fD$$

where:
$F$ = focal length in the test medium
$c$ = sound velocity
$f$ = frequency
$D$ = element diameter or aperture

In the case of the most commonly used linear and square phased arrays with rectangular elements, the beam will be focused in the steering direction (parallel to the sides of the individual elements) and unfocused in the passive direction (parallel to the ends of the individual elements). Increasing the aperture size increases the sharpness of the focused beam, as can be seen in the beam profiles below. Axial and transverse beam profiles show the difference in relative sound pressure for different apertures (see Fig. 12-11).

### 12.2.6 Factors Affecting Beam Shaping and Steering

The response of any ultrasonic test system is a combination of factors: the transducer being used, the type of instrument used and its settings, and the acoustic properties of the test material. The responses produced by phased array transducers, like those from any other ultrasonic transducers for NDT, will be relative to transducer design parameters, like frequency, size, and mechanical damping, and to the parameters of the excitation pulse that is used to drive it.

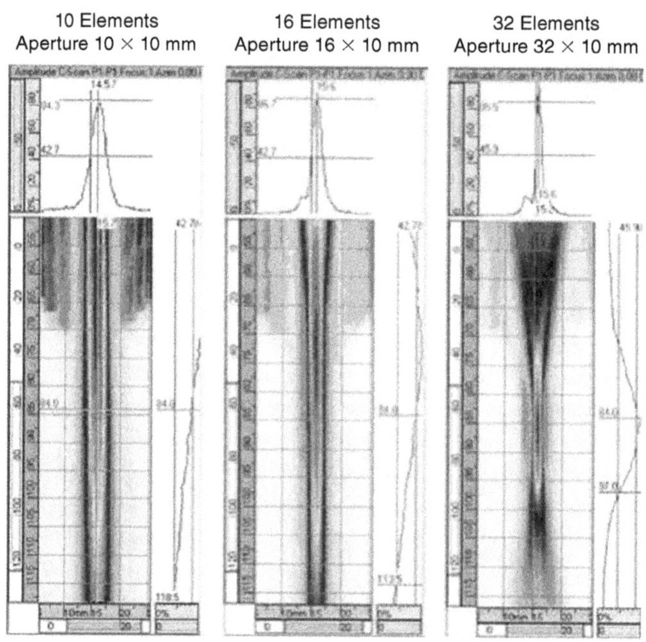


**FIGURE 12-11** Beam focus with different apertures.

Four important transducer parameters will have a number of interrelated effects on performance.

**Frequency:** As noted in Chapter 7, the test frequency has a significant effect on near field length and beam spreading. In certain instances, higher frequencies can provide better signal-to-noise ratio than lower frequencies since they offer potentially sharper focusing and thus a tighter, more optimized focal spot. At the same time, penetration in any test material will decrease because of increasing material attenuation and scattering as frequency increases. Applications involving very long sound paths or test materials that exhibit high attenuation or scattering will require the use of lower frequencies. Most industrial phased array probes have frequencies between 2 and 10 MHz, although higher and lower frequencies are also available.

**Element size:** As the size of individual elements in an array decreases, its beam steering capability increases. For manufacturing reasons, the minimum practical element size in commercial probes is typically around 0.2 mm. However, if the element size is less than one wavelength, strong unwanted side lobes will occur, so element size must be larger at low frequencies.

**Number of elements:** As the number of elements in an array increases, so can the physical coverage area of the probe and its sensitivity, focusing capability, and steering capability. At the same time, use of large arrays must often be balanced against issues of system complexity and cost.

**Pitch and aperture:** Pitch is the distance between individual elements, while aperture is the effective size of a pulsing element that usually comprises a group of individual elements that are pulsed simultaneously (virtual aperture). For phased array transducers, a selected number of elements are grouped together to form the effective aperture for which beam spreading can be approximated from conventional transducer models. To optimize steering range, pitch must be small. For optimum sensitivity, minimum unwanted beam spreading, and strong focusing, the aperture must be large. Setups for a given test will represent a compromise between these considerations. Current phased array instruments most commonly support focal laws for up to 16 element apertures. More advanced systems allow up to 32 or even 64 element apertures with large arrays.

The key concepts for a general understanding of phased array beam generation can be summarized as follows: A group of elements is fired according to programmed focal laws. This creates the desired transducer aperture and beam characteristics. The table below lists the effects associated with each change in focal law programming.

| Change | Effect |
|---|---|
| Decreasing pitch and element width while keeping the number of elements constant | Increases beam steering capability |
| Increasing pitch or frequency for a given aperture | Eventually creates unwanted grating lobes |
| Increasing element width | Can create side lobes (as in conventional UT), reduces beam steering |
| Increasing active aperture by using many small elements with small pitch | Increases focusing factor (sharpness of beam) |

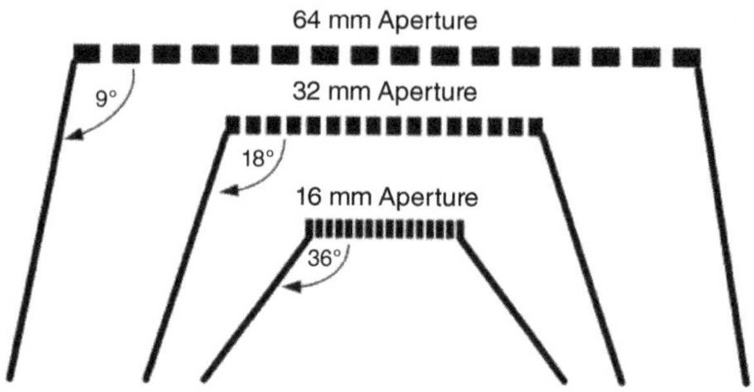


**FIGURE 12-12** Beam steering limits.

For phased array transducers, the maximum steering angle (at –6 dB sensitivity) in a given case is derived from the beam spread equation below and demonstrated in Fig. 12-12. It can be easily seen that small elements have more beam spreading and hence higher angular energy content, which can be combined to maximize steering. When the number of elements is constant, the maximum beam steering angle increases as the aperture size decreases. As element size decreases, more elements must be pulsed together to maintain sensitivity.

$$\sin \theta_{st} = 0.514\lambda/e$$

where:

$\sin \theta_{st}$ = sine of maximum steering angle
$\lambda$ = wavelength in test material
$e$ = element width

### 12.2.7 Phased Array Probe Selection Summary

Selecting phased array probes for a given inspection is always a compromise between selecting the optimum pitch, element width, and aperture. Using a large number of small elements increases steering, reduces side lobes, and provides focusing, but it can be limited by the cost of manufacturing and instrument complexity. Most standard phased array instruments will support apertures of up to 16 elements. Separating elements by greater distances within the probe may seem to be an easy way to gaining aperture size, but that creates unwanted grating lobes.

It is important to note that manufacturers of phased array transducers offer standard probes that have been designed with these compromises in mind, resulting in optimized performance for the intended use. Actual transducer selection in a given case will ultimately be governed by the specific application needs. In some cases, multi-angle steering will be required over small metal paths; therefore, large aperture sizes are not needed or desired. In other cases, the application may require coverage of large areas for laminar defects, requiring large apertures and a linear scan format with multiple grouped elements but no angular steering at all. In general, the user should apply best practices from his or her conventional UT experience for frequency and aperture selection.

### 12.2.8 Focal Law Sequencing (Image Generation)

As noted previously, the essence of phased array testing is an ultrasonic beam whose direction (refracted angle) and focus can be steered electronically by varying

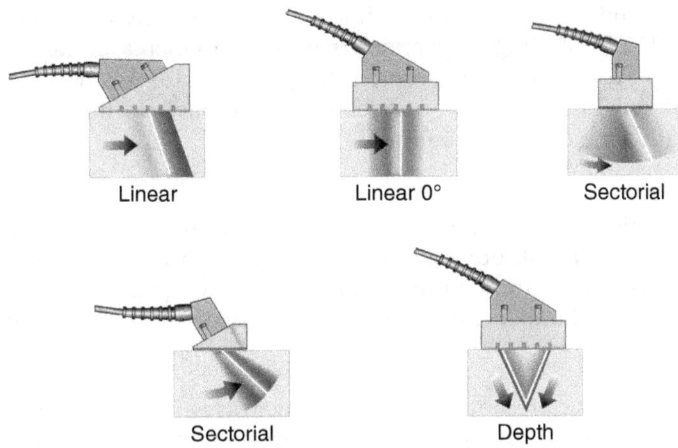


**Figure 12-13** Beam sequencing examples.

the excitation delay of individual elements or groups of elements. Sequencing the beam steering permits multiple angles and/or multiple point inspection from a single probe and a single probe position. The ability to sweep the beam from a fixed location greatly increases detection capability. Common patterns are seen in Fig. 12-13.

### 12.2.9 Imaging Modes

Both conventional and phased array ultrasonic techniques utilize high-frequency sound waves to examine the internal structure of a test piece or measure its thickness, and both rely on the same basic laws of physics that govern sound wave propagation. Similar concepts are employed in both ultrasonic technologies to present ultrasonic data.

Conventional ultrasonic instruments for NDT commonly utilize transducers consisting of either a single active element that both generates and receives high-frequency sound waves, or two paired elements, one for transmitting and one for receiving. A typical conventional instrument consists of a single-channel pulser and receiver that generates and receives ultrasonic energy, with an integrated digital acquisition system which is coordinated with an onboard display and measurement module. In more advanced systems, multiple pulser receiver channels can be used with a group of transducers to increase zone of coverage for evaluating different depths or flaw orientations and can further provide alarm outputs. Conventional ultrasonics can also be integrated with positional encoders, controllers, and software as part of an imaging system.

Phased array instruments, on the other hand, are naturally multichannel as they need to provide excitation patterns (focal laws) to transducer assemblies with from 16 to as many as 256 elements. Unlike conventional flaw detectors, phased array systems can sweep a sound beam from a single probe through a range of refracted angles, along a linear path, or dynamically focus at a number of different depths, thus increasing both flexibility and capability in inspection setups. This added ability to generate multiple transducer paths within one probe adds a powerful advantage in detection and naturally adds the ability to "visualize" an inspection by creating an image of the inspection zone. Phased array imaging provides the user with the ability to see relative point to point changes and multi-angular defect responses, which can assist in flaw discrimination and sizing. While this may seem inherently complex, it can actually simplify expanding inspection coverage with increased detection ability by eliminating the complex fixtures and multiple transducers that are often required by conventional UT inspection applications. The basic formats for phased array data presentation are discussed below.

### A-Scan Data

All ultrasonic flaw detectors typically record two fundamental parameters of an echo: how large it is (amplitude) and where it occurs in time with respect to a zero point (pulse transit time). Transit time in turn is usually correlated to reflector depth or distance, based on the sound velocity of the test material and the simple relationship:

$$\text{Distance} = \text{velocity} \times \text{time}$$

The most basic presentation of ultrasonic waveform data is in the form of an A-scan, or waveform display, in which echo amplitude and transit time are plotted on a simple grid with the vertical axis representing amplitude and the horizontal axis representing time. The example in Fig. 12-14 shows a version with a rectified waveform; unrectified RF displays are also used. The horizontal bar on the screen is a gate that selects a portion of the wave train for analysis, typically for measurement of echo amplitude and/or distance.

When using phased array probes, there are multiple A-scans that are stacked to create an image from a single probe position. Typically the user selects which particular A-scan or finite group of A-scans to view from the image. In Fig. 12-14, a 64-element linear array probe is used to sweep the aperture over four side drill holes. The A-scan on the left is a selected A-scan from the aperture over the shallowest side drill hole. The A-scan on the right is a composite of all the A-scans showing the reflections across the length of the entire array.

### Cross-Sectional B-Scans

A cross-sectional B-scan provides a detailed end view of a test piece along a single axis (see Fig. 12-15). Successive A-scans are plotted over elapsed time or actual encoded transducer position so as to draw full cross sections of the test piece along

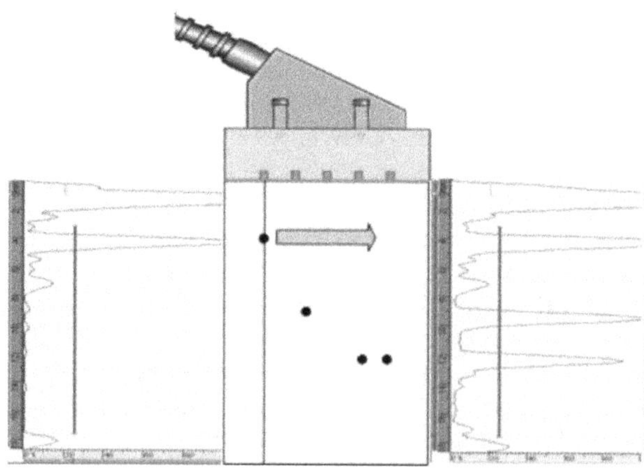

**Figure 12-14** Single and composite A-scan from a linear array probe.

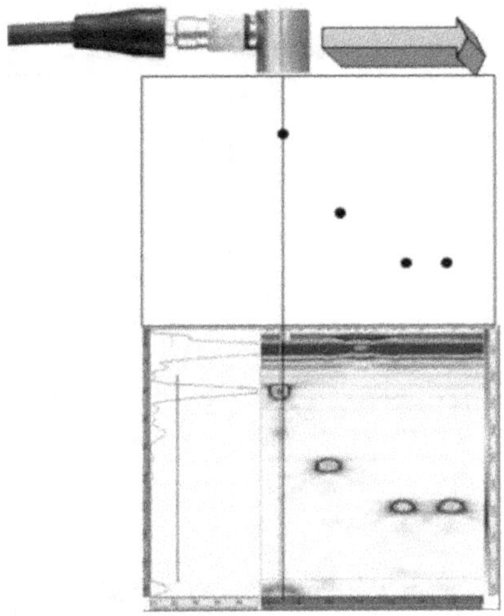

**FIGURE 12-15** Cross-sectional B-scan.

the scanned line. This allows visualization of both near and far surface reflectors within the part. With this technique, the full waveform data is often stored at each location and may be recalled from the image for further evaluation or verification.

To accomplish this, each digitized point of the waveform is plotted so that color, representing signal amplitude, appears at the proper depth. Successive A-scans are digitized, related to color, and "stacked" at user-defined intervals (elapsed time or position) to form a true cross-sectional image.

### Linear Scans

A phased array system can use electronic scanning along the length of a linear array probe to create a cross-sectional profile without moving the transducer. As each focal law is sequenced, the associated A-scan is digitized and plotted. Successive apertures are "stacked" creating a live cross-sectional view. In practice this electronic sweeping is done in real time so that a live cross section can be continually viewed as the transducer is physically moved. Below is an image made with a 64-element linear phased array probe, programmed to use 16 elements to form an aperture and sequenced the starting increments by one element. This results in 49 individual waveforms that are stacked to create the real-time cross-sectional view, seen in Fig. 12-16, across the transducer's 1.5 inch length.

It is also possible to scan at a fixed angle across elements, as seen in Fig. 12-17. As discussed later, this is very useful for automated weld inspection. Using a 64-element linear phased array probe with wedge, shear waves can be generated at a user-defined angle (often 45, 60, or 70 degrees). With aperture sequencing through the length of the probe, full volumetric weld data can be collected without the need for physically moving the probe perpendicular to the weld centerline while scanning. This provides for single-pass inspection along the weld length.

### Sectorial Scans (S-Scan)

In a linear scan, all focal laws employ a fixed angle with sequencing apertures. Sectorial scans, on the other hand, use fixed apertures and steer through a defined sequence

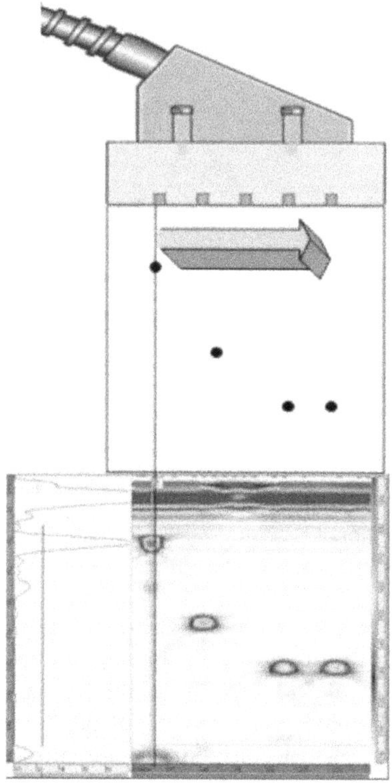

**Figure 12-16** Normal beam linear scan.

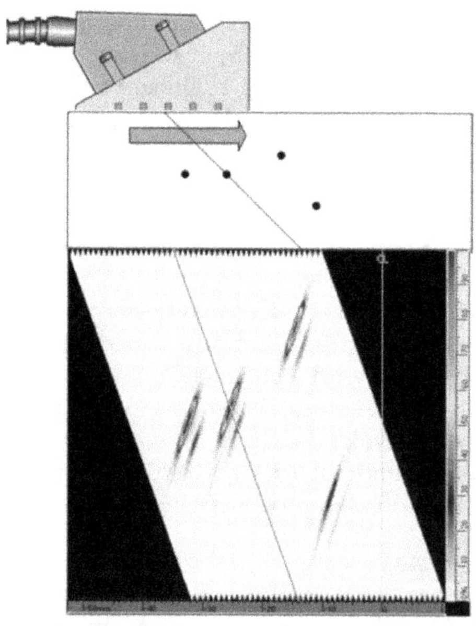

**Figure 12-17** Fixed-angle linear scan.

of angles. Sectorial scans are unique to phased array testing because of their multiple-angle nature and are the most commonly used phased array scanning mode. As with the linear scan, the image presentation is a cross-sectional picture of the inspected area of the test piece.

Two main S-scan formats are typically used. The most familiar, very common in medical imaging, uses a 0 degree (flat) plastic wedge to steer longitudinal waves to create a pie-shaped image showing laminar and slightly angled defects. The beam commonly sweeps from −30 degrees to +30 degrees (see Fig. 12-18).

The second format, seen in Fig. 12-19, employs an angled plastic wedge to increase the incident beam angle for generation of shear waves, most commonly in the refracted angle range of 30 to 70 degrees. This technique is similar to conventional angle beam inspection, except that the beam sweeps through a range of angles rather than a just single fixed angle determined by a wedge.

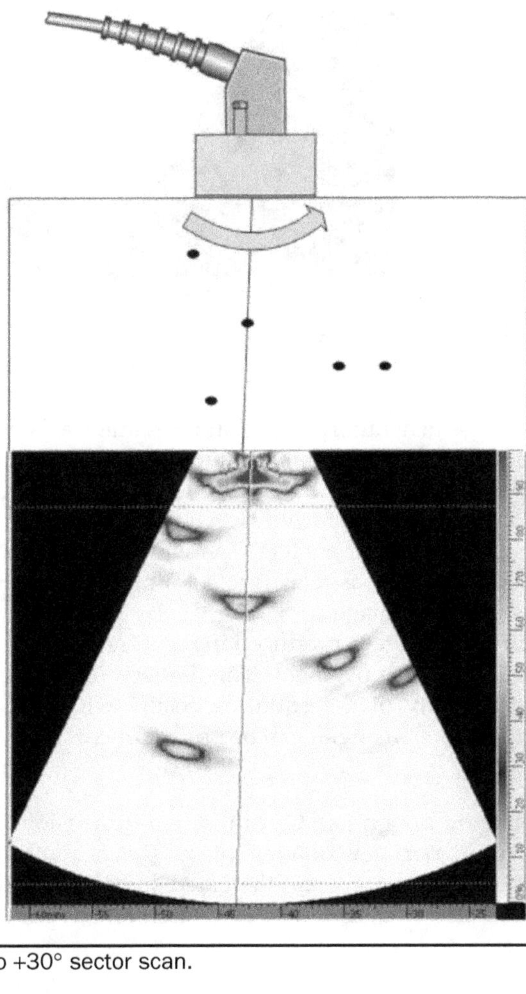

**Figure 12-18** −30° to +30° sector scan.

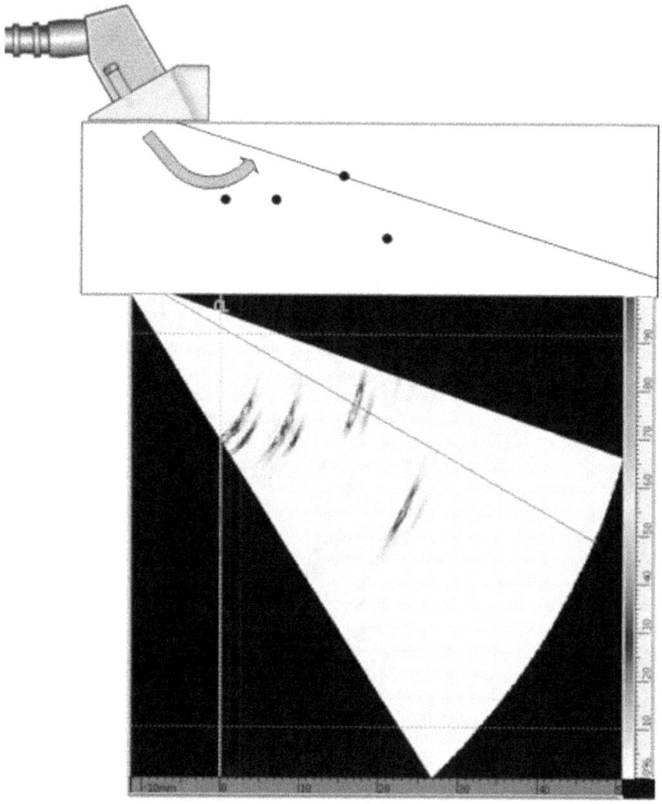

FIGURE 12-19 +40° to +70° sector scan.

The actual image generation works on the same stacked A-scan principle that was discussed previously. The user defines the angle start, end, and step resolution to generate the sectorial image. Note that the aperture remains constant, with each defined angle generating a corresponding beam with characteristics defined by aperture, frequency, damping, and other setup parameters. The waveform response from each angle (focal law) is digitized and plotted as related to color at the appropriate corresponding angle, building a cross-sectional image.

The sectorial scan image is produced in real time, so as to continually offer dynamic imaging with transducer movement. This is very useful for defect visualization and increases the probability of detection, especially with respect to randomly oriented defects, as many inspection angles can be used at once.

### One-Line Scan

Another presentation option is a C-scan, a two-dimensional presentation of data displayed as a top or planar view of a test piece. This is similar in its graphic perspective to an X-ray image. Colors represent the gated signal amplitude or depth at each point in the test piece. Planar images can be generated on flat parts by associating amplitude and/or depth data to *X-Y* position, or on cylindrical parts by tracking axial and angular position. For conventional ultrasound, a mechanical scanner with encoders is used to track the transducer's coordinates to the desired index resolution.

A C-scan from a phased array system is very similar to one from a conventional probe. With phased array systems, however, a C-scan may be made by physically moving along just one axis while the beam electronically scans along the other according to the focal law sequence. Signal amplitude or depth data is collected within a gated region of interest, just as with conventional C-scans. In the case of phased arrays, data is plotted with each focal law progression, using the programmed beam aperture. This is often referred to as a one-line scan.

The image in Fig. 12-20 is a scan of a test block using a 5 MHz, 64-element linear array probe with a straight wedge or shoe. Each focal law uses 16 elements to form the aperture, and with each pulse the starting element increments by one. This results in 49 data points that are plotted (horizontally in the image below) across the transducer's 37 mm (1.5 in.) length. As the transducer is moved in a straight line forward, a planar C-scan view emerges. Encoders will normally be used whenever precise geometrical correspondence of the scan image to the part must be maintained, although non-encoded manual scans can also provide useful information in many cases. Figure 12-20 shows a linear array probe with arrows indicating the generalized electronic beam steering and direction of physical motion. The C-scan image shows reference holes in a test bock. The test block is shown with the flat bottom holes up for conceptual purpose.

While the graphic resolution may not be as detailed as in a conventional C-scan using sharply focused immersion transducers because of the larger effective beam size, there are advantages. The phased array system is field portable, which the conventional

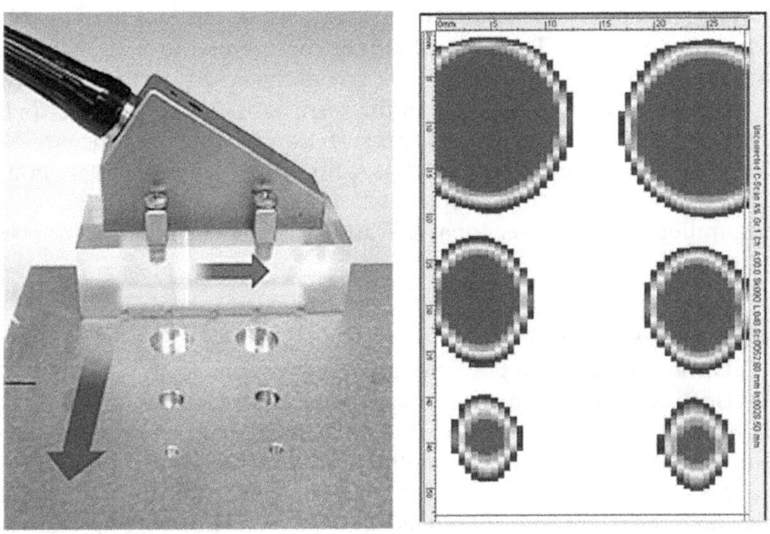

**FIGURE 12-20** C-scan generation with linear array probe.

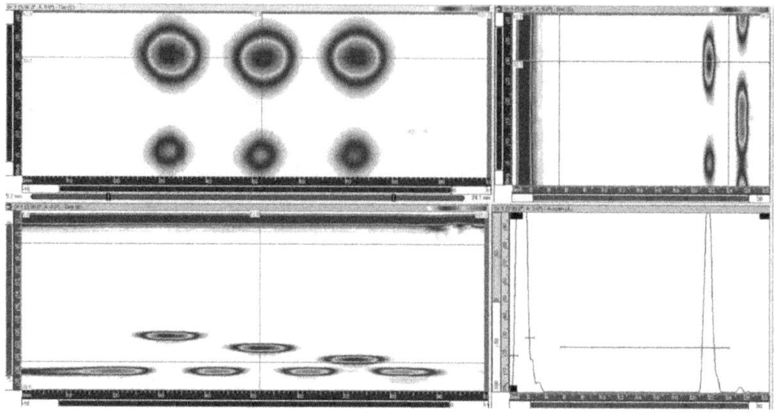

FIGURE 12-21 One-line scan shown with sourced linear scan, cross-sectional B-scan and A-scan.

system is not, and it can cost significantly less. Additionally, a phased array image can often be made in a few seconds, whereas a conventional immersion scan will typically take several minutes. Wider area C-scans can be made by using a two-axis encoded scanner using a large index (90 percent transducer aperture for overlap) and stitching each one-line scan together.

### *Cross-Sectional B-scans versus Linear Scans*

Once an encoded scan is produced via a one-line scan, cross-sectional data is available in two planes as shown in Fig. 12-21. In the one-line scan (upper left corner), the horizontal cursor represents the distance traveled along the length of the plate at a selected aperture or focal law. The vertical cursor represents the data collected through electronic steering of the virtual aperture across the width of the probe (linear scan) at a selected position along the length of the scan. Cross-sectional data can be sourced with the vertical cursor to display the linear scan (upper right) and the cross-sectional B-scan (lower left). The A-scan (lower right) is from the cursor intersection point one-line scan. While the linear scan cross section is bound by the transducer's width and element configuration, the cross-sectional B-scan is limited by the encoder resolution and the total length scanned.

### *Combined Image Display*

Phased array images are a powerful analysis tool, thanks to their ability to provide real-time visualization of volumetric data. Through the electronic scanning process, imaging truly becomes real time and is used in both manual and automated systems to increase the probability of detection. The ability to display multiple image types and store complete raw waveform information for the entire inspection allows post-scanning analysis of the data, especially in automated and more advanced phased array instruments. Because all the ultrasonic waveform data is collected, this post-analysis allows reconstruction of sectorial scans, C-scans and/or B-scans with corresponding A-scan information at any inspection location. For example, the screen in Fig. 12-22 simultaneously

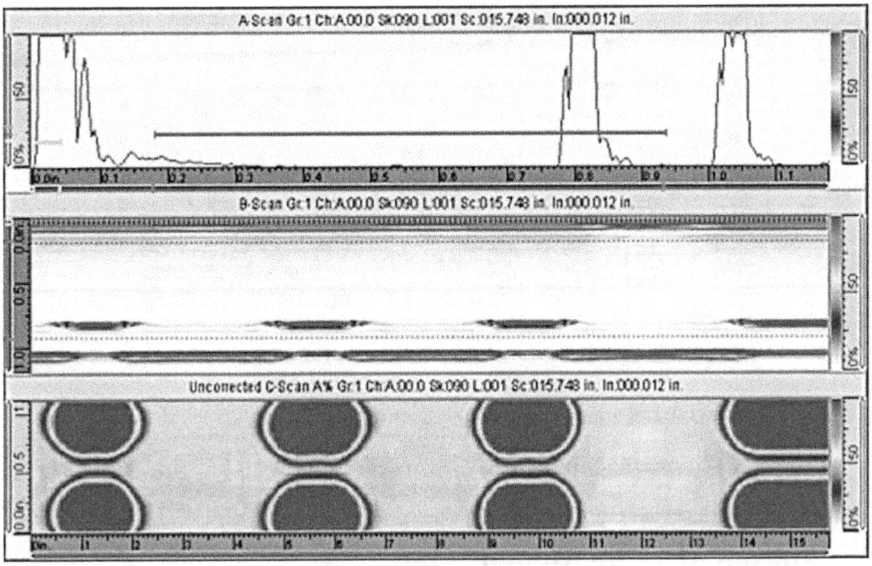


**FIGURE 12-22** Simultaneous A-scan, cross-sectional B-scan, and C-scan display.

displays the rectified A-scan waveform, a cross-sectional B-scan profile, and a C-scan image of a set of reference holes in a steel test block.

### *Scan Rate and Data Acquisition*

When generating B-scans or C-scans, a phased array probe may be moved either by hand or by an automated scanning fixture. In either case, data acquisition may be either free-running, based solely on the instrument's update rate, or correlated to probe position through the use of electromechanical encoders. As noted previously, correlating ultrasonic data with actual transducer position allows plotting a proportional view and matches data to specific areas of the part being inspected. The encoder records the location of each data acquisition with respect to a desired user-defined scan pattern and index resolution.

In order to avoid gaps in data acquisition, it is important to consider the speed at which the probe is moving and the distance resolution of the encoder. In short, the instrument's data acquisition rate must be greater than the scanning speed divided by the encoder resolution. The acquisition rate is determined by instrument design and setup. Pulse repetition frequency (PRF) and the number of focal laws being generated for each acquisition are key factors affecting data acquisition rate. The PRF divided by the number of focal laws represents the fastest possible acquisition rate for a phased array system. However, that number may be further adjusted by factors like averaging, digital sampling rate, and processing time. Consult the instrument manufacturer for details.

Once the acquisition rate has been established, the maximum scan speed may be calculated based on the desired encoder resolution, or vice versa. The effect of excessive scanning speed for a given encoder resolution is loss of signal and image gaps, as seen in the scan images in Fig. 12-23.

$$\text{Acquisition rate} > \frac{\text{scanning speed}}{\text{encoder resolutions}}$$

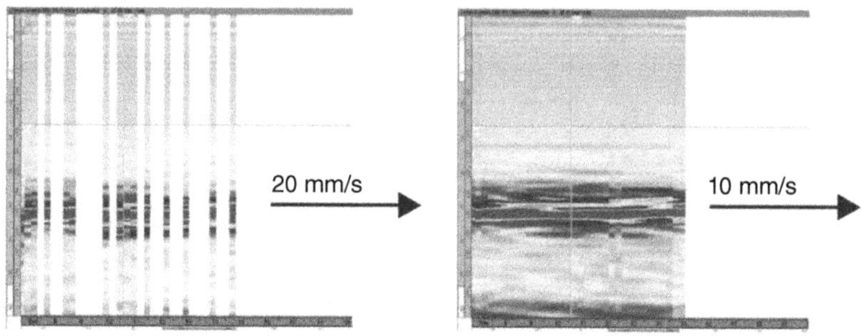


**FIGURE 12-23** Data loss due to excessive scan speed.

## 12.3 Phased Array Instrumentation

As discussed previously, there are numerous configurations of phased array probes. While the linear array probe is certainly the most commonly used configuration, customized probes with high element counts and varying element placements are also available. These probes are often designed to meet demanding application needs requiring high-speed full volumetric coverage and/or complex beam steering. As a result, there are varying levels of phased array instrumentation now commercially available in three general classifications: field portable manual, field portable automated, and rack instruments for in-line inspection.

### 12.3.1 Basic Concepts

The fundamental requirement of all phased array instruments is the ability to configure a defined group of elements with the programmed pulser and receiver delays commonly referred to as a focal law. The instrument forms images by sequencing focal laws with different pulser and receiver delays across the same aperture, or multiplexing the same law across stepped apertures.

During pulsing, a trigger is sent to a bank of pulsers with the delay sequence required to generate the desired beam in the material. The received signals are digitized and delayed according to a focal law and summed to form a single RF response. This waveform is then amplified, filtered as required, processed, and stored. A conceptual block diagram of a focal law as applied is shown in Fig. 12-24.

### 12.3.2 Important Specifications for Phased Array Instruments

Many phased array instruments include a conventional ultrasonic (UT) capability. This allows an operator to check phased array inspection results with manual UT techniques as required. In some cases, simultaneous UT and phased array testing can increase coverage. Inclusion of multiple NDT methods increases the instrument's flexibility to maximize its use. It is important to verify that the performance and features of the UT mode in a phased array instrument are not compromised.

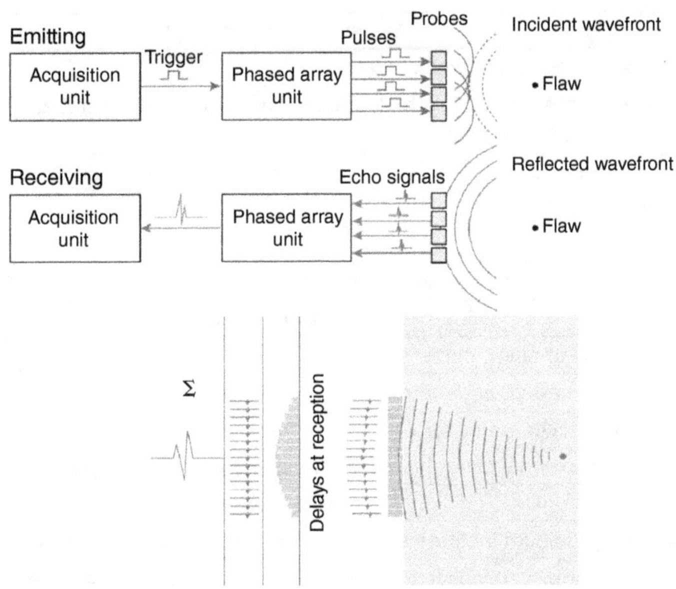


**Figure 12-24** Beam forming and time delay for pulsing and receiving multiple beams (same phase and amplitude).

When evaluating UT instruments, users often specify a number of functional characteristics including receiver bandwidth, filtering, pulser performance, measurement gate control, readings format, and available inputs and outputs. In addition, many flaw detection standards and codes have been developed and are employed for sizing a variety of defects. Certain inspections require that a specific code be followed. These codes can apply to the inspection of welds, volumetric inspection of forgings, and many other metallic and composite structures. As a result, a variety of analysis and measurement tools have become available in conventional digital flaw detectors to automate and record information required by these codes. A summary of typical UT specifications show in the following table.

| Pulser |
|---|
| Type: Spike or Square Wave |
| Energy: Higher voltage increases penetration |
| Tuning frequency |
| PRF and measurement rate |
| **Receiver** |
| Bandwidth |
| Available filters |
| Time varied gain |
| Overall dynamic range |
| Rectification modes |

| Measurement and Display |
|---|
| Number of alarm and measurement gates |
| Measurement modes: sound path, depth, distance from front of probe, dB, dB to curve, leg, echo to echo, interface gating |
| A-scan display modes: minimum, maximum, composite, hollow, filled, peak memory |
| Measurement resolution |
| Range |
| Single-value B-scan (not available on most flaw detectors) |
| **Inputs and Outputs** |
| Number and type of alarm outputs |
| USB for printing, saving, or data transfer |
| Availability of encoder inputs for linking data to position |
| Trigger input for external control of pulser firing and acquisition cycle |
| **Sizing Options** |
| API American Petroleum Institute standard practice for ID crack sizing in piping |
| AWS D1.1, 1.2 American Welding Society D Rating for qualification of weldment |
| DAC Distance Amplitude Correction |
| DGS Distance Gain-Sizing Technique |

Because of the multi-element nature of phased array instruments, there are additional key specifications that need further consideration and review.

**Pulser Voltage:** Phased array does not require as much pulser voltage to drive the elements of probes as does conventional UT instruments. Typically, phased array instruments allow low- and high-voltage selection of a tunable square wave of ≤100 volts.

**Number of Pulsers:** Will define the maximum number of elements that can be grouped to form an active aperture, also referred to as virtual probe aperture.

**Number of Receivers:** Will define the total number of elements that can be used for sequencing apertures that lead to the potential increase in coverage from a single probe footprint.

***XX:YY:*** Naming convention used where *XX* = number of pulsers and *YY* = number of receiver paths. The number of receivers is always greater than or equal to the number of pulsers. Instruments from 16:16 to 32:128 are available in field portable packaging. Higher pulser and receiver combinations are available for in-line inspection and/or systems that use larger element count probes.

**Focal Law Number:** The maximum number of focal laws that can be combined to form an image is often specified. In general, higher *XX:YY* designated configurations can support more focal laws. However, increasing the number of focal laws does not always mean more functionality. Take the example shown in Fig. 12-25 that uses a 16-element probe to produce an S-scan of three side-drilled holes from 40 to 70 degrees with steering resolutions of 0.5 degree (64 laws) and 4 degree (8 laws). While the image will be slightly better defined with finer angle increments, detection at coarser resolution is adequate.

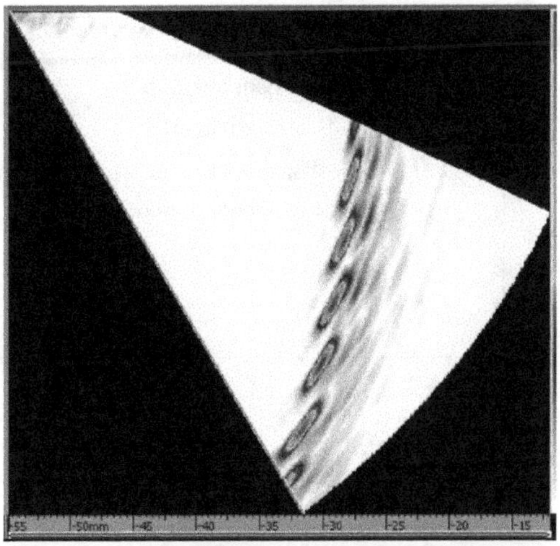

0.5 degree increment focal law

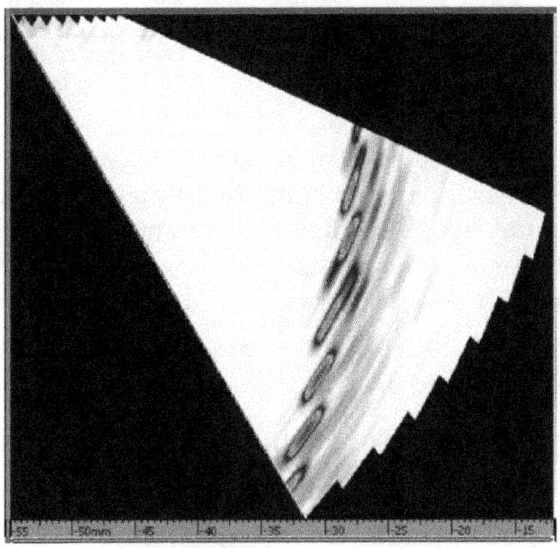

4 degree increment focal law

**FIGURE 12-25** Changes in image resolution due to finer and coarser angular increment.

The number of focal laws required for a particular inspection is dependent on a number of factors. These include aperture of probe, number of elements used, and type of scan; S-scan or linear scan. For example, Fig. 12-26 shows the number of focal laws required for performing linear scans with varying combinations of virtual probe apertures and total number of elements assuming a single-element step.

**Image Display Update Rate:** Instruments can vary greatly in display update in various image modes.

For phased array imaging modes:

**Maximum Image Display Rate:** This equals PRF ÷ Number of Focal Laws, where maximum image display rate is limited to 60 Hz.

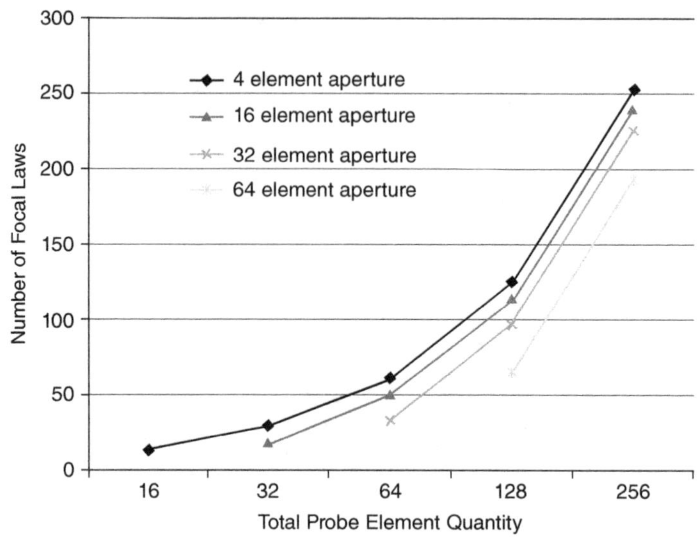


**FIGURE 12-26** Relationship between elements and number of focal laws.

**A-scan Display Update Rate:** The A-scan display update rate of a single focal law will vary between instruments. In some instruments, the A-scan update rate is limited by the maximum image display update and may vary if the A-scan is displayed independently or with an image. For this reason, in some applications, it may be important to verify A-scan display update in various image display modes.

**A-scan Display Types:** All phased array instruments allow the ability to select a single A-scan from the phased array image. Additional capabilities may include:

- Simultaneous display of multiple A-scans from a PA image
- Highest amplitude A-scan from a gated region of a phased array image
- A composite waveform constructed from all the steered A-scans used to form the image

**Maximum Measurement and Acquisition Rate:** This rate typically matches the pulse repetition frequency for generation of the entire focal law sequence used to create the desired image. The PRF can be limited by range, spurious wrap-around signals, and the number of A-scan points to be stored per focal law. As in conventional flaw detectors, some phased array instruments allow a PRF greater than the display update rate, whereas others are bound by a display update limit of 60 Hz or less.

**Probe Recognition:** This is the ability to recognize phased array probes that reduces operator setup time and errors by automatically configuring an instrument setup with the proper number of elements and probe geometry.

**Image Types:** Sectorial and linear scans are typically available in phased array instruments. The ability to stack these image modes to create amplitude and depth C-scans allows planar images to be formed and provides expanded means for sizing defects.

**Image Sourcing:** The ability to extract one type of image from another. For example:

- An A-scan from S-scan, linear scan, C-scan, and/or cross-sectional B-scan
- An S-scan, linear scan, and/or cross-sectional B-scan from a C-scan

**Image Linking:** The ability to link and synchronize the displays of image-sourcing hierarchy.

**Image Display:** The number, layout, and definition of the windows that can be used to simultaneously display an image and A-scan data may vary between instruments.

**Data Storage:** Some instruments allow the ability to store raw data, including all RF waveforms. This is particularly useful when collecting data over large areas and when 100 percent inspection record needs to be maintained. Full data sets of encoded scans also allow the ability to analyze the full volume of the inspected parts by sectioning the data as required. The number of waveform points to be stored is often user selectable. More basic instruments may save data from the image, but not the A-scan, whereas others store only a snapshot of an image and/or A-scan.

**Data Reprocessing:** This is the ability to change gate positions, gate types, positions, or focal law to reprocess and display updated image data and measurements from a stored data set. This typically requires an instrument with full waveform storage capabilities.

**Multigroup Support:** More advanced phased array instruments allow multiple focal law sequence groups to be sequenced on one or more connected transducers. Each group is commonly referred to as an inspection channel. This is especially useful in cases in which it is important to collect volumetric data which will be analyzed offline.

For example, a 5-MHz, 16-element probe can be programmed for a 40- to 70-degree sector scan, while a second group can be defined for a 5-MHz 64-element probe to perform a 60-degree linear scan with an aperture of 16 elements, stepping by 1 element over the entire 64-element length. To summarize the possibilities:

- Multiple channels set up with the same phased array probe
- Multiple channels set up with different phased array probe
- Multiple channels defined across NDT modalities, such as phased array with conventional UT and/or TOFD

The number and types of groups supported vary by instrument.

**Encoding:** There are two classes of instruments generally available: manual and encoded. A manual phased array instrument works much like a conventional flaw detector as it provides real-time data. Along with an A-scan, the instrument also shows real-time S-scan or linear scan images, which can aid in detection and discontinuity analysis. The ability to use and visualize more than one angle at a time in a test would be the main reason for using this type of instrument. In some cases, such as crack sizing, the image can be used as a tool to help size crack depth.

A phased instrument with encoding capabilities merges programmed focal law sequences and probe positional data to accumulate planar and cross-sectional images of the test specimen. These encoded images allow for planar sizing of defects. In instruments that also store full waveform data, images can be reconstructed to provide cross-sectional views along the length of the scan or to regenerate planar C-scans at various levels.

**Reference Cursors:** Instruments will provide various cursors that can be used on an image as aids for interpretation, sizing, and depth measurement. In a sectorial scan, it is possible to use cursors for measurement of crack height. Approximate defect size can be measured as area in encoded C-scans and in length or width with cross-sectional B-scans or linear scans as well.

## 12.4 Basic Techniques for Instrument Setup

### 12.4.1 Initial Instrument Setup

As in the case of conventional ultrasonic testing, any phased array inspection begins with a basic instrument setup, which includes parameters related to both the probe being used and the test piece that is being inspected. Proper setup is essential for proper performance. This section summarizes the initial setup procedure for typical phased array instruments.

#### *Probe-Related Parameters*

As in conventional ultrasonic inspection, knowledge of probe frequency and the application allows proper setting of receiver filtering, rectification, and gain, as well as pulser frequency and voltage. Refracted wedge angle, when utilized, allows proper geometrical measurements to be made when the material velocity has been established. In order to correctly set up and use any phased array instrument and probe combination, a number of additional key parameters must be input in order to enable proper focal law calculation and dimensional mapping.

For a phased array probe, the total number of elements, element pitch, and element width must be known. The coupling method must be specified, whether it be through direct contact, a 0 degree wedge, an angled wedge, or immersion. While the method and terminology that describe delay line and wedge parameters may vary slightly with different manufactures, they all target geometrical and acoustic descriptions required for focal law and image generation. Key parameters shown in Fig. 12-27 include incident angle, wedge/delay velocity, height offset (z), and distance from

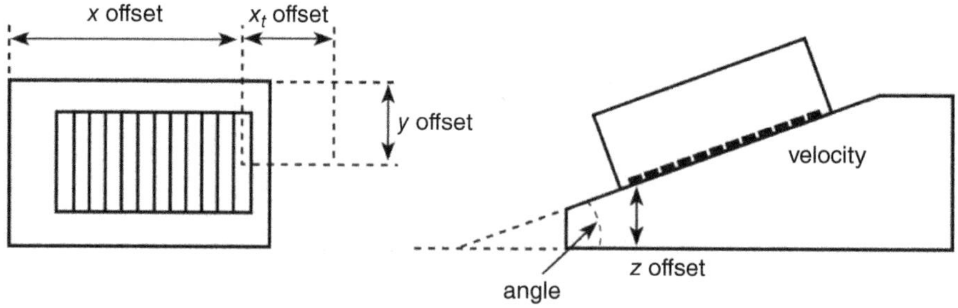


**FIGURE 12-27** Wedge/probe dimensional parameters.

front of wedge ($x$) to the center of probe's first element. Additionally, the distance from the side of the wedge to the center of the elements (scan offset $y$) is useful when stitching together a scan where the probe movement is tracked in both $X$- and $Y$-axes mechanical motions.

With many instruments, the important pulser/receiver and probe parameters can be set automatically by probe recognition software that identifies the transducer being used and calls up a programmed default setup. Similarly, wedge selection menus allow the operator to choose a wedge type, from which the instrument automatically recalls the necessary setup values. In cases in which probe recognition software is not available, the user must enter these parameters manually.

#### Material-Related Parameters

Sound velocity in the test material must be entered or calibrated (as discussed in Section 12.4.2) in order to properly calculate delays to produce desired refracted beams and to measure resulting sound paths. The geometric parameters of the test piece thickness and diameter can also be specified for compensating for the effect of curvature or depth and distance measurements when performing circumferentially oriented tests on pipes.

#### Focal Law Sequence Parameters

Once all the necessary parameters have been entered for calculation of intended beams, the end user must specify the sequence of focal laws that will be created by the probe and plotted as an image in real time. For the purpose of this overview we will concentrate on linear arrays with the understanding that many different probe configurations exist with more complex steering sequences (such as 1.5d, 2d, dual linear, annular, etc.). Examples of advanced focal laws include dynamic depth focusing, lateral skewing, or compound focusing in combination with angle steering.

As already discussed in detail, for linear array probes there are two major fundamental imaging techniques, sectorial and linear scans. In both scan types, the number of elements forming the virtual probe aperture and focus depth to be used for each focal law is defined. For S-scans, the start angle, end angle, and angular step size are entered to generate the desired angular sweep and resolution. For linear scans the user needs to input the total number of elements over which the aperture can be swept as well as the element step from one virtual aperture to the next. This defines the electronic scanning range of the aperture.

### 12.4.2 Calibration Methods

Once the initial setup values have been recalled or entered, it is necessary to further calibrate the phased array instrument for optimum performance and accuracy with a given combination of probe and material. Typically, velocity, wedge delay (zero offset), and sensitivity (gain) will be calibrated across all angles with the aid of appropriate reference standards. Most instruments incorporate calibration software that will prompt the user through the necessary steps of each process. Consult the operating manual of your instrument for specific procedures. The reference standard used for setup may be an IIW-type test block made of the test material, or other appropriate blocks, if required by specific inspection procedures. Note that sound velocity calibration should always be performed before wedge delay calibration, since material velocity information directly impacts calculation of beam steering parameters.

### *Velocity Calibration*

Velocity calibration is used to precisely calibrate the speed of sound in a test material. A default value for sound velocity is usually entered during the initial setup process. Velocity calibration then establishes a more precise value for the specific test material. This process requires a reference block with two reflectors at a precisely known depth or distance, commonly the arcs or back wall of a test block. Velocity calibration is commonly performed by means of an echo-to-echo measurement between two known reflectors in order to obtain a velocity measurement independent of wedge delay or refracted angle.

For setups using angle wedges, velocity calibration is commonly performed with either the two radii of an IIW-type block (sound path) or two side-drilled holes (depth). In Fig. 12-28, the probe is positioned on an IIW block at the 0 mark, aimed at the 100 mm and 50 mm radii. In the first step, the calibration gate is placed on the echo representing the 50 mm sound path, the echo is maximized, and then the sound path distance represented by the echo is entered into the calibration menu. The gate is then repositioned around 100 mm for input of the second depth required for calibration and the process is repeated.

The instrument then calculates the sound velocity corresponding to the measured pulse transit time across the known sound path represented by the distance between the two reflectors. This measurement is independent of zero offset.

In the case of setups using straight wedges or no wedge, the calibration is performed with two echoes from flat plate reflectors, such as a step block or the sidewall of an IIW-type calibration block as shown in Fig. 12-29. These echoes may represent either two different thicknesses, or two multiple back wall echoes from a single thickness.

### *Wedge Delay Calibration*

Wedge delay calibration corrects for the delay between the electronic firing of the transducer elements and the point in time at which the steered sound beam enters the test piece after passing through the wedge. This parameter is known as "zero offset" in conventional ultrasonic testing. In phased array testing, this is a more complex calibration

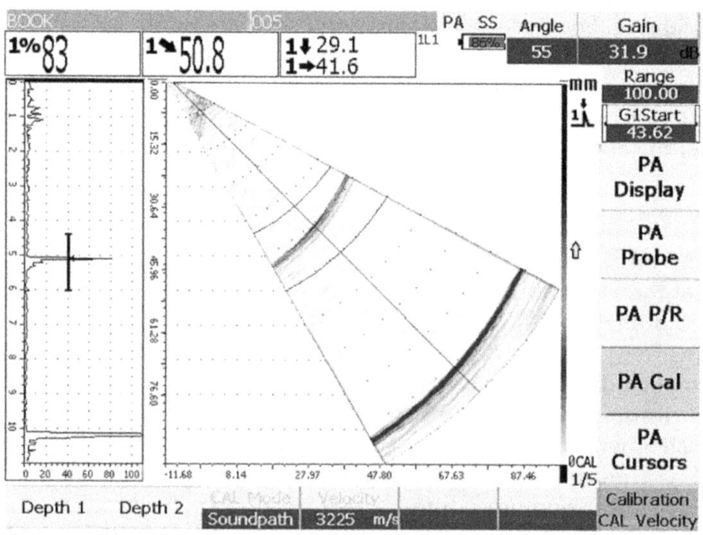


**FIGURE 12-28** Velocity calibration method using IIW block radii.

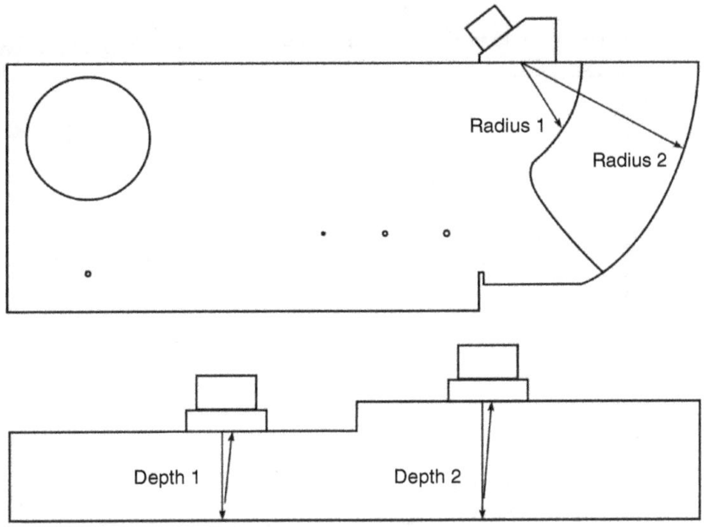


FIGURE 12-29 Common test blocks for velocity calibration.

because each focal law in an angle scan has a different wedge delay. Each angular component of the scan exits the wedge at a different point and therefore travels through a different amount of wedge material before reaching the test piece. Thus, the instrument must apply a different time correction to each focal law. While this is pre-calculated along with the delay laws when a wedge type is entered from a setup menu, this value can be further adjusted to correct for small measurement errors.

This calibration can be performed from a single radius or side-drilled hole. One common reference for S-scans is the 15-mm-deep side-drilled hole in an IIW-type reference block. In Fig. 12-30, after the appropriate calibration mode was selected, the echo from this hole was peaked at a middle angle (55 degrees) and the known depth of the hole was entered. The probe was then scanned across the hole so that a reflection is obtained at all angles in the scan, in this case 40 to 70 degrees. Any measurement variations between angles are automatically normalized via wedge delay adjustment, thus maximizing the accuracy of depth and distance measurements. In this case, a zero offset error at the higher angles that initially led to erroneous depth measurement of a 15-mm-deep hole was corrected following wedge delay calibration.

As can be seen in Fig. 12-30, some instruments have the capability to display the image corrected, so as to display the image at the true depth of the reflector at each calibrated focal law. While this apparently "turns" the defect, it does allow true post-acquisition analysis from the image data. Other instruments calculate the live measurement without change or correction of image data, limiting post-analysis capabilities.

A wedge delay calibration is also required when using a straight wedge. In the case of a normal beam linear scan, this can simply be performed on a flat test block where all parts of the reflector are an equal distance from the probe. In the case of low-angle sector scans utilizing straight wedges, calibration is performed by sweeping the beam across a reflector of known depth, typically a side-drilled hole as found in an IIW block. An exception occurs during inspection of anisotropic materials. For this specialized refracted shear inspection, it is common to input an average material velocity. A matrix

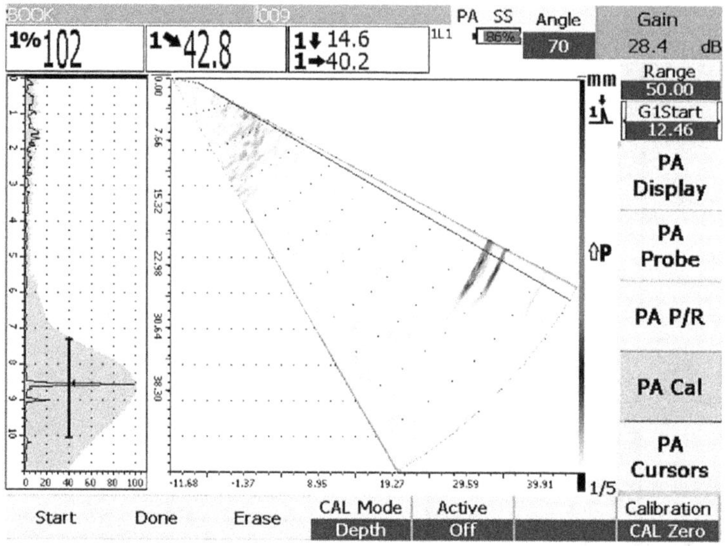


Uncalibrated S-scan

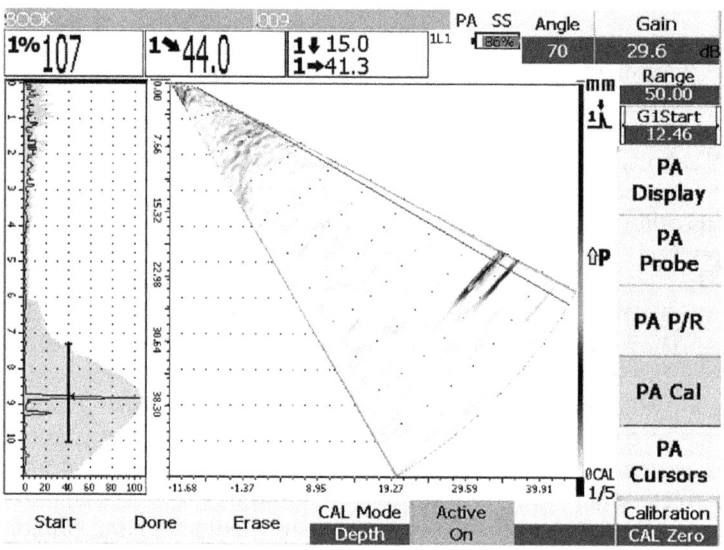


Calibrated S-scan

**Figure 12-30** Sector scans shown before and after wedge delay calibrations.

of wedge delay adjustments can be created using a reference block of the anisotropic material. The appropriate wedge delay is then individually modified at the angle of detection during the inspection.

### *Sensitivity Calibration*

Just as wedge delay varies across the angles of a sector scan, so does sensitivity. This variation comes from the combination of three factors. These are: changes in the length of the sound path and, thus, the attenuation within the wedge, the fact that the mechanical refraction used along with electronic steering to generate a beam is less efficient for sound coupling at higher angles approaching 70 degrees, and element-to-element sensitivity variations within the probe. Depending on the type of instrument being used, some manufacturers provide one-step sensitivity normalization across all angles, whereas others require each focal law to be calibrated separately. Figure 12-31 shows the former.

In Fig. 12-31, a test block with a row of equally sized horizontal holes was used as the reference reflector. As with wedge delay calibration, the probe is scanned across the target so that a reflection is captured from all angles of the beam. As shown in the image, sensitivity drops off as the angle increases, so the holes captured by higher angle components of the beam appear lighter than those in the lower angles. Following normalization, sensitivity is equalized across all focal laws and all holes are displayed at the same amplitude.

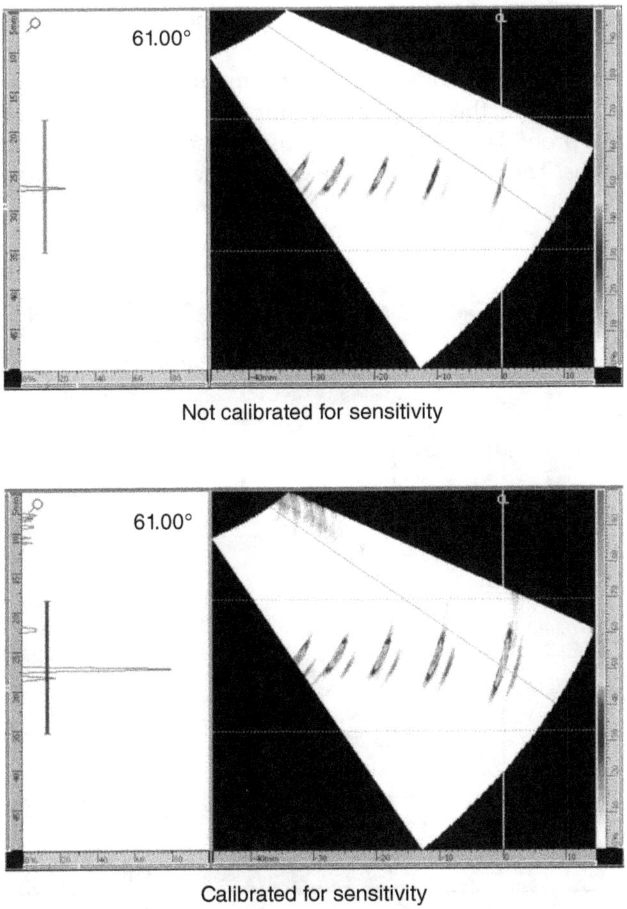


**FIGURE 12-31** S-scans before and after sensitivity calibration.

## 12.5 Evaluation of Test Results

It must first be noted that proper interpretation of the images generated by phased array instruments requires a good understanding of the principles of conventional ultrasonic flaw detection. While phased array instruments are characterized by both the flexibility offered by beam steering and the visualization aid offered by graphic displays, the physics factors related to wave propagation and reflection from flaws are exactly the same as in conventional ultrasonic testing. Notably, these include beam spreading effects. Additionally, as with any application of ultrasonic NDT, appropriate operator training in instrument operation and data interpretation is essential for reliable results.

### 12.5.1 Depth and Distance Measurement

Like most contemporary conventional flaw detectors, phased array instruments calculate and display a variety of depth and distance measurements based on instrument calibration, measured sound path length, and trigonometry, using echoes that have been selected by means of flaw gates. These measurements must take into account the sweeping of beam angle and beam index point that can be found in phased array applications.

The simplest case involves straight beam linear scans, in which the beam advances across the length of the array probe but the angle remains constant. In Fig. 12-32, a 64-element probe is used to scan across a test block that contains reference holes at varying depths. The vertical cursor selects an aperture from which a gated A-scan and its

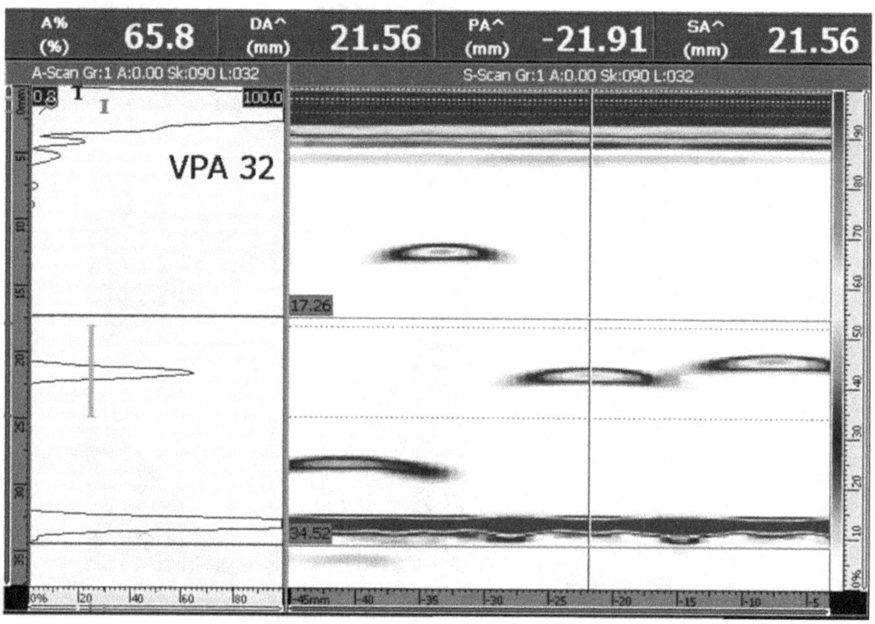


**FIGURE 12-32** Depth measurement from linear scan.

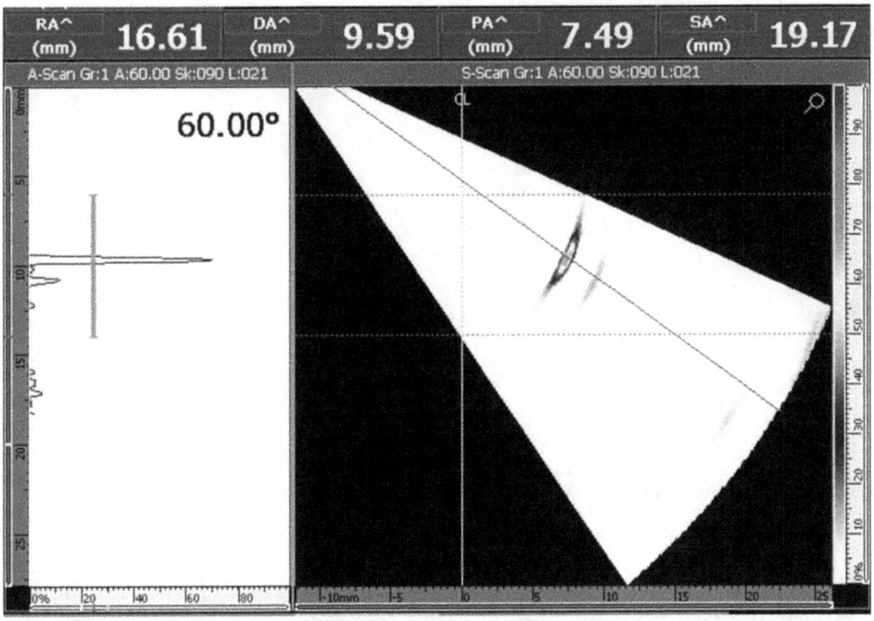


**FIGURE 12-33** Dimensional measurements from the 60-degree angle of the S-scan.

associated measurements are displayed. From this A-scan, the depth is measured as 21.56 mm. Since this is a straight beam scan, reflector depth and sound path distance are equivalent, and depth measurement is straightforward.

Angled scans require trigonometric interpretation for calculation of reflector position, and in the case of sector scans, require calculation across multiple angles as well. Figure 12-33 shows an example of a simple A-scan/S-scan display with measurements of the echo from a reference hole in a test block, with the 60-degree beam component selected by the cursor.

The scale on the left side of the screen shows calibrated depth for both the A-scan and the S-scan images, and the scale at the bottom of the S-scan shows surface distance relative to the front of the wedge. The vertical cursor marks the relative position of the front edge of the wedge on the test piece. The angled cursor on the S-scan identifies which angular component of the beam is being measured, in this case, 60 degrees. The horizontal lines indicate the limits of the flaw gate within the image.

The readouts across the top of the screen show the most commonly measured dimensions as outlined in Fig. 12-34. These are: forward distance from the sound exit point of the wedge to the reflector (RA), reflector depth (DA), forward distance from the front of the wedge (PA) to the reflector, and distance to the reflector along the sound path (SA). Because the beam exit point in phased array ultrasonics typically moves between focal laws, knowing the reflector distance from the front of the wedge is typically more useful than the distance from the variable beam index point.

As with conventional flaw detection, when using angled beams in phased array tests, reflectors will commonly be found in skip paths beyond the first leg of a sound beam. S-scans normally represent this by flipping the second leg of the sound path into

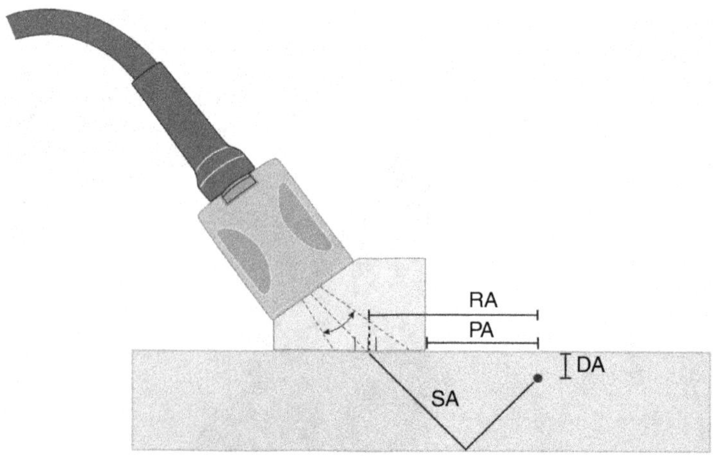


**Figure 12-34** Designation of dimensions measured at each angle.

the lower part of the cone-shaped display, as seen in Fig. 12-35. While the beam actually reflects upward from the bottom of the test piece, the image displays it as if it were continuing along its path for twice the thickness of the test piece.

In the case of swept angle scans, interpretation requires operator understanding of the fact that an S-scan will commonly display indications at different angles in multiple legs, as seen in this screen. Thus, the cursors marking skip legs are very important. Figure 12-36 illustrates a test block with a side-drill reflector that produces three indications from a 40- to 70-degree S-scan. In Fig. 12-37, the dotted horizontal white cursors

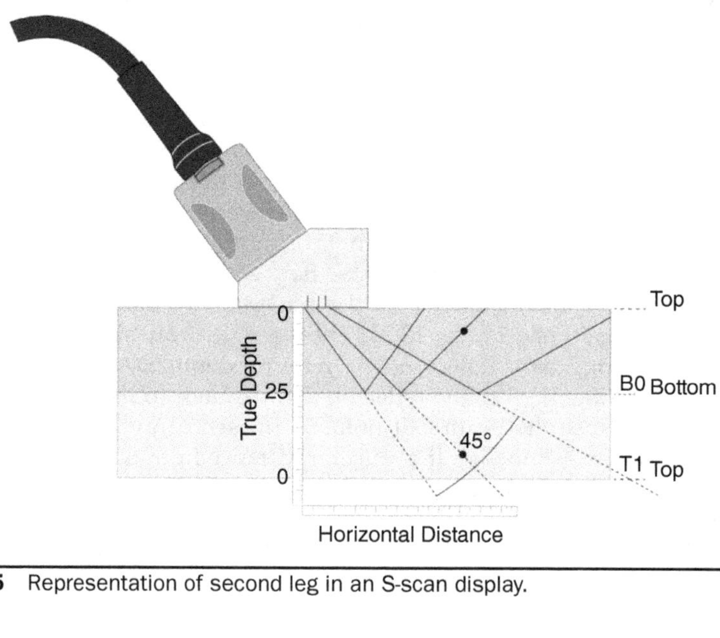


**Figure 12-35** Representation of second leg in an S-scan display.

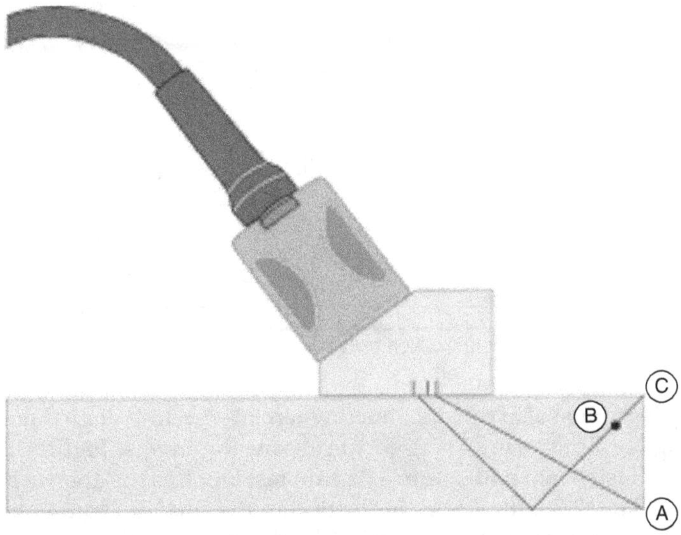


**FIGURE 12-36** Angular beam paths in test block.

represent the boundaries between the first and second legs, and second and third legs. Three relevant indications are seen in the scan. Indication A, in the first leg of the 65-degree beam component, comes from the lower corner of the test block. Indication B, in the second leg at 45 degrees, comes from the reference hole. Indication C, peaking at a slightly higher angle, comes from the upper corner of the block at the transition between the second and third legs.

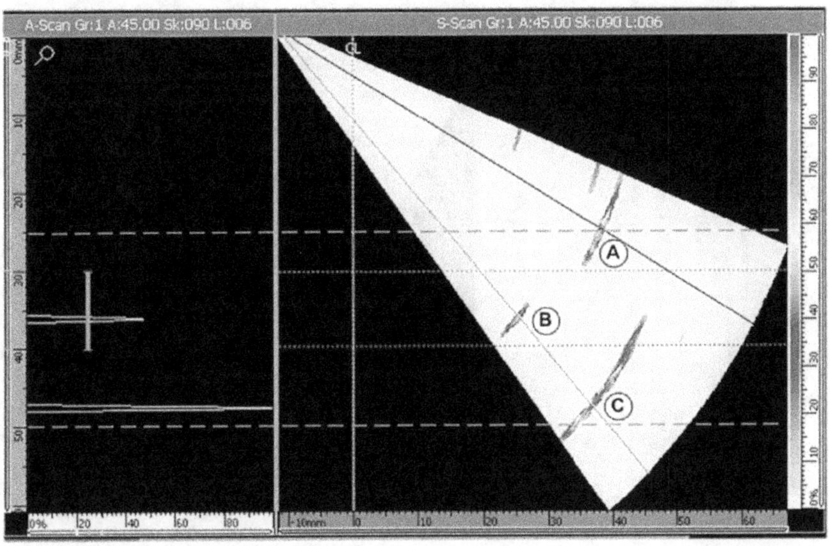


**FIGURE 12-37** S-scan showing multiple reflectors appearing in the first or second leg.

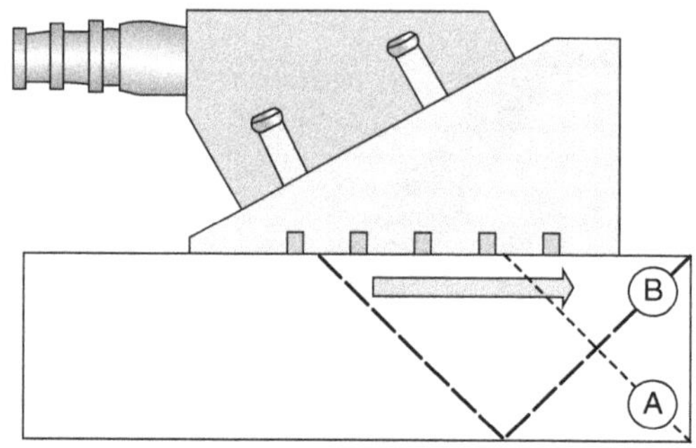


**Figure 12-38** Beam paths in 25-mm test block.

Skip legs must also be considered when interpreting angled linear scans. The same "flipping" or "mirroring" is used to present the image. Figure 12-38 shows the first and second leg beam paths from a 25-mm test block as the aperture is moved across the probe. In Fig. 12-39, the upper indication in the images represents the reflection from the bottom corner of the block, marked by the dotted horizontal cursor as the end of the first leg, and the lower indication represents a second leg indication off the upper corner, coming from the opposite end of the linear array.

### 12.5.2 Sizing Methods

Phased array instruments commonly incorporate flaw-sizing software that is derived from conventional flaw detection techniques. These include distance amplitude correction (DAC) and time-varied gain (TVG), DGS/AVG, and AWS D1.1 sizing algorithms, all of which are applied to the A-scans from which scan images are derived.

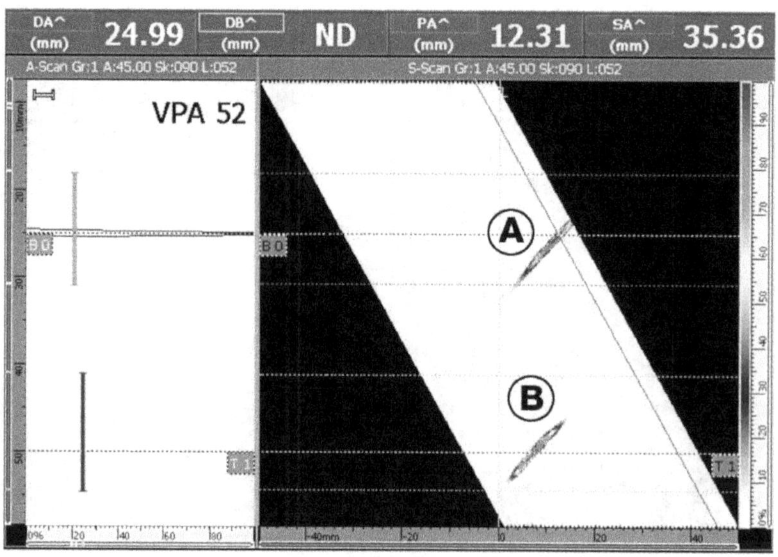


**Figure 12-39** First and second leg indications from linear scan.

Phased array instruments also provide $X$–$Y$ cursors that can be used to estimate defect size from an image; however, these must be used with care since the information they offer is typically imprecise.

### *Direct Cursor Measurement*

As noted previously, cursors placed over an indication on a scan display can be used to accurately identify the location of a reflector with respect to its depth and/or position with relationship to the probe. However, flaw sizing from an image as opposed to depth/distance measurement from an A-scan must be approached with caution and awareness of the uncertainties created by beam shape, probe element size, scan index, and the interactions of beams and reflectors. All of these have a modifying effect on the shape of a display image. This is especially true in angle scans where reflector indications can be stretched out because they are seen at multiple angles within the scan. Measurements from a scan image are typically approximations only.

Sizing by cursor measurement from an image typically works best in linear scans that identify relatively large laminar defects, utilizing measurement of indication width at the 50 percent (–6 dB) level with $X$-axis cursors. Figure 12-40 shows the indication from a 10-mm-diameter flat bottom hole in a test block. In this case, the cursors measure flaw width as 8.5 mm. The accuracy of sizing depends on aperture and beam spread as in conventional UT. In addition, the element pitch establishes the increment of each focal law. Smaller pitch will allow better definition of the edges of the defect.

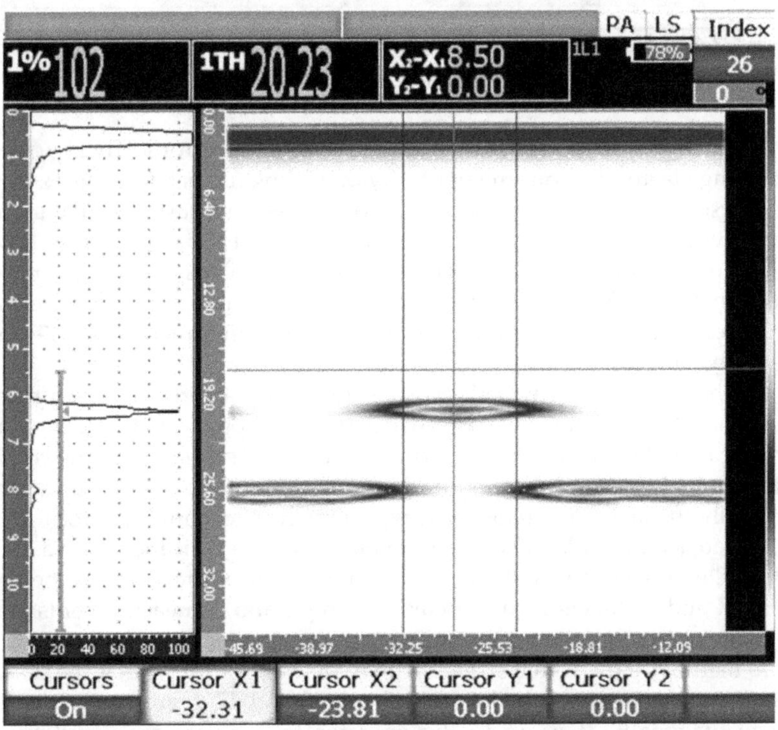


**Figure 12-40** Sizing defect from linear scan using image cursors.

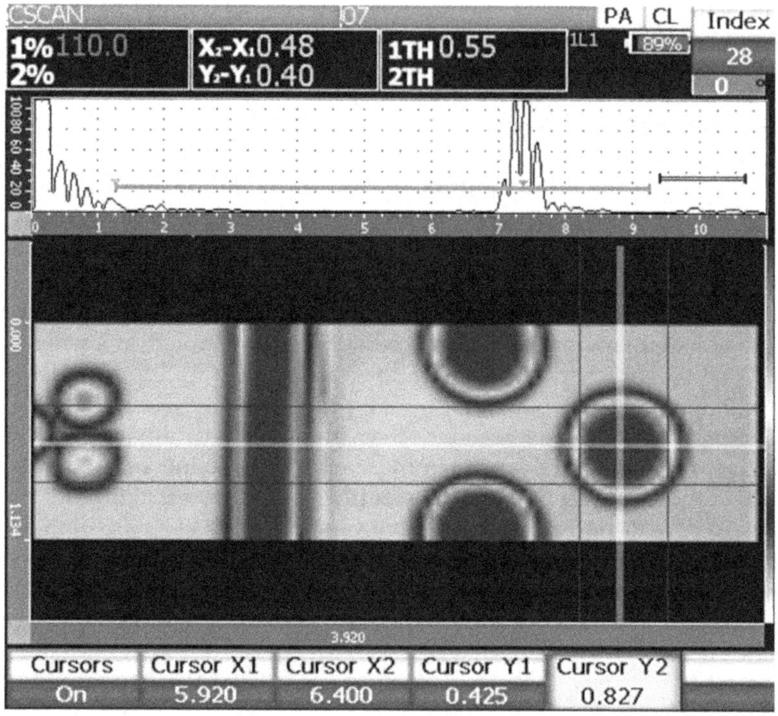


**FIGURE 12-41** Sizing indication with image cursors on C-scan.

Phased array systems offer the further ability to electronically sharpen a beam focus for sizing clarity without the necessity of adding a mechanical lens.

Cursors can also be applied to encoded C-scan images to size indications seen in planar view. In Fig. 12-41, pairs of horizontal and vertical cursors have been moved over the indication from a reference hole. The instrument measures the separation between the two $X$-axis cursors and the two $Y$-axis cursors to calculate width and height in the scan plane. The $X$-axis cursors size in the scan direction, the $Y$-axis cursors along the probe length.

In the case of S-scans, reflectors are frequently oversized when measured from an image and thus this technique provides only an approximation of actual reflector size, which can be better determined from the A-scan. This is due to the combined effects of beam spreading, the inherent fuzziness of the outer edges of ultrasonic beams, and especially the fact that a reflector is typically seen from multiple focal laws representing several adjacent angular increments. In Fig. 12-42, the indication from a 1.25-mm-diameter side-drilled hole in an IIW block is measured at the approximate 50 percent (–6 db) level in the $X$ and $Y$ directions using cursors. The $X$ and $Y$ measurements are 2 mm, which, following a trigonometric calculation at 45 degrees, suggest a reflector width of 2.8 mm, more than twice the actual size. A primary reason for this is that the hole is reflecting all angular components of the beam between 43 and 47 degrees, effectively stretching the image diagonally. It is up to the operator to evaluate the significance of flaw size measurements from screen displays, based on knowledge of the application.

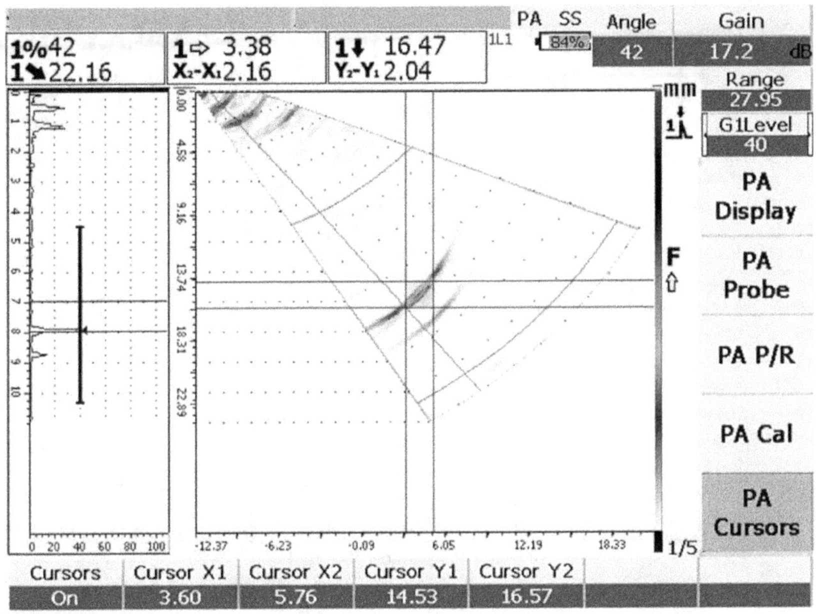


**Figure 12-42** Measurement error of cursor measurements on S-scan.

This general approach can also be used as part of a tip diffraction test to measure approximate crack height or depth with a calibrated instrument. In many cases, the base and tip reflections from a crack can be seen from a single probe position in a multi-angle scan (see Fig. 12-43), allowing dimensional measurement via either of two techniques.

Figure 12-44 shows the A-scan and S-scan response to shallow and deep reference notches in a test block. The large lower indication represents the reflection from the

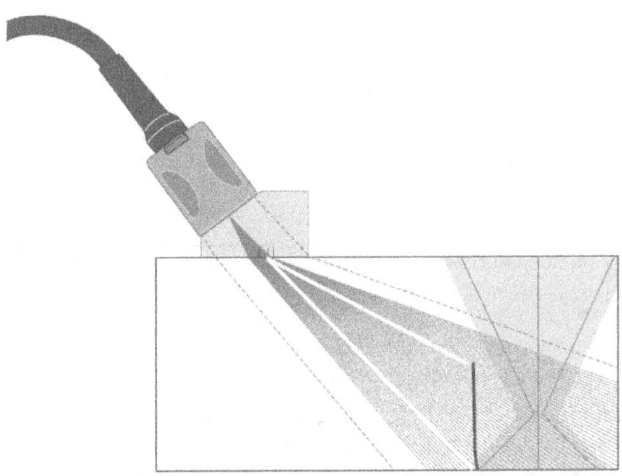

**Figure 12-43** Representation of tip diffraction technique.

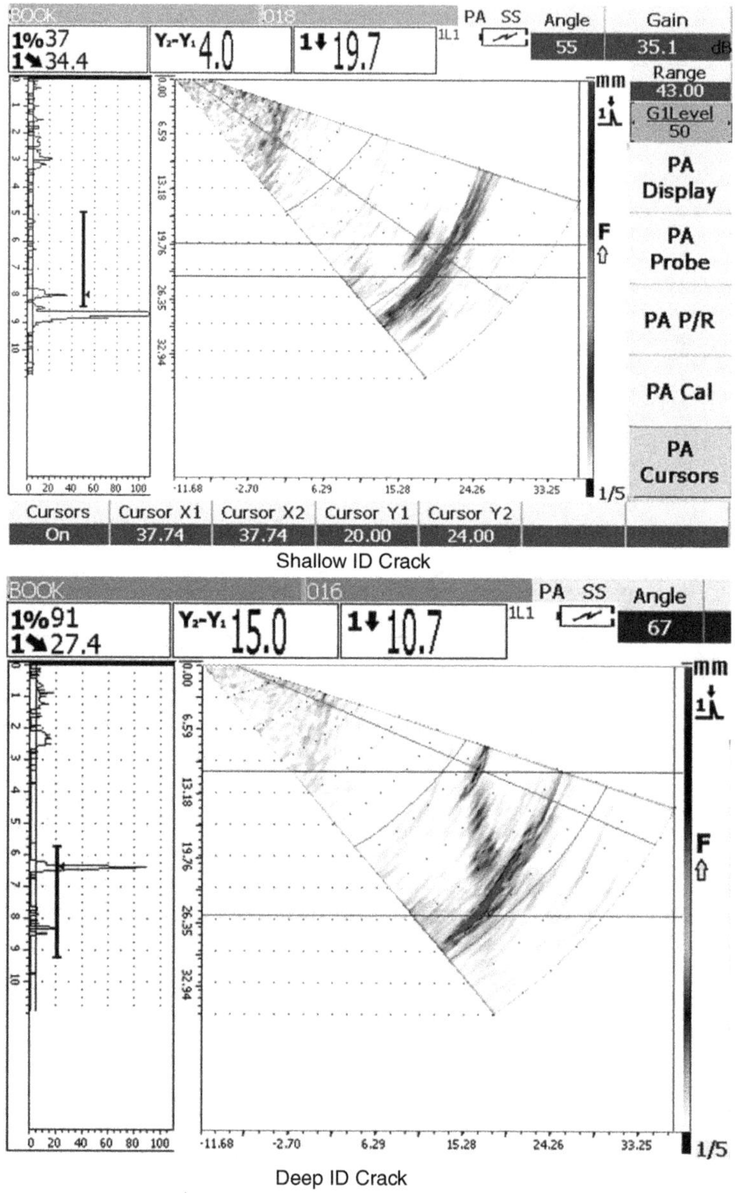


**Figure 12-44** S-scans showing tip diffraction from shallow and deep ID connected cracks.

base of the notch, seen across multiple angles, and the smaller indication represents the tip diffraction signal seen in higher angles. The angular cursor placed on the tip diffraction signal measures the depth of that reflector directly. Additionally, the two horizontal cursors placed at the center of both the notch base and tips are used to measure the notch height from the bottom of the test block. Note that in the case of the very large indication from the base, it is necessary to lower instrument gain when positioning the cursor to bring the indication out of saturation.

### *Time-Varied Gain and Distance Amplitude Correction*

In both conventional and phased array ultrasonic testing, echoes from reflectors of the same size decrease in amplitude with distance or depth because of material attenuation and beam spreading. Distance amplitude correction (DAC) and time-varied gain (TVG) are closely related techniques used to compensate for this effect when echoes are displayed. In phased array instruments, both DAC and TVG are set up by first normalizing gain across all focal laws in a scan using a single reference reflector, and then using multiple reference standards with equal size reflectors at different distances to create a matrix of echo amplitude versus sound path length through a range of distances, again across all focal laws.

Time-varied gain (also known as time-corrected gain, TCG) employs software to normalize the displayed echo amplitude from targets of the same size at varying distances or depths. It does this by using the amplitude/distance calibration matrix to increase receiver gain as reflector distance, and hence pulse transit time, increases. Three equal-sized holes at different depths in a test block appear at different amplitudes in a linear scan image (Fig. 12-45) due to attenuation and beam-spreading effects.

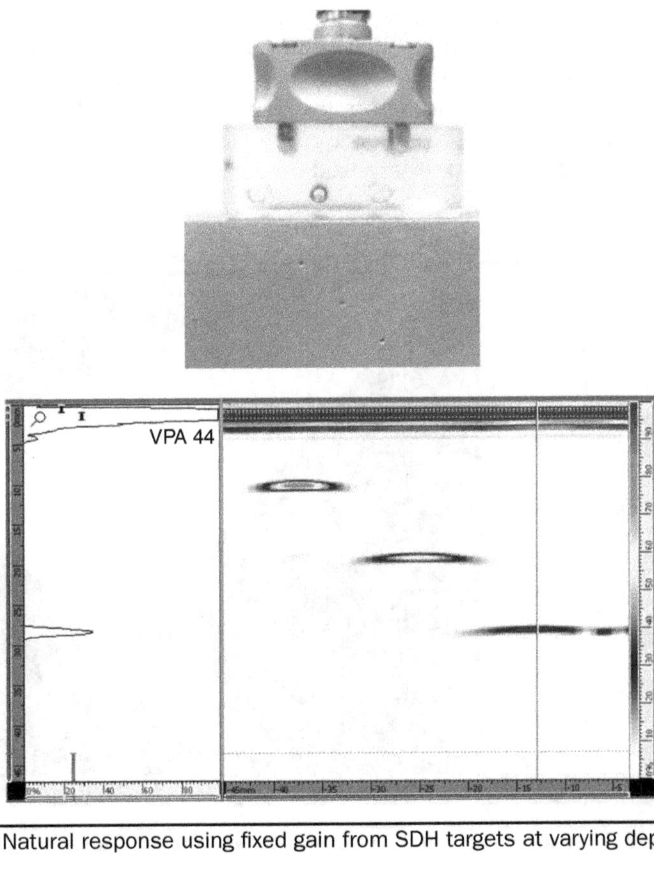


**FIGURE 12-45** Natural response using fixed gain from SDH targets at varying depths.

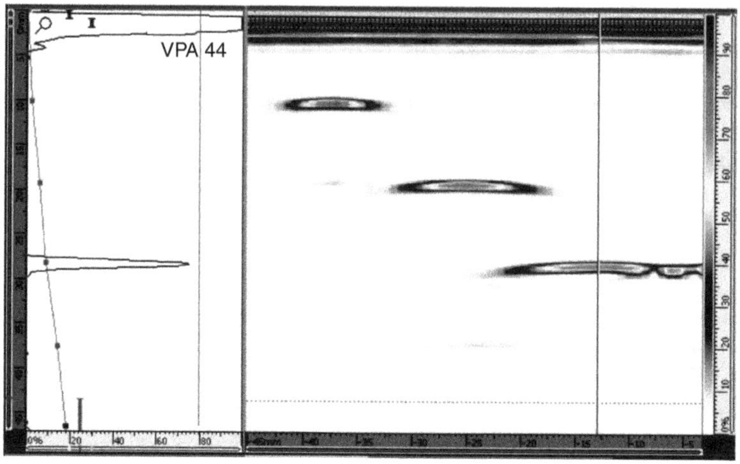


**Figure 12-46** Response using time-varied gain to equalize reflector amplitude.

Following the application of time-varied gain, the reflector amplitude at all three depths is essentially equal (Fig. 12-46). The sloping line superimposed on the A-scan shows how relative gain now increases with depth.

While TVG varies gain, distance amplitude correction maintains constant gain across a distance range while plotting the amplitude of echoes from reflectors of the same size at different distances from the transducer in graphic form. In a phased array instrument, DAC provides a reference curve that is applied to the selected angle A-scan. In Fig. 12-47, each curve superimposed on the A-scan marks an equivalent signal level

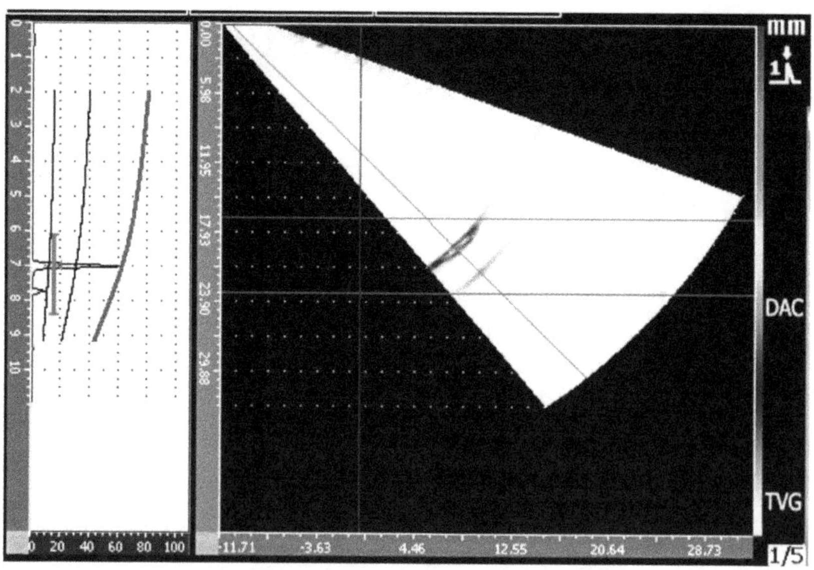


**Figure 12-47** S-scan and selected 45-degree A-scan with DAC curve.

following the ASME-3 format, with the upper line representing the reference calibration and the next two lines representing equivalent reflectors at 6 dB and 14 dB below the reference level. Echoes at the same amplitude with respect to a curve represent equivalent reflectors regardless of their actual height. Indications can be identified from the S-scan image and then echoes at each angle can be evaluated from the A-scan curves.

### DGS (AVG)

The DGS technique (distance gain size) method for flaw sizing was developed in Germany, where it is known as AVG (Abstand Verstarkung Grosse). It is a mathematical technique for estimating reflector size at different distances from a probe, based on an initial reference calibration and a calculation of relative probe sensitivity that is based on near field, beam spreading, and material attenuation effects. In DGS calibration, plotted curves relate A-scan echo amplitude to the size of an equivalent flat bottom hole (ERS, or equivalent reflector size) at increasing depth. DGS is valid only for reflectors that are smaller than the cross section of the sound beam.

In the case of swept-angle S-scans, DGS values are typically plotted for selected single angles, commonly 45, 60, and 70 degrees. A gated indication seen at 45 degrees in an S-scan image (Fig. 12-48) is marked with a cursor and DGS software calculates that its amplitude is equal to that of a 2.8-mm-diameter flat-bottom hole. In the customary DGS display, the upper curve superimposed on the A-scan indicates the selected DGS reference level, in this case 2 mm, and the lower curve marks an amplitude 50 percent below that.

### AWS D1.1

The American Welding Society D1.1 procedure is used in angle beam testing to size defects with respect to a reference reflector (commonly a side-drilled hole), comparing

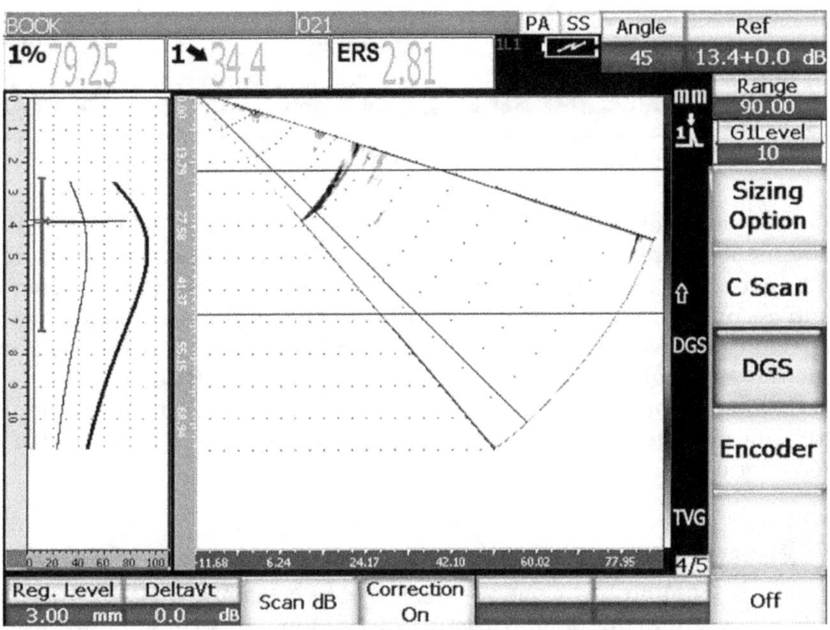


**FIGURE 12-48** S-scan with DGS curve on selected 45-degree A-scan.

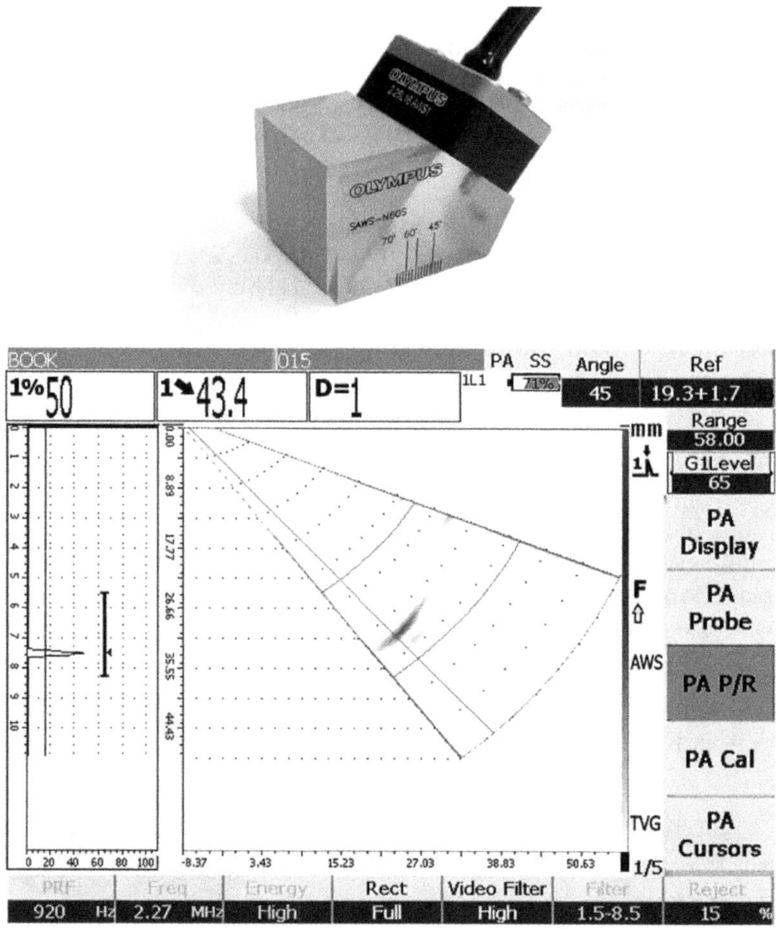


**Figure 12-49** Phased array AWS probe and AWS D1.1 display.

the amplitude of the indication to that from the reference reflector and adjusting for attenuation as required across the sound path length (see Fig. 12-49). The indication rating, or "D value," can then be classified with respect to tables generated by AWS and recorded in a report. The phased array probe is designed with frequency, aperture, and wedge footprint per the AWS D1.1 code. Phased array affords the ability to use a single probe and calibrated setup for 45-, 60-, and 70-degree inspections. Following calibration, an indication is detected in the 45-degree component of a sector scan and rated as D = 1, meaning its amplitude is greater than that of the reference reflector used for setup.

## 12.6 Phased Array Applications

Through its ability to generate steered sound beams, phased array offers unique advantages in many nondestructive inspections. It is common for users to employ software tools which provide ray tracing for the probes being used relative to the part being inspected in order to verify the part coverage through the steering range. This greatly assists in developing an appropriate scan plan as shown in Fig. 12-50. The following applications provide some examples of how phased array technology can be used to simplify, improve, and enhance many inspections currently done with conventional UT.

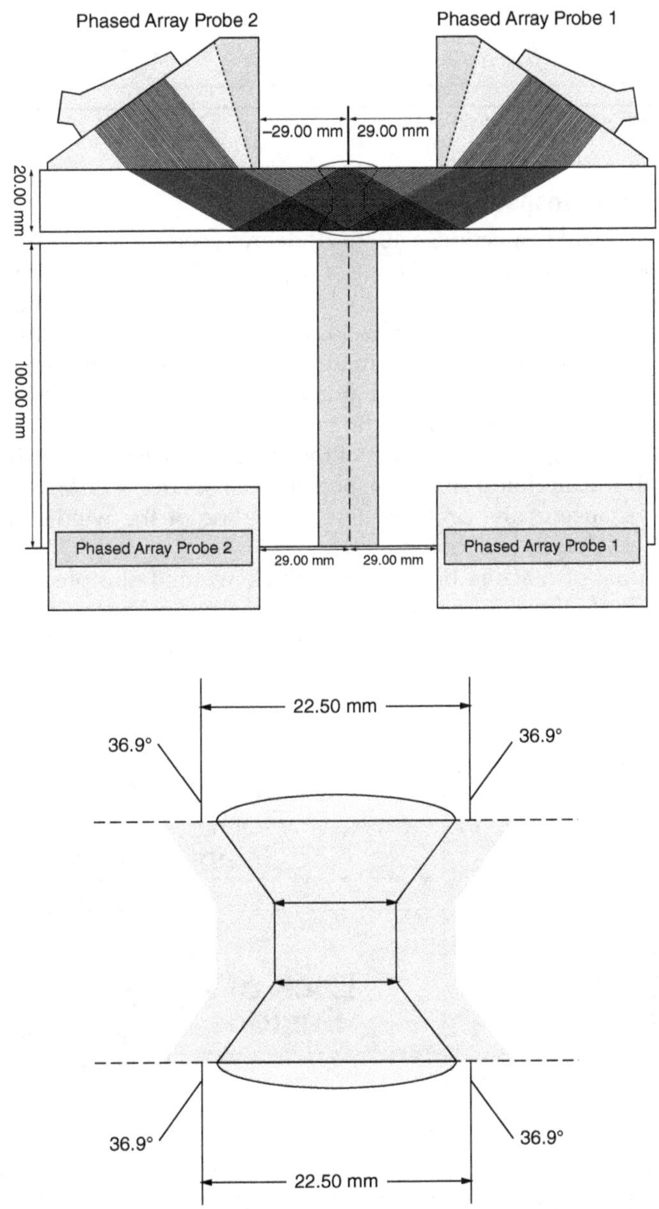


**FIGURE 12-50** Ray tracing used to develop a scan plan of 20-mm-thick weld using two 5-MHz, 64-element linear array probes at 60 degrees.

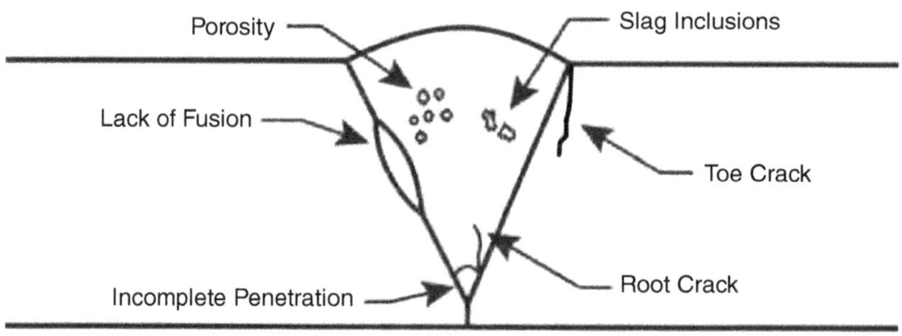


**Figure 12-51** Typical weld defects.

### 12.6.1 Weld Inspection

For manual weld inspection, a 40- to 70-degree sector scan is commonly used. Simultaneous live inspection across a range of angles, inclusive of the primary angles specified for conventional UT (45, 60, and 70 degrees), enhances the probability of detection. In order to cover the volume of the weld, it is necessary to manually scan in a raster fashion while skewing the probe. During this process, the sector scan aids in differentiation of flaws from weld geometry and aids identification of flaw types. By using a tool on board the instrument or by importing a CAD drawing, the geometry of the weld being inspected can be generated as part of the image to assist in identifying flaws within the weld's volume. In doing so, the inspector must set the weld overlay position by inputting the distance of the probe to the center line of the weld at the point of interest. Typical weld defects are shown in Fig. 12-51.

Examples of S-scans on a 16-mm-thick welded sample with weld overlay are shown in Figs. 12-52, 12-53, and 12-54. These images were generated with a 5-MHz 16-element probe and angle beam wedge sweeping from 40 to 70 degrees across the weld. In Fig. 12-52, lack of fusion is clearly visible in the S-scan in the second leg,

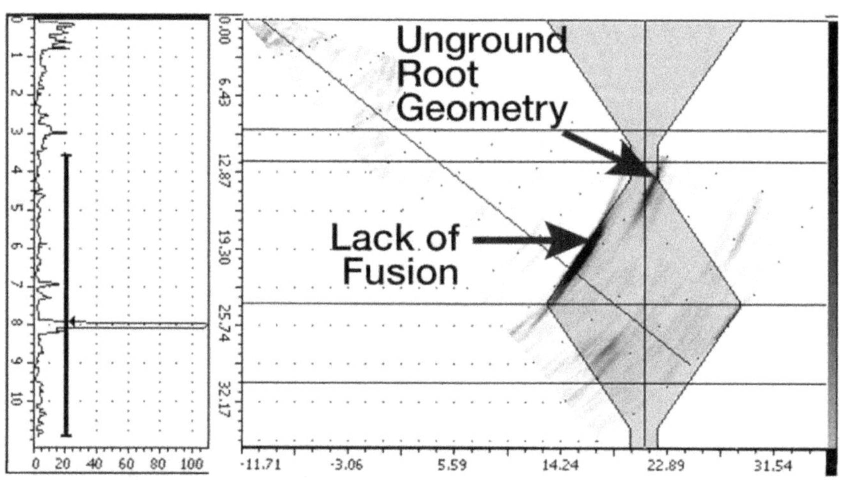


**Figure 12-52** Lack of fusion.

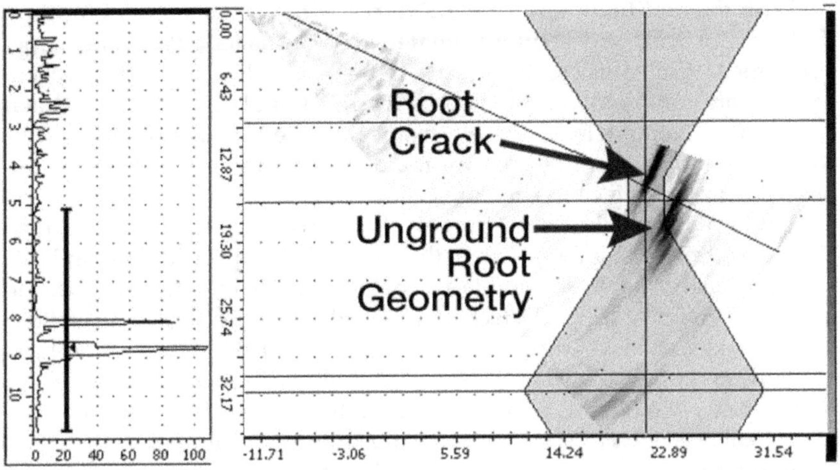


**FIGURE 12-53** Root crack.

across multiple angles, at the high amplitude expected from a planar defect. The sourced A-scan shows a sharp rise and fall as would be expected. A root crack is visualized with the S-scan at high refracted angles within the first leg. Echo dynamic of the 60 degree A-scan shows the echo walking up the crack as the transducer is moved toward and away from the weld. Further confirmation is achieved by scanning in a similar fashion from the opposite side of the weld. In both of these examples, the unground root geometry is clearly imaged. Porosity appears in the S-scan of Fig. 12-54 as multiple smaller amplitude reflections that "dance" with probe movement. The sourced A-scan exhibits multiple facets.

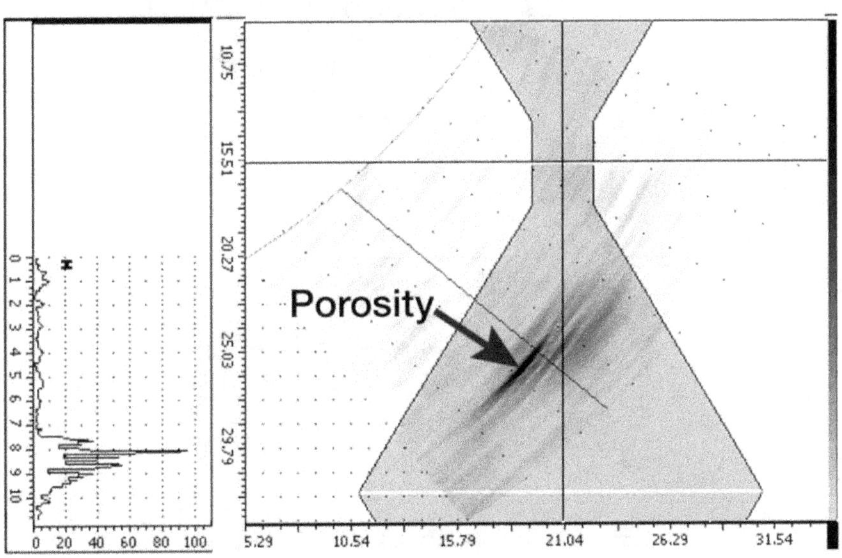


**FIGURE 12-54** Porosity cluster.

With the addition of scanning fixtures with encoders to a PA probe, automated (fully motorized) and semi-automated (hand-pushed) PA weld inspection can be performed. While this can be done with S-scans, full volumetric inspection of thicker welds more commonly employs primary fixed-angle linear scans. When using a linear array, a fixed angle can be electronically swept forward and backward from the weld thus eliminating the need for manual rastering or the requirement for multiple probes with varying centerline offsets.

During this "one-line scan," full waveform data is captured at encoded positions as the probe is scanned along the length of the weld, to allow post analysis and full record of the inspection. An example of a simple encoder attached to linear array probe for weld inspection is shown in Fig. 12-55. Defects can be analyzed by software "dissection" of edge and length cross-sections. C-scan data can be used for lengthwise sizing. Instruments with multigroup support allow linear scans with multiple fixed angles and/or a probe on either side of the weld. With this technique, probability of detection, repeatability, and speed are greatly enhanced. It is also possible to combine TOFD with phased array (see Fig. 12-57), leveraging the sensitivity of TOFD to skewed defects for increased detection reliability. In addition to allowing multiple inspection groups, advanced software allows

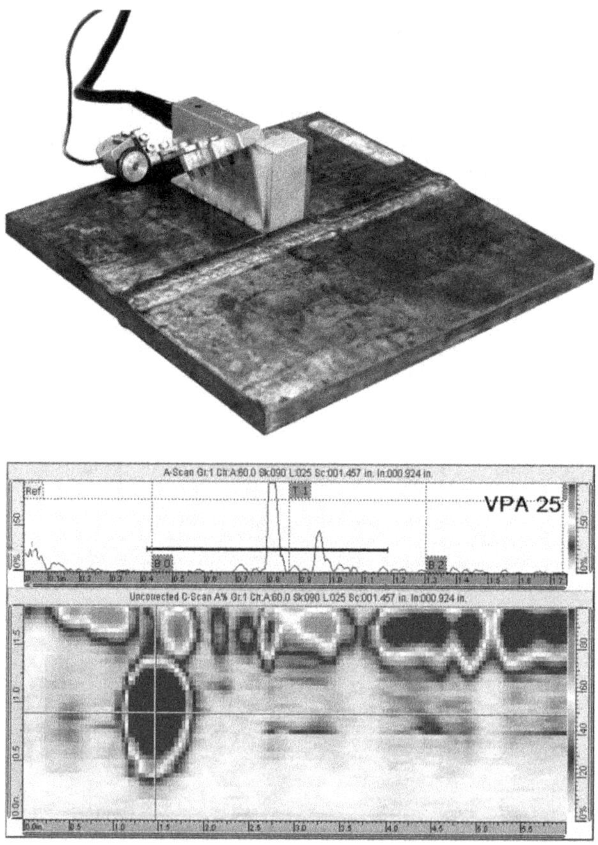


**FIGURE 12-55** Scanning fixture on weld.

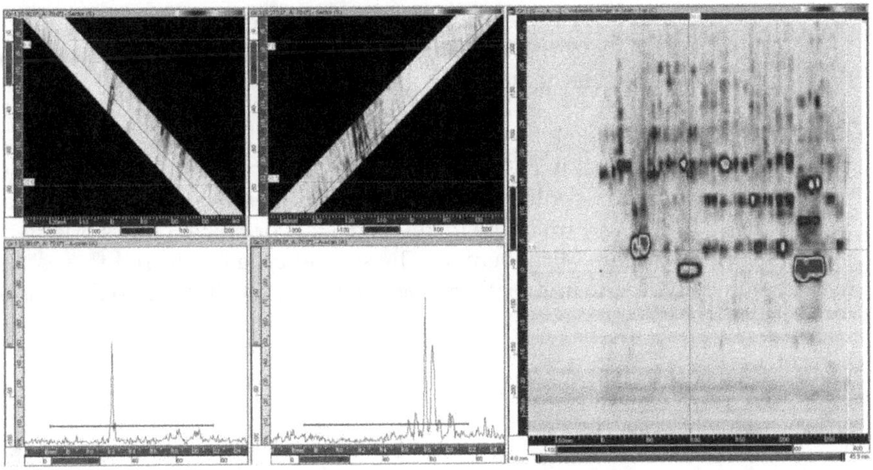

**Figure 12-56** Combined image of multiple linear scans and volumetric C-scan data using two arrays on opposite side of weld.

for all of the data accumulated from each group to be merged volumetrically (see Fig. 12-56). This allows data to be reviewed more efficiently and with better probability of detection.

### *Intergranular Stress Corrosion Cracking*

Intergranular stress corrosion cracking (IGSCC) is the phenomenon of multifaceted cracking occurring along grain boundaries of susceptible materials in an aggressive environment under load. As the cracks penetrate further into the material, they reduce the supporting structure of the material to the point of failure from overload. As the crack follows trans-granular paths, there is a tendency for crack branching.

**Figure 12-57** Multimodality weld inspection PA S-scan and linear scan simultaneous with conventional UT-TOFD.

This branching can be readily visualized using a 45- to 70-degree S-scan. Images in Fig. 12-58 of two 20-mm-thick stainless parts show shallow (10 percent) and deep (65 percent) ID-connected cracking.

### Hydrogen-Induced Cracking

Hydrogen-induced cracking (HIC) occurs in susceptible steels that are exposed to aqueous environments containing hydrogen sulfides (see Fig. 12-59). Manifestation occurs in two forms. The first type forms as blisters or blister cracks oriented parallel to the part surface and are fairly benign. The second type is referred to as stress-oriented hydrogen-induced cracking, or stepwise cracking (SOHIC). SOHIC forms as cracking

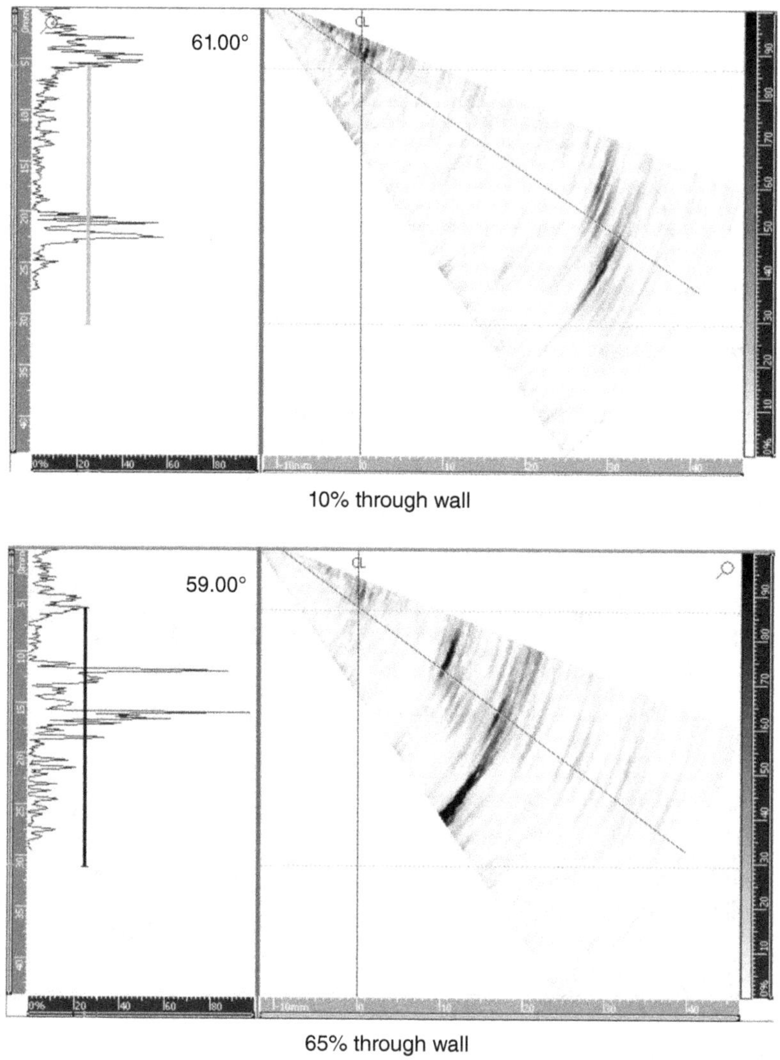


**FIGURE 12-58** S-scans of IGSCC multifaceted ID-connected cracking.

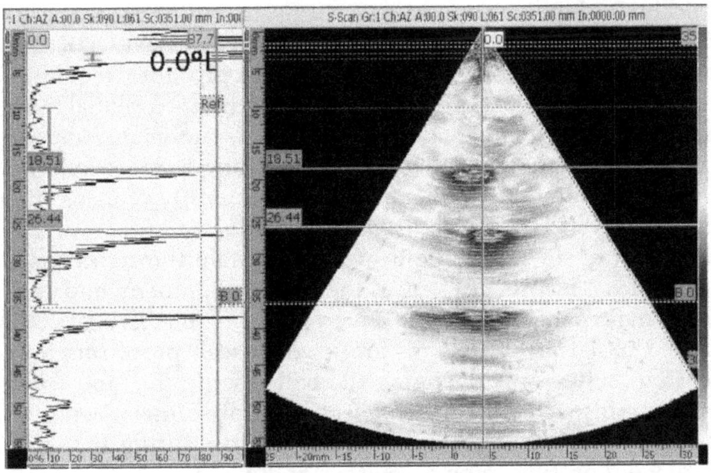

HIC with no stepwise cracking visible (no SOHIC)

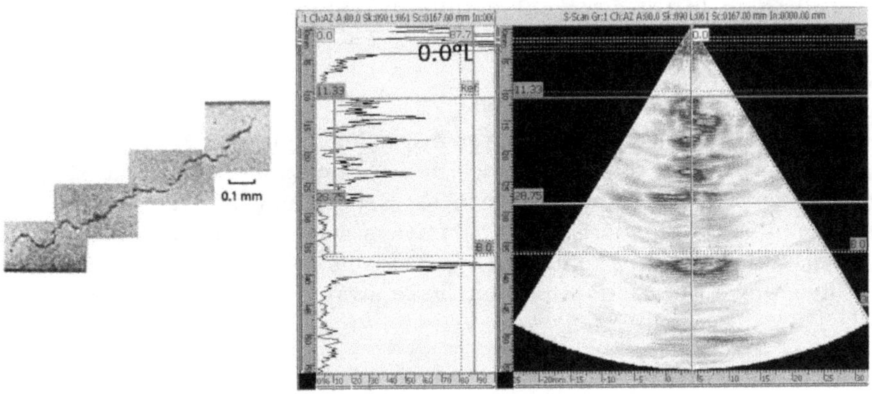

HIC with SOHIC visible

**Figure 12-59** S-scans of hydrogen-induced cracking with and without stepwise cracking.

between HIC blisters, at an angle to the surface through the thickness of the plate. Because SOHIC can reduce load carrying capabilities to a greater degree than HIC, detection is more critical. For detection of HIC and SOHIC, an S-scan is set up for −30 to +30 degree L-wave propagation. The array is skewed back and forth to optimize signals. Typically, the beam is focused at the midwall, as most HIC and SOHIC occur at one-third to two-thirds depth. The operator looks for additional signals between HIC reflections to identify SOHIC.

### *Field-Inspected Girth Welds*

Gas pipelines are made of high-strength steel and operate at a significant percentage of their yield strength. Pipes are girth-welded on site, typically using automated welding systems. They are then rapidly inspected, coated, and buried. Due to the demanding construction cycle, it is important that any weld defects be detected and analyzed very quickly. Onshore, the inspection cycle for mounting the scanner, scanning, evaluating the data, removing the scanner, and getting to the next weld is under four minutes. Developed codes, such as DNV and API 1104 for pipelines, require zone discrimination with inspection focus set at every 1–3 mm of depth. It is essential to store the data during the inspection cycle for regulatory purposes. As automated ultrasonic testing (AUT) offers better detection and sizing accuracy leading to lower rejection rates, it is often preferred over traditional radiography. The complexity of using multichannel conventional UT, however, had limited its implementation.

Phased array simplifies the inspection system through consolidation of probes and mechanics. This not only eases the setup, alignment, and calibration processes, but also provides increased zone coverage for better vertical sizing while reducing scan time. Phased array systems are used to scan pipes ranging in diameter from 2 in. to 56 in. while only changing the band, setup file, and wedge. In Fig. 12-60 a weld is being inspected as the next section of pipe line is being welded. Rapid data acquisition and analysis ensures welded quality during the construction phase. A variety of weld profiles, pipe diameters, and wall thicknesses can be accommodated by recalling the appropriate setup files without the need for changing probe positioning. Instruments with multigroup support allow multiple primary angles to be used within a single setup.

### 12.6.2 Composite Inspection

Full C-scan records are becoming an inspection requirement as CFRP (carbon fiber reinforced plastic) use continues to increase in commercial aircraft. Phased array inspection has been increasingly adopted to increase speed and simplify inspections for both flat and radiused CFRP parts. Beam parameters, such as aperture (element group size) and focus distance, can be controlled to provide flexible setups and superior results with a single probe. For CFRP inspection, linear array probes are used to produce a series of

**Figure 12-60** PipeWizard system for field automated girth weld inspection.

compressional waves by moving the acoustic beam along the major axis of an array without any mechanical movement. Typical linear phased array probes for inspection of CFRP with thicknesses of 3 to 25 mm are 3.5 or 5 MHz with 24 to 128 total individual elements.

In addition to the ability to select and display A-scans across various apertures along the probe, a live cross-sectional image (linear scan) can be displayed, dimensionally accurate to the probe's length. This cross-sectional view allows visualization of ply-to-ply interface abnormalities, especially delamination depth measurement and single axis length sizing. A typical linear scan may consist of the user defining an 8-mm aperture (8 elements grouped together), stepping one element at a time over a 128-element span, resulting in a 120-mm cross-sectional view. While larger numbers of elements allow a wider inspection area within a single pass, smaller element count linear arrays allow better access and coupling to curved and restricted access areas while still providing cross-sectional visualization. Phased array probes can be designed to minimize the dead zone (approximately 1.5 mm) from the side of the probe to the last element to allow a maximum inspection area as close to the edges as possible. As with single-element UT transducers, linear array probes are often used with plastic delay lines or with water irrigated wedges to provide more consistent coupling while scanning large areas. In Fig. 12-61 a live cross-sectional image allows clear visualization of ply delamination deep in a CRFP aerospace component. The probe configuration shown minimizes water flow and use. It also provides an acoustic impedance match to the material effectively reducing the front surface interface reflection to improve the ability to see defects in the first ply layers.

A one-line scan C-scan format can be used to rapidly accumulate swatches of data from a variety of part shapes. As the footprint of the probe and encoder can be quite small, this method is very versatile. When inspecting large, flat or near-flat CFRP panels, a two-axis mechanical scanner can be used. More capable instruments will automatically stitch together linear scans that are position encoded in two axes, dramatically increasing the speed of inspection as compared to conventional single-element C-scanning. Scanner designs, such as shown in Fig. 12-62, are lightweight, water resistant, and have mounting options to allow horizontal, vertical, or upside-down operation.

### 12.6.3 Radius Inspection Using Phased Array

Inspection of radiused CRFP with conventional UT can be challenging. However, use of a linear curved array allows the acoustic beam to be stepped around either an ID or OD while maintaining normality to the part radius, producing live cross-sectional visualization. As described earlier, adding a single encoder produces a C-scan image of delaminations in the radius over the entire length of the part. This method is fast, provides full coverage, improves probability of detection, and results in an inspection with full traceability. The choice of the appropriate probe depends on the geometry of the part. The main factors that need to be considered are the inspection type (ID or OD), the radius of the corner, and the angle of the corner. Once the choice of probes has been narrowed, the thickness of the part, as well as the footprint of the probe and wedge, must be taken into consideration to ensure access to restricted areas that may exist. Probes are available to accommodate a variety of radii, as seen in Fig. 12-63. Fixturing is required to enable adjustment of probe to material radius alignment and maintenance of alignment during scanning.

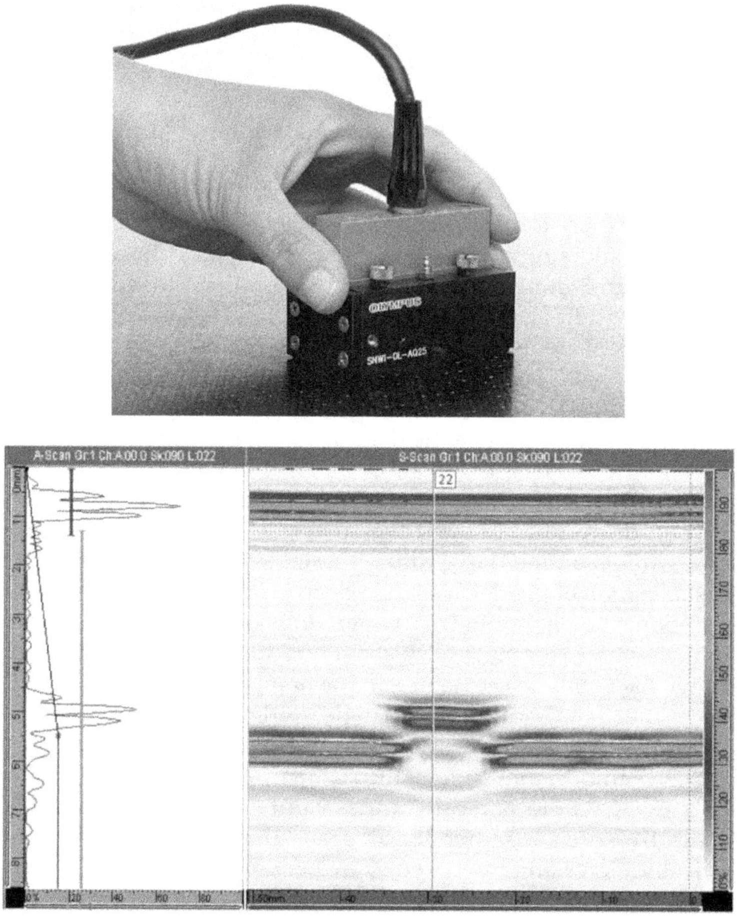

**Figure 12-61** Linear array probe with irrigated wedge (top). Live linear scan with A-scan showing delamination near bottom of CRFP (bottom).

With complete storage of inspection data, test results can be reviewed and analyzed after the inspection. When collected, fully linked A-scan, linear scan, B-scan, and C-scan data provide a permanent record of inspection that can be recalled, reviewed, and analyzed. As with linear scans on flat panels, indication size for length of radius defects in the scan direction can be directly measured. Determining the length of the defect from the linear scan following the radius, however, requires a geometric conversion. This sizing takes into account the radius of the phased array probe, the radius of the corner, the depth of the indication, the thickness of the part, and the uncorrected indication size. This conversion is also dependent on the type of inspection (ID or OD). Along with manual indication sizing, PC software is often used to automate the analysis process. Defects can be clustered and sized based on inspection requirements. A common scenario is to determine statistical variation of a "clean area" to establish a reference point or noise level. Defects are then automatically characterized on similar areas via a user specified signal-to-noise allowance, minimum defect size, and percent area restrictions.

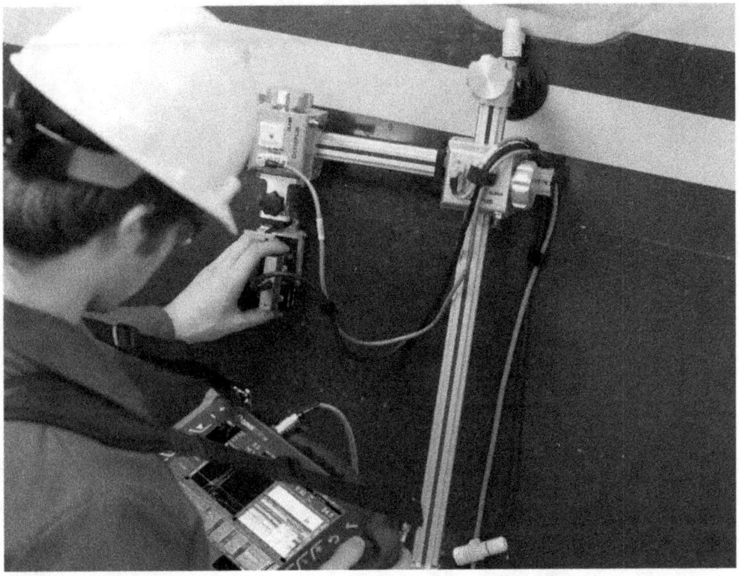

**Figure 12-62** Portable phased array instrument and two-axis scanner for flat/near-flat, large-area CFRP inspection.

### *Inline Phased Array Inspection*

Phased array can be used for inline inspection of tubes, bars, and billets. Phased array offers advantages over conventional ultrasonics in detection, setup, consolidation of probes, and throughput. While it is possible to generate cross-sectional (B-scan or linear scan) and C-scan views, it is not necessarily practical to do so. The power of phased array for these inspections is the ability to perform consolidated multichannel inspections with beam overlap and control for consistent detection, increased coverage, as well as simplified mechanics, setup, and calibration. As such, it is typical to use real-time alarm outputs and view the inline process in real time with multichannel strip chart views.

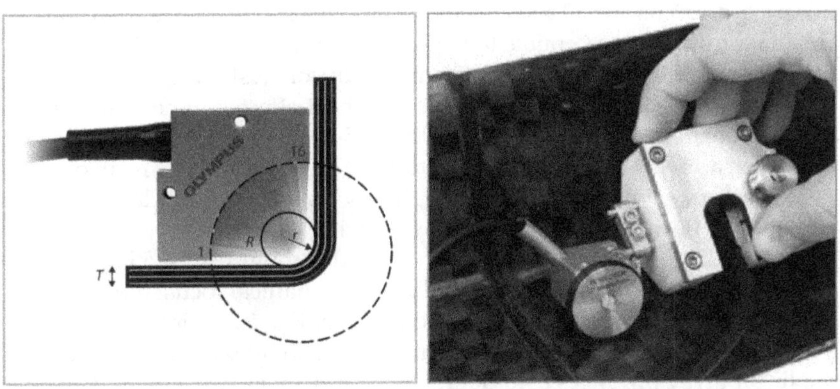


**Figure 12-63** Conceptual acoustic beam for ID inspection (left), probe fixture and single encoder for ID inspection (right).

**Figure 12-64** Full-body inspection using phased array.

In Fig. 12-64, a full volumetric rotating bar inspection system is shown with four surface riding array probes. This configuration combines an eddy current array probe for surface defects with one longitudinal and two shear wave phased array probes to detect internal and near surface defects.

### *Seamless and Welded Pipe*

Common products for oil country tubular goods (OCTG) demand inspection. Seamless products include drill pipes that rotate the drill bit and circulate the drilling fluid, casing pipe used to line the hole, and tubing which the oil or gas is produced from the well bore. Electric resistive welded pipe (ERW) has a singular vertical seam along its length. It is popular in the OCTG industry and is also used in other applications, including fencing, water and gas delivery, and structural purposes.

ERW inspection is often done in-line. As seen in Fig. 12-65, conventional UT requires two probes, one for the ID and one for the OD on each side of the weld zone, and they cover only a narrow zone susceptible to large detection variance, especially when the weld positioning relative to the probes varies. In comparison, just one PA probe on each side of the weld provides wide sector coverage of the heat-affected zone with constant amplitude. Through beam steering, constant amplitude detection can be maintained within the entire inspection area, even when there is significant mechanical movement of the weld line.

When inspecting seamless piping, full-body inspection of laminar, axial, transverse, and oblique flaws can be optimized by utilizing electronic focusing for improved detection capability and linear scans for increased throughput (see Fig. 12-66). Curved passive element linear array probes produce a mechanical focus in the radial axis. Laminar defects are detected using a 0-degree L-wave. For axial defect detection, the probe is mechanically tilted to generate a shear wave at about 40 degrees in steel. Beam steering the linear array produces a 40-degree refracted shear wave for traverse flaw detection. To detect oblique flaws, mechanical offset used to generate a ~40-degree shear wave in steel in a radial direction is combined with electronic steering and focusing.

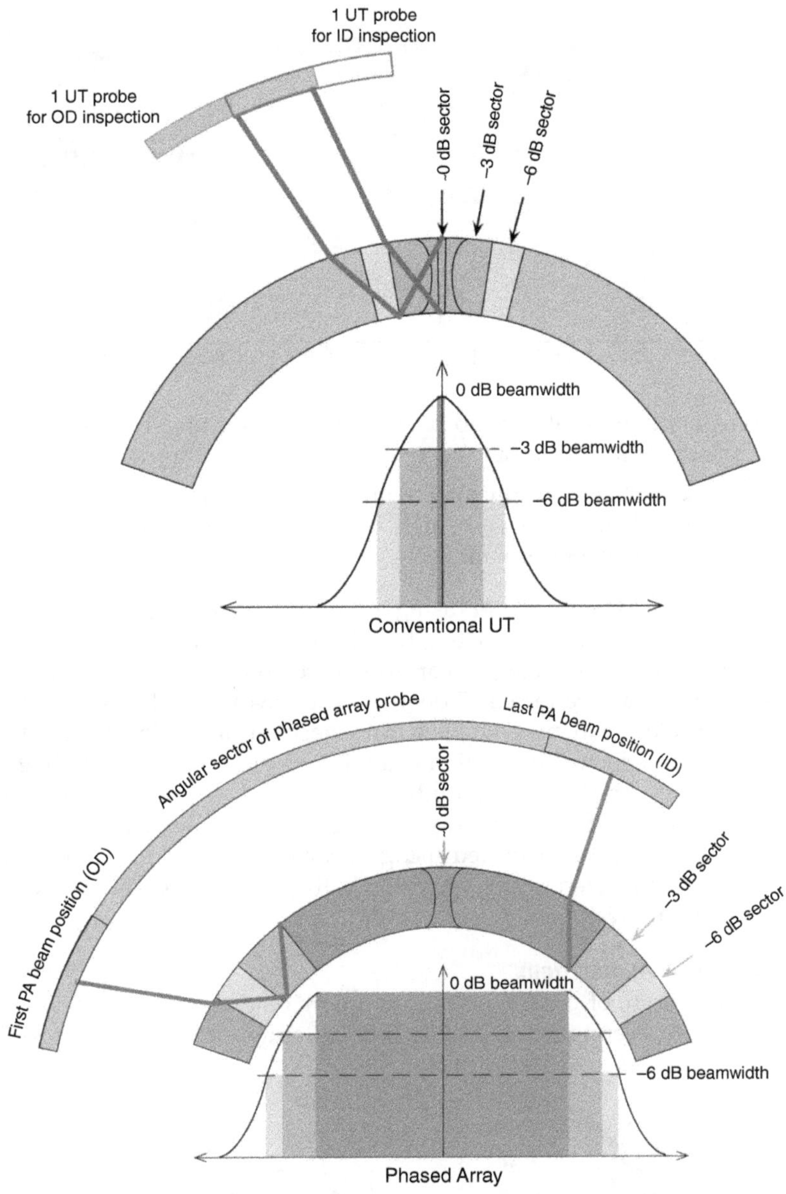


**FIGURE 12-65** Conventional versus PA beams for ERW inspection.

### Round and Square Billets

Phased array inspection of round and square billets for end use in auto, forgings, or finished products provides benefits in:

- A large helicoidally pitch covered by only one PA probe rather than several mechanical-interlacing inspection heads

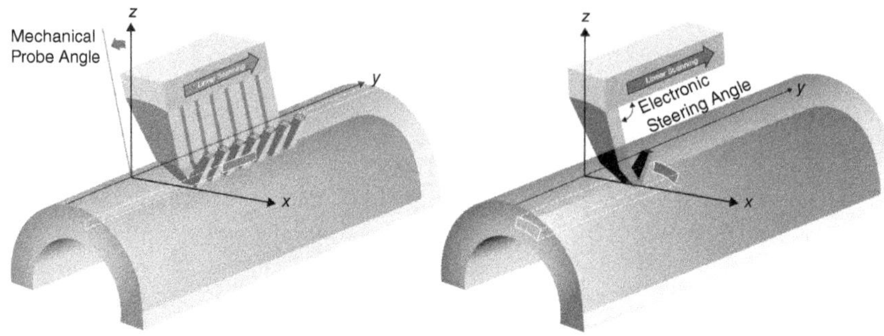


**Figure 12-66** Common probe and beam orientations for full-body piping inspection.

- A large physical overlap between two consecutive beams (focal laws) that are electronically adjustable depending on the size of the defects being detected
- An optimized beam profile with less than 2-dB drop between apertures (focal laws), obtaining a very high density of energy in the pipe and improving repeatability performance

This is an improvement over using multiple single-element probes or paintbrush probes, which have fixed focal length and can have up to 6-dB loss between beams.

For rotating bar systems, linear arrays are configured for a 0-degree longitudinal wave inspection and refracted 45-degree shear waves in the billet as shown in Fig. 12-67.

For nonrotating inspection, there is no mechanical rotation, rather the bar is pushed through a phased array inspection head. For round billets, curved linear array probes are used for generation of longitudinal waves with depth focalization for volumetric inspection and refracted shear wave testing of the perimeter, as shown in Fig. 12-68.

Square bar can be inspected using linear probes for longitudinal wave full-body inspection and steered shear wave for both near-surface and far-corner inspection, as shown in Fig. 12-69.

### *Miscellaneous Applications*

Phased array can, of course, be used in many other inspections. Numerous industries leverage beam steering for defect visualization and restrictive area access. Transportation

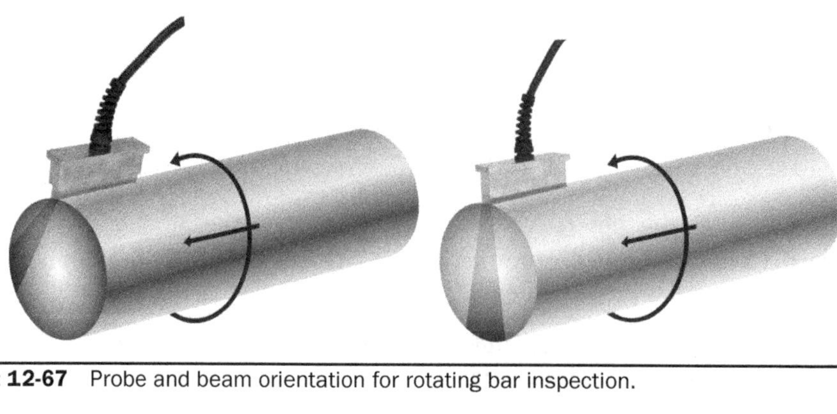

**Figure 12-67** Probe and beam orientation for rotating bar inspection.

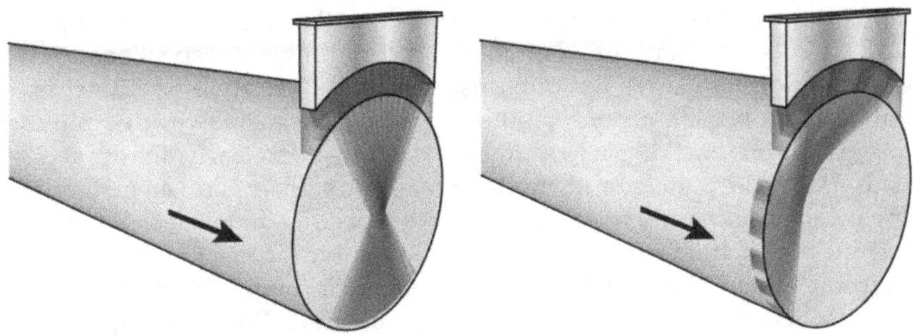

**FIGURE 12-68** Probe and beam orientation for nonrotating bar inspection.

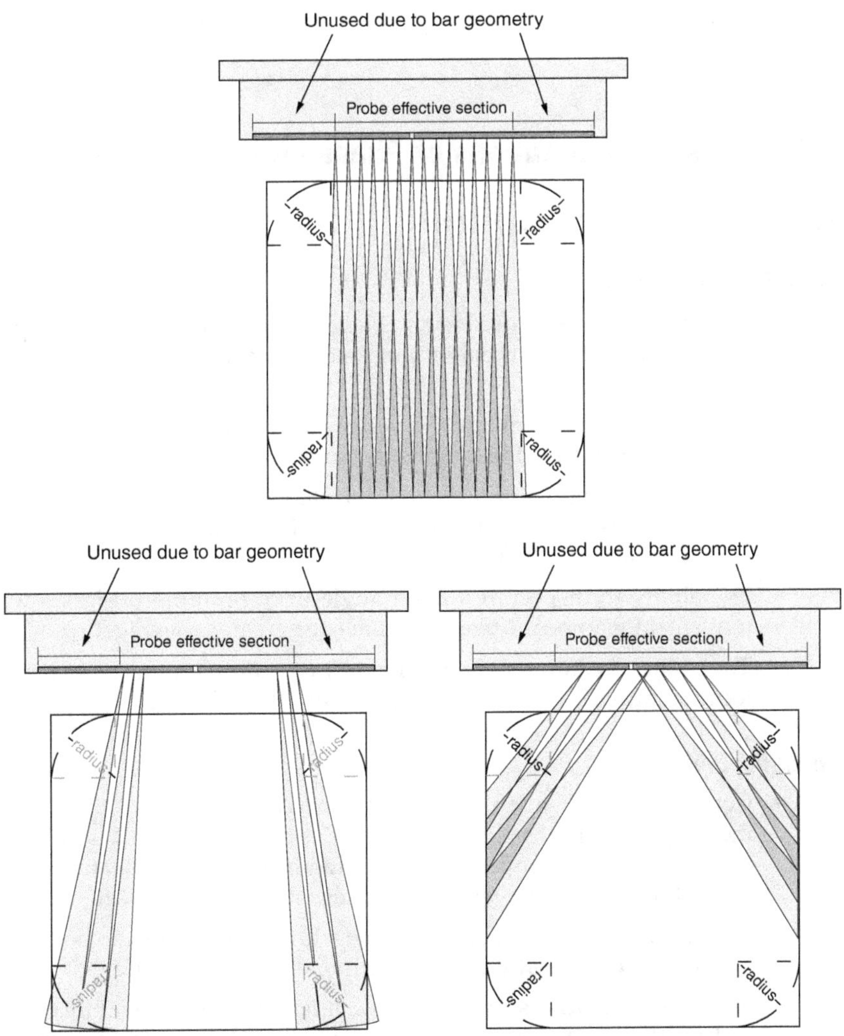


**FIGURE 12-69** Probe and beam orientation for square billet inspection.

industries use this technology for inspections that include landing gears, scribe line surface cuts at aluminum lap joints, rotors, engine inspections, and rail testing. Construction industries not only inspect a wide variety of weldment configurations but also use phased array for other applications, such as bridge pin inspection and bolt corrosion/cracking where image visualization clearly shows geometry features such as thread pattern and mode converted signals. Power generation and petrochemical industries use phased array for inspection of nozzles, dovetails, rotors, wind turbines, and high-throughput corrosion monitoring. Quality control of general components and raw materials is enhanced through volumetric inspection of bar stock, whether it be polymer, metallic, or ceramic. Bond testing of joined materials, weld processes such as friction stir welds and laser welds and high-speed component testing also benefit from the use of phased arrays.

As the technology continues to develop and becomes more and more affordable, opportunities for its use on a variety of applications, both simple and complex, will continue to increase.

## 12.7 Benefits and Limitations of Phased Array Inspection

### 12.7.1 Benefits of Phased Arrays

#### *Probability of Detection*

- The visual images provided by linear and S-scans from a single probe location can offer increased probability of detection and better data interpretation than a single A-scan.
- Reflections from geometrical reflectors are more readily distinguishable and are less apt to be falsely interpreted as flaws.
- As with all imaging, point data presented simultaneously with its "neighbors" provides perspective between critical defects and variations that are caused by native material characteristics.
- The ability to inspect at a fixed angle from multiple positions without probe movement via linear scanning allows for greater coverage.
- The ability to simultaneously inspect at multiple angles from a single probe location provides increased detection of off-angled flaws with a single setup.

#### *Simplification*

- Conventional UT often requires the use of more probes than phased array to meet volumetric inspection needs. The ability to use a single phased array probe to generate multiple angles, depth of focus, or beam positions reduces inspection setup complexity and eases mechanical fixturing requirements.
- A wide spectrum of problems can be addressed, setups can be changed in a few minutes, and, typically, greater component dimensional flexibility is available.
- Small matrix arrays can offer significantly more flexibility for inspecting restricted areas than conventional transducers by providing greater geometrical coverage from a single probe position.

### Speed

- Scanning with phased arrays to create an accumulated image of the part volume can be much faster than with conventional single-element transducers using mechanical scanners.

### Data Storage

- Whether from a single location or accumulated encoded scan, phased array images provide a more intuitive representation of defects for reporting.
- Availability of full data storage in advanced instruments permits detailed analysis and archiving.
- Post analysis can be performed on collected data.
- Image and A-scan data can be sourced from the data.
- Various cross-sectional images across width or length of part (in the case of encoded scans) can be viewed.
- Volumetric coverage of accumulated C-scans can be changed by adjusting gating parameters.

## 12.7.2 Limitations of Phased Arrays

Phased array systems are still bound by the same acoustic limitations that affect conventional ultrasonic testing. Phased arrays still require the use of liquid or gel couplant and are affected by material properties such as surface roughness, attenuation, grain structure, acoustic impedance, etc. Beam parameters such as near field, focusing limits, and beam spread are still governed by aperture and frequency. In addition, phased arrays have some specific limitations.

- **Grating lobes:** Phased array probes can generate unwanted grating lobes or side lobes, two closely related phenomena caused by sound energy that spreads out from the transducer at angles other than the primary path. This phenomenon is not limited to phased array systems, as unwanted lobes also occur with conventional transducers as element size increases, but the phenomenon can be complicated by multi-angle sound paths. These unwanted ray paths can reflect off surfaces in the test piece and cause spurious indications on an image. The amplitude of grating lobes is affected by pitch size, the number of elements, frequency, and bandwidth. Grating lobes will occur whenever the size of individual elements in an array is equal to or greater than the wavelength, and there will be no grating lobes when element size is smaller than half a wavelength. (For element sizes between one-half and one wavelength, the generation of grating lobes will depend on the steering angle.) Thus, the simplest way to minimize grating lobes in a given application is to use a transducer with a small pitch. Specialized transducer design incorporating subdicing (cutting elements into smaller elements) and varying element spacing will also reduce unwanted lobes. In practice, controlling PA probe design largely eliminates the problem.
- **Beam profiles:** The beam profiles in Fig. 12-70 compare two situations for a 5-MHz probe where the aperture is approximately the same 15.5 mm but the element count is reduced. The beam shown in the upper graph is generated by

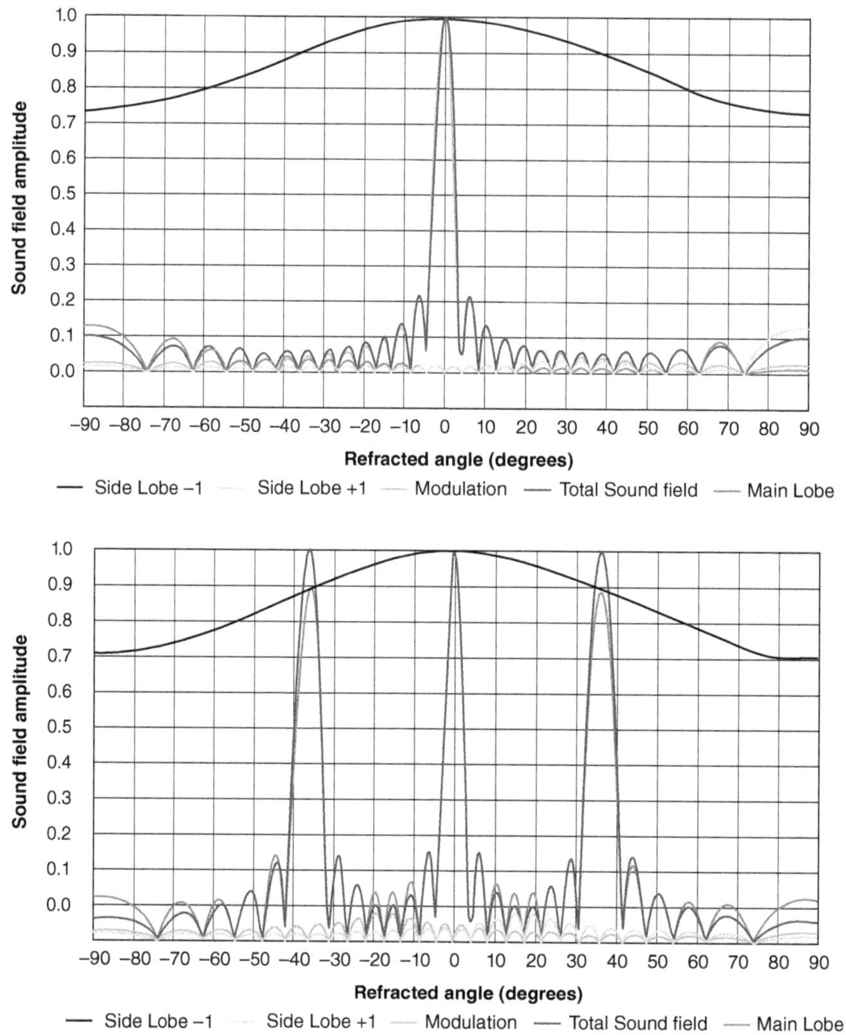


**FIGURE 12-70** Grating lobe from fine and coarse pitch array.

16 elements at a 1-mm pitch and the beam in the lower graph by 8 elements at a 2-mm pitch. The beam model in the upper graph has a clear main lobe, whereas the beam in the lower graph has two spurious lobes at approximately a +/−30 degree angle to the center axis of the beam.

- **Dead elements:** There is the potential for individual elements to fail (as in conventional probes). Typically, up to one-quarter of the elements in an array must be non-functional before serious beam-forming deficiencies occur. Guidelines are available for identifying and treating dead elements. Checking each element before inspection is possible. Performing calibrations against reference standards will highlight any performance degradations.

- **Sweep restrictions:** The range of angular sweep possible with S-scans has a practical limitation. As the desired steering angle increases away from the natural refraction angle of the wedge, reflected signal amplitudes drop significantly. Many inspection codes for weldments and volumetric material outline procedures for sensitivity calibrations that are appropriate for defect detection and sizing. The ability to calibrate across all the angles with good signal-to-noise response dictates the practical sweep range.
- **Displays:** With many different display formats possible, it is important for the operator to know which display shows what, whether it is geometrically corrected or uncorrected, and how the displays interact.
- **Cost:** Phased array instruments and probes are higher priced than the more conventional UT A-scan flaw detectors. This is to be expected, as phased arrays provide more detailed image data. When compared to conventional UT systems used with one or more probes and mechanical scanners to form images, the reduction in complexity offered by phased array often will provide more favorable costing.
- **Training:** Proper training and qualification of skilled operators for nondestructive testing modalities is always an ongoing effort. As the use of industrial phased array is relatively new, this qualification is even more challenging. Increased acceptance and use has driven the development and availability for specialized phased array training courses.
- **Codes:** Nondestructive inspection methods, detection, sizing, as well as critical defect classifications, are governed by specific industrial codes. These codes have been developed by quality agencies, industrial societies or via internal company requirements. Use of phased array may be restricted if inspection needs to be performed in accordance with codes intended for conventional single channel A-scan devices. As the use of phased array technology has mainstreamed, more codes and code cases have been developed specifically for the use of phased array inspection.

## 12.8 Appendix: Full Matrix Capture and Total Focusing Method

### 12.8.1 History and Development

Just like phased array ultrasonic testing (PAUT), full matrix capture (FMC) technology was borrowed from the medical field in the late 1970s. At that time, multiple nondestructive testing (NDT) research groups and manufacturers began to exploit and promote the value of a data visualization method known today as the total focusing method (TFM). Around 2015, devices had started to make their entry into the inspection world. However, several years after, these methods remain unclear to some practitioners and confusion still exists between conventional PAUT and FMC acquisition as well as the multiple ways to process the acquired data.

This appendix will cover the fundamentals of FMC and the differentiation between FMC and TFM, as well as between TFM and conventional PAUT. It will also touch base on the most common modes of propagation and their common applications. This appendix should provide a basic understanding of TFM imaging for inspectors who already have a knowledge of PAUT imaging.

### 12.8.2 Differences between Conventional Phased Array Ultrasound and Full Matrix Capture

In both PAUT and TFM, a multielement probe is used to transmit pulsed ultrasound waves in the component and to record the waveform (time based). These waveforms are then cumulated to generate an image of scatterers in the inspected piece. The way an ultrasound image can be visualized is by stacking together sub-images called *frames*. For example, in PAUT, a sectorial scan (S-scan) is composed of an arrangement of A-scans (amplitude vs. time) acquired at different angles and then stacked together. In this case, the individual A-Scans would be considered as "frames" in the definition of a PAUT S-scan. The strategy is the same for FMC and TFM and is done in processing these frames in the fastest way possible for displaying and refreshing the global image in real time. The fundamental difference between conventional PAUT and FMC/TFM resides in the approach used for the A-Scans acquisition and "frame" processing.

### 12.8.3 Differences between FMC and TFM

These two acronyms are often mistakenly used interchangeably. The important distinction between FMC and TFM is that FMC is an *acquisition strategy* only. It describes the way the entire matrix of A-scan is collected for a given probe. All the possibilities of sound propagation exist in the FMC. It would be extremely difficult for an operator to make sense of a raw FMC data set (see FMC image in Fig. 12-72b). TFM on the other hand is an imaging method used to visualize the relevant components of the FMC acquisition. It is an image generated from the FMC acquisition that uses every summed focal point to display them for every pixel of the image. Therefore, a TFM image cannot be reprocessed. The FMC needs to be kept in order to process a new image in a different mode of propagation. Note that the acquisition of FMC can also be visualised or imaged in other ways than using TFM.

### 12.8.4 Conventional Phased Array Imaging

For PAUT, the acquisition and imaging process are presented in Fig. 12-71 using again an S-scan as an example for imaging. The flow chart in Fig. 12-71a describes the steps from scan plan definition by the user to the image displayed on the screen. Figure 12-71b illustrates the frame acquisition process in the case of a S-scan. The S-scan is composed of individual frames, which correspond to A-scans captured at different incremental angles. During the acquisition, a group of $N_{el}$ elements (known as the aperture) fire and record at the same time. A delay is applied to each element to steer the ultrasound beam at the required angle, and to focus it at a desired depth in the piece. A sound beam is created in the component and the echoes from within the component are recorded in a single A-scan signal. In Fig. 12-71b, the dotted line crosses in the test piece represent the desired focal depth of the ultrasound beam for each frame acquisition. Each frame is then parametered by the refracted angle $\theta$ and the focus depth $d$. Therefore, the total number of frames to acquire is the number of discrete angles composing the global image.

An advantage of PAUT is that the technique requires a limited amount of acquisitions. Images obtained through PAUT are therefore displayed very fast. The drawback of PAUT is that the frames are focused only at one constant depth. Scatterers located outside of the focal region will appear blurry and somewhat larger than an identical scatterer appearing in the focal zone: The physical resolution of the resulting imaging system may therefore vary strongly over the displayed image.

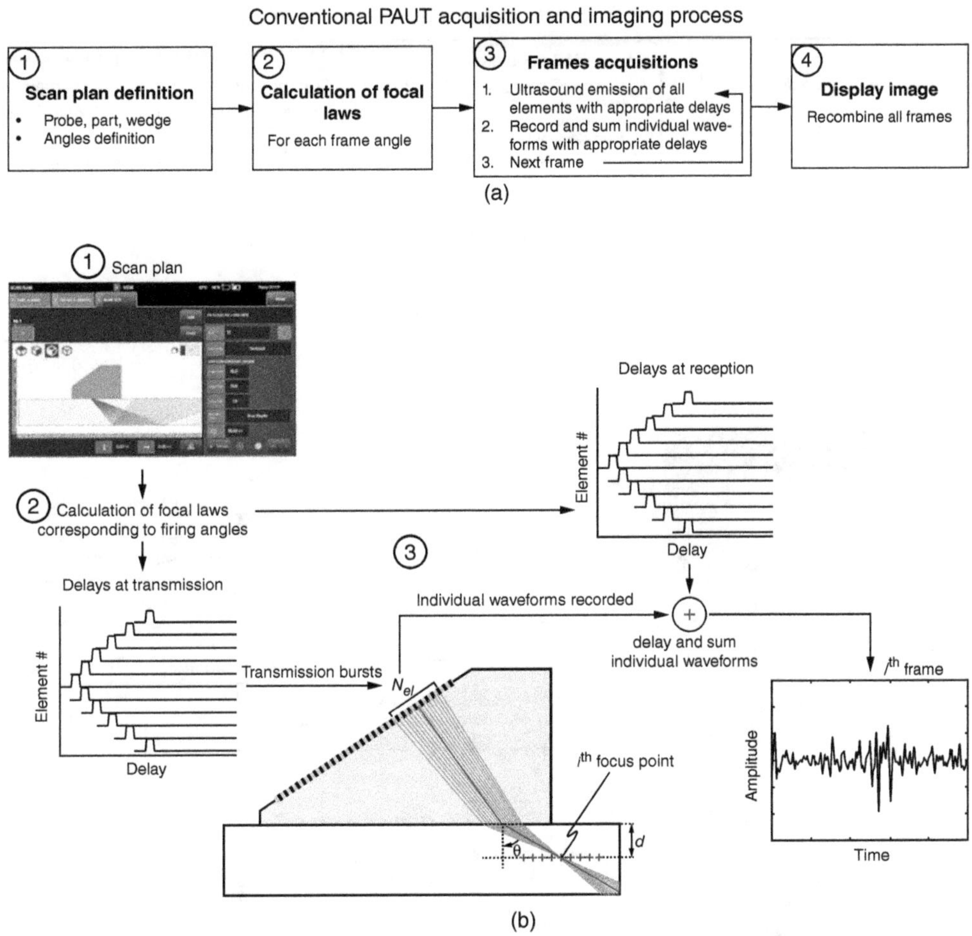


**Figure 12-71** Acquisition and imaging process of conventional PAUT. (*a*) Flow chart. (*b*) Illustrated chart of frame acquisition.

## 12.8.5 Phased Array Imaging with Total Focusing Method Imaging

As mentioned above in the conventional PAUT section, although frames are very fast, the focus can only occur at a given location within the test piece. With TFM the goal is to have an imaging method which displays the amplitude over focal lines at multiple depths, hence producing a highly resolved image everywhere and not over a single depth line as in the conventional PAUT approach presented above.

In TFM, instead of decomposing the image in sectorial angles and keep the same focal depth for each frame, the image is decomposed in pixels and the beam is focused at the pixel position. An overview of the process is illustrated in Fig. 12-72.

In the case of an acquisition with PAUT, each frame would require an acquisition, the time to generate a TFM image would increase dramatically because the number of pixels constituting an image is much higher than the number of discrete angles

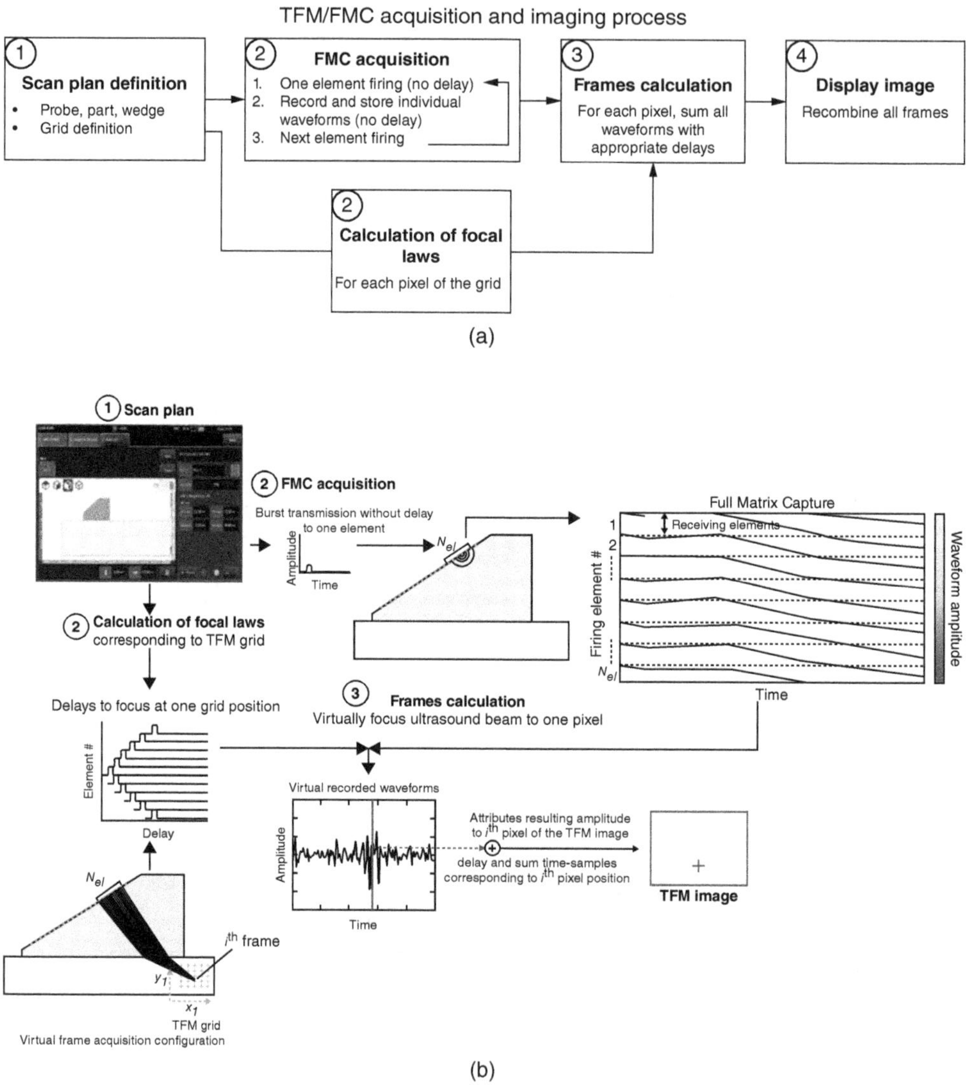


**FIGURE 12-72** Acquisition and imaging process of TFM imaging. (a) Flow chart. (b) Illustrated chart.

required to generate an S-scan. For example, an S-scan swept through 100 angles requires 100 acquisition, whereas a TFM image of 100 × 100 pixels would require 10,000 acquisitions. Therefore, another acquisition strategy is considered, in which frames are calculated in post-processing, as illustrated in Fig. 12-72. This strategy requires a set of focal laws corresponding to each pixel position, and a set of raw fundamental waveforms called FMC. With these two elements, fundamental waveforms can be delayed and summed appropriately to synthetically generate ultrasound beams, both in transmission and reception, and focused at each pixel location. Therefore, the image is considered to be "focused everywhere."

For FMC, the full aperture of the probe is most commonly used as it gives the best focusing result for a given probe design. The acquisition consists of acquiring all the waveforms between all individual pairs of probe elements. The number of acquisitions required to build the FMC is then equal to the number of elements of the probe (if the full aperture of the probe is used). This means that the FMC acquisition contains all the information regarding sound propagation between each element of the probe including reflections at interfaces and scattering by flaws. It is worth mentioning that any type of PAUT acquisition can be reconstructed using the FMC, including sectorial scan, plane wave imaging (PWI), dynamic depth focusing (DDF), etc.

Using the FMC acquisition process, the number of acquisitions required to generate an image is of the same order than PAUT but storing the individual FMC data sets require significant storage capacity, transfer bandwidth and processing power. For a given electronics, the TFM/FMC is therefore slower than conventional PAUT.

### 12.8.6 Differences between PAUT and TFM Images

Both FMC/TFM and PAUT require an array transducer where each element on the array is independently pulse. The probe selected should have an aperture where the far field will encompass the entirety of the area to be inspected. In comparison to PAUT, the FMC approach usually requires the entire array in order to achieve the best results.

In order to illustrate differences in PAUT and TFM imaging, the setup presented in Fig. 12-73 was considered. For this experimental case, a linear phased array (PA) probe is used to image identical side drilled holes (SDHs) distributed in an elliptical way in a steel block.

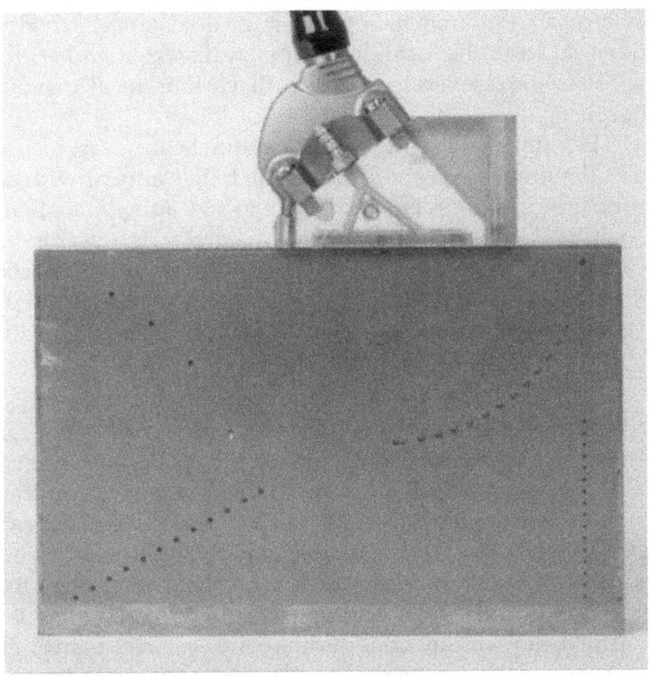

**Figure 12-73** Experimental setup for PAUT and TFM inspection.

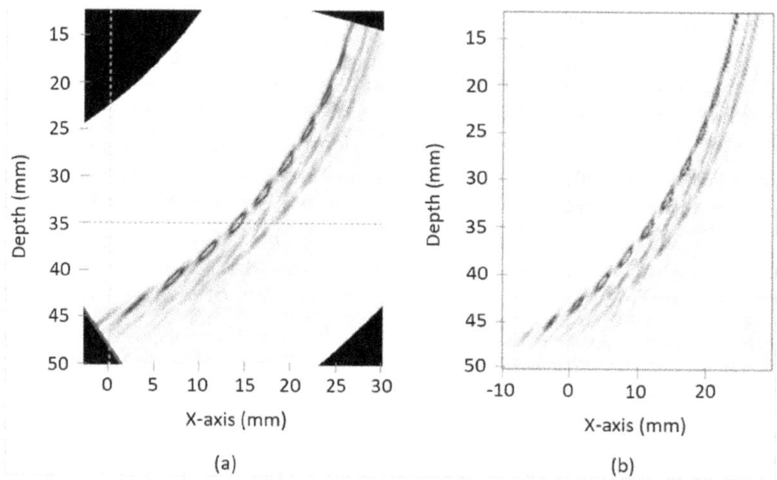


**Figure 12-74** Comparison of the amplitude data collected on a steel block with elliptical series of evenly spaced identical side-drilled-holes (SDHs). Image obtained using (*a*) fixed-depth PAUT S-scan and (*b*) TFM method. The horizontal dashed line in (*a*) represents the focus depth at 35 mm.

An S-scan (a) and a TFM image (b) obtained with the same inspection configuration are presented in Fig. 12-74.

In the S-scan, each frame is acquired using a unique focal distance of 35 mm (represented by the dashed line). SDHs located in the focal area appear with similar amplitude and size. For such a focal distance, the area where the image resolution is optimal is larger compared to a shorter focal distance, which explains this result. SDHs located far from the focal depth appear distorted and with a significantly lower amplitude. Therefore, several images with various focal distances are needed for a more uniform sizing of all the SDHs.

In the TFM image Fig. 12-74, ultrasound beams are focused at each pixel, as explained in the previous section. Hence, all SDHs appear with an optimal resolution, and only one image is necessary to be able to size adequately SDHs located in a larger range of depths. Nevertheless, a lateral distortion is observed for SDHs located at extremity of the electronic focalization capabilities. This distortion is intrinsic to the physics of ultrasonic beam for a given probe and is therefore still present in TFM image. To image these areas appropriately, a different mode of propagation would need to be used.

## 12.8.7 Modes of Propagations and Their Applications

Modes of propagations (also known as imaging paths, wave sets, ultrasonic paths, or ultrasonic modes) are the paths where ultrasound waves can travel for a given probe and taken into account by the instrument calculations. Modes of propagations are defined by their transmitting and receiving path pattern. For example, a direct path (or pulse echo modes) would describe a wave set that comes from the probe to the datum and directly back to the probe again (see Fig. 12-75). This can be done using a shear (transverse) or compression (longitudinal) waves. The names given to these modes commonly represent the pattern used. For example, for a direct mode where the

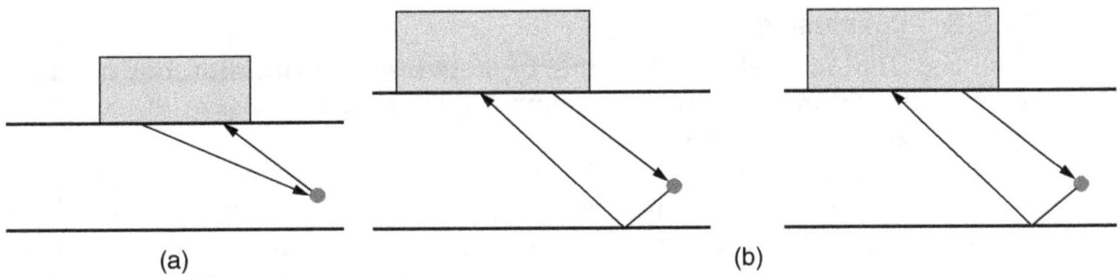


**Figure 12-75** (*a*) Direct path; with sound going from the probe to the datum point and directly back to the probe. (*b*) Indirect path; with sound going from the probe to the datum point and returning to the probe by reflecting off the back wall back to the probe.

transmission will use a shear (transverse) wave and where the reception will also use a shear (transverse) wave, the imaging path will be called T-T (transverse-transverse). Waves can also be called indirect (or self tandem modes), where one of the transmitting paths (either transmit or receive) will be reflected off the back wall before returning to the probe or before going to the datum (see Fig. 12-75a and b).

Selecting the right mode of propagation for a given inspection can be a challenge. The main factor that determines the selection of a mode of propagation is the expected orientation of the discontinuity. Like in any UT inspection, the perpendicularity of the signal with the flaw is key in detection. This is not different from TFM.

Some simulation tools are available to help the inspector in the selection of the proper modes (see Fig. 12-76).

A vertical linear reflector (example flat bottomed hole [FBH]) will commonly require an indirect (i.e., self tandem) mode to have the proper perpendicularity of the signal, where volumetric discontinuities (example SDH) are better detected with direct (pulse echo) modes. Note that the angle of a linear reflector will also affect the selection. For example, a FBH or crack at a 5-degree angle will have a better reflectivity with a TTT mode than one at a −45 degree angle.

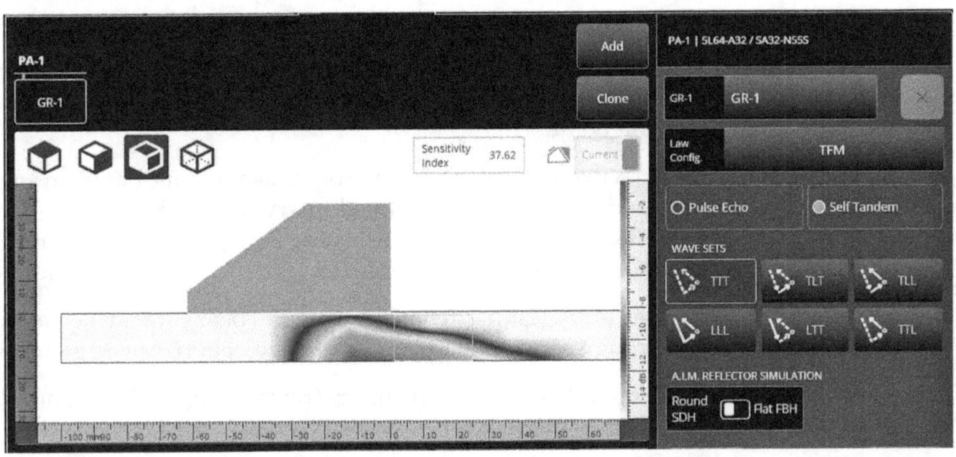


**Figure 12-76** Simulation tool to visualize the amplitude response for a TTT mode of propagation on a volumetric reflector.

### 12.8.8 Conclusion

FMC and TFM still rely on the same basic ultrasonic principles, but the differences reside in the nature of the frames composing the image as well as the selection of the proper mode of propagations for a given inspection.

The principal advantage of TFM is that, for the proper mode of propagation, the image has an optimal resolution everywhere, compared to an image produced with PAUT which is highly resolved only in the focal area of the beam. However, TFM conveys the same general limitations of PAUT processing. The image resolution is still constrained to the limitations of electronic focusing, that in turn is also limited by the physical attribute of the probe (frequency, pitch, elevation). These limitations are important to keep in mind, when an inspection to be constantly "focused everywhere" is expected.

## 12.9 Phased Array Glossary

**A-Scan** An ultrasonic waveform plotted as amplitude with respect to time. It may be either rectified or unrectified (RF). When using phased array probes, there are multiple A-scans that are stacked to create an image from a single probe position. Typically, the user selects which particular A-scan or finite group of A-scans to view from the image.

**Aperture** In phased array testing, the width of the transducer element or group of elements that are pulsed together with a delay law to create the intended beam.

**Apodization** A computer-controlled function that applies lower excitation voltage to the outside elements of an array in order to reduce the amplitude of unwanted side lobes.

**Azimuthal Scan** An alternate term for S-scan. It is a two-dimensional view of all amplitude and time or depth data from all focal laws of a phased array probe, corrected for delay and refracted angle. An S-scan also refers to the action of sweeping the beam through a range of angles.

**B-Scan** A two-dimensional image of ultrasonic data plotted as reflector depth or distance with respect to beam position. B-scans may be either single value or cross-sectional.

**B-Scan, Single Value** A two-dimensional image based on plotting the first or largest reflector within a gate as a function of time or transducer position. This format is commonly used in ultrasonic flaw detectors and advanced thickness gages, and it shows one reflector at each data point.

**B-Scan, Cross-Sectional** A two-dimensional image of ultrasonic data, based on amplitude color encoding of the entire A-scan at each timed interval or transducer position. The end image shows all reflectors in a cross section, rather than just the first or largest. This allows visualization of both near and far surface reflectors within the sample.

**Beam Forming** In phased array testing, generating a sound beam at a particular position, angle and/or focus, through sequential pulsing of the elements of an array transducer.

**Beam Steering** The capability to modify the refracted angle of the sound beam generated by a phased array probe.

**C-Scan** A two-dimensional view of ultrasonic amplitude or time/depth data displayed as a top view of the test piece.

**Calibration, Sensitivity** A procedure that electronically equalizes amplitude response across all beam components in a phased array scan. This typically compensates for both element-to-element sensitivity variations, varying energy transfer at different refracted angles, and changes in wedge sound path length that may occur from one focal law to the next.

**Calibration, Wedge Delay** A procedure that electronically compensates for the different sound paths taken by different beam components in a wedge, used to normalize the measured sound path length to a reflector.

**E-Scan** Also termed an electronic scan. In some industries, E-scan is referred to as a "linear scan" or "linear electronic scan," due to the ability to move the acoustic beam along a phased array probe without any mechanical movement. A constant aperture is multiplexed across a group of active elements. E-scans are performed at a constant angle and along the phased array probe length. For angle beam inspections, changes in sound path through the wedge are compensated at each focal law. From this, an image is formed similarly to a cross-sectional B-scan; however, stacked A-scans are the result of the focal law sequence, not time interval or position.

**Focal Laws** The programmed pattern of time delays and/or element groups and wedge delays applied to pulsing and receiving from the individual elements that comprise the defined aperture of a phased array probe, in order to move, steer, and/or focus the resulting sound beam and echo response.

**Grating Lobes** Spurious components of a sound beam diverging to the sides of the center of energy, caused by even sampling across the probe elements. Grating lobes occur only with phased array transducers and are caused by ray components associated with the regular, periodic spacing of the small individual elements. See also Side Lobes.

**Linear Scan** Sometimes referred to as E-scan, the ability to move the acoustic beam along the major axis of the array without any mechanical movement. A constant aperture is multiplexed across a group of active elements. Linear scans are performed at a constant angle and along the phased array probe length. For angle beam inspections, changes in sound path through the wedge are compensated at each focal law. From this, an image similar to a cross-sectional B-scan is formed; however, stacked A-scans are the result of the focal law sequence, not time interval or position. In some industries, this term is also used to describe a one-line scan.

**Multigroup Inspection** The ability to program multiple channels using one or more probes and/or modalities simultaneously during an inspection. For phased array probes, this includes the ability to use multiple probes and/or to use a single PA probe with different sequences.

**One-Line Scan** A single pass mechanical scan of a phased array probe parallel to a weld or region to be inspected. Typically done with a linear array probe to create a C-scan line image of amplitude or depth data as a function of electronic steering versus mechanical position. These scans can be produced with either E-scans or S-scans.

**Phased Array Probe** A multi-element ultrasonic probe (typically with 16, 32, or 64 elements) used to generate steered beams by means of phased pulsing and receiving.

**Phasing** The interaction of two or more waves of the same frequency, but with different time delays, which may result in either constructive or destructive interference.

**Pitch** The separation between individual elements in a phased array transducer.

**Plane, Active** The orientation parallel to the phased array probe axis consisting of multiple elements. Beam steering occurs along the active plane.

**Plane, Passive** The orientation parallel to the individual transducer element length or probe width.

**Plane, Steering** The orientation in which the beam direction is varied for a phased array probe.

**Resolution, Angular** In phased array systems, the angular resolution is the minimum angular value between two A-scans where adjacent discontinuities located at the same depth are individually resolvable.

**Resolution, Lateral** In phased array systems, the minimum lateral separation between two specified reflectors that permits discrete identification of each. This is related to both the design of the array transducer and the selected focal law programming.

**Sector Scan (S-Scan)** A two-dimensional view of all amplitude and time or depth data from all focal laws of a phased array probe corrected for delay and refracted angle.

**Side Lobes** Spurious components of a sound beam diverging to the sides of the center of energy, produced by acoustic pressure leaking from transducer elements at different angles from the main lobe. Side lobes are generated by all types of ultrasonic transducers. See also Grating Lobes.

**Virtual Aperture** The combined width of a group of phased array elements that are pulsed together with a delay law.

## References

American Society for Nondestructive Testing. *Nondestructive Testing Handbook,* 3rd ed., Vol. 7, Ultrasonic Testing, American Society for Nondestructive Testing, Columbus, Ohio, 2007.

Davis, J. M., and Michael Moles, "Resolving Capabilities of Phased Array Sectorial Scans (S-scans) on Diffracted Tip Signals," *Insight* (Northampton, UK), Vol. 48, No. 4, April 2006.

Davis, J. Mark, and Michael Moles, "Back to Basics: Beam Sweeping and Encoded Phased Array Units," *Materials Evaluation*, Vol. 65, No. 6, June 2007, pp. 777–778.

Granillo, Jesse, and Michael Moles, "Portable Phased Array Applications," *Materials Evaluation*, Vol. 63, No. 4, April 2005, pp. 394–404.

*Introduction to Phased Array Ultrasonic Technology Applications*, RD Tech Corp., Quebec, QC, 2004.

Krautkramer, J., and H. Krautkramer. *Ultrasonic Testing of Materials*, 4th ed., Springer-Verlag, Berlin and New York, 1990.

*Phased Array Testing Basic Theory for Industrial Applications*, 1st ed., Olympus NDT, Waltham, Mass., November 2010.

"Phased Array Tutorial," www.olympus-ims.com/en/ndt-tutorials/phased-array/, Olympus NDT, Waltham, MA, 2012.

# CHAPTER 13

# An Introduction to Ultrasonic Guided Wave Inspection

## 13.1 What Is an Ultrasonic Guided Wave?

Sonic waves are associated with hearing audible sounds including musical instruments and talking. The frequency of these sounds is less than 20 kHz. All other sound vibrations above 20 kHz are called ultrasonic waves. As an example, if we strike a large steel object with a hammer, or other force via a piezoelectric or electromagnetic device, the wave velocity in steel is about 6,000 meters per second for a longitudinal wave. A longitudinal wave is one in which the actual particle velocity of the particles being considered in the material is actually parallel or in the same direction as the wave vector or wave propagation direction. Since shear stresses can be present in solid materials, an additional wave can be generated that is called a shear wave. In this case, the particle velocity is at 90 degrees to the wave propagation direction with a speed of about 3,000 meters per second in steel. These two waves, longitudinal and shear, are called bulk waves, since they travel in semi-infinite or infinite solid media with no boundary resistance. No boundaries are present to interact with the propagating waves. These longitudinal and shear wave velocities are a function of the material being considered and are almost the same value for sonic and ultrasonic wave propagation, assuming that the waves are propagating in homogenous, isotropic material. These are the values popularly found in the literature, presented by such equipment manufacturers as General Electric Inspection Technologies, Olympus NDT, and others.

Let us now go beyond bulk waves to an explanation of ultrasonic guided waves. Boundaries of a structure are required to have ultrasonic guided wave propagation. Combinations of longitudinal and shear waves are present in the structure. The waves interact with boundaries in a very special way. The boundaries could be the surface of a very thick structure where the thick structure is considered as a half-space or a semi-infinite media. In this case, Rayleigh surface waves can propagate over the surface of a thick steel plate, for example or over any thick structure where the frequency is such that the wave length is small compared to the thickness of the structure. Rayleigh waves can also travel over the surface of the earth. The Rayleigh surface wave velocity in metals can be estimated as a function of Poisson's ratio, which for steel, as an example, is around 2,900 meters per second. Guided waves can also propagate in many different kinds of waveguides including thin plates, rods, tubes, and multilayered structures. In this case, the ultrasonic waves bounce back and forth inside the waveguide, leading to all sorts of interference phenomenon. In fact, most work in ultrasonic guided waves is a study of interference and wave superposition. Imagine, pumping ultrasonic energy into a plate with an initial starting angle and a specific frequency. As the waves bounce back and forth, mode conversion occurs, whereby each time an interface is struck, both longitudinal and shear waves are reflected and/or refracted as in the case of multilayered media. For the particular angle and frequency chosen, the interference phenomena could be totally constructive, destructive, or intermediate in nature. We could consider a very complex geometric problem of wave interference to see what is actually happening. There will be certainly hundreds of solutions of constructive interference points leading to a whole set of incident angles and frequencies that could represent solutions to the problem. It turns out now that we could look into a solution to a problem in wave mechanics in two different ways: one geometrics and the second utilizing classical wave mechanics. We could consider a governing wave equation in solid media, Navier's equation, subjected to specific boundary conditions, for example, in a plate using stress-free boundary conditions. Utilizing the theory of elasticity in wave mechanics along with Navier's equation and subsequent strain-displacement equations and a constitutive equation such as generalized Hooke's law, with assumed harmonic solutions. To satisfy the boundary conditions, one could come up with all of the constructive interference points, hence possible solutions to the problem via a solution to a system of homogeneous equations. These constructive interference points can be plotted to produce a phase velocity dispersion curve of phase velocity versus frequency. Note that this advanced wave mechanics solution leads to results that are identical to the geometric approach of finding suitable incident angles and frequencies leading to constructive interference points. The relationship between incident angle and phase velocity is simply Snell's law, so incident angle or phase velocity could be plotted against frequency. As a consequence, each natural waveguide, plate, tube, etc. has its own unique phase velocity dispersion curve.

An interesting turn of events now comes up. There are virtually hundreds of solutions to an inspection problem available from the phase velocity dispersion curves. How do we pick the best solution? Every point on a dispersion curve has a different particle velocity vibration characteristic across the thickness of the structure. As an example, maximum in-plane vibration could occur on the surface of a structure at a particular phase velocity and frequency value. If this point were selected as a solution and the structure were placed under water, the water would have almost no influence on the wave propagation characteristics as energy leakage into the fluid would not take place. A fluid cannot support shear wave propagation which comes about when considering in-plane vibration on the surface of a structure. Let's consider one additional example, suppose we wanted to examine a weak interface in a multilayer structure, we would have to search the phase velocity dispersion curve space to seek out a special wave structure across a thickness of a multilayer structure in such a way that we would obtain a maximum shear stress or in-plane displacement at the interface being considered. This phase velocity and frequency value would then have excellent sensitivity to the weak interface situation at the particular layer being designated. Every guided wave problem is approached in a similar fashion. Upon selection of a particular point in the phase velocity dispersion curve space it becomes possible to design an ultrasonic transducer that excites that particular point. Precise excitation is difficult, however, because of the existence of a phase velocity spectrum and a frequency spectrum. These are concerns though for another day. The sensor design could be an angle beam transducer with an excitation line at a constant phase velocity value on the dispersion curve. As frequency is swept across the frequency axis, the specific modes and frequencies will be generated. An alternate probe design could be a comb transducer or an inter-digital transducer design, where the excitation line goes from the origin of the phase velocity dispersion curve at an angle of the wavelength as the excitation line crosses many modes in the phase velocity dispersion curve space. Again, as the frequency is swept along the frequency axis, the modes crossed by the excitation lines will be generated. The comb spacing will be wavelength, which is the actual slope of the excitation line in the phase velocity dispersion curve space. Note that it becomes possible to move freely over the entire phase velocity dispersion curve space by changing angle in the horizontal excitation line approach or by changing element spacing in the sloped line of slope wavelength from the origin in the phase velocity dispersion curve space.

Let's now consider the long-time solution to a wave propagation problem. In the bulk wave case, since waves are traveling in infinite space, there is no vibration aspect of the problem to be considered since there are no wave reflection and transmission factors to consider. On the other hand, for guided wave propagation, the long time solution in many cases leads to a vibrations problem. This may not occur if no boundary exists in a particular direction as the wave is transmitted to infinity. In examining a closed structure, as a finite plate or tube, the reflection and transmission factors for each entry of a wave onto a boundary leads to a variety of constructive and destructive interference situations. The long time solution therefore leads to a vibrations or modal analysis problem. There will be specific resonant frequency values for a structure as well as specific vibrational modal patterns at both on and off resonance. Some work is being done today to examine the transition from the initial transient response onto a long time vibrations solution. As a consequence, ultrasonic nondestructive evaluation and structural health monitoring is being developed today by considering many aspects of ultrasonic bulk wave analysis, ultrasonic guided wave analysis, and also modal vibration analysis.

## 13.2 Market Forces

Ultrasonic guided wave integration into NDE and structural health monitoring (SHM) is growing rapidly for a variety of reasons. Quite often it is the only way to solve a problem because of limited access to a structure and the ability to inspect a large area from a sensor at only a single location on the structure. Ultrasonic guided waves are very different from the more standard ultrasonic bulk wave inspection where hundreds of test modes become possible, whereas for bulk waves only two modes of inspection are possible, longitudinal and shear. With the initiation of guided wave inspection taking off around 15 years ago, there were high expectations on use which was subsequently held back because of a lack of theoretical understanding and weakness in computational power required in modeling analysis. In technology transfer from laboratory to field, we often encountered many challenges of coatings, hidden, buried structures, geometrical intricacies in environment, etc. Many of their problems have now been overcome and technology transfer and product development are moving rapidly forward. Guided wave innovation is amazing in application, sensitivity, and penetration power. Some of these advancements are discussed on these pages. Taking place is a paradigm shift from bulk wave ultrasonics to guided waves because of reduced cost, less inspection time, and greater coverage and the ability to solve new problems with no prior solution potential. There is also a paradigm shift from NDE to SHM because of improved reliability, continuous screening, and early warning of defect growth and structural failure. As a result, there is baked-in potential for prefab sensor installation. There are also many cost and safety benefits associated with SHM. Expectations for market growth are also focused on new markets and not necessarily replacement in the bulk wave market.

From discussions with many users, funded research, and number of publications, up to 20 percent of the current bulk wave market could be replaced with an expansion in new SHM and NDE markets which could expand by 35 percent or more over the next decade compared to the current ultrasonic bulk wave market. In fact, the anticipated growth rate for NDE, SHM, and ultrasonic guided wave services could be about 10 percent per year starting in 2020. Reasons for growth can be associated with improvements in guided wave application and innovation in new product development programs with integration into wireless and energy harvesting systems. Instead of tens of specialized guided wave products, guided wave product trends are directed toward custom to configurable, easier to use, lightweight, speed, and improved coverage. More transparency is also becoming possible with improved software so that training becomes simpler and straightforward.

The future looks bright for ultrasonic guided waves in NDE and SHM, with applications in pipe, rail, bridge, aircraft, power plants, processing plants, highways, buildings, and in almost any government or industrial facility.

## 13.3 Literature Survey

### 13.3.1 General

Many textbooks are available on the subject of ultrasonic guided waves. For example, Graff's book [1975] covers waves in infinite media and in various waveguides. Rose's book [1999] covers fundamentals including dispersion curves, wave structure, source influence, and excitability. Fundamentals of wave propagation are presented nicely by Achenbach [1973], Auld [1973], and Datta [2008]. In Achenbach [2003], some interesting reciprocity concepts associated with guided and Lamb waves are treated. A recent book that highlights some interesting aspects of guided waves in composite materials is presented by Rokhlin, Chimenti, and Nagy [2011]. Zhu and Rose [1999] employ boundary element methods (BEM) and simulation for guided wave analysis. Rose [1994] also covers wave mechanics principles and Gao et al. [2010] introduce the concept of goodness curves to assist in mode and frequency selection in studying such features as skew, a variety of wave structure variables, excitability, etc. Yan et al. [2010a] use ultrasonic guided wave mode and frequency selection for defect detection and characterization in a multilayer hybrid laminate. Fromme and Sayir [2002] studied the detection of cracks at rivet holes using guided waves. Rizzo et al. [2009] studied fatigue crack detection in waveguides. Ing and Fink [1998] introduced time reversal techniques. Wilcox [2003a] employed rapid signal processing techniques to remove the effect of dispersion from guided wave signals.

### 13.3.2 Pipe

Ultrasonic guided waves were first introduced in tubing inspection by Rose et al. [1993]. Following the research results of Ditri and Rose [1992], Li and Rose developed an analytical method to simulate the angular profiles in hollow cylinders [2001a]. In 2002, Li and Rose presented the phased array focusing technique for longitudinal guided waves propagating in cylindrical shells. Rose and Mudge [2002] introduced a commercial application of the guided wave focusing technique with a pipeline inspection tool, the Teletest-focus instrument. Sun et al. [2003a] carried out the angular profile calculations and the phased array focusing technique for torsional guided wave propagation in hollow cylinders. If the angular profiles show that ultrasonic energy is naturally concentrated at a circumferential location at a particular propagation distance, this phenomenon, which is called "natural focusing," can be used to improve pipeline inspection. See Sun et al. [2003b]. Real-time phased array technology for pipe was introduced to the world by Rose et al. [2002, 2003].

Circumferential waves were studied by Thompson et al. [1972], who used Lamb waves for steel plate and pipe inspections. Van Velsor et al. [2009] developed enhanced circumferential guided wave coating disbond detection methods with guided wave physically based features.

Defect imaging and synthetic focusing methods were introduced by Hayashi et al. [2005], Mu and Rose [2008a] by focusing and with coatings by Mu et al. [2008b], Davies and Cawley [2009], Mu et al. for synthetic focusing [2010], and Velichko and Wilcox [2009] with post-processing.

Pavlakovic et al. [1997] introduced a commercial software package for dispersion curve calculation: DISPERSE. Demma et al. [2001] studied mode conversion of longitudinal and torsional guided modes due to pipe bends. Ma et al. [2007] studied sludge and blockage detection inside pipes.

SAFEM-efficient computational techniques were introduced for guided wave problems by Hayashi et al. [2002]. Magnetostrictive methods were pioneered by Kwun et al. [1998] covering guided wave inspection of pipe using magnetostrictive sensors. Kim et al. [2005] used magnetostrictive transducers with ferromagnetic strips 45 degrees from the axis of a pipe.

### 13.3.3 Rail

Grewal [1996] used Rayleigh waves in rail. Hayashi et al. [2003] computed guided wave dispersion curves for a bar with an arbitrary cross section with a rod and rail example. Wilcox et al. [2003b] also introduced some interesting aspects of rail inspection. Gunawan and Hirose [2005] used boundary element analysis of guided waves in a bar with an arbitrary cross section. Bartoli et al. (including Lanza di Scalea) [2006] employed modeling aspects of wave propagation in damped waveguides of arbitrary cross section. Lee et al. [2009] used a guided wave approach for defect detection under shelling in rail with wave structure selection examples for sensor design to achieve special effects.

### 13.3.4 Bonding and Composites

Mal et al. [1988] studied guided wave evaluation of adhesive bonds. Chimenti et al. [1988] made use of guided waves reflection and propagation in fluid-coupled composite laminates. Datta et al. [1990] studied edge and layering effects of guided waves in a multilayered composite plate. Na [2002] researched ultrasonic guided waves for steel bar concrete interface testing. Hosten and Castaings [2005] employed finite elements method (FEM) for modeling the guided waves propagation in structures with weak interfaces. Puthillath and Rose [2010] introduced higher-order ultrasonic guided wave inspection of a titanium patch that was bonded to an aluminum aircraft skin.

### 13.3.5 Phased Array for Plates

Li and Rose [2001b] implemented guided wave mode control by use of a phased transducer array. Giurgiutiu et al. [2002] invented a virtual beam steering technique by using PZT transducer arrays. Wilcox [2003c] developed omni-directional guided wave transducer arrays for rapid inspections of large areas in plates. Yan and Rose in 2007 used beam steering in composites and in 2009 with a phased array time delay comb. Salas and Cesnik [2010] presented guided wave structural health monitoring (SHM) by using composite long-range variable-length emitting radar (CLoVER) transducers in composite materials. Senesi et al. [2010] developed guided wave frequency-steerable arrays for SHM.

### 13.3.6 Guided Wave Imaging and Tomography

Hinders et al. [1999] studied Lamb wave tomographic scanning and Gao et al. [2005] introduced guided wave mode and frequency selection for tomography for aircraft structures. Breon et al. [2007] did guided wave damage detection tomography for SHM in pipeline elbows, and Van Velsor et al. [2007] studied pipe defects by employing guided wave tomography technique. Yan et al. [2010b] also used ultrasonic guided wave imaging techniques in SHM.

Some research has been carried out to improve the guided wave imaging techniques. Ruzzene [2007] employed frequency wave number domain filtering for improved damage visualization. Hall and Michaels [2011] studied computational efficiency of ultrasonic guided wave imaging algorithms.

### 13.3.7 Ultrasonic Vibration

Rose et al. [2010b] introduced an ultrasonic guided wave modal analysis technique (UMAT) for defect detection, followed by work by Yan and Rose [2010c].

### 13.3.8 Deicing

Venna et al. [2007], Zhu et al. [2008], and Gao et al. [2009] utilized ultrasonic guided waves for deicing experiments.

### 13.3.9 SHM

Michaels et al. [2004] used feature-based classifiers for ultrasonic SHM. Raghavan and Cesnik [2004] modeled use of piezoelectric-based Lamb wave generation and sensing for SHM. Many other investigators are also using guided waves in SHM.

### 13.3.10 Nonlinear

Nonlinear methods are now finding applications in ultrasonic guided wave analysis. See Cantrell et al. [2001], Kim et al. [2006], and Pruell et al. [2009].

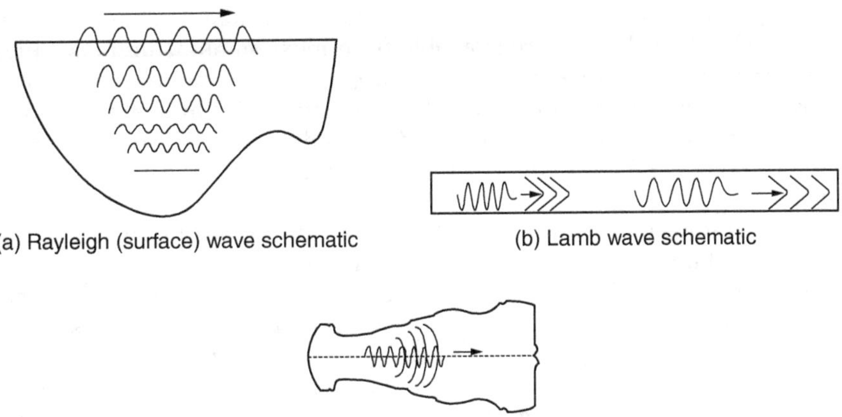

(a) Rayleigh (surface) wave schematic

(b) Lamb wave schematic

(c) Stonely wave schematic

**FIGURE 13-1** Guided wave possibilities.

## 13.4 Fundamentals

Compared to ultrasonic bulk waves that travel in infinite media with no boundary influence, guided waves require a structural boundary for propagation. As an example, some guided wave possibilities are illustrated in Fig. 13-1 for a Rayleigh surface wave, a Lamb wave, and a Stonely wave at an interface between two materials.

There are many other guided wave possibilities, of course, beyond that presented in Fig. 13-1, as long as a boundary on either one or two sides of the wave is considered. Let's look further into a variety of different natural waveguides as outlined in Table 13-1.

When you think about it, most structures are really natural waveguides provided the wavelengths are large enough with respect to some of the key dimensions in the waveguide. If the wavelengths are very small, then bulk wave propagation can be considered, those waves that were used traditionally for many years in ultrasonic nondestructive pulse echo and through transmission testing. One very interesting difference, of many, associated with guided waves, is that many different wave velocity values can be obtained as a function of frequency, whereas for most practical bulk

| List of Natural Waveguides |
|---|
| • Plates (aircraft skin) |
| • Rods (cylindrical, square, rail, etc.) |
| • Hollow cylinder (pipes, tubing) |
| • Multilayer structures |
| • Curved or flat surfaces on a half-space |
| • Layer or multiple layers on a half-space |
| • An interface |

**TABLE 13-1** List of Natural Waveguides

wave propagation purposes, the wave velocity is independent of frequency. In fact, tables of wave velocities are available from most manufacturers of ultrasonic equipment that are applicable to bulk wave propagation in materials, showing just a single wave velocity value for longitudinal waves and one additional value for shear waves.

A comparison of bulk wave and guided wave ultrasonic inspection is illustrated in Fig. 13-2. Note the coverage volume increases as a result of guided wave inspection.

To get some idea of how guided waves are developed in a waveguide, just imagine a bunch of bulk waves bouncing back and forth inside a waveguide with mode conversions between longitudinal and shear constantly taking place at each boundary. The resulting superimposed wave form traveling along the waveguide is just a sum of all of these waves including amplitude and phase information. You can visualize the outcome being strongly dependent on frequency and impinging wave angles of propagation inside the structure. The strongly superimposed results are actually points that end up on the phase velocity dispersion curve for the structure. Elsewhere there is strong cancellation. To solve for the points on the dispersion curve either you could consider a partial wave summation process accounting for all reflections and mode conversions or you could solve an appropriate boundary value problem in wave propagation.

Information on ultrasonic guided waves can be found in many excellent textbooks, see for example, Graff [1975], Rose [1999], Achenbach [1973], Auld [1990], Datta [2008], and Achenbach [2003]. A very reasonable up-to-date state-of-the-art literature review and evaluation of ultrasonic guided waves can be easily found in the literature.

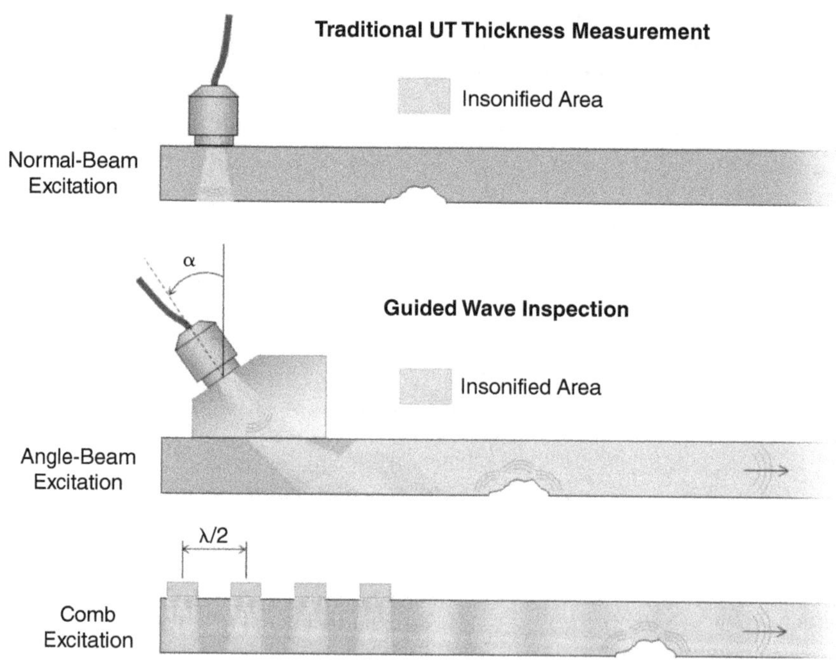


**FIGURE 13-2** Comparison of bulk wave and guided wave inspection methods.

- 100 percent coverage.
- Inspection over long distances from a single probe position.
- Ability to inspect hidden structures and structures under water, coatings, soil insulations, and concrete.
- Beam focusing capability is possible for improved probability of detection, reduce false alarm rate, penetration power, and inspection confidence.
- Often greater sensitivity than that obtained in other NDT techniques.
- Cost-effectiveness because of inspection simplicity and speed.
- Avoidance of removal/reinstallment of insulation or coating, except at location of transducer tool.
- Excellent overall defect circumferential length in pipe.
- Lateral and depth sizing potential.

**TABLE 13-2** Principal Benefits of Guided Waves

The use of ultrasonic guided waves is increasing tremendously over the past decade because of a variety of different reasons, notably understanding and improved computational efficiency for complex problem solving. The principal benefits of guided waves can be summarized as follows. Inspection over long distances from a single probe position is possible giving complete volumetric coverage of the item to be inspected as presented in Fig. 13-2. There's no need for scanning; all of the data is acquired from the single probe position. Quite often, greater sensitivity than that obtained in standard normal beam ultrasonic inspection or other NDT techniques can be obtained, even with low frequency ultrasonic guided wave inspection techniques. There is also an ability to inspect hidden structures, structures under water, coatings, insulations, and concrete because of the inspection capability from a single probe position via wave structure change and controlled mode sensitivity along with an ability to propagate over long distances. There is also a tremendous cost-effectiveness associated with guided wave propagation and inspection because of the inspection simplicity and speed. See Table 13-2 for a brief summary of the benefits.

## 13.5 Dispersion

The subject of dispersion and the propagation of either dispersive or nondispersive modes is a very critical one to understand when dealing with ultrasonic guided waves. In Figs. 13-3 and 13-4, for example, you can see an example of dispersive and nondispersive guided wave propagation. For nondispersive wave propagation, the pulse duration remains constant as the wave travels through the structure. On the other hand, for dispersive wave propagation, since wave velocity is a function of frequency, the pulse duration changes from point to point inside the structure. This is because each harmonic of the particular input pulse packet travels at a different wave velocity. There's a decrease in amplitude of the waveform and an increase in pulse duration, but energy is still conserved, unless of course, lossy media is considered.

### 13.5.1 Dispersion Curves

Let's now consider the development of a phase velocity computation in a waveguide. A particular waveguide such as a plate or pipe can be used as an example. The appropriate

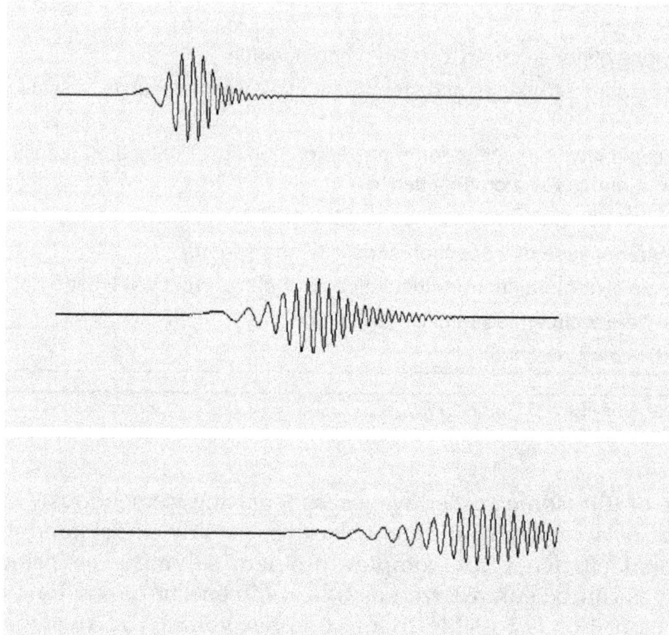

**Figure 13-3** Dispersive wave propagation as it travels along a structure.

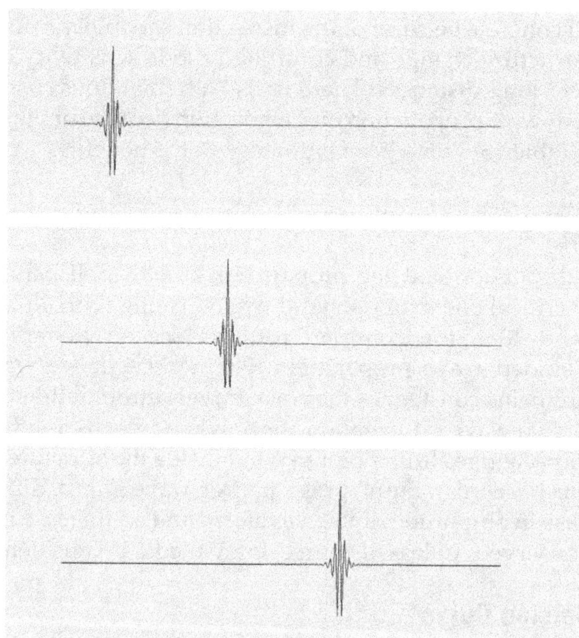

**Figure 13-4** Nondispersive wave propagation as it travels along a structure.

boundary conditions in these items call for a traction-free upper and lower surface, or outer and inside surface respectively. If we now consider some form of the governing Navier's wave equation in rectangular or cylindrical coordinates and an assumed harmonic solution for displacement, we can through elasticity theory derive the equations to satisfy the boundary conditions of the problem being studied. This leads to a transcendental equation, or a characteristic equation that most often requires a numerical solution. In extracting the roots from the characteristic equation, usually associated with a system of homogeneous equations, the determinant of the coefficient matrix must be set equal to zero. The resulting characteristic equations for a plate are as follows:

$$\frac{\tan(qh)}{\tan(ph)} = -\frac{4k^2pq}{(q^2-k^2)^2} \quad \text{for symmetric modes} \tag{13-1}$$

$$\frac{\tan(qh)}{\tan(ph)} = -\frac{(q^2-k^2)^2}{4k^2pq} \quad \text{for anti symmetric modes} \tag{13-2}$$

where:

$$p^2 = \left(\frac{\omega}{c_L}\right)^2 - k^2 \quad \text{and} \quad q^2 = \left(\frac{\omega}{c_T}\right)^2 - k^2 \tag{13-3}$$

In this case, the roots extracted determine the phase velocity versus frequency values that can be plotted, as illustrated in Fig. 13-5. In Fig. 13-5 are shown the phase velocity dispersion curves for a particular traction-free steel plate. The modes are labeled as antisymmetric A0, A1, A2, A3, etc. or symmetric as S0, S1, S2, S3, etc. The S0 mode intersects the zero frequency axis at the plate velocity value for the plate. The plate velocity formula is as follows:

$$C_{\text{plate}} = \left[\frac{E}{\rho(1-\upsilon^2)}\right]^{1/2} \tag{13-4}$$

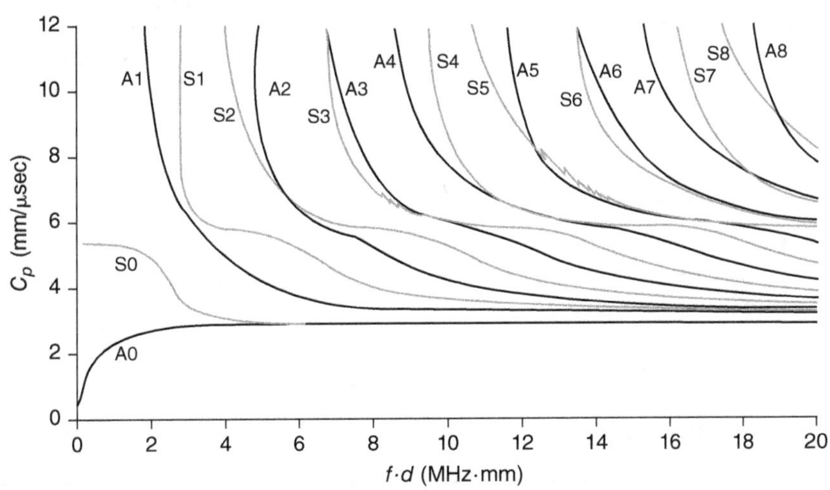


**FIGURE 13-5** Phase velocity dispersion curve for a carbon steel plate.

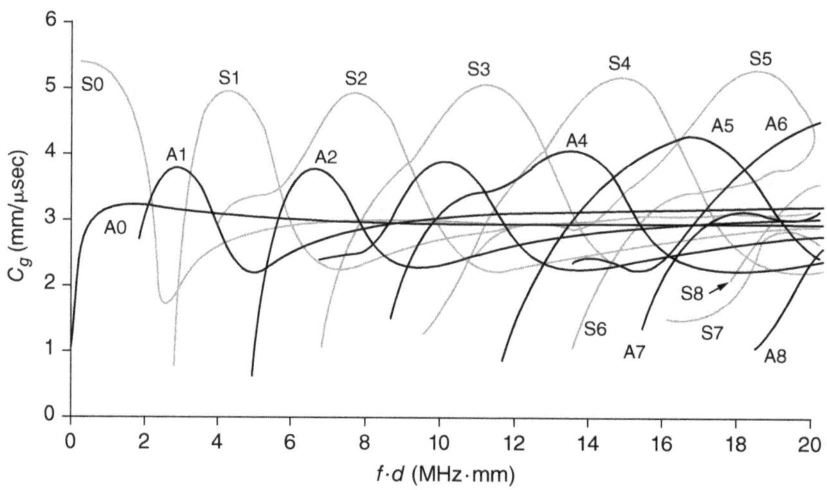


**FIGURE 13-6** Group velocity dispersion curves for a carbon steel plate.

The A0 mode at high frequency approaches the Rayleigh surface wave velocity. All other modes at new high frequency converge to the shear wave velocity for the plate. Notice the cut-off frequencies at high phase velocities $C_p$. The group velocity curves are derived from the phase velocity curves and represent the velocity of a packet of waves of similar frequency. See Fig. 13-6. The group velocity formula is as follows:

$$C_g = C_p^2 \left[ C_p - (fd) \frac{dc_p}{d(fd)} \right]^{-1} \tag{13-5}$$

Note that wave structure across the thickness of a waveguide changes as you move along frequencies for a particular mode. Sample results are shown in Fig. 13-7. Note that for an *fd* valve of .037 the in-plane displacement on the outer surface of the plate is a maximum, whereas for an *fd* valve of .245 the in-plane displacement is maximum on the outer surface. Wave structure considerations and mode and frequency choice have

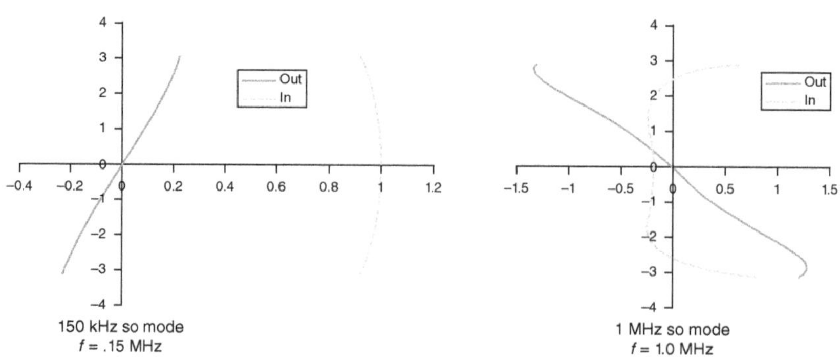


**FIGURE 13-7** Sample normalized wave structures of the SO mode in a 0.245-in.-thick steel plate. Solid line: Out-of-plane displacement; Dotted line: In-plane displacement.

serious consequences in defect detection sensitivity, penetration power, and influence to water loading situations as an example.

Let us now consider dispersion curves for a pipe and also consider longitudinal and torsional loading. The modes are labeled as L(0,1), L(0,2), L(0,3), etc. or torsional T(0,1), T(0,2), etc. The particular limits in the diagram are surface wave velocity, shear wave velocity, and cut-off frequencies as shown in Fig. 13-8. Group velocity dispersion curves are shown in Fig. 13-9. Details on the development and the nomenclature considered here

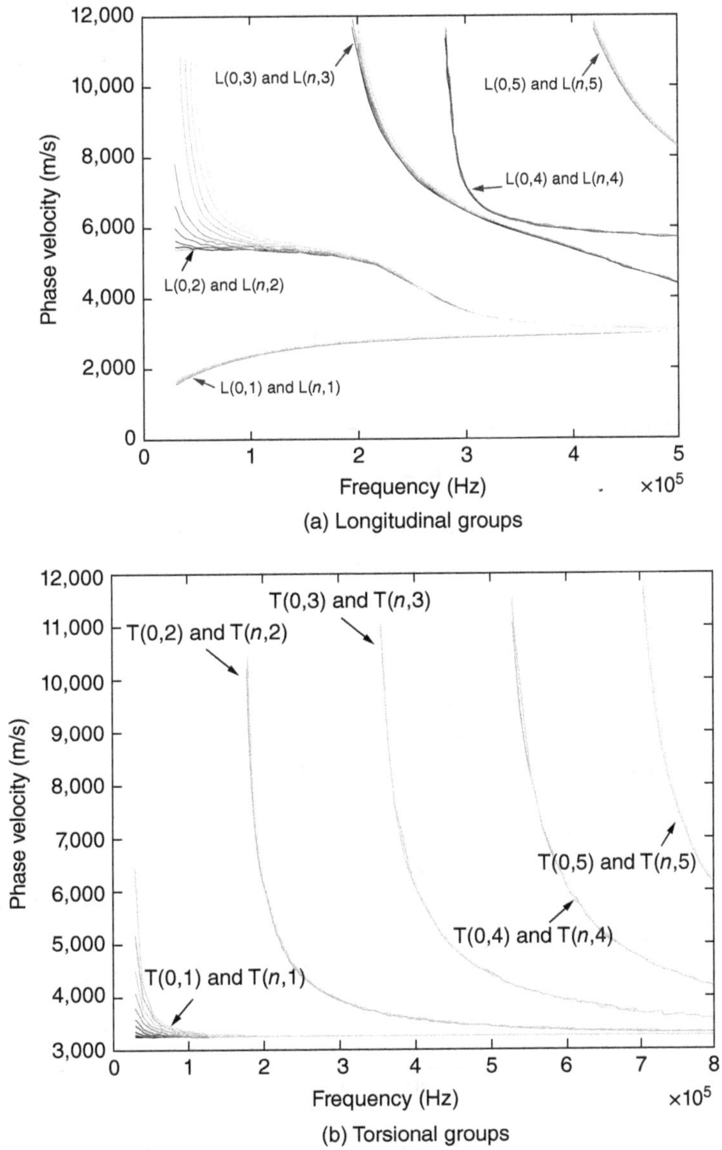


**FIGURE 13-8** Sample PHASE velocity dispersion curves for a 16-in. schedule 30 elastic steel uncoated pipe.

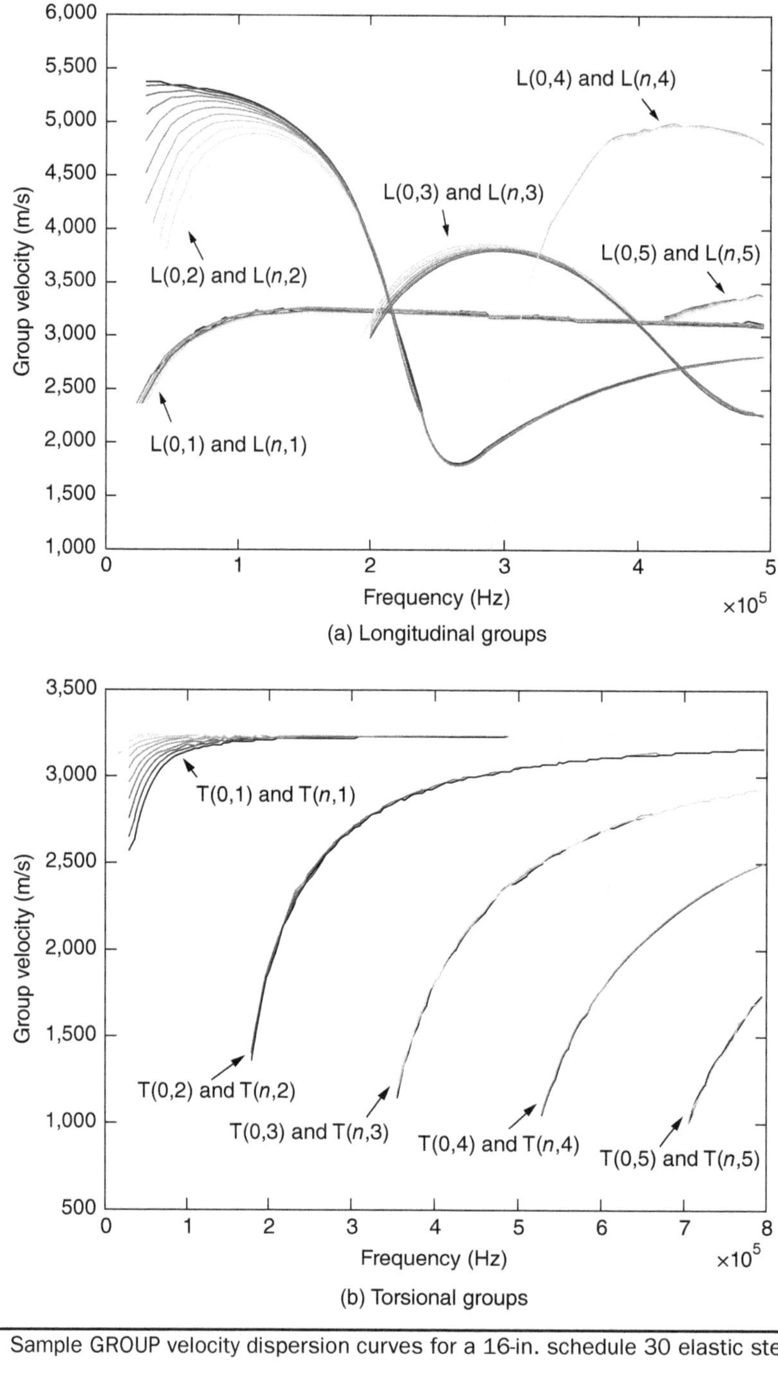


**Figure 13-9** Sample GROUP velocity dispersion curves for a 16-in. schedule 30 elastic steel bare pipe.

can be found in such references as Rose [1999]. Flexural modes are shown in Figs. 13-8 and 13-9 as well as the axisymmetric modes. Flexural modes are associated with partial loading around the circumference of a pipe. Derivable from the phase velocity dispersion curves are sets of the group velocity dispersion curves. The values of the group velocity dispersion curves depend on the ordinate and slope values of the phase velocity dispersion curves as indicated in the group velocity formula. The group velocity is what you actually measure in an experiment. See Rose [1999] for details.

If a steel pipe is filled with or is under water, there will be energy leakage as the wave travels along the pipe, because of an out-of-plane displacement component that would load the liquid. The in-plane displacement components would not travel into the liquid media since this would be like shear loading on the fluid. If you solve this wave propagation problem, or as another example the wave propagation associated with bitumen coating on a pipe, there would also be leakage of ultrasonic energy as the wave propagates along the plate. Following the phase and group velocity dispersion curves, the complex roots from the characteristic equation would then lead to a set of attenuation dispersion curves. A sample attenuation dispersion curve is shown in Fig. 13-9.

A pipe sample problem is used here. For the common plate problem in the literature, the modes are labeled as A0, A1, A2, …, S0, S1, S2, … because of symmetric and antisymmetric character. In the case of guided wave in pipes, the axisymmetric longitudinal waves are labeled as L(0,1), L(0,2), L(0,3), … and the axisymmetric torsional waves as T(0,1), T(0,2), T(0,3), …. Flexural modes are also possible due to partial loading around the circumference of a pipe. See Sun et al. [2003a, 2003b] for example for more details on flexural modes. The flexural or antisymmetric modes are shown in Fig. 13-10 whenever $n \neq 0$. Note in Fig. 13-10 that attenuation does not always increase as frequency is increased as in a usual bulk wave problem. Some modes are also attenuated more quickly than others.

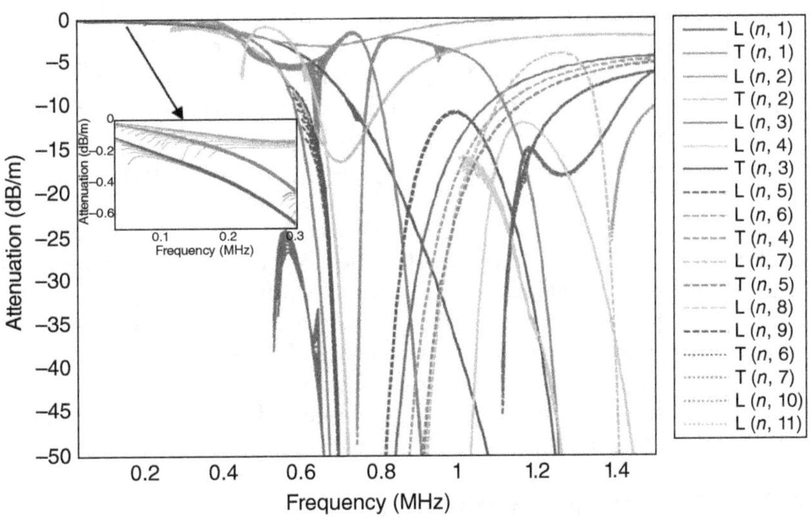


Attenuation dispersion curves for guided wave modes with circumferential order $n$ from 0 to 10. Pipe size: a 4-in. schedule 40 steel hollow cylinder coated with 0.02 in. thick E&C 2057 Cat 9 epoxy.

**FIGURE 13-10** A sample complete solution of guided wave attenuation dispersion curves.

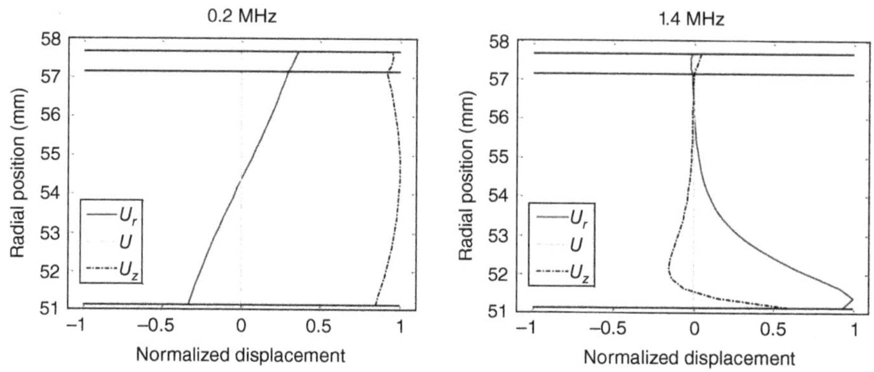


**FIGURE 13-11** Sample wave structures L(*n*, 2). *Note:* The energy is mostly concentrated at the inner surface of the coated pipe at high frequency.

All guided wave problems have associated with them the development of appropriate dispersion curves and corresponding wave structures. Of thousands of points on a dispersion curve, only certain ones lead to a successful inspection, i.e., those with greatest penetration power, maximum displacement on the outer, center, or inner surface, with only in-plane vibration on the surface to avoid leakage into a fluid, or with minimum power at an interface between a pipe and a coating, etc. A sample set of wave structure curves is illustrated in Fig. 13-11 for a pipe. In plane $u_z$, out-of-plane $u_r$ and shear horizontal $u_\theta$ are shown. Notice the in-plane vibration behavior across the thickness compared to the out-of-plane motion. The wave structure changes from point to point along every mode on a dispersion curve. The characteristics of every point on each dispersion curve are different, primarily with respect to wave structure, a critical feature for the development of an efficient test technique for a particular structure. For a particular mode any frequency value, for example, the in-plane displacement could be totally dominant across the thickness with almost no out-of-plane vibration. This mode, as an example, would travel very far even if the pipe were full of water.

In addition to the Lamb-type longitudinal waves illustrated here so for a pipe, there could be torsional, guided wave propagation in the pipe as well, depending on the sensor loading situation. In this case, the shear horizontal-type wave produces an in-plane component that's perpendicular to the wave propagation or wave vector direction but in the circumferential direction of the pipe. For the torsional mode, there is no out-of-plane displacement. The leakage into a nonviscous fluid media from a pipe would be nonexistent as far as wave propagation is concerned. Keep in mind, however, that mode conversion at a defect could create some leaky reflected waves.

Note that dispersion curve examples for other waveguides can be found in Rose [1999]. Of particular interest might be the closed-form solution possibility for shear horizontal waves in a plate.

### 13.5.2 Bulk versus Guided Waves

Further comparison of bulk and guided waves can be seen in Table 13-3. Key elements of the differences between isotropic and anisotropic media are listed in Table 13-4. Methods of determining characteristic equations for anisotropic waveguides can be found in the literature. See also Rose [1999].

| | Bulk | Guided |
|---|---|---|
| Phase Velocities | Constant | Function of frequency |
| Group Velocities | Same as phase velocities | Generally not equal to phase velocity |
| Pulse Shape | Nondispersive | Generally dispersive |

**Table 13-3** Ultrasonic Bulk versus Guided Wave Propagation Considerations

| | Isotropic | Anisotropic |
|---|---|---|
| Wave Velocities | Not function of launch direction | Function of launch direction |
| Skew Angles | No | Yes |

*Note:* All metals are not isotropic. For example, columnar dendritic centrifugally cast stainless steel is anisotropic. This must be considered in any wave propagation studies.

**Table 13-4** Ultrasonic Wave Considerations for Isotropic versus Anisotropic Media

| Bulk Wave | Guided Wave |
|---|---|
| Tedious and time-consuming | Fast |
| Point by point scan (accurate rectangular grid scan) | Global in nature (approximate line scan) |
| Unreliable (can miss points) | Reliable (volumetric coverage) |
| High-level training required for inspection | Minimal training |
| Fixed distance from reflector required | Any reasonable distance from reflector acceptable |
| Reflector must be accessible and seen | Reflector can be hidden |

**Table 13-5** A Comparison of the Currently Used Ultrasonic Bulk Wave Technique and the Proposed Ultrasonic Guided Wave Procedure for Plate and Pipe Inspection

A further practical comparison of the use of bulk and guided waves is presented in Table 13-5, in particular for plate and pipe inspections.

### 13.5.3 Aspects of NDT and SHM

Major differences between NDT and SHM are highlighted in Table 13-6, the most crucial of which is the requirement of a baseline in SHM. Very complex structures can be studied with SHM that are not even approachable with NDT. Baseline information can

| NDT | SHM |
|---|---|
| Off-line evaluation | On-line evaluation |
| Time-based maintenance | Condition-based maintenance |
| Finds existing damage | Determines fitness-for-service and remaining useful life |
| Baseline generally not available | Starts with baseline on initial data collection |

**Table 13-6** The Major Differences between NDT and SHM

change, however, and such items as temperature and boundary condition compensation are being studied. Utilization of guided waves in an SHM environment has great potential.

### 13.5.4 Source Influence

The development of the dispersion curves discussed so far employs a harmonic plane wave excitation in the waveguide. Because of a bounded transducer problem, though, we must study a source influence problem for a particular size sensor. The finite size of a transducer and various vibration characteristics give rise to a phase velocity spectrum. Therefore, in addition to the ordinary frequency spectrum, there is a phase velocity spectrum, and because of these two spectral bandwidths of frequency and phase velocity, it makes it difficult to excite a specific point on a dispersion curve. Some interesting discussion on the source influence problem can be found in Rose [1999], Ditri et al. [1992], and Rose et al. [1994]. Mode separation in the dispersion curve space then becomes useful for single-mode excitation potential.

Guided wave energy can be induced into a waveguide by a variety of different techniques including piezoelectric, magnetostrictive, electromagnetic acoustic means (EMAT), laser, or physically controlled impact. The challenge is to excite a particular mode at a specific frequency. Normal piezoelectric beam probes can be used. Angle beam piezoelectric sensors can also be used to impart beams that lead to desired kinds of guided waves in a pipe or plate. A comb transducer can also be used, either piezoelectric, magnetostrictive, EMAT, etc., that is, considering a number of different elements at a specific spacing, that together pump ultrasonic energy into the plate, hence causing wave propagation of a certain wavelength in the waveguide. The excitation zones in the phase velocity dispersion curve can be evaluated by the source being considered in the problem. Again Rose [1999], Ditri et al. [1992], and Rose et al. [1994] can provide details in this exercise. A comb transducer, as an example, could be wrapped completely around a pipe or laid out as fingers or an interdigital transducer design on a plate. See Figs. 13-12 and 13-13. Activation lines on a dispersion curve are horizontal in the case of Fig. 13-12 in using Snell's law and angled from the origin of a slope to $\lambda$, the spacing of the elements shown in Fig. 13-13.

Because of a phase velocity spectrum and a frequency spectrum, the excitation zone is not a single point in the phase velocity dispersion curve; it is a region as illustrated in Figs. 13-14 and 13-15. See Rose [1999] for details.

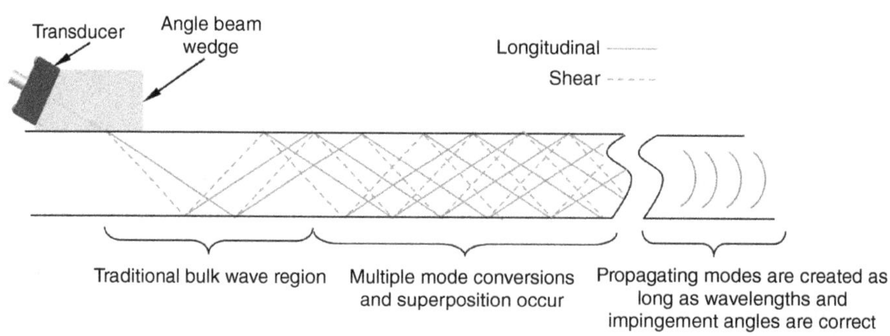


**FIGURE 13-12** Angle beam guided wave generation.

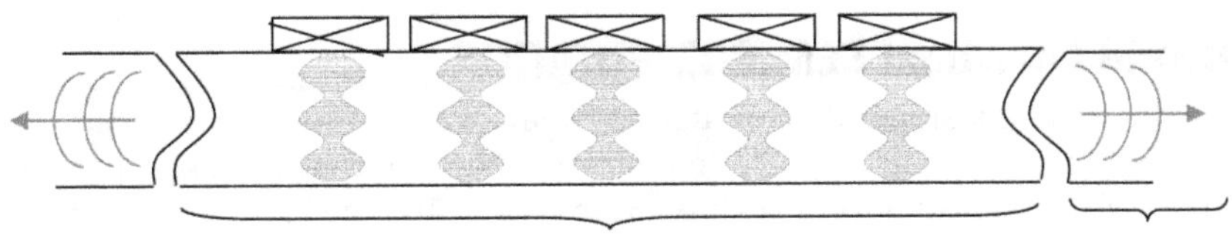

**FIGURE 13-13** Guided wave generation with a comb transducer.

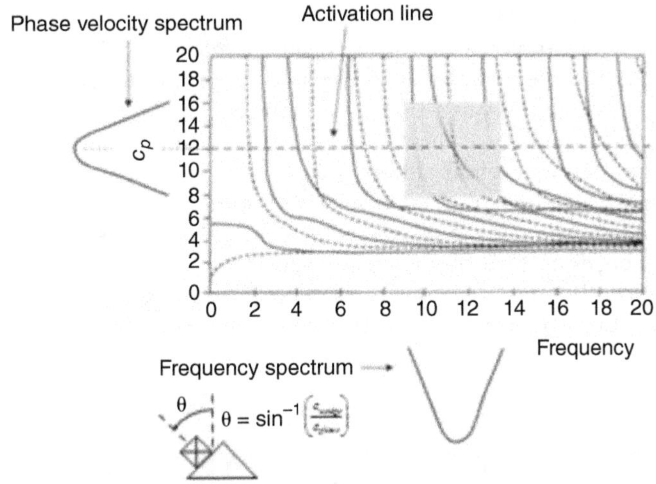


**FIGURE 13-14** Source for a typical angle beam excitation or an ability to generate a specific mode and frequency.

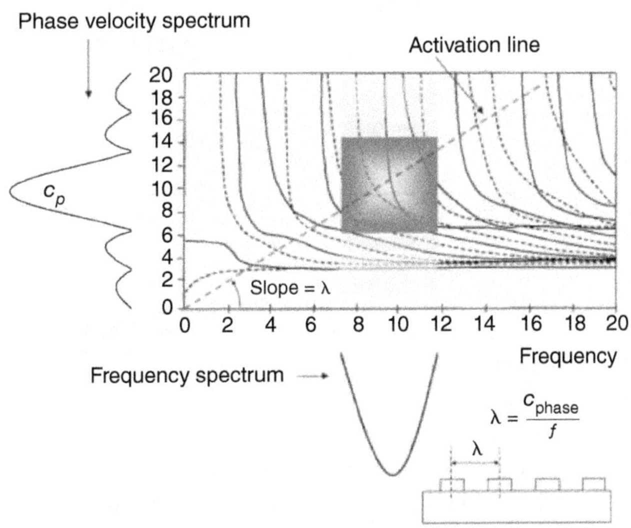


**FIGURE 13-15** Source influence for a typical comb transducer excitation or an ability to generate a specific mode and frequency.

## 13.6 Hybrid Analytical FEM COMPUTATION

We will now consider the modeling analysis component associated with the understanding of ultrasonic guided waves. The approach makes use of a hybrid analytical FEM computation as exhibited in Fig. 13-16. The FEM computation process is useless unless proper boundary conditions are considered in a simulation of an ultrasonic guided wave inspection. The boundary condition comes from a sensor design that comes from a choice of appropriate wave structure from the phase velocity dispersion curve space, hence an appropriate mode and frequency choice to solve a particular problem in NDE or SHM.

A few sample problems that have been solved using the hybrid analytical FEM computation are listed in Table 13-7 along with appropriate references.

Phased array systems have risen to ecstatic heights in the NDT field for ultrasonic bulk waves because of electronic beam scanning with one transducer array compared to manual scanning with many different angle beam probes. Laid out in Fig. 13-17 is the historical progression of phased array development from the 1970s to current-day developments. Just recently, phased arrays for real time and synthetic focusing in practical pipe inspection applications have been introduced by Rose and Mudge [2002]. Even more recently phased arrays for plate inspection were introduced by Li and Rose [2001b] and Giurgiutiu and Bao [2002].

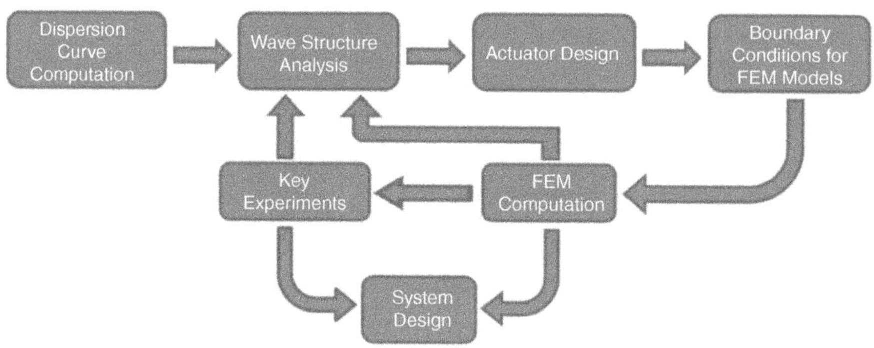


**FIGURE 13-16** The hybrid analytical FEM approach for solving guided wave problems.

- Rail—transverse cracking under shelling, defect in base
- Adhesive bonding—weak interface
- Water-loaded structure—to avoid false alarms
- Gas entrapment—leaky and nonleaky waves
- De-icing—maximum shear stress at the interface
- Piping—flexural mode best selection for focusing

**TABLE 13-7** Examples of the Use of the Hybrid Analytical FEM Approach for Solving Guided Wave Problems

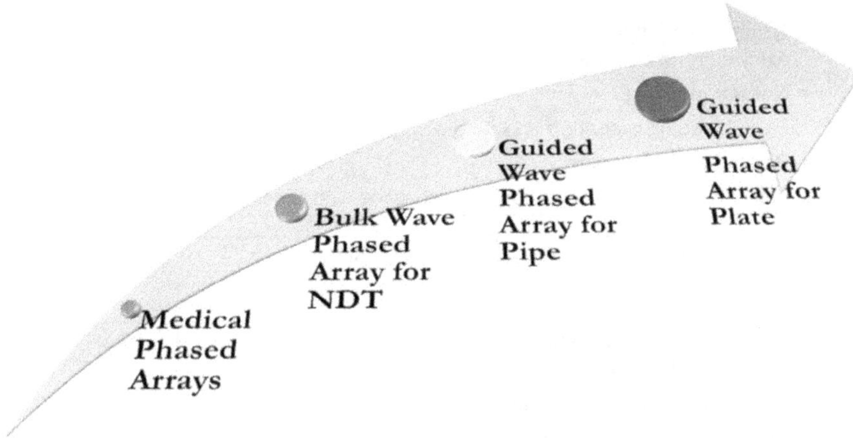


**FIGURE 13-17** Guided wave phased array technologies.

### 13.6.1 Applications

#### *Pipe*

Static shot FEM simulations at different times of axisymmetric guided wave propagation in a pipe are demonstrated in Fig. 13-18. To go beyond the axisymmetric wave to focusing considerations, the methods of focusing are listed in Table 13-8. A static shot example of a real-time phased array result is illustrated in Fig. 13-19. Notice the achievement of the focal spot in the 5th frame. The focal spot can be changed in size by changing the probe and instrument design parameters. The focal spot can then be moved anywhere axially and circumferentially in the pipe. The benefits of focusing are described in Table 13-9.

A sample experiment is presented here in Figs. 13-20 and 13-21 where all three of the defects shown in Fig. 13-20 are found and presented in Fig. 13-21. The upper diagram shows an FFA plot over a frequency range from 20 to 80 kHz showing that defect

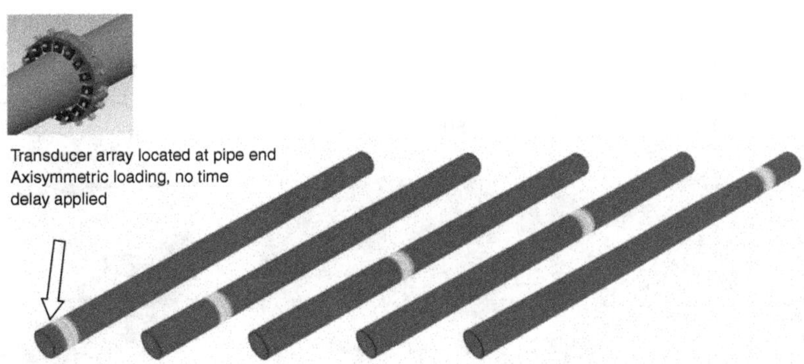


**FIGURE 13-18** Axisymmetric guided wave propagation along the pipe.

| |
|---|
| • **Frequency tuning**<br>Axisymmetric excitation and receiving |
| • **Natural focusing**<br>Partial loading excitation and receiving |
| • **Phased array focusing**<br>Multi-element array excitation and receiving with time delay and amplitude tuning |

**Table 13-8** Guided Wave Pipe Focusing Techniques

| |
|---|
| • Improved defect probability of detection (approximately 3 percent CSA, compared to 9 percent for axisymmetric) |
| • Decreased defect false alarm rate |
| • Increased inspection confidence |
| • Improved penetration power for coated pipe |
| • Excellent defect circumferential location analysis |
| • Improved SNR compared to axisymmetric |
| • 6 to infinite dB improvement compared to axisymmetric |

**Table 13-9** Focused Guided Wave Inspection Benefits

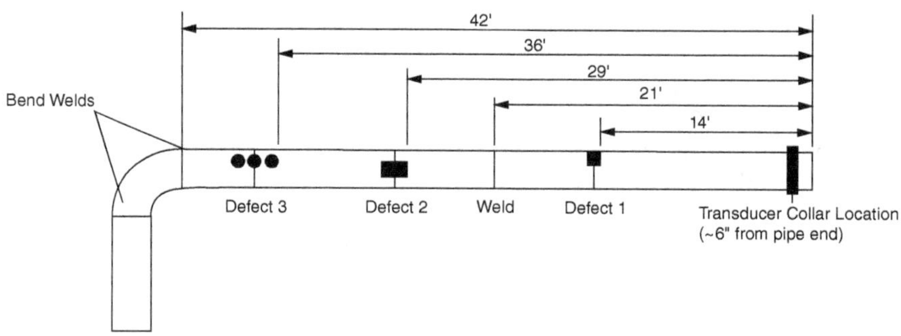


**Figure 13-19** Guided wave active focusing in pipe FE simulation results.

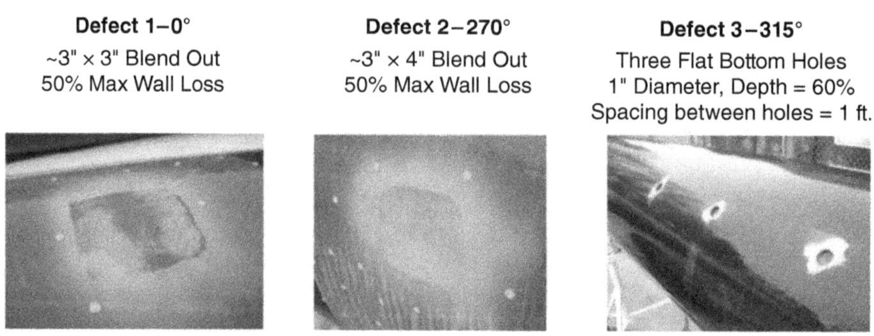


**Figure 13-20** Details of the pipe section that was studied.

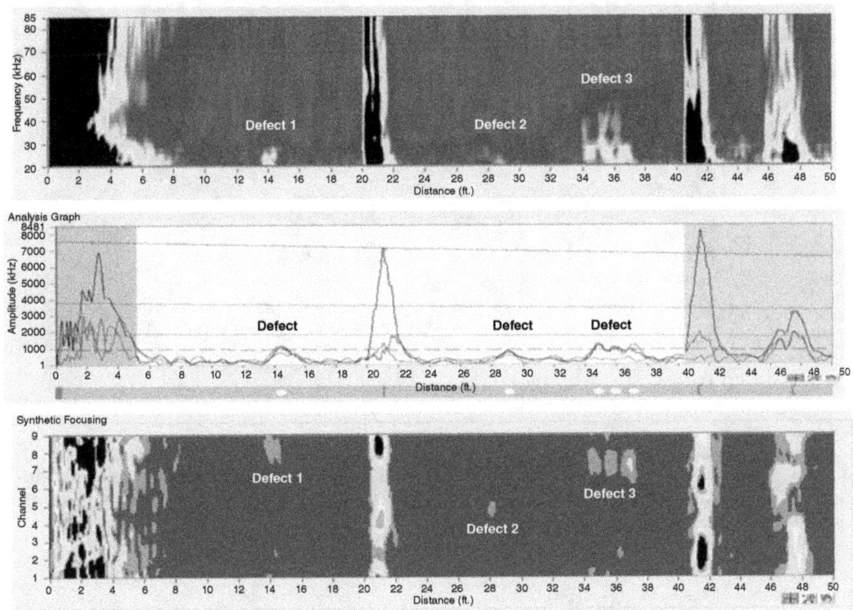


**FIGURE 13-21** Fast frequency analysis [FFA], envelope, and synthetic focusing result after examining the pipe section in Fig. 13-20.

detection sensibility is a function of frequency. The FFA plot shows the best frequency to see each defect. Each defect could have a different preferred frequency in its reflection characteristics. The second trace is a video envelope of the RF waveform clearly showing the defects and welds in the structure. The third trace shows a pipe laid out flat with an image of the defect locations both axially and circumferentially. Excellent results were obtained.

To illustrate the focusing principle in a practical sense, consider the partial loaded field natural focusing result as calculated in Fig. 13-22. The loading transducer is centered at zero degrees. With the pipe laid out flat it can be seen that the natural focal spot in this example occurs at the opposite side of the pipe at 180 degrees at an axial location of around 10 meters. If we were to have an angular profile due to some partial loading around the circumference at some axial position as shown in Fig. 13-23a, and if we were to arrange a series of partial loadings around the circumference and phase them properly, we could end up with a superimposed result at the same axial distance with a strong focal point as plotted in Fig. 13-23b. Two focused experimental results are diagrammed in Fig. 13-24. These profiles provide confidence in a call that a defect is indeed there and its circumferential position can be clearly determined.

Wave propagation into a pipe elbow and beyond can create blind spots at a specific frequency due to mode conversion. Pipe coating and buried pipes can also seriously reduce penetration distances.

Presented in Figs. 13-25 and 13-26 are some interesting aspects of pipe elbow inspection. Frequency tuning can help but a challenge remains to inspect pipelines beyond elbow regions.

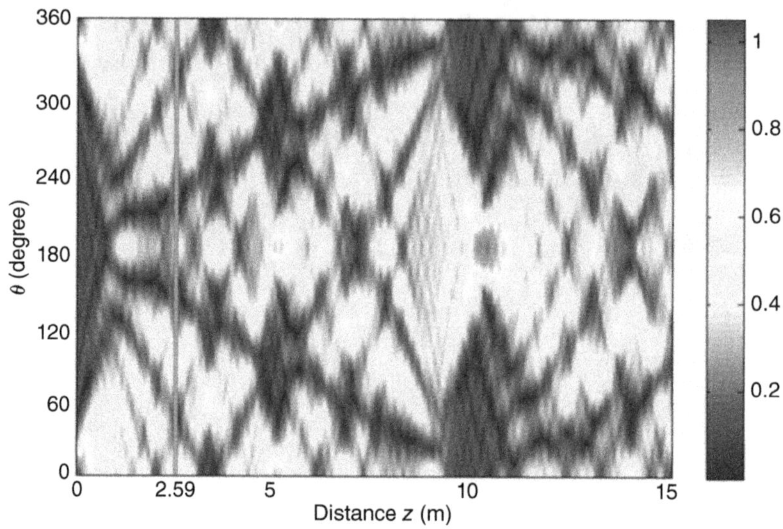


**FIGURE 13-22** Guided wave displacement interference pattern in an 8-in. schedule 40 steel pipe (unwrapped). The waves were generated by a 45° source loading vibrating in the circumferential direction at 80 kHz.

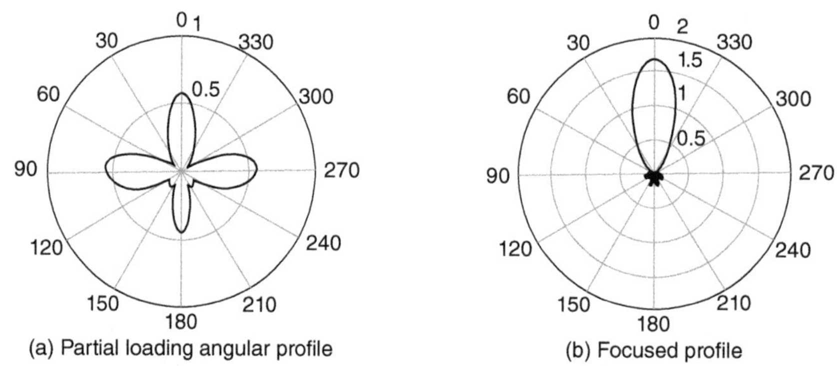


**FIGURE 13-23** Pipe active focusing concept.

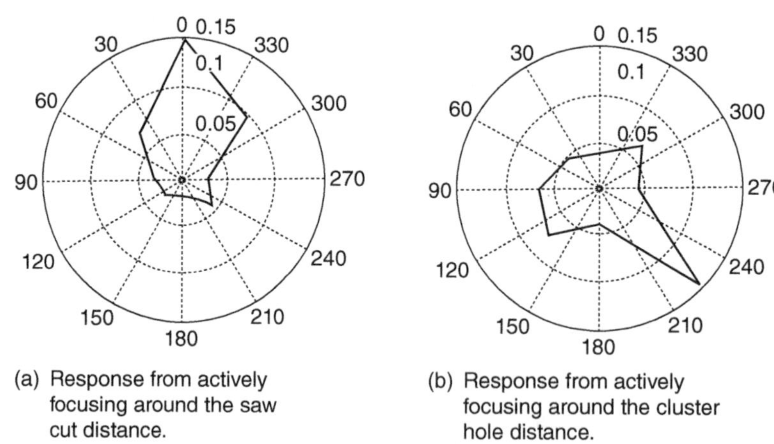


**FIGURE 13-24** Active focusing results from pipe end at 65 kHz.

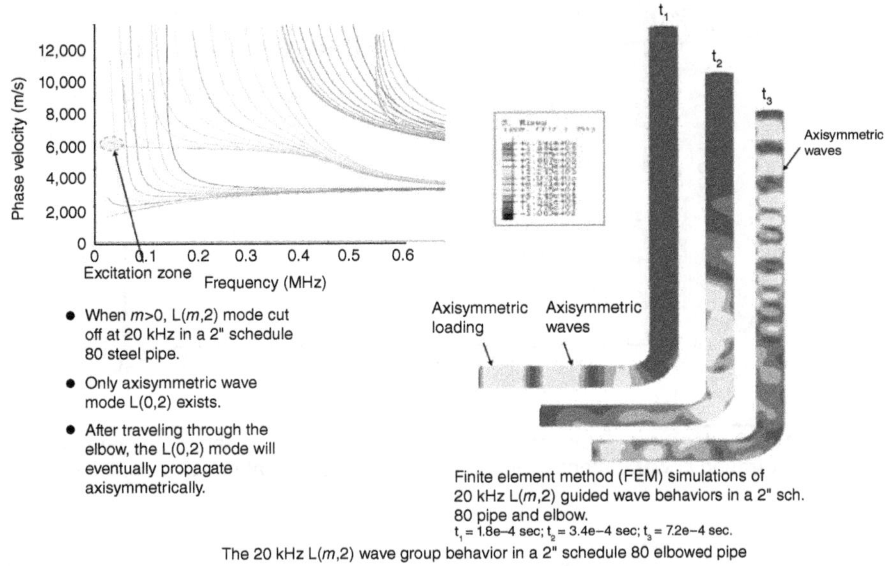


**FIGURE 13-25** Inspection beyond some elbow sections is often easy and straightforward. *Note:* Returned axisymmetric behavior beyond elbow.

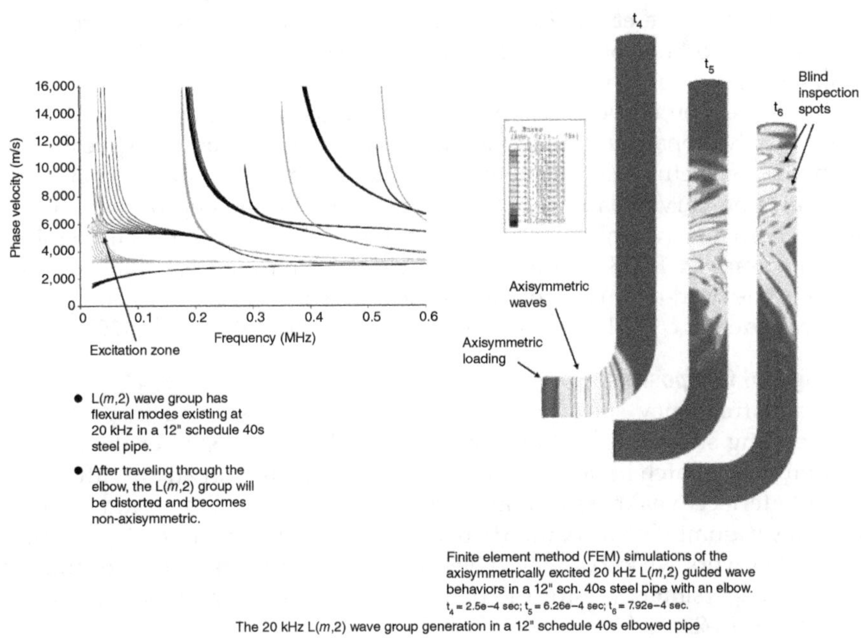


**FIGURE 13-26** Inspection beyond some elbow sections is difficult and incomplete. *Note:* Non-axisymmetric behavior beyond elbow.

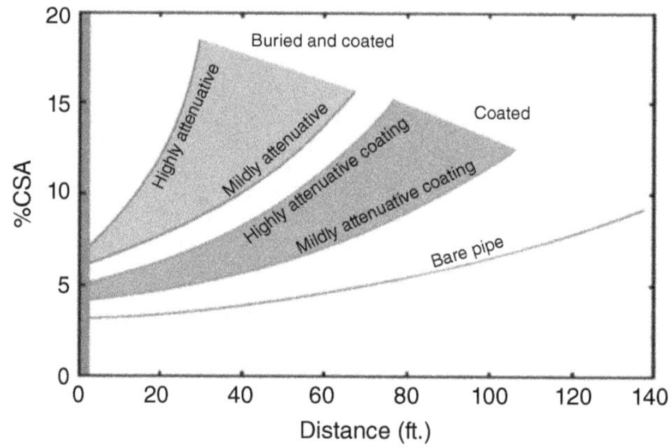


**Figure 13-27** Guided wave detection sensitivity concept curves (coated pipe buried in clay) showing penetration power distance reduction due to various coatings and soil influence.

Another challenge is depicted in Fig. 13-27. Unfortunately, coated and buried pipe decreases ultrasonic guided wave penetration power significantly.

### Rail

Guided waves are also finding increased use in a variety of different rail inspection applications. See Lee et al. [2009], for an example. We'll consider here the ability to cover various portions of a rail, from head, to web, to base in this brief discussion. The dispersion curves for a rail are close together, and in fact almost cover the entire phase velocity dispersion curve space except for bounds, as an example on the lower end where we only propagate a pseudo-Rayleigh surface wave mode. Despite the complexity, the curves are useful. Wave structure variations occur from one region to another in the dispersion curve space. It becomes possible with the appropriate mode and frequency choice to have energy propagate in the head only, or in the entire rail. See a few static shots in Fig. 13-28. If transverse crack detection under shelling was a goal, the energy in the head-only example would work well. If you were seeking a solution to find a base defect, complete rail coverage would be useful. See Lee [2009] for details.

### Bonding and Composites

Mode and frequency selection are also useful when solving inspection problems with bonding situations for composite materials. Let's consider the inspection of a titanium repair patch bonded to an aluminum structure on an aircraft. If we were to find an interface weakness or a kissing bond, guided waves would be useful as we could get a dominant in-plane displacement component into the interface which from earlier work we know would be sensitive to the weak interface. To do this, consider the dispersion curve for the titanium to aluminum bonded structure as calculated in Fig. 13-29.

Superimposed on the curves is the magnitude of the in-plane displacement component. Red areas, for example, point to good portions of the curve to carry out an inspection. Of the possibilities, consider the source influence study depicted in Fig. 13-30. The mode selected is somewhat separated from the others, and can easily

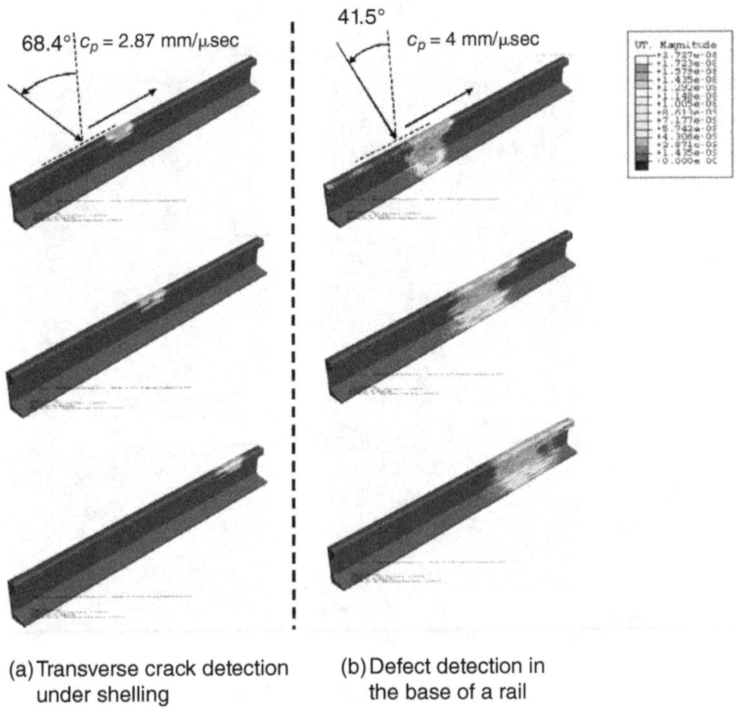


**FIGURE 13-28** Rail coverage as a function of mode selection.

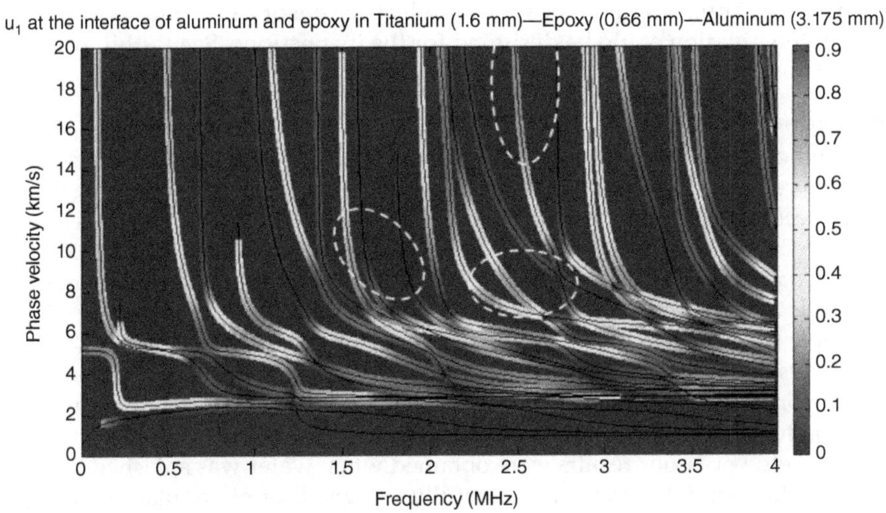


**FIGURE 13-29** Interfacial displacement (in-plane) at Al-epoxy interface in Ti-epoxy-Al specimen.

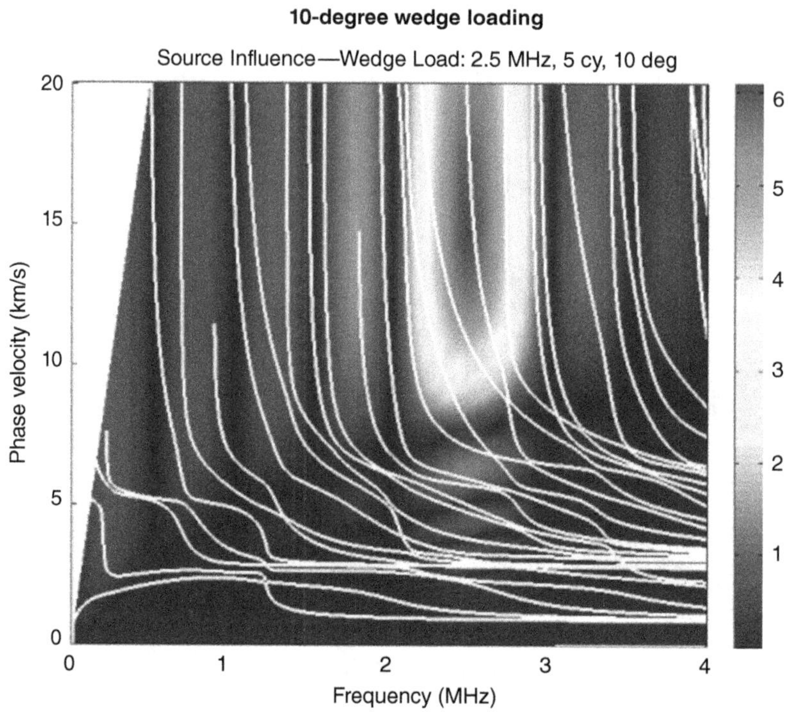


**FIGURE 13-30** Source influence study of guided wave loading sources.

be generated with an actuator having a narrow phase velocity spectrum and frequency spectrum. An angle beam piezoelectric sensor or a comb-type piezoelectric or EMAT activation could be designed for the inspections. See Puthillath et al. [2010] for details.

### *Imaging and Tomography*

Ultrasonic guided wave tomography with a variety of different sensor distributions on a structure can become a valuable tool in SHM. The basic guided wave tomographic image reconstruction process is illustrated in Figs. 13-31 and 13-32. More sensors can produce better images, but sparse array utilization can be very useful. Keep in mind that the data acquisition process should be for a particular guided wave mode, perhaps the mode and frequency most sensitive to the kind of defect that you are looking for. A sample result of corrosion on a pipe elbow is illustrated in Fig. 13-33.

Success depends, however, on using the right actuators. For example, in the plate with a corrosion defect shown in Fig. 13-34, a guided wave tomography experiment was conducted with a set of cylindrically shaped transducers commercially available off the shelf; very poor results were obtained when water was splashed onto the plate. On the other hand, when using a carefully designed set of annular array transducers, excellent results were obtained whether the plate was dry or wet. See Fig. 13-35. This is possible by selecting a point on the phase velocity dispersion curve space with dominant in-plane displacement so that the ultrasonic wave would not leak into the water. The annular array sensor is similar to a comb-type transducer, and its activation line in

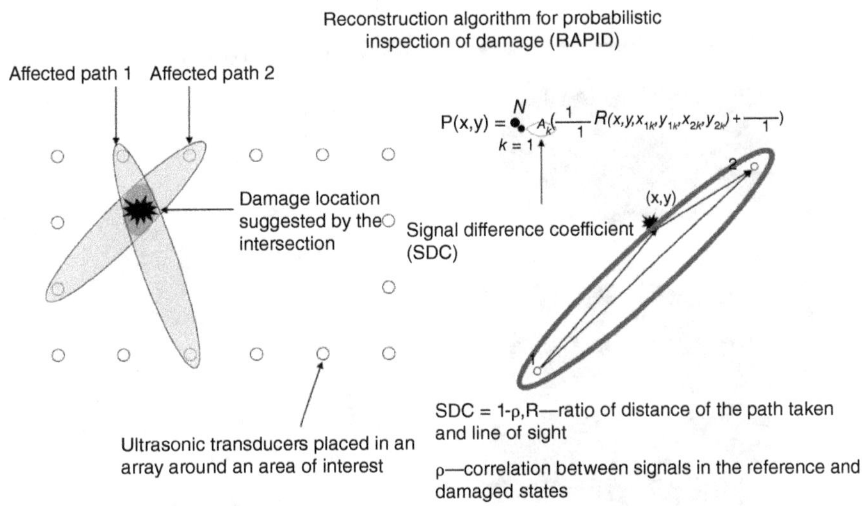


**FIGURE 13-31** The basic tomographic image reconstruction process.

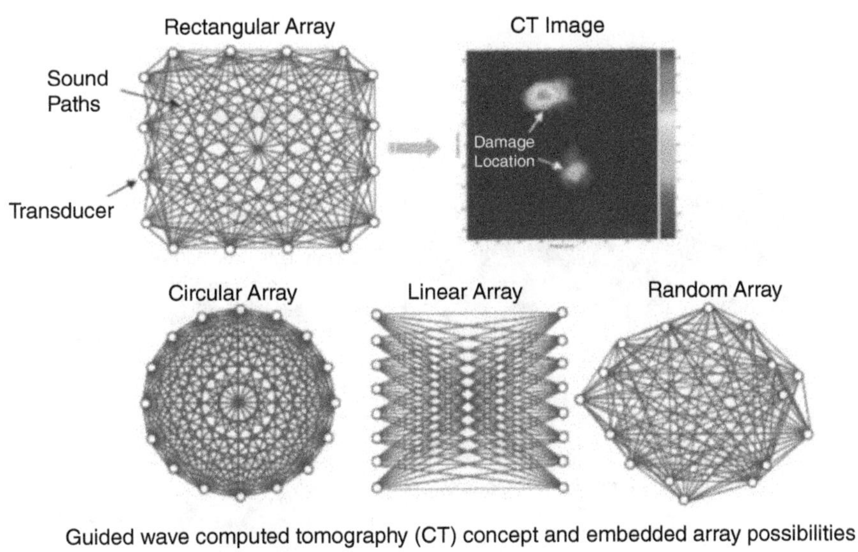


**FIGURE 13-32** Guided wave computed tomography (CT) concept and embedded array possibilities.

the phase velocity dispersion curve space can intersect the optimal test point. The sensor design process is based on selecting an appropriate wave structure as shown in Fig. 13-36. For emphasis on the concept, see Fig. 13-37. The annular array can be designed with different spacing between the elements or application of phasing or the use of time delays from one element to the next.

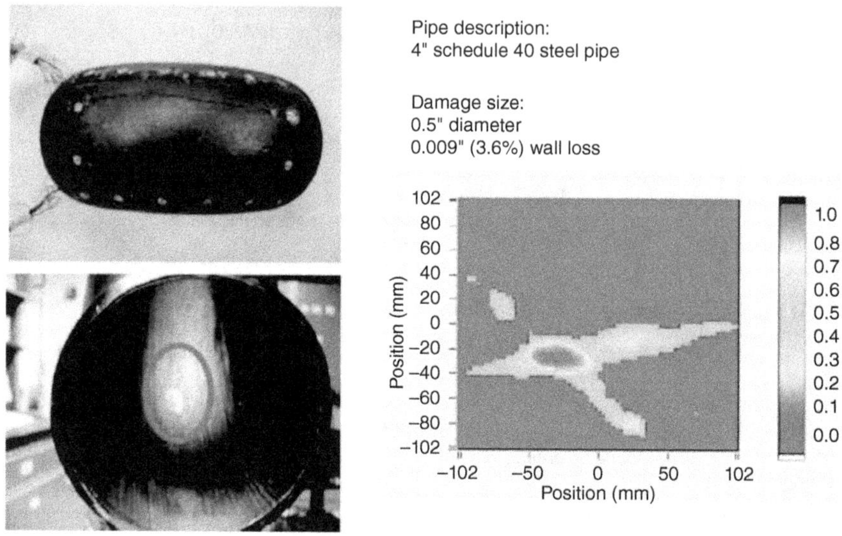


**FIGURE 13-33** Tomographic pipe elbow SHM showing interior erosion simulation via acid etch.

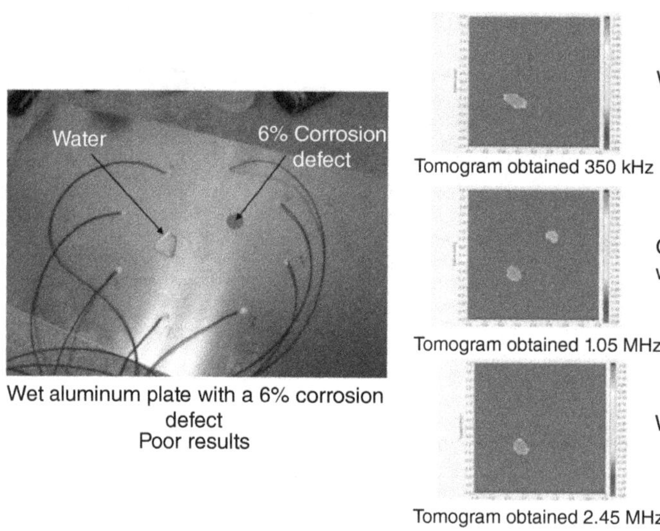


**FIGURE 13-34** The water-loaded plate influence using piezoelectric disk sensors for corrosion detection.

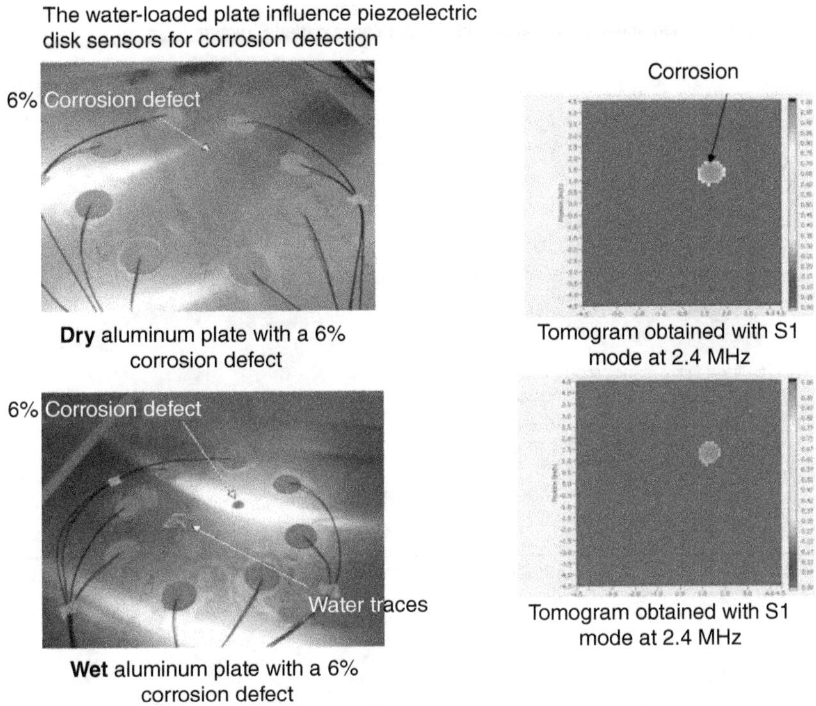


**FIGURE 13-35** Corrosion detection with good annular array sensors.

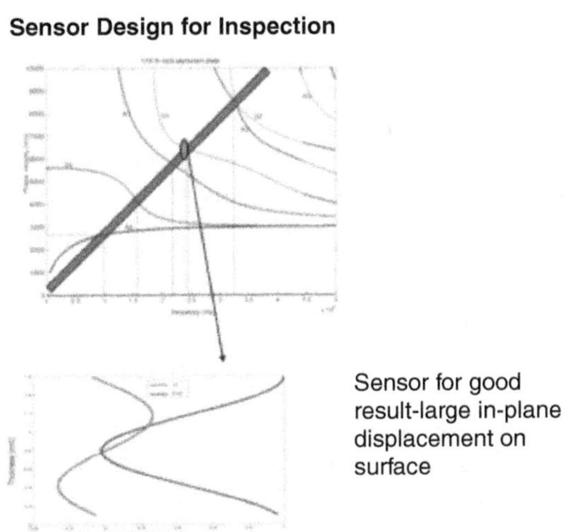


**FIGURE 13-36** Wave structure selection for achieving a document in-plane displacement component to avoid water influence.

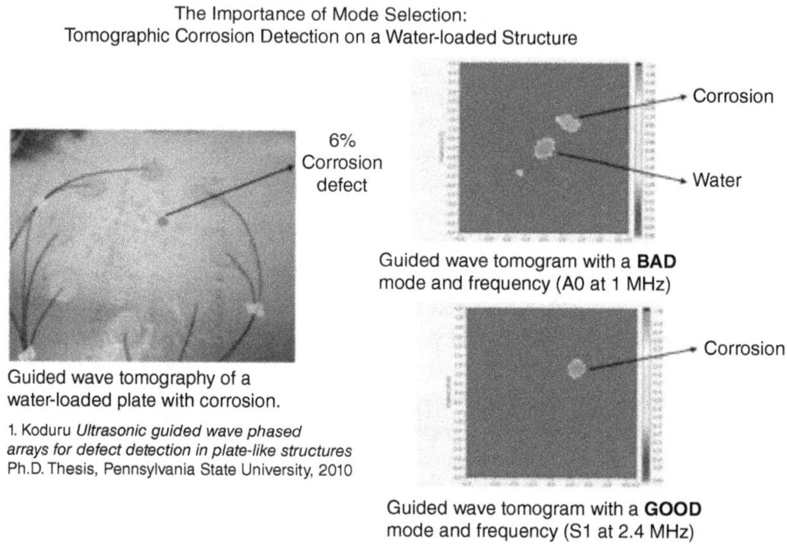


**FIGURE 13-37** Use of an annular array transducer with a bad and good wave structure selection.

## 13.7 Phased Array for Plate

Phased array application is now going beyond medical and ultrasonic bulk wave phased arrays, to guided wave phased arrays for pipe, and now to guided wave phased arrays for plates. A simulation of a radar-like scan is simulated in Fig. 13-38 with static shot shown at seven positions. Proper sensor design is required to do this from a small circular array of sensors at the center of a plate. Keep in mind here that appropriate mode and frequency control is also necessary in order to optimize defect detection sensitivity, to scan a composite plate, and to overcome skew angle and group velocity variation in the plate. Multiple scans might be necessary. A sample result is presented in Fig. 13-39. Some of the challenges in moving this method forward are outlined in Table 13-10.

Phased array possibilities for an anisotropic plate are much more complex compared to an isotropic plate. The dispersion curve becomes a function of direction and skew angles can lead to missed portions of a structure. Multimode and frequency sensors are often required for complete evaluation of a structure. See Fig. 13-40 for a description of the situation. The directional dependence of the wave velocities and energy distributions of guided waves in multilayer anisotropic composite plates sometimes make it impossible to steer the guided wave beam into a certain direction with a phased array. An example of failure is shown to the left, by choosing guided wave modes and frequencies that are insensitive to wave launching directions in the composite plates; successful beam steering can be achieved for any direction, good example to the right.

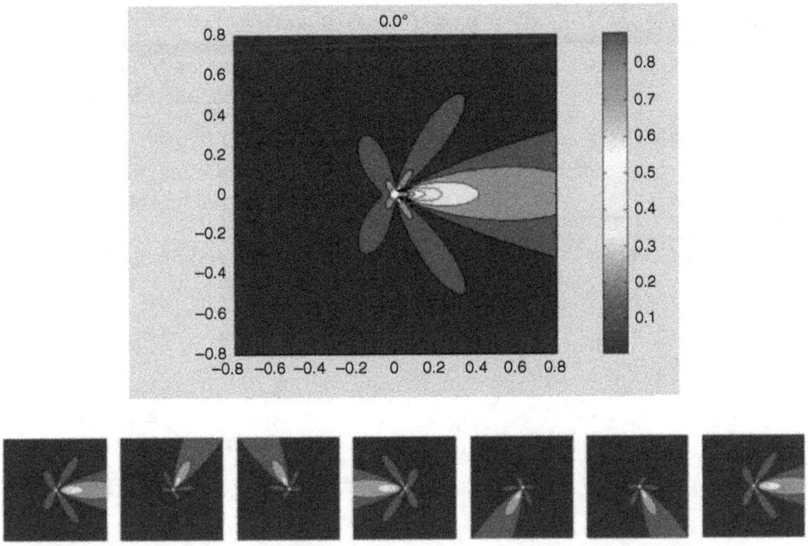


**FIGURE 13-38** Ultrasonic guided wave beam steering concept using phased arrays. *Notes:* Guided wave beam can be steered into different directions by applying phase delays to the elements of an array, circular array example.

**Product Description:** Phased array guided wave inspection of plate-like structures. From a single probe position, a guided wave beam is steered throughout the structure via time delay control of elements within an array.
**Competition:** Point by point bulk wave inspection
**Advantage:** Allow 100% coverage, ability to look at hidden structures, rapid screening possible
**Product Availability:** Under Development

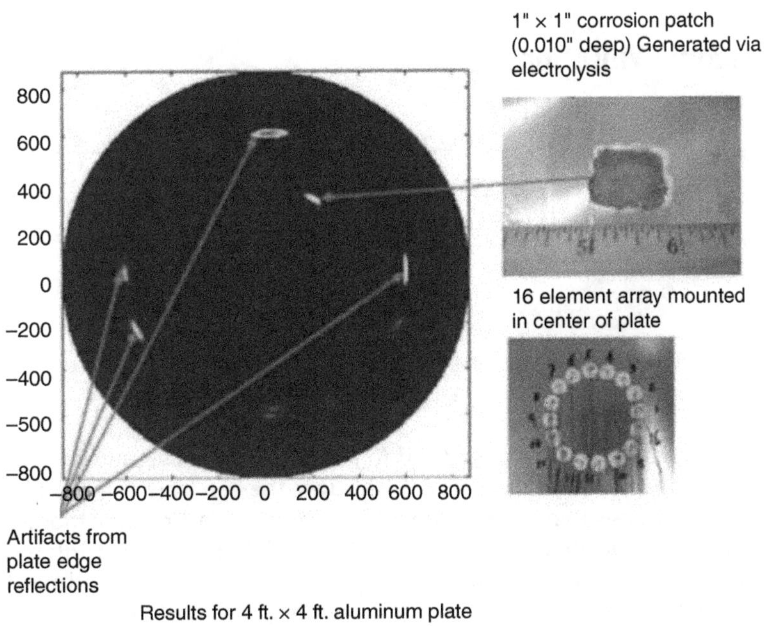


**FIGURE 13-39** Phased array guided wave inspection for plate.

| Technical Challenges | Potential Solution |
|---|---|
| • Multimodes application environment | • Guided wave transducer design and mode tuning to achieve single mode excitation and detection |
| • Guided wave dispersion | • Wave number/frequency domain beam steering |
| • Skew effects in anisotropic plates | • Select a guide wave mode and frequency that is insensitive to the material anisotropy |

**TABLE 13-10** Guided Wave Phased Array Technical Challenges and Solutions

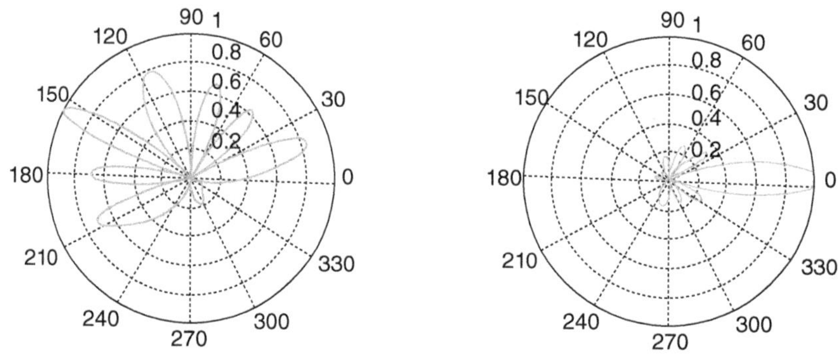


**FIGURE 13-40** Phased array for multilayer anisotropic composite plates.

## 13.8 Ultrasonic Vibrations

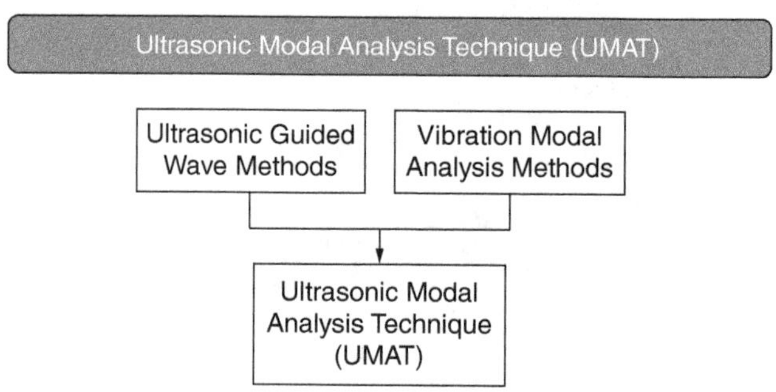


A new ultrasonic guided wave method to provide a robust signature of a structure has recently been introduced. This technique bridges the gap from transient high frequency guided wave analysis to more traditional low frequency ultrasonic modal analysis. The characteristics of the methods are presented in Table 13-11. An interesting result is shown in Fig. 13-41. The high frequency ultrasonic resonances and modal pattern or ODS depends on the ultrasonic loading function. Sensitivity optimization is possible

| | Defect Detection Size | Sender Positions | Speed |
|---|---|---|---|
| Ultrasonic Guided Wave | Very small ✓ | Many × | Slow × |
| Vibration Modal Analysis | Large × | One ✓ | Fast ✓ |
| UMAT | Small to intermediate ✓ | One ✓ | Fast ✓ |

✓ good news, × bad news.

**Table 13-11** Method Comparisons

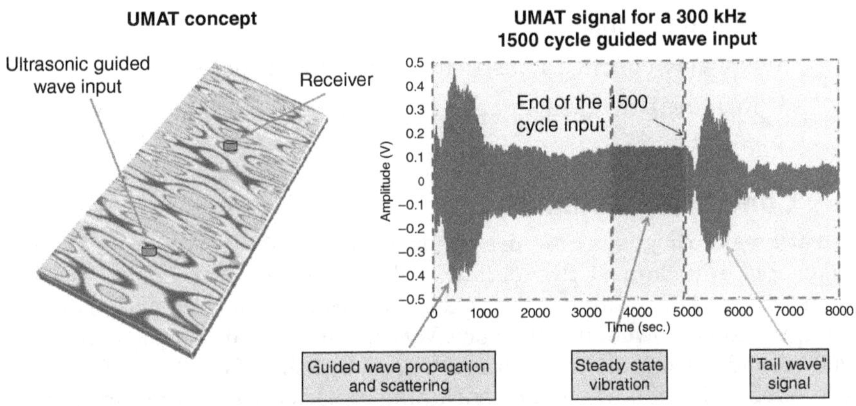


- Steady-state controlled vibrations induced by specific guided wave mode excitation
- Selective guided wave mode excitation achieved by phased annular array transducer activation
- Vibration sensitivity related to guided wave sensitivity, based on stress/displacement wave structure
- Analysis of vibration patterns and resonant frequencies under guided wave mode and frequency tuning to detect defects or changes in boundary condition

**Figure 13-41** The importance of mode selection in the ultrasonic vibration concept.

here, again, by selecting the proper mode and frequency for the phase velocity dispersion curve space. This approach is particularly useful in manufacturing hundreds of a complex shaped part and follow-up later, in an in-service environment. Structural and boundary condition variability poses serious challenges to date for this method but some specific applications are working out quite well.

## 13.9 Advantages and Limitations

### 13.9.1 Advantages of Ultrasonic Guided Wave Inspection

1. Provides rapid screening for in-service degradation (long-range inspection)—potential to achieve hundreds of feet (meters) of inspection area.
2. Capable of detecting internal and external metal loss.
3. Reduced time and cost to gain access.

4. Insulated line can be inspected with minimal insulation removal.
5. Corrosion under supports can be examined without a need for lifting.
6. Inspection at elevated locations with minimal need for scaffolding, and inspection of road crossings and buried pipes.
7. Data can be fully recorded.
8. Fully automated data collection protocols.

### 13.9.2 Limitations of Ultrasonic Guided Wave Inspection

1. Interpretation of data is highly operator dependent.
2. Difficult to find small pitting conditions.
3. Not very effective at inspecting areas close to accessories.
4. Requires a detailed, qualified procedure

## 13.10 Concluding Remarks

Great breakthroughs on the use of ultrasonic guided waves in NDT and SHM have taken place in the last two decades. The advances were made possible because of increased understanding, interpretation of results, and tremendous leaps forward in computational power and efficiency. Very few investigators were involved from 1985 to 2000 as indicated in literature survey work, but since 2000 the work and interest have exploded. What follows in the concluding remarks is a brief outline of a few successes and challenges associated with practical application of ultrasonic guided waves.

Ultrasonic guided waves for aircraft and composite material inspections have come a long way in the past decade or so. Many successes have come about but many challenges remain, some of which are outlined below. The same is true for pipeline inspection.

Successes in guided waves, in general, over the past decade, are listed in Table 13-12, in composite materials in Table 13-13, in aircraft applications in Table 13-14, and in pipeline inspection in Table 13-15. The topics speak for themselves.

- Computational developments and understanding basic principles
- Phased array and focusing developments in plates and pipes (animations and practice)
- Demonstration of optimal mode and frequency selections for penetration power, fluid loading influences, and defect detection sensitivities

**Table 13-12** Successes—Guided Waves in General

- Understanding guided wave behavior in anisotropic media (slowness profiles and skew angle influence)
- Development of ultrasonic guided wave tomographic imaging methods
- Comb sensor designs for optimal mode and frequency selection (linear comb and annular arrays)
- Implementation of time-delay comb-type transducers for phased array composite plate inspection

**Table 13-13** Successes—Composite Materials

| |
|---|
| • Demonstration of feasibility studies in lap splice, tear strap, skin to core delamination, corrosion detection, and other applications |
| • Introduction of ultrasonic modal analysis testing (linking guided waves to modal vibration analysis) |
| • Guided wave ice detection and deicing studies initiated |

**Table 13-14** Successes—Aircraft Applications

| |
|---|
| • Understanding of both axisymmetric and non-axisymmetric modes |
| • Achieving excellent penetration power with special sensors, focusing, and mode and frequency choices |
| • Handling fluid loading with torsional modes |
| • Defect sizing accomplishments to less than 5 percent cross-sectional area |
| • Reduced false alarm calls in inspection due to focusing for confirmation |
| • Circumferential location and length of defect estimations with focusing |
| • Testing of pipe under insulation, coatings, and/or soil |

**Table 13-15** Successes—Pipe Inspection

Finally, although many promising methods are evolving into promising inspection tools, numerous challenges remain. Many of the challenges are focused on technology transfer work tasks to a realistic practical environment. Challenges are listed in Table 13-16 for guided waves in general, in Table 13-17 for composite materials, in Table 13-18 for aircraft applications, and in Table 13-19 for pipeline inspection. The topics are self explanatory and can be used to define future work efforts.

| |
|---|
| • Long numerical solution computation times often necessary with FEM (especially for high frequency, and solutions at long distances from the actuator) |
| • Modeling accuracy is critically dependent on accurate input parameters (especially for anisotropic and viscoelastic properties, interface conditions, and defect characteristics) |
| • Signal interpretations often difficult (due to multimode propagation and mode conversion, along with special test structure geometric features) |
| • Mode and frequency selection for optimal penetration power and defect detection sensitivity |
| • Low-frequency limitation for thick structures |
| • Equipment cost and reliability |
| • Sensor robustness to environmental situations like temperature and humidity |
| • Sensor robustness to high stress, mechanical vibrations, shock, and radiation |
| • Adhesive bonding challenges for mounting sensors and sustainability |
| • Sensor type and development for achieving the desired guided wave mode and frequency (especially considering source influence, side lobes, and multimode generation potential) |
| • Guided wave propagation in structures with variations in cross-sectional areas and thicknesses |
| • Development of regulatory codes |
| • Educational level and inspector training |
| • Lack of realistic mock-up tests specimens for training and test development |

**Table 13-16** Challenges—Guided Waves in General

| |
|---|
| • Dealing with complex anisotropy and wave velocity and skew angle as a function of direction<br>• Viscoelastic influences<br>• Penetration power due to anisotropy, viscoelasticity, and inhomogeneity<br>• Optimal mode and frequency selection and appropriate transducer designs<br>• Differentiating critical composite damage such as delamination defects from structural variability during fabrication (including minor fiber misalignments, ply-drops, inaccurate fiber volume fraction, and so on)<br>• Guided wave inspection of composites with unknown material properties |

**Table 13-17** Challenges—Composite Materials

| |
|---|
| • Sensor and instrumentation weight considerations<br>• Cabling and wireless potential and energy harvesting<br>• Adhesive bonding inspection reliability (especially kissing bonds)<br>• Robustness of guided wave sensors under in-flight conditions<br>• High temperature environment possibilities<br>• Many wave scattering sources such as rivets, fasteners, stiffeners, and joints<br>• Influences of aircraft paint and embedded metallic mesh in composite airframes for lightning protection |

**Table 13-18** Challenges—Aircraft Applications

| |
|---|
| • Penetration power distances for plastic pipe and coated and buried steel pipe<br>• Tees, elbows, bends, and number of elbows and inspection beyond elbows<br>• Quantification in defect location, characterization, sizing, especially depth determination<br>• Inspection reliability and false alarms (due to multimode propagation, mode conversions, and so many pipe features like welds, branches, etc.)<br>• Reducers, expanders, unknown layout drawings, cased pipes, and sleeves<br>• Inspection time for a complete real-time focusing scan of a pipe<br>• Temperature effects on coating<br>• Improved axial and lateral resolution<br>• Reduction of dead zone |

**Table 13-19** Challenges—Pipe Inspection

New directions point to NDE and SHM with inexpensive distributive sensor networks and sparse arrays for a line of sight analysis, tomography, phased array work, and ultrasonic vibrations. Besides defect detection and location analysis for screening, new work also points to more detailed quantitative characterization analysis. Integration with wireless and energy harvesting technologies are underway. Emphasis on lightweight easy-to-use are of great interest and are being developed by several manufacturers.

## Acknowledgments

Thanks are given to my M.S. and Ph.D. students over the years, as well as all of the employees at FBS, Inc. for their contributions to the field of ultrasonic guided waves. Cooperation has been a fun and rewarding experience. Thanks also to all of my sponsors over the years and many worldwide friends and colleagues in ultrasonics.

*Note:* Portions of this chapter are taken from Rose [2011], Workman et al. [2007], and Rose [2010a].

## References

Achenbach, J.D., *Wave Propagation in Elastic Solids*, Elsevier Science, Amsterdam, 1973.

Achenbach, J.D., *Reciprocity in Elastodynamics*, Cambridge University Press Cambridge, United Kingdom, 2003.

Auld, B.A., *Acoustic Fields and Waves in Solids*, John Wiley and Sons, 1973, Hoboken, NJ (Reprinted R.E. Krieger Publ. Co., Malabar, Florida, 1990).

Bartoli, I., Marzani, A., Lanza di Scalea, F., and Viola, E., "Modeling wave propagation in damped waveguides of arbitrary cross-section," *Journal of Sound and Vibration*, 295(3–5), 685–707, August 22, 2006.

Breon, L.K. Van Velsor, J.K., and Rose, J.L., "Guided wave damage detection tomography for structural health monitoring in critical zones of pipelines," *Materials Evaluation*, 65, 1215–1219, 2007.

Cantrell, J.H., and Yost, W.T., "Nonlinear ultrasonic characterization of fatigue microstructures," *International Journal of Fatigue*, 23(1), 487–490, 2001.

Chimenti, D.E., and Nayfeh, A.H., "Experimental ultrasonic reflection and guided wave propagation in fibrous composite laminates," *American Society of Mechanical Engineers, Applied Mechanics Division, AMD*, 90, 29–37, 1988.

Datta, S.K., and Shah, A.H., *Elastic Waves in Composite Media and Structures: With Applications to Ultrasonic Nondestructive Evaluation*, CRC Press, Boca Raton, FL, 2008.

Datta, S.K., Shah, A.H., and Karunasena, W., "Edge and layering effects in a multilayered composite plate," *Computers and Structures*, 37(2), 151–162, 1990.

Davies, J., and Cawley, P., "The application of synthetic focusing for imaging crack-like defects in pipelines using guided waves," *IEEE Transactions on Ultrasonics, Ferroelectrics, and Frequency Control*, 56(4), 759–771, 2009.

Demma, A., Cawley, P., and Lowe, M.J.S., "Mode conversion of longitudinal and torsional guided modes due to pipe bends," *AIP Conference Proceedings*, 557A, 172–179, 2001.

Ditri, J.J., and J.L. Rose, "Excitation of guided elastic wave modes in hollow cylinders by applied surface tractions," *Journal of Applied Physics*, 72(7), 2589–2597, 1992.

Fromme, P., and Sayir, M.B., "Detection of cracks at rivet holes using guided waves," *Ultrasonics*, 40(1–8), 199–203, May 2002.

Gao, H., and Rose, J.L., "Ice detection and classification on an aircraft wing with ultrasonic shear horizontal guided waves," *IEEE Transactions on Ultrasonics, Ferroelectrics, and Frequency Control*, 56(2), 334–344, 2009.

Gao, H., and Rose, J.L., "Goodness dispersion curves for ultrasonic guided wave based SHM: a sample problem in corrosion monitoring," *The Aeronautical Journal of the Royal Aeronautical Society*, 114, January 2010.

Gao, H., Shi, Y., and Rose, J.L., "Guided wave tomography on an aircraft wing with leave in place sensors," *Review of Progress in Quantitative Nondestructive Evaluation*, 760, 1788–1794, 2005.

Giurgiutiu, V., and Bao, J., "Embedded ultrasonic structural radar with piezoelectric wafer active sensors for the NDE of thin-wall structures," *ASME International Mechanical Engineering Congress and Exposition, Proceedings*, 31–38, 2002.

Graff, K.F., *Wave Motion in Elastic Solid*, Oxford University Press, New York, NY, 1975.

Grewal, D.S., "Improved ultrasonic testing of railroad rail for transverse discontinuities in the rail head using higher order Rayleigh (M21) waves," *Materials Evaluation*, 54, 983–986, 1996.

Gunawan, A., and Hirose, S., "Boundary element analysis of guided waves in a bar with an arbitrary crosssection," *Engineering Analysis with Boundary Elements*, 29(10), 913–924, October 2005.

Hall, J.S., and Michaels, J.E., "Computational efficiency of ultrasonic guided wave imaging algorithms," *IEEE Transactions on Ultrasonics, Ferroelectrics, and Frequency Control*, 58(1), 244–248, January 2011.

Hayashi, T., Kawashima, K., Sun, Z., and Rose, J.L., "Analysis of flexural mode focusing by a semianalytical finite element method," *Journal of Acoustical Society of America*, 113, 1241–1248, 2002.

Hayashi, T., and Murase, M., "Defect imaging with guided waves in a pipe," *The Journal of the Acoustical Society of America*, 117, 2134–2140, 2005.

Hayashi, T., Song, W., and Rose, J.L., "Guided wave dispersion curves for a bar with an arbitrary crosssection, a rod and rail example," *Ultrasonics*, 41(3), 175–183, May 2003.

Hinders, M.K., Malyarenko, E.V., and McKeon, J.C.P., "Ultrasonic Lamb wave tomographic scanning," *Proceedings of SPIE—The International Society for Optical Engineering*, 3586, 279–291, 1999.

Hosten, B., and Castaings, M., "Finite elements methods for modeling the guided waves propagation in structures with weak interfaces," *Journal of the Acoustical Society of America*, 117(3), 1108–1113, March 2005.

Ing, R.K., and Fink, M., "Time-reversed Lamb waves," *IEEE Transactions on Ultrasonics, Ferroelectrics and Frequency Control*, 45(4), 1032–1043, July 1998.

Kim, Y.Y., Park, C.I., Cho, S.H., and Man, S.W., "Torsional wave experiments with a new magnetostrictive transducer configuration," *Journal of the Acoustical Society of America*, 117 (6), 3459–3468, June 2005.

Kwun, H., and Dynes, C., "Long-range guided wave inspection of pipe using the magnetostrictive sensor technology—Feasibility of defect characterization," *Proceedings of SPIE—The International Society for Optical Engineering*, 3398, 28–34, 1998.

Lee, C.Y., Rose, J.L., and Cho, Y., "A guided wave approach to defect detection under shelling in rail," *NDT & E International*, 42(3), 174–180, April 2009.

Li, J., and Rose, J.L., "Excitation and propagation of non-axisymmetric guided waves in a hollow cylinder," *Journal of the Acoustical Society of America*, 109(2), 457–464, 2001a.

Li, J., and Rose, J.L., "Implementing guided wave mode control by use of a phased transducer array," *Ultrasonics, Ferroelectrics & Frequency Control*, 48(3), 761–768, May 2001b.

Li, J., and Rose, J.L., "Angular-profile tuning of guided waves in hollow cylinders using a circumferential phased array," *IEEE Transactions on Ultrasonics, Ferroelectrics and Frequency Control*, 49(12), 1720–1729, 2002.

Ma, J., Lowe, M.J.S., and Simonetti, F., "Feasibility study of sludge and blockage detection inside pipes using guided torsional waves," *Measurement Science and Technology*, 18(8), 2629–2641, August 1, 2007.

Mal, A.K., Xu, P., and Bar-Cohen, Y., "Ultrasonic nondestructive evaluation of adhesive bonds," *American Society of Mechanical Engineers, Materials Division (Publication) MD*, 9, 85–89, 1988.

Michaels, J.E., Cobb, A.C., and Michaels, T.E., "A comparison of feature-based classifiers for ultrasonic structural health monitoring," *Proceedings of SPIE–The International Society for Optical Engineering*, 5394, 363–374, 2004.

Mu, J., and Rose, J.L., "Long range pipe imaging with a guided wave focal scan," *Materials Evaluation*, 66(6), 663–666, 2008a.

Mu, J., and Rose, J.L., "Guided wave propagation and mode differentiation in hollow cylinders with viscoelastic coatings," *Journal of the Acoustical Society of America*, 124(2), 866–874, 2008b.

Mu, J., Zhang, L., and Rose, J.L., "Pipe inspection with guided wave synthetic focusing technique," *Materials Evaluation*, 68(10), 1171–1176, 2010.

Na, W., Kundu, T., and Ehsani, M.R., "Ultrasonic guided waves for steel bar concrete interface testing," *Materials Evaluation*, 60(3), 437–444, 2002.

Pavlakovic, B., Lowe, M., Alleyne, D., and Cawley, P., "DISPERSE: a general purpose program for creating dispersion curves," *Review of Progress in Quantitative Nondestructive Evaluation*, 1185–1192, 1997.

Pruell, C., Kim, J.-Y., Qu, J., and Jacobs, L.J., "A nonlinear-guided wave technique for evaluating plasticitydriven material damage in a metal plate," *NDT&E International*, 42(3), 199–203, April 2009.

Puthillath, P., and Rose, J.L., "Ultrasonic guided wave inspection of a titanium patch bonded to an aluminum aircraft skin," *International Journal of Adhesion and Adhesives*, 30, 566–573, 2010.

Raghavan, A., and Cesnik, C.E.S., "Modeling of piezoelectric-based Lamb wave generation and sensing for structural health monitoring," *Proceedings of SPIE*, 5391, 419–430, 2004.

Rizzo, P., Cammarata, M., Dutta, D., Sohn, H., and Harries, K., "An unsupervised learning algorithm for fatigue crack detection in waveguides," *Smart Materials and Structures*, 18(2), 025016 (11 pp.), February 2009.

Rokhlin, S.I., Chimenti, S.E., and Nagy, P.B., *Physical Ultrasonics of Composites*, Oxford University Press New York, NY CB2 8BS, United Kingdom, 2011.

Rose, J.L., "Guided waves successes and challenges for ultrasonic testing in NDT and SHM," *Materials Evaluation*, 68(5), 495–500, 2010a.

Rose, J.L., "The upcoming revolution in ultrasonic guided waves," *Proceedings of SPIE*, 7983, 798302 (2011); 10.1117/12.897025, [2011].

Rose, J. L., *Ultrasonic Waves in Solid Media*, Cambridge University Press, Cambridge, UK, 1999.

Rose, J.L., Ditri, J., and Pilarski, A., "Wave mechanics principles in acousto-ultrasonic NDE," *Journal of Acoustic Emission*, 12(1/2), 23–26, 1994.

Rose, J.L., Ditri, J.J., Pilarski, A., Rajana, K.M., and Carr, F.T., A guided wave inspection technique for nuclear steam generator tubing," *NDT&E International*, 27(6), 307–310, 1993.

Rose, J.L., and Mudge, P.J., "Flexural Mode Focusing in Pipe," *8th European Conference on Nondestructive Testing*, Barcelona, Spain, June 17–21, 2002.

Rose, J.L., Sun, Zongqi, Mudge, Peter J., and Avioli, Michael J., "Guided wave flexural mode tuning and focusing for pipe testing," *Materials Evaluation*, 61(2), 162–167, February 2003.

Rose, J.L., Yan, F., and Zhu, Y., "Ultrasonic guided wave modal analysis technique (UMAT) for defect detection," *Proceedings of IMAC XVIII, Jacksonville*, FL, February 2010b.

Ruzzene, M., "Frequency wavenumber domain filtering for improved damage visualization," *Smart Materials and Structures*, 16(6), 2116–2129, 2007.

Salas, K.I., and Cesnik, C.E.S. "Clover: An alternative concept for damage interrogation in structural health monitoring systems," *Aeronautical Journal of the Royal Aeronautical Society*, 113(1144), 339–356, 2009.

Senesi, M., Xu, B., and Ruzzene, M., "Experimental characterization of periodic frequency-steerable arrays for structural health monitoring," *Smart Materials and Structures*, 19, 055026 (11 pp), April 2010.

Sun, Z., Zhang, L., Gavigan, B.J., Hayashi, T., and Rose, J.L., "Ultrasonic flexural torsional guided wave pipe inspection potential," *ASME Proceedings of Pressure Vessel and Piping Division Conference*, PVP 456, 29–34, 2003a.

Sun, Z., Zhang, L., and Rose, J.L., "Flexural longitudinal and torsional modes natural focusing phenomena in a pipe," *Proceedings of QNDE*, 23, 193–197, 2003b.

Thompson, R.B., Alers, G.A., and Tennison, M.A., "Application of direct electromagnetic lamb wave generation to gas pipeline inspection," *Conference: IEEE Ultrason Symp, Proc*, 91–94, October 4, 1972–October 7, 1972.

Van Velsor, J.K., Gao, H., and Rose, J.L., "Guided-wave tomographic imaging of defects in pipe using a probabilistic reconstruction algorithm," *Insight: Non-Destructive Testing and Condition Monitoring*, 49(9), 532–537, 2007.

Van Velsor, J.K., Rose, J.L., and Nestleroth, J.B., "Enhanced coating disbound detection capabilities in pipe using circumferential shear horizontal guided waves," *Materials Evaluation*, 67 (10), 1179–1188, 2009.

Velichko, A., and Wilcox, P.D., "Post-processing of guided wave array data for high resolution pipe inspection," *Journal of the Acoustical Society of America*, 126(6), 2973–2982, 2009.

Venna, S., Lin, Y., and Botura, G., "Piezoelectric transducer actuated leading edge de-icing with simultaneous shear and impulse forces," *Journal of Aircraft*, 44(2), 509–515, 2007.

Wilcox, P.D., "A rapid signal processing technique to remove the effect of dispersion from guided wave signals," *IEEE Transactions on Ultrasonics, Ferroelectrics and Frequency Control*, 50(4), 419–427, April 2003a.

Wilcox, P.D., "Omni-directional guided wave transducer arrays for the rapid inspection of large areas of plate structures," *IEEE Transactions on Ultrasonics, Ferroelectrics and Frequency Corntrol*, 50(6), 699–709, 2003c.

Wilcox, P.D., Pavlakovic, B.N., Evans, M.J., Vine, K.A., Cawley, P., Lowe, M.J.S., and Alleyne, D.N., "Long Range Inspection of Rail Using Guided Waves" in Review of Progress in Quantitative NDE, 22, pp. 236–243, 2003b.

Workman, Gary L., and Kishoni, Doron (eds.), Ultrasonic Testing, Volume 7 in *Nondestructive Testing Handbook*, 3rd ed. (J. Rose contribution), American Society for Nondestructive Testing, Columbus, OH, 2007.

Yan, F., Qi, X., Rose, J.L., and Weiland, H., "Ultrasonic guided wave mode and frequency selection for defect detection and characterization in a multilayer hybrid laminate," *Materials Evaluation*, 68(2), 169–175, 2010a.

Yan, F., and Rose, J.L., "Defect detection using a ultrasonic guided wave modal analysis technique (UMAT)," *Proceedings of SPIE*, Vol. 7650, 76500R, 2010c.

Yan, F., and Rose, J.L., "Guided wave phased array beam steering in composite plates," *Proceedings of SPIE* 6532:65320G, 2007.

Yan, F., and Rose, J.L., "Time delay comb transducers for aircraft inspection," *The Aeronautical Journal of the Royal Aeronautical Society*, 113(1144), 417–427, 2009.

Yan, F., Rose, J.L., and Royer, R.L., Jr., "Ultrasonic guided wave imaging techniques in structural health monitoring," *Journal of Intelligent Material Systems and Structures*, 21(3), 377–384, 2010b.

Zhu, Y., Palacios, J.L., Rose, J.L., and Smith, E.C., "De-icing of multilayer composite plates using ultrasonic guided waves," *Collection of Technical Papers - 49th AIAA/ASME/ASCE/AHS/ASC Structures, Structural Dynamics and Materials Conference*, 2008.

Zhu, W., and Rose, J.L., "Lamb wave generation and reception with time-delay periodic linear arrays: ABEM simulation and experimental study," *IEEE Transactions on Ultrasonics, Ferroelectrics and Frequency Control*, 46(3), 654–664, May 1999.

# CHAPTER 14
# Shearography Nondestructive Testing

## 14.1 History and Development

Since it was first introduced in 1986, the shearography nondestructive testing (NDT) technology has been providing large productivity increases in manufacturing and overhaul of metal and composite structures in aerospace, marine, and rail industries. A high-throughput optical inspection technique, shearography uses a common path, laser-based imaging interferometer to detect, measure, and locate anomalies by imaging submicroscopic changes to a test part surface when an appropriate stress is applied. The shearography camera inspects a full field at a time, rather than scanning a single-point sensor or an array of sensors as with phased array ultrasonic testing. It is non-contact (except for portable vacuum shearography), non-contaminating, and near real time. While laser light is a non-penetrating radiation, shearography systems are capable of inspecting composite structures for both surface and subsurface discontinuities such as impact damage, disbonds, delaminations, near surface porosity, wrinkled fibers, fiber bridging, foreign objects (FOs), heat damage, and cracks.

Shearography is an evolutionary development from holographic interferometry first developed by Karl Stetson in 1964 [1]. Initial shearography techniques invented by Y. Y. Hung in 1982 [2] were further enhanced with the development of electronic phase detection by S. Nakadate [3] in 1988. Commercialization of electronic shearography NDT began in 1986 with the first integration of the technology into a major aircraft manufacturing program by J. Newman [4].

Shearography NDT methods are mature and can be highly effective solutions for a wide range of composite NDT applications. Common successful applications include metal and composite honeycomb, foam-core panels, bonded elastomers or cork, solid composite laminates, fiber-wound structures such as composite over-wrap pressure vessels (COPVs), solid-propellant rocket motors, and liquid-propellant rocket engines. Test objects ranging from less than 1 in. in width to very large composite structures have been successfully inspected. Figure 14-1 shows the composite (fiberglass/foam-core) wind fairings added to the entire span of the Bronx Whitestone Bridge in 2003 [5]. These foam-core panels were inspected during manufacturing and in situ using portable vacuum shearography.

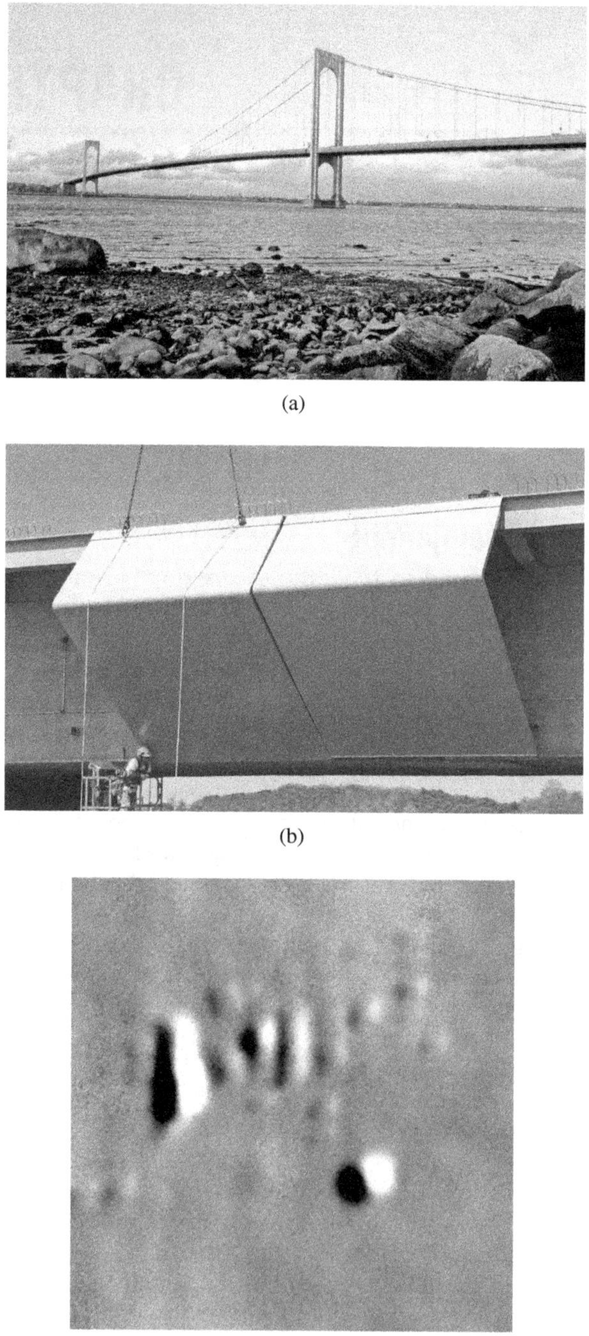

(a)

(b)

(c)

**Figure 14-1** In 2003, triangular-shaped fiberglass and foam-core wind fairings, similar to those shown in (*b*), were added to stiffen and improve the wind aerodynamics of the Bronx Whitestone Bridge. Weighting a total of 890,000 lb, these 390 FRP structures were the largest composite project in the world at that time. All panels were inspected for disbonds and delamination using portable vacuum shearography equipment. In (*a*), is a shearography result with multiple disbonds in the $12 \times 12$-in. FOV of the instrument.

Benefits of shearography include very high throughput. Large production systems easily achieve throughput rates of 300 to 700 $ft^2/h$ (27 to 65 $m^2/h$) compared to typically only 10 $ft^2/h$ (0.93 $m^2/h$) with ultrasonic C-scan. In many cases shearography NDT methods are noncontact, ideal for space applications. Production shearography cameras do not require precision contour following and can inspect a structure at an offset angle, eliminating the costs associated with precision gantries such as required for UT C-scan systems. In addition, because many shearography NDT methods are noncontact composite structures manufactured with multiple cure cycles can be inspected in-process allowing repairs or scrapping of the part with the lowest possible cost implications. The technology lends itself to a wide range of configurations to meet the discontinuity detection and logistical requirements.

As with all NDT methods in use, a critical element is operator training and certification to recognized standards. Shearography NDT operators are trained and tested to the requirements of the American Society of Nondestructive Testing (ASNT) SNT-TC1A and CP105, National Aerospace Standard NAS410, or European Standard EN 4179. Shearography operator training courses meeting these requirements are commercially available.

Shearography NDT is based on the phenomenon that when an object is subjected to a change in load, subsurface anomalies can produce slight local deformations on the surface (Figs. 14.2 and 14.3). These out-of-plane deformations may be as small as several nanometers (nm) but are easily imaged by a shearography camera. For example, when a carbon-fiber laminate panel is exposed to infrared radiation the change in temperature causes primarily in-plane thermal expansion. In a panel with a delamination, the laminate on the heated side of the panel expands more and sooner than the laminate on the cooler side of the panel away from the heat. This differential expansion coupled with the discontinuity boundary causes the material over the discontinuity to deform out of plane. A shearography camera is used to capture a reference image before the application of the thermal stress to the panel and then a second image as the heat is flowing through the laminate plies. These images are processed to produce an image of the delamination that can be measured and located on the part. In addition to thermal stress, the range of shearography techniques include pressure changes, vacuum stress, vibration stress using both sonic and ultrasonic signals, mechanical load changes, varying magnetic field loads, and microwave stress changes. Figure 14-2 shows the setup for shearography inspection of a COPV and in Fig. 14-3 an aircraft radome.

## 14.2 Theory and Principles of Operation

A shearography NDT system consists of a laser light source, a shearing image interferometer, an image processing computer, display monitor, and a means to provide a controlled and repeatable stress to the test object [2, 4, 6–13]. Shearography cameras are relatively resistant to environmental vibration and motion. Large production systems with 30-ft (9.14-m) gantry scanners have been built as well as portable systems for on-vehicle inspection. Shearography is relatively insensitive to test part bending or deformation due to the applied stress but still highly sensitive to local deformation caused by a discontinuity.

Shearography cameras are sensitive to changes in the distance from the object surface to the camera and in practice these $z$-axis surface deformations may be as small as 2 nm, depending on environmental noise. Test parts can be inspected with a few images

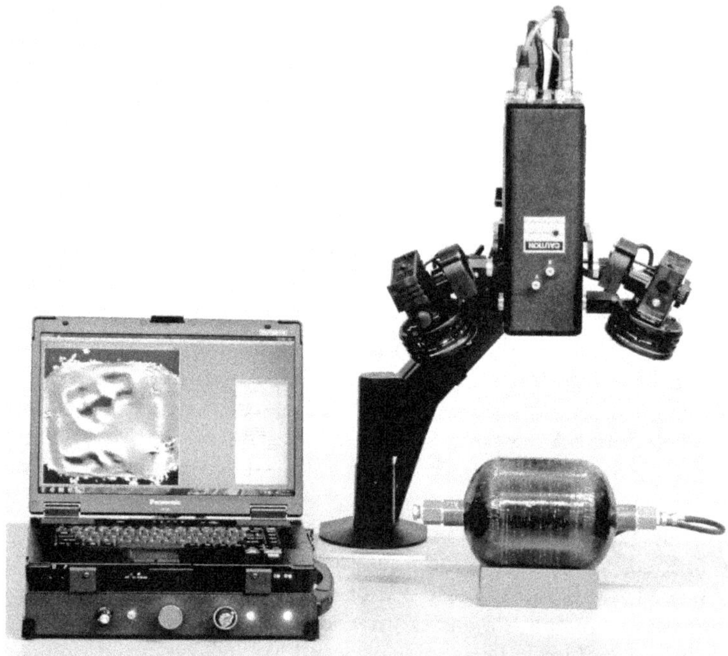

(a)

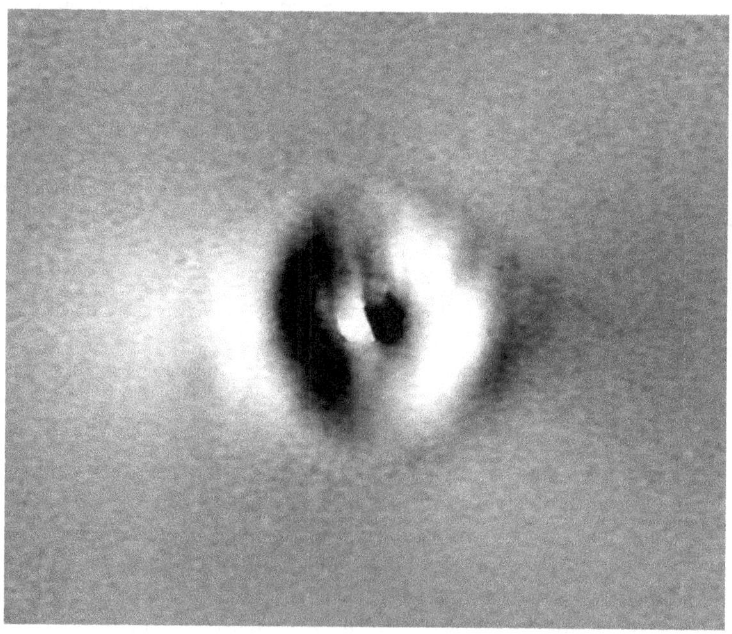

(b)

**Figure 14-2** A shearography test setup (*a*) for a small carbon-fiber COPV is shown. The tank is rated for 7250 psi (50 MPa) using small pressure changes of only 5 psi (0.03 MPa) to image non-visible impact damage as shown in (*b*). The small impact strike is in the center of the circular delamination which measures 3.2 in. (81.3 cm) in diameter. No visual evidence of damage was seen on this COPV.

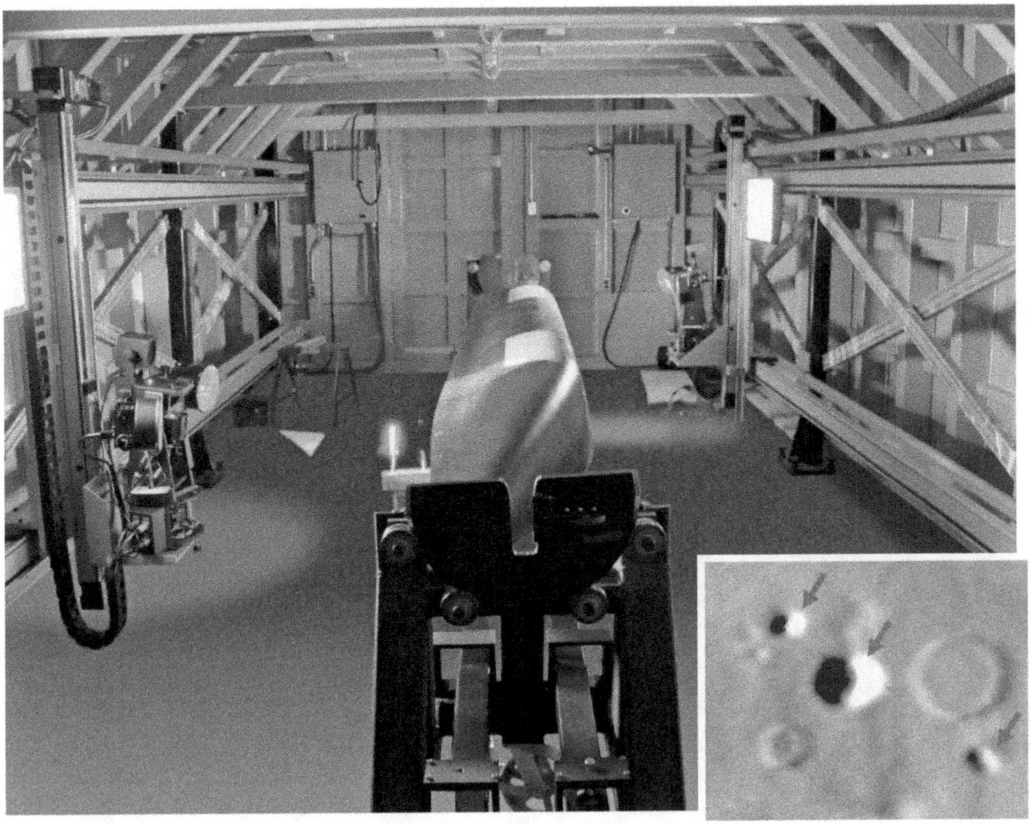

FIGURE 14-3 At left, an aircraft radome is shown undergoing shearography testing in a test chamber using a partial vacuum change from ambient to –2 psi (–0.014 MPa). Unlike older impedance bond testing methods routinely used for radomes, shearography can distinguish defects like impact damage and disbonds from repairs thus eliminating needless rework. The result at bottom right shows three indications of impact damage which appear as crater-like circles (arrows) with darker gray levels on the left and white on the right. Two repairs with the reverse-edge gray levels (referred to as phase) are also seen. The flat appearance of the repairs is due to the added stiffness of the epoxy repair reducing the deformation of the radome face sheet when the test was performed.

using a large field of view (FOV) camera lens setting, or for greater resolution many images with a smaller FOV can be recorded and automatically stitched together. The optimal FOV for a shearography test depends on the maximum allowable discontinuity size, camera resolution, laser illumination power, the ability to uniformly apply a stress change to the test part over the FOV, and background noise. Figure 14-4 shows a schematic diagram of a thermal shearography system that includes the laser and optical elements for test part illumination, an imaging shearing optical system consisting of a beam splitter with a two-axis tilting mirror, a second mirror with a piezoelectric phase stepper to translate the mirror in half wavelength steps at video frame rates and finally the CCD camera. The laser light is expanded through lenses to illuminate the test area on the panel. Light from point P1 is reflected from the panel surface, where it is well bonded to the core. Light from point P2 is reflected from the surface above a

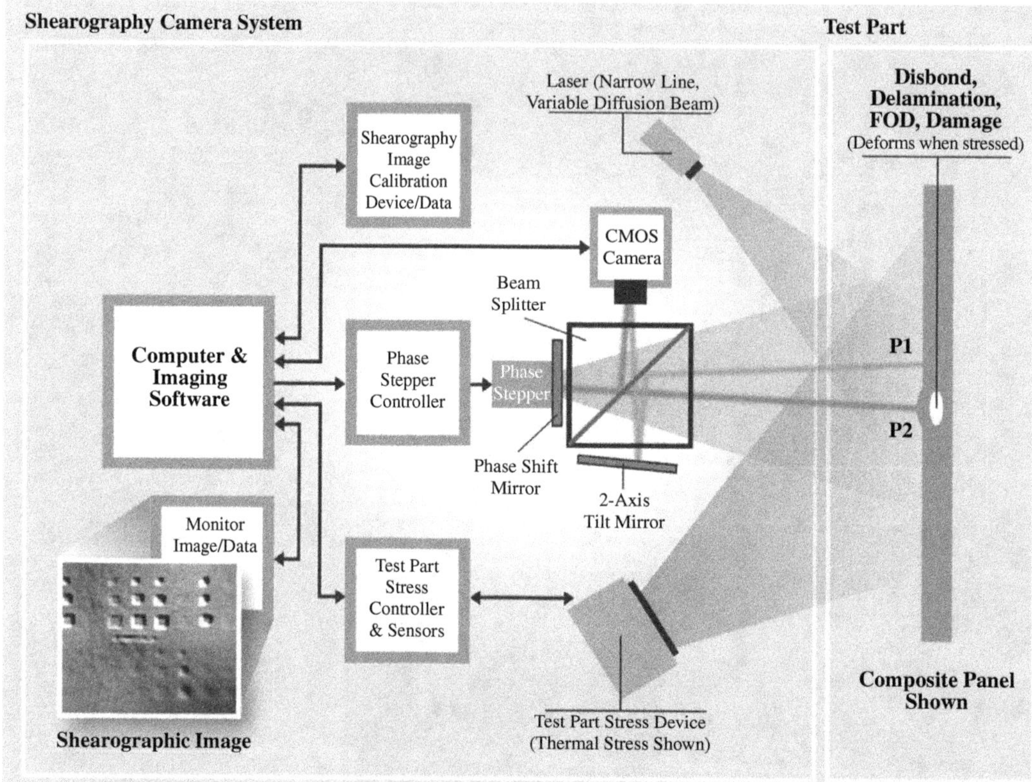


(a)

**Figure 14-4** (*a*) Schematic diagram of a digital shearography camera and thermal stress system. The separation distance and direction, or shear vector, between the paired points P1 and P2 on the test part surface determine the camera sensitivity to surface deformation, and the light intensity detected by each pixel in the CMOS camera is determined by the complex addition of the light from these two points on the target. (*b*) An example of a shearography camera with all of the elements in (*a*). Instruments with this configuration are used on benchtop mounts, tripods, gantrys, or robots, depending on the application. All the shearography electro-optical components and the single-frequency laser are built into the sealed housing. Thermal stress is achieved with the 2 to 1-kW quartz thermal lamps.

skin-to-core disbond. If the panel is stressed with a small temperature change or a partial vacuum, the panel face sheet above the disbond will deform out-of-plane toward the shearography camera. This shorter distance traveled by light from point P2 causes a phase shift with respect to light from point P1. Light from both P1 and P2 are combined by the shearing interferometer at a single pixel in the CMOS array as depicted in Fig. 14-5.

The phase stepper applies a $\pi/2$ phase step to a mirror in the interferometer at video frame rates (typically 30 to 120 frames per second) to generate a phase map and subsequent quantitative determination of the deformation derivatives between two strain states [3, 4, 11, 14]. As the applied load on the test object is changed, two sets of phase-stepped images, (I1-I4) and (I5-I6), are captured and the phase calculation is

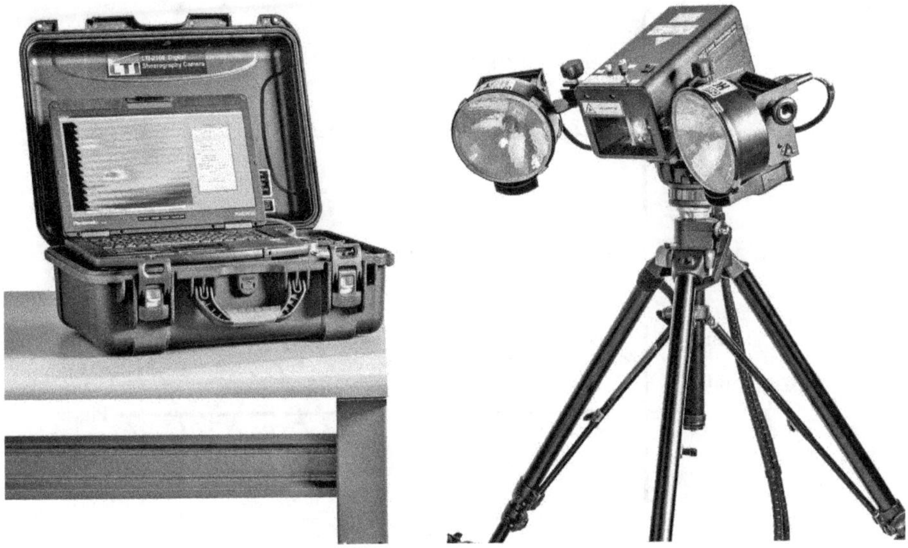

(b)

**Figure 14-4** *(Continued)*

performed for each pixel in the image, using the following equation for the four-phase-step method:

$$\Delta(x, y) = \tan^{-1}[I_8(x, y) - I_6(x, y)/I_5(x, y) - I_7(x, y)] - \tan^{-1}[I_4(x, y) - I_2(x, y)/[I_1(x, y) - I_3(x, y)]$$

where $I_1$ through $I_8$ are eight sequentially phase-stepped captured images, described by:

$$I_1(x, y) = I'(x, y) + I''(x, y) \cos[\Phi(x, y)]$$
$$I_2(x, y) = I'(x, y) + I''(x, y) \cos[\Phi(x, y) + \pi/2]$$
$$I_3(x, y) = I'(x, y) + I''(x, y) \cos[\Phi(x, y) + \pi]$$
$$I_4(x, y) = I'(x, y) + I''(x, y) \cos[\Phi(x, y) + 3\pi/2]$$
$$I_5(x, y) = I'(x, y) + I''(x, y) \cos[\Phi(x, y) + \Delta(x, y)]$$
$$I_6(x, y) = I'(x, y) + I''(x, y) \cos[\Phi(x, y) + \Delta(x, y) + \pi/2]$$
$$I_7(x, y) = I'(x, y) + I''(x, y) \cos[\Phi(x, y) + \Delta(x, y) + \pi]$$
$$I_8(x, y) = I'(x, y) + I''(x, y) \cos[\Phi(x, y) + \Delta(x, y) + 3\pi/2]$$

where:

$I'$ = bias intensity

$I''$ = modulation intensity

$\Phi$ = random phase variable due to reflection of the laser light from test object

$\Delta$ = quantity directly proportional to the differential displacement due to the test part deformation from the applied load change

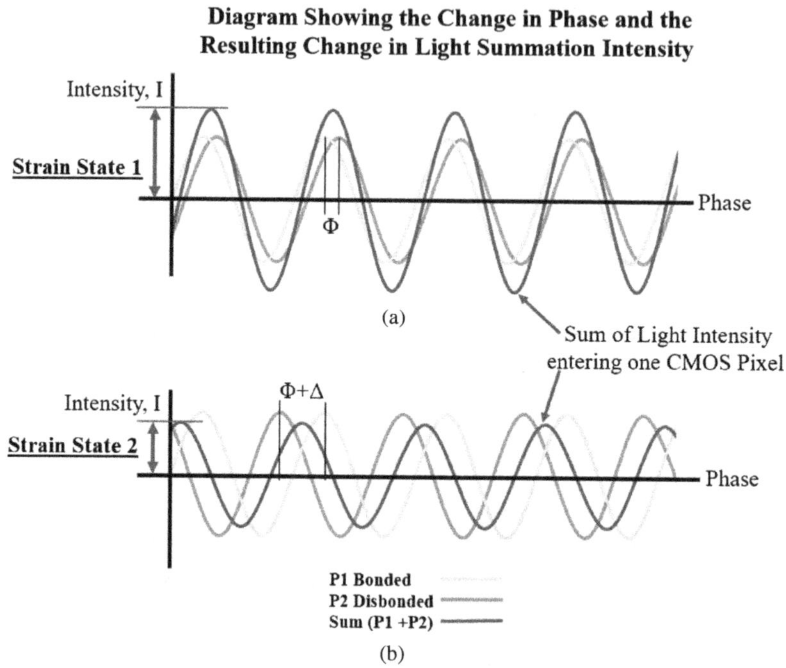


**FIGURE 14-5** (*a*) shows how the coherent, single-frequency light, from adjacent points on the part, is combined in each pixel in the shearography camera. The random phase difference Φ results from the random surface roughness on a diffusely reflecting test part surface. Stressing the part causes a relative phase shift Δ between light from well-bonded homogeneous material and light from the surface above discontinuity or non-homogeneous material. In two dimensions, this is expressed as $\Delta(x, y)$. (*b*) shows a plot of the light intensity (I) at each pixel in the camera detector and how I changes due to a phase shift when the strain state changes and the light paths are reflected from areas on the test part with different compliance or stiffness caused by the presence of an anomaly.

The result of the phase calculation yields a wrapped phase map image that shows the surface deformation derivative resulting from the load change. A flat metal plate deformed with a point load and imaged with a shearography camera reveals the surface deformation, as shown in Fig. 14-6.

Shearography images also include quantitative calibration data for the image scale and the image offset or shear vector. These data are used to obtain accurate discontinuity dimensions $D$ by subtracting the shear vector component from the measured dimension $D_m$. For example, if the shear vector, $S_v$, is in the horizontal (or 0°) direction with a shear of 0.5 in. (5 mm), the total horizontal uncorrected measured dimension $D_m$ of any indication will be greater than $S_v$. The measured dimension $D$ in the vertical direction, perpendicular to the direction of $S_v$, will equal the measured dimension $D_m$. For any measured dimension at any angle α to the shear vector $D$ and be calculated a $D = (D_m - \cos \alpha)$. The calibration data is also used during calculation of the integrated shearogram to extract the $z$-axis deformation of the part surface anywhere in the field of view.

Finally, the calibration data can be combined with the stress data to measure discontinuity deformation rates as a function of the applied stress change. This technique

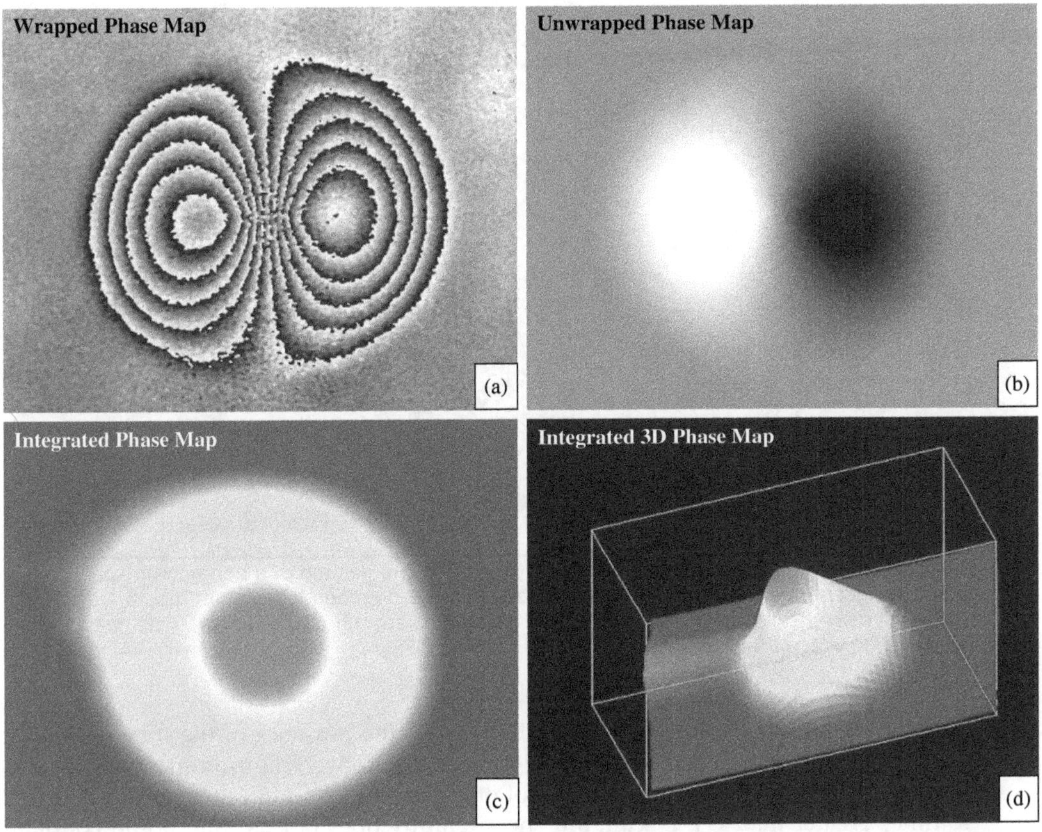


**FIGURE 14-6** Shearography images. (*a*) Phase map of the deformation derivative. (*b*) Unwrapped phase map. (*c*) Integrated unwrapped phase map. (*d*) 3D representation of the plate deformation. Numerical values of deformation over the entire field of view can be exported for analysis and comparison to finite element models.

has been used for example to establish a model for grading impact damage to honeycomb distinguishing between crushed core and disbonds. Applications using thermal stress can us a noncontact remote temperature sensor to measure the test part temperature change and the resulting discontinuity thermal deformation.

Real-time phase maps and unwrapping algorithms can be performed at video frame rates and quantitative measurement of the $z$-axis deformation can be computed. For NDT, shearography images can provide details of surface damage and subsurface discontinuities and structural features. Figure 14-7 shows a vacuum stress shearogram of a $14 \times 24$-in. ($36 \times 61$ cm) section of a metal spacecraft honeycomb radiator panel with face sheets 0.025 in. (0.63 mm) thick. The cells, disbonds, and voids are detected. Two shearograms were made in 14 s in a shearography test chamber with a pressure reduction of only 0.5 psi (3.4 kPa) below ambient air pressure. The metal face sheet over the honeycomb cells and discontinuities develop submicroscopic dimples as the air pressure in the test chamber is reduced. One indication in particular (large white arrow, left side) is a disbond on top of a coolant tube. This discontinuity was not detectable

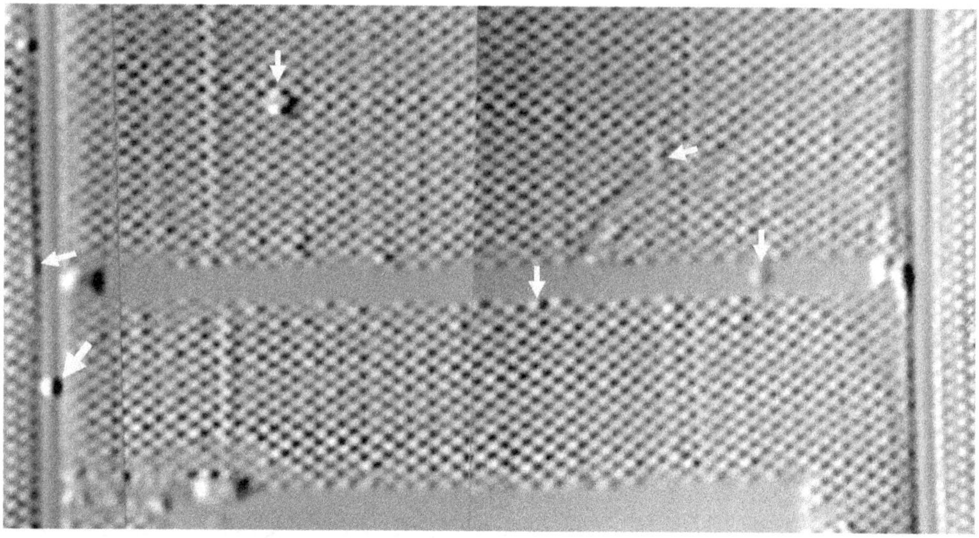

**Figure 14-7** A shearogram of a 24 × 14 in. (61 × 35.6 cm) section of a honeycomb spacecraft radiator panel showing the honeycomb cells, disbonds, and embedded coolant tubes. A disbond is seen at near the left edge on top of an embedded coolant tube (large arrow). This disbond was not detected with ultrasonic C-scan.

with through-transmission ultrasonic NDT, due to the presence of the tube beneath it blocking the ultrasonic signal. In addition, shearography NDT was specified for this spacecraft program, and many like it, not only because of a high throughput that was 10 times greater than UT C-scan, but shearography does not use any liquid couplant required to induce uniform ultrasonic signals into the panel. With shearography so the test parts remain dry and uncontaminated. In addition, shearography NDT in many cases, is able to detect anomalies in areas with complex geometry or conditions that prevent the use of traditional NDT techniques.

## 14.3 Shearography Equipment and Accessories

Shearography cameras usually are built with integrated stressing equipment such as thermal or vacuum stress. The probability of detection (PoD) for each type of camera and stressing mechanism must be determined for the structures and discontinuity types expected for a given application. The development and production of accurate representative NDT standards is a part of every shearography NDT process. Shearography systems are divided into two main types: First, shearography cameras and accessory subsystems for scanning the test part and equipment for stressing the test part. The scanning equipment ranges from manually moved tripod, integrated pan/tilt mechanisms, robots to robotic $x/y$ gantries. Stressing equipment can include thermal lamps, sound compression drivers, vacuum chambers as seen in Fig. 14-3, and ultrasonic exciters. Second, shearography cameras are available with integrated stressing equipment, such as the portable vacuum/thermal stress system shown in Fig. 14-8. Systems such as these are used for composite sandwich structures on composite launch vehicle fuel tanks, aircraft, naval ships, yachts, and even rail cars (Fig. 14-9).

(a)

(b)

**Figure 14-8** (*a*) Several classes of composite naval vessels are built using a portable vacuum/thermal stress shearography NDT system throughout for the inspection of the FRP face sheet to foam-core bonds, as seen in (*b*). The disbond (*c*) measured $9 \times 3.2$ in. in size. The typical throughput for a portable vacuum shearography system is about 150 $ft^2/h$.

(c)

**FIGURE 14-8** (*Continued*)

(a)

**FIGURE 14-9** (*a*) A small thermal shearography camera is shown being used to inspect a mass transit rail car interior for disbonds. The fiberglass/Nomex® cored sandwich structure is easily inspected for face sheet to core disbonds as shown in (*b*). The camera includes tools for measuring size and area of indications.

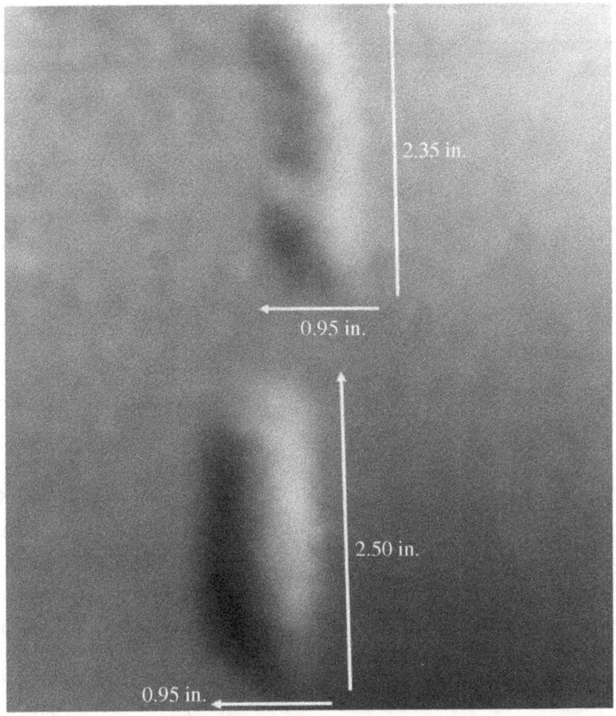


(b)

**FIGURE 14-9** *(Continued)*

## 14.4 Applications and Techniques

Shearography NDT uses a wide range of stressing and load timing techniques to detect discontinuities in materials and structures. These stress techniques include heating or cooling the test part, vacuum, pressure, mechanical bending, sound, ultrasonic signals, and microwaves [15]. The applied load changes are typically very low, compared to normal operating conditions of the test piece. For example, composite over-wrapped pressure vessels (COPVs) are typically pressurized to 1/1000 of the nominal working pressure or less to detect cracks, impact damage, and fiber bridging [16]. Refer to Table 14-1 for stressing methods and applications.

### 14.4.1 Pressure Shearography

Pressure shearography has provided good results on a wide variety of pressure-containing structures, including COPVs (see Fig. 14-10), composite drive shafts, tubular struts, and solid-propellant rocket motors (see Fig. 14-11). Pressure shearography uses a gas to pressurize the test object to between 0.01 and 5 percent working pressure [16]. Anomalies that affect the hoop strain or create strain concentrations generate indications of the surface. Test articles are usually placed on a support frame, allowing part rotation. Pressure shearography proceeds by capturing reference images and a bias pressure, increasing the pressure to a test level and capturing

| Method | Units | Range | Applications |
|---|---|---|---|
| Thermal | Degrees °F, °C<br>Surface temperature change | 1–30°F<br>(–17 to –1°C) | *Carbon-fiber laminate panels:* Impact damage, delamination, wrinkled fibers, porosity, embedded foreign material, repairs, foreign objects (FO)<br>*Sandwich panels—honeycomb, foam core:* Impact damage, skin-to-core disbond, damaged core, foam-foam disbonds, metal core-skin disbonds, repairs, kissing disbonds, skin delamination, water<br>*Resin transfer molded composites:* Resin lean areas, porosity, damage<br>*Engine stators, vanes, composite fan blades:* Erosion strip bond, voids, resin lean areas, damage, FO<br>*Steel, aluminum, ceramics, composites:* Surface breaking or near surface breaking cracks<br>*COPV with metal liners:* Disbonds at the liner/composite bond, fiber bridging |
| Partial vacuum | psi, $kg/cm^2$ pressure reduction change | –0.02 to –7.0 psid (–0.00141 to –0.49215 $kg/cm^2$) | *Elastomers:* Coatings, rubber and plastic components and structures, tires, solid rocket motor liners, rubber to substrate bond, cork to substrate bond: voids, disbonds<br>*Sandwich panels—honeycomb, foam cores:* Such as radomes, aircraft control surfaces, flaps, air brakes, helicopter blades, turbine engine ducts, laminated wood structures: impact damage, voids, disbonds<br>*COPV:* Fiber bridging, liner to composite wrap disbonds |
| Pressure | psi, $kg/cm^2$ pressure change | 0.01 to 5,000+ psid (0.0007 to –351.5348 $kg/cm^2$) | *COPV and composite rocket motors:* Impact damage, composite cracks, broken fibers, fiber bridging, liner wrinkles, porosity<br>*Pressure vessels and heat exchangers/radiators:* Metal pressure tanks, actively cooled liquid-propellant rocket engine thrust chambers, nozzles, exit cones, thrust ramps, piping, vented core honeycomb |
| Vibration | Air coupled 0.5–20 kHz; mechanically coupled 4.0–500 kHz | 90–125 dB | *Foam rocket thermal protection systems:* Damage, disbonds, delamination, cracks<br>*Light-weight honeycomb:* Spacecraft solar panels and structures<br>*Metal honeycomb:* Face sheet to core disbonds<br>*Brazed metallic bonded and plasma sprayed engine compressor seals:* Disbonds |

**Table 14-1** Shearography NDT Stressing Methods and Applications [17]

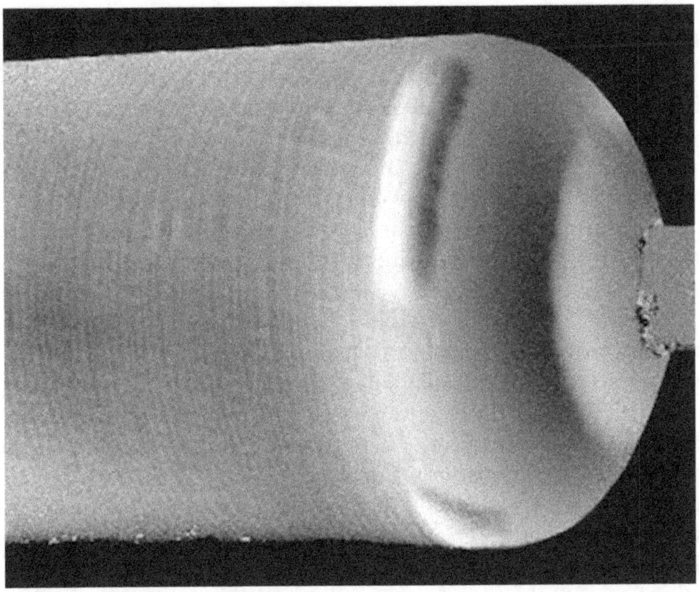

**FIGURE 14-10** Shearography image of the end of a 6.3-in.-diameter (160-mm) COPV showing fiber bridging areas on the barrel transition area. With a maximum allowable working pressure of 8,560 psi (59 MPa), the shearography test pressure with the result shown above was made with only a 5.0-psi (0.034-MPa) pressure change.

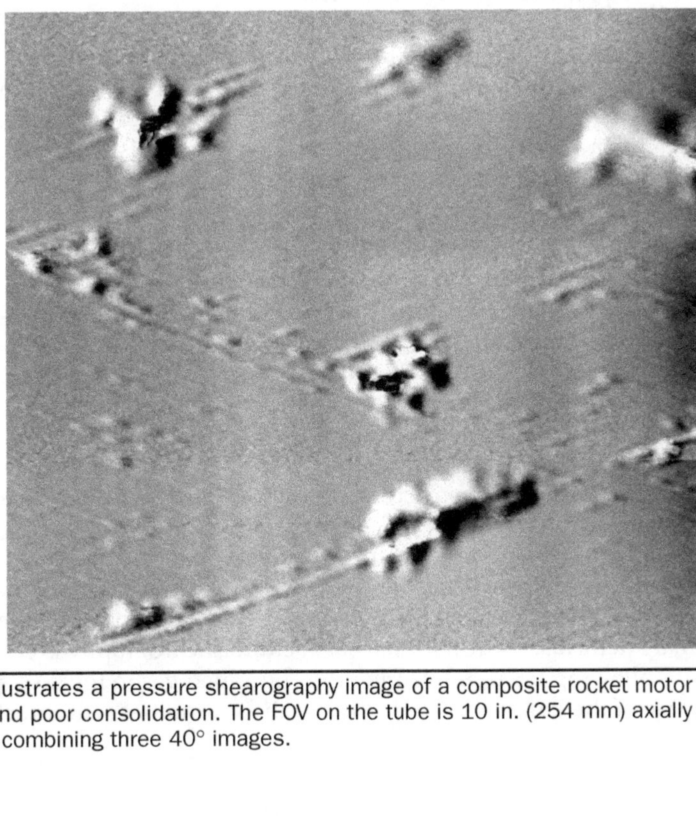

**FIGURE 14-11** Illustrates a pressure shearography image of a composite rocket motor tube showing porosity, voids, and poor consolidation. The FOV on the tube is 10 in. (254 mm) axially and covers 120° circumferentially combining three 40° images.

the stressed data. The vessel can be vented to prepare for the next test. Shearography data can be taken both during pressurization and venting, if the phase of the data taken during venting is inverted to match that taken during pressurization.

Flight hardware, depending on contamination requirements, may require specific cleanliness precautions such as using ultrapure GN2 with oil-free valves, fittings, and piping. Each area on the tank is tested sequentially with the COPV rotated 30° to 45° after each section is completed. The end domes are also inspected at a 45° angle. The FOV on the tank is only as large as can be fully illuminated with the shear camera laser. Controlling the test pressure requires valves to pressurize and vent the test object. Commercial pressure shearography systems are available that allow rapid, preset pressurization and venting of the COPV and can set a minimum bias pressure to increase testing throughput and preload of the COPV.

Pressure shearography has also been used to image porosity and poor consolidation of composite tubes and pressure vessels [15]. The tensile loads on the fibers in the hoop and longitudinal directions on tank create strain concentrations at the site of porosity, voids, and poor consolidation.

Pressurizing the COPV increases the hoop strain on the aluminum liner and increases the compressive force on the Teflon insert between the liner and the composite over-wrap material. Pressure changes produce only a small strain change at the site of the insert, but a large strain change at the fiber bridging discontinuity as seen in Figs. 14-10 and 14-12a. Due to the slight insulating properties of the insert and delamination and fiber bridging discontinuities, the insert affects the thermal strain response of the COPV in several ways. First, the Teflon has a different thermal diffusivity and thermal expansion coefficient from the virgin structure. The result is a local high-strain state and a highly detectable shearography response as seen in Fig. 14-12b.

Representative NDT standards are critical to selecting the best technique, which may also involve multiple tests with different methods to capture all possible discontinuity types. Again, in the example of the COPV in Fig. 14-12, thermal stress is good for detecting liner to composite disbonds and foreign material (FO) as well as fiber bridging. For a COPV, pressure stress shearography is best for impact damage, cracks, and wrinkled or buckled liners, as shown in Fig. 14-13.

### 14.4.2 Partial Vacuum Shearography

Partial vacuum shearography can be accomplished in several ways. For portable testing, a vacuum window may be used consisting of a Lexan plate with edge seals and a variable, controllable vacuum blower. Often a small shear camera is built into a portable inspection head as shown in Fig. 14-14. Shearography inspection rates of 150 $ft^2/h$ (13.9 $m^2/h$) are easily achieved. Honeycomb and foam-core aircraft control surfaces, rudder as shown in Fig. 14-14, stabilizers, flaps as shown in Figs. 14-15 and 14-16, and wing panels can be test in situ. Naval composites, yachts, railcars, and wind turbine blades are other composite structures tested extensively with portable vacuum shearography.

Partial vacuum stressing reduces the outside ambient air pressure on the test part. Internal disbonds or delaminations in the material contain small quantities of air and when the air pressure on the part surface is reduced, the test part surface at the location of the anomaly deforms. Critical for partial vacuum shearography is ensuring there are no leak paths allowing the rapid venting the entrapped gas in the discontinuity within the time frame of the data acquisition. Vented discontinuities will come to equilibrium

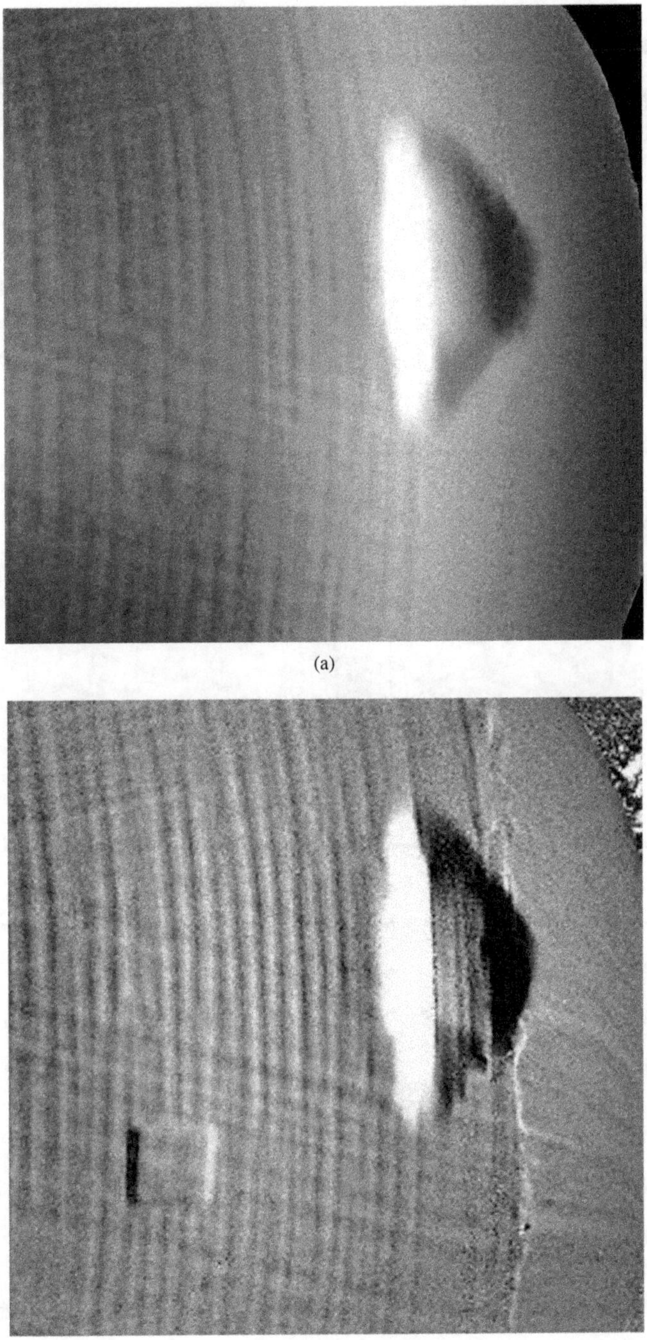

(a)

(b)

**FIGURE 14-12** A 6.3-in.-diameter (16-cm) COPV tested with shearography methods. (*a*) Pressure shearography shows the fiber bridging discontinuity only; (*b*) thermal shearography reveals both fiber bridging and a square Teflon® insert between the 0.35-in.-thick (8.9-mm) composite and the aluminum tank liner.

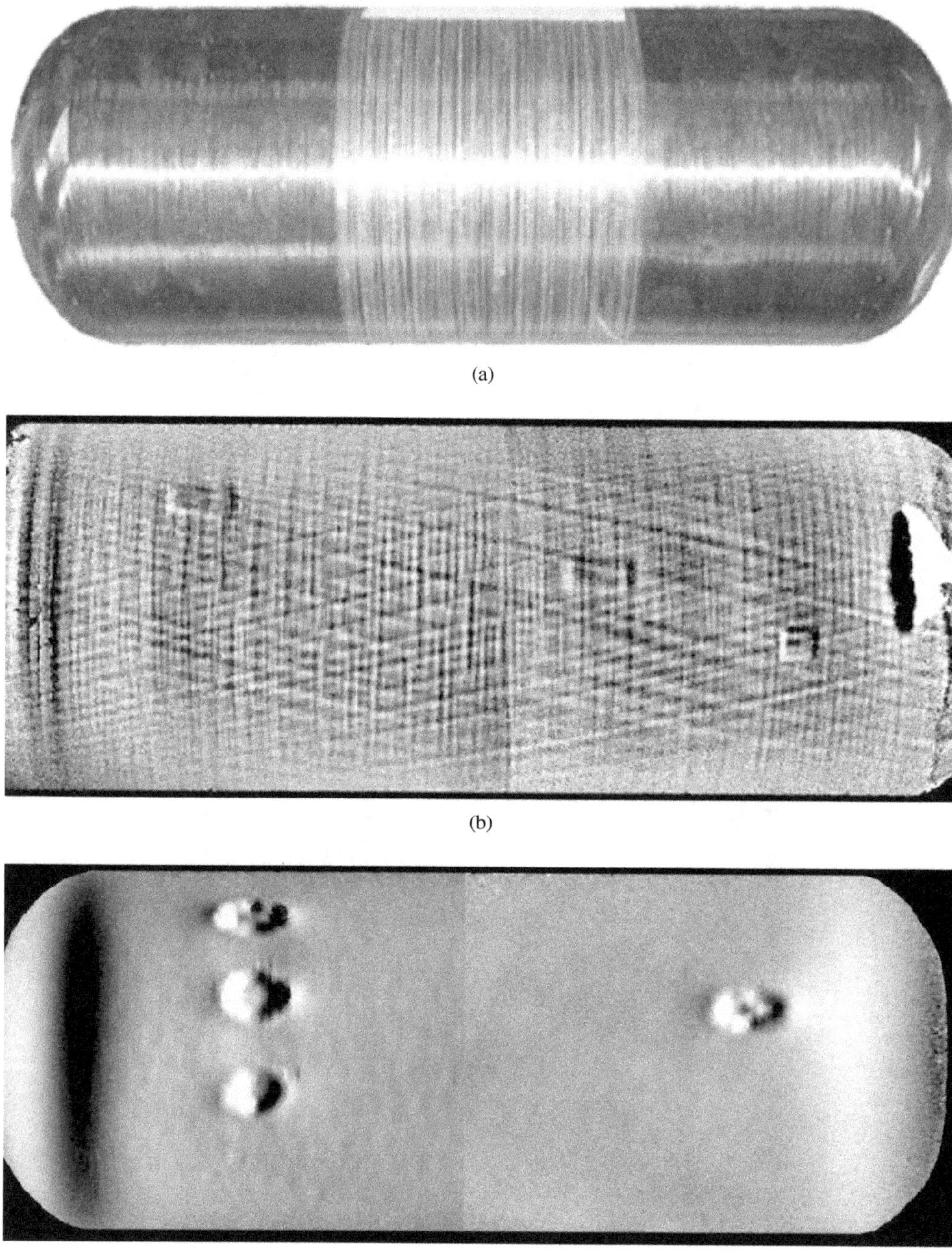

(a)

(b)

(c)

**FIGURE 14-13** (*a*) A 6.3-in.-diameter carbon-fiber COPV is shown. This tank has a maximum operating pressure of 8,560 psi (59 MPa). (*b*) Shearography tests with thermal stress show three Teflon inserts placed on the aluminum liner then over-wrapped with the carbon-fiber wrap. (*c*) The shearography result with a pressure change of 2.5 psi (0.017 MPa). Non-visible impact damage is easily detected, producing very strong crater-like indications. These indications are areas of delamination and fracturing of the composite matrix.

(a)

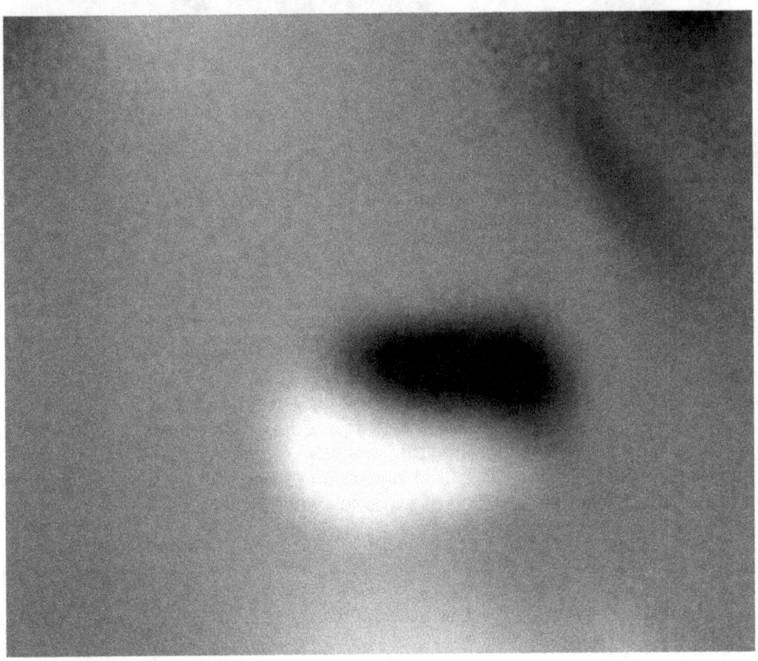

(b)

**Figure 14-14** (*a*) An aircraft rudder inspection with a portable vacuum shearography system. (*b*) A shearography image is seen of disbond measuring 3.5 × 2.0 in. (8.9 × 5.1 cm).

**Figure 14-15** A shearography camera is shown mounted on a large *x/y* scan gantry in a large vacuum test chamber. This system is used for shearography NDT of large honeycomb aircraft components. The air pressure inside this chamber can be reduced from ambient to –5 psi (–0.035 kg/cm$^2$) in less than 15 s, causing subsurface anomalies to produce surface deformations detected with a shearography camera. The shearography test results are automatically stitched, allowing both the indication measurement and location. Test part scan speeds of 400 ft$^2$/h are easily achieved by these systems. The author is standing at the back wall for scale.

with pressure changes allowing the surface deformation to relax and the shearography indications to disappear.

Several important characteristics of partial vacuum shearography NDT can be gleamed from the equation for calculation the displacement $W$ of a circular disbond with a diameter $D$ and depth $H$ in an isotropic material with an elastic modulus $E$. A uniform partial vacuum is applied yielding a pressure change of $P$.

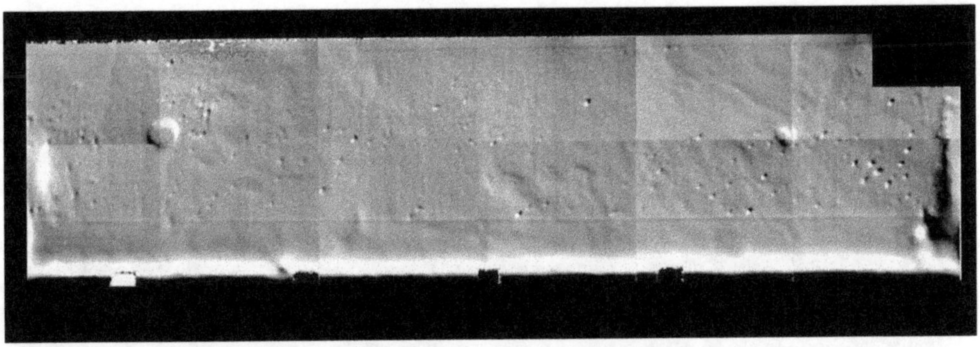

**Figure 14-16** A 14-ft-long (304-cm) aircraft flap tested in a partial vacuum shearography chamber with automatic image stitching. Over 90 disbonds are detected requiring repair. This flap was inspected in less than 15 min including placement in the test chamber. Shearography NDT technology is providing large productivity increases in aerospace manufacturing and overhaul.

The *z*-axis displacement for the partial vacuum stressed disbond may be calculated as [16–20]:

$$W = k\left(\frac{PD^4}{EH^3}\right)$$

Notice first that the *z*-axis displacement increases with the 4th power of the disbond diameter *D*. For discontinuities at the same depth, doubling the diameter increases the displacement 16 times. As a result, small changes in the discontinuity diameter can have a large effect on the resulting displacement, which roughly corresponds to the signal strength and the signal-to-noise ratio. To double the displacement for given size disbond $D_1$ at the same depth, a disbond $D_2$ would have to be only 18.9 percent larger:

$$D_2 = D_1\sqrt[4]{2} = 1.189\,D_1$$

The following examples of vacuum shearograms illustrate some of the remarkable effects of this method for both metal and composite honeycomb. In Fig. 14-17, a metal honeycomb leading-edge flap that had been removed from an operational jet hours earlier has a large face sheet to core disbond the full width of the 18-in. FOV. In addition, a cracked core is seen, which is a very dangerous condition and a sign of imminent failure. In Fig. 14-18, a honeycomb test panel with carbon laminate face sheets and Nomex core is shown to have numerous programmed disbonds even on the ramp areas of the panel. Figure 14-19 shows a panel with imbedded foreign object debris (FOD). In this panel, shearogram detected all the FOD detected with through transmission C-scan.

### 14.4.3 Thermal Shearography

Thermal shearography has been applied successfully to a wide range of structures and material types, including solid laminates, metal or composite sandwich panels (limited to near side inspections only), COPV, and even for surface breaking fine crack detection in metal aircraft and nuclear reactor components.

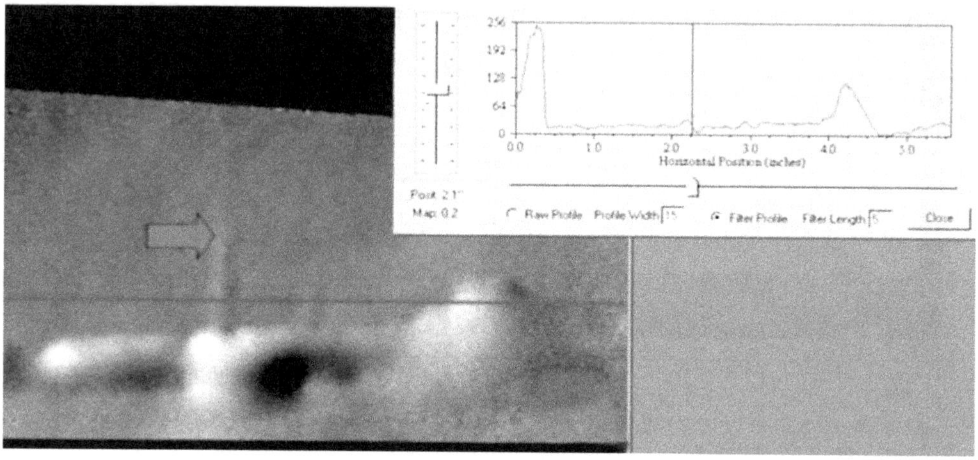


**FIGURE 14-17** This shearogram shows a vacuum inspection of an aircraft leading-edge flap revealing a large honeycomb disbond. Repeated stress loading in operation is causing failure of the core adhesive bond (red arrow), a situation that could have led to structural failure. The full chord width of the flap is shown. The insert shows a line scan with the waveform of the signal at the failing core splice joint (black line) and at the disbond.

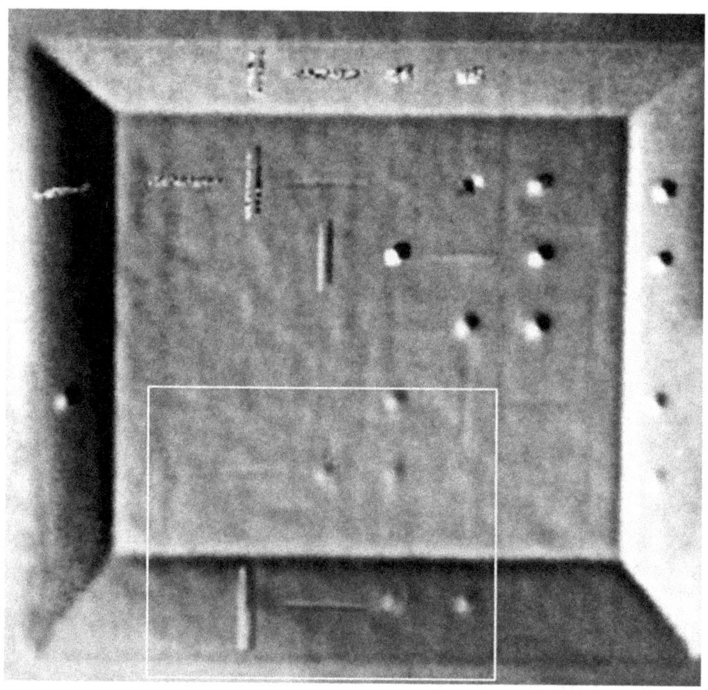

(a)

**FIGURE 14-18** (*a*) A vacuum shearography image of a carbon-fiber sandwich panel with Nomex core measuring 14 × 14 in. (36 × 36 cm) inspected using a −1.0-psi (13.7-kPa) pressure reduction change. Indications are Teflon inserts at plies 4 and 6 as well as between the eighth ply and the adhesive on top of the core. (*b*) A close-up of the bottom edge ramp area.

(b)

**Figure 14-18** *(Continued)*

Unlike thermography which images surface temperature changes (or derivatives) over time, thermal shearography uses a temperature change to reveal local changes in the coefficient of thermal expansion (CTE) due to the presence of an anomaly. Thermal shearography is not affected by the test part emissivity and is equally able to inspect polished aluminum or painted surfaces without any surface preparation except to remove any grease or moisture.

For materials with a positive CTE, which includes most materials such as carbon laminate, aluminum, steel, and fiberglass, the test part surface is heated. For materials with a negative CTE, such as Kevlar, the component part must be tested by cooling the surface. Thermal stress is typically accomplished using radiant light with a high IR component such as shown in Fig. 14-20. These 1-kW quartz lamps with adjustable reflectors are mounted to the shearography camera and heat at distances up to 10 ft any target in the FOV of the shear camera.

Thermal shearography performed using hot air guns to heat the part is discouraged but it can produce adequate results in materials with a low thermal emissivity, such as polished aluminum. The user must be careful to heat the inspection area evenly. Manual heating of the test part with a hot air gun is also highly subjective and not quantifiable. For materials with a negative CTE, cooling the test part can be accomplished using compressed $CO_2$ or a vortex cold air gun, again though with the same limitations.

The thermal lamps are computer controlled from the shear camera software to provide consistent repeatable heating for each test. The reference images may be captured

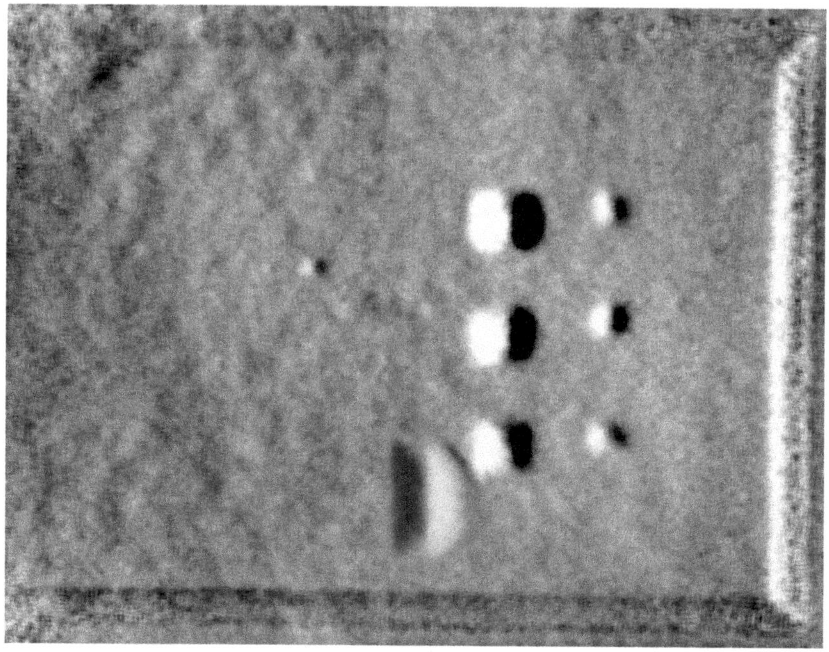

(a)

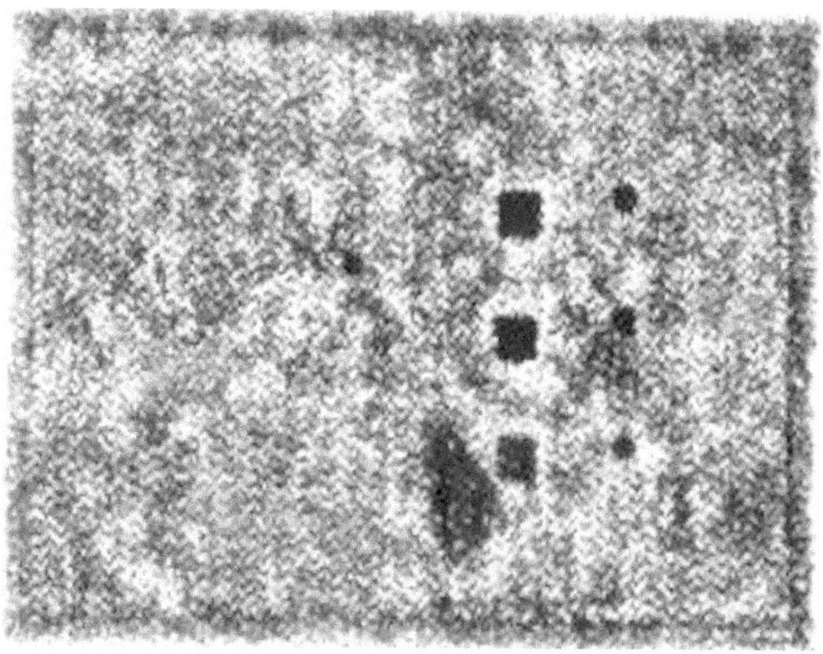

(b)

**Figure 14-19** (*a*) FOD panel shearography result. A partial vacuum shearography test reveals various types of FOD including a razor blade, backing material, Kapton®, paper, and polyethylene. This 20 × 16-in. (51 × 41-cm) panel was inspected in two tests lasting 15 s. (*b*) Through-transmission UT result. An ultrasonic C-scan of this panel shows substantially identical results, with a considerably slower inspection rate and a lower signal-to-noise ratio.

(a)

(b)

**Figure 14-20** (*a*) A remote-controlled shearography camera with two 1-kW thermal lamps mounted on the camera to direct the IR energy wherever the camera is pointed, here an aircraft radome. (*b*) A thermal shearography camera on an automated gantry for scanning a large aluminum sandwich panel for a commercial aircraft.

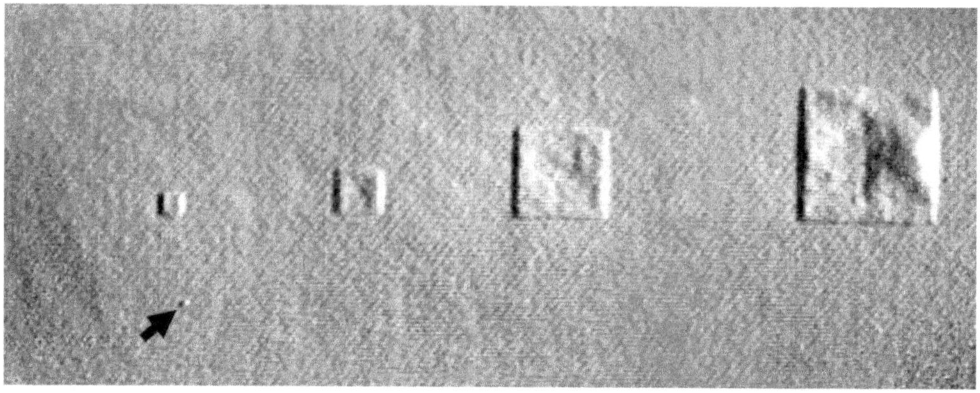

**Figure 14-21** NDT standard panel with five-ply carbon-fiber face sheets and Nomex honeycomb manufactured with machined core and Teflon inserts in sizes starting with 0.25 in. (6.35 mm). A small natural discontinuity indication is seen lower left (arrow). The small shear vector, $S_v = 0.1$ in. at 0°, provides excellent detail and skin-to-core disbond sensitivity.

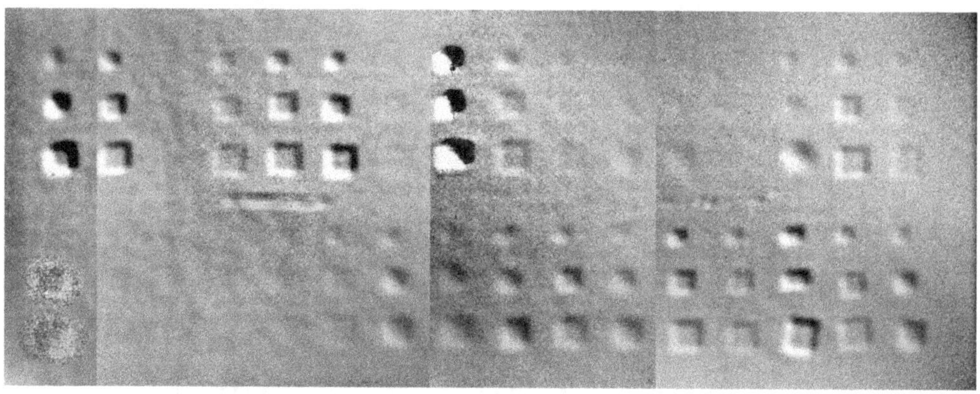

**Figure 14-22** Four thermal shearograms stitched to image a 30-in.-long, 20-ply carbon-fiber laminate. Pieces of foreign material, peel ply, and Teflon were placed in the ply lay-up throughout the full thickness of the panel. The difference in the coefficient of thermal expansion for the FOD as compared to the carbon-fiber laminate produces the slight out of plane surface changes when thermally stressed allowing detection with the shearography camera.

either before [referred to as refresh before heat (RBH)] or after [referred to as refresh after heat (RAH)] the thermal stress is applied. Thermal stress parameters may be time-based or temperature-change-based using a remote sensor to measure the real-time test part surface temperature change during heating.

Thermal shearography can provide excellent disbond detection in sandwich panels. (As illustrated in Figs. 14-21 and 14-22). Defects typically detectable in sandwich panels with thermal shearography include face sheet to core disbonds, damaged core, disbonded core splice joints, crushed core, and impact damage. Defect sensitivity varies with the sandwich construction. Materials with high coefficients of thermal expansion $C_t$ such as metallic and carbon-fiber laminate face sheets and metal cores yield the best shearography results, often detecting disbonds on both near and far sides from one side. Figure14-23 is one such example of a thermal shearography test of NDT STD for a satellite dish antenna made with carbon laminate face sheets and aluminum core. Programmed disbonds on both sides are detected. Nomex and foam-core sandwich panels can be effectively inspected but only on a single side due to low thermal conductivity; for these materials, vacuum shearography is highly effective.

The detection of fiber wrinkles in composite structures can be extremely important and challenging for traditional NDT methods. Figure 14-24 shows an example of a very serious tooling induced fiber wrinkle in a large aircraft fuselage, detected with thermal shearography. Fiber wrinkles can often also occur at corners and curves. Being a non-contact NDT method, where contour following and being normal to the part surface is not a requirement, shearography is excellent for detecting wrinkled fibers.

Thermal shearography has been used successfully to inspect composite repairs, as shown in Fig. 14-25. Engineered repairs can contain porosity, disbonds, and FO, all of which have been detected. Impacts can leave no visible trace on a composite surface.

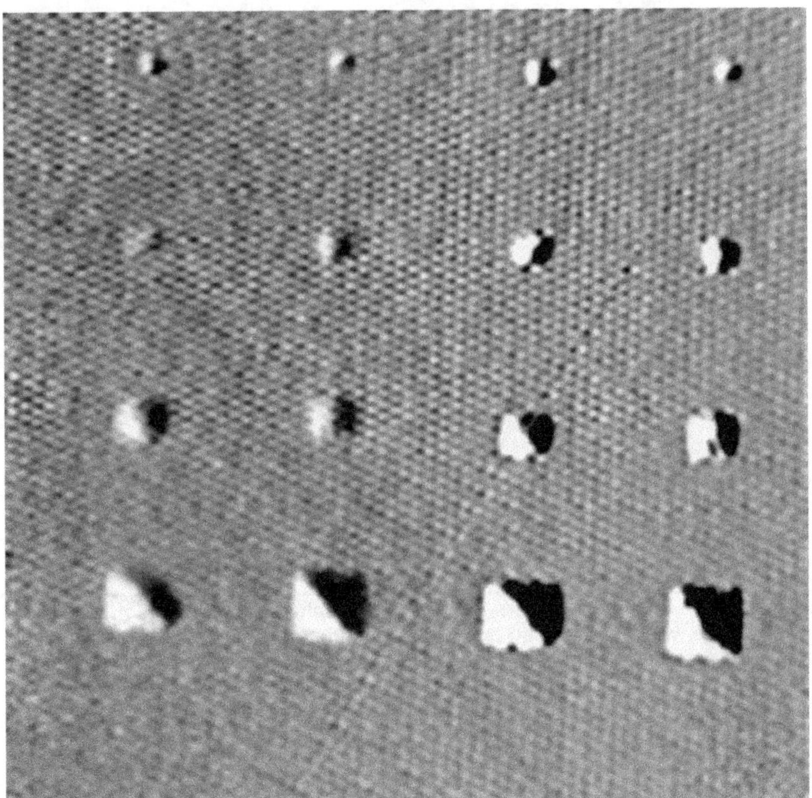

**FIGURE 14-23** A thermal shearography image of a panel with thin carbon-fiber face sheets and aluminum honeycomb core. This NDT STD, measuring 16 × 16 in. (40 × 40 cm), was manufactured with Teflon inserts between the core and the adhesive to simulate disbonds in a spacecraft antenna. Eight near-side skin-to-core inserts are detected in the right side of the panel and eight far-side inserts on the left side.

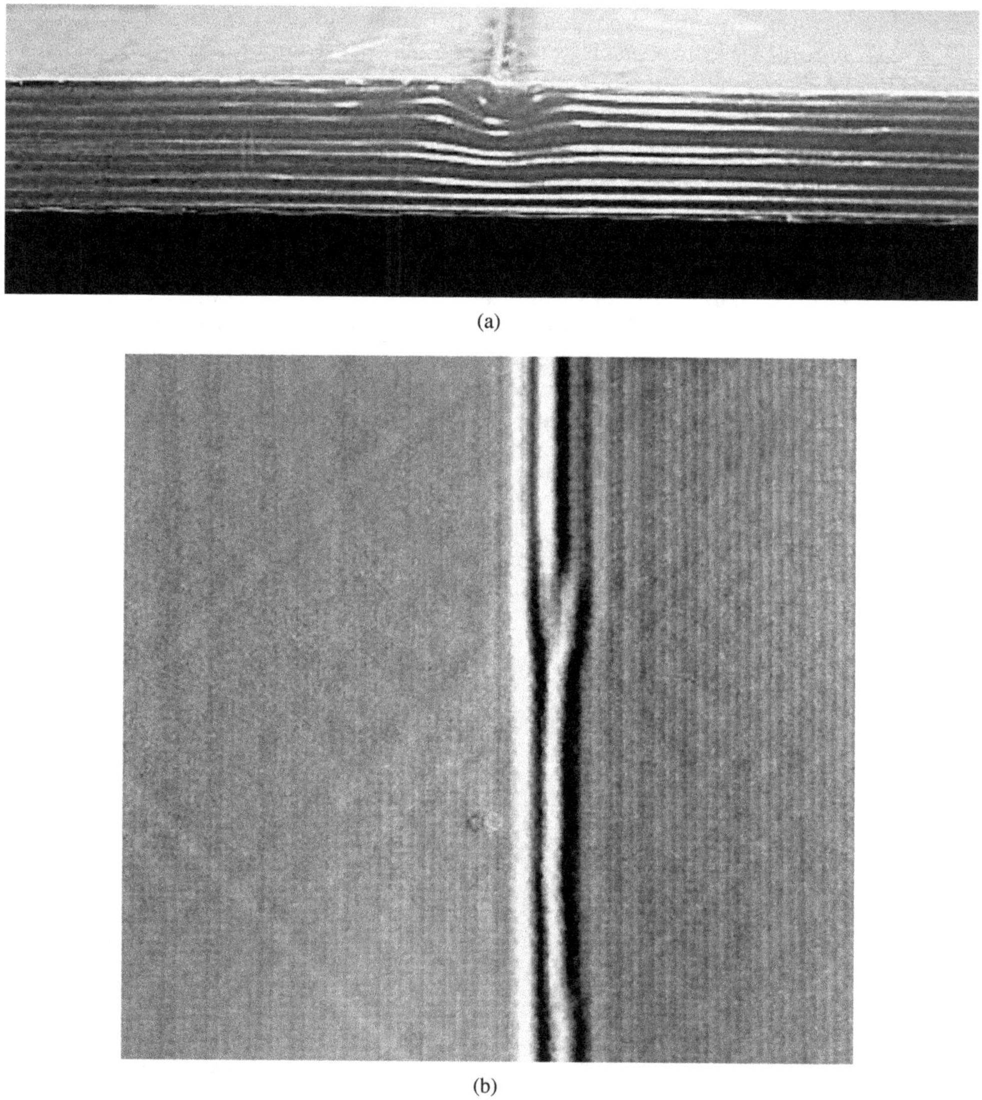

**Figure 14-24** (*a*) Wrinkled fibers are seen in this photo of a section of a composite aircraft fuselage. The wrinkles run in an axial direction substantially weakening the ability of the fuselage to support pressurization loads. (*b*) The thermal shearography result taken from what would be the outside surface of the fuselage showing the wrinkle in the fibers running from top to bottom. The FOV is 12 × 12 in. (30.5 × 30.5 cm).

The thermal shearography provides outstanding sensitivity for impact damage to carbon-fiber laminates.

### 14.4.4 Vibration Shearography

Vibration shearography uses either an air coupled or a mechanically coupled ultrasonic source to induced vibrations into the test part in the frequency ranges for the resonance

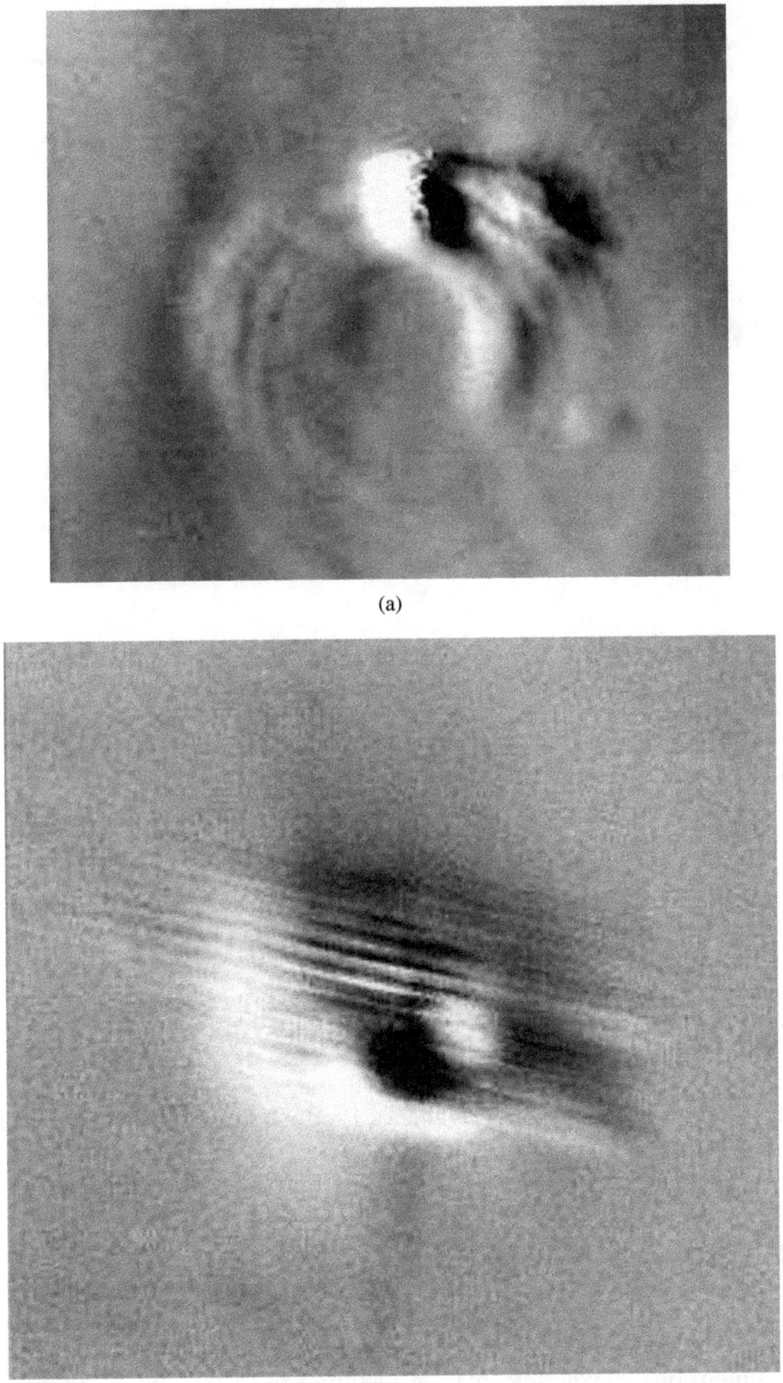

(a)

(b)

**Figure 14-25** (*a*) A thermal shearography test of an 8-in.-diameter (203-mm) composite repair on an aircraft vertical stabilizer shows a large 1.5-in. disbond and areas of porosity. (*b*) Non-visible impact damage to a solid carbon laminate wing panel. The actual impact site appears like a crater with broken fibers and a deeper under-laying delamination.

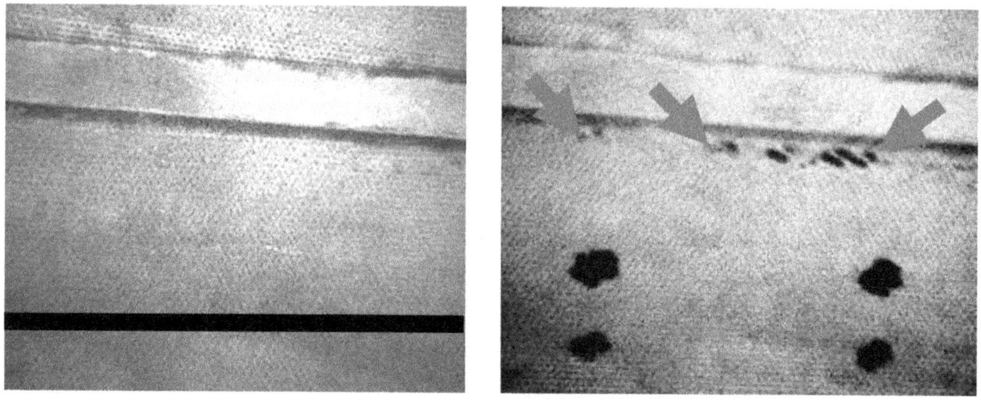

(a) (b)

**Figure 14-26** (*a*) A live image as seen through the shearography camera of a large aircraft engine inlet duct with perforated aluminum face sheets and metallic honeycomb. (*b*) The shearography results show natural disbonds (arrows) and four mill flat bottomed holes from the far side, measuring 0.5 and 0.8 in. in diameter. Testing is real time.

modes of anomalies. Air-coupled acoustic drive (ACAD) should be considered for light weight structures or insulation materials such as honeycomb with thin face sheets and foam insulation used for launch vehicle cryogenic fuel tanks. Mechanically coupled acoustic drive (MECAD) should be considered for metal to metal bonding, metal honeycomb and is especially good for structures with perforated face sheets such as aircraft engine inlet ducts such as shown in Fig. 14-26. Vibration shearography is fully real time and the system will operate continuously during data acquisition as the camera scans and briefly stops to acquire data.

Acoustic stress shearography (ACAD) testing is performed using a speaker with a directional horn. Production systems such as shown in Fig. 14-27 use an acoustic horn or waveguide to focus the acoustic energy over a 1-m$^2$ area for inspecting foam isolation of a launch vehicle cryogenic fuel tank. Power levels for ACAD typically produce acoustic pressure levels sometimes as high as 110 dB (decibels). Both ACAD and MECAD can produce sound levels well above safe exposure limits without hearing protection. Because of the very short inspection time required, acoustic shearography is ideal for large structures or launch vehicles. Production systems in current use are easily capable of an inspection rate of 1,500 ft$^2$/h.

## 14.5 Evaluation and Reporting of Test Results

The protocol for evaluating shearography images derives for the feasibility studies to determine the efficacy of the shearography stress method for a given application. Shearography instruments have a range of tools for measuring indication sizes, area, linear dimensions, *z*-axis deformation resulting from the stress change. These measurements compensate for the shear vector in every image (except image integration) and are used to accept or reject indications against the test part discontinuity accept/reject criteria.

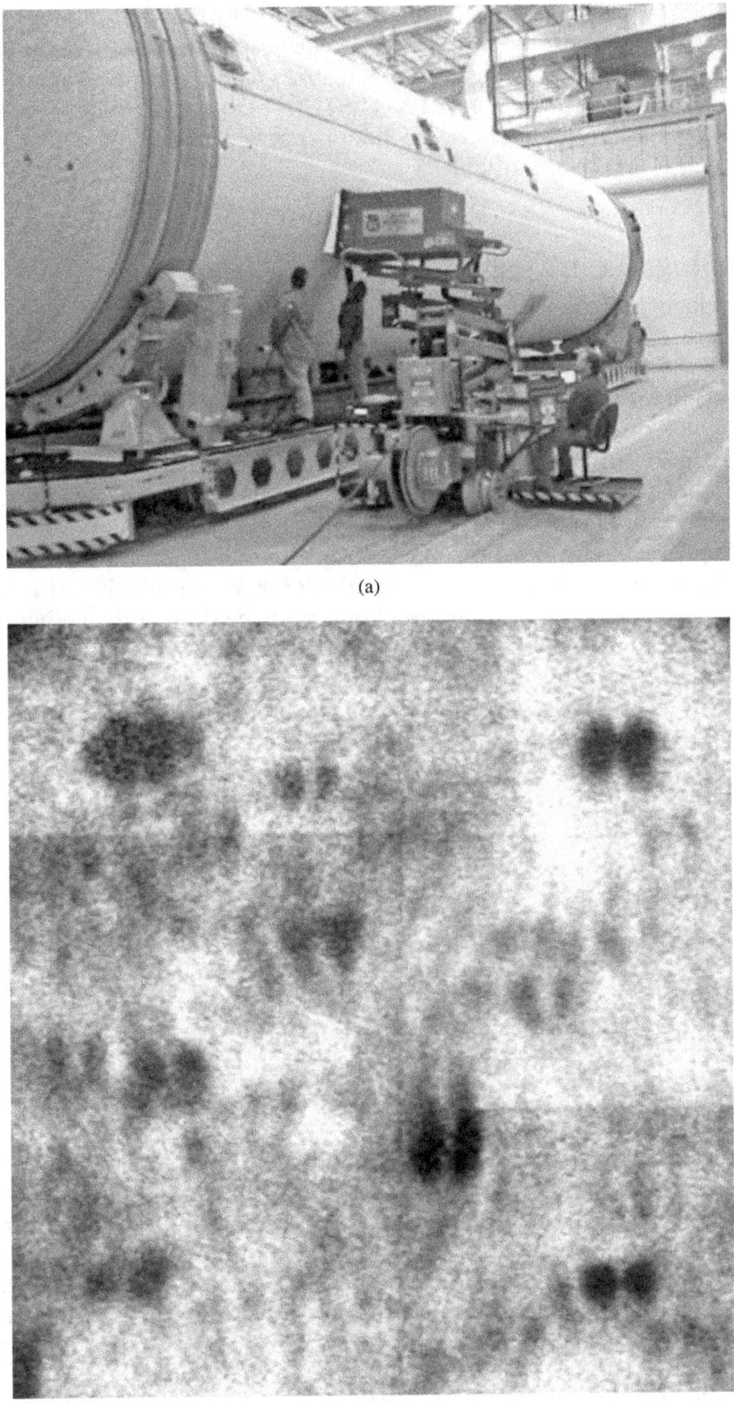

(a)

(b)

**Figure 14-27** (*a*) An acoustic shearography system testing foam insulation on a large cryogenic rocket fuel tank. (*b*) A typical shearography test result. The black areas are foam to substrate disbonds. The FOV is 3 × 3 ft, with a test time of only 60 ms.

Shearography test reports include detailed description and photographs of the test setup and the shearography procedure. Also included are a listing of the test parameters including the shear vector, range to target, stress levels and rates, and data acquisition timing. Shearography test images of NDT STD(s) before and after every test run, images of all indications with dimensions. Secondary supplemental testing with UT is sometimes used for depth, where applicable.

## 14.6 Advantages and Limitations

The advantages of shearography NDT of conventional methods include high through-put. The time required to gather the raw data is typically determined by the time required to induce the load change be it thermal flow in a carbon-fiber laminate or pressurization of a rocket engine thrust chamber. The data processing is generally accomplished in less than 60 to 200 ms. A second advantage is no requirement for part contour following. Tilted test part surfaces can be inspected, greatly reducing the cost of a scanning shearography system. Third, shearography methods can employ many different test part stress methods often several in the same test. This diversity of techniques often leads to a positive NDT solution. Shearography can test a number of materials such as bond cork and foam that are extremely difficult for conventional NDT methods. Finally, shearography techniques and interpretation tend to be simpler than other NDT methods. For example, in NAS 410 the recommended training hours for level 2 shearography is 60 h and for ultrasonic testing is 80 h.

Limitations of the shearography method include reduced discontinuity sensitivity at depth. The deeper the flaw is from the surface the larger it must be to be detected. Extensive testing of matrix NDT STDs for common solid laminates have proven detection of 0.25-in.-diameter void through 15 plies. Imaging black glossy curved surfaces can be challenging. The use of penetrant developer to increase the surface reflectivity can help, if allowed. Finally, shearography images can be degraded by environmental vibration and noise.

## 14.7 Laser Safety

Laser shearography NDT systems utilize laser light to illuminate the surface of a test article being inspected. The laser provides a convenient source of monochromatic-coherent light that makes the implementation of shearography NDT possible. With the exception of extremely low powered laser systems, virtually all laser products pose some form of hazard [20]. The most common hazards associated with lasers come from the direct exposure of the eyes and skin. Within the United States, laser systems are classified in accordance with the regulations set forth by the Center for Devices and Radiological Health (CDRH) division of the Food and Drug Administration (FDA). Additional federal, state, and local regulations may also apply to further regulate the use of a laser product for a given application. Many of these secondary regulations are based on classification data provided by the American National Standards Institute (ANSI) [15, 16]. In the European community, laser system standards are overseen primarily by the International Electrotechnical Commission (IEC) and the British Standards Institution (BSI).

## 14.8 Glossary of Key Terms

**Coherent Light Source** A light source that converts electrical energy to a coherent beam of light.

**Composite Material** A substance consisting of two or more materials, insoluble in one another, which are combined to form a useful engineering material possessing certain properties not possessed by the constituents.

**Composite Component** A finished part containing composite material(s) that is in its end use application configuration, and which has undergone processing, fabrication, and assembly to the extent specified by the drawing, purchase order, or contract.

**Displacement Derivatives ($\partial w/\partial x$)** Rate of spatial displacement change, where $w$ is the surface displacement and $x$ is the surface coordinates.

**Filament Wound Pressure Vessel** An inner shell over-wrapped with composite layers that form a *composite shell*. The inner shell or liner may consist of an impervious metallic or nonmetallic material. The vessel may be cylindrical or spherical and will have at least one penetration with valve attachments for introducing and holding pressurized liquids or gases. Also referred to as a composite over-wrapped pressure vessel of COPV.

**Shear Vector** The image separation direction angle and distance formed in an imaging shearing interferometer used for NDT. Shear vector is expressed as the angle from the $x$ axis and distance between identical points in each image in inches or mm.

**Shearogram** The speckle pattern observed in laser shearography where a displacement derivative is detected.

**Stress** The stress change resulting from the application of an engineered force applied to the test object during shearography inspection. Stress most frequently used are vacuum, pressure, heat, mechanical and acoustic vibration in the sonic or ultrasonic ranges, magnetic fields, electric fields, microwave or mechanical load(s).

## References


1. Stetson, K. A., 50 Years of holographic interferometry, *SPIE Optics and Measurement Conference Proceedings*, Vol. 9442, 2014.
2. Hung, Y. Y., "Shearography: A new optical method for strain measurement and nondestructive testing," *Optical Engineering*, Vol. 21, No. 3, pp. 391–395, 1982.
3. Nakadata, S., "Phase detection of equidistant fringes for highly sensitive optical sensing," *Journal of the Optical Society of America A*, Vol. 5, No. 8, pp. 1265–1269, 1988.
4. Newman, J. W., "Laser Testing: Shearography & Holography," P.O. Moore, editor, *The Nondestructive Testing Overview*, Columbus Ohio, American Society for Nondestructive Testing, 2012.
5. "New York's Bronx-Whitestone Bridge Span to Feature 390 Composite Wind Fairings," NetComposites, http://netcomposites.com/news/2003/october/22/new-york-s-bronx-whitestone-bridge-span-to-feature-390-composite-wind-fairings/, 2003.

6. Caufield, H. J. and Lu, S., *Applications of Holography*, New York, NY, Wiley-Interscience, 1970.
7. Chatters, T., Pouet, B., and Krishnaswamy, S., "Shearography With Synchronized Pressure Stressing," D. O. Thompson and D. E. Chimenti, editors, *Review of Progress in Quantitative NDE*, La Jolla, CA, Plenum Press, pp. 19–426, 1992.
8. Deaton, J. B., Jr. and Rogowhi, R. S., "Optimization of Image Shearing Parameters in Electronic Shearography for Nondestructive Evaluation," D. O. Thompson and D. E. Chimenti, editors, *NDE: An Answer for an Aging World*, Columbus OH, American Society for Nondestructive Testing, pp. 204–206, 1994.
9. Gryzagoridis, J., Oliver, G., and Findeis, D., "Modal frequency versus shearography in detecting and locating voids/delaminations in sandwich composites," *Insight*, Vol. 55, No. 5, pp. 249–252, 2013.
10. Post, D., "Optical Interference for Deformation Measurements—Classical Holographic and Moire' Interferometry," W. W. Stinchcomb, editor, *Mechanics of Nondestructive Testing*, New York, NY, Plenum Press, 1980.
11. Spicer, J. B., Champion, J. L., Osiander, R., and Spicer, J. W. M., "Time Resolved Shearography and Thermographic NDE Methods for Graphite Epoxy/Honeycomb Composites," *Materials Evaluation*, Vol. 54, No. 10, pp. 1210–1213, 1996.
12. Vest, V. M., *Holographic Interferometry*, New York NY, John Wiley, 1979.
13. Gryzagoridis, J., "Laser based nondestructive inspection techniques," *Journal of Nondestructive Evaluation*, Vol. 31, No. 4, pp. 295–302, 2012.
14. "Standard Practice for Shearography of Polymer Matrix Composites and Sandwich Core Materials in Aerospace Applications," American Society for Testing and Materials, ASTM E2581-14, web page: https://www.astm.org/Standards/E2581.htm.
15. "Standard Guide for Nondestructive Testing of the Composite Overwraps in Filament Wound Pressure Vessels Used in Aerospace Applications," American Society for Testing and Materials, ASTM E2981, 2015, web page: https://www.astm.org/Standards/E2981.htm.
16. Timoshenko, S. P. and Goodier, A. N., *Theory of Elasticity*, 3rd Edition, New York, McGraw-Hill Classic Textbook Reissue Series, 1970.
17. Newman, J. W., "Shearographic & Holographic Testing," R. H. Bossi, editor, *ASNT Industry Handbook Aerospace NDT*, Columbus Ohio, American Society for Nondestructive Testing, 2012, https://www.asnt.org/MajorSiteSections/Publications/NDT_Handbooks/Industry_Handbook_Aerospace.aspx.
18. Timoshenko, S. P. and Woinowsky-Krieger, S., *Theory of Plates and Shells*, 2nd Edition, New York, McGraw-Hill Book Company, 1959.
19. Prescott, J., *Applied Elasticity*, New York, Dover, 1946.
20. Sliney, D. H., Safety with Lasers and Other Optical Sources: A Comprehensive Handbook, New York, Springer Science, 1980.

# APPENDIX

# Basic Metric–U.S. Customary Conversions

| To Convert | Into | Multiply by |
|---|---|---|
| Celsius | Fahrenheit | 9/5 and add 32 |
| Centimeters | Inches | 0.3937 |
| Cubic centimeters | Cubic inches | 0.06102 |
| Cubic meters | Cubic feet | 35.31 |
| Grams | Ounces | 0.03527 |
| Kilograms | Pounds | 2.205 |
| Kilometers | Miles | 0.6214 |
| Liters | Gallons (U.S.) | 0.2642 |
| Meters | Feet | 3.281 |
| Meters | Yards | 1.094 |
| Millimeters | Inches | 0.0393 |
| Square centimeters | Square inches | 0.1550 |
| Square kilometers | Square miles | 0.3861 |
| Square meters | Square feet | 10.76 |
| Square meters | Square yards | 1.196 |

Metric to U.S. Customary

| To Convert | Into | Multiply by |
|---|---|---|
| Cubic feet | Cubic meters | 0.02832 |
| Cubic inches | Cubic centimeters | 16.39 |
| Fahrenheit | Celsius | 5/9 after subtracting 32 |
| Feet | Meters | 0.3048 |
| Gallons (U.S.) | Liters | 3.785 |

*(Continued)*

| To Convert | Into | Multiply by |
|---|---|---|
| Inches | Millimeters | 25.40 |
| Inches | Centimeters | 2.540 |
| Miles | Kilometers | 1.609 |
| Ounces | Grams | 28.35 |
| Pounds | Kilograms | 0.4536 |
| Square feet | Square meters | 0.0929 |
| Square inches | Square centimeters | 6.452 |
| Square miles | Square kilometers | 2.590 |
| Square yards | Square meters | 0.8361 |
| Yards | Meters | 0.9144 |

U.S. Customary to Metric

# Index

*Note:* An *f* follows page numbers referencing figures; a *t*, tables; an *n*, notes.

## A

## B

## C

## D

## Q

## T

## U

## V

## W



## X

ngramcontent.com/pod-product-compliance
ng Source LLC
ne TN
62159200926
LV00006B/1286